THE DUXBURY SERIES IN STATISTICS AND DECISION SCIENCES

THE DUXBURY ADVANCED SERIES IN STATISTICS AND DECISION SCIENCES

Student's t (single mean)

$$t = \frac{\bar{x} - \mu_0}{s/\sqrt{n}}$$

Student's t (comparing two means)

$$t = \frac{(\bar{x}_1 - \bar{x}_2) - D_0}{s\sqrt{\frac{1}{n_1} + \frac{1}{n_2}}}$$

Pooled Estimate of σ^2

$$s^2 = \frac{\sum_{i=1}^{n_1}(x_{1i} - \bar{x}_1)^2 + \sum_{i=1}^{n_2}(x_{2i} - \bar{x}_2)^2}{(n_1 - 1) + (n_2 - 1)}$$

Correlation Coefficient

$$r = \frac{S_{xy}}{\sqrt{S_{xx}S_{yy}}}$$

where

$$S_{yy} = \sum_{i=1}^{n}(y_i - \bar{y})^2 = \sum_{i=1}^{n} y_i^2 - \frac{\left(\sum_{i=1}^{n} y_i\right)^2}{n}$$

$$S_{xx} = \sum_{i=1}^{n}(x_i - \bar{x})^2 = \sum_{i=1}^{n} x_i^2 - \frac{\left(\sum_{i=1}^{n} x_i\right)^2}{n}$$

$$S_{xy} = \sum_{i=1}^{n}(x_i - \bar{x})(y_i - \bar{y}) = \sum_{i=1}^{n} x_i y_i - \frac{\left(\sum_{i=1}^{n} x_i\right)\left(\sum_{i=1}^{n} y_i\right)}{n}$$

Least Squares Estimators of β_0 and β_1

$$\hat{\beta}_1 = \frac{S_{xy}}{S_{xx}} \quad \text{and} \quad \hat{\beta}_0 = \bar{y} - \hat{\beta}_1 \bar{x}$$

Least Squares Line (single independent variable)

$$\hat{y} = \hat{\beta}_0 + \hat{\beta}_1 x$$

Mann-Whitney U

$$U_1 = n_1 n_2 + \frac{n_1(n_1 + 1)}{2} - T_1 \quad \text{and} \quad U_2 = n_1 n_2 + \frac{n_2(n_2 + 1)}{2} - T_2$$

Chi-Square Statistic for Contingency Tables

$$X^2 = \sum_{i=1}^{k} \frac{(n_i - np_i)^2}{np_i}$$

Spearman's Rank Correlation Coefficient

$$r_s = 1 - \frac{6\sum_{i=1}^{n} d_i^2}{n(n^2 - 1)}$$

INTRODUCTION
TO PROBABILITY
AND STATISTICS

EIGHTH EDITION

INTRODUCTION TO PROBABILITY AND STATISTICS

WILLIAM MENDENHALL
University of Florida, Emeritus

ROBERT J. BEAVER
University of California, Riverside

PWS-KENT PUBLISHING COMPANY   **BOSTON**

PWS-KENT
Publishing Company

20 Park Plaza
Boston, Massachusetts 02116

Library of Congress Cataloging-in-Publication Data

Mendenhall, William.
    Introduction to probability and statistics/William Mendenhall,
Robert Beaver.—8th ed.
        p.   cm.
    Includes index.
    ISBN 0–534–92409–3
    1. Mathematical statistics.   2. Probabilities.
    I. Beaver, Robert
J. II. Title.
QA276.M425   1990   90–38775
519.5—dc20   CIP

International Student Edition ISBN: 0–534–98264–6

Printed in the United States of America.

91 92 93 94 95—10 9 8 7 6 5 4 3 2

Sponsoring Editor: Michael Payne
Assistant Editor: Marcia Cole
Editorial Assistant: Tricia Schumacher
Production Editor: Susan M. C. Caffey
Manufacturing Coordinator: Marcia A. Locke
Interior/Cover Designer: Susan M. C. Caffey
Cover Art: *Le Pacifique* by Masahiro Kasai. A silkscreen print used with permission of the artist.
Typesetter: Polyglot Pte. Ltd.
Cover Printer: Henry N. Sawyer Company, Inc.
Text Printer/Binder: R. R. Donnelley & Sons Co., Inc.

Credits

Exercises 1.1 (p. 7), 2.23 (p. 39), 2.35 (p. 45), 2.60 (p. 60), 3.37 (p. 92), 3.44 (p. 96), 4.22 (p. 143), 5.73 (p. 194), 7.38 (p. 264), 10.71 (p. 436), and the case studies on pages 10 and 280 are reprinted by permission of the *New York Times*. Copyright ©1987/88/89 by the *New York Times* Company.

Exercises 2.50 (p. 53), 4.14 (p. 142), 4.42 (p. 150), 4.67 (p. 160), 4.79 (p. 162), 6.26 (p. 221), 6.52 (p. 231), 8.26 (p. 304), 9.82 (p. 379), and 10.10 (p. 403) are reprinted by permission of the *Wall Street Journal*, © Dow Jones & Company, Inc. (1987/88/89). All Rights Reserved Worldwide.

Example 3.16 on pages 94 and 95 contains information excerpted from material originally appearing in *Discover* magazine, in an article entitled, "AIDS Diary," by Robert Gallo, ©1988, p. 38.

Exercises 5.33 (p. 189), 6.56 (p. 232), 7.12 (p. 253), 14.7 (p. 589), and 14.70 (p. 633) are reprinted with permission from *Science News*, the weekly newsmagazine of science, copyright 1989 by Science Service, Inc.

Exercise 9.122 on page 389 is taken from the *Journal of Psychology*, 123 (4), pp. 397–404, 1989. Reprinted with permission of the Helen Dwight Reid Educational Foundation. Published by Heldref Publications, 4000 Albermarle St., N.W., Washington, D.C. 20016. Copyright ©1989.

Data from "Selected Mechanical Factors Associated with Acceleration in Ice Skating" used in Exercises 7.14 (p. 253), 7.57 (p. 272), and 7.58 (p. 272) is reprinted with permission from the *Research Quarterly for Exercise and Sport*, December, 1983. Data from "Ventilatory Threshold, Running Economy and Distance Running Performance of Trained Athletes" used in Exercises 10.9 (p. 402) and 14.46 (p. 624) are reprinted from the *Research Quarterly for Exercise and Sport*, September, 1983. Data from "The Relationship Between Game and Sport Involvement in Later Childhood: A Preliminary Investigation," used in Exercises 12.20 (p. 507) and 12.50 (p. 525) is reprinted from the *Research Quarterly for Exercise and Sport*, June, 1983. *Research Quarterly* is a publication of the American Alliance for Health, Physical Education, Recreation and Dance, 1990 Association Drive, Reston, VA 22091.

CONTENTS

► PREFACE

In the eighth edition of *Introduction to Probability and Statistics*, we have attempted to revise the text to improve the students' understanding of statistics and its use in everyday life, as well as its use in scientific research and experimental situations. This edition, like its predecessors, presents statistical inference, the **objective of statistics**, as its theme. Throughout the preparation of the eighth edition, our primary thrust has been to make the material more relevant and to increase its pedagogical effectiveness. To this end, we have revised exercise sets to provide a broad range of practical, relevant exercises, many of which are drawn from current periodicals and journals. These exercises will be meaningful both for those whose careers will utilize the results of statistical analysis of data, and for others who merely need to make sense of the reports and claims that appear daily in newspapers, magazines, and television presentations.

Following are some specific changes in the eighth edition:

1. In discussions involving univariate data (in Chapters 1–9) the response variable is x instead of y.

2. In Chapter 2, z-scores have been included in Section 2.9, "Measures of Relative Standing."

3. Chapters 3 and 4 of the previous edition have been combined into one chapter entitled "Probability and Probability Distributions." Combinatorial counting rules and Bayes' Rule remain as optional sections in this combined chapter.

4. The revised Chapter 4, "Several Useful Discrete Distributions," now includes the Poisson, hypergeometric, uniform, and geometric distributions as well as the binomial probability distribution. A discussion of lot acceptance sampling appears as a subtopic in Section 4.2. However, the concepts and ideas related to testing hypotheses that appeared in the binomial chapter in the seventh edition have been moved to Chapter 8, "Large Sample Tests of Hypotheses."

5. The table of binomial probabilities has been expanded to include sample sizes $n = 2, 3, \ldots, 12, 15, 20,$ and 25 for values of $p = .01, .05, .10, .20, \ldots, .90, .95,$ and .99. A new table of cumulative Poisson probabilities for various values of μ is also included in Appendix III.

6.   Chapter 5, "The Normal and Other Continuous Distributions," now includes a brief discussion of the uniform and exponential distributions.

7.   Chapter 6, "Sampling Distributions," includes a new introductory section, which explains the concept of a sampling distribution based on a random sample of size $n = 3$ from a population of $N = 5$ observations.

8.   Chapters 14 and 15 of the seventh edition, dealing with experimental design and the analysis of variance, have been combined into the new Chapter 13, and the Latin Square design has been deleted from this chapter.

9.   MINITAB has been incorporated throughout the text as representative of one of the many available interactive statistical packages. The MINITAB commands used in the text are summarized at the end of chapters in which they are used. SAS and SPSS printouts are also provided in the chapters dealing with multiple regression analysis and the analysis of enumerative data.

10.   The case studies in Chapters 1, 4, 7, 8, and 13 are either new or have been updated.

11.   Exercises throughout the text have been revised and their number reduced by elimination, revision, and replacement of dated exercises. Nonetheless, there are still a large number and variety of exercises in the text.

12.   Finally, we have attempted to reduce the length of the book — not only by revising exercise sets, but also by careful editing of the text, and the condensation of four chapters into two.

Exercises are graduated in level of difficulty so that all students can solve some of them, a substantial number can solve most, and a few of the best students are challenged to solve all without error. Symbols are used to identify areas of application such as business, industry, education, the social and biological sciences, and so on. The coding used to indicate the area of application appears in the inside front cover.

A color coding of certain exercise numbers has been provided to indicate that the solutions to those exercises appear in the *Partial Solutions Manual*, which is available to students. Also, starred exercises (*) signal those problems that are optional.

The authors are grateful to Michael Payne and the editorial staff of PWS-KENT for their patience, assistance, and cooperation in the preparation of this edition. Special thanks go to Barbara Beaver for reading, editing, and typing manuscript material, and for her diligence in preparing the solutions manual for this and previous editions. Thanks are also due to

William H. Beyer, *University of Akron*; Curtis K. Church, *Middle Tennessee State University*; Eugene A. Enneking, *Portland State University*; Michael Eurgubian, *Santa Rosa Junior College*; Dale Everson, *University of Idaho*; Iris B. Fetta, *Clemson University*; Wei-Min Huang, *Lehigh University*; David D. Krueger, *St. Cloud State University*; Robin H. Lock, *St. Lawrence University*; Rhonda Magel, *North Dakota State University*; Barbara Treadwell McKinney; *Western Michigan University*; James L. Mullenex, *James Madison University*; Glenn Rock, *Clarion University of Pennsylvania*; and Galen R. Shorack, *University of Washington* for their helpful reviews of the manuscript. We wish to thank authors and organizations for allowing us to reprint selected material; acknowledgments are made wherever such material appears in the text. Finally, we acknowledge the assistance of our families and their partnership in this writing endeavor.

William Mendenhall
Robert J. Beaver

WHAT IS STATISTICS?

Case Study

Are all children in the United States receiving the same quality of education, or are some merely being "shuffled along" until they finally drop out of the system? The case study at the end of Chapter 1 discusses a study involving 35,000 children and provides some interesting observations with regard to this question.

General Objective

The purpose of this chapter is to identify the nature of statistics, its objective, and how it plays an important role in the sciences, in industry, and, ultimately, in our daily lives.

Specific Topics

1 Use of statistics in practical situations (1.1)
2 Population versus sample (1.2)
3 Elements of a statistical problem (1.3)
4 Role of the statistician in inference making (1.4)

▷ I.I   ILLUSTRATIVE STATISTICAL PROBLEMS

What is statistics? How does it function? How does it help solve certain practical problems? Rather than attempt a definition at this point, let us examine several problems that might come to the attention of the statistician. From these examples we can then select the essential elements of a statistical problem.

In predicting the outcome of a national election, pollsters interview a predetermined number of people throughout the country and record their preference. On the basis of this information a prediction is made. Similar problems are encountered in market research (What fraction of smokers prefer cigarette brand A?), in sociology (What fraction of rural homes have electricity?), and in industry (What fraction of items purchased or produced are defective?).

The yield (production) of a chemical plant is dependent upon many factors unique to the chemical plant under consideration. By observing these factors and the yield over a period of time, we can construct a prediction equation relating yield to the observed factors. The plant's future production for a given set of factor conditions can be predicted by using this equation. One encounters similar problems in the fields of education, sociology, psychology, and the physical sciences. For instance, colleges wish to predict the grade point average of college students based on information obtained *prior* to college entrance. Here, we might relate grade point average to rank in high school class, the type of high school, College Board examination scores, socioeconomic status, or any number of similar factors.

In addition to prediction, the statistician is concerned with decision making based on observed data. Consider the problems of determining the effectiveness of a new cold vaccine. For simplification, let us assume that ten people have received the new cold vaccine and are observed over a winter season. Of these ten, eight survive the winter without acquiring a cold. Is the vaccine effective?

Two physically different teaching techniques are used to present a subject to two groups of students of comparable ability. At the end of the instructional period, a measure of achievement is obtained for each group. On the basis of this information, we ask: Do the data present sufficient evidence to indicate that one method produces, on the average, higher student achievement?

Consider the sampling inspection of purchased items in a manufacturing plant. On the basis of such an inspection, each lot of incoming goods must be either accepted or rejected and returned to the supplier. The inspection might involve drawing a sample of ten items from each lot and recording the number of defective items. The decision to accept or reject the lot could then be based on the number of defective items observed.

▷ I.2   THE POPULATION AND THE SAMPLE

The examples we have cited are varied in nature and complexity, but each involves prediction or decision making. In addition, each of these examples involves sampling. A specified number of items (objects or bits of information)—a **sample**—is drawn from a much larger body of data called the **population**. Note that the word *population* refers to data and not to people. The pollster draws a sample of opinion (those interviewed) from the statistical population, which is the set of opinions corresponding to all eligible voters in the country. For the cold vaccine experiment, the sample consists of observations made of the ten individuals receiving the vaccine. We assume that the sample is representative of a much larger body of data—the **population**—based on all people who could have received the vaccine.

Which is of primary interest, the sample or the population? In all the examples given above, we are primarily interested in the population. Since we cannot interview all the people in the United States, we must predict their behavior on the basis of information obtained from a representative sample. Similarly, it is not possible to give all possible users a cold vaccine. The manufacturer of the drug must predict the effectiveness of the vaccine in the purchasing public from information extracted from the sample. Hence, the sample may be of immediate interest, but we are primarily interested in describing the population from which the sample is drawn.

Definition

> A **population** is the set representing all measurements of interest to the sample collector.

Definition

> A **sample** is a subset of measurements selected from the population of interest.

As used by most people, the word *sample* has two meanings. It can refer to the set of objects on which measurements are to be taken or it can refer to the measurements themselves. A similar double use could be made of the word *population*.

For example, you read in the newspapers that a Gallup poll was based on a sample of 1823 people. In this use of the word *sample*, the objects selected for the sample are clearly people. Presumably each person interviewed is asked a particular question, and that person's response represents a single item of data. The collection of data corresponding to the people represents a sample of data.

In a study of sample survey methods, it is important to distinguish between the objects measured and the measurements themselves. To experimenters the objects measured are called **experimental units**. The sample survey statistician calls them **elements of the sample**.

▷ **1.3   THE ESSENTIAL ELEMENTS OF A STATISTICAL PROBLEM**

The objective of statistics is to make inferences (predictions, decisions) about a population based on information contained in a sample. How do we achieve this objective? Every statistical problem contains five elements. The first and most important element is a clear specification of the question to be answered and the population of data that is related to the question.

The second element of a statistical problem is deciding how to select the sample. This is called the design of the experiment, or the sampling procedure. This element is important because gathering data costs money and time. In fact, it is not unusual for an experiment or a statistical survey to cost $50,000 or $500,000, and the costs of many biological or technological experiments can run into the millions. And what do these experiments and surveys produce? Numbers on a sheet of paper or, in brief, information. So planning the experiment is important. Including too many observations in the sample is often costly and wasteful; including too few observations results in little useful information. In addition, the way the sample is selected will often affect the amount of information per observation and thereby the cost. A good sampling design can sometimes reduce the cost of data collection to one-tenth or as little as one-hundredth of the cost of another sampling design.

The third element of a statistical problem is analyzing the sample data. No matter how much information the data contain about the practical question, you must use an appropriate method of data analysis to extract the desired information.

The fourth element of a statistical problem is using the sample data to make an inference about the population. As you will learn, many different procedures can be employed to make an estimate or decision about some characteristic of a population or to predict the value of some member of the population. For example, ten different methods might be available to estimate human response to a new drug, but one procedure might be much more accurate than another. Therefore, you will want to employ the best inference-making procedure when you use sample data to make an estimate or decision about a population or a prediction about some member of a population.

The final element of a statistical problem identifies what is perhaps the most important contribution of statistics to inference making. It answers the question, How good is the inference? To illustrate, suppose an agency conducts a statistical survey for you and estimates that your company's product will gain 34% of the market this year. How much confidence can you place in this estimate? You will quickly realize that you are lacking some important information. Of what value is the estimate without a measure of its reliability? Is the estimate accurate to within 1%, 5%, or 20%? Is it reliable enough to be used in setting production goals? As you will learn, statistical estimation, decision-making, and prediction procedures enable you to calculate a measure of goodness for every inference. Consequently, in a practical inference-making situation, every inference should be accompanied by a measure that tells you how much confidence you can place in the inference.

To summarize, a statistical problem involves the following:

1. A clear specification of the question and the population of data related to the question.
2. The design of the experiment or sampling procedure.
3. The collection and analysis of data.
4. The procedure for making inferences about the population based on sample information.
5. The provision of a measure of goodness (reliability) for the inference.

It is extremely important to note that the steps in the solution of a statistical problem are sequential; that is, you must identify the population of interest and plan how you will collect the data before you can collect and analyze it. All these operations must precede the ultimate goal, making inferences about the population based on information contained in the sample. These steps—carefully identifying the pertinent population and designing the experiment or sampling procedure—are often omitted. The experimenter may select the sample from the wrong population or may plan the data collection in a manner that intuitively seems reasonable or logical but that may be an extremely poor plan from a statistical point of view. The resulting data may be difficult or impossible to analyze, the data may contain little or no pertinent information, or, inadvertently, the sample may not be representative of the population of interest. This means that every experimenter should either be knowledgeable in the statistical design of experiments and sample surveys or should consult an applied statistician for the appropriate design *before* the data are collected.

▷ 1.4   THE ROLE OF STATISTICS AND THE STATISTICIAN

Statistics is an area of science concerned with the extraction of information from numerical data and its use in making inferences about a population from which the data are obtained. What is the role of the statistician in making inferences about a population based on sample data? One of the major contributions of the statistician is designing experiments and surveys, thereby reducing their cost and size. The statistician quantifies information and studies various designs and sampling procedures, searching for the procedure that yields a specified amount of information in a given situation at a minimum cost.

The other major contribution is in the inference making itself. The statistician studies various inferential procedures, looking for the best predictor or decision-making process for a given situation. Even more important, the statistician provides information concerning the goodness of an inferential procedure. When we make a prediction, we want to know something about the error in our prediction. When we make a decision, we want to know the chance that our decision is correct. Our built-in individual prediction and decision-making systems do not provide immediate answers to these important questions. They can be evaluated only by observation over a long period of time. In contrast, statistical procedures provide verifiable answers to these questions.

▷ 1.5   SUMMARY

Statistics is an area of science concerned with the **design of experiments** or sampling procedures, the **analysis of data**, and the **making of inferences** about a **population** of measurements from information contained in a **sample**. The statistician is concerned with developing and using procedures for design, analysis, and inference making that will provide the best inference at a minimum cost. In addition to making the best inference, the statistician is concerned with providing a quantitative measure of the goodness of the inference-making procedure.

A careful identification of the target population and careful design of the sampling procedure are often omitted, but they are essential steps in drawing inferences from experimental data. Poorly designed sampling procedures often produce data that are of little or no value (although this may not be obvious to the experimenter). After you make an inference, cast a critical eye upon it. Be sure you acquire a measure of its reliability.

▷ 1.6   A NOTE TO THE READER

We have stated the objective of statistics and we hope that we have answered the question, What is statistics? The remainder of this text is devoted to the development of the basic concepts involved in statistical methodology. In other words, we want to explain how statistical techniques actually work and why.

As you proceed through this text, you may wonder why we devote so much space to the making of inferences about populations from sample data and so little to the design of experiments. The reason is that the principles of good design do not become apparent until you know how to make inferences. As a consequence, your first contact with a design that may greatly increase the information in an experiment is in Section 9.5. An elementary discussion of experimental design follows in Chapter 13. The purpose of this text is to introduce you to the concepts and some of the elementary methods of statistics. For assistance in designing real-life experiments or surveys, you will want to consult a professional statistician.

Statistics involves very heavy use of applied mathematics. Most of the fundamental rules (called theorems in mathematics) are developed and based on a knowledge of calculus or higher mathematics. Inasmuch as this is meant to be an introductory text, we omit proofs except where they can be easily derived. Where concepts or theorems can be shown to be intuitively reasonable, we will attempt to give a logical explanation. We will attempt to convince you with the aid of examples and intuitive arguments rather than with rigorous mathematical derivations.

You should refer occasionally to Chapter 1 and review the objective of statistics and the elements of a statistical problem. Each of the following chapters should, in some way, be directed toward answering the questions posed here. Each is essential to completing the overall picture of statistics.

REFERENCES

Careers in Statistics. 3d ed. Washington, D.C.: American Statistical Association, 1980.

Tanur, J. M. et al., eds. *Statistics: A Guide to the Unknown.* 3d ed. Belmont, California: Wadsworth and Brooks-Cole, 1989.

EXERCISES

Understanding the Concepts

 1.1 Prineville, Oregon, a city of 5445 people, is the seat of the only one of the nation's 3106 counties that has voted for the winner in every Presidential election in which it has taken part. An NBC poll conducted in Prineville was reported in the *New York Times* (November 2, 1988), one week prior to the 1988 presidential election. In a sample of 1086 of the county's 13,500 residents, Vice President Bush was favored by 52%, Dukakis by 39%, and 9% were undecided.

    a. If the NBC pollsters were planning to use these results to predict the outcome of the 1988 election, describe the population of interest to them.

    b. Describe the actual population from which they have drawn a sample.

    c. Is the sample selected by the pollsters representative of the population described in part (a)? Explain.

 *1.2 You are a candidate for your state legislature and you want to survey voter attitudes regarding your chances of winning. Identify the population of interest to you and from which you would like to select your sample. How will this population be dependent on time?

 1.3 A medical researcher wants to estimate the survival time of a patient after the onset of a particular type of cancer and after a particular regimen of radiotherapy. Identify the population of interest to the medical researcher. Can you perceive some problems in sampling this population?

 1.4 An educational researcher wants to evaluate the effectiveness of a new method for teaching reading to deaf students. Achievement at the end of a period of teaching is to be measured by a student's score on a reading test. Discuss the population (or populations) that might be of interest to the researcher.

 1.5 A substantial amount of emission control tampering in cars and light trucks is cited in a federal study (*New York Times*, November 4, 1985). Emission tampering is defined as having occurred if an emission control system is either removed from a vehicle or is inoperable. In a 14-city survey, a sample of cars and light trucks manufactured between 1974 and 1984 was taken, and deliberate tampering was found in 22% of the vehicles. Another 29% of the vehicles showed evidence of tampering, but there was no evidence to confirm that the tampering was deliberate.

    a. Describe the population from which this sample was selected.

    b. Describe the sample.

    c. Does the sample selected from the population appear to be representative of all cars and light trucks operating in the United States that were manufactured between 1974 and 1984?

    d. Is the sample in part (b) representative of all cars and light trucks operating in the United States today? Why or why not?

 1.6 An article in *USA Today* (March 21, 1985) emphasizes an important point: If a statistical inference is to be valid, the sample must be representative of the population of interest to the researcher. The article describes a controversy over a study conducted by Dr. Robert Case and his wife, Nan, at New York's St. Luke's–Roosevelt Hospital Medical Center. The Cases

---

* Throughout the text, those exercise numbers in color indicate exercises whose solutions appear in the Student's Partial Solutions Manual.

concluded that harried, impatient, aggressive Type A individuals who have had a heart attack are no more prone to a new heart attack or death following a heart attack than easy-going Type B individuals who have had a heart attack. To reach this conclusion, the researchers examined a group of 516 heart attack victims. Half were Type A's and half were Type B's. In both groups, approximately the same percentage, 10%, died in the three years following their heart attack. Dr. Meyer Friedman, the originator of the Type A theory—a theory that Type A heart attack victims are much more prone to follow-up heart attacks and death—was quoted as saying, "This study relied solely on a questionnaire to distinguish Type A's and Type B's. That has been a discredited technique for 15 years." In response, the Cases claim that the questionnaire method for classifying individuals as Type A or Type B is 75–90% reliable.

Explain the basis of the controversy in terms of samples and populations. Do you think that the sample of Type A and Type B patients employed in the case study are representative of the corresponding populations of all Type A and Type B heart attack victims in the United States? Explain.

1.7 A nationwide television survey of 1200 people found that "95.1 percent of the respondents thought that their local school districts should require a student to pass reading and math exams before receiving a high school diploma" (*Florida Times Union*, April 1, 1985). According to the *Times Union* article, "The survey's sample closely reflected the nation's makeup in terms of age, geography, race, and family income." However, the article noted that "nearly 53 percent of the respondents attended college or graduated from one compared with 33 percent for the nation as a whole."

   a. Based on these quotes, do you think that the 1200 respondents are a representative sample of all adults in the United States? Explain.

   b. If the sample is nonrepresentative, how would you interpret the survey results?

1.8 A study in the *New England Journal of Medicine* (February 16, 1989) reports that long-term heavy alcohol consumption has a major toxic effect on the strength of both the heart and the skeletal muscles. The team of investigators studied 50 white males, age 25 to 59, who had *voluntarily* entered an alcohol treatment facility, all of whom were from emotionally stable environments and had no signs of ill health. An age-matched group of 50 white male physicians who were not heavy drinkers was also studied.

   a. Describe the two populations of interest to the experimenters.

   b. Describe the samples.

   c. Are the two samples representative of the populations in which the experimenters are interested?

   d. Why did the experimenters choose to isolate heavy drinkers who were from emotionally stable environments and who had no signs of ill health?

1.9 Jeffries, Voris, and Yang give data on the numbers of two types of barnacles found on 10 *T. orientalis* lobsters caught in the seas near Singapore ("Diversity and Distribution of the Pedunculate Barnacle *Octolasmis* Gray, 1825 Epizoic on the Scyllarid Lobster, *Thenus orientalis* [Lund, 1793]," *Crustaceana* 46, no. 3 [1984]). The data shown in the following table give the carapace length in millimeters (mm) and the numbers of two types of barnacles on each of the lobsters. The 10 lobsters were selected from among a total of 43 lobsters acquired from fishermen and fish markets and collected from the seas in the vicinity of Singapore. Suppose that we were interested in the number of *O. tridens* barnacles that one would find on a *T. orientalis* lobster.

   a. Describe the population that characterizes the number of *O. tridens* barnacles on a *T. orientalis* lobster.

b.   What do the numbers in the table under the *O. tridens* heading represent?

| | | Number of barnacles | | |
Field number	Carapace length (mm)	O. tridens	O. lowei	Total
AO61	78	645	6	651
AO62	66	320	23	343
AO66	65	401	40	441
AO70	63	364	9	373
AO67	60	327	24	351
AO69	60	73	5	78
AO64	58	20	86	106
AO68	56	221	0	221
AO65	52	3	109	112
AO63	50	5	350	355

1.10   The California bar examination, described as one of the hardest in the nation, is given twice each year. Of the 4555 students who took the exam in February of 1989, 2283 passed (*Press Enterprise*, Riverside, Calif., May 31, 1989). This pass rate of 50.2% was a 15-year high; however, state bar officials insist that the difficulty level of the exam has remained the same during this decade.

a.   Do the 4555 students who took the exam in February, 1989 represent a population or a sample?

b.   If the 4555 students are a sample, describe the population from which the sample was drawn.

c.   What assumptions must we make about the examination in order to be certain that the sample was representative of the population of interest?

1.11   Scientists are beginning to make some progress in the fight against the deadly AIDS virus. The head of the National Cancer Institute reports that the chance of living 18 months or longer after being diagnosed with AIDS has risen from 30% in 1982 to 60% today (*Press Enterprise*, Riverside, Calif., June 8, 1989).

a.   Describe the population of interest.

b.   How do you think the head of the National Cancer Institute obtained the 18-month survival rates quoted in the newspaper?

1.12   Researchers from California State University and the University of California will soon begin a series of experiments in an attempt to determine whether there is a relationship between nutrients and violent behavior (*Press Enterprise*, Riverside, Calif., October 4, 1989). The subjects of the experiment are violent inmates from the California Youth Authority. Preliminary studies have shown that many violent offenders have substandard levels of such substances as zinc, cadmium, manganese, or lead, which could be provided in the form of vitamin supplements.

a.   Describe the population of interest to the experimenters.

b.   Describe the sample.

c.   Do the experimenters intend to prove that lack of essential vitamins *causes* violent behavior? Explain.

WHO READS BEST?

The level of education in the United States has become a subject of much interest in many segments of our society. Politicians, administrators, company executives, teachers, and parents are all concerned that our schools are not providing the highest possible levels of education in training our children to enter the highly technical work force of the future. Are all of the children receiving the same quality of education, or are some merely being "shuffled along" until they finally drop out of the system?

To assess the progress of our educational system, Congress now requires a continuous testing project every two years, run by the Educational Testing Service of Lawrence Township, New Jersey, under contract with the Federal Department of Education. A report entitled "Who Reads Best?" was produced in 1986 and was based on a reading test given to 35,000 children nationwide in grades 4, 7, and 11 in public and private schools (*New York Times*, February 26, 1988). The report claims that the teaching techniques used for good readers and poor readers are not the same.

According to the report, all children, regardless of grade level, have difficulty evaluating and interpreting what they read. However, poor readers have more difficulty than good readers. Teaching strategies that can be be used to emphasize comprehension and critical thinking include pointing out hard words, reading new parts aloud, previewing passages, giving out lists of questions, explaining how to find the main idea, asking for opinions, and organizing discussions. When the surveyed students were asked how often their teachers used these techniques, younger students and those who scored in the bottom quarter of the test indicated less exposure to these activities. On the other hand, older students and those in the top 25% reported more time spent on "before- and after- reading activities."

An interesting sidelight to the report is that the testing service was slow to release the 1986 results because of unbelievable "precipitous declines in average reading proficiency" among the fourth and eleventh grade students in the test group. The testing service was concerned that there might be a "glitch in their test procedures." Hence, they were not willing to make comparisons between the 1984 and 1986 results, switching the scales of the new results to avoid comparisons with 1984. The conclusions of the report, however, were not in doubt.

1. Describe the population of interest to the experimenter.
2. Is the experimenter interested in making inferences about any subpopulations within the main population of interest? If so, describe some of these subpopulations.

3. Describe the sample. Is the sample representative of the population of interest?

4. Describe the experimental units. Describe the measurements collected by the experimenter.

5. Can you think of any reasons that might explain the "glitch" in the testing service's procedure in 1986?

CHAPTER 2

DESCRIBING SETS OF DATA

Case Study

Is your blood pressure normal, or is it too high or too low? The case study at the end of this chapter examines a large set of blood pressure data. You will use the methods of Chapter 2 to describe this set of data and compare your blood pressure with others of your sex and age group.

General Objective

Some of the data that we collect represent a sample selected from a population. Other data may represent the entire population, as in a national census. In either case, we need to be able to describe the data set. The objective of this chapter is to present two methods for describing data sets:

1. Graphical descriptive methods
2. Numerical descriptive methods

Graphical descriptive methods describe the data using charts and graphs. Numerical descriptive methods use numbers to help construct a mental picture of the data.

Specific Topics

1 Relative frequency distributions (2.1)

2 Stem and leaf displays (2.2)

3 Measures of central tendency (2.4)

4 Measures of variability (2.5)

5 Tchebysheff's Theorem and the Empirical Rule (2.6)

6   Shortcut calculation of s^2 (2.7)

7   Percentiles, quartiles, and z-scores (2.9)

8   Box plots (2.10)

2.1   A GRAPHICAL METHOD FOR DESCRIBING A SET OF DATA: RELATIVE FREQUENCY DISTRIBUTIONS

When confronted with a set of measurements, it is often difficult to make sense out of the data. One solution to this problem is to present the data in a pictorial or graphical display. One such statistical display, called a **relative frequency histogram**, can be explained by example.

The data presented in Table 2.1 represent the grade point averages of 30 Bucknell University freshmen recorded at the end of the freshmen year. By examining the table, you can quickly see that the largest and smallest grade point averages are 3.4 and 1.9, respectively, but how are the remaining 28 GPAs distributed? To answer this question, we divide the interval into an arbitrary number of subintervals of equal length. As a rule of thumb, the number of subintervals chosen should range from 5 to 20; the more data available, the more subintervals. These subintervals, or **classes**, are chosen so that each measurement can fall in *one and only one* subinterval. The classes form cells or pockets similar to the pockets of a billiard table. Each measurement, like a billiard ball, will fall into one and only one pocket. The measurements are categorized according to the class or pocket into which they fall, and the resulting tabulation is presented graphically in the form of a **frequency histogram**.

Table 2.1
Grade point averages of 30 Bucknell University freshmen

2.0	3.1	1.9	2.5	1.9
2.3	2.6	3.1	2.5	2.1
2.9	3.0	2.7	2.5	2.4
2.7	2.5	2.4	3.0	3.4
2.6	2.8	2.5	2.7	2.9
2.7	2.8	2.2	2.7	2.1

For the grade point averages in Table 2.1, we choose to use eight intervals of equal length. Since the length of the total grade point average span is $(3.4 - 1.9) = 1.5$, a convenient choice of interval length is $(1.5 \div 8) = .1875$, rounded off to .2. Rather than begin the first interval at the lowest value, 1.9, we choose a starting value

Table 2.2

Tabulation of relative frequencies for data of Table 2.1

Class i	Class boundaries	Tally	Class frequency f_i	Class relative frequency
1	1.85–2.05	111	3	3/30
2	2.05–2.25	111	3	3/30
3	2.25–2.45	111	3	3/30
4	2.45–2.65	~~1111~~ 11	7	7/30
5	2.65–2.85	~~1111~~ 11	7	7/30
6	2.85–3.05	1111	4	4/30
7	3.05–3.25	11	2	2/30
8	3.25–3.45	1	1	1/30

of 1.85, and form subintervals from 1.85 to 2.05, 2.05 to 2.25, 2.25 to 2.45, and so on. By choosing 1.85, we make it impossible for any measurement to fall on the class boundaries and eliminate any ambiguity regarding the disposition of a particular measurement.

The 30 measurements are now categorized according to the class into which they fall, as shown in Table 2.2. The eight classes are labeled from 1 to 8 for identification purposes. The boundaries for the eight classes, along with a tally of the number of measurements falling in each class, are given in the second and third columns of the table. The fourth column gives the number of measurements falling into a particular class, say class i, called the **class frequency** and designated by f_i. The last column of the table presents the fraction or proportion of the total number of measurements falling into each class. We call this proportion on the **class relative frequency**. If we let n represent the total number of measurements—for instance, in our example, $n = 30$—then the relative frequency for the ith class is f_i divided by n:

$$\text{Relative frequency} = \frac{f_i}{n}$$

Figure 2.1

Relative frequency histogram

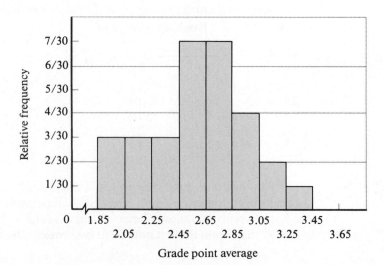

The resulting tabulation can be presented graphically in the form of a **relative frequency histogram**, as in Figure 2.1. Rectangles are constructed over each class interval, their height being proportional to the proportion of the total number of measurements (class relative frequency) falling in each class interval. When we look at the relative frequency histogram, we can see at a glance how the grade point averages are distributed over the interval 1.9 to 3.4.

In recent years, computers and microcomputers have become readily available to many students, providing them with an invaluable tool. In the study of statistics, even the beginning student can use packaged programs to perform statistical analyses with a high degree of speed and accuracy. Some of the more common statistical packages available at computer facilities are MINITAB™*, SAS (Statistical Analysis System), SPSS (Statistical Package for the Social Sciences), and BMPD (Biomedical Package).

These programs, called **statistical software**, differ in terms of the types of analyses available, the options within the programs, and the forms of printed results (called **output**). However, they are all similar. In this text, we will primarily use MINITAB as a statistical tool; understanding the basic output of this package will help you interpret the output from other software systems. However, a word of caution is necessary regarding the use of the computer for statistical analysis. Inaccurate entry of data, improper procedure commands, or incorrect interpretation of output will result in improper conclusions, even though you did use a computer!

The MINITAB software package can be used to create a relative frequency histogram for data that has been stored in a column format. Within the MINITAB package, the histogram is generated using the command HISTOGRAM for data that has been stored in a particular column (say C1). The program automatically chooses classes with conveniently rounded midpoints. If a measurement falls on a class boundary, it is assigned to the class with the larger midpoint. The printout gives the midpoints of the class intervals and the number of measurements per interval; it uses a star (∗) to represent a measurement in the histogram display. The MINITAB histogram for the grade point average data is presented in Table 2.3(a), and the SAS histogram in Table 2.3(b).

Table 2.3
(a) MINITAB histogram for the grade point average data

MTB > HISTOGRAM C1

Histogram of C1 N = 30

Midpoint	Count	
2.0	3	***
2.2	3	***
2.4	3	***
2.6	7	*******
2.8	7	*******
3.0	4	****
3.2	2	**
3.4	1	*

* MINITAB is the trademark of MINITAB, Inc., 215 Pond Lab., University Park, PA 16802.

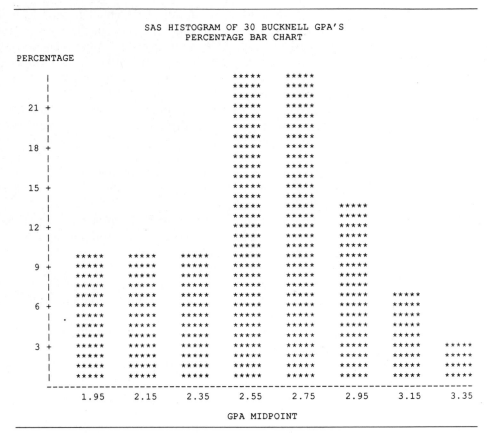

Consider the relative frequency histogram in Figure 2.1 in greater detail. What proportion of the students had grade point averages equal to 2.7 or greater? Checking the relative frequency histogram, we see that the proportion involves all classes to the right of 2.65. Using Table 2.2, we see that 14 students had grade point averages greater than or equal to 2.7. Hence, the proportion is 14/30, or approximately 47%. We note that this value is also the percentage of the total area of the histogram in Figure 2.1 lying to the right of 2.65.

Suppose that we write each of the 30 grade point averages on a piece of paper, place them in a hat, and then draw one piece of paper from the hat. What is the chance that this paper contains a grade point average greater than or equal to 2.7? Since 14 of the 30 pieces of paper are marked with numbers greater than or equal to 2.7, we say that we have 14 chances out of 30 or that the probability is 14/30. You have undoubtedly encountered the word *probability* in ordinary conversation; we will defer a definition of it and a discussion of its significance until Chapter 3.

Although we are interested in describing the set of $n = 30$ measurements, we might also be interested in the population from which the sample was drawn, which consists of grade point averages of all freshmen currently in attendance at Bucknell

University. Or, if we are interested in the academic achievement of college freshmen in general, we might consider our sample as representative of the population of freshmen attending Bucknell or colleges *similar* to Bucknell. What proportion of the grade point averages were greater than or equal to 2.7? If we possessed the relative frequency histogram for the population, we could give the exact answer to this question. Since we do not have this information, we are forced to make an **inference** using our sample information. Our estimate for the true population proportion, based on the sample information, would likely be 14/30, or 47%. Without knowledge of the population relative frequency histogram, we would infer that the population histogram is similar to the sample histogram and that approximately 47% of the grade point averages in the population are greater than or equal to 2.7. Most likely this estimate would differ from the true population percentage. We will examine the magnitude of this error in Chapter 7.

The relative frequency histogram is often called a *relative frequency distribution* because it shows the manner in which the data are distributed along the horizontal axis of the graph. The rectangles constructed above each class are subject to two interpretations. They represent the proportion of observations falling in a given class. Also, if a measurement is drawn from the data, a particular class relative frequency is the chance or probability that the measurement will fall into that class. The most significant feature of the sample frequency histogram is that it provides information on the population frequency histogram that describes the population. An important point to note here is that different samples from the same population will result in different sample histograms, even when the class boundaries remain fixed. However, we would expect the sample and the population frequency histograms to be similar. The degree of resemblance will increase as more and more data are added to the sample. If the sample were enlarged to include the entire population, the sample and population would be the same and the histograms would be identical.

In the preceding discussion we showed you how to construct a frequency distribution for the grade point average data in Table 2.1, and we explained how such a distribution could be interpreted. Before concluding this topic, we summarize the principles that you should employ in constructing a frequency distribution for a set of data.

PRINCIPLES TO EMPLOY IN CONSTRUCTING A FREQUENCY DISTRIBUTION

1. *Determine the number of classes.* It is usually best to have from 5 to 20 classes. The larger the amount of data available, the more classes should be employed. If you have too few classes, you might be concealing important characteristics of the data by grouping. If you have too many classes, empty classes may result and the distribution will be meaningless. The number of classes should be determined from the amount of data present and the unformity of the data. A small sample would require fewer classes.

2. *Determine the class width.* As a general rule for finding the class width, divide the difference between the largest and smallest measurements by the number of classes desired and add enough to the quotient to arrive at a convenient figure for class

width. All classes should be of equal width. This allows you to make uniform comparisons of the class frequencies.

3.   *Locate the class boundaries.* The lowest class must include the smallest measurement. Then add the remaining classes. Class boundaries should be chosen so that it will be impossible for a measurement to fall on a boundary.

EXERCISES   Basic Techniques

2.1   Construct a relative frequency histogram for the following set of data.

3.1	4.9	2.8	3.6	2.5
4.5	3.5	3.7	4.1	4.9
2.9	2.1	3.5	4.0	3.7
2.7	4.0	4.4	3.7	4.2
3.8	6.2	2.5	2.9	2.8
5.1	1.8	5.6	2.2	3.4
2.5	3.6	5.1	4.8	1.6
3.6	6.1	4.7	3.9	3.9
4.3	5.7	3.7	4.6	4.0
5.6	4.9	4.2	3.1	3.9

a.   Approximately how many class intervals should you use?
b.   Suppose you decide to use classes starting at 1.55 with a class width of .5 (i.e., 1.55 to 2.05, 2.05 to 2.55). Construct the relative frequency histogram for the data.
c.   What fraction of the measurements are less than 5.05?
d.   What fraction of the measurements are larger than 3.55?

2.2   The length of time (in months) between the onset of a particular illness and its recurrence was recorded for $n = 50$ patients. The times of recurrence are as follows:

2.1	4.4	2.7	32.3	9.9
9.0	2.0	6.6	3.9	1.6
14.7	9.6	16.7	7.4	8.2
19.2	6.9	4.3	3.3	1.2
4.1	18.4	.2	6.1	13.5
7.4	.2	8.3	.3	1.3
14.1	1.0	2.4	2.4	18.0
8.7	24.0	1.4	8.2	5.8
1.6	3.5	11.4	18.0	26.7
3.7	12.6	23.1	5.6	.4

a.   Construct a relative frequency histogram for the data.
b.   Give the fraction of recurrence times less than or equal to 10.

Applications

2.3 In the following table, the United States Census Bureau details the federal government's income through federal income taxes (on a per capita basis) as well as the amount spent per capita in federal aid for each of the 50 states in fiscal year 1986.

State	Taxes	U.S. Aid	State	Taxes	U.S. Aid
AL	$ 740	$ 434	MT	$ 755	$ 723
AK	3490	1244	NE	700	413
AZ	975	367	NV	1084	434
AR	770	474	NH	472	394
CA	1144	419	NJ	1096	440
CO	718	374	NM	989	579
CT	1202	471	NY	1278	697
DE	1343	495	NC	881	360
FL	780	278	ND	907	638
GA	806	448	OH	843	443
HI	1400	446	OK	895	423
ID	743	434	OR	715	497
IL	848	434	PA	898	481
IN	810	363	RI	908	585
IA	863	434	SC	863	391
KS	777	359	SD	570	646
KY	863	479	TN	682	443
LA	807	453	TX	667	313
ME	940	573	UT	820	485
MD	1047	439	VT	923	617
MA	1314	528	VA	836	345
MI	1019	476	WA	1169	427
MN	1163	501	WV	964	554
MS	731	512	WI	1148	483
MO	712	391	WY	1569	929

Source: Data from U.S. Commerce Department, Bureau of the Census, *The World Almanac & Books of Facts,* 1989 edition, copyright © Newspaper Enterprise Association, Inc. 1988, New York, NY 10166.

a. Construct a relative frequency histogram to describe the data on per capita federal taxes for the 50 states.

b. Construct a relative frequency histogram to describe the data on federal aid to the 50 states.

c. Compare the shapes of the relative frequency histograms in parts (a) and (b). Are they similar or different?

2.4 The following data provide the rushing statistics for the top rushers in the National Football League through the games of October 8, 1989 (*Sporting News,* October 16, 1989). Statistics include the number of attempts, the total number of rushing yards, and the average number of yards per carry.

| National Football Conference | | | |
Name/Team	Att	Yds	Avg
Anderson, Chicago	100	530	5.3
Bell, Los Angeles	100	512	5.1
Riggs, Washington	108	468	4.3
Fullwood, Green Bay	87	434	5.0
B. Sanders, Detroit	67	354	5.3
Anderson, New York	98	346	3.5
Craig, San Francisco	84	318	3.8
Tate, Tampa Bay	64	282	4.4
Hilliard, New Orleans	84	267	3.2
Settle, Atlanta	71	254	3.6
Walker, Dallas	81	246	3.0
Howard, Tampa Bay	56	185	3.3
S. Mitchell, Phoenix	43	165	3.8
Fenney, Minnesota	40	157	3.9
Cunningham, Philadelphia	32	149	4.7
Byner, Washington	19	148	7.8
Toney, Philadelphia	39	142	3.6
Majkowski, Green Bay	29	128	4.4
Morris, Washington	37	126	3.4

| American Football Conference | | | |
Name/Team	Att	Yds	Avg
Okoye, Kansas City	103	487	4.7
Dickerson, Indianapolis	102	446	4.4
Brooks, Cincinnati	70	429	6.1
Thomas, Buffalo	69	348	5.0
Humphrey, Denver	68	294	4.3
Warner, Seattle	64	235	3.7
Allen, Los Angeles	52	219	4.2
Worley, Pittsburgh	56	199	3.6
Highsmith, Houston	45	192	4.3
Williams, Seattle	55	189	3.4
Hoge, Pittsburgh	62	185	3.0
Ball, Cincinnati	45	184	4.1
Winder, Denver	56	182	3.3
Butts, San Diego	44	180	4.1
Pinkett, Houston	45	179	4.0
Stephens, New England	58	179	4.0
Perryman, New England	46	177	3.8
Metcalf, Cleveland	45	169	3.8
Spencer, San Diego	41	167	4.1
White, Houston	50	156	3.1

    a.   Construct two relative frequency histograms to describe the total rushing yards for the top rushers in the NFC and the AFC. Use the same class intervals for both histograms.

    b.   Compare the two sets of data by comparing their relative frequency histograms.

2.5   Refer to Exercise 2.4.

    a.   Construct a relative frequency histogram describing the average number of yards per carry for the top rushers in the National Football League.

    b.   What percentage of the top rushers in the NFL average more than 5 yards per carry?

2.6   In order to decide on the number of service counters needed for stores to be built in the future, a supermarket chain wanted to obtain information on the length of time (in minutes) required to service customers. To obtain information on the distribution of customer service times, a sample of 1000 customers' service times was recorded. Sixty of these are shown in the accompanying tabulation.

3.6	1.9	2.1	.3	.8	.2
1.0	1.4	1.8	1.6	1.1	1.8
.3	1.1	.5	1.2	.6	1.1
.8	1.7	1.4	.2	1.3	3.1
.4	2.3	1.8	4.5	.9	.7
.6	2.8	2.5	1.1	.4	1.2
.4	1.3	.8	1.3	1.1	1.2
.8	1.0	.9	.7	3.1	1.7
1.1	2.2	1.6	1.9	5.2	.5
1.8	.3	1.1	.6	.7	.6

a. Construct a relative frequency histogram for the data.

b. What fraction of the service times are less than or equal to 1 minute?

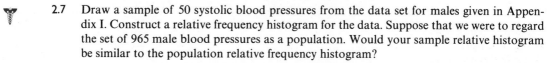

2.7 Draw a sample of 50 systolic blood pressures from the data set for males given in Appendix I. Construct a relative frequency histogram for the data. Suppose that we were to regard the set of 965 male blood pressures as a population. Would your sample relative histogram be similar to the population relative frequency histogram?

2.8 Draw a sample of 50 systolic blood pressures from the data set for females given in Appendix I. Construct a relative frequency histogram for the data using the same class intervals as used for the data set in Exercise 2.7. Does the histogram for females resemble the histogram for males constructed in Exercise 2.7?

2.2 STEM AND LEAF DISPLAYS

The **stem and leaf display** is an alternative method for describing a set of data. This method was proposed by John Tukey as part of a newly emerging area of statistics called **exploratory data analysis** (EDA). The objective of EDA is to provide the experimenter with simple techniques that allow him to look more effectively at his data. The stem and leaf display presents a histogram-like picture of the data, while allowing the experimenter to retain the actual observed values of each data point. Hence, the stem and leaf display is partly tabular and partly graphical in nature.

Table 2.4 gives the top 40 stocks on the over-the-counter (OTC) market ranked by percentage of outstanding shares traded on a particular day. In creating a stem and leaf display for these data, we divide each observation into two parts: the stem and the leaf. For example, we could divide each observation at the decimal point. The portion to the left of the point of division is the stem; the portion to the right is the leaf. Thus, the stem and leaf for the observation 7.15 are

Stem	Leaf
7	15

Table 2.4
Top 40 stocks on the OTC market

22.88	5.49	4.40	3.44	2.88
7.99	5.26	4.05	3.36	2.74
7.15	5.07	3.94	3.26	2.74
7.13	4.94	3.93	3.20	2.69
6.27	4.81	3.78	3.11	2.68
6.07	4.79	3.69	3.03	2.63
5.98	4.55	3.62	2.99	2.62
5.91	4.43	3.48	2.89	2.61

Alternatively, we could choose the point of division between the tenths and hundredths decimal places, whereby

Stem	Leaf
71	5

The choice of the stem and leaf coding depends on the nature of the data set. The stem and leaf display is constructed by using the following steps.

CONSTRUCTING A STEM AND LEAF DISPLAY

1. List the stem values, in order, in a vertical column.
2. Draw a vertical line to the right of the stem values.
3. For each observation, record the leaf portion of that observation in the row corresponding to the appropriate stem.
4. Reorder the leaves from lowest to highest within each stem row.
5. If the number of leaves appearing in each stem row is too large, divide the stems into two groups, the first corresponding to leaves beginning with digits 0 through 4, and the second corresponding to leaves beginning with digits 5 through 9. [This subdivision can be increased to five groups if necessary.]
6. Provide a key to your stem and leaf coding, so that the reader can recreate the actual measurements from your display.

The stem and leaf display for the data in Table 2.4 is shown in Figure 2.2. If we were to choose the point of division between the tenths and hundredths decimal places, there would be a large number of stems, from 26 to 228, and the display would not provide a good visual description of the data. Hence, we choose to use the decimal point as the point of division. As an aid to decoding, we write

Leaf unit $= 0.10$

5 00 represents 5.00

How does the stem and leaf display in Figure 2.2 describe the 40 OTC stocks? If you turn the display sideways, you can see that the stocks are displayed in a histogram-like picture. *The data are not symmetric*, with most of the observations

Figure 2.2
Stem and leaf display for the data in Table 2.4

2	61	62	63	68	69	74	74	88	89	99	
3	03	11	20	26	36	44	48	62	69	78	93	94
4	05	40	43	55	79	81	94					
5	07	26	49	91	98							
6	07	27										
7	13	15	99									
HI	22.88											

falling between 2.00% and 5.00%. Notice that one extremely large observation (22.88) is listed as HI rather than extending the stem and leaf display to accommodate the one large stem.

The stem and leaf display has an advantage over the relative frequency histogram in that it allows you to reconstruct the actual data set; also, it lists the observations in order of magnitude. Hence, we can tell from the display that the lowest observation is 2.61 and the highest is 22.88. If we wish to find the fifth smallest measurement, we can count up from the smallest measurement and identify the fifth smallest as 2.69.

Sometimes the available stem choices result in a display containing too many stems (and very few leaves within a stem) or too few stems (and many leaves within a stem). This is the case with the Bucknell University grade point data (Table 2.1). In this situation, we may divide the too few stems by stretching them into two or more lines, depending on the leaf values with which they will be associated. Two options are available:

1. The stem is divided into two parts. The first is associated with leaves having 0, 1, 2, 3, or 4 as their first digit.

2. The stem is divided into five parts. The first is associated with leaves having 0 or 1 as their first digit; the second is associated with leaves having 2 or 3 as their first digit; and so on. The last stem part is associated with leaves having 8 or 9 as their first digit.

Since the data in Table 2.1 vary from 1.9 to 3.4, using the decimal point as the point of division would produce only three stem rows. Such a stem and leaf display would not produce a good descriptive picture of the data. We could divide each stem into two parts; this division would result in four stems (since the lower portion of the first stem and the upper portion of the last stem are unnecessary). Dividing each stem into five parts would result in nine stems, and would produce a more visually descriptive display. The stem and leaf display for this option is shown in Figure 2.3. Notice the similarity between the stem and leaf display in Figure 2.3 and the relative histogram in Figure 2.1. However, the stem and leaf display allows the reproduction of the actual data set if necessary.

A stem and leaf display can be generated by using the STEM AND LEAF command in the MINITAB package. When the MINITAB command STEM AND

Figure 2.3
Stem and leaf display
for the data in Table 2.1

```
1 | 9   9
2 | 0   1   1
2 | 2   3
2 | 4   4   5   5   5   5   5
2 | 6   6   7   7   7   7   7
2 | 8   8   9   9
3 | 0   0   1   1
3 |
3 | 4
```

Leaf unit = 0.1
1 2 represents 1.2

```
MTB  > STEM AND LEAF C1;
SUBC > TRIM.

Stem-and-leaf  of C1          N = 40
Leaf  Unit = 0.10

       10     2   6666677889
       17     3   0122344
      (5)     3   66799
       18     4   044
       15     4   5789
       11     5   024
        8     5   99
        6     6   02
        4     6
        4     7   11
        2     7   9

              HI   228,
```

LEAF was implemented for the OTC stock data that was stored in column C1 of a data array, the stem and leaf display in Table 2.5 was produced. The MINITAB program chooses the decimal point as the point of division, and divides each stem into two parts. The first column in the display is the number of leaves on that stem or on a stem closer to the nearer end of the display. For example, there are 17 data values less than or equal to 3.4, while there are 6 data values greater than or equal to 6.0. For the stem class containing the data value midway from each end, the number in parentheses gives the number of leaves for that stem. Notice that the computer truncates. the second decimal place in the data, sacrificing some accuracy in the display in order to retain its simplicity and clarity.

To summarize, a stem and leaf display is easy to construct and the display creates the same sort of figure produced by a relative frequency histogram. In addition, it permits the user to reconstruct the data set and to identify observations ordered by their relative magnitude.

EXERCISES

Basic Techniques

2.9   Construct a stem and leaf display for the data in Exercise 2.1. The data are reproduced below for your convenience.

3.1	4.9	2.8	3.6	2.5
4.5	3.5	3.7	4.1	4.9
2.9	2.1	3.5	4.0	3.7
2.7	4.0	4.4	3.7	4.2
3.8	6.2	2.5	2.9	2.8
5.1	1.8	5.6	2.2	3.4
2.5	3.6	5.1	4.8	1.6
3.6	6.1	4.7	3.9	3.9
4.3	5.7	3.7	4.6	4.0
5.6	4.9	4.2	3.1	3.9

a. Compare the stem and leaf display with the relative frequency histogram constructed in Exercise 2.1.

b. Use the stem and leaf display to find the smallest observation and to find the eighth and ninth largest observations.

2.10 Use the following set of data to answer questions (a) and (b).

4.5	3.2	3.5	3.9	3.5	3.9
4.3	4.8	3.6	3.3	4.3	4.2
3.9	3.7	4.3	4.4	3.4	4.2
4.4	4.0	3.6	3.5	3.9	4.0

a. Construct a stem and leaf display by using the leading digit as the stem.

b. Construct a stem and leaf display by using each leading digit twice. Has this technique improved the presentation of the data?

Applications

2.11 Construct a stem and leaf display for the supermarket service times given in Exercise 2.6. Compare the stem and leaf display with the relative frequency histogram constructed in that exercise. Do the two graphical descriptions of the data seem to convey the same information?

2.12 The federal government has begun cracking down on colleges and trade schools whose former students have defaulted on federally backed student loans. Defaults on these loans will cost the taxpayers about $1.8 billion this year, the fourth highest item in the Education Department's budget (*Press Enterprise*, Riverside, Calif., June 2, 1989). The data below are default rates for 24 colleges or trade schools in Southern California. The default rate represents the percentage of students who defaulted in 1986 or 1987 who are required to start repaying loans in 1986.

41.9	31.0	22.0	11.0	5.8
41.1	27.1	21.1	9.8	5.7
34.6	25.4	17.8	8.7	4.0
32.6	24.3	17.8	8.1	3.6
32.0	24.2	14.5	8.0	

Source: Data from U.S. Education Department.

Construct a stem and leaf display to describe the data.

2.13 Refer to Exercise 2.12. The MINITAB command STEM was used to produce the stem and leaf display on the top of page 26 for the default rates in federally backed student loans for 24 Southern California schools. Explain the choice of stem and leaf used in the MINITAB analysis. Does this display differ from the display you constructed in Exercise 2.12?

2.14 Construct a stem and leaf display for your sample of 50 male systolic blood pressures given in Exercise 2.7. Compare it with the relative frequency histogram you constructed in Exercise 2.7.

```
Stem-and-leaf of C1        N = 24
Leaf Unit = 1.0

        2     0   34
        8     0   558889
       10     1   14
       12     1   77
       12     2   1244
        8     2   57
        6     3   1224
        2     3
        2     4   11
```

 2.15   Construct a stem and leaf display for the sample of 50 female systolic blood pressures drawn in Exercise 2.8. Compare it with the relative frequency histogram you constructed in Exercise 2.8.

▷ 2.3 NUMERICAL METHODS FOR DESCRIBING A SET OF DATA

Graphical methods are extremely useful for conveying a rapid general description of collected data and for presenting data. In many respects they support the saying that a picture is worth a thousand words. There are, however, limitations to the use of graphical techniques for describing and analyzing data. For instance, suppose that we want to describe our data verbally before a group of people. Unable to present the histogram visually, we would be forced to use other descriptive measures that would convey to the listeners a mental picture of the histogram.

A second and not so obvious limitation of the histogram and other graphical techniques is that they are difficult to use for purposes of statistical inference. Presumably, we use the sample histogram to make inferences about the shape and location of the population histogram, which describes the population and is unknown to us. Our inference is based upon the correct assumption that some degree of similarity will exist between the two histograms, but we are then faced with the problem of measuring the degree of similarity. We know when two figures are identical, but this situation will not likely occur in practice. Therefore, if the sample and population histograms differ, how can we measure the degree of difference or, expressing it positively, the degree of similarity? To be more specific, we might wonder about the degree of similarity between the histogram in Figure 2.1 and the frequency histogram for the population of grade point averages from which the sample was drawn. Although the difficulties we would face in trying to answer this question are not insurmountable, it is preferable to seek other descriptive measures that readily lend themselves to use as predictors of the shape of the population frequency distribution.

The limitation of the graphical method of describing data can be overcome by the use of numerical descriptive measures. Thus, we would like to use the sample data to calculate a set of numbers that will convey a good mental picture of the frequency distribution and will be useful in making inferences concerning the population.

> Numerical descriptive measures computed from population measurements are called **parameters**; those computed from sample measurements are called **statistics**.

▷ 2.4 MEASURES OF CENTRAL TENDENCY

In constructing a mental picture of the frequency distribution for a set of measurements, we would likely envision a histogram similar to that shown in Figure 2.1 for the data on grade point averages. One of the first descriptive measures of interest would be a **measure of central tendency**—that is, a measure that locates the center of the distribution. The grade point data ranged from a low of 1.9 to a high of 3.4, with the center of the histogram being located in the vicinity of 2.6. Let us now consider some definite rules for locating the center of a distribution of data.

One of the most common and useful measures of central tendency is the arithmetic average of a set of measurements. This measure is also often referred to as the **arithmetic mean**, or simply the **mean**, of a set of measurements. Since we will wish to distinguish between the means for the sample and for the population, we will use the symbol $\bar{x}$ (x bar) to represent the sample mean and μ (Greek lowercase letter mu) to represent the mean of the population.

Definition

> The **arithmetic mean** of a set of n measurements $x_1, x_2, x_3, \ldots, x_n$ is equal to the sum of the measurements divided by n.

The procedures for calculating a sample mean and many other statistics are conveniently expressed as formulas. Consequently, we will need a symbol to represent the process of summation. If we denote the n quantities that are to be summed as $x_1, x_2, \ldots, x_n$, then their sum is denoted by the symbol

$$\sum_{i=1}^{n} x_i$$

The symbol $\sum_{i=1}^{n}$ ($\sum$ is the Greek capital letter sigma) tells us to sum the elements that appear to the right of the $\sum$ sign, commencing with x_1 (i.e., $i = 1$) and proceeding in order to x_n. Thus,

$$\sum_{i=1}^{3} x_i = x_1 + x_2 + x_3$$

and

$$\sum_{i=1}^{n} x_i = x_1 + x_2 + \cdots + x_n$$

Using this notation, we can express the formula for the sample mean as shown in the following display.

Sample mean: $\bar{x} = \dfrac{\sum\limits_{i=1}^{n} x_i}{n}$

Population mean: μ

EXAMPLE 2.1   Find the mean of the measurements 2, 9, 11, 5, 6.

Solution   Substituting the measurements into the formula, we have

$$\bar{x} = \frac{\sum\limits_{i=1}^{n}}{n} = \frac{2 + 9 + 11 + 5 + 6}{5} = 6.6$$

Note that this value, $\bar{x} = 6.6$, falls near the middle of the set of sample measurements 2, 9, 11, 5, and 6.

We have seen that $\bar{x}$ is used to locate the center of a set of sample measurements. A more important use of $\bar{x}$ is as an estimator (predictor) of the value of the unknown population mean μ. For example, the mean of the sample given in Table 2.1 is

$$\bar{x} = \frac{\sum\limits_{i=1}^{n} x_i}{n} = \frac{77.5}{30} = 2.58$$

Note that this value falls approximately in the center of the set of sample measurements. The mean of the entire population of grade point averages is unknown to us, but if we were to estimate its value, our estimate of μ would be 2.58. Although the value of $\bar{x}$ varies from sample to sample, the population mean μ remains the same.

A second measure of central tendency is the **median**, which is the value in the middle position.

Definition

> The **median** m of a set of n measurements $x_1, x_2, x_3, \ldots, x_n$ is the value of x that falls in the middle position when the measurements are ranked in order from the smallest to the largest.

Hence, the median divides a set of measurements into two equal parts. If the number n of measurements is odd, this number will be the measurement with rank equal to $(n + 1)/2$. If the number of measurements is even, the median is chosen as the value of x halfway between the two middle measurements, that is, halfway between the measurement ranked $n/2$ and the one ranked $(n/2) + 1$.

RULE FOR CALCULATING A MEDIAN

Rank the n measurements from the smallest to the largest.

1. If n is odd, the median m is the measurement with rank $(n + 1)/2$.
2. If n is even, the median m is the value of x that is halfway between the measurement with rank $n/2$ and the measurement with rank $(n/2) + 1$. Alternatively, the median is the average of the two middle measurements.

EXAMPLE 2.2   Find the median for the set of measurements 9, 2, 7, 11, 14.

Solution   We first rank the $n = 5$ measurements from the smallest to the largest: 2, 7, 9, 11, 14. Since $n = 5$ is odd, the median has rank $(n + 1)/2 = 3$ and is equal to 9.   ◁

EXAMPLE 2.3   Find the median for the set of measurements 9, 2, 7, 11, 14, 6.

Solution   Since $n = 6$ is even, we rank the measurements 2, 6, 7, 9, 11, 14, and choose the median halfway between the two middle measurements, 7 and 9. Therefore, the median is equal to 8.   ◁

Although both the mean and the median are good locations of the center of a distribution of measurements, the median is less sensitive to extreme values. For example, if the distribution is symmetrical about its mean (see Figure 2.4(a)), the mean and the median are equal. In contrast, if a distribution is skewed to the left or to the right, the mean shifts toward the direction of skewness. Figure 2.4(b) shows a distribution skewed to the right. Since the large extreme values in the upper tail of the distribution increase the sum of the measurements, the mean shifts toward the direction of skewness. The median is not affected by these extreme values because the numerical values of the measurements are not used in its computation.

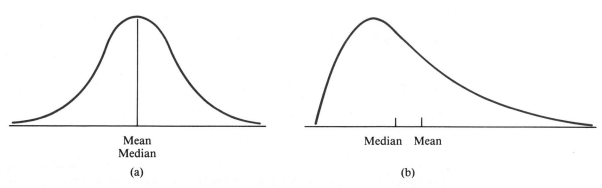

Figure 2.4   Relative frequency distributions showing the effect of extreme values on the mean and median

▷ **2.5** MEASURES OF VARIABILITY

Once we have located the center of a distribution of data, the next step is to provide a measure of the **variability**, or **dispersion**, of the data. Consider the two distributions shown in Figure 2.5. Both distributions are located with a center at $x = 4$, but there is a vast difference in the variability of the measurements about the mean for the two distributions. The measurements in Figure 2.5(a) vary from 3 to 5; in Figure 2.5(b) the measurements vary from 0 to 8.

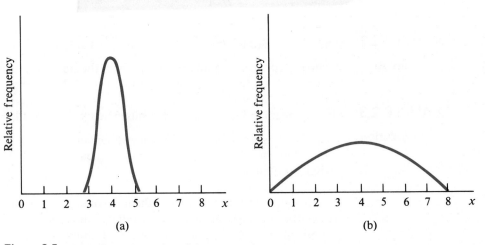

(a)   (b)

Figure 2.5   Variability or dispersion of data

Variation is a very important characteristic of data. For example, if we are manufacturing bolts, excessive variation in the bolt diameter would imply a high percentage of defective product. On the other hand, if we are using an examination to discriminate between good and poor accountants, we would be most unhappy if the examination always produced test grades with little variation, since this would make discrimination very difficult.

In addition to the practical importance of variation in data, a measure of this characteristic is necessary to the construction of a mental image of the frequency distribution. Numerous measures of variability exist, and we will discuss a few of the most important ones.

The simplest measure of variation is the **range**.

Definition

> The **range** of a set of n measurements $x_1, x_2, x_3, \ldots, x_n$ is defined as the difference between the largest and smallest measurements.

For the grade point data in Table 2.1, the measurements vary from 1.9 to 3.4. Hence, the range is $(3.4 - 1.9) = 1.5$. The range is easy to calculate, easy to interpret, and quite adequate as a measure of variation for small sets of data. But for large sets, it can be insensitive to substantial differences in data variation. This insensitivity is illustrated by the two relative frequency distributions shown in Figure 2.6. Both distributions have the same range, but the data of Figure 2.6(b) are more variable than the data of Figure 2.6(a).

Figure 2.6
Distributions with equal ranges and unequal variability

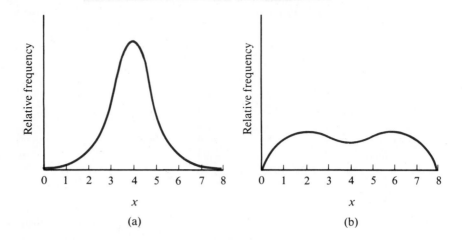

(a)   (b)

Can we find a measure of variability that is more sensitive than the range? Consider, as an example, the sample measurements 5, 7, 1, 2, 4. We can depict these data graphically, as in Figure 2.7, by showing the measurements as dots falling along the x axis. Figure 2.7 is called a **dot diagram**.

Figure 2.7
Dot diagram

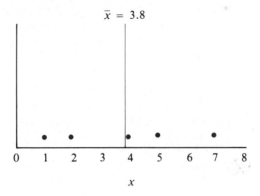

Calculating the mean as the measure of central tendency, we obtain

$$\bar{x} = \frac{\sum\limits_{i=1}^{n} x_i}{n} = \frac{19}{5} = 3.8$$

and we locate $\bar{x}$ on the dot diagram. We can now view variability in terms of distance between each dot (measurement) and the mean $\bar{x}$. If the distances are large, we can say that the data are more variable than if the distances are small. More explicitly, we define the **deviation** of a measurement from its mean to be the quantity $(x_i - \bar{x})$. Note that measurements to the right of the mean produce positive deviations, and those to the left produce negative deviations. The values of x and the deviations for our example are shown in the first and second columns of Table 2.6.

Table 2.6
Computation of
$$\sum_{i=1}^{n} (x_i - \bar{x})^2$$

x_i	$(x_i - \bar{x})$	$(x_i - \bar{x})^2$
5	1.2	1.44
7	3.2	10.24
1	−2.8	7.84
2	−1.8	3.24
4	.2	.04
19	0.0	22.80

If we now agree that deviations contain information on variation, our next step is to construct a formula based on the deviations that will provide a good measure of variation. As a first possibility, we might choose the average of the deviations. Unfortunately, the average will not work, because some of the deviations are positive, some are negative, and the sum is always zero (unless round-off errors have been introduced into the calculations). Note that the deviations in the second column of Table 2.6 sum to zero.

You may have observed an easy solution to this problem. Why not calculate the average of the absolute values* of the deviations? This method has, in fact, been employed as a measure of variability in exploratory data analysis. We prefer to overcome the difficulty caused by the sign of the deviations by working with the sum of their squares,

$$\sum_{i=1}^{n} (x_i - \bar{x})^2$$

For a fixed number of measurements this quantity will be relatively large for highly variable data and relatively small for less variable data.

Definition

The **variance of a population** of N measurements $x_1, x_2, \ldots, x_N$ is defined to be the average of the squares of the deviations of the measurements about their mean μ. The population variance is denoted by σ^2 (σ is the Greek lowercase letter sigma) and is given by the formula

$$\sigma^2 = \frac{1}{N} \sum_{i=1}^{N} (x_i - \mu)^2$$

* The absolute value of a number is its magnitude, ignoring its sign. For example, the absolute value of −2, represented by the symbol $|-2|$, is 2. The absolute value of 2, that is, $|2|$ is 2.

The letter N is used to denote the number of measurements in the population and n to denote the number of measurements in the sample.

Typically, we do not have all the population measurements available and must be satisfied with sample measurements selected from the population. Thus, we must use the **variance of a sample**, as defined next.

Definition ✪

> The **variance of a sample** of n measurements $x_1, x_2, \ldots, x_n$ is defined to be the sum of the squared deviations of the measurements about their mean $\bar{x}$ divided by $(n - 1)$. The sample variance is denoted by s^2 and is given by the formula
>
> $$s^2 = \frac{1}{n-1} \sum_{i=1}^{n} (x_i - \bar{x})^2$$

For example, we may calculate the variance for the set of $n = 5$ sample measurements presented in Table 2.6. The square of the deviation of each measurement is recorded in the third column of Table 2.6. Adding, we obtain

$$\sum_{i=1}^{5} (x_i - \bar{x})^2 = 22.80$$

The sample variance is

$$s^2 = \frac{1}{n-1} \sum_{i=1}^{n} (x_i - \bar{x})^2 = \frac{22.80}{4} = 5.70$$

You may wonder about the apparent inconsistency in the definitions of the population and sample variances. The sample mean $\bar{x}$ is used as an estimator of the population mean μ. Although it was not specifically stated, it is implied that the sample mean provides a good estimate of μ. In the same vein, we might reasonably assume that

$$s'^2 = \frac{1}{n} \sum_{i=1}^{n} (x_i - \bar{x})^2$$

would provide a good estimate of the population variance σ^2, based on a set of sample measurements. However, for small samples (n small) s'^2 tends to underestimate σ^2, and the sample variance s^2 provides better estimates of σ^2 than does s'^2. Note that s^2 and s'^2 differ only in the denominators and that when n is large, s'^2 and s^2 will be approximately equal. In later chapters we will have occasion to use an estimator of the population variance σ^2. **In all our calculations we will use s^2 rather than s'^2 and refer to s^2 as the sample variance.***

At this point you may be understandably disappointed with the practical significance attached to variance as a measure of variability. Large variances imply a large amount of variation, but this statement only permits comparison of several sets of data. When we attempt to say something specific concerning a single set of data, we are at a loss. For example, what can be said about the variability of a set

---

* The quantity s^2 is properly referred to as the *estimator* of the population variance.

of data with a variance equal to 100? The question cannot be answered with the facts we have. We remedy this situation by introducing a new definition and, in Section 2.6, a theorem and a rule.

Definition

> The **standard deviation** of a set of n sample measurements $x_1, x_2, \ldots, x_n$ is equal to the positive square root of the variance.

The variance is measured in terms of the square of the original units of measurement. If the original measurements are in inches, the variance is expressed in square inches. Taking the square root of the variance, we obtain the standard deviation, which conveniently returns the measure of variability to the original units of measurement.

SYMBOLS

Sample Standard Deviation:　$s = \sqrt{\dfrac{\sum\limits_{i=1}^{n} (x_i - \bar{x})^2}{n - 1}}$

Population Standard Deviation:　$\sigma = \sqrt{\sigma^2}$

The population standard deviation is denoted by σ and is calculated as the positive square root of the population variance σ^2.

Now that we have defined the standard deviation, you might wonder why we bothered to define the variance in the first place. Actually, both the variance and the standard deviation play an important role in statistical inference.

▷ 2.6　ON THE PRACTICAL SIGNIFICANCE OF THE STANDARD DEVIATION

We now introduce a useful theorem developed by the Russian mathematician Tchebysheff. Proof of the theorem is not difficult, but we omit it from our discussion.

Theorem 2.1　**Tchebysheff's Theorem**
Given a number k greater than or equal to 1 and a set of n measurements $x_1, x_2, \ldots, x_n$, at least $(1 - 1/k^2)$ of the measurements will lie within k standard deviations of their mean.

Tchebysheff's Theorem applies to any set of measurements and can be used to describe either a sample or a population. We will use the notation appropriate for populations, but you should realize that we could just as easily use the mean and the standard deviation for the sample.

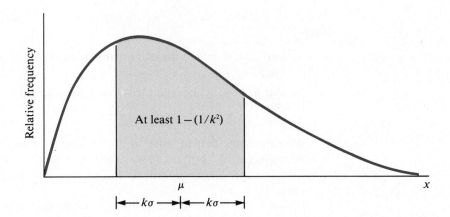

Figure 2.8
Illustrating
Tchebysheff's
Theorem

At least $1 - (1/k^2)$

The idea involved in Tchebysheff's Theorem is illustrated in Figure 2.8. An interval is constructed by measuring a distance $k\sigma$ on either side of the mean μ. Note that the theorem is true for any number we choose for k as long as it is greater than or equal to 1. Then, computing the fraction $[1 - (1/k^2)]$, we see that Tchebysheff's Theorem states that at least that fraction of the total number n of measurements lies in the constructed interval.

Let us choose a few numerical values for k and compute $[1 - (1/k^2)]$ (see Table 2.7). When $k = 1$, the theorem states that at least $1 - [1/(1)^2] = 0$ of the measurements lie in the interval from $(\mu - \sigma)$ to $(\mu + \sigma)$, a most unhelpful and uninformative result. However, when $k = 2$, we observe that at least $1 - [1/(2)^2] = 3/4$ of the measurements lie in the interval from $(\mu - 2\sigma)$ to $(\mu + 2\sigma)$. At least $8/9$ of the measurements lie within three standard deviations of the mean, that is, in the interval from $(\mu - 3\sigma)$ to $(\mu + 3\sigma)$. Although $k = 2$ and $k = 3$ are very useful in practice, k need not be an integer. For example, the fraction of measurements falling within $k = 2.5$ standard deviations of the mean is at least $1 - [1/(2.5)^2] = .84$.

Table 2.7
Illustrative values
of $[1 - (1/k^2)]$

k	$1 - (1/k^2)$
1	0
2	3/4
3	8/9

EXAMPLE 2.4   The mean and variance of a sample of $n = 25$ measurements are 75 and 100, respectively, Use Tchebysheff's Theorem to describe the distribution of measurements.

Solution   We are given $\bar{x} = 75$ and $s^2 = 100$. The standard deviation is $s = \sqrt{100} = 10$. The distribution of measurements is centered about $\bar{x} = 75$, and Tchebysheff's Theorem states:

a. *At least* 3/4 of the 25 measurements lie in the interval $\bar{x} \pm 2s = 75 \pm 2(10)$, that is, 55 to 95.

b. *At least* 8/9 of the measurements lie in the interval $\bar{x} \pm 3s = 75 \pm 3(10)$, that is, 45 to 105.

We emphasize the "at least" in Tchebysheff's Theorem because the theorem is very conservative, applying to *any* distribution of measurements. In most situations the fraction of measurements falling into the specified interval will exceed $[1 - (1/k^2)]$.

We now state a rule that describes accurately the variability of a particular bell-shaped distribution and describes reasonably well the variability of other mound-shaped distributions of data. The frequent occurrence of mound-shaped and bell-shaped distributions of data in nature—hence, the applicability of our rule—leads us to call it the Empirical Rule.

EMPIRICAL RULE

Given a distribution of measurements that is approximately bell-shaped (see Figure 2.9), the interval

- $(\mu \pm \sigma)$ contains approximately 68% of the measurements
- $(\mu \pm 2\sigma)$ contains approximately 95% of the measurements
- $(\mu \pm 3\sigma)$ contains all or almost all of the measurements

The bell-shaped distribution shown in Figure 2.9 is commonly known as the **normal distribution** and will be discussed in detail in Chapter 5. The Empirical Rule applies exactly to data that possess a normal distribution, but it also provides an excellent description of variation for many other types of data.

Figure 2.9
Normal (bell-shaped)
distribution

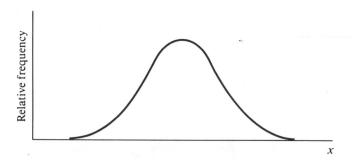

EXAMPLE 2.5   A time study is conducted to determine the length of time necessary to perform a specified operation in a manufacturing plant. The length of time necessary to

complete the operation is measured for each of $n = 40$ workers. The mean and standard deviation are found to be 12.8 and 1.7, respectively. Describe the sample data by using the Empirical Rule.

Solution   To describe the data, we calculate the intervals

$$(\bar{x} \pm s) = 12.8 \pm 1.7 \qquad \text{or} \qquad 11.1 \text{ to } 14.5$$

$$(\bar{x} \pm 2s) = 12.8 \pm 2(1.7) \qquad \text{or} \qquad 9.4 \text{ to } 16.2$$

$$(\bar{x} \pm 3s) = 12.8 \pm 3(1.7) \qquad \text{or} \qquad 7.7 \text{ to } 17.9$$

According to the Empirical Rule, we expect approximately 68% of the measurements to fall into the interval from 11.1 to 14.5, approximately 95% to fall into the interval from 9.4 to 16.2, and all or almost all to fall into the interval from 7.7 to 17.9.

 If we doubt that the distribution of measurements is mound-shaped, or wish for some other reason to be conservative, we can apply Tchebysheff's Theorem and be absolutely certain of our statements. Tchebysheff's Theorem tells us that at least 3/4 of the measurements fall into the interval from 9.4 to 16.2 and at least 8/9 into the interval from 7.7 to 17.9.

Before leaving this topic, we might ask how well the Empirical Rule applies to the grade point data of Table 2.1. We have already shown that the mean is $\bar{x} = 2.58$, and in Section 2.7 we will show that the standard deviation for the $n = 30$ measurements is $s = .37$. The appropriate intervals are calculated and the number of measurements falling in each interval are recorded. The results are shown in Table 2.8 with k in the first column and the interval $\bar{x} \pm ks$ in the second column, using $\bar{x} = 2.58$ and $s = .37$. The frequency, or number of measurements falling in each interval, is given in the third column, and the relative frequency in the fourth column. Note that the observed relative frequencies are quite close to the relative frequencies specified in the Empirical Rule.

Table 2.8

Frequency of measurements lying within k standard deviations of the mean for the data from Table 2.1

k	Interval $\bar{x} \pm ks$	Frequency in interval	Relative frequency
1	2.21–2.95	19	.63
2	1.84–3.32	29	.97
3	1.47–3.69	30	1.0

Another way to see how well the Empirical Rule and Tchebysheff's Theorem apply to the grade point data is to mark off the intervals $\bar{x} \pm s, \bar{x} \pm 2s$, and $\bar{x} \pm 3s$ on the relative frequency histogram for the data. This is shown in Figure 2.10. Now recall that the area under the histogram over an interval is proportional to the

Figure 2.10

Histogram for the grade point data (Figure 2.1) with intervals $\bar{x} \pm s$, $\bar{x} \pm 2s$, and $\bar{x} \pm 3s$ superimposed

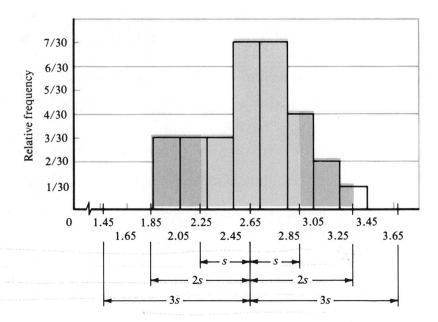

number of measurements falling in the interval and visually observe the proportion of the area above the interval $\bar{x} \pm s$. You will observe that this proportion is near the .68 specified by the Empirical Rule. Similarly, you will note that almost all of the area lies above the interval $\bar{x} \pm 2s$. Clearly, both the Empirical Rule and Tchebysheff's Theorem, using $\bar{x}$ and s, provide a good description for the grade point data.

To conclude, note that Tchebysheff's Theorem can be proved mathematically, and that it applies to any set of data. It gives a *lower* bound to the fraction of measurements to be found in an interval $\bar{x} \pm ks$, where k is some number greater than or equal to 1. In contrast, the Empirical Rule is an arbitrary statement about the behavior of data, a "rule of thumb." Although the percentages contained in the rule come from the area under the normal curve (Figure 2.9), the same percentages hold approximately for distributions with varying shapes as long as they tend to be roughly mound-shaped (the data tend to pile up near the center of the distribution).

EXERCISES

Basic Techniques

2.16　Given $n = 5$ measurements, 0, 5, 1, 1, 3, find
　　a.　$\bar{x}$　　　　　　　　b.　m　　　　　　　　c.　s

2.17　Given $n = 8$ measurements, 3, 1, 5, 6, 4, 4, 3, 5, find
　　a.　the range　　　　　b.　$\bar{x}$　　　　　　　c.　s^2

2.18　Given $n = 10$ measurements, 3, 5, 4, 6, 10, 5, 6, 9, 2, 8, find
　　a.　$\bar{x}$　　　　　　　　b.　m　　　　　　　　c.　s

2.19 Suppose that you want to create a mental picture of the relative frequency histogram for a large data set consisting of 1000 observations and that you know that the mean and standard deviation of the data set are equal to 36 and 3, respectively.

a. If you are fairly certain that the relative frequency distribution of the data is mound-shaped, how might you picture the relative frequency distribution? (*Hint:* Use the Empirical Rule.)

b. If you have no prior information concerning the shape of the relative frequency distribution, what can you say about the relative frequency histogram? (*Hint:* Construct intervals $\bar{x} \pm ks$ for several choices of k.)

Applications

2.20 The average price of a new car has nearly doubled in the past ten years, reaching a high of $14,387 in 1988 (*USA Today*, July 6, 1989). Suppose that you were interested in purchasing a new car in 1988, and were concerned with the distribution of new car costs. If the standard deviation of the distribution were approximately $4000, within what limits would most car costs fall?

2.21 In Exercise 2.20, we stated that the average price of a new car in 1988 was $14,387. Visualize the distribution of new car costs in 1988 and answer the following questions.

a. Do you think that the distribution of car costs is symmetrical about the mean, or is it skewed? Explain.

b. If you were choosing a measure of central tendency to locate the center of the distribution, would you choose the mean or the median? Explain.

2.22 The typical American household is shrinking in size, according to a recent report by the Census Bureau (*Press Enterprise*, May 5, 1989). This decline reflects the fact that families are having fewer children, young people are setting up housekeeping on their own rather than living with their parents, and the growing number of elderly in the population are now maintaining their own households. As of July 1, 1988, the average population in the nation's 91.5 million households was 2.62 people, down from 2.75 people in 1980 and 5.79 people in 1790!

a. Based on these statistics and your general knowledge about the number of people in a household, describe the shape of the distribution of American household sizes.

b. Based on your answer to part (a), would you expect the mean household size to be greater than, less than, or equal to the median household size? Explain.

2.23 The cost of educating our children has gone up again. Although graduation rates and college entrance exam scores have stayed roughly the same, the average expenditure per student has increased from $3470 in 1985 to $3752 in 1986 (*New York Times*, February 26, 1988). Visualize the national distribution of expenditures per student in 1986 and suppose that the standard deviation of these expenditures is $1200.

a. Within what limits would you expect at least 3/4 of the expenditures to lie?

b. Within what limits would you expect at least 8/9 of the expenditures to lie?

2.24 Refer to Exercise 2.23.

a. Do you think that the distribution of expenditures per student is symmetric about the mean of $3752, or skewed? Explain.

b. If the distribution in part (a) is approximately mound-shaped, describe the distribution of expenditures per student using the Empirical Rule.

2.25 In an article entitled "You Aren't Paranoid If You Think Someone Eyes Your Every Move," the *Wall Street Journal* (March 19, 1985) notes that big business collects detailed statistics on your behavior. Jockey International knows how many undershorts you own; Frito-Lay, Inc. knows which you eat first—the broken pretzels in a pack or the whole ones; and, to get to specifics, Coca-Cola knows that you put 3.2 ice cubes in a glass. Have you ever put 3.2 ice cubes in a glass? What did the *Wall Street Journal* article mean by that statement?

2.26   Is your breathing rate normal? Actually, there is no standard breathing rate for humans. It can vary from as low as 4 breaths per minute to as high as 70 or 75 for a person engaged in strenuous exercise. Suppose that the resting breathing rates for college-age students have a relative frequency distribution that is mound-shaped with a mean equal to 12 and a standard deviation of 2.3 breaths per minute. What fraction of all students would have breathing rates in the interval

a.   9.7 to 14.3 breaths per minute
b.   7.4 to 16.6 breaths per minute
c.   more than 18.9 or less than 5.1 breaths per minute?

2.27   An analytical chemist wanted to determine the number of moles of cupric ions in a given volume of solution by electrolysis. The solution was partitioned into $n = 30$ portions of .2 milliliter each. Each of the $n = 30$ unknown portions was tested. The average number of moles of cupric ions for the $n = 30$ portions was found to be .17 mole; the standard deviation was .01 mole.

a.   Describe the distribution of the measurements for the $n = 30$ portions of the solution, using Tchebysheff's Theorem.
b.   Describe the distribution of the measurements for the $n = 30$ portions of the solution, using the Empirical Rule. (Would you expect the Empirical Rule to be suitable for describing these data?)
c.   Suppose that the chemist had only employed $n = 4$ portions of the solution for the experiment and obtained the readings .15, .19, .17, and .15. Would the Empirical Rule be suitable for describing the $n = 4$ measurements? Why?

2.28   According to the Environmental Protection Agency (EPA), chloroform, which in its gaseous form is suspected of being a cancer-causing agent, is present in small quantities in all of the country's 240,000 public water sources. If the mean and standard deviation of the amounts of chloroform present in the water sources are 34 and 53 micrograms per liter, respectively, describe the distribution for the population of all public water sources.

2.29   The table in Exercise 2.3 gives the per capita federal tax for each of the 50 states in fiscal year 1986.

a.   Find the average per capita federal tax for the entire United States.
b.   Find the median per capita federal tax for the 50 states and compare it with the mean calculated in part (a).
c.   Based on your comparison in part (b), would you conclude that the distribution of per capita federal taxes is skewed? Explain.

2.7   A SHORT METHOD FOR CALCULATING THE VARIANCE

The calculation of the variance and standard deviation of a set of measurements is difficult, regardless of the method employed, but it is particularly tedious if we proceed according to the definition, by calculating each deviation individually as shown in Table 2.6. However, the sum of squares of the deviations can always be obtained by using the alternative formula given in the following display.

SHORTCUT FORMULA FOR CALCULATING THE SUM OF SQUARES OF DEVIATIONS

$$\sum_{i=1}^{n} (x_i - \bar{x})^2 = \sum_{i=1}^{n} x_i^2 - \frac{\left(\sum_{i=1}^{n} x_i \right)^2}{n}$$

Table 2.9 presents the individual observations of Table 2.6 in the first column and the squares of these observations in the second column. It follows that

$$\sum_{i=1}^{n} (x_i - \bar{x})^2 = \sum_{i=1}^{n} x_i^2 - \frac{\left(\sum_{i=1}^{n} x_i \right)^2}{n}$$

$$= 95 - \frac{(19)^2}{5}$$

$$= 95 - \frac{361}{5}$$

$$= 95 - 72.2 = 22.8$$

Table 2.9
Table for simplified calculations: $\sum_{i=1}^{n} (x_i - \bar{x})^2$

x_i	x_i^2
5	25
7	49
1	1
2	4
4	16
19	95

Substituting this expression for the sum of squares of the deviations into the formula for s^2, we obtain an easy computational method for calculating s^2.

SHORTCUT FORMULA FOR CALCULATING s^2

$$s^2 = \frac{\sum_{i=1}^{n} (x_i - \bar{x})^2}{n - 1} = \frac{\sum_{i=1}^{n} x_i^2 - \frac{\left(\sum_{i=1}^{n} x_i \right)^2}{n}}{n - 1}$$

Most calculators with statistical capabilities have built-in programs that will calculate $\bar{x}$ and s or μ and σ by accumulating $\sum_{i=1}^{n} x_i$ and $\sum_{i=1}^{n} x_i^2$ and then using the shortcut formulas given in this section. The sample standard deviation key is usually marked with s or σ_{n-1}, and the population standard deviation key is usually marked with σ or σ_N. In using any calculator with these built-in functions, be sure that you know which calculation is being carried out by each function key. With some data, like the data in the next example, for which $\bar{x}$, s, and $\sum_{i=1}^{n} (x_i - \bar{x})^2$ are known, verifying which keys are used to calculate $\bar{x}$, s, and σ is a simple exercise.

EXAMPLE 2.6   Calculate the sample standard deviation for the $n = 30$ grade point averages in Table 2.1.

Solution   Using a calculator with built-in statistical functions, you can verify the following:

$$\sum_{i=1}^{n} x_i = 77.5$$

$$\sum_{i=1}^{n} x_i^2 = 204.19$$

Using the shortcut formula,

$$\sum_{i=1}^{n} (x_i - \bar{x})^2 = \sum_{i=1}^{n} x_i^2 - \frac{\left(\sum_{i=1}^{n} x_i\right)^2}{n}$$

$$= 204.19 - \frac{(77.5)^2}{30} = 204.19 - 200.21$$

$$= 3.98$$

It follows that the standard deviation is

$$s = \sqrt{\frac{\sum_{i=1}^{n} (x_i - \bar{x})^2}{n-1}} = \sqrt{\frac{3.98}{29}} = .37$$ ◁

▷   **2.8**   A CHECK ON THE CALCULATION OF s

Tchebysheff's Theorem and the Empirical Rule can be used to detect gross errors in the calculation of s. For example, we know that at least three-fourths or, in the case of a mound-shaped distribution, nearly 95% of a set of measurements will lie within two standard deviations of their mean. Consequently, most of the sample measurements will lie in the interval $\bar{x} \pm 2s$, and the range will approximately equal $4s$. This is, of course, a very rough approximation, but it can serve as a useful check that will detect large errors in the calculation of s. If we let R equal the range,

$$R \approx 4s$$

then s is approximately equal to $R/4$; that is,

$$s \approx \frac{R}{4}$$

The computed value of s using the shortcut formula should be of roughly the same order as the approximation.

EXAMPLE 2.7   Use the approximation above to check the calculation of s for Table 2.7.

Solution   The range of the five measurements is

$$R = 7 - 1 = 6$$

Then,

$$s \approx \frac{R}{4} = \frac{6}{4} = 1.5$$

This is of the same order as the calculated value $s = \sqrt{\dfrac{22.8}{4}} = 2.4$.   ◁

Note that the range approximation is not intended to provide an accurate value for s.* Rather, its purpose is to detect gross errors in calculating, such as the failure to divide the sum of squares of deviations by $(n - 1)$ or the failure to take the square root of s^2. Both errors yield solutions that are many times larger than the range approximation of s.

EXAMPLE 2.8   Use the range approximation to determine an approximate value for the standard deviation for the data in Table 2.1.

Solution   The range is $R = 3.4 - 1.9 = 1.5$. Then

$$s \approx \frac{R}{4} = \frac{1.5}{4} = .375$$

We have shown $s = .37$ for the data in Table 2.1. The approximation is very close to the actual value of s.   ◁

TIPS ON PROBLEM SOLVING

1. Be careful about rounding numbers. Carry your calculations of $\sum\limits_{i=1}^{n} (x_i - \bar{x})^2$ to at least six significant figures.

---

* The range for a sample of n measurements will depend on the sample size n. The larger the value of n, the more likely you will observe extremely large or small values of x. The range for large samples, say, $n = 50$ or more observations, may be as large as $6s$, while the range for small samples, say, $n = 5$ or less, may be as small or smaller than $2s$.

2. After you have calculated the standard deviation s for a set of data, compare its value with the range of the data. The Empirical Rule tells you that approximately 95% of the data should fall in the interval $\bar{x} \pm 2s$; that is, a very approximate value for the range will be $4s$. Consequently, a very rough rule of thumb is that

$$s \approx \frac{\text{range}}{4}$$

This crude check will help you to detect large errors—for example, failure to divide the sum of squares of deviations by $(n - 1)$ or failure to take the square root of s^2.

EXERCISES

Basic Techniques

2.30 Given $n = 8$ measurements, 4, 1, 3, 1, 3, 1, 2, 2:
   a. Calculate $\bar{x}$.
   b. Calculate s^2 and s using the formula given by the definition in Section 2.5.
   c. Use the shortcut formula in Section 2.7 to calculate s^2 and s. Compare the results with those found in part (b).

2.31 Given $n = 5$ measurements, 2, 1, 1, 3, 5:
   a. Calculate $\bar{x}$.
   b. Use the shortcut formula in Section 2.7 to calculate s^2 and s.
   c. Calculate a range estimate of s (Section 2.8) as a rough check on your calculations in part (b).

2.32 Calculate $\bar{x}$, m, and s for the data in Exercise 2.1. Then examine the relative frequency histogram that you constructed for those data. Do these numerical descriptive measures provide a good description of the histogram?

Applications

2.33 The length of time required for an automobile driver to respond to a particular emergency situation was recorded for $n = 10$ drivers. The times, in seconds, were .5, .8, 1.1, .7, .6, .9, .7, .8, .7, .8.
   a. Scan the data and use the procedure in Section 2.8 to find an approximate value for s. Use this value to check your calculations in part (b).
   b. Calculate the sample mean $\bar{x}$ and the standard deviation s. Compare with part (a).

2.34 Refer to Exercise 2.4. The average number of yards per carry for the top rushers in the National Football League are shown here in a different format.

5.3	3.2	3.6	3.7	4.1
5.1	3.6	4.4	4.2	4.0
4.3	3.0	3.4	3.6	4.0
5.0	3.3	4.7	4.3	3.8
5.3	3.8	4.4	3.4	3.8
3.5	3.9	6.1	3.0	4.1
3.8	4.7	5.0	4.1	3.1
4.4	7.8	4.3	3.3	

a. Find the mean and standard deviation of the data set.
b. Find the percentage of measurements in the intervals $\bar{x} \pm s$, $\bar{x} \pm 2s$, and $\bar{x} \pm 3s$.
c. How do the percentages obtained in part (b) compare with those given by the Empirical Rule? Explain.

2.35 A report by the Department of Education (*New York Times*, February 26, 1988) revealed that teacher salaries in the United States rose 5.3% in 1987, to an average of $26,551. The highest statewide average, $43,970 per year, was reported in Alaska, while the lowest average, $18,871 per year, was reported in South Dakota.
a. Use the range approximation to estimate the standard deviation of the average salaries by state.
b. If the distribution of statewide average salaries is approximately mound-shaped, what proportion of the states have average salaries between $20,276 and $39,101?

2.36 The number of television viewing hours per household and the prime viewing times are two factors that affect television advertising income. A random sample of 25 households in a particular viewing area produced the following estimates of viewing hours per household:

3.0	6.0	7.5	15.0	12.0
6.5	8.0	4.0	5.5	6.0
5.0	12.0	1.0	3.5	3.0
7.5	5.0	10.0	8.0	3.5
9.0	2.0	6.5	1.0	5.0

a. Scan the data and use the procedure in Section 2.8 to find an approximate value for *s*. Use this value to check your calculations in part (b).
b. Calculate the sample mean $\bar{x}$ and the sample standard deviation *s*. Compare *s* with the approximate value obtained in part (a).
c. Find the percentage of the viewing hours per household that fall in the interval $\bar{x} \pm 2s$. Compare with the corresponding percentage given by the Empirical Rule.

2.37 To estimate the amount of lumber in a tract of timber, an owner decided to count the number of trees with diameters exceeding 12 inches in randomly selected 50-by-50-foot squares. Seventy 50-by-50-foot squares were chosen and the selected trees were counted in each tract. The data are as follows:

7	8	7	10	4	8	6
9	6	4	9	10	9	8
3	9	5	9	9	8	7
10	2	7	4	8	5	10
9	6	8	8	8	7	8
6	11	9	11	7	7	11
10	8	8	5	9	9	8
8	9	10	7	7	7	5
8	7	9	9	6	8	9
5	8	8	7	9	13	8

    a.   Construct a relative frequency histogram to describe the data.

    b.   Calculate the sample mean $\bar{x}$ as an estimate of μ, the mean number of timber trees for all 50-by-50-foot squares in the tract.

    c.   Calculate s for the data. Construct the intervals $\bar{x} \pm s$, $\bar{x} \pm 2s$, and $\bar{x} \pm 3s$. Count the percentage of squares falling in each of the three intervals and compare with the corresponding percentages given by the Empirical Rule and Tchebysheff's Theorem.

2.38   Refer to Exercise 2.2. The length of time (in months) between the onset of a particular illness and its recurrence are reproduced below.

2.1	4.4	2.7	32.3	9.9
9.0	2.0	6.6	3.9	1.6
14.7	9.6	16.7	7.4	8.2
19.2	6.9	4.3	3.3	1.2
4.1	18.4	.2	6.1	13.5
7.4	.2	8.3	.3	1.3
14.1	1.0	2.4	2.4	18.0
8.7	24.0	1.4	8.2	5.8
1.6	3.5	11.4	18.0	26.7
3.7	12.6	23.1	5.6	.4

    a.   Find the range.

    b.   Use the range approximation to find an approximate value for s.

    c.   Compute s for the data and compare with your approximation from part (b).

2.39   Refer to Exercise 2.38.

    a.   Examine the data and count the number of observations falling in the interval $\bar{x} \pm s$, $\bar{x} \pm 2s$, and $\bar{x} \pm 3s$.

    b.   Do the percentages falling in these intervals agree with Tchebysheff's Rule? the Empirical Rule?

    c.   Why might the Empirical Rule be unsuitable for describing these data?

2.40   Suppose that some measurements occur more than once and that the data $x_1, x_2, \ldots, x_k$ are arranged in a frequency table as shown here.

Observations	Frequency f_i
x_1	f_1
x_2	f_2
$\vdots$	$\vdots$
x_k	f_k
	n

Then,

$$\bar{x} = \frac{\sum\limits_{i=1}^{k} x_i f_i}{n}, \quad \text{where} \quad n = \sum_{i=1}^{k} f_i$$

and

$$s^2 = \frac{\displaystyle\sum_{i=1}^{k} x_i^2 f_i - \frac{\left\{\displaystyle\sum_{i=1}^{k} x_i f_i\right\}^2}{n}}{n-1}$$

Although these formulas for grouped data are primarily of value when you have a large number of measurements, demonstrate their use for the sample 1, 0, 0, 1, 3, 1, 3, 2, 3, 0, 0, 1, 1, 3, 2.

a.   Calculate $\bar{x}$ and s^2 directly, using the formulas for ungrouped data.
b.   The frequency table for the $n = 15$ measurements is

x	f(x)
0	4
1	5
2	2
3	4
	$n = 15$

Calculate $\bar{x}$ and s^2 using the formulas for grouped data. Compare with your answers to part (a).

2.41   The International Baccalaureate (IB) program is an accelerated academic program offered at a growing number of high schools throughout the country. Students enrolled in this program are enrolled in accelerated or advanced placement courses, and must take IB examinations in each of six subject areas at the end of their junior or senior year. Students are scored on a scale of 1–7, with 1–2 being poor, 3 mediocre, 4 average, and 5–7 excellent. During its first year of operation at John W. North High School in Riverside, California, 17 juniors attempted the IB economics exam with the following results:

Exam grade	Number of students
7	1
6	4
5	4
4	4
3	4

Calculate the mean and standard deviation for these scores.

2.42   To illustrate the utility of the Empirical Rule, consider a distribution that is heavily skewed to the right, as shown in the figure on page 48.

a.   Calculate $\bar{x}$ and s for the data shown. (*Note:* There are 10 zeros, 5 ones, and so on.)
b.   Construct the intervals $\bar{x} \pm s$, $\bar{x} \pm 2s$, and $\bar{x} \pm 3s$ and locate them on the frequency distribution.
c.   Calculate the proportion of the $n = 25$ measurements falling in each of the three intervals. Compare with Tchebysheff's Theorem and the Empirical Rule. Note that while

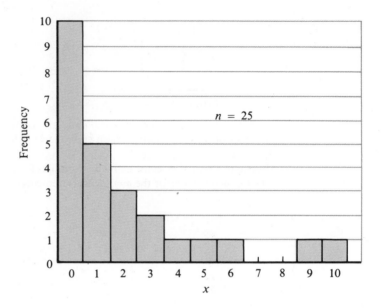

the proportion falling in the interval $\bar{x} \pm s$ does not agree closely with the Empirical Rule, the proportions falling in the intervals $\bar{x} \pm 2s$ and $\bar{x} \pm 3s$ agree very well. Many times this is true, even for non-mound-shaped distributions of data.

2.43   Refer to your sample of 50 male systolic blood pressures given in Exercise 2.7.
     a.   Calculate a range estimate of s to use as a check against the calculation of s.
     b.   Calculate $\bar{x}$ and s for the data. Check the value of s against part (a).
     c.   Calculate the percentages of observations in the intervals $\bar{x} \pm s$, $\bar{x} \pm 2s$, and $\bar{x} \pm 3s$. Compare with the Empirical Rule.

▷ **2.9** MEASURES OF RELATIVE STANDING

Sometimes we want to know the position of one observation relative to others in a set of data. For example, if you took an examination based on a total of 35 points, you might want to know how your score of 30 compared to the scores of the other students in the class. The mean and standard deviation of the scores can be used to calculate a z-score, which will measure the relative standing of a measurement in a data set.

Definition   The **sample z-score** is a measure of relative standing defined by

$$z\text{-score} = \frac{x - \bar{x}}{s}$$

A **z-score measures the distance between an observation and the mean, measured in units of standard deviation.** For example, suppose we know that the mean and standard deviation of the test scores (based on a total of 35 points) are 25 and 4, respectively. The z-score for your score of 30 is calculated as follows:

$$z\text{-score} = \frac{x - \bar{x}}{s} = \frac{30 - 25}{4} = 1.25$$

Your score of 30 lies 1.25 standard deviations above the mean ($30 = \bar{x} + 1.25s$).

Sample z-scores by themselves may not convey a strong sense of relative position other than to indicate that a score is either above or below the mean of the data set. However, for any data set, the z-score used in conjunction with Tchebysheff's Theorem allows us to make some conservative statements about the relative standing of an observation. Moreover, for mound-shaped data, the Empirical Rule can be used in conjunction with the z-score to make more powerful statements. Since at least 75%, and more likely 95%, of the observations lie within two standard deviations of their mean, z-scores between -2 and $+2$ are highly likely. Therefore, the exam score of 30 with $z = 1.25$ is not unusually high or unusually low. At least 89%, and more likely all, of the observations lie within three standard deviations of their mean, so that z-scores exceeding 3 in absolute value are very unlikely. The experimenter should closely examine any observation that has a z-score exceeding 3 in absolute value. Perhaps the measurement was recorded incorrectly or does not belong to the population being sampled. Perhaps it is just a highly unlikely observation. Such an unusually large or small observation is called an **outlier**.

EXAMPLE 2.9 Consider a sample of $n = 10$ measurements:

   1, 1, 0, 15, 2, 3, 4, 0, 1, 3

The measurement $x = 15$ appears to be unusually large. Calculate the z-score for this observation and state your conclusions.

Solution For the sample,

$$\sum_{i=1}^{10} x_i = 30 \quad \text{and} \quad \sum_{i=1}^{10} x_i^2 = 266$$

Then,

$$\bar{x} = \frac{\sum_{i=1}^{10} x_i}{n} = \frac{30}{10} = 3.0$$

$$\sum_{i=1}^{n} (x_i - \bar{x})^2 = \sum_{i=1}^{n} x_i^2 - \frac{\left(\sum_{i=1}^{n} x_i\right)^2}{n}$$

$$= 266 - \frac{(30)^2}{10} = 176$$

$$s^2 = \frac{\sum_{i=1}^{n} (x_i - \bar{x})^2}{n - 1} = \frac{176}{9} = 19.5556 \quad \text{and} \quad s = 4.42$$

The z-score for the suspected outlier, $x = 15$, is calculated as

$$z\text{-score} = \frac{x - \bar{x}}{s} = \frac{15 - 3}{4.42} = 2.71$$

Hence, the measurement $x = 15$ lies 2.71 standard deviations above the sample mean, $\bar{x} = 3.0$. Although the z-score does not exceed 3, it is close enough so that we suspect that $x = 15$ is an outlier. We should examine our sampling procedure to see whether $x = 15$ is a faulty observation.

A **percentile** is another measure of relative standing that is most often used for large data sets. (Percentiles are not very useful for small data sets.)

Definition

> Let $x_1, x_2, \ldots, x_n$ be a set of n measurements arranged in order of magnitude. The **pth percentile** is the value of x that exceeds $p\%$ of the measurements and is less than the remaining $(100 - p)\%$.

EXAMPLE 2.10

Suppose that you have been notified that your score of 610 on the Verbal Graduate Record Examination placed you at the 60th percentile in the distribution of scores. Where does your score of 610 stand in relation to the scores of others who took the examination?

Solution

Scoring at the 60th percentile means that 60% of all the examination scores were lower than your score and 40% were higher.

Viewed graphically, a particular percentile—say the 60th percentile—is a point on the x axis located so that 60% of the area under the relative frequency histogram for the data lies to the left of the 60th percentile (see Figure 2.11) and 40% of the area lies to the right. Thus, by our definition, the median of a set of data is the 50th percentile because half of the measurements in a data set are smaller than the median and half are larger.

The 25th and 75th percentiles, called the **lower** and **upper quartiles**, along with the median (the 50th percentile), locate points that divide the data into four sets of

Figure 2.11
The 60th percentile shown on the relative frequency histogram for a data set

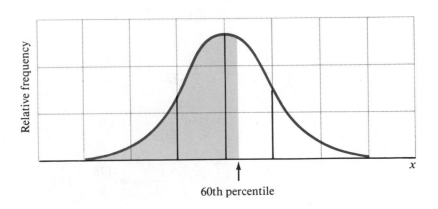

60th percentile

Figure 2.12
Location of quartiles

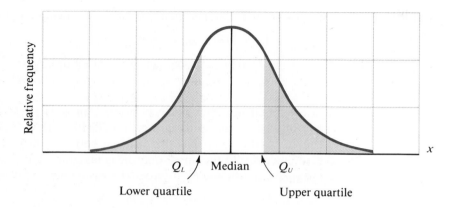

equal number. Twenty-five percent of the measurements will be less than the lower (first) quartile, 50% will be less than the median (the second quartile), and 75% will be less than the upper (third) quartile. Thus, the median and the lower and upper quartiles are located at points on the x axis so that the area under the relative frequency histogram for the data is partitioned into four equal areas, as shown in Figure 2.12.

Definition

> Let $x_1, x_2, \ldots, x_n$ be a set of n measurements arranged in order of magnitude. The **lower quartile (first quartile)**, Q_L, is the value of x that exceeds one-fourth of the measurements and is less than the remaining three-fourths. The **second quartile** is the median. The **upper quartile (third quartile)**, Q_U, is the value of x that exceeds three-fourths of the measurements and is less than one-fourth.

For small sets of data, a quartile may fall between two observations, in which case many numbers may satisfy the preceding definition. To avoid this ambiguity, we will use the following rules to locate the sample quartiles.

RULES FOR CALCULATING SAMPLE QUARTILES

1. When the measurements $x_1, x_2, \ldots, x_n$ are arranged in order of magnitude, the **lower quartile**, Q_L, is the value of x in position $.25(n + 1)$, and the **upper quartile**, Q_U, is the value of x in position $.75(n + 1)$.

2. When $.25(n + 1)$ and $.75(n + 1)$ are not integers, the quartiles are found by interpolation, using the values in the two adjacent positions.*

We demonstrate the procedure with the next example.

---

* This definition of quartiles is consistent with the one used in the MINITAB package. However, some texts use ordinary rounding when finding quartile positions, while others compute a sample quartile as the average of the values in adjacent positions only when $.25(n + 1)$ and $.75(n + 1)$ end in the fraction .5.

EXAMPLE 2.11   Find the lower and upper quartiles for the following set of measurements:

16, 25, 4, 18, 11, 13, 20, 8, 11, 9

Solution   Rank the $n = 10$ measurements from smallest to largest.

4, 8, 9, 11, 11, 13, 16, 18, 20, 25

The lower quartile is the value in position $.25(n + 1) = .25(10 + 1) = 2.75$, and the upper quartile is the value in position $.75(10 + 1) = 8.25$. The lower quartile is taken to be the value 3/4 the distance between the second and third ordered measurements, and the upper quartile is taken to be the value 1/4 of the distance between the eighth and ninth ordered measurements. Therefore,

$$Q_L = 8 + .75(9 - 8) = 8 + .75 = 8.75$$

and

$$Q_U = 18 + .25(20 - 18) = 18 + .5 = 18.5$$ ◁

Many of the numerical descriptive measures that we have discussed are easily found by using the command DESCRIBE in the MINITAB package. As part of its output, this program produces the values of the mean, the standard deviation, the median, and the lower and upper quartiles, as well as the values of some other statistics that we have not discussed. The **trimmed mean** (given as TMEAN) is the mean of the middle 90% of the measurements after excluding the smallest 5% and the largest 5%. Unlike the ordinary arithmetic mean, the trimmed mean is not sensitive to extremely large or extremely small values in the data set. Using the command DESCRIBE to summarize the data in Example 2.11 produced the computer output shown in Table 2.10. Notice that the quartiles are identical to those calculated in that example.

Table 2.10
MINITAB output using the command DESCRIBE for the data in Example 2.11

MTB > DESCRIBE C1

	N	MEAN	MEDIAN	TRMEAN	STDEV	SEMEAN
C1	10	13.50	12.00	13.25	6.28	1.98

	MIN	MAX	Q1	Q3
C1	4.00	25.00	8.75	18.50

EXERCISES

Basic Techniques

2.44   Find the median and the lower and upper quartiles for the following data:

8, 7, 1, 4, 6, 6, 4, 5, 7, 6, 3, 0

2.45   Refer to the data in Exercise 2.44.
a.   Calculate $\bar{x}$ and s.

b. Calculate the z-score for the smallest and largest observations. Are either of these observations unusually large or unusually small?

2.46 Find the median and the lower and upper quartiles for the following data:

19, 12, 16, 0, 14, 9, 6, 1, 12, 13, 10, 19, 7, 5, 8

Applications

2.47 The table in Exercise 2.3 gives the per capita federal tax for each of the 50 states in fiscal year 1986.
  a. Find the median (see Exercise 2.29) and the lower and upper quartiles for the data set.
  b. Are there any states that pay unusually high or unusually low per capita taxes? What are the z-scores associated with these states' taxes?

2.48 Find the median and the lower and upper quartiles for the data on times until the recurrence of an illness in Exercise 2.2. Use these descriptive measures to construct a mental image of the relative frequency histogram for the data. Compare your visualization with the relative frequency histogram that you constructed in Exercise 2.2.

2.49 If you scored at the 69th percentile on a placement test, how would your score stand in relation to the others?

2.50 According to researchers Edward Lazear and Robert Michael, the allocation of family income among all household members is far from uniform. In fact, about 2 1/2 times as much is spent on adult household members as on children (*Wall Street Journal*, December 8, 1988). The article states: "About one in 10 households probably spends less than $20 a child for every $100 spent on each adult; another 10% spend more than $55 a child." Identify any percentiles for the distribution of amounts spent per child for every $100 spent per adult.

2.51 An article in *American Demographics* (Schreiner, T. and Hamel, R., "Upstate, Downstate," September, 1989, p. 60) provides some interesting statistics on differences between residents of San Francisco and residents of Los Angeles. A survey of shoppers in the two areas showed that Los Angeles residents have a median income of $35,000 and that more than 10% have an income of $75,000 or more. The average income of San Franciscans is estimated at $32,000, while 7% have incomes in the $75,000- plus range. Identify any percentiles that can be determined from this information.

2.52 Refer to Exercise 2.51. The survey of shoppers also revealed that 16% of San Franciscans are age 65 or older, compared with 13% in Los Angeles. In the distribution of ages for residents of San Francisco, at what percentile does age 65 lie? For the residents of Los Angeles, at what percentile does age 65 lie?

2.53 In recent years, increasing concern has begun to surface over the amount of toxic chemical emissions that are presently being pumped into the environment. All emissions data must be carefully examined, however, to determine whether the figures are biased by the inclusion or exclusion of certain information. The table on page 54 gives the Environmental Protection Agency's (EPA's) top ten counties for toxic waste in 1987, along with a *USA Today* computer analysis of the same data. The EPA reports all toxic emissions including sodium sulfate, while the *USA Today* reports only toxic emissions into the environment (not waste treated or hauled away) excluding sodium sulfate, which is now considered nontoxic. Emissions are reported in millions of pounds.
  a. Find the mean and standard deviation of the EPA emissions for the ten counties.
  b. Calculate the z-score for the emissions reported by the EPA for San Bernardino County. Is this an unusually high amount?

County	EPA	USA Today
San Bernardino, CA	5200	22.5
Harris, TX	612	78.3
Calhoun, TX	581	579.2
Wayne, MI	512	17.7
Harrison, MS	423	53.0
Brazoria, TX	404	197.9
St. James, LA	349	312.5
Milam, TX	329	329.1
Jefferson, TX	309	198.0
St. Charles, LA	284	209.1

Source: Larry Sanders, *USA Today* database research, "EPA report spurs confusion." Copyright 1989, *USA Today*. Excerpted with permission.

c. If you were told that an industrial plant in San Bernardino County emitted a large amount of sodium sulfate, would this explain the difference in the emissions amounts for San Bernardino County as reported by the EPA and *USA Today*?

2.10   THE BOX PLOT

The **box plot** is another technique used in explanatory data analysis (EDA), which can be used to describe not only the behavior of the measurements in the middle of the distribution but also their behavior at the ends or tails of the distribution. Values that lie very far from the middle of the distribution in either direction are called **outliers**. An outlier may result from transposing digits when recording a measurement, or from incorrectly reading an instrument dial, or perhaps from a malfunctioning piece of equipment, and so on. Even when there are no recording or observational errors, a data set may contain one or more valid measurements that, for one reason or another, differ markedly from the others in the set. These outliers can cause a marked distortion in the values of commonly used numerical measures such as $\bar{x}$ and s. In fact, outliers may themselves contain important information not shared with the other measurements in the set. Therefore, isolating outliers, if they are present, is an important step in any preliminary analysis of a data set. The box plot is designed expressly for this purpose.

A box plot is constructed by using the median and two other measures, known as **hinges**. The median divides the ordered data set into two halves; the hinges are the values in the middle of each half of the data. Hinges are very similar to quartiles and, in effect, serve the same purpose. The actual difference between the value of a hinge and a quartile is quite small and decreases as the number of measurements increases. Nevertheless, we retain the distinction between these two measures and use hinges in constructing a box plot. *However, we can think of a hinge as playing the role of a quartile.* Hinges are calculated using the following procedure.

CALCULATING HINGES

1. Calculate the position of the median, $(n + 1)/2$, and drop the fraction $1/2$ if there is one. This quantity is called $d(M)$ and measures the "depth" of the median from each end of the ordered measurements.

2. Calculate the position of the hinges as

$$\frac{d(M) + 1}{2}$$

The hinges are the values in position $[d(M) + 1]/2$ as measured from each end of the ordered data set.

For example, for a data set with $n = 10$, the position of the median is $(10 + 1)/2 = 5.5$, and $d(M) = 5$. The position of the hinges is then

$$\frac{d(M) + 1}{2} = \frac{5 + 1}{2} = 3$$

The hinges are those values in the third position, as measured from each end of the ten ordered measurements. If the position of the hinges ends in $1/2$, a hinge will be the average of the two adjacent values in the ordered data set.

The dispersion of the measurements is now measured in terms of the difference between the hinges, called the **H-spread**, which is approximately equal to $(Q_U - Q_L)$, the **interquartile range**. A data value will be identified as an outlier depending upon its relative position with respect to boundary points called inner and outer fences. The **inner fences** are defined as follows:

lower inner fence = lower hinge − 1.5(H-spread)

upper inner fence = upper hinge + 1.5(H-spread)

The **outer fences** are defined as follows:

lower outer fence = lower hinge − 3(H-spread)

upper outer fence = upper hinge + 3(H-spread)

The data values in each tail closest to, but still inside, the inner fences are called the **adjacent values**. Values lying between an inner fence and its neighboring outer fence are termed "outside" and are considered to be mild outliers. Values outside the outer fences are termed "far outside" and are considered to be extreme outliers. A box plot combines all this information in a pictorial display as follows.

CONSTRUCTING A BOX PLOT

1. Calculate the median and the upper and lower hinges, and locate them on a horizontal line representing the scale of measurement.

2. Draw a box whose ends are the upper and lower hinges. Draw a line through the box at the value of the median.

3. Calculate the **H-spread** = upper hinge − lower hinge. Then calculate the **inner and outer fences**. Determine suspect and extreme outliers, and locate them on the box plot.

4. Locate the **adjacent values** on the box plot. Draw a dashed line from the box to the corresponding adjacent values.

EXAMPLE 2.12 Construct a box plot for the data in Example 2.9.

Solution The $n = 10$ measurements in the sample, ranked from the smallest to the largest, are

$$0, 0, 1, 1, 1, 2, 3, 3, 4, 15$$

The position of the median is $(n + 1)/2 = 11/2 = 5.5$. Hence, the median is $(1 + 2)/2 = 1.5$ and the depth of the median is

$$d(M) = 5$$

The position of the hinges, then, is

$$\frac{d(M) + 1}{2} = \frac{5 + 1}{2} = 3$$

and the hinges are 1 and 3, the values in the third position as measured from each end of the ordered data set. The H-spread is $3 − 1 = 2$, and the outer and inner fences are calculated as

upper inner fence = $3 + 1.5(2) = 6$

lower inner fence = $1 − 1.5(2) = −2$

upper outer fence = $3 + 3(2) = 9$

lower outer fence = $1 − 3(2) = −5$

The value $x = 15$ lies outside the outer fences, and hence is considered an extreme outlier.

To construct a box plot for the data, a box is drawn whose ends are the upper and lower hinges, as shown in Figure 2.13. A line is drawn through the box at the value of the median. The outlier, $x = 15$, is plotted using a capital O. The adjacent values, lying just inside the inner fences are 0 and 4, and are connected to the box with a dashed line.

The box plot emphasizes the fact that the outlier lies far from the central 50% of the measurements lying between the hinges. The box plot also indicates that these

Figure 2.13
Box plot for the data in
Example 2.9

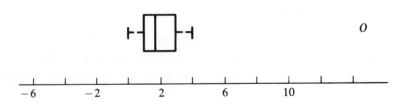

Table 2.11
MINITAB printout of
the box plot for the
data in Example 2.9

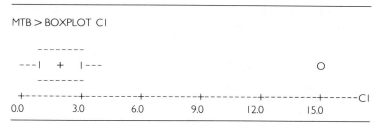

MTB > BOXPLOT C1

data are positively skewed (skewed to the right) since the median is not equally spaced between the two hinges, but rather, lies closer to the lower hinge.

For these same data the command BOXPLOT in the MINITAB package produced the box plot in Table 2.11. Notice that, except for scale, the box plots are identical and lead to the same conclusions concerning the outlier, $x = 15$. ◁

EXERCISES

Basic Techniques

2.54   Construct a box plot for the following data and identify any outliers.

25, 22, 26, 23, 27, 26, 28, 18, 25, 24, 12

2.55   Construct a box plot for the following data and identify any outliers.

3, 9, 10, 2, 6, 7, 5, 8, 6, 6, 4, 9, 22

Applications

 2.56   In Exercises 2.29 and 2.47 we found the median and the lower and upper quartiles for the per capita federal tax data of Exercise 2.3.
a.   Find the upper and lower hinges for the data. Compare the hinges to the upper and lower quartiles found in Exercise 2.47.
b.   Construct a box plot for the data.
c.   Are there any outliers? If so, how would you explain them?

 2.57   Environmental scientists are increasingly concerned with the accumulation of toxic elements in marine mammals and the transfer of such elements to the animals' offspring. The striped dolphin (*Stenella coeruleoalba*), considered to be the top predator in the marine food chain, was the subject of one such study. The mercury concentration (micrograms/gram) in the livers of 28 male striped dolphins are shown below.

1.70	183.00	221.00	286.00
1.72	168.00	406.00	315.00
8.80	218.00	252.00	241.00
5.90	180.00	329.00	397.00
101.00	264.00	316.00	209.00
85.40	481.00	445.00	314.00
118.00	485.00	278.00	318.00

a. Construct a box plot for the data.
b. Are there any outliers?
c. If you knew that the first four dolphins were all less than three years old, while all of the others were more than eight years old, would this information help explain the difference in the magnitude of those four observations? Explain.

 2.58 The total toxic emissions reported by the EPA for ten counties in the United States (Exercise 2.53) are reproduced below. Data are reported in millions of pounds.

5200, 612, 581, 512, 423, 404, 349, 329, 309, 284

a. Construct a box plot for the data and identify any outliers.
b. Do the results of part (a) agree with the results obtained using the z-score in Exercise 2.53?

2.11   SUMMARY

Methods for describing sets of measurements fall into one of two categories: graphical methods or numerical methods. The relative frequency histogram and the stem and leaf display are extremely useful graphical methods for characterizing a set of measurements. Numerical descriptive measures are numbers that attempt to create a mental image of the frequency histogram (or frequency distribution). We have restricted the discussion to measures of central tendency and variation, the most useful of which are the mean and standard deviation. While the mean possesses intuitive descriptive significance, the standard deviation is significant only when used in conjunction with Tchebysheff's Theorem and the Empirical Rule. The objective of sampling is the description of the population from which the sample was obtained. This objective is accomplished by using the sample mean $\bar{x}$ and the quantity s^2 as estimators of the population mean μ and variance σ^2.

Many descriptive methods and numerical measures have been presented in this chapter, but these constitute only a small percentage of those that might have been discussed. Many special computational techniques usually found in elementary texts have been omitted. Because the advent and common use of calculators and computers have minimized the importance of special computational formulas, we have chosen to focus on the main objective of modern statistics and this text—statistical inference.

2.12   MINITAB COMMANDS

At the end of this and other chapters, we will supply the MINITAB commands and subcommands used in implementing procedures and analyses introduced in the chapter.

In general, when one or more subcommands are to be used, the main command and subsequent subcommands are followed by a *semicolon*, and *the last subcommand is followed by a period*. In the presentation of a command and its admissible subcommands, the letter C represents a column number, K represents a constant,

M represents a matrix, and E represents one or more of the three preceding letters. For example, the command DESCRIBE is used to calculate descriptive measures for data stored in a column, whereas the command PRINT can be used to print a column of data, one or more stored constants, a matrix, or some combination of all three.

```
BOXPLOT C

    INCREMENT = K
    START      = K [end = K]
    BY C

    LINES      = K
    NOTCH      [K%] (sign confidence interval)
    LEVELS     K . . . K
```

```
HISTOGRAM   C . . . C

    INCREMENT = K
    START      = K [end = K]
    BY C
    SAME scales for all columns
```

```
STEM-AND-LEAF display of C . . . C

    TRIM outliers
    INCREMENT = K
    BY C
```

```
DESCRIBE C . . . C

    BY C
```

REFERENCES

Devore, J., and Peck, R. *Statistics: The Exploration and Analysis of Data.* St. Paul: West, 1986.

Dixon, W. J., et al. *BMDP Statistical Software Manual.* Vol. 1. Berkeley: University of California, 1988.

Freund, J. E. *Modern Elementary Statistics.* 6th ed. Englewood Cliffs, N.J.: Prentice-Hall, 1984.

Koopmans, L. H. *An Introduction to Contemporary Statistics.* 2d ed. Boston: Duxbury Press, 1987.

Neter, J., Wasserman, W., and Whitmore, G. A. *Applied Statistics.* 3d ed. Boston: Allyn and Bacon, 1987.

Ott, L. *An Introduction to Statistical Methods and Data Analysis.* 3d ed. Boston: PWS-KENT, 1988.

Rosner, B. *Fundamentals of Biostatistics.* 3d ed. Boston: PWS-KENT, 1990.

Ryan, T. A., Joiner, B. L., and Ryan, B. F. *Minitab Student Handbook.* 2d ed. Boston: Duxbury Press, 1985.

SPSS, Inc. Staff. *SPSS-X User's Guide.* 3d ed. New York: McGraw-Hill, 1987.

Tukey, J. W. *Exploratory Data Analysis.* Reading, Mass.: Addison-Wesley, 1977.

Velleman, P. F., and Hoaglin, D. C. *Applications, Basics, and Computing of Exploratory Data Analysis.* Boston: PWS-KENT, 1981.

SUPPLEMENTARY EXERCISES

2.59 Conduct the following experiment: toss ten coins and record x, the number of heads observed. Repeat this process $n = 50$ times, thus providing 50 values of x.

a. Construct a relative frequency histogram for these measurements.

b. Calculate $\bar{x}$, s^2, and s.

c. Find the proportion of measurements lying in the interval $\bar{x} \pm s$.

d. Find the proportion of measurements lying in the interval $\bar{x} \pm 2s$. Are these results consistent with Tchebysheff's Theorem?

e. Is the frequency histogram relatively mound-shaped? Does the Empirical Rule adequately describe the variability of the data?

2.60 During the 1988 presidential campaign, many different news agencies conducted weekly polls to determine the percentage of voters considered likely to vote for one candidate or another. During the early part of October, 1988, the following results were reported (*New York Times*, October 18, 1988):

News agency	Bush/Quayle (%)	Dukakis/Bentsen (%)
NBC/*Wall Street Journal*	55	38
ABC/*Washington Post*	50	47
Gallup	49	43
Los Angeles Times	44	41
New York Times/CBS News	47	42

a. Calculate $\bar{x}$, s^2, and s for the percent of voters likely to vote for the Bush/Quayle ticket based on these five polls.

b. Find the range of the Bush/Quayle percentages. Find the ratio of range to s. Does the range approximation provide a good estimate of s in this case?

c. Does the NBC/*Wall Street Journal* percentage appear to be unusually high? Calculate the z-score for this observation.

2.61 The following data represent the median sale price for existing single-family homes during the first quarters of 1987 and 1988 for 26 selected metropolitan areas in the United States.

City	1987	1988	City	1987	1988
Akron, OH	$ 54,700	$ 57,100	Milwaukee, WI	$ 67,800	$ 72,600
Anaheim, CA	158,400	183,800	New York, NY	169,400	186,600
Birmingham, AL	67,500	73,100	Omaha, NE	57,800	58,300
Chattanooga, TN	57,500	61,200	Philadelphia, PA	79,800	81,400
Cleveland, OH	62,900	65,300	Portland, OR	62,700	62,800
Detroit, MI	64,300	71,500	Providence, RI	101,300	123,300
El Paso, TX	58,000	57,800	St. Louis, MO	71,900	74,100
Grand Rapids, MI	51,700	55,400	Salt Lake City, UT	68,100	65,300
Honolulu, HI	174,000	198,400	San Antonio, TX	67,600	63,400
Jacksonville, FL	62,700	68,900	San Francisco, CA	159,000	178,800
Las Vegas, NV	76,100	75,700	Syracuse, NY	65,200	68,100
Los Angeles, CA	130,100	159,900	Toledo, OH	53,700	56,500
Louisville, KY	52,100	51,100	West Palm Beach, FL	96,900	94,500

Source: The World Almanac & Book of Facts, 1989 edition, copyright ©Newspaper Enterprise Association, Inc. 1988, New York, NY 10166.

a. Construct a relative frequency histogram for the median home price during the first quarter of 1988.

b. Calculate the range for the 1988 prices. Calculate the range approximation for s.

c. Calculate $\bar{x}$, s^2, and s. Compare the approximation to s that was obtained in part (b).

d. Find the proportion of the 1988 prices falling in the intervals $\bar{x} \pm s$, $\bar{x} \pm 2s$, and $\bar{x} \pm 3s$. Do these results agree with Tchebysheff's Theorem and the Empirical Rule?

2.62 Refer to Exercise 2.61. The **MINITAB** commands DESCRIBE and BOXPLOT produced the output shown below for the 1988 median prices.

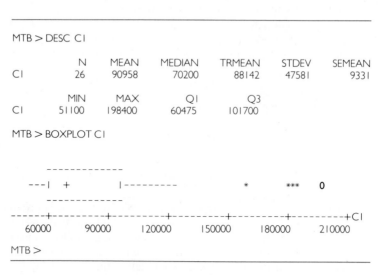

```
MTB > DESC C1

            N      MEAN    MEDIAN    TRMEAN    STDEV    SEMEAN
C1         26     90958     70200     88142    47581      9331

          MIN       MAX        Q1        Q3
C1      51100    198400     60475    101700

MTB > BOXPLOT C1


            -------------
    ---I  +            I---------            *       ***    0
            -------------

------+---------+---------+---------+---------+---------+C1
    60000     90000    120000    150000    180000    210000

MTB >
```

a. Compare the values for $\bar{x}$ and s with those obtained in Exercise 2.61.
b. What are the upper and lower quartiles for the data?
c. Are there any outliers in the data set? If so, how would you explain them?

2.63 Refer to Exercise 2.61. The **MINITAB** commands DESCRIBE and HISTOGRAM produced the output shown below for the 1987 median home prices.

```
MTB > DESC C3

            N      MEAN    MEDIAN    TRMEAN    STDEV    SEMEAN
C3         26     84277     67550     81896    39203      7688

          MIN       MAX        Q1        Q3
C3      51700    174000     57950     98000

MTB > HIST C3

Histogram of C3    N = 26

Midpoint    Count
   50000       4    ****
   60000       7    *******
   70000       6    ******
   80000       2    **
   90000       0
  100000       2    **
  110000       0
  120000       0
  130000       1    *
  140000       0
  150000       0
  160000       2    **
  170000       2    **
```

Compare the histogram for the 1987 prices generated by MINITAB with the relative frequency histogram that you constructed for the 1988 median home prices. Describe and compare the median home prices for these two years.

2.64 The range approximation for s can be improved if it is known that the sample is drawn from a bell-shaped distribution of data. Thus, the calculated s should not differ substantially from the range divided by the appropriate ratio given in the following table:

Number of measurements	5	10	25
Expected ratio of range to s	2.5	3	4

For the data in Exercise 2.60, estimate s as suggested. Compare this estimate with the calculated s.

2.65 In contrast to aptitude tests, which are predictive measures of what one can accomplish with training, achievement tests tell what an individual can do at the time of the test. Mathematics achievement test scores for 400 students were found to have a mean and a variance equal to 600 and 4900, respectively. If the distribution of test scores was mound-shaped, approximately how many of the scores would fall in the interval 530 to 670? Approximately how many scores would be expected to fall in the interval 460 to 740?

2.66 Petroleum pollution in seas and oceans stimulates the growth of some types of bacteria. A count of petroleumlytic microorganisms (bacteria per 100 milliliters) in ten portions of seawater gave the following readings:

49, 70, 54, 67, 59, 40, 61, 69, 71, 52

a.   Observe the data and guess the value for s by use of the range approximation.
b.   Calculate $\bar{x}$ and s and compare with the range approximation of part (a).

2.67 Why do statisticians generally prefer to divide the sum of squares of deviations of the sample measurements by $(n - 1)$ rather than n when estimating a population variance σ^2?

2.68 The percentage of city telephone subscribers who use unlisted numbers is on the increase. The distribution of the percentage of unlisted numbers in cities has a mean and standard deviation that are near 14% and 6%, respectively. If you were to pick a city at random, is it likely that the percentage of unlisted numbers would exceed 20%? Explain. Within what limits would you expect the percentage to fall?

2.69 An industrial concern uses an employee screening test with average score μ and standard deviation $\sigma = 10$. Assume that the test-score distribution is mound-shaped and that a score of 65 qualifies an applicant for further consideration. What is the value of μ such that approximately 2.5% of the applicants qualify for further consideration?

2.70 Attendances at a high school's basketball games were recorded and found to have a sample mean and variance of 420 and 25, respectively. Calculate $\bar{x} \pm s, \bar{x} \pm 2s$, and $\bar{x} \pm 3s$, and state the approximate fraction of measurements you would expect to fall in these intervals according to the Empirical Rule.

$\bar{x} \pm s$ _____ fraction_____

$\bar{x} \pm 2s$ _____ fraction_____

$\bar{x} \pm 3s$ _____ fraction_____

2.71 A study conducted by the National Center for Health Statistics (*Press Enterprise*, June 7, 1989) reports that the average length of all the marriages that ended in divorce in 1986 was 9.6 years. The study also found that "divorces were most likely to occur early in marriages, with 33.6 percent taking place between the first and fourth years of wedlock."

    a. Based on the information given above, describe the distribution of length of marriages for those marriages that ended in 1986. Is the distribution likely to be symmetric about the mean, or is it skewed?

    b. Suppose that the standard deviation of the marriage lengths is approximately 4.2 years. What proportion of marriages ending in 1986 lasted at least 18 years?

 2.72 The College Board's verbal and mathematics scholastic aptitude tests are scored on a scale of 200 to 800. Although originally designed to produce mean scores approximately equal to 500, the mean verbal and math scores in recent years have been as low as 463 and 493, respectively, and have been trending downward. It seems reasonable to assume that a distribution of all GRE scores, either verbal or math, is mound-shaped. If σ is the standard deviation of one of these distributions, what is the largest value (approximately) that σ might assume? Explain.

 2.73 The mean duration of television commercials on a given network is 75 seconds, with a standard deviation of 20 seconds. Assuming that duration times are approximately normally distributed:

    a. What is the approximate probability that a commercial will last less than 35 seconds?

    b. What is the approximate probability that a commercial will last longer than 55 seconds?

 2.74 A random sample of 100 foxes were examined by a team of veterinarians to determine the prevalence of a particular type of parasite. Counting the number of parasites per fox, the veterinarians found that 69 foxes had no parasites, 17 had one parasite, and so on. The following is a frequency tabulation of the data:

Number of parasites, x	0	1	2	3	4	5	6	7	8
Number of foxes, f	69	17	6	3	1	2	1	0	1

    a. Construct a relative frequency histogram for x, the number of parasites per fox.

    b. Calculate $\bar{x}$ and s for the sample.

    c. What fraction of the parasite counts fall within two standard deviations of the mean? three standard deviations? Do these results agree with Tchebysheff's Theorem? the Empirical Rule?

 2.75 Consider a population consisting of the number of teachers per college at small two-year colleges. Suppose that the number of teachers per college has an average $\mu = 175$ and a standard deviation $\sigma = 15$.

    a. Use Tchebysheff's Theorem to make a statement about the percentage of colleges that have between 145 and 205 teachers.

    b. Assume that the population is normally distributed. What fraction of the colleges have more than 190 teachers?

2.76 From the following data, a student calculated s to be .263. On what grounds might we doubt his accuracy? What is the correct value (to the nearest hundredth)?

17.2	17.1	17.0	17.1	16.9
17.0	17.1	17.0	17.3	17.2
17.1	17.0	17.1	16.9	17.0
17.1	17.3	17.2	17.4	17.1

2.77   In an article entitled "If Fido Takes a Piece of the Postman, Feds May Take a Chunk Out of You," the *Wall Street Journal* (July 10, 1981) notes that the Postal Service is helping letter carriers assemble evidence needed to seek damages from dog owners for the harmful antipostman behavior of their pets. The article notes that the mean cost in medical bills and lost time per dog bite is $300; but, of course, the actual cost can run much higher. Suppose that the distribution of cost per bite has a mean equal to $300 and a standard deviation equal to $200.

a.   Why might the relative frequency distribution of costs per dog bite be skewed to the right as shown below?

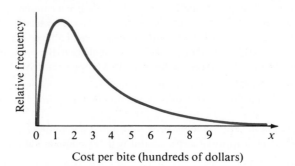

b.   Approximately what percentage of the costs per bite are less than $700?

HOW IS YOUR BLOOD PRESSURE?

Blood pressure is the pressure that the blood exerts against the walls of the arteries. When physicians or nurses measure your blood pressure, they take two readings. The **systolic** pressure is the pressure when the heart is contracting and therefore pumping. The **diastolic** blood pressure is the pressure in the arteries when the heart is relaxing. The diastolic blood pressure is always the lower of the two readings. Blood pressure varies from one person to another. It will also vary for a single individual from day to day and even within a given day.

If your blood pressure is too high, it can lead to a stroke or a heart attack. If it is too low, blood will not get to your extremities and you may feel dizzy. Low blood pressure is usually not serious.

So, what should *your* blood pressure be? A systolic blood pressure of 120 would be considered normal. One of 150 would be high. But since blood pressure varies with sex and increases with age, a better gauge of the relative standing of your blood pressure would be obtained by comparing it with the population of blood pressures of all persons of your sex and age in the United States. Of course, we cannot supply you with that data set, but we can show you a very large sample selected from it. Appendix I provides blood pressure data on 1911 persons, 966 males and 945 females between the ages of 15 and 20. The data are part of a health survey conducted by the National Institutes of Health (NIH) from 1976 to 1980.

Examine Appendix I and you will see that the persons in the survey are numbered from 1 to 1911. Entries for each person include the person's age and systolic and diastolic blood pressures at the time the blood pressure was recorded.*

1. Using the systolic blood pressure data set, construct a relative frequency histogram for the 965 males and another for the 945 females. Use a statistical software package if you have access to one. Compare the two histograms.

2. Describe the two data sets using numerical descriptive measures. Use these measures to compare the two sets of data.

3. How does your blood pressure compare to those in a comparable sex and age group? Use a measure of relative standing to determine whether your blood pressure is "normal" or whether it is too high or too low.

* A few of the readings are missing.

CHAPTER 3

◁ ◀ # PROBABILITY AND PROBABILITY DISTRIBUTIONS

Case Study

In his exciting novel *Congo* (New York: Alfred A. Knopf, 1980), author Michael Crichton describes an expedition racing to find boron-coated blue diamonds in the rain forests of eastern Zaire. Can probability help the heroine, Karen Ross, in her search for the Lost City of Zinj? The case study at the end of Chapter 3 involves Ross's use of probability in decision-making situations.

General Objectives

Now that you have learned to describe a data set, how can you use sample data to infer the nature of the sampled population? The technique is based on probability. Consequently, one objective of this chapter is to present the basic concepts of the theory of probability, including the following:

1. Experiments and events.
2. The methods for calculating the probability of an event, such as observing a particular sample outcome.

Most importantly, you will gain some insight into how to use the probabilities of sample outcomes to infer the nature of a sampled population. Most samples are collections of numerical measurements on a variable of interest. The variable measured is called a random variable because its values depend upon the chance selection of the elements in the sample. A second objective of this chapter is to identify types of random variables and to develop probability distributions that serve as models for random variables whose possible values constitute a countable set.

Specific Topics

▷ 3.1   THE ROLE OF PROBABILITY IN STATISTICS

Probability and statistics are related in a most curious way. **In essence, probability is the vehicle that enables the statistician to use information in a sample to make inferences or to describe the population from which the sample was obtained.** We can illustrate this relationship with a simple example.

Consider a balanced die with its familiar six faces. By balanced we mean that the chance of observing any one of the six sides on a single toss of the die is just as likely as observing any other side. Tossing the die can be viewed as an experiment that could conceivably be repeated a large number of times, thus generating a population of numbers where the measurement x would be either 1, 2, 3, 4, 5, or 6. Assume that the population is so large that each value of x occurs with equal frequency. Note that we do not actually generate the population; rather, it exists conceptually. Now let us toss the die once and observe the value of x. This one measurement represents a sample of $n = 1$ drawn from the population. What is the probability that the sample value of x will equal 2? Knowing the structure of the population, we realize that each value of x has an equal chance of occurring and hence the probability that $x = 2$ is $1/6$. This example illustrates the type of problem considered in probability theory. The population is assumed to be known, and we are concerned with calculating the probability of observing a particular sample. Exactly the opposite is true in statistical problems, where we assume the population to be unknown and the sample to be known, and we wish to make inferences about the population. **Thus probability reasons from the population to the sample, while statistics acts in reverse, moving from the sample to the population.**

To illustrate how probability is used in statistical inference, consider the following example. Suppose that a die is tossed $n = 10$ times and the number of dots x that appear is recorded after each toss. This represents a sample of $n = 10$ measurements drawn from a much larger body of tosses, the population, which could be generated if we wished. Suppose that all ten measurements resulted in $x = 1$. We wish to use this information to make an inference concerning the population of tosses; specifically, we wish to infer that the die is or is not balanced. Having observed ten tosses, each resulting in $x = 1$, we would be somewhat suspicious of the die and would likely reject the theory that the die is balanced. We reason as follows: If the die is balanced as we hypothesize, observing ten identical measurements is most improbable. Hence, either we have observed a rare event or our hypothesis is false. We would likely be inclined toward the latter conclusion. Notice that the decision was based on the probability of observing the sample, assuming our theory to be true.

This example emphasizes the importance of probability in making statistical inferences. In the following discussion of the theory of probability, we will assume the population to be known and calculate the probability of drawing various samples. In doing so, we are really choosing a **model** for a physical situation because the actual composition of a population is rarely known in practice. Thus the probabilist models a physical situation (the population) with probability much as the sculptor models with clay.

In the sections that follow, we present some basic concepts of probability that will aid you in understanding references to probabilities as they relate to statistical inference. To help you grasp these concepts, we will use simple experiments involving the toss of one or more dice or coins. Practical applications follow these simple examples.

3.2 PROBABILITY AND THE SAMPLE SPACE

Data are obtained either by observation of uncontrolled events in nature or by controlled experimentation in the laboratory. To simplify our terminology, we seek a word that will apply to either method of data collection, and hence we define the term *experiment*.

Definition

> An **experiment** is the process by which an observation (or measurement) is obtained.

Note that the observation need not produce a numerical value. Here are some typical examples of experiments.

1. Recording a test grade.
2. Making a measurement of daily rainfall.
3. Interviewing a householder to obtain his or her opinion on a greenbelt zoning ordinance.

4. Testing a printed circuit board to determine whether it is a defective product or an acceptable product.

5. Tossing a coin and observing the face that appears.

Let us now direct our attention to a careful analysis of an experiment and the construction of a mathematical model for a population. A by-product of our development will be a systematic and direct approach to the solution of probability problems.

We begin by noting that each experiment may result in one or more outcomes, which we will call **events** and denote by capital letters.

Definition   An **event** is the outcome of an experiment.

EXAMPLE 3.1   Experiment: Toss a die and observe the number appearing on the upper face. Some events would be as follows:

1. Event A: observe an odd number.
2. Event B: observe a number less than 4.
3. Event E_1: observe a 1.
4. Event E_2: observe a 2.
5. Event E_3: observe a 3.
6. Event E_4: observe a 4.
7. Event E_5: observe a 5.
8. Event E_6: observe a 6.

The eight events described in Example 3.1 do not represent a complete listing of all possible events associated with the experiment. However, you can see that there is a difference between events A and B and events E_1, E_2, E_3, E_4, E_5, and E_6. Event A will occur if event E_1, E_3, or E_5 occurs, that is, if we observe a 1, 3, or 5. Thus A could be decomposed into a collection of simpler events, namely, E_1, E_3, and E_5. Likewise, event B will occur if E_1, E_2, or E_3 occurs, and could be viewed as a collection of smaller or simpler events. In contrast, it is impossible to decompose events $E_1, E_2, E_3, \ldots, E_6$. These events are called **simple events**.

Definition   An event that cannot be decomposed is called a **simple event**. Simple events will be denoted by the symbol E with a subscript.

The events $E_1, E_2, \ldots, E_6$ represent a complete listing of all simple events associated with the experiment in Example 3.1. **An experiment will result in one and only one of the simple events.** For instance, if a die is tossed, we will observe either a 1, 2, 3, 4, 5, or 6, but we cannot possibly observe more than one of the simple events at the same time. Hence, a list of simple events provides a breakdown of all possible indecomposable outcomes of the experiment.

EXAMPLE 3.2   Experiment: Toss a coin. The simple events are:

E_1: observe a head.

E_2: observe a tail.

EXAMPLE 3.3   Experiment: Toss two coins. The simple events are:

Event	Coin 1	Coin 2
E_1	Head	Head
E_2	Head	Tail
E_3	Tail	Head
E_4	Tail	Tail

We can now define the outcomes of an experiment in terms of the associated simple events.

Definition   An **event** is a collection of one or more simple events.

The set of all simple events associated with an experiment is called the **sample space**. In order to visualize a problem, an experiment can be described graphically using a **Venn diagram**, in which events are depicted as circular portions of the sample space. For example, the sample space for the toss of a single die consists of the six simple events $E_1, E_2, \ldots, E_6$, while the event A consists of the simple events E_1, E_3, and E_5.

In Figure 3.1, the large outer box represents the sample space. The circle inside the box divides the sample space into two portions: the portion inside the circle, which represents the event A (an odd number is observed), and the portion outside the circle, which represents the event that A did not occur (an even number is observed). Keep in mind that a single repetition of this experiment will result in one and only one simple event. However, an arbitrary event such as A (an odd number is observed) will occur if any simple event within the circle (E_1, E_3, or E_5) occurs on that single repetition.

Figure 3.1
Venn diagram for die tossing

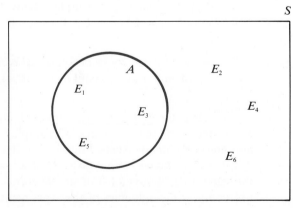

▷ **3.3   THE PROBABILITY OF AN EVENT**

The probability associated with an event is a measure of belief that the event will occur on the next repetition of the experiment. One way of assessing the probability of an event is to repeat the experiment a very large number of times N and to record n, the number of times that A occurs in these N repetitions. Then the **probability of the event A** is estimated by

$$P(A) = \frac{n}{N}$$

More properly, $P(A)$ is the limiting value of the fraction n/N as N becomes infinitely large. This interpretation of the meaning of probability, an interpretation held by many laypersons, is called the **relative frequency concept of probability**. In tossing a balanced die in a fair manner, we would expect that in a large number of tosses, the numbers 1, 2, 3, 4, 5, and 6 would appear with approximately the same frequency. Hence, it would be reasonable to take

$$P(E_1) = P(E_2) = \cdots = P(E_6) = \frac{1}{6}$$

Using the relative frequency definition of probability, it is apparent that for any event A, $P(A)$ is a fraction lying between 0 and 1 with $P(A) = 0$ if A never occurs, and $P(A) = 1$ if A always occurs. Therefore, the closer its value to 1, the more likely A is to occur.

Definition ⊗   Two events, A and B, are **mutually exclusive** if, when one event occurs, the other cannot, and vice versa.

For example, in the toss of a fair die, the events

   A: observe an odd number.

   C: observe an even number.

are mutually exclusive, since if A occurs, C cannot, and vice versa. Simple events are mutually exclusive, and therefore the probabilities associated with simple events satisfy the following conditions.

REQUIREMENTS FOR SIMPLE EVENT PROBABILITIES

   1. $0 \le P(E_i) \le 1$ for all E_i
   2. $\Sigma P(E_i) = 1$

When it is possible to write down the simple events associated with an experiment and assess their respective probabilities, we can find the probability of an event A by summing the probabilities for the simple events contained in the event A.

Definition

> The **probability of an event** A is equal to the sum of the probabilities of the simple events contained in A.

EXAMPLE 3.4   Calculate $P(A)$ and $P(B)$ for the die-tossing experiment in Example 3.1.

Solution   Event A occurs if E_1, E_3, or E_5 occurs. Therefore,

$$P(A) = P(E_1) + P(E_3) + P(E_5)$$

$$= \frac{1}{6} + \frac{1}{6} + \frac{1}{6}$$

$$= \frac{1}{2}$$

Event B will occur if E_1, E_2, or E_3 occurs. Therefore,

$$P(B) = P(E_1) + P(E_2) + P(E_3)$$

$$= \frac{3}{6}$$

$$= \frac{1}{2}$$ ◁

EXAMPLE 3.5   Calculate the probability of observing exactly one head in a toss of two coins.

Solution   Construct the sample space letting H represent a head and T a tail.

Event	First coin	Second coin	$P(E_i)$
E_1	H	H	1/4
E_2	H	T	1/4
E_3	T	H	1/4
E_4	T	T	1/4

It would seem reasonable to assign a probability of 1/4 to each of the simple events. We are interested in

Event A: observe exactly one head.

Simple events E_2 and E_3 are in A. Hence,

$$P(A) = P(E_2) + P(E_3)$$

$$= \frac{1}{4} + \frac{1}{4}$$

$$= \frac{1}{2}$$ ◁

EXAMPLE 3.6   Consider the following experiment involving two urns. Urn 1 contains two white balls and one black ball. Urn 2 contains one white ball. A ball is drawn from urn 1 and placed in urn 2. Then a ball is drawn from urn 2. What is the probability that the ball drawn from urn 2 will be white? (See Figure 3.2.)

Figure 3.2
Representation of
the experiment in
Example 3.6

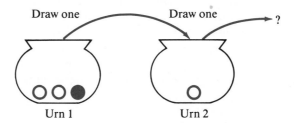

Solution   The problem is easily solved once we have listed the simple events. For convenience, number the white balls 1, 2, and 3, with ball 3 residing in urn 2. The simple events are listed below using W_i to represent the ith white ball and B to represent the black ball. For example, E_1 is the event that white ball 1 is drawn from urn 1, placed in urn 2, and then drawn from urn 2.

| | Drawn from || |
Event	Urn I	Urn 2	$P(E_i)$
E_1	W_1	W_1	1/6
E_2	W_1	W_3	1/6
E_3	W_2	W_2	1/6
E_4	W_2	W_3	1/6
E_5	B	B	1/6
E_6	B	W_3	1/6

Event A, drawing a white ball from urn 2, occurs if $E_1, E_2, E_3, E_4,$ or E_6 occurs. Once again, it would seem reasonable to assume that the simple events are equally likely and to assign a probability of 1/6 to each of the six simple events. Hence,

$$P(A) = P(E_1) + P(E_2) + P(E_3) + P(E_4) + P(E_6)$$

$$= \frac{1}{6} + \frac{1}{6} + \frac{1}{6} + \frac{1}{6} + \frac{1}{6}$$

$$= \frac{5}{6}$$

◁

The sample space provides a probabilistic model for a population that has a great deal of utility in addition to elegance. The model provides a simple, logical, direct method for calculating the probability of an event or, if you like, the probability of a sample drawn from a theoretical population. A practical limitation to this or any other method for calculating probabilities is that the initial assignment of probabilities to the simple events is somewhat subjective. Consequently, a valid solution would require that the subjective probability assignments provide a good measure of the likelihood of occurrence of the simple events.

A second limitation, which specifically applies to the simple event procedure, becomes evident when the method is applied to certain probability problems. Listing the simple events can become tedious, and it is essential that none are omitted. Since the total number of simple events in S may run into the millions, it is convenient to have counting rules to simplify the counting of simple events. Although not necessary for understanding the basic concepts of probability, counting rules are included in optional Section 3.7. If you wish to develop an ability to solve complex probability problems using the simple event approach, you should move directly to that section.

TIPS ON PROBLEM SOLVING

Calculating the probability of an event: the simple event approach

1. Use the following steps for calculating the probability of an event by summing the probabilities of the simple events.

   a. Define the experiment.

   b. Identify a typical simple event. List the simple events associated with the experiment and test each one to make certain that it cannot be decomposed. This defines the sample space S.

   c. Assign reasonable probabilities to the simple events in S, making certain that $0 \leq P(E_i) \leq 1$ and $\sum_S P(E_i) = 1$.

   d. Define the event of interest A as a specific collection of simple events. (A simple event is in A if A occurs when the simple event occurs. Test *all* simple events in S to locate those in A.)

   e. Find $P(A)$ by summing the probabilities of simple events in A.

2. When the simple events are equiprobable, obtain the sum of the probabilities of the simple events in A [step (e)] by counting the points in A and multiplying by the probability per simple event.

3. Calculating the probability of an event by using the five-step procedure described in part 1 is systematic and will lead to the correct solution if all steps are followed correctly. Major sources of error include

   a. failure to define the experiment clearly [step (a)]

   b. failure to specify simple events [step (b)]

   c. failure to list all the simple events

   d. failure to assign valid probabilities to the simple events

EXERCISES   Basic Techniques

3.1   An experiment involves tossing a single die. Specify which of the following are simple events:

   A: observe a 2.

   B: observe an odd number.

   C: observe a number less than 4.

   D: observe both A and B.

   E: observe A or B or both.

   F: observe both A and C.

Calculate the probabilities of the events D, E, and F by summing the probabilities of the appropriate simple events.

3.2   A sample space contains five simple events, E_1, E_2, E_3, E_4, and E_5. If $P(E_3) = .4$, $P(E_4) = 2P(E_5)$, and $P(E_1) = P(E_2) = .15$, find the probabilities of E_4 and E_5.

3.3   A sample space contains ten simple events, $E_1, E_2, \ldots, E_{10}$. If $P(E_1) = 3P(E_2) = .45$ and the remaining simple events are equiprobable, find the probabilities of these remaining simple events.

3.4   A particular basketball player hits 70% of her free throws. When tossing a pair of free throws, the four possible simple events and three of their associated probabilities are

Simple event	Outcome of first free throw	Outcome of second free throw	Probability
1	Hit	Hit	.49
2	Hit	Miss	?
3	Miss	Hit	.21
4	Miss	Miss	.09

   a.   Find the probability that the player will hit on the first free throw and miss on the second.
   b.   Find the probability that the player will hit on at least one of the two free throws.

3.5   The game of roulette uses a wheel containing 38 pockets. Thirty-six pockets are numbered 1, 2, . . . , 36 and the remaining two are marked 0 and 00. The wheel is spun and a pocket is identified as the "winner." Assume that the observance of any one pocket is just as likely as any other.
   a.   Identify the simple events in a single spin of the roulette wheel.
   b.   Assign probabilities to the simple events.
   c.   Let A be the event that you observe either a 0 or a 00. List the simple events in the event A and find P(A).
   d.   Suppose that you were to place bets on the numbers 1 through 18. What is the probability that one of your numbers would be the winner?

3.6   A jar contains four coins: a nickel, a dime, a quarter, and a half-dollar. Three coins are randomly selected from the jar.
   a.   List the simple events in S.
   b.   What is the probability that the selection will contain the half-dollar?
   c.   What is the probability that the total amount drawn will equal 60 cents or more?

3.7   Two dice are tossed. What is the probability that the sum of the numbers shown on the dice is equal to 7? to 11?

Applications

3.8   A survey classified a large number of adults according to whether they were judged to need eyeglasses to correct their reading vision and whether they used eyeglasses when reading. The proportions falling in the four categories are shown below. (Note that a small proportion, .02. of adults used eyeglasses when, in fact, they were judged not to need them.)

Judged to need eyeglasses	Use eyeglasses for reading	
	Yes	No
Yes	.44	.14
No	.02	.40

If a single adult is selected from this large group, find the probability that
a.   The adult is judged to need eyeglasses.
b.   The adult needs eyeglasses for reading but does not use them.
c.   The adult uses eyeglasses for reading whether he or she needs them or not.

3.9   According to *Webster's New Collegiate Dictionary*, a divining rod is "a forked rod believed to indicate [divine] the presence of water or minerals by dipping downward when held over a vein." To test the claims of success by a divining rod expert, four cans are buried in the ground, two empty and two filled with water. The expert will use the divining rod to test each of the four cans and will decide which two contain water.
a.   Define the experiment.
b.   List the simple events in S.
c.   If the rod is completely useless in locating water, what is the probability that the expert correctly identifies (by guessing) the two cans containing water?

3.10   Refer to Exercise 3.9. Suppose the experiment were conducted using five cans: three empty and two containing water. Answer parts (a), (b), and (c) of the exercise.

3.11   A tea taster is required to taste and rank three varieties of tea, A, B, and C, according to the taster's preference.
a.   Define the experiment.
b.   List the simple events in S.
c.   If the taster has no ability to distinguish difference in taste between teas, what is the probability that the taster will rank tea type A as the most desirable? as the least desirable?

3.12   Four union men, two from a minority group, are assigned to four distinctly different one-man jobs.
a.   Define the experiment.
b.   List the simple events in S.
c.   If the assignment to the jobs is unbiased, that is, if any one ordering of assignments is as probable as any other, what is the probability that the two men from the minority group are assigned to the least desirable jobs?

3.13   The odds are 2:1 that when players A and B play racquetball, A wins. Suppose that A and B play three matches and that the winners of the matches are recorded. Using the letters A and B to denote the winner of each match, the eight simple events are listed in the accompanying table. As you will subsequently learn, under certain conditions it is reasonable to assume that the simple event probabilities are as listed in the table.

Winner of match			Sample point	
1	2	3	i	$P(E_i)$
A	A	A	1	8/27
A	A	B	2	4/27
A	B	A	3	4/27
A	B	B	4	2/27
B	A	A	5	4/27
B	A	B	6	?
B	B	A	7	2/27
B	B	B	8	1/27

Using these probabilities,
a.   Find $P(E_6)$.
b.   Find the probability that A wins at least two of the three matches.

3.14   An investor has the option of investing in three of five recommended stocks. Unknown to the investor, only two will show a substantial profit within the next five years. If the investor selects the three stocks at random (giving every combination of three stocks an equal chance of selection), what is the probability that the investor selects the two profitable stocks? What is the probability that he will select only one of the two profitable stocks?

3.15   Although much progress has been made in bridging the gap between blacks and whites in American society, there is still much to be accomplished. In the past decade, the number of black managers, professionals, technicians, and government officials has risen by 52%. However, the gap between black and white median incomes is wider now than it was in the late 1970s (Lacayo, Richard, "Between Two Worlds," *Time*, March 13, 1989, p. 58). In particular, the median income of blacks in 1987 was $18,500, compared to $35,000 for whites. Suppose that a group of four black Americans is surveyed, and that each one reveals whether his or her income exceeded $18,500 in 1987.
a.   How many simple events will be in the sample space?
b.   List the simple events.
c.   Identify the simple events in the list of events below:

   A: at least two had income exceeding $18,500.

   B: exactly two had income exceeding $18,500.

   C: exactly one had income less than or equal to $18,500.

d.   Recall that, by definition, half of the black population has an income exceeding the median income. Using this information, assign reasonable probabilities to the simple events and find $P(A)$, $P(B)$, and $P(C)$.

3.16   Two city commissioners are to be selected from a total of five to form a subcommittee to study the city's traffic problems.
a.   Define the experiment.
b.   List the simple events in S.
c.   If all possible pairs of commissioners have an equal probability of selection, what is the probability that commissioners Jones and Smith will be selected?

3.4 EVENT COMPOSITION AND EVENT RELATIONS

Frequently we are interested in experimental outcomes that can be described as being formed by some composition of two or more events. **Compound events**, as the name suggests, can be formed in one of two ways or by some combination of the two, namely, a **union** or an **intersection**.

Let A and B be two events defined on the sample space S.

Definition

The **intersection** of events A and B, denoted by AB^*, is the event that both A and B occur.

Definition

The **union** of events A and B, denoted by the symbol $A \cup B$, is the event that A or B or both occur.

Figure 3.3
Venn diagram AB

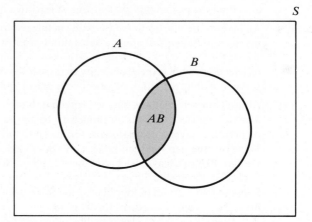

Figure 3.4
Venn diagram $A \cup B$

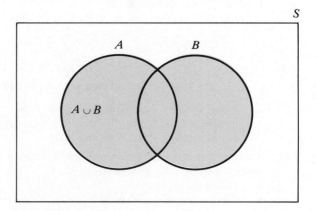

* Some authors use the symbol $A \cap B$.

A Venn diagram representation of AB, the intersection of events A and B, is given in Figure 3.3, while Figure 3.4 gives the Venn diagram representation of $A \cup B$. If we can enumerate the simple events and their probabilities, then $P(AB)$ and $P(A \cup B)$ can be found by summing the probabilities associated with the simple events comprising AB and $A \cup B$.

EXAMPLE 3.7   Refer to the experiment in Example 3.3, where two coins are tossed, and define

Event A: at least one head

Event B: at least one tail

Define events A, B, AB, and $(A \cup B)$ as collections of simple events.

Solution   Recall that the simple events for this experiment were

E_1: HH (head on first coin, head on second)

E_2: HT

E_3: TH

E_4: TT

The occurrence of simple events E_1, E_2, and E_3 implies and hence defines event A. The other events could similarly be defined

Event B: E_2, E_3, E_4

Event AB: E_2, E_3

Event $(A \cup B)$: E_1, E_2, E_3, E_4

Note that $(A \cup B) = S$, the sample space, and is thus certain to occur.

When the two events A and B are mutually exclusive, it means that when A occurs, B cannot, and vice versa. Figure 3.5 is a Venn diagram representation of two such events with no simple events in common. Mutually exclusive events are also

Figure 3.5
Two disjoint events

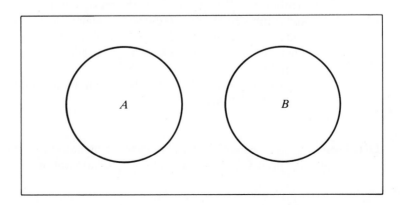

referred to as **disjoint events**. When A and B are mutually exclusive

1. $P(AB) = 0$
2. $P(A \cup B) = P(A) + P(B)$

That is, if $P(A)$ and $P(B)$ are known, we do not need to enumerate the simple events comprising $A \cup B$ and sum their respective probabilities; rather, we can simply sum $P(A)$ and $P(B)$.

EXAMPLE 3.8   An oil-prospecting firm plans to drill two exploratory wells. Past evidence is used to assess the following possible outcomes.

Event	Description	Probability
A	Neither well produces oil or gas	.80
B	Exactly one well produces oil or gas	.18
C	Both wells produce oil or gas	.02

Find $P(A \cup B)$ and $P(B \cup C)$.

Solution   By their definition, events A, B, and C are jointly mutually exclusive, since the occurrence of one event precludes the occurrence of either of the other two. Therefore,

$$P(A \cup B) = P(A) + P(B) = .80 + .18 = .98$$

while

$$P(B \cup C) = P(B) + P(C) = .18 + .02 = .20$$

The event $A \cup B$ can be described as the event that *at most* one well produces oil or gas, while $B \cup C$ describes the event that *at least* one well produces gas or oil.

The concept of unions and intersections can be extended to more than two events. For example, the union of three events A, B, and C would be the event that A, B, C, or any combination of the three occurs. This event, which is denoted by the symbol $A \cup B \cup C$, would be the set of simple events that are in A or B or C or in any combination of those events. Similarly, the intersection of the three events A, B, and C, denoted as ABC, is the event that all three of the events A, B, and C occur. This event is the collection of simple events that are common to the three events A, B, and C.

Complementation is another event relationship that often simplifies probability calculations.

Definition   The **complement** of an event A, denoted by $\bar{A}$, consists of all the simple events in the sample space S that are not in A.

The complement of A is the event that A does not occur. Therefore, A and $\bar{A}$ are mutually exclusive, and $A \cup \bar{A} = S$, the sample space. It follows that $P(A) + P(\bar{A}) = 1$ and

$$P(A) = 1 - P(\bar{A})$$

For example, if the event A is the occurrence of at least one head in the toss of three fair coins, then the event $\bar{A}$ is the occurrence of no heads in the toss, and

$$P(A) = 1 - P(\bar{A}) = 1 - (.5)^3 = 1 - .125 = .875$$

▷ 3.5 CONDITIONAL PROBABILITY AND INDEPENDENCE

Two events are often related so that the probability of the occurrence of one depends on whether the second has or has not occurred. For instance, suppose that one experiment consists in observing the weather on a specific day. Let A be the event "observe rain" and B be the event "observe an overcast sky." Events A and B are obviously related. The probability of rain, $P(A)$, is not the same as the probability of rain given prior information that the day is cloudy. The probability of A, $P(A)$, would be the fraction of the entire population of observations that result in rain. Now let us look only at the subpopulation of observations that result in B, a cloudy day, and the fraction of these that result in A. This fraction, called the **conditional probability of A given B**, may equal $P(A)$, but we would expect the chance of rain, given that the day is cloudy, to be larger. **The conditional probability of A, given that B has occurred, is denoted as**

$$P(A \mid B)$$

where the vertical bar in the parentheses is read "given" and events appearing to the right of the bar are the events that have occurred.

We will define the conditional probabilities of B given A and A given B as follows.

Definition

> The **conditional probability of B**, given that A has occurred, is
>
> $$P(B \mid A) = \frac{P(AB)}{P(A)} \quad \text{if } P(A) \neq 0$$
>
> The **conditional probability of A**, given that B has occurred, is
>
> $$P(A \mid B) = \frac{P(AB)}{P(B)} \quad \text{if } P(B) \neq 0$$

For example, consider a toss of a fair die. Let B be the outcome for which the number is less than 4 (that is, a 1, 2, or 3), and let A be the outcome for which the number is odd. If the event B has occurred, and the outcome is a 1, 2, or 3, then the chance that the number is odd is 2/3. We have just described the conditional probability of A given B.

By attaching some numbers to the probabilities in the weather example, we can see that this definition of conditional probability is consistent with the relative frequency concept of probability. Recall that A denotes rain on a given day; B denotes the day is cloudy. Now suppose that 10% of all days are rainy and cloudy [that is, $P(AB) = .10$] and 30% of all days are cloudy [$P(B) = .30$].

This situation is graphically portrayed in Figure 3.6. Each simple event in event B, denoted by the large egg-shaped area, is associated with a single cloudy day. Since 30% of all days will be cloudy, we can regard this area as .30. Ten percent of all days, or 1/3 of all cloudy days, will also be rainy. These days are included in the color-shaded event AB. If a single day is selected from the set of all days representing the population, what is the probability that we will select a rainy day, given that we know the day is cloudy; that is, what is $P(A\,|\,B)$?

Figure 3.6
Events A and B

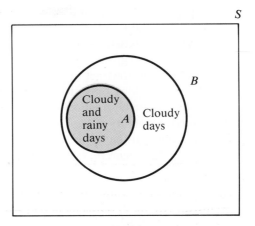

Since we already know that the day is cloudy, we know that the simple event to be selected must fall in event B (Figure 3.6). One-third of these days will result in rain. Hence, the probability that we will select a rainy day is

$$P(A\,|\,B) = \frac{1}{3}$$

We can see that this result agrees with our definition for $P(A\,|\,B)$; that is:

$$P(A\,|\,B) = \frac{P(AB)}{P(B)} = \frac{.10}{.30} = \frac{1}{3}$$

EXAMPLE 3.9   Calculate $P(A\,|\,B)$ for the die-tossing experiment described in Example 3.1.

Solution   Given that event B (a number less than 4) has occurred, we have observed either a 1, 2, or 3, which occur with equal frequency. Of these three simple events, exactly two

(1 and 3) result in event A (an odd number). Hence,

$$P(A \mid B) = \frac{2}{3}$$

Or, we could obtain $P(A \mid B)$ by substituting into the equation

$$P(A \mid B) = \frac{P(AB)}{P(B)} = \frac{1/3}{1/2} = \frac{2}{3}$$

Note that $P(A \mid B) = 2/3$ while $P(A) = 1/2$, indicating that A and B are dependent on each other.

Definition

Two events A and B are said to be **independent** if and only if either

$$P(A \mid B) = P(A)$$

or

$$P(B \mid A) = P(B)$$

Otherwise, the events are said to be **dependent**.

Translating this definition into words: Two events are independent if the occurrence or nonoccurrence of one of the events does not change the probability of the occurrence of the other event. If $P(A \mid B) = P(A)$, then $P(B \mid A)$ will also equal $P(B)$. Similarly, if $P(A \mid B)$ and $P(A)$ are unequal, then $P(B \mid A)$ and $P(B)$ will also be unequal.

EXAMPLE 3.10   Refer to the die-tossing experiment in Example 3.1. Are events A and B mutually exclusive? Are they complementary? Are they independent?

Solution   Event A: E_1, E_3, E_5

Event B: E_1, E_2, E_3

Event AB is the set of simple events in both A and B. Since AB includes events E_1 and E_3, A and B are not mutually exclusive. They are not complementary because B is not the set of all outcomes in S that are not in A. The test for independence lies in the definition; that is, we will check to see if $P(A \mid B) = P(A)$. From Example 3.9, $P(A \mid B) = 2/3$. Then, since $P(A) = 1/2$, $P(A \mid B) \neq P(A)$ and, by definition, events A and B are dependent.

A second approach to the solution of probability problems is based on the classification of compound events, event relations, and two probability laws, which we will now state and illustrate. The "laws" can be simply stated and taken as fact as long as they are consistent with our model and with reality. The first is called the Additive Law of Probability and applies to unions.

THE ADDITIVE LAW OF PROBABILITY

Given two events A and B, the probability of the union $(A \cup B)$ is equal to

$$P(A \cup B) = P(A) + P(B) - P(AB)$$

If A and B are mutually exclusive, $P(AB) = 0$ and

$$P(A \cup B) = P(A) + P(B)$$

The Additive Law conforms to reality and our model. Note in Figure 3.7 that the sum $P(A) + P(B)$ contains the sum of the probabilities of all simple events in $(A \cup B)$ but includes a double counting of the probabilities of all outcomes in the intersection AB. Subtracting $P(AB)$ gives the correct result.

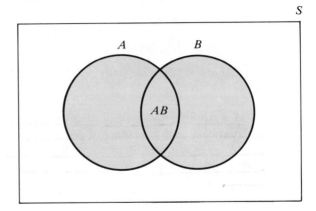

Figure 3.7
The union of two events, A and B $(A \cup B$ is shaded)

The second law of probability is called the Multiplicative Law and applies to intersections.

THE MULTIPLICATIVE LAW OF PROBABILITY

Given two events A and B, the probability of the intersection AB is

$$P(AB) = P(A)P(B \mid A)$$
$$= P(B)P(A \mid B)$$

If A and B are independent, $P(AB) = P(A)P(B)$.

The Multiplicative Law follows from the definition of conditional probability.

Using probability laws to calculate the probability of a compound event is less direct than listing simple events and requires some experience and ingenuity. The approach involves expressing the event of interest as a union or intersection (or combination of both) of two or more events whose probabilities are known or easily calculated. This can often be done in many ways; the trick is to find the right combination, a task requiring no small amount of creativity in some cases. The usefulness of event relations is now apparent. If the event of interest is expressed as a union of mutually exclusive events, the probabilities of the intersections will equal zero. If the events are independent, we can use the unconditional probabilities to calculate the probability of an intersection. Examples 3.11 through 3.15 illustrate the use of the probability laws and the technique just described.

EXAMPLE 3.11 Calculate $P(AB)$ and $P(A \cup B)$ for Example 3.1.

Solution Recall that $P(A) = P(B) = 1/2$ and $P(A|B) = 2/3$. Then

$$P(AB) = P(B)P(A|B)$$

$$= \left(\frac{1}{2}\right)\left(\frac{2}{3}\right)$$

$$= \frac{1}{3}$$

and

$$P(A \cup B) = P(A) + P(B) - P(AB)$$

$$= \frac{1}{2} + \frac{1}{2} - \frac{1}{3}$$

$$= \frac{2}{3}$$

These solutions will agree with those obtained using the simple event approach.

EXAMPLE 3.12 Consider the experiment in which two coins are tossed. Let A be the event that the toss results in at least one head. Find $P(A)$.

Solution $\bar{A}$ is the collection of simple events implying the event "two tails." Because $\bar{A}$ is the complement of A,

$$P(A) = 1 - P(\bar{A})$$

The event $\bar{A}$ will occur if both of two independent events occur: "tail on the first coin," T_1, and "tail on the second coin," T_2. Then $\bar{A}$ is the intersection of T_1 and T_2, or

$$\bar{A} = T_1 T_2$$

Applying the Multiplicative Law and noting that T_1 and T_2 are independent events,

$$P(\bar{A}) = P(T_1)P(T_2) = \left(\frac{1}{2}\right)\left(\frac{1}{2}\right) = \frac{1}{4}$$

Then $P(A) = 1 - P(\bar{A}) = 1 - 1/4 = 3/4$.

EXAMPLE 3.13   Refer to Example 3.12 and find the probability of tossing exactly one head.

Solution   Event B, "exactly one head," is the union of two mutually exclusive events B_1 and B_2, where

B_1:   head on the first coin, tail on the second

B_2:   head on the second coin, tail on the first

Then B_1 and B_2 are both intersections of independent events:

$$B_1 = H_1 T_2$$

$$B_2 = H_2 T_1$$

Applying the Multiplicative Law for independent events,

$$P(B_1) = P(H_1)P(T_2)$$

$$= \left(\frac{1}{2}\right)\left(\frac{1}{2}\right) = \frac{1}{4}$$

and

$$P(B_2) = P(H_2)P(T_1)$$

$$= \left(\frac{1}{2}\right)\left(\frac{1}{2}\right) = \frac{1}{4}$$

Then $P(B) = P(B_1) + P(B_2)$, because B_1 and B_2 are mutually exclusive events, and

$$P(B) = \frac{1}{4} + \frac{1}{4} = \frac{1}{2}$$

This problem was solved in Example 3.5 using the simple event approach.

EXAMPLE 3.14   A city council contains eight members, two of whom are local contractors. If two councilmen are selected at random to fill vacancies on the zoning committee, what is the probability that both contractors will be selected?

Solution   Let D_1 be the event that the first councilman selected is a contractor, and correspondingly, let D_2 be the event that the second is a contractor. The event that both councilmen are contractors is A. Then A will occur if both D_1 and D_2 occur or, equivalently, A is the intersection of D_1 and D_2.

$$A = D_1 D_2$$

Applying the Multiplicative Law,

$$P(A) = P(D_1)P(D_2 | D_1)$$

The probability of selecting a contractor on the first draw is 2/8, or 1/4. Similarly, the probability that the second councilman will be a contractor, given that the first was a contractor, is 1/7. Then

$$P(A) = P(D_1)P(D_2|D_1) = \left(\frac{1}{4}\right)\left(\frac{1}{7}\right) = \frac{1}{28}$$

(*Note:* This solution can be easily obtained using the simple event approach.)

EXAMPLE 3.15 Two cards are drawn from a deck of 52 cards. Calculate the probability that the draw will include an ace and a ten.

Solution   Event A: draw an ace and a ten.

Then $A = B \cup C$, where

B: draw the ace on the first draw and the ten on the second.

C: draw the ten on the first draw and the ace on the second.

B and C were chosen to be mutually exclusive and also to be intersections of events with known probabilities. Thus,

$$B = B_1 B_2 \quad \text{and} \quad C = C_1 C_2$$

where

B_1:   draw an ace on the first draw.

B_2:   draw a ten on the second draw.

C_1:   draw a ten on the first draw.

C_2:   draw an ace on the second draw.

Applying the Multiplicative Law,

$$P(B_1 B_2) = P(B_1)P(B_2|B_1)$$

$$= \left(\frac{4}{52}\right)\left(\frac{4}{51}\right)$$

and

$$P(C_1 C_2) = \left(\frac{4}{52}\right)\left(\frac{4}{51}\right)$$

Then, applying the Additive Law,

$$P(A) = P(B) + P(C)$$

$$= \left(\frac{4}{52}\right)\left(\frac{4}{51}\right) + \left(\frac{4}{52}\right)\left(\frac{4}{51}\right) = \frac{8}{663}$$

Check each composition carefully to be certain that it is actually equal to the event of interest.

TIPS ON PROBLEM SOLVING

Calculating the probability of an event: event composition approach

1. Use the following steps to calculate the probability of an event using the event composition approach:
    a. Define the experiment.
    b. Clearly visualize the nature of the simple events. Identify a few to clarify your thinking.
    c. Write an equation expressing the event of interest, say A, as a composition of two or more events using either or both of the two forms of composition (unions and intersections). Note that this equates point sets. Make certain that the event implied by the composition and event A represent the same set of simple events.
    d. Apply the Additive and Multiplicative Laws of Probability to step (c) and find $P(A)$.

2. Be careful with step (c). You can often form many compositions that will be equivalent to event A. The trick is to form a composition in which all the probabilities appearing in step (d) will be known. Visualize the results of step (d) for any composition and select the one for which the component probabilities are known.

3. Always write down letters to represent events described in an exercise. Then write down the probabilities that are given and assign them to events. Identify the probability that is requested in the exercise. This can help you arrive at the appropriate event composition.

EXERCISES Basic Techniques

3.17   An experiment can result in one of five equally likely simple events, $E_1, E_2, \ldots, E_5$. Events A, B, and C are defined as follows:

A: E_1, E_3

B: E_1, E_2, E_4, E_5

C: E_3, E_4

Find the probabilities associated with the following compound events by listing the simple events in each.

a.	$\bar{A}$	e.	BC	i.	$A \mid B$
b.	$\bar{B}$	f.	$A \cup B$	j.	$A \cup B \cup C$
c.	AB	g.	$A \cup C$	k.	$\overline{AB}$
d.	AC	h.	$B \mid C$	l.	$\bar{A}\bar{B}$

3.18   Refer to Exercise 3.17. Use the definition of a complementary event to find:
   a. $P(\bar{A})$ b. $P(\overline{AB})$ c. $P(\bar{A}\bar{B})$
   Do the results agree with those obtained in Exercise 3.17?

3.19 Refer to Exercise 3.17. Use the definition of conditional probability to find:

a. $P(A|B)$ b. $P(B|C)$

Do the results agree with those obtained in Exercise 3.17?

3.20 Refer to Exercise 3.17. Use the Additive and Multiplicative Laws of Probability to find:

a. $P(A \cup B)$ b. $P(AB)$ c. $P(BC)$

Do the results agree with those obtained in Exercise 3.17?

3.21 Refer to Exercise 3.17. Determine whether events A and B are:

a. Independent   b. Mutually exclusive

3.22 An experiment consists of tossing a single die and observing the number of dots shown on the upper face. Events A, B, and C are defined as follows:

A: observe a number less than 4.

B: observe a number less than or equal to 2.

C: observe a number greater than 3.

Find the probabilities associated with the following compound events using either the simple event approach or the event composition approach.

a. S d. ABC g. BC

b. $A|B$ e. AB h. $A \cup C$

c. B f. AC i. $B \cup C$

3.23 Refer to Exercise 3.22.

a. Are events A and B independent? mutually exclusive?

b. Are events A and C independent? mutually exclusive?

Applications

3.24 According to a study by the Bureau of Justice Statistics, the odds of a United States citizen being murdered are 1 in 133 (*Gainesville Sun*, May 6, 1985). The odds are 1 in 131 for white males, 1 in 21 for black males, 1 in 369 for white females, and 1 in 104 for black females. If a large community consists of 34% white males, 36% white females, 14% black males, and 16% black females, what is the probability of a member of the community being murdered?

3.25 In an article entitled "Drug Tests for Jobs Spreads, Raises Questions," the *St. Augustine Record* (November 5, 1985) quotes a Miami drug-testing consultant as saying, "Testing of prospective employees for drug use has become so prevalent that detectable users may soon find it impossible to get work." Proponents of the procedure point to the improvement in worker efficiency and the reduction of absenteeism, accidents, and theft that can be achieved by eliminating drug users from the work force. Opponents claim that the procedure is creating a class of unhirables and that some persons may be placed in this class because the tests themselves are not 100% reliable. The article notes that EMIT, the most widely used test, is 97–98% accurate. Suppose that a company uses a test procedure that is 98% accurate—that is, it correctly identifies a person as a drug user or nonuser with probability .98—and that, to reduce the chance of error, each job applicant is required to take two tests. If the outcomes of the two tests on the same person are independent events, what is the probability that

a. A nondrug user will fail both tests?

b. A drug user will be detected; that is, he or she will fail at least one test?

c. A drug user will pass both tests?

3.26 A study of the behavior of a large number of drug offenders after treatment for drug abuse suggests that the likelihood of conviction within a two-year period after treatment may depend on the offender's education. The proportions of the total number of cases falling in four education-conviction categories are shown in the accompanying table.

| | Status within two years after treatment | | |
Education	Convicted	Not convicted	Totals
Ten years or more	.10	.30	.40
Nine years or less	.27	.33	.60
Totals	.37	.63	1.00

Suppose that a single offender is selected from the treatment program. Define the events.

A: the offender has ten or more years of education.

B: the offender is convicted within two years after completion of treatment.

Find the approximate probabilities for these events.

a. A

b. B

c. AB

d. $A \cup B$

e. $\bar{A}$

f. $\overline{A \cup B}$

g. $\overline{AB}$

h. A given that B has occurred

i. B given that A has occurred

3.27 Use the probabilities of Exercise 3.26 to show that

a. $P(AB) = P(A)P(B \mid A)$

b. $P(AB) = P(B)P(A \mid B)$

c. $P(A \cup B) = P(A) + P(B) - P(AB)$

3.28 When an American marriage ends in divorce, it is most likely that the woman is the partner who is dissatisfied. A study conducted by the National Center for Health Statistics (*Press Enterprise*, June 7, 1989) indicates that in 1986, 61.5% of all divorce petitions in the United States were filed by wives, 32.6% by husbands, and the remaining petitions were filed jointly.

a. What percentage of all divorce petitions were filed jointly?

b. If a couple were divorced in 1986, what is the probability that the wife did not unilaterally file the divorce petition?

If two couples were divorced in 1986, what is the probability that the wife unilaterally filed the divorce petition in both cases? in at least one case?

3.29 Refer to Exercise 3.28. The study reported that the filing rates for divorce petitions have changed within the past decade, as indicated in the following table:

	1975	1980	1986
Filed by wife	67.2%	63.4%	61.5%
Filed by husband	29.4%	30.2%	32.6%
Filed jointly	3.4%	6.4%	5.9%

Suppose that two women are interviewed, one of whom was divorced in 1975, and the other in 1986.

a. What is the probability that both women unilaterally filed their divorce petitions?

b. What is the probability that one woman unilaterally filed the divorce petition, while the other petition was filed by the woman's husband?

c. What is the probability that at least one woman filed the divorce petition jointly with her husband?

3.30 In an article entitled "Experts Agree Nobody's Opinion Is a Sure Thing," the *Orlando Sentinel* (April 20, 1985) cites a number of cases where expert assessments of a situation varied greatly. One example concerned the "peer review" system for granting foundation money to support university faculty research. Each of a group of university faculty research proposals was evaluated by a group of experts in the research field, and the group then decided whether the research was worthy of funding. When the same proposals were submitted to new groups of equally qualified experts, the decision to fund was reversed in 25–30% of the cases. Suppose that the probability that a project is judged suitable for funding by a peer review group is .2 and that the probability of a reversal in the decision by a second peer review group is .3. What is the probability that a project worthy of funding will be

a. approved by both groups?
b. disapproved by both groups?
c. approved by one group?

3.31 An article discussing a National Education Association plan to support programs aimed at reducing public school dropouts notes that in some cities the dropout rate is 60% for blacks and 75% for Hispanics. Suppose that the school population in a large city is 54% white, 30% black, and 16% Hispanic. If the overall city dropout rate is 42%, what is the dropout rate for whites?

3.32 A certain article is visually inspected by two successive inspectors. When a defective article comes through, the probability that it gets by the first inspector is .1. Of those that get past the first inspector, the second inspector will "miss" five out of ten. What fraction of the defectives get by both inspectors?

3.33 A survey of people in a given region showed that 20% were smokers. The probability of death due to lung cancer, given that a person smoked, was roughly ten times the probability of death due to lung cancer, given that a person did not smoke. If the probability of death due to lung cancer in the region is .006, what is the probability of death due to lung cancer given that a person is a smoker?

3.34 A smoke detector system uses two devices, *A* and *B*. If smoke is present, the probability that it will be detected by device *A* is .95; by device *B*, .98; and by both devices, .94.

a. If smoke is present, find the probability that the smoke will be detected by device *A* or device *B* or both devices.
b. Find the probability that the smoke will not be detected.

3.35 The "no pass, no play" rule is making life difficult for Texas public school students. Figures released in January 1986 by the Dallas Independent School District show that 1 in 4 of the 4300 students participating in sports have been prohibited from playing because they are failing one or more subjects (*Gainesville Sun*, January 31, 1986). Under the statewide "no pass, no play" rule, a student who fails a course is disqualified from participating in extracurricular activities for six weeks. Suppose that the probability that an athlete who has not previously been disqualified will be disqualified for the next six weeks is .15 and that the probability that an athlete who has been disqualified will be disqualified again in the next six weeks is .5. If 30% of the athletes have been disqualified before, what is the probability that an athlete will be disqualified during the next six-week period?

3.36 A new prenatal test has been developed to detect Tay-Sachs disease, a genetic disorder that is usually fatal in early childhood. If both parents are carriers of the disease, the probability that their offspring will develop the disease is approximately .25 (*Wall Street Journal*, July 22, 1985). Suppose that a husband and wife are both carriers of the disease and that the wife is pregnant on three different occasions. If the occurrence of Tay-Sachs in any one offspring is independent of the occurrence in any other, what is the probability that

a. All three children will develop Tay-Sachs disease?
b. Only one child will develop Tay-Sachs disease?
c. The third child will develop Tay-Sachs disease, given that the first two did not?

3.37 When Brian Bosworth announced that he would become a professional football player in the supplemental draft of 1987, National Football League officials explained the rules for determining the drafting order for NFL teams as follows (*New York Times*, May 12, 1987):

> "The team with the worst record, Tampa Bay, puts 28 of its logos into a barrel, the Indianapolis Colts put 27 in, and so on down to the Jets with eight and the [Super Bowl Champion] Giants with one. Twenty-eight logos are then picked to set the order. Once a team is picked, its remaining logos in the barrel are ignored."

a. How many logos will be put in the barrel?
b. What is the probability that the Super Bowl Champion Giants will obtain the first pick?
c. What is the probability that the Tampa Bay Buccaneers will obtain the first pick?
d. What is the probability that the Indianapolis Colts and the New York Jets will have the first two picks?

3.38 Two people enter a room and their birthdays (ignoring years) are recorded.
a. Identify the nature of the simple events in S.
b. What is the probability that the two people have a specific pair of birth dates?
c. Identify the simple events in event A, "both persons have the same birthday."
d. Find $P(A)$.
e. Find $P(\bar{A})$.

3.39 If n people enter a room, find the probability that

A: none of the persons have the same birthday.

B: at least two of the persons have the same birthday.

Solve for
a. $n = 3$ b. $n = 4$
[*Note:* Surprisingly, $P(B)$ increases rapidly as n increases. For example, for $n = 20$, $P(B) = .411$; for $n = 40$, $P(B) = .891$.]

 3.40 A worker-operated machine produces a defective item with probability .01 if the worker follows the machine's operating instructions exactly, and with probability .03 if he does not. If the worker follows the instructions 90% of the time, what proportion of all items produced by the machine will be defective?

3.6 BAYES' RULE (Optional)

Frequently, we wish to find the conditional probability of an event A, given that an event B has already occurred. For example, we might wish to know the probability of rain tomorrow, given that it has rained during the preceding seven days. Hence, we assume that some state of nature exists, and we wish to calculate the probability of some event that will occur in the future.

Equally interesting is the probability that a certain state of nature exists given that a certain sample is observed. One such problem occurs in screening tests, which used to be associated primarily with medical diagnostic tests, but are now finding application in a variety of fields. Camera technology and automatic test equipment are routinely used in inspecting parts in high-volume production processes. Corporate drug testing of employees, steroid testing of athletes, home pregnancy tests, and AIDS (acquired immune deficiency syndrome) testing are some other familiar applications.

Very few, if any, screening tests are perfect. There is always the risk that defective parts, diseased persons, or contaminated products will go undetected (a false negative). On the other hand, good parts, healthy people, or safe products may be classified as defective, sick, or unsafe (a false positive). The effectiveness of a screening test is evaluated by assessing the probability of a false negative or a false positive. In this section we present an application of the Multiplicative Law, together with conditional probabilities, in a form derived by the probabilist Thomas Bayes, which allows for the determination of the probability of false positives and false negatives, both of which are conditional probabilities.

Consider an experiment that involves selection of a sample from one of k mutually exclusive and only possible populations, referred to as **states of nature**, and denoted by $S_1, S_2, \ldots, S_k$ with *prior* probabilities $P(S_1), P(S_2), \ldots, P(S_k)$, respectively. The sample is selected and results in the event A, but the population (or state of nature) giving rise to the sample is unknown. The problem is to determine the population from which the sample was selected using the conditional probabilities $P(S_i \mid A)$ for $i = 1, 2, \ldots, k$. These probabilities are known as **posterior probabilities**, that is, probabilities that result when prior probabilities are updated using sample information.

To find the conditional probability that the sample was selected from population i given that event A was observed, $P(S_i \mid A)$, $i = 1, 2, \ldots, k$, note that A could have been observed if the sample was selected from population 1, population 2, or any one of the k populations $S_1, S_2, \ldots, S_k$. The probability that population i was selected *and* that event A occurred is the intersection of the events S_i and A, or (AS_i). These events, $(AS_1), (AS_2), \ldots, (AS_k)$, are mutually exclusive and hence

$$P(A) = P(AS_1) + P(AS_2) + \cdots + P(AS_k)$$

This relationship is shown in Figure 3.8 for $k = 3$ subpopulations. Then the probability that the sample came from population i is

$$P(S_i \mid A) = \frac{P(AS_i)}{P(A)} = \frac{P(S_i)P(A \mid S_i)}{\sum_{j=1}^{k} P(AS_j)} = \frac{P(S_i)P(A \mid S_i)}{\sum_{j=1}^{k} P(S_j)P(A \mid S_j)}$$

The expression for $P(S_i \mid A)$ is known as Bayes' Rule for the probability of causes. As you can see, it follows easily from the definition of conditional probability.

Bayes suggested that if the prior probabilities, $P(S_i)$, were unknown, they be taken as equally probable—that is, $P(S_i) = 1/k$, for $i = 1, 2, \ldots, k$. In many instances, however, the experimenter does have at least some approximation to the prior probabilities for $S_1, S_2, \ldots, S_k$.

Figure 3.8
Representation of the
event A for $k = 3$
subpopulations

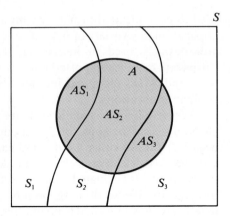

BAYES' RULE

Let $S_1, S_2, \ldots, S_k$ represent the k mutually exclusive, only possible states of nature with prior probabilities $P(S_1), P(S_2), \ldots, P(S_k)$. If an event A occurs, the posterior probability of S_i given A is the conditional probability

$$P(S_i \mid A) = \frac{P(S_i)P(A \mid S_i)}{\sum_{j=1}^{k} P(S_j)P(A \mid S_j)}$$

for $i = 1, 2, \ldots, k$.

EXAMPLE 3.16   To evaluate the effectiveness of a screening procedure, we will evaluate the probabilities of a false negative or a false positive using the following notation.

T^+:   the test is positive and indicates that the person has the disease

T^-:   the test is negative and indicates that the person does not have the disease

D: the person tested has the disease

$\bar{D}$:   the person tested does not have the disease

In a 1987 assessment of the HIV test for AIDS (MacNeil/Lehrer Newshour segment entitled "Lax Labs," August 12, 1987), the "sensitivity" of the test, defined as the probability of a positive result given the person has the disease, was reported as

$$P[T^* \mid D] = .98$$

while the "specificity" of the test, defined as the probability that the test is negative given the person does not have the disease, was reported as

$$P[T^- \mid \bar{D}] = .99$$

If the proportion of the general population in the United States infected with the AIDS virus is 45,000/242,200,000 = .000018580, or approximately 19 per million

("Aids Diary," *Discover*, January 1988), find

a. $P[\bar{D} \mid T^+]$, the probability of a false positive

b. $P[D \mid T^-]$, the probability of a false negative

Solution   From the given information, we know the following:

$$P(D) = .0001858 \qquad P(\bar{D}) = .9998142$$

$$P[T^+ \mid D] = .98 \qquad P[T^- \mid D] = .02$$

$$P[T^+ \mid \bar{D}] = .01 \qquad P[T^- \mid \bar{D}] = .99$$

a.   From Bayes' Rule,

$$P[\bar{D} \mid T^+] = \frac{P(\bar{D}T^+)}{P(T^+)}$$

with

$$\begin{aligned}
P(T^+) &= P(T^+D) + P(T^+\bar{D}) \\
&= P(D)P(T^+ \mid D) + P(\bar{D})P(T^+ \mid \bar{D}) \\
&= (.0001858)(.98) + (.9998142)(.01) \\
&= .00018208 + .00999814 \\
&= .01018022
\end{aligned}$$

Therefore, the probability of a false positive is

$$P[\bar{D} \mid T^+] = \frac{.00999814}{.01018022} = .98211434$$

b.   Using a similar calculation,

$$P[D \mid T^-] = \frac{P(DT^-)}{P(T^-)}$$

with

$$\begin{aligned}
P(T^-) &= P(T^-D) + P(T^-\bar{D}) \\
&= P(D)P(T^- \mid D) + P(\bar{D})P(T^- \mid \bar{D}) \\
&= (.0001858)(.02) + (.9998142)(.99) \\
&= .0000037 + .98981606 \\
&= .9898198
\end{aligned}$$

Therefore, the probability of a false negative is

$$P[D \mid T^-] = \frac{.0000037}{.9898198} = .0000037$$

Hence, the probability of a false positive is near 1 and very likely, while the probability of a false negative is quite small and very unlikely.   ◁

Another way to view Bayes' Rule is as a method of incorporating the information from sample observations to adjust the probability of some event. For example, if you had no information on the result of a person's diagnostic test, you would regard the probability that he or she is infected with the AIDS virus as .0001858 (since .0186% of all people are infected with the AIDS virus). However, given the added information that the person's test was positive, the probability that he or she is infected with the AIDS virus is

$$P[D \mid T^+] = .0179$$

Thus, based on the test information, the probability that the person is infected is adjusted from a very small probability, .000186, to .0179—a 100-fold increase.

EXERCISES   Basic Techniques

3.41 A sample is selected from one of two populations A_1 and A_2, with probabilities $P(A_1) = .7$ and $P(A_2) = .3$. If the sample has been selected from A_1, the probability of observing an event B is $P(B \mid A_1) = .2$. Similarly, if the sample has been selected from A_2, the probability of observing B is $P(B \mid A_2) = .3$. If a sample is selected and event B is observed, what is the probability that the sample was selected from population A_1? from A_2?

3.42 If an experiment is conducted, one and only one of three mutually exclusive events A_1, A_2, and A_3, can occur with these probabilities.

$$P(A_1) = .2 \qquad P(A_2) = .5 \qquad P(A_3) = .3$$

The probabilities of a fourth event B occurring, given that events A_1, A_2, or A_3 occur, are

$$P(B \mid A_1) = .2 \qquad P(B \mid A_2) = .1 \qquad P(B \mid A_3) = .3$$

If event B is observed, find $P(A_1 \mid B)$, $P(A_2 \mid B)$, and $P(A_3 \mid B)$.

Applications

3.43 City crime records show that 20% of all crimes are violent, and 80% are nonviolent, involving theft, forgery, and so on. Ninety percent of violent crimes are reported versus 70% of nonviolent crimes. If a crime in progress is reported to the police, what is the probability that the crime is violent?

3.44 A recent test conducted by the Federal Aviation Administration found that guards hired to screen passengers for weapons at the boarding gate had a very poor rate of detection. The detection rates for weapons carried by F.A.A. inspectors or placed in their carry-on luggage averaged 80%, but the rates varied from 34% to 99% for the airports tested (*New York Times*, June 18, 1987). Suppose that in a particular city, airport A handles 50% of all airline traffic, while airports B and C handle 30% and 20%, respectively. The detection rates at the three airports are .9, .5, and .4, respectively. If a passenger at one of the airports is found to be carrying a weapon through the boarding gate, what is the probability that the passenger is using airport A? airport C?

3.45 How often are convicted felons freed on parole, only to commit another serious crime? A study conducted by the Bureau of Justice Statistics reports that the overall recidivism rate, the probability that a parolee will be rearrested for a new felony or serious misdemeanor within six years, is 69% (*Florida Times-Union/Jacksonville Journal*, Jacksonville, Fla.,

May 25, 1987). Furthermore, the recidivism rate for male parolees is 70%, compared to 52% for female parolees. Define the following events.

   R: parolee is rearrested within six years.

   M: parolee is male.

a.   Using the information given above, find the probability that a parolee is male. [*Hint:* Use the fact that $P(R) = P(RM) + P(R\bar{M})$.]

b.   If a parolee is rearrested within six years, what is the probability that the parolee is male? What is the probability that the parolee is female?

3.46   As items come to the end of a production line, an inspector chooses items to undergo a complete inspection. Of all items produced, 10% are defective. Sixty percent of all defective items go through a complete inspection, and 20% of all good items go through a complete inspection. Given that an item is completely inspected, what is the probability that it is defective?

3.47   Medical case histories indicate that different illnesses may produce identical symptoms. Suppose that a particular set of symptoms, which we will denote as event *H*, occurs only when any one of three illnesses — *A*, *B*, or *C* — occurs (for the sake of simplicity, we will assume that illnesses *A*, *B*, and *C* are mutually exclusive). Studies show that the probabilities of getting the three illnesses are

$$P(A) = .01$$

$$P(B) = .005$$

$$P(C) = .02$$

The probabilities of developing the symptoms *H*, given a specific illness, are

$$P(H \mid A) = .90$$

$$P(H \mid B) = .95$$

$$P(H \mid C) = .75$$

Assuming that an ill person shows the symptoms *H*, what is the probability that the person has illness *A*?

3.48   A student answers a multiple-choice examination question that has four possible answers. Suppose that the probability that the student knows the answer to the question is .80 and the probability that the student guesses the answer is .20. Assume that if the student guesses, the probability of selecting the correct answer is .25. If the student correctly answers a question, what is the probability that the student really knew the correct answer?

3.49   Suppose that 5% of all people filing the long income tax form seek deductions that they know are illegal, and an additional 2% incorrectly list deductions because they are unfamiliar with income tax regulations. Of the 5% who are guilty of cheating, 80% will deny knowledge of the error if confronted by an investigator. If the filer of the long form is confronted with an unwarranted deduction and he or she denies knowledge of the error, what is the probability that he or she is guilty?

3.50   A particular football team is known to run 30% of its plays to the left and 70% to the right. A linebacker on an opposing team notes that the right guard shifts his stance most of the time (80%) when plays go to the right and that he uses a balanced stance the remainder of the time. When plays go to the left, the guard takes a balanced stance 90% of the time and the shift stance the remaining 10%. On a particular play, the linebacker notes that the guard takes a balanced stance. What is the probability that the play will go to the left?

▷ 3.7 USEFUL COUNTING RULES (Optional)

This section presents three rules that fall into the realm of combinatorial mathematics and that can be used to solve probability problems involving a large number of simple events. For example, suppose that you are interested in the probability of an event A and you know that the simple events in S are equiprobable. Then

$$P(A) = \frac{n_A}{N}$$

where

n_A = number of simple events in A

N = number of simple events in S

Often we can use counting rules to find the values of n_A and N and thereby eliminate the necessity of listing the simple events in S.

The first rule, called the *mn* rule, is as follows:

mn RULE

With m elements $a_1, a_2, a_3, \ldots, a_m$ and n elements $b_1, b_2, b_3, \ldots, b_n$, it is possible to form mn pairs that contain one element from each group.

To illustrate, suppose that four companies have job openings in each of three areas: sales, manufacturing, and personnel. How many job opportunities are available? This situation contains two sets of "things": companies (four) and types of jobs (three). Therefore, as shown in Figure 3.9, there are three jobs for each of

Figure 3.9
Company-job
combinations

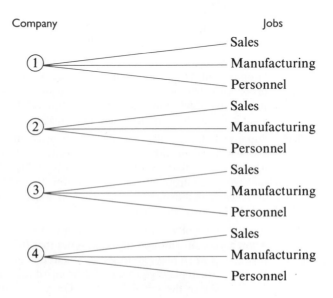

the four companies, or $(4)(3) = 12$ possible pairings of companies and jobs. This example illustrates a use of the *mn* rule.

EXAMPLE 3.17   Two dice are tossed. How many simple events are associated with the experiment?

Solution   The first die can fall in one of six ways; that is, $m = 6$. Likewise, the second die can fall in $n = 6$ ways. Since an outcome of this experiment involves pairing of the numbers showing on the faces of the two dice, the total number N of simple events is

$$N = mn = 6(6) = 36$$

EXAMPLE 3.18   Urn 1 contains two white balls and one black ball, while urn 2 contains one white ball. A ball is drawn from urn 1 and placed in urn 2. A ball is then drawn from urn 2. How many simple events are associated with the experiment?

Solution   A ball can be chosen from urn 1 in one of $m = 3$ ways. After one of these ways has been chosen, a ball may be drawn from urn 2 in $n = 2$ ways. The total number of simple events is

$$N = mn = 3(2) = 6$$

A COUNTING RULE FOR FORMING PAIRS, TRIPLETS, AND SO ON

Given k groups of elements, n_1 elements in the first group, n_2 in the second, . . . , and n_k in the kth group, the number of ways of selecting one element from each of the k groups is

$$n_1 n_2 n_3 \cdots n_k$$

EXAMPLE 3.19   How many simple events are in the sample space when three coins are tossed?

Solution   Each coin can land in one of two ways. Hence,

$$N = 2(2)(2) = 8$$

EXAMPLE 3.20   A truck driver can take three routes from city A to city B, four from city B to city C, and three from city C to city D. If, when traveling from A to D, the driver must proceed from A to B to C to D, how many possible A-to-D routes are available?

Solution   Let

$$m = \text{number of routes from } A \text{ to } B = 3$$

$$n = \text{number of routes from } B \text{ to } C = 4$$

$$t = \text{number of routes from } C \text{ to } D = 3$$

Then the total number of ways to construct a complete route, taking one subroute from each of the three groups (A to B), (B to C), and (C to D), is

$$mnt = (3)(4)(3) = 36$$

A second useful mathematical rule is associated with orderings or permutations. For instance, suppose that we have three books, b_1, b_2, and b_3. In how many ways can the books be arranged on a shelf, taking them two at a time? Listing all combinations of two in the first column and a reordering of each in the second column, we get the following enumeration:

Combinations of two	Reordering of combinations
$b_1 b_2$	$b_2 b_1$
$b_1 b_3$	$b_3 b_1$
$b_2 b_3$	$b_3 b_2$

The number of permutations is six, a result easily obtained from the *mn* rule. The first book can be chosen in $m = 3$ ways, and once selected, the second book can be chosen in $n = 2$ ways. The result is $mn = 6$.

In how many ways can three books be arranged on a shelf, taking three at a time? Enumerating, we obtain

$b_1 b_2 b_3$	$b_2 b_1 b_3$	$b_3 b_1 b_2$
$b_1 b_3 b_2$	$b_2 b_3 b_1$	$b_3 b_2 b_1$

a total of six. This, again, could be obtained easily by the extension of the *mn* rule. The first book can be chosen and placed in $m = 3$ ways. After choosing the first, the second can be chosen in $n = 2$ ways, and finally, the third in $t = 1$ way. Hence, the total number of ways is

$$N = mnt = 3 \cdot 2 \cdot 1 = 6$$

Definition   An ordered arrangement of r distinct objects is called a **permutation**. The number of ways of ordering n distinct (different) objects taken r at a time will be designated by the symbol P_r^n.

A COUNTING RULE FOR PERMUTATIONS

The number of ways that you can arrange n distinct objects taking them r at a time is

$$P_r^n = n(n - 1)(n - 2) \cdots (n - r + 1)$$

$$= \frac{n!}{(n - r)!}$$

where $n! = n(n - 1)(n - 2) \cdots (3)(2)(1)$ and $0! = 1$.

We are concerned with the number of ways of filling r positions with n distinct objects. Applying the extension of the *mn* rule, the first object can be chosen in one of n ways. After choosing the first, the second can be chosen in $(n - 1)$ ways, the third in

$(n - 2)$ ways, and the rth in $(n - r + 1)$ ways. Hence, the total number of ways is

$$P^n_r = n(n - 1)(n - 2) \cdots (n - r + 1)$$

EXAMPLE 3.21 Three lottery tickets are drawn from a total of 50. Assume that order is of importance. How many simple events are associated with the experiment?

Solution The total number of simple events is

$$P^{50}_3 = \frac{50!}{47!} = 50(49)(48) = 117{,}600 \qquad \lhd$$

EXAMPLE 3.22 A piece of equipment is composed of five parts that can be assembled in any order. A test is to be conducted to determine the time necessary for each order of assembly. If each order is to be tested once, how many tests must be conducted?

Solution The total number of tests would equal

$$P^5_5 = \frac{5!}{0!} = 5(4)(3)(2)(1) = 120 \qquad \lhd$$

The enumeration of the permutations of books in the previous discussion was performed in a systematic manner, first writing the combinations of n books taken r at a time, and then writing the rearrangements of each combination. In many situations, ordering is unimportant and we are interested solely in the number of possible combinations. For instance, suppose that an experiment involves the selection of five men, a committee, from a total of 20 candidates. Then the simple events associated with this experiment correspond to the different combinations of men selected from the group of 20. How many simple events (different combinations) are associated with this experiment? Since order in a single selection is unimportant, permutations are irrelevant. Thus we are interested in the number of combinations of $n = 20$ things taken $r = 5$ at a time.

Definition The **number of combinations** of n objects taken r at a time will be denoted by the symbol C^n_r. [*Note:* Some authors prefer the symbol $\binom{n}{r}$.]

A COUNTING RULE FOR COMBINATIONS

The number of distinct combinations of n distinct objects that can be formed, taking them r at a time, is

$$C^n_r = \frac{n!}{r!(n - r)!}$$

The relationship between the number of **combinations** and the number of **permutations** of n things taken r at a time is given by

$$C^n_r = \frac{P^n_r}{r!}$$

Expressed in this way, we see that C_r^n results from dividing the number of permutations by $r!$, the number of ways of rearranging each distinct selection of r items from n.

EXAMPLE 3.23   A radio tube may be purchased from five suppliers. In how many ways can three suppliers be chosen from the five?

Solution $$C_3^5 = \frac{5!}{3!2!} = \frac{(5)(4)}{2} = 10$$ ◁

The following example illustrates the use of the counting rules in the solution of a probability problem.

EXAMPLE 3.24   Five manufacturers, of varying but unknown quality, produce a certain type of electronic device. If we were to select three manufacturers at random, what is the chance that the selection would contain exactly two of the best three?

Solution   Without enumerating the simple events, we would likely agree that each point, that is, any combination of three, would be assigned equal probability. If N points are in S, then each event receives probability

$$P(E_i) = \frac{1}{N}$$

Let n be the number of events in which two of the best three manufacturers are selected. Then the probability of including two of the best three manufacturers in a selection of three is

$$P = \frac{n}{N}$$

Our problem is to use the counting rules to find n and N. Since order within a selection is not important and is not recorded, each selection is a combination and hence,

$$N = C_3^5 = \frac{5!}{3!2!} = 10$$

Determination of n is more difficult, but it can be obtained using the mn rule. Let a be the number of ways of selecting exactly two from the best three, or

$$C_2^3 = \frac{3!}{2!1!} = 3$$

and let b be the number of ways of choosing the remaining manufacturer from the two poorest, or

$$C_1^2 = \frac{2!}{1!1!} = 2$$

Then the total number of ways of choosing two of the best three in a selection of three is $n = ab = 6$.

Hence, the probability p is equal to

$$p = \frac{6}{10}$$

Many other counting rules are available in addition to the three presented in this section. If you are interested in this topic, you should consult one of the many texts on combinatorial mathematics.

TIPS ON PROBLEM SOLVING

Many students have difficulty deciding which (if any) of the three counting rules to apply in a given problem. The following tips may help.

1. Look at the problem and note whether a simple event is formed by
    a. selecting elements from each of *two* (*or more*) sets (a situation that suggests the use of the *mn* rule), or
    b. selecting *r* elements from a single set of *n* elements (a situation that suggests the use of either combinations or permutations).
2. If the situation is 1(b), decide whether combinations or permutations should be used. If every different ordering of the elements in the group of *r* leads to a different simple event, use permutations. If ordering does not produce a new simple event, use combinations. Your diagnostic thought process should perform the checks indicated in the decision tree shown in the accompanying diagram.

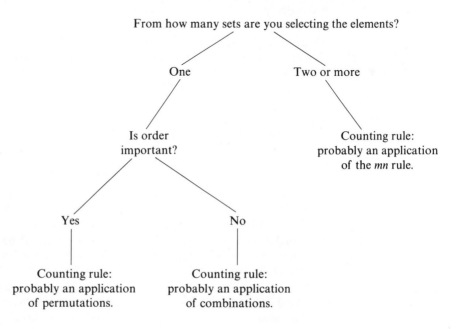

3.   If visualizing the appropriate counting rule (or rules) to use for a problem involving a large number of simple events is difficult, construct a miniature version of the problem so that the points can be counted manually. This may lead to a solution for the more complex version.

EXERCISES   Basic Techniques

3.51   You have *two* sets of distinctly different elements, ten in the first group and eight in the second. If you select one element from each group, how many different pairs can you form?

3.52   You have *three* sets of distinctly different elements, four in the first set, seven in the second, and three in the third. If you select one element from each set, how many pairs of triplets can you select?

3.53   You have *one* set of ten distinctly different elements. If you select two elements from among the ten, how many different pairs can you select?

3.54   Refer to Exercise 3.53. Suppose that the two elements are to fill two positions. How many different ways can you fill the two positions by selecting two elements from the set of ten?

3.55   You have *one* set of twelve distinctly different elements. If you select three elements to fill three different positions, how many different ways can you fill the positions?

3.56   You have *two* sets of distinctly different elements, ten in the first set and eight in the second. If you select two elements from each group, how many different sets of four elements can you select?

Applications

Use the Tips on Problem Solving to help you diagnose and solve the following problems.

3.57   A salesperson in New York is preparing an itinerary for a visit to six major cities. The distance traveled, and hence the cost of the trip, will depend on which city is visited first, second, . . . , sixth. How many different itineraries (and hence trip costs) are possible?

3.58   A company executive wishes to fill three vice-presidential positions—sales, manufacturing, and finance—from among 24 lower-echelon company managers. How many different options does the executive have for filling the positions?

3.59   Probability played a role in the rigging of the April 24, 1980, Pennsylvania state lottery (*Los Angeles Times*, September 8, 1980). To determine each digit of the three-digit winning number, each of the numbers 0, 1, 2, . . . , 9 is placed on a Ping-Pong ball; the ten balls are blown into a compartment; and the number selected for the digit is the one on the ball that floats to the top of the machine. To alter the odds, the conspirators injected a liquid into all balls used in the game except those numbered 4 and 6, making it almost certain that the lighter balls would be selected and determine the digits in the winning number. They then proceeded to buy lottery tickets bearing the potential winning numbers. How many potential winning numbers were there (666 was the eventual winner)?

3.60   Refer to Exercise 3.59. Hours after the rigging of the Pennsylvania state lottery was announced, Connecticut state lottery officials were stunned to learn that *their* winning number for the day was 666 (*Los Angeles Times*, September 21, 1980).

   a.   All evidence indicates that the Connecticut selection of 666 was due to pure chance. What is the probability that a 666 would be drawn in Connecticut, given that a 666 had been selected in the April 24, 1980, Pennsylvania lottery?

b. What is the probability of drawing a 666 in the April 24, 1980, Pennsylvania lottery (remember, this drawing was rigged) *and* a 666 on the September 19, 1980, Connecticut lottery?

3.61 A study is to be conducted to determine the attitudes of nurses in a hospital towards various administrative procedures that are currently employed. If a sample of ten nurses is to be selected from a total of 90, how many different samples could be selected? (Note that order within a sample is unimportant.)

3.62 If five cards are to be selected, one after the other, in sequence, from a 52-card deck, and each card is to be replaced in the deck before the next draw, how many different selections are possible?

3.63 Refer to Exercise 3.62. Suppose that the five cards are drawn from the 52-card deck simultaneously and without replacement. How many different hands could be selected?

3.64 The following case occurred in the city of Gainesville, Florida, in 1976. The eight-member Human Relations Advisory Board considered the complaint of a woman who claimed discrimination, based on her sex, on the part of a local surveying company. The Board, composed of five women and three men, voted 5–3 in favor of the plaintiff, the five women voting in favor of the plaintiff, the three men against. The attorney representing the company appealed the Board's decision by claiming sex bias on the part of the Board members. If the vote in favor of the plaintiff was 5–3 and the Board members were not biased by sex, what is the probability that the vote would split along sex lines (five women for, three men against)?

3.65 For each of 20 questions on a multiple-choice test, a student can choose one of five possible answers.
a. How many completely different sets of answers are possible for the test?
b. If a person guessed on all of the questions, what is the probability that all of the questions would be answered correctly?

3.66 A lineup of ten men is conducted to test the ability of a witness to identify three burglary suspects. Suppose that the three burglary suspects who committed the crime are in the lineup. If the witness is actually unable to identify the suspects but feels compelled to make a choice, what is the probability that the three guilty men will be selected by chance? What is the probability that the witness will select three innocent men?

3.8 RANDOM VARIABLES

Observations generated by an experiment fall into one of two categories: quantitative or qualitative. For example, the daily production in a manufacturing plant would be a quantitative or numerically measurable observation, whereas weather descriptions, such as rainy, cloudy, or sunny, would be qualitative. Statisticians are concerned with both quantitative and qualitative data, although the former are perhaps more common.

In some instances it is possible to convert qualitative data to quantitative data by assigning a numerical value to each category to form a scale. Industrial production is often scaled according to first grade, second grade, and so on. In general, however, the number of items or individuals in each category becomes the quantitative measurement of interest. The events of interest associated with the experiment are the values that the data can take and are, therefore, numerical events.

Suppose that the variable measured in the experiment is denoted by the symbol x. For example, consider the experiment that consists of tossing two coins. Suppose that we are interested in the variable $x =$ number of heads. The variable x can take three values: $x = 0$, $x = 1$, or $x = 2$. The simple events for this experiment are

$$E_1 : HH \qquad E_2 : HT \qquad E_3 : TH \qquad E_4 : TT$$

The numerical event $x = 0$ includes the simple event E_4; $x = 1$ includes simple events E_2 and E_3; and $x = 2$ includes the simple event E_1. The variable x is called a **random variable** because the value that x assumes depends upon the random outcome of an experiment. Furthermore, notice that one and only one value of x corresponds to each simple event, but the converse is not true. Simple events E_2 and E_3 are both associated with the same value of x. Two conclusions can be drawn from this example. First, the relationship between the random variable x and the simple events satisfies the definition of a functional relation. Second, for each simple event in the sample space there corresponds one and only one value of x. Therefore, we define a random variable as follows:

Definition

A **random variable** is a numerically valued function defined over a sample space.

The value of a random variable will vary from trial to trial as the experiment is repeated. However, for a given experiment, the random variable x will be designated as being either **discrete** or **continuous**, according to the values that x may assume.

Definition

A **discrete random variable** is one that can assume a countable* number of values.

A discrete random variable is easily identified by examining the number of values it can assume. **If the number of values that the random variable may assume can be counted, it must be discrete.**

Some typical examples of discrete random variables are:

1. The number of defective bolts in a sample of ten drawn from industrial production.
2. The number of building permits issued in a city during the last six months.
3. The number of malfunctions of an airplane engine during a six-month period.
4. The number of people in the waiting room in a doctor's office.

---

* Countable means that the values that the random variable can assume can be associated with the integers 1, 2, 3, 4, . . . ; that is, they can be counted.

Definition

> A **continuous random variable** is one that can assume the infinitely large number of values corresponding to the points on a line interval.

The word *continuous*, an adjective, means proceeding without interruption. It, in itself, provides the key for identifying continuous random variables. **Look for a measurement with a set of values that form points on a line with no interruptions or intervening spaces between them.**

Typical examples of continuous random variables are:

1. The height of a human.

2. The length of life of a human cell.

3. The amount of sugar in an orange.

4. The length of time required to complete an assembly operation in a manufacturing process.

3.9 PROBABILITY DISTRIBUTIONS FOR DISCRETE RANDOM VARIABLES

The distinction between discrete and continuous random variables is important since different probability models are required for each type of variable. We will focus our attention on discrete random variables in the remainder of this chapter. Continuous random variables are the subject of Chapter 5.

Since each value of the random variable x is a numerical event, we can apply the methods of this chapter to obtain $p(x)$, the probability associated with each of the values of x.

Definition

> The **probability distribution** for a discrete random variable is a formula, table, or graph that provides $p(x)$, the probability associated with each of the values of x.

The events associated with different values of x cannot overlap because one and only one value of x is assigned to each simple event, and hence the values of x represent mutually exclusive numerical events. Summing $p(x)$ over all values of x equals the sum of the probabilities of all simple events and hence equals 1. We can therefore state two requirements for a discrete probability distribution.

REQUIREMENTS FOR A DISCRETE PROBABILITY DISTRIBUTION

1. $0 \le p(x) \le 1$

2. $\sum\limits_{\text{all } x} p(x) = 1$

EXAMPLE 3.25 Consider an experiment that consists of tossing two coins and let x equal the number of heads observed. Find the probability distribution for x.

Solution The simple events for this experiment with their respective probabilities are as follows:

Simple event	Coin 1	Coin 2	$P(E_i)$	x
E_1	H	H	1/4	2
E_2	H	T	1/4	1
E_3	T	H	1/4	1
E_4	T	T	1/4	0

Because E_1 is associated with the simple event "observe a head on coin 1 and a head on coin 2," we assign it the value $x = 2$. Similarly, we assign $x = 1$ to event E_2, and so on. The probability of each value of x can be calculated by adding the probabilities of the simple events in that numerical event. The numerical event $x = 0$ contains one simple event, E_4; $x = 1$ contains two simple events, E_2 and E_3; and $x = 2$ contains one event, E_1. The values of x with respective probabilities are given in Table 3.1. Observe that $\sum_{x=0}^{2} p(x) = 1$.

Table 3.1
Probability distribution for x (x = number of heads)

x	Simple events in x	$p(x)$
0	E_4	1/4
1	E_2, E_3	1/2
2	E_1	1/4

$$\sum_{x=0}^{2} p(x) = 1$$

The probability distribution in Table 3.1 can be presented graphically in the form of the relative frequency histogram (see Section 2.1).* The histogram for the random variable x would contain three classes, corresponding to $x = 0$, $x = 1$, and $x = 2$. Since $p(0) = 1/4$, the theoretical relative frequency for $x = 0$ is 1/4; $p(1) = 1/2$, and hence the theoretical frequency for $x = 1$ is 1/2. The histogram is given in Figure 3.10.

If you were to draw a sample from this population—that is, if you were to throw two balanced coins, say $n = 100$ times, and each time record the number of heads observed, x, and then construct a histogram using the 100 measurements on

* The probability distribution in Table 3.1 can also be presented using a formula, which is presented in Section 4.2.

Figure 3.10
Probability histogram
showing $p(x)$ for
Example 3.25

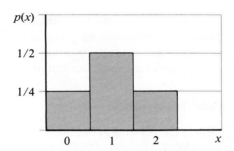

x—you would find that the histogram for the sample would appear very similar to that for $p(x)$ in Figure 3.10. If you were to repeat the experiment $n = 1000$ times, the similarity would be much more pronounced. ◁

EXAMPLE 3.26   Let x equal the number observed on the throw of a single balanced die. Find $p(x)$.

Solution   The simple events for this experiment are given in Table 3.2. Assign $x = 1$ to E_1, $x = 2$ to E_2, and so on. Since each value of x contains only one simple event, $p(x)$, the probability distribution for x would appear as shown in the fifth column of Table 3.2. Note that

$$p(x) = \frac{1}{6}, \quad x = 1, 2, 3, 4, 5, 6$$

gives the probability distribution as a formula. The corresponding histogram is given in Figure 3.11.

Table 3.2
Tossing a die:
probability distribution
for x

Simple event	Number on upper face	$p(E_i)$	x	$p(x)$
E_1	1	1/6	1	1/6
E_2	2	1/6	2	1/6
E_3	3	1/6	3	1/6
E_4	4	1/6	4	1/6
E_5	5	1/6	5	1/6
E_6	6	1/6	6	1/6

Figure 3.11
Probability histogram
for $p(x) = 1/6$
in Example 3.26

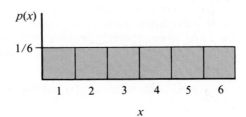

There are a number of useful discrete random variables—useful because their probability distributions, given as formulas, provide good models for the relative frequency distributions of many types of data observed in real life. Formulas for several useful discrete distributions are given in Chapter 5.

EXERCISES   Basic Techniques

3.67   Identify the following as discrete or continuous random variables:
   a.   The total number of points scored in a football game.
   b.   The shelf life of a particular drug.
   c.   The height of the ocean's tide at a given location.
   d.   The length of a two-year-old black bass.
   e.   The number of aircraft near-collisions in a year.

3.68   Identify the following as discrete or continuous random variables:
   a.   The increase in length of life achieved by a cancer patient as a result of surgery.
   b.   The tensile breaking strength, in pounds per square inch, of one-inch-diameter steel cable.
   c.   The number of deer killed per year in a state wildlife preserve.
   d.   The number of overdue accounts in a department store at a particular time.
   e.   Your blood pressure.

3.69   A random variable x has the following probability distribution:

x	0	1	2	3	4	5
$p(x)$	.1	.3	.4	.1	?	.05

   a.   Find $p(4)$.
   b.   Construct a probability histogram to describe $p(x)$.

3.70   A random variable x can assume five values, 0, 1, 2, 3, and 4. A portion of the probability distribution is shown below.

x	0	1	2	3	4
$p(x)$	.1	.3	.3	?	.1

   a.   Find $p(3)$.
   b.   Construct a probability histogram for $p(x)$.
   c.   Simulate the experiment by marking ten poker chips (or coins)—one with a 0, three with a 1, three with a 2, and so on. Mix the chips thoroughly, draw one, and record the observed value of x. Repeat the process 100 times. Construct a relative frequency histogram for the 100 values of x and compare with the probability histogram in part (b).

3.71   A jar contains two black balls and two white balls. Suppose that the balls have been thoroughly mixed and two are randomly selected from the jar.
   a.   List all simple events for this experiment and assign appropriate probabilities to each.
   b.   Let x equal the number of white balls in the selection. Then assign the appropriate value of x to each simple event.
   c.   Calculate the values of $p(x)$ and display them in tabular form. Show that $\sum_{x=0}^{2} p(x) = 1$.

d. Construct a probability histogram for $p(x)$.

e. Simulate the experiment by actually drawing two balls from a jar that contains two black balls and two white balls. (This experiment can also be conducted using coins instead of balls.) Repeat the drawing process 100 times, each time recording the value of x that was observed. Construct a relative frequency histogram for the 100 observed values of x and compare with the probability histogram in part (d).

Applications

3.72 If you toss a pair of dice, the sum T of the number of dots appearing on the upper faces of the dice can assume the value of an integer in the interval $2 \leq T \leq 12$.

a. Find the probability distribution for T and display it in tabular form.

b. Construct a probability histogram for $p(T)$.

3.73 A key ring contains four office keys that are identical in appearance. Only one will open your office door. Suppose you randomly select one key and try it. If it does not fit, you randomly select one of the three remaining keys. If it does not fit, you randomly select one of the last two. Each different sequence that could occur in selecting the keys represents one of a set of equiprobable simple events.

a. List the simple events in S and assign probabilities to the simple events.

b. Let x equal the number of keys that you try before you find the one that opens the door $(x = 1, 2, 3, 4)$. Then assign the appropriate value of x to each simple event.

c. Calculate the values of $p(x)$ and display them in tabular form.

d. Construct a probability histogram for $p(x)$.

3.74 What are the financial concerns that most often worry Americans? A survey conducted by the Gallup Organization (*Good Housekeeping*, October, 1989) indicates that paying the mortgage is a concern for 19% of all those surveyed, while paying for children's education is mentioned by 11%. Suppose that we randomly select three people and ask them to list their financial concerns.

a. Find the probability distribution for x, the number of people listing "paying the mortgage" as a concern.

b. Construct the probability histogram for $p(x)$.

c. Find the probability that more than one person will list "paying the mortgage" as a financial concern.

3.75 A company has five applicants for two positions: two women and three men. Suppose that the five applicants are equally qualified and that no preference is given for choosing either sex. Let x equal the number of women chosen to fill the two positions.

a. Find $p(x)$.

b. Construct a probability histogram for x.

3.76 A piece of electronic equipment contains six transistors, two of which are defective. Three transistors are selected at random, removed from the piece of equipment, and inspected. Let x equal the number of defectives observed, where $x = 0, 1,$ or 2. Find the probability distribution for x. Express the results graphically as a probability histogram.

3.77 Past experience has shown that on the average only one in ten wells drilled hits oil. Let x be the number of drillings until the first success (oil is struck). Assume that the drillings represent independent events.

a. Find $p(1)$, $p(2)$, and $p(3)$.

b. Give a formula for $p(x)$.

c. Graph $p(x)$.

3.78 Two tennis professionals A and B are scheduled to play a match in which the winner is determined by the first player to win three sets in a total that cannot exceed five sets. The event that A wins any one set is independent of the event that A wins any other, and the probability that A wins any one set is equal to .6. Let x equal the total number of sets in the match; that is, $x = 3, 4,$ or 5. Find $p(x)$.

3.79 Shortly after the October 1989 earthquake in Northern California, a California poll indicated that six out of ten Californians have already stored or are now likely to store emergency water and food in preparation for a major earthquake (*Press Enterprise*, October 31, 1989). Suppose that we conduct a telephone survey of California residents to determine whether or not they are going to store emergency supplies in case of an earthquake. Find the probability distribution for x, the number of calls until the first person is found who is *not* going to store emergency supplies.

3.10 MATHEMATICAL EXPECTATION FOR DISCRETE RANDOM VARIABLES

The probability distribution provides a model for the theoretical frequency distribution of a random variable and hence must have a mean, a standard deviation, and other descriptive measures associated with the theoretical population that it represents. Since both the mean and the variance are averages, we will calculate the mean value of a random variable that is defined over a theoretical population. The population mean is called the **expected value** of the random variable.

The method for calculating the population mean or expected value of a random variable can be more easily understood by considering an example. Let x equal the number of heads observed in the toss of two coins. For convenience, $p(x)$ is given as

x	0	1	2
$p(x)$	1/4	1/2	1/4

Suppose that the experiment is repeated a large number of times, say $n = 4{,}000{,}000$ times. Intuitively, we would expect to observe approximately 1 million zeros, 2 million ones, and 1 million twos. Then the average value of x would equal

$$\frac{\text{sum of measurements}}{n} = \frac{1{,}000{,}000(0) + 2{,}000{,}000(1) + 1{,}000{,}000(2)}{4{,}000{,}000}$$

$$= \frac{1{,}000{,}000(0)}{4{,}000{,}000} + \frac{2{,}000{,}000(1)}{4{,}000{,}000} + \frac{1{,}000{,}000(2)}{4{,}000{,}000}$$

$$= (1/4)(0) + (1/2)(1) + (1/4)(2)$$

Note that the first term in this sum is equal to $(0)p(0)$, the second is equal to $(1)p(1)$, and the third is equal to $(2)p(2)$. The average value of x, then, is

$$\sum_{x=0}^{2} xp(x) = 1$$

This result is not an accident, and it provides some intuitive justification for the definition of the expected value of a discrete random variable x.

Definition

> Let x be a discrete random variable with probability distribution $p(x)$ and let $E(x)$ represent the **expected value of x**. Then
>
> $$E(x) = \sum_x xp(x)$$
>
> where the elements are summed over all values of the random variable x.

Note that if $p(x)$ is an accurate description of the relative frequencies for a real population of data, then $E(x)$ is equal to μ, the mean of the population. We will assume this to be true and let $E(x)$ be synonymous with μ; that is, we will assume that $\mu = E(x)$.

EXAMPLE 3.27   Consider the random variable x representing the number observed on the toss of a single die. The probability distribution for x is given in Example 3.26. Then the expected value of x would be

$$E(x) = \sum_{x=1}^{6} xp(x) = (1)p(1) + (2)p(2) + \cdots + (6)p(6)$$

$$= (1)(1/6) + (2)(1/6) + \cdots + (6)(1/6)$$

$$= \frac{1}{6} \sum_{x=1}^{6} x = \frac{21}{6} = 3.5$$

Note that this value $\mu = E(x) = 3.5$ locates the center of the probability distribution exactly (see Figure 3.11).

EXAMPLE 3.28   Eight thousand tickets are to be sold at $5 each in a lottery conducted to benefit the local fire company. The prize is a $12,000 automobile. If you purchase two tickets, what is your expected gain?

Solution   Your gain x may take one of two values. Either you will lose $10 (that is, your gain will be $-\$10$) or win $11,990 with probabilities 7998/8000 and 2/8000, respectively. The probability distribution for the gain x is as follows:

x	$p(x)$
$-\$10$	7998/8000
$\$11,900$	2/8000

The expected gain will be

$$E(x) = \sum_x xp(x)$$

$$= (-\$10)\left(\frac{7998}{8000}\right) + (\$11,990)\left(\frac{2}{8000}\right) = -\$7$$

Recall that the expected value of x is the average of the theoretical population that would result if the lottery were repeated an infinitely large number of times. If this were done, your average or expected gain per lottery would be a loss of $7. ◁

EXAMPLE 3.29   Determine the yearly premium for a $1000 insurance policy covering an event that, over a long period of time, has occurred at the rate of two times in 100. Let x equal the yearly financial gain to the insurance company resulting from the sale of the policy, and let C equal the unknown yearly premium. Calculate the value of C such that the expected gain $E(x)$ will equal zero. Then C is the premium required to break even. To this, the company would add administrative costs and profit.

Solution   The first step in the solution is to determine the values that the gain x may take and then to determine $p(x)$. If the event does not occur during the year, the insurance company will gain the premium of $x = C$ dollars. If the event does occur, the gain will be negative; that is, the company will lose $1000 less the premium of C dollars already collected. Then $x = -(1000 - C)$ dollars. The probabilities associated with these two values of x are 98/100 and 2/100, respectively. The probability distribution for the gain would be

$x = $ gain	$p(x)$
C	98/100
$-(1000 - C)$	2/100

Since we want the insurance premium C such that, in the long run (for many similar policies), the mean gain will equal zero, we will set the expected value of x equal to zero and solve for c. Then

$$E(x) = \sum_x xp(x)$$

$$= C\left(\frac{98}{100}\right) + [-(1000 - C)]\left(\frac{2}{100}\right) = 0$$

or

$$\frac{98}{100}C + \left(\frac{2}{100}\right)C - 20 = 0$$

Solving this equation for C, we obtain $C = \$20$. Therefore, if the insurance company were to charge a yearly premium of $20, the average gain calculated for a large number of similar policies would equal zero. The actual premium would equal $20 plus administrative costs and profit. ◁

Just as we used numerical descriptive measures to describe a relative frequency distribution (Chapter 2), we wish to use the mean and variance (ultimately, the standard deviation) of a random variable to describe its probability distribution. Knowing μ and σ, we could use Tchebysheff's Theorem or the Empirical Rule to describe $p(x)$. The variance σ^2 of a random variable is defined to be the mean value of the square of the deviation of x from its mean, that is,

$$\sigma^2 = E[(x - \mu)^2]$$

This leads to the problem of finding the mean value of a function of a random variable, namely, the expected value of $(x - \mu)^2$. The rule for finding the expected value of any function of x, say $g(x)$, is given as follows:

Definition

Let x be a discrete random variable with probability distribution $p(x)$, and let $g(x)$ be a numerical valued function of x. Then the **expected value of $g(x)$** is

$$E[g(x)] = \sum_x g(x)p(x)$$

In Chapter 2 we defined the population variance to be the average of the squares of the deviations of the measurements from their mean. Because taking an expectation is equivalent to "averaging," we define the **variance** and the **standard deviation** as shown in the displays.

Definition

Let x be a discrete random variable with probability distribution $p(x)$ and expected value $E(x) = \mu$. The **variance of x** is

$$\sigma^2 = E[(x - \mu)^2] = \sum_x (x - \mu)^2 p(x)$$

where the elements are summed over all values of the random variable x.*

Definition

The **standard deviation σ of a random variable x** is equal to the square root of its variance.

EXAMPLE 3.30   Find the variance σ^2 for the population associated with Example 3.25, the coin-tossing problem. In Section 3.10 the expected value of x was shown to equal 1.

Solution   The variance is equal to the expected value of $(x - \mu)^2$, or

$$\sigma^2 = E[(x - \mu)^2] = \sum_x (x - \mu)^2 p(x)$$

$$= (0 - 1)^2 p(0) + (1 - 1)^2 p(1) + (2 - 1)^2 p(2)$$

$$= (1)(1/4) + (0)(1/2) + (1)(1/4) = 1/2$$

Then $\sigma = \sqrt{1/2} = .707$. The values $\mu = 1$ and $\sigma = .707$ can be used to describe the probability distribution shown in Figure 3.10.   ◁

---

* It can be shown (proof omitted) that

$$\sigma^2 = \sum_x (x - \mu)^2 p(x) = \sum_x x^2 p(x) - \mu^2 = E(x^2) - \mu^2$$

This result is analogous to the shortcut formula for the sum of squares of deviations given in Chapter 2.

EXAMPLE 3.31   Let x be a random variable with the probability distribution given in the following table.

x	-1	0	1	2	3	4	5
$p(x)$	.05	.10	.40	.20	.10	.10	.05

Find μ, σ^2, and σ. Graph $p(x)$ and locate the interval $\mu \pm 2\sigma$ on the graph. What is the probability that x will fall in the interval $\mu \pm 2\sigma$?

Solution
$$\mu = E(x) = \sum_{x=-1}^{5} xp(x)$$

$$= (-1)(.05) + (0)(.10) + (1)(.40) + \cdots + (4)(.10) + (5)(.05)$$

$$= 1.70$$

$$\sigma^2 = E[(x - \mu)^2] = \sum_{x=-1}^{5} (x - \mu)^2 p(y)$$

$$= (-1 - 1.70)^2(.05) + (0 - 1.70)^2(.10) + \cdots + (5 - 1.70)^2(.05)$$

$$= 2.11$$

and

$$\sigma = \sqrt{\sigma^2} = \sqrt{2.11} = 1.45$$

The interval $\mu \pm 2\sigma$ is $1.70 \pm (2)(1.45)$ or -1.20 to 4.60.

The graph of $p(x)$ and the interval $\mu \pm 2\sigma$ are shown in Figure 3.12. Note that $x = -1, 0, 1, 2, 3$, and 4 fall in the interval. Therefore,

$$P[\mu - 2\sigma < x < \mu + 2\sigma] = p(-1) + p(1) + p(2) + \cdots + p(4)$$

$$= (.05) + (.10) + (.40) + (.20) + (.10) + (.10)$$

$$= .95$$

Figure 3.12
The probability histogram for $p(x)$ in Example 3.31

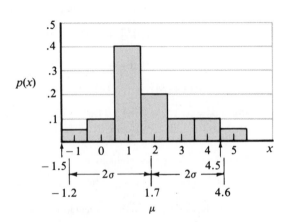

The method for calculating the expected value of x or some function of x for a continuous random variable is similar to what we have done, but in practice it involves the use of calculus. However, the basic results concerning expectations are the same for continuous and discrete random variables. For example, regardless of whether x is continuous or discrete, $E(x) = \mu$, $\sigma^2 = E(x - \mu)^2$, and $E(x - \mu)^2 = E(x^2) - [E(x)]^2$.

EXERCISES   Basic Techniques

3.80   Let x be a discrete random variable with the probability distribution given in the following table.

x	0	1	2	3	4	5
$p(x)$	.05	.10	.20	.40	.20	.05

a.   Find μ, σ^2, and σ.
b.   Construct a probability histogram for $p(x)$.
c.   Locate the interval $\mu \pm 2\sigma$ on the x axis of the histogram. What is the probability that x will fall in this interval?
d.   If you were to select a very large number of values of x from the population, would most fall in the interval $\mu \pm 2\sigma$? Explain.

3.81   Let x be a discrete random variable with the probability distribution given in the following table.

x	1	2	3	4	5	6	7
$p(x)$	.05	.20	.35	.20	.10	.05	.05

a.   Find μ, σ^2, and σ.
b.   Construct a probability histogram for $p(x)$.
c.   Locate the interval $\mu \pm 2\sigma$ on the x axis of the histogram. What is the probability that x will fall in this interval?
d.   If you were to select a very large number of values of x from the population, would most fall in the interval $\mu \pm 2\sigma$? Explain.

3.82   Let x represent the number of times a customer visits a grocery store in a one-week period. Assume that the following is the probability distribution of x.

x	0	1	2	3
$p(x)$	.1	.4	.4	.1

Find the expected value of x. This is the average number of times a customer visits the store.

Applications

3.83   Exercise 3.5 described the game of roulette. Suppose that you were to bet $5 on a single number, say, the number 18. The payoff on this type of bet is usually 35 to 1. What is your expected gain?

3.84   In Exercise 2.25, we learned that big business knows a lot about our habits. One of the statistics quoted by the *Wall Street Journal*, that the number of ice cubes placed in a cold drink is 3.2, affects the amount of beverage consumed per drink. Since the number of ice cubes used per drink will vary, not only from person to person but from glass to glass, suppose that a study of a large number of drinks produced the following probability distribution for the number x of ice cubes used per glass.

x	1	2	3	4	5	6
$p(x)$	.05	.20	.40	.20	.10	.05

a.   Find the expected number of ice cubes used per glass.
b.   Find the variance and standard deviation of x.
c.   What did the Coca-Cola Company mean when it stated that we use 3.2 ice cubes per glass?

3.85   The 1985 Sea Pines Heritage Classic professional golf tournament was held at the Harbour Town Golf Links in Hilton Head, South Carolina. The length and the par score (score expected on a hole if a player makes no errors) for each of the 18 holes is shown below (*USA Today*, April 19, 1985).

Hole	Yards	Par	Hole	Yards	Par
1	415	4	10	443	4
2	497	5	11	423	4
3	412	4	12	420	4
4	188	3	13	358	4
5	528	5	14	153	3
6	404	4	15	556	5
7	172	3	16	398	4
8	455	4	17	169	3
9	332	4	18	465	4
	3403	36		3385	35

One professional golfer plays better on short-distance holes. Experience has shown that the number x of shots required for 3, 4, and 5 par holes has the probability distributions shown below.

Par 3 holes		Par 4 holes		Par 5 holes	
x	$p(x)$	x	$p(x)$	x	$p(x)$
2	.12	3	.14	4	.04
3	.80	4	.80	5	.80
4	.06	5	.04	6	.12
5	.02	6	.02	7	.04

What is the golfer's expected score
a.   on a par 3 hole?
b.   on a par 4 hole?
c.   on a par 5 hole?

3.86 A $50,000 diamond is insured for its total value by paying a premium of D dollars. If the probability of theft in a given year is estimated to be .01, what premium should the insurance company charge if it wants the expected gain to equal $1000?

3.87 The maximum patent life for a new drug is 17 years. Subtracting the length of time required by the FDA for testing and approval of the drug provides the actual patent life of the drug, that is, the length of time that a company has to recover research and development costs and make a profit. Suppose that the distribution of the length of patent life for new drugs is as shown below.

Years x	3	4	5	6	7	8	9	10	11	12	13
$p(x)$	.03	.05	.07	.10	.14	.20	.18	.12	.07	.03	.01

a. Find the expected number of years of patent life for a new drug.
b. Find the standard deviation of x.
c. Find the probability that x falls in the interval $\mu \pm 2\sigma$.

3.88 Experience has shown a shipping company that the cost of delivering a small package within 24 hours is $14.80. The company charges $15.50 for shipment but guarantees to refund the charge if delivery is not made within 24 hours. If the company fails to deliver only 2% of its packages within the 24-hour period, what is the expected gain per package?

3.89 A manufacturing representative is considering taking out an insurance policy to cover possible losses incurred by marketing a new product. If the product is a complete failure, the representative feels that a loss of $80,000 would be incurred; if it is only moderately successful, a loss of $25,000 would be incurred. Insurance actuaries have determined from market surveys and other available information that the probabilities that the product will be a failure or only moderately successful are .01 and .05, respectively. Assuming that the manufacturing representative is willing to ignore all other possible losses, what premium should the insurance company charge for the policy in order to break even?

3.90 Refer to the analysis of the professional golfer's expected scores given in Exercise 3.85. It can be shown (proof omitted) that the expected value of a sum of a set of random variables is equal to the sum of their expected values. Thus if $x_1, x_2, \ldots, x_{18}$ represent the golfer's scores on holes 1, 2, ..., 18, respectively, then the golfer's expected score for the entire 18 holes is equal to the sum of the expected scores for each of the 18 holes. Find the golfer's expected score for one round of 18 holes at the Harbour Town Golf Links.

3.91 The probability that a tennis player A can win a set from tennis player B is one measure of the comparative abilities of the two players. In Exercise 3.78 you found the probability distribution for x, the number of sets required to play a best-of-five-sets match, given that the probability that A wins any one set—call this $P(A)$—is .6.
a. Find the expected number of sets required to complete the match for $P(A) = .6$.
b. Find the expected number of sets required to complete the match when the players are of equal ability, that is, $P(A) = .5$.
c. Find the expected number of sets required to complete the match when the players differ greatly in ability, that is, say, $P(A) = .9$.

▷ 3.11 RANDOM SAMPLING

Since probability distributions are theoretical models for population relative frequency distributions, samples selected from populations can be viewed as observations on random variables. As noted in earlier chapters, the way a sample is selected from a population affects the quantity of information in the sample (Chapter 13) as well as the probabilities of observing particular samples. Since the probability of observed sample outcomes is the basis for inference making (a topic introduced in Chapters 7 and 8), it is important that we now define one of the most commonly employed sampling procedures—**simple random sampling**.

Simple random sampling gives every different sample in the population an equal chance of being selected. To illustrate, suppose that we wish to select a sample of $n = 2$ from a population containing $N = 4$ elements (we are choosing a small value of N to simplify our discussion). If the four elements are identified by the symbols x_1, x_2, x_3, and x_4, then there are six different samples that could be selected from the population, namely,

Sample	Observations in sample
1	x_1, x_2
2	x_1, x_3
3	x_1, x_4
4	x_2, x_3
5	x_2, x_4
6	x_3, x_4

If the sample were selected so that each of these six samples had an equal chance of selection (probability equal to 1/6), the sample would be called a simple random (or, simply, "random") sample.

It can be shown* that the number of ways of selecting $n = 2$ elements from a set of $N = 4$, denoted by the symbol C_2^4, is

$$C_2^4 = \frac{4!}{2!2!} = \frac{4 \cdot 3 \cdot 2 \cdot 1}{(2 \cdot 1)(2 \cdot 1)} = 6$$

The symbol $n!$ (read "n factorial"), defined in Optional Section 3.7, is used to denote the product $n(n - 1)(n - 2) \cdots (3)(2)(1)$. Thus, $5! = 5(4)(3)(2)(1) = 120$. The quantity 0! is defined to be equal to 1. In general, the number of ways of selecting n elements from a set of N is

$$C_n^N = \frac{N!}{n!(N - n)!}$$

For example, if you wish to conduct an opinion poll of 5000 people based on a sample of $n = 100$, there are C_{100}^{5000} different combinations of people who could be selected in the sample. If the sampling is conducted so that each of these combinations has an equal probability of being selected, the sample is called a simple random sample.

---

* An explanation of this result along with examples and applications is given in Section 3.7.

Definition

> Let N and n represent the numbers of elements in the population and sample, respectively. If the sampling is conducted so that each of the C_n^N samples has an equal probability of being selected, the sampling is said to be random and the result is said to be a **simple random sample**.

Perfect random sampling is difficult to achieve in practice. If the population is not too large, we might write each of the N numbers on a poker chip, mix the total, and select a sample of n chips. The numbers on the poker chips would specify the measurements to appear in the sample. In many situations, the population is conceptual, as in an observation made during a laboratory experiment. Here the population is envisioned to be the infinitely large number of measurements obtained when the experiment is repeated over and over again. If we want a sample of $n = 10$ measurements from this population, we repeat the experiment ten times and hope that the results represent, to a reasonable degree of approximation, a random sample.

The simplest and most reliable way to select a random sample of n elements from a large population is to use a table of random numbers (see Table 14, Appendix III). Random-number tables are constructed so that integers occur randomly and with equal frequency. For example, suppose that the population contains $N = 1000$ elements. Number the elements in sequence, 1 to 1000. Then turn to a table of random numbers such as the excerpt shown in Table 3.3.

Table 3.3
Portion of a table of random numbers

15574	35026	98924
45045	36933	28630
03225	78812	50856
88292	26053	21121

Select n of the random numbers in order. The population elements to be included in the random sample will be given by the first three digits of the random numbers (unless the first four digits are 1000). Thus if $n = 5$, we would include elements numbered 155, 450, 32, 882, and 350. To ensure against using the same sequence of random numbers over and over, select different starting points in Table 14 to begin selecting random numbers for different samples.

Not all samples selected in the real world are simple random samples. Some sampling techniques are partly systematic and partly random. For instance, if we wish to determine the voting preference of the nation in a presidential election, we would not chose a random sample from the population of voters. By pure chance, all the voters appearing in the sample might be drawn from a single city, say San Francisco, which might not be representative of the population. We would prefer a random selection of voters from smaller political districts, perhaps states, and allot a specified number of voters for each state. The information from the randomly selected subsamples drawn from the respective states would be combined to form a prediction concerning the entire population of voters in the country. The purpose of systematic sampling, as in the design of experiments in general, is to obtain a maximum of information for a fixed sample size. This, we recall, was one of the five elements of a statistical problem discussed in Chapter 1.

EXERCISES   Basic Techniques

3.92   Find C_n^N for
a. $N = 5, n = 3$ b. $N = 10, n = 4$ c. $N = 15, n = 5$

3.93   Find C_n^N for
a. $N = 7, n = 2$ b. $N = 12, n = 5$ c. $N = 100, n = 3$

3.94   How many different samples of $n = 5$ observations can be selected from a population containing 500 observations?

3.95   If a random sample of four observations is selected from a population containing 100 observations, what is the probability that one particular sample of four observations will be selected?

Applications

3.96   Suppose that a telephone company executive wishes to select a random sample of $n = 20$ (a small number is used to simplify the exercise) out of 7000 customers for a survey of customer attitudes concerning service. If the customers are numbered for identification purposes, indicate the customers whom you will include in your sample. Use the random-number table and explain how you selected your sample.

3.97   A small city contains 20,000 voters. Use the random-number table to identify the voters to be included in a random sample of $n = 15$.

3.98   A random sample of the public opinion for a small town was obtained by selecting every tenth person to pass by the busiest corner in the downtown area. Will this sample have the characteristics of a random sample selected from the town's citizenry? Explain.

3.99   The following notice, appearing in *USA Today* (April 19, 1985), refers to one of the many telephone call-in opinion polls conducted by the major television networks. "NBC will conduct a phone vote via 900 numbers (50 cents a call) during its 1:00 P.M. EST baseball telecasts Saturday on whether a designated hitter should be used in both leagues. Bill Macatee will update the results until 4:00 P.M. from the studio."

   Many television viewers will interpret the proportion of telephone calls responding "yes" to NBC's question as an estimate of the proportion of all baseball fans in favor of using a designated hitter in both leagues. Explain why the respondents will not represent a random sample of the opinions of all baseball fans. Explain the types of distortions that could creep into a call-in opinion poll.

▷ 3.12 SUMMARY

The theories of both probability and statistics are concerned with samples drawn from populations. Probability assumes that the population is known and calculates the probability of observing a particular sample. Statistics assumes the sample to be known and, with the aid of probability, attempts to describe the unknown population frequency distribution.

A random variable is a numerically valued outcome of an experiment. Random variables can be classified as discrete or continuous, depending on whether the number of simple events in the sample space is or is not countable. The theoretical frequency distribution for a discrete random variable is called a probability distribution and often can be derived using the techniques in this chapter.

The expected value of a random variable is the average of the random variable calculated for the theoretical population defined by its probability distribution. Mathematical expectation can be used to find the mean and variance, and hence the standard deviation of a random variable. These quantities can be used to describe a probability distribution in the same way that the mean and standard deviation were used to describe distribution of data in Chapter 2.

Inferences about a population will be based on the probability of an observed sample, and this probability will depend on how the sample was selected from the population. Simple random sampling is the most common method for selecting a sample from a population. A simple random sample is selected from the population in such a way that every possible different sample has an equal probability of being selected. The inferential techniques described in Chapter 6 and later chapters are derived assuming that simple random sampling has been used.

REFERENCES

Feller, W. *An Introduction to Probability Theory and Its Applications*, Vol. 1. 3d ed. New York: Wiley, 1968.

Freund, J. E., and Walpole, R. E. *Mathematical Statistics*. 4th ed. Englewood Cliffs, N.J.: Prentice Hall, 1987.

Hoel, P. G.; Port, S. C.; and Stone, C. J. *Introduction to Probability Theory*. Boston: Houghton Mifflin, 1971.

Hogg, R. V. and Craig, A. T. *Introduction to Mathematical Statistics*. 4th ed. New York: Macmillan, 1986.

Mendenhall, W.; Wackerly, D; and Scheaffer, R. L. *Mathematical Statistics with Applications*. 4th ed. Boston: PWS-KENT, 1990.

Meyer, P. L. *Introductory Probability and Statistical Applications*. New York: Wiley, 1960.

Mood, A. M., et al. *Introduction to the Theory of Statistics*. 3d ed. New York: McGraw-Hill, 1973.

Riordan, J. *An Introduction to Combinatorial Analysis*. Princeton, N.J.: Princeton University Press, 1980.

Ross, S. *A First Course in Probability*. 3d ed. New York: Macmillan, 1988.

Triola, M. F. *Elementary Statistics*. 4th ed. Redwood City, Calif.: Benjamin/Cummings, 1989.

SUPPLEMENTARY EXERCISES

[Starred (*) exercises are optional]

3.100 Identify the following as continuous or discrete random variables:
a. The number of homicides in Detroit during a one-month period.
b. The length of time between arrivals at an outpatient clinic.

    c.   The number of typing errors on a page of typing.

    d.   The number of defective light bulbs in a packet containing four bulbs.

    e.   The time required to finish an examination.

3.101   Identify the following as discrete or continuous random variables:

    a.   The weight of two dozen shrimp.

    b.   A person's body temperature.

    c.   The number of people waiting for treatment at a hospital emergency room.

    d.   The number of properties for sale by a real estate agency.

    e.   The number of claims received by an insurance company during one day.

3.102   Identify the following as discrete or continuous random variables:

    a.   The number of people in line at a supermarket checkout counter.

    b.   The depth of a snowfall.

    c.   The length of time for an automobile driver to respond when faced with an impending collision.

    d.   The number of aircraft arriving at the Atlanta airport in a given hour.

3.103   The discrete random variable x and its probability distribution $p(x)$ are as follows:

x	0	2	3	4
$p(x)$	1/8	1/4	1/2	1/8

    a.   Find $\mu = E(x)$.   b.   Find $E(x^2)$.   c.   Find σ^2 and σ.

3.104   Let x equal the number of dots observed when a die is tossed, and let $p(x)$ equal 1/6, $x = 1, 2, 3, \ldots, 6$. We found in Example 3.28 that $\mu = E(x) = 3.5$. Find σ^2 and show that $\sigma = 1.71$. Then find the probability that x will fall in the interval $\mu \pm 2\sigma$.

3.105   Draw a sample of $n = 50$ measurements from the die-throwing population of Example 3.26 by tossing a die 50 times and recording x after each toss. Calculate $\bar{x}$ and s^2 for the sample. Compare $\bar{x}$ with the expected value of x in Example 3.28 and s^2 with the variance of x obtained in Exercise 3.104. Do $\bar{x}$ and s^2 provide good estimates of μ and σ^2?

3.106   A survey conducted by the Merit Systems Protection Board found that 69% of workers who knew firsthand of some example of government waste failed to report it. Approximately 23% of those who had reported a case of fraud said that they had suffered some reprisal, such as demotion, poor performance rating, and so on (*Orlando Sentinel*, January 20, 1985). Assume that the probability that a worker will fail to report a case of fraud is the same as the probability of failing to report waste, .69. Find the probability that a worker who observes a case of fraud will report it and will subsequently suffer some form of reprisal.

3.107   Two cold tablets are accidentally placed in a box containing two aspirin tablets. The four tablets are identical in appearance. One tablet is selected at random from the box and is swallowed by patient A. A tablet is then selected at random from the three remaining tablets and is swallowed by patient B. Define the following events as specific collections of simple events:

    a.   The sample space S.

    b.   The event A that patient A obtained a cold tablet.

    c.   The event B that exactly one of the two patients obtained a cold tablet.

    d.   The event C that neither patient obtained a cold tablet.

3.108   Refer to Exercise 3.107. By summing probabilities of simple events find $P(A)$, $P(B)$, $P(AB)$, $P(A \cup B)$, $P(C)$, $P(AC)$, and $P(A \cup C)$.

3.109　A coin is tossed four times and the outcome is recorded for each toss.
　　　a.　List the simple events for the experiment.
　　　b.　Let A be the event that the experiment yields exactly three heads. List the simple events in A.
　　　c.　Assign reasonable probabilities to the simple events and find $P(A)$.

3.110　A retailer sells two styles of high-priced, compact disk players that experience indicates are in equal demand. (Fifty percent of all potential customers prefer style 1, and 50% favor style 2.) If the retailer stocks four of each, what is the probability that the first four customers seeking a CD player all purchase the same style?
　　　a.　Define the experiment.
　　　b.　List the simple events.
　　　c.　Define the event of interest A as a specific collection of simple events.
　　　d.　Assign probabilities to the simple events and find $P(A)$.

3.111　A boxcar contains seven complex electronic systems. Unknown to the purchaser, three are defective. Two of the seven are selected for thorough testing and are then classified as defective or nondefective.
　　　a.　List the simple events for this experiment.
　　　b.　Let A be the event that the selection includes no defectives. List the simple events in A.
　　　c.　Assign probabilities to the simple events and find $P(A)$.

3.112　A heavy-equipment salesman can contact either one or two customers per day with probability 1/3 and 2/3, respectively. Each contact will result in either no sale or a $50,000 sale with probability 9/10 and 1/10, respectively. What is the expected value of his daily sales?

3.113　A county containing a large number of rural homes is thought to have 60% of those homes insured against a fire. Four rural home owners are chosen at random from the entire population, and x are found to be insured against a fire. Find the probability distribution for x. What is the probability that at least three of the four will be insured?

3.114　A fire-detection device uses three temperature-sensitive cells acting independently of each other in such a manner that any one or more can actuate the alarm. Each cell has a probability of $p = .8$ of actuating the alarm when the temperature reaches 100 degrees or more. Let x equal the number of cells actuating the alarm when the temperature reaches 100 degrees. Find the probability distribution for x. Find the probability that the alarm will function when the temperature reaches 100 degrees.

3.115　Find the expected value and variance for the random variable x defined in Exercise 3.114.

3.116　If you toss a pair of dice, the sum T of the numbers of dots appearing on the upper faces of the dice can assume the value of an integer in the interval $2 \leq T \leq 12$.
　　　a.　Find $E(T)$.
　　　b.　Find σ^2.
　　　c.　Use your graph of $p(T)$, found in Exercise 3.72, and locate the interval $\mu \pm 2\sigma$.
　　　d.　What is the probability that T will assume a value in the interval $\mu \pm 2\sigma$?

3.117　A die is tossed twice. What is the probability that the sum of the numbers observed will be greater than nine?

3.118　Toss a die and a coin. If event A is the occurrence of a head and an even number, and event B is the occurrence of a head and a 1, find $P(A)$, $P(B)$, $P(AB)$, and $P(A \cup B)$. (Solve by listing the simple events.)

3.119　Refer to Exercise 3.118 and calculate $P(AB)$ and $P(A \cup B)$. Use the laws of probability.

3.120　A salesman figures that the probability of his consummating a sale during the first contact with a client is .4, but improves to .55 on the second contact if the client did not buy during the

first contact. Suppose that the salesman makes one and only one callback to any client. If the salesman contacts a client, calculate
a.   the probability that the client will buy.
b.   the probability that the client will not buy.

3.121   A man takes either a bus or the subway to work with probabilities .3 and .7, respectively. When he takes the bus, he is late 30% of the days. When he takes the subway, he is late 20% of the days. If the man is late for work on a particular day, what is the probability that he took the bus?

3.122   Suppose that independent events A and B have nonzero probabilities. Show that A and B cannot be mutually exclusive.

3.123   The failure rate for a guided missile control system is 1 in 1000. Suppose that a duplicate, but completely independent, control system is installed in each missile so that if the first fails, the second can still take over. The reliability of a missile is the probability that it does not fail. What is the reliability of the modified missile?

3.124   Suppose that at a particular supermarket the possibility of waiting five minutes or longer for checkout at the cashier's counter is .2. On a given day, a man and his wife decide to shop individually at the market, each checking out at different cashier counters. If they both reach cashier counters at the same time, answer the following questions:
a.   What is the probability that the man will wait less than five minutes for checkout?
b.   What is the probability that both the man and his wife will be checked out in less than five minutes? (Assume that the checkout times for the two are independent events.)
c.   What is the probability that one or the other or both will wait five minutes or more?

3.125   A quality control plan calls for accepting a large lot of crankshaft bearings if a sample of seven is drawn and none is defective. What is the probability of accepting the lot if none in the lot is defective? if 1/10 is defective? if 1/2 is defective?

3.126   It is said that only 40% of all people in a community favor the development of a mass transit system. If four citizens are selected at random from the community, what is the probability that all four favor the mass transit system? that none favors the mass transit system?

3.127   The weatherman forecasts rain with probability .6 today and .4 tomorrow. Experience has shown that in this particular locale, it rains one day in four and the probability of rain on two successive days is .15. If it is raining outside when we hear his forecast (as so often happens), what is the probability of rain tomorrow?

3.128   A research physician compared the effectiveness of two blood pressure drugs A and B by administering the two drugs to each of four pairs of identical twins. Drug A was given to one member of a pair, drug B to the other. If, in fact, there is no difference in the effect of the drugs, what is the probability that the drop in the blood pressure reading for drug A would exceed the corresponding drop in the reading for drug B for all four pairs of twins? Suppose that drug B created a greater drop in blood pressure than drug A for each of the four pairs of twins. Do you think this provides sufficient evidence to indicate that drug B is more effective in lowering blood pressure than drug A?

3.129   To reduce the cost of detecting a disease, blood tests are conducted on a pooled sample of blood collected from a group of n people. If no indication of the disease is present in the pooled blood sample (as is usually the case), none have the disease. If analysis of the pooled blood sample indicates that the disease is present, each individual must submit to a blood test. The individual tests are conducted in sequence. If among a group of five people, one person has the disease, what is the probability that six blood tests (including the pooled test) are required to detect the single diseased person? If two people have the disease, what is the probability that six tests are required to locate both diseased people?

3.130 How many times should a coin be tossed to obtain a probability equal to or greater than .9 of observing at least one head?

3.131 An oil prospector will drill a succession of holes in a given area to find a productive well. The probability that she is successful on a given trial is .2.
    a. What is the probability that the third hole drilled is the first hole to locate a productive well?
    b. If her total resources allow the drilling of only three holes, what is the probability that she locates a productive well?

3.132 Suppose that two defective refrigerators are included in a shipment of six refrigerators. The buyer begins to test the six refrigerators one at a time.
    a. What is the probability that the last defective refrigerator is found on the fourth test?
    b. What is the probability that no more than four refrigerators need be tested before both defective refrigerators are located?
    c. Given that one defective refrigerator has been located in the first two tests, what is the probability that the remaining defective refrigerator is found on the third or fourth test?

3.133 Two men each toss a coin. They obtain a "match" if either both coins are heads or both are tails. Suppose the tossing is repeated three times.
    a. What is the probability of three matches?
    b. What is the probability that all six tosses (three for each man) result in tails?
    c. Coin tossing provides a model for many practical experiments. Suppose that the coin tosses represented the answers given by two students for three specific true–false questions on an examination. If the two students gave three matches for answers, would the low probability found in (a) suggest collusion?

3.134 Experience has shown that 50% of the time, a particular union-management contract negotiation led to a contract settlement within a two-week period, 60% of the time the union strike fund was adequate to support a strike, and 30% of the time both conditions were satisfied. What is the probability of a contract settlement given that the union strike fund is adequate to support a strike? Is settlement of a contract within a two-week period dependent on whether the union strike fund is adequate to support a strike?

3.135 Suppose the probability of remaining with a particular company ten years or more is 1/6. A man and a woman start work at the company on the same day.
    a. What is the probability that the man will work there less than ten years?
    b. What is the probability that both the man and the woman will work there less than ten years? (Assume that they are unrelated and their lengths of service are independent of each other.)
    c. What is the probability that one or the other or both will work ten years or more?

3.136 Accident records collected by an automobile insurance company give the following information: The probability that an insured driver has an automobile accident is .15; if an accident has occurred, the damage to the vehicle amounts to 20% of its market value with probability .80, 60% of its market value with probability .12, and a total loss with probability .08. What premium should the company charge on a $12,000 car so that the expected gain by the company is zero?

3.137 If you own common stock in a corporation, you can sometimes increase the return on your investment by selling an option. For a stated price, the purchaser of the option gains the right to buy your stock at any time up to a specified expiration date. Suppose you purchased 200 shares of a stock at $25 per share and you sell an option for the option purchaser to buy the stock at any time within the next six months for $30 per share. If the stock reaches $30 per share within the next six months, your stock will be sold and you will gain $5 per share plus $2 per share (from the dividends and the sale of the option) less $109 commission to the broker for selling your stock. If you do not sell, you will gain $2 per share. If the probability of the stock reaching $30 per share within the next six months is .7, what is the expected return (in dollars) from your 200 shares of stock? What is the expected annual rate of return, in percentage, on your investment? (*Note:* The $15 commission on the sale of the option has been ignored.)

3.138   A political analyst wishes to select a sample of $n = 20$ people from a population of 2000. Use the random-number table to identify the people to be included in the sample.

3.139   A population contains 50,000 voters. Use the random-number table to identify the voters to be included in a random sample of $n = 15$.

*3.140   How many different telephone numbers of five digits can be formed if the first digit is a 3 or a 4?

*3.141   A piece of equipment can be assembled in three operations that can occur in any sequence.
a.   Give the total number of ways in which the equipment can be assembled.
b.   Comparative tests are to be conducted to determine the best assembly procedure. If each assembly procedure is to be tested and compared with every other procedure exactly once, how many tests must be conducted?

*3.142   Electronic fuses are produced by five production lines in a manufacturing operation. The fuses are costly and reliable, and are shipped to suppliers in 100-unit lots. Because testing is destructive, most buyers test only a small number of fuses before deciding to accept or reject incoming lots. All five production lines produce fuses at the same rate and normally produce only 2% defective fuses, which are randomly dispersed in the output. Unfortunately, production line 1 suffered mechanical difficulty and produced 5% defectives during the month of March. The manufacturer discovered the problem only after the fuses had been shipped. A customer received a lot produced in March and tested three fuses. One failed. What is the probability that the lot came from one of the four other lines?

*3.143   A student prepares for an exam by studying a list of ten problems. He can solve six of them. For the exam, the instructor selects five questions at random from the list of ten. What is the probability that the student can solve all five problems on the exam?

*3.144   A monkey is given twelve blocks—three shaped like squares, three like rectangles, three like triangles, and three like circles. If he draws three of each kind in order—say, three triangles, then three squares, and so on—would you suspect that the monkey associates identically shaped figures? Calculate the probability of this event.

PROBABILITY AND DECISION MAKING IN THE CONGO

In his exciting novel *Congo* (New York: Alfred A. Knopf, 1980), Michael Crichton describes a search by Earth Resources Technology Services (ERTS), a geological survey company, for deposits of boron-coated blue diamonds, diamonds that ERTS believes to be the key to a new generation of optical computers. In the novel, ERTS is racing against an international consortium to find the Lost City of Zinj, a city that thrived on diamond mining and existed several thousand years ago (according to African fable), deep in the rain forests of eastern Zaire.

After the mysterious destruction of its first expedition, ERTS launches a second expedition under the leadership of Karen Ross, a 24-year-old computer genius who is accompanied by Professor Peter Elliot, an anthropologist; Amy, a talking gorilla; and the famed mercenary and expedition leader, "Captain" Charles Munro. Ross's efforts to find the city are blocked by the consortium's offensive actions, by the deadly rain forest, and by hordes of "talking" killer gorillas whose perceived mission is to defend the diamond mines. Ross overcomes these obstacles by using space-age computers to evaluate the probabilities of success for all possible circumstances and all possible actions that the expedition might take. At each stage of the expedition, she is able to quickly evaluate the chances of success.

At one stage in the expedition, Ross is informed by her Houston headquarters that their computers estimate that she is 18 hours, 20 minutes behind the competing Euro–Japanese team, instead of 40 hours ahead. She changes plans and decides to have the 12 members of her team—Ross, Elliot, Munro, Amy, and eight native porters—parachute into a volcanic region near the estimated location of Zinj. As Crichton relates, "Ross had double-checked outcome probabilities from the Houston computer, and the results were unequivocal. The probability of a successful jump was .7980, meaning that there was approximately one chance in five that someone would be badly hurt. However, given a successful jump, the probability of expedition success was .9943, making it virtually certain thay they would beat the consortium to the site."

Keeping in mind that this is an excerpt from a novel, let us examine the probability, .7980, of a successful jump. If you were one of the 12-member team, what is the probability that you would successfully complete your jump? In other words, if the probability of a successful jump by all 12 team members is .7980, what is the probability that a single member could successfully complete the jump?

CHAPTER 4

SEVERAL USEFUL DISCRETE DISTRIBUTIONS

Case Study

Is the Pilgrim I nuclear reactor responsible for an increase in cancer cases in the surrounding area? A political controversy was set off when the Massachusetts Department of Public Health found an unusually large number of cases in a four-mile-wide coastal strip just north of the nuclear reactor in Plymouth, Massachusetts. The case study at the end of Chapter 4 examines how this question can be answered using one of the discrete probability distributions presented in this chapter.

General Objectives

In Chapter 3 you learned the difference between discrete and continuous random variables. In this chapter, you will study several useful discrete random variables, and you will learn how the formulas for their probability distributions, means, and standard deviations are acquired. An introduction to the subject of inference making is provided in the form of lot acceptance sampling for defectives.

Specific Topics

1 The binomial probability distribution (4.2)
2 The mean and variance for the binomial random variable (4.2)
3 The Poisson probability distribution (4.3)
4 The hypergeometric probability distribution (4.4)
5 The uniform and geometric probability distributions (4.5)

> ## 4.1   INTRODUCTION

In Chapter 3 we found that random variables defined over a finite or countably infinite number of simple events are called **discrete random variables**. Examples of discrete random variables abound in the physical and social sciences, as well as in business and economics, but three discrete probability distributions serve as **models** for a large number of these applications. These three distributions are **the binomial, the Poisson, and the hypergeometric probability distributions**. In this chapter we will study these distributions, discussing their development as logical models for discrete processes observed in different physical settings.

> ## 4.2   THE BINOMIAL PROBABILITY DISTRIBUTION

One of the most elementary, useful, and interesting discrete random variables, the binomial random variable, is associated with the coin-tossing experiment described in Examples 3.5 and 3.25. As an illustration, consider a sample survey conducted to predict voter preference in a political election. Interviewing a single voter bears a similarity, in many respects, to tossing a single coin, because the voter's response may be in favor of our candidate—a "head"—or may be against (or indicate indecision)—a "tail." In most cases, the fraction of voters favoring a particular candidate does not equal one-half, but in most national presidential elections, the fraction of the total vote favoring the winning presidential candidate is *very near* one-half.

Similar polls are conducted in the social sciences, in industry, and in education. The sociologist is interested in the fraction of army recruits who are white females; the marketer of soft drinks desires knowledge concerning the fraction of cola drinkers who prefer his or her brand; the teacher is interested in the fraction of high school seniors who pass basic proficiency tests in reading, language, and mathematics. Each person sampled is analogous to the toss of an unbalanced coin, since the probability of a "head" is usually not one-half. Although dissimilar in some respects, the surveys described above often exhibit, to a reasonable degree of approximation, the characteristics of a **binomial experiment**.

Definition

> **A binomial experiment** is one that has the following properties:
>
> 1. The experiment consists of n identical trials.
> 2. Each trial results in one of two outcomes. For lack of a better nomenclature, the one outcome is called a success, S, and the other a failure, F.
> 3. The probability of success on a single trial is equal to p and remains the same from trial to trial. The probability of a failure is equal to $(1 - p) = q$.
> 4. The trials are independent.
> 5. We are interested in x, the number of successes observed during the n trials.

EXAMPLE 4.1   Suppose that there are approximately 1,000,000 adults in a county and that an unknown proportion p favor the Equal Rights Amendment (ERA). A sample of 1000 adults will be chosen in such a way that every one of the 1,000,000 adults has an equal chance of being selected, and each adult is asked whether he or she favors the ERA. (The ultimate objective of this survey is to estimate the unknown proportion p, a problem that we will discuss in Chapter 7.) Is this a binomial experiment?

Solution   To decide whether this is a binomial experiment, we must see whether the sampling satisfies the five characteristics described in the preceding definition.

1. The sampling consists of $n = 1000$ identical trials. One trial represents the selection of a single adult from the 1,000,000 adults in the county.

2. Each trial will result in one of two outcomes. A person will either favor the amendment or will not. These two outcomes could be associated with the "success" and "failure" of a binomial experiment.*

3. The probability of a success will equal the proportion of adults favoring the ERA. For example, if 500,000 of the 1,000,000 adults in the county favor the ERA, then the probability of selecting an adult favoring the ERA out of the 1,000,000 in the county is $p = .5$. For all practical purposes, this probability will remain the same from trial to trial even though adults selected in the earlier trials are not replaced as the sampling continues.

4. For all practical purposes, the probability of a success on any one trial will be unaffected by the outcome on any of the others (it will remain very close to p).

5. We are interested in the number x of adults in the sample of 1000 who favor the ERA.

Because the survey satisfies the five characteristics reasonably well, for all practical purposes it can be viewed as a binomial experiment.   ◁

EXAMPLE 4.2   A purchaser who has received a shipment containing 20 personal computers (PCs) wants to sample three of the PCs to see if they are in working order before accepting the shipment. The nearest three PCs are selected for testing and, afterward, are declared either defective or nondefective. Unknown to the purchaser, two of the PCs in the shipment of 20 are defective. Is this a binomial experiment?

Solution   Again, we check the sampling procedure against the characteristics of a binomial experiment.

1. The experiment consists of $n = 3$ identical trials. Each trial represents the selection and testing of one PC from the total of 20.

2. Each trial results in one of two outcomes. Either a PC is defective (call this a "success") or it is not (a "failure").

---

* Although it is traditional to call the two possible outcomes of a trial "success" and "failure," they could have been called "head" and "tail," "red" and "white," or any other pair of words. Consequently, the outcome called a success does not need to be viewed as a success in the ordinary usage of the word.

3. Suppose that the PCs were randomly loaded into a boxcar so that any one of the 20 PCs could have been placed near the boxcar door. Then the unconditional probability of drawing a defective PC on a given trial would be 2/20.

4. The condition of independence between trials is *not* satisfied because the probability of drawing a defective PC on the second and third trials will be dependent on the outcome of the first trial. For example, if the first trial results in a defective PC, then there is only one defective left in the remaining 19 in the boxcar. Therefore, the conditional probability of success on trial 2, given a success on trial 1, is 1/19. This differs from the unconditional probability of a success on the second trial (which is 2/20). Thus the trials are dependent and the sampling does not represent a binomial experiment.   ◁▭

Example 4.2 illustrates an important point. If the sample size n is large relative to the population size N, then the probability of success p will not remain constant from trial to trial. Hence, the trial outcomes will be dependent and the resulting experiment will not be a binomial experiment. **As a rule of thumb, if $n/N \geq .05$, the resulting experiment will not be binomial.**

The probability distribution for a simple binomial random variable (the number of heads in the tosses of two coins) was derived in Example 3.25. The probability distribution for a binomial experiment consisting of n tosses is derived in exactly the same way, but the procedure is much more complex when the number n of trials is large. We will omit this derivation and will simply present the **binomial probability distribution** and its mean, variance, and standard deviation, as shown in the following display.

BINOMIAL PROBABILITY DISTRIBUTION

$$p(x) = C_x^n p^x q^{n-x} = \frac{n!}{x!(n-x)!} p^x q^{n-x}$$

where x, the number of successes in n trials, may take values $0, 1, 2, \ldots, n$; p is the probability of success on a single trial; and C_x^n is defined as

$$\frac{n!}{x!(n-x)!}$$

Mean: $\mu = np$
Variance: $\sigma^2 = npq$
Standard Deviation: $\sigma = \sqrt{npq}$

In the formula for $p(x)$ the quantity $p^x q^{n-x}$ represents the probability of observing a simple event with x successes and $(n-x)$ failures; the term C_x^n, defined as $n!/x!(n-x)!$, counts the number of such simple events. The probability $p(x) = C_x^n p^x q^{n-x}$ associated with a particular value of x in n independent trials is the term involving p to the power x in the series expansion of the binomial $(p+q)^n$—hence, the name *binomial probability*.

Figure 4.1
Binomial probability
distributions

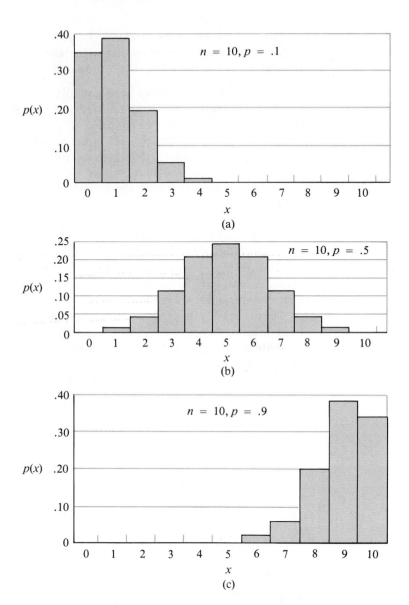

Graphs of three binomial probability distributions are shown in Figure 4.1—the first for $n = 10, p = .1$; the second for $n = 10, p = .5$; and the third for $n = 10, p = .9$.

EXAMPLE 4.3   Over a long period of time it has been observed that a given rifleman can hit a target on a single trial with probability equal to .8. Suppose that he fires four shots at the target.

a. What is the probability that he will hit the target exactly two times?
b. What is the probability that he will hit the target at least two times?
c. What is the probability that he will hit the target exactly four times?

Solution   Assume that the trials are independent and that the probability p of hitting the target remains constant from trial to trial, $n = 4$ and $p = .8$. Let x denote the number of shots that hit the target. Then, for $x = 0, 1, 2, 3, 4$, we have

$$p(x) = C_x^4(.8)^x(.2)^{4-x}$$

a. $p(2) = C_2^4(.8)^2(.2)^{4-2}$

$$= \frac{4!}{2!2!}(.64)(.04)$$

$$= \frac{4(3)(2)(1)}{2(1)(2)(1)}(.64)(.04)$$

$$= .1536$$

The probability is .1536 that he will hit the target exactly two times.

b. $P(\text{at least two}) = p(2) + p(3) + p(4)$

$$= 1 - p(0) - p(1)$$

$$= 1 - C_0^4(.8)^0(.2)^4 - C_1^4(.8)(.2)^3$$

$$= 1 - .0016 - .0256$$

$$= .9728$$

The probability is .9728 that he will hit the target at least two times.

c. $p(4) = C_4^4(.8)^4(.2)^0$

$$= \frac{4!}{4!0!}(.8)^4(1)$$

$$= .4096$$

The probability is .4096 that he will hit the target exactly four times.

Note that these probabilities would be incorrect if the rifleman could observe the location of each hit on the target and thereby adjust his aim. In that case, the trials would be dependent and p would likely increase from trial to trial.   ◁

LOT ACCEPTANCE SAMPLING

EXAMPLE 4.4   Large lots of incoming product at a manufacturing plant are inspected for defectives by using a **sampling plan**, in which a random sample of size n is selected from each lot and inspected, and the number x of defectives is recorded. If x is less than or equal to a specified acceptance number, a, the lot is accepted. If x exceeds a, the lot is rejected. Suppose that a manufacturer employs a sampling plan with $n = 10$ and $a = 1$. If the lot contains exactly 5% defectives, what is the probability that the lot will be accepted? rejected? Assume that successive draws are independent.

Solution   Let x be the number of defectives observed. Then $n = 10$ and the probability of observing a defective on a single trial is $p = .05$. Hence, $p(x) = C_x^{10}(.05)^x(.95)^{10-x}$, and

$$P(\text{accept}) = p(0) + p(1) = C_0^{10}(.05)^0(.95)^{10} + C_1^{10}(.05)^1(.95)^9$$

$$= .914$$

$$P(\text{reject}) = 1 - P(\text{accept})$$

$$= 1 - .914 = .086$$

Although in a practical situation we would not know the exact value of p, we would want to know the probability of accepting bad lots (lots for which p is large) and good lots (lots for which p is small). This example shows how to calculate this probability of acceptance for various values of p.

A graph of the probability of lot acceptance for various lot fractions that are defective is called the **operating characteristic curve** for the sampling plan. A satisfactory lot acceptance sampling plan is one for which **the probability of accepting lots with a low fraction defective should be high and the probability of accepting lots with a high fraction defective should be low**. The probability of acceptance always decreases as the fraction defective increases, a result that is in agreement with our intuition.

EXAMPLE 4.5   Calculate the probability of lot acceptance for a sampling plan with sample size $n = 5$ and acceptance number $a = 0$ for lot fraction defectives $p = .1, .3,$ and $.5$. Sketch the operating characteristic curve for the plan.

Solution   The lot will be accepted if $n = 5$ items are sampled and $a = 0$ defectives are observed.

$$P(\text{accept}) = p(0) = C_0^5 p^0 q^5 = q^5$$

$$P(\text{accept given that } p = .1) = (.9)^5 = .590$$

$$P(\text{accept given that } p = .3) = (.7)^5 = .168$$

$$P(\text{accept given that } p = .5) = (.5)^5 = .031$$

A sketch of the operating characteristic curve can be obtained by plotting the three calculated points and connecting the points with a smooth curve. In addition, the probability of acceptance must equal 1 when $p = 0$ and must equal 0 when $p = 1$. The operating characteristic curve is given in Figure 4.2.

Acceptance sampling, which operates in an objective manner, is an example of statistical inference because the procedure implies a decision concerning the lot fraction defective p. If you accept the lot, you infer that the true lot fraction defective p is some relatively small, acceptable value. If you reject a lot, it is clear that you think p is too large. Consequently, lot acceptance sampling is an inference-making procedure concerning the lot fraction defective. The operating characteristic curve for the sampling plan provides a measure of the goodness of this inferential procedure.

Figure 4.2
Operating characteristic
curve, $n = 5$, $a = 0$

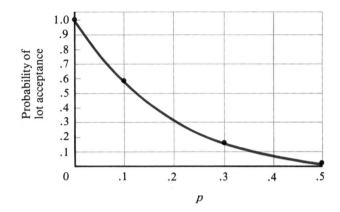

Calculating binomial probabilities is a tedious task when n is large. To simplify our calculations, the sum of the binomial probabilities from $x = 0$ to $x = a$ is presented in Table 1 of Appendix III for $n = 2, 3, \ldots, 12, 15, 20,$ and 25. More extensive tables, which give the sums of the binomial probabilities for many different values of n and p, are listed in the references at the end of this chapter.

To illustrate the use of Table 1, suppose we wish to find the sum of the binomial probabilities from $x = 0$ to $x = 3$ for $n = 5$ trials and $p = .6$. That is, we wish to find

$$P(x \leq 3) = \sum_{x=0}^{3} p(x) = p(0) + p(1) + p(2) + p(3)$$

where

$$p(x) = C_x^5(.6)^x(.4)^{5-x}$$

Since the table values give

$$P(x \leq a) = \sum_{x=0}^{a} p(x)$$

we seek the table value in the row corresponding to $a = 3$ and the column for $p = .6$. The table value, .663, is shown in Table 4.1 on page 138 as it appears in Table 1, Appendix III. Therefore, the sum of the binomial probabilities from $x = 0$ to $x = a = 3$ (for $n = 5$, $p = .6$) is .663.

Table 1 can also be used to find individual binomial probabilities. For example, suppose we wish to find $p(3)$ when $n = 5$ and $p = .6$. Since $P(x = 3) = P(x \leq 3) - P(x \leq 2)$, we write

$$p(3) = \sum_{x=0}^{3} p(x) - \sum_{x=0}^{2} p(x) = .663 - .317 = .346$$

The values $\sum_{x=0}^{3} p(x)$ and $\sum_{x=0}^{2} p(x)$ are found directly from Table 1, indexing $n = 5$ with $p = .6$. In general, an individual binomial probability may be found by subtracting successive entries in the table for a given value of p.

Table 4.1 Portion of Table 1, Appendix III, for $n = 5$

a	0.01	0.05	0.10	0.20	0.30	0.40	*p* 0.50	0.60	0.70	0.80	0.90	0.95	0.99	a
0	—	—	—	—	—	—	—	—	—	—	—	—	—	0
1	—	—	—	—	—	—	—	—	—	—	—	—	—	1
2	—	—	—	—	—	—	—	—	—	—	—	—	—	2
3	—	—	—	—	—	—	—	.663	—	—	—	—	—	3
4	—	—	—	—	—	—	—	—	—	—	—	—	—	4

EXAMPLE 4.6   A regimen consisting of a daily dose of vitamin C was tested to determine its effectiveness in preventing the common cold. Ten people who were following the prescribed regimen were observed for a period of one year. Eight survived the winter without a cold. Suppose that the probability of surviving the winter without a cold is .5 when the vitamin C regimen is not followed. What is the probability of observing eight or more survivors, given that the regimen is ineffective in increasing resistance to colds?

Solution   Assuming that the vitamin C regimen is ineffective, the probability p of surviving the winter without a cold is .5. The probability distribution for x, the number of survivors, is

$$p(x) = C_x^{10}(.5)^x(.5)^{10-x}$$

Then, by direct calculation,

$$P(8 \text{ or more}) = p(8) + p(9) + p(10)$$
$$= C_x^{10}(.5)^{10} + C_x^{10}(.5)^{10} + C_x^{10}(.5)^{10}$$
$$= .055$$

If we consult Table 1 in Appendix III for $n = 10$, we have

$$P(x \geq 8) = \sum_{x=8}^{10} p(x) = 1 - \sum_{x=0}^{7} p(x)$$

The quantity

$$\sum_{x=0}^{7} p(x)$$

can be found by moving across the top of Table 1 to the column for $p = .5$ and down that column to the row corresponding to $a = 7$. We read

$$\sum_{x=0}^{7} p(x) = .945$$

Then, as before,

$$P(x \geq 8) = \sum_{x=8}^{10} p(x) = 1 - \sum_{x=0}^{7} p(x) = 1 - .945 = .055 \qquad \triangleleft$$

Both individual and cumulative binomial probabilities are available through the MINITAB package. Individual binomial probabilities associated with each value of x for any combination of n and p can be found by using the probability density function command PDF followed by a semicolon (;) and then the subcommand BINOMIAL N P followed by a period. The variable N is the given sample size and P is the probability of success. Cumulative binomial probabilities can be found by using the cumulative distribution function command CDF followed by a semicolon (;) and then the subcommand BINOMIAL N P followed by a period.

The MINITAB output for both the PDF and CDF commands when $n = 10$ and $p = .5$ is given in Table 4.2. The PDF command gives rise to the individual probabilities $P(x = K)$; the CDF command gives rise to the cumulative probabilities $P(x \leq K)$. (The letter K plays the role of the letter a in Table 1 of Appendix III.) Notice that in the MINITAB output $P(x \leq 7) = .9453$, so that $P(x \geq 8) = (1 - .9453) = .0547$, which, to three-decimal accuracy, agrees with our earlier results using Table 1.

Table 4.2 MINITAB output of binomial probabilities when $n = 10$ and $p = .5$

MTB > PDF;		MTB > CDF;	
SUBC > BINOMIAL 10 .5.		SUBC > BINOMIAL 10 .5.	

BINOMIAL WITH N = 10 P = 0.500000		BINOMIAL WITH N = 10 P = 0.500000	
K	P(X = K)	K P(X LESS OR = K)	
0	0.0010	0	0.0010
1	0.0098	1	0.0107
2	0.0439	2	0.0547
3	0.1172	3	0.1719
4	0.2051	4	0.3770
5	0.2461	5	0.6230
6	0.2051	6	0.8281
7	0.1172	7	0.9453
8	0.0439	8	0.9893
9	0.0098	9	0.9990
10	0.0010	10	1.0000

The MINITAB output may not list the probabilities for all values of $x = 0, 1, 2, \ldots, n$ for various combinations of n and p, since the MINITAB package has an internal check that stops the calculations when $P(x = K) = 0$ [or equivalently, when $P(x \leq K) = 1$] to within a preassigned level of accuracy.

EXAMPLE 4.7 Find the mean and standard deviation for a binomial probability distribution with $n = 10$ and $p = .5$. Find the probability that x falls in the interval $\mu \pm 2\sigma$.

Solution   The mean and standard deviation are

$$\mu = np = 10(.5) = 5$$

$$\sigma = \sqrt{npq} = \sqrt{10(.5)(.5)} = 1.58$$

Therefore, the interval $\mu \pm 2\sigma$ is given by

$$5 \pm 2(1.58) = 5 \pm 3.2$$

or from 1.8 to 8.2. This interval includes the values $x = 2, 3, \ldots, 8$. Therefore,

$$P[2 \le x \le 8] = P[x \le 8] - P[x \le 1]$$

$$= .989 - .011 = .978$$

a result that agrees with Tchebysheff's Theorem and agrees approximately with the Empirical Rule.

EXAMPLE 4.8   How are scores on a multiple-choice test evaluated? A score of 0 on an objective test (questions requiring complete recall of the material) indicates that the person was unable to recall the test material at the time the test was given. In contrast, a person with little or no recall knowledge of the test material can achieve a higher score on a multiple-choice test because the person only needs to recognize (in contrast to recall) the correct answer and because some questions will be answered correctly just by chance, even if the person does not know the correct answers. Consequently, the no-knowledge score for a multiple-choice test may be well above 0. If a multiple-choice test contains 100 questions, each with six possible answers, what is the expected score for a person who has no knowledge of the test material? Within what limits would a no-knowledge score fall?

Solution   Let p equal the probability of a correct choice on a single question and let x equal the number of correct responses out of the $n = 100$ questions. Assume that "no-knowledge" means that a student will randomly select one of the six possible answers for each question and hence that $p = 1/6$. Then for $n = 100$ questions, the expected score for a student with no knowledge would be $E(x)$, where

$$E(x) = np = 100\left(\frac{1}{6}\right) = 16.7 \text{ correct questions}$$

To evaluate the variation of no-knowledge scores, we need to know σ, where

$$\sigma = \sqrt{npq} = \sqrt{(100)(1/6)(5/6)} = 3.7$$

Based on Tchebysheff's Theorem and the Empirical Rule, we would expect x to fall within the interval $(\mu \pm 2\sigma)$ with a high probability and almost certainly within the interval $(\mu \pm 3\sigma)$. The intervals are

$$(\mu \pm 2\sigma) = (16.7 \pm 7.4) \qquad \text{or} \qquad 9.3 \text{ to } 24.1$$

$$(\mu \pm 3\sigma) = (16.7 \pm 11.1) \qquad \text{or} \qquad 5.6 \text{ to } 27.8$$

This compares with a score of 0 for a no-knowledge student taking an objective recall test. (*Comment:* A histogram of the binomial probability distribution will be very mound-shaped for $n = 100$ and $p = 1/6$. Hence, we would expect the Empirical Rule to work very well. The justification for this statement will be given in Chapter 6.) ◁

EXERCISES   Basic Techniques

4.1   A jar contains five balls: three red and two white. Two balls are randomly selected without replacement from the jar, and the number x of red balls is recorded. Explain why x is or is not a binomial random variable. (*Hint:* Compare the characteristics of this experiment with the characteristics of a binomial experiment given in Section 4.1.) If the experiment is binomial, give the values of n and p.

4.2   Refer to Exercise 4.1. Assume that the sampling was conducted with replacement. That is, assume that the first ball was selected from the jar, observed, and then replaced, and that the balls were then mixed before the second ball was selected. Explain why x, the number of red balls observed, is or is not a binomial random variable. If the experiment is binomial, give the values of n and p.

4.3   Evaluate the following binomial probabilities:
   a. $C_2^8(.3)^2(.7)^6$ b. $C_0^4(.05)^0(.95)^4$
   c. $C_3^{10}(.5)^3(.5)^7$ d. $C_1^7(.2)^1(.8)^6$

4.4   Use the formula for the binomial probability distribution to calculate the values of $p(x)$ and construct the probability histogram for
   a. $n = 7, p = .2$ b. $n = 7, p = .5$ c. $n = 7, p = .8$

4.5   Refer to Exercise 4.4. For each of the binomial random variables given in that exercise, calculate
   a. $P(x = 1)$ b. $P(x \geq 1)$ c. $P(x > 1)$
   d. $P(x \leq 1)$ e. $\mu = np$ f. $\sigma = \sqrt{npq}$

4.6   Use Table 1 in Appendix III to find the sum of the binomial probabilities from $x = 0$ to $x = a$ for
   a. $n = 10, p = .1, a = 3$ b. $n = 15, p = .6, a = 7$ c. $n = 25, p = .5, a = 14$

4.7   Use Table 1 in Appendix III to evaluate the following probabilities for $n = 7$ and $p = .8$:
   a. $P(x \geq 4)$ b. $P(x = 2)$ c. $P(x < 2)$ d. $P(x > 1)$
   Verify these answers using the values of $p(x)$ calculated in part (c), Exercise 4.4.

4.8   Find $\sum_{x=0}^{a} p(x)$ for
   a. $n = 20, p = .05, a = 2$ b. $n = 15, p = .7, a = 8$ c. $n = 10, p = .9, a = 9$

4.9   Use Table 1, Appendix III, to find
   a. $P\{x < 12\}$ for $n = 20, p = .5$ b. $P\{x \leq 6\}$ for $n = 15, p = .4$
   c. $P\{x > 4\}$ for $n = 10, p = .4$ d. $P\{x \geq 6\}$ for $n = 15, p = .6$
   e. $P\{3 < x < 7\}$ for $n = 10, p = .5$

4.10   Find the mean and standard deviation for a binomial distribution with
   a. $n = 1000, p = .3$ b. $n = 400, p = .01$
   c. $n = 500, p = .5$ d. $n = 1600, p = .8$

4.11   Find the mean and standard deviation for a binomial distribution with $n = 100$ and
   a. $p = .01$ b. $p = .9$ c. $p = .3$
   d. $p = .7$ e. $p = .5$

4.12 In Exercise 4.11 the mean and standard deviation for a binomial random variable were calculated for a fixed sample size, $n = 100$, and for different values of p. Graph the values of the standard deviation for the five values of p given in Exercise 4.11. For what value of p does the standard deviation seem to be a maximum?

4.13 Let x be a binomial random variable with $n = 20$ and $p = .1$.
  a. Calculate $P(x \leq 4)$ using the binomial formula.
  b. Calculate $P(x \leq 4)$ using Table 1 in Appendix III.
  c. Use the MINITAB output given below to calculate $P(x \leq 4)$. Compare the results of parts (a), (b), and (c).

```
MTB  > PDF;
SUBC > BINOMIAL   20   .1.

     BINOMIAL WITH N = 20   P = 0.100000
        K              P(X = K)
        0               0.1216
        1               0.2702
        2               0.2852
        3               0.1901
        4               0.0898
        5               0.0319
        6               0.0089
        7               0.0020
        8               0.0004
        9               0.0001
       10               0.0000
```

  d. Calculate the mean and standard deviation of the random variable x.
  e. Use the results of part (d) to calculate the intervals $\mu \pm \sigma$, $\mu \pm 2\sigma$, and $\mu \pm 3\sigma$. Find the probability that an observation will fall in each of these intervals.
  f. Are the results of part (e) consistent with Tchebysheff's Theorem? with the Empirical Rule? Why or why not?

Applications

4.14 It is not always easy to complete business transactions by telephone. A survey by Adia Personnel Services (*Wall Street Journal*, November 29, 1988) indicates that only one in six business calls is completed to the intended party on the first try. Suppose that seven businessmen are chosen at random and that each tries to make a business call. Let x be the number of calls completed to the intended party on the first try. Explain why x is or is not a binomial random variable. If x is binomial, give the values of n and p.

4.15 The fear of air travel is becoming more and more prevalent among American adults. In a survey of 1162 American adults conducted by Media General-Associated Press (*New York Times*, February 3, 1989), 56% of those surveyed said that airline security on international flights was inadequate. Does this sampling represent a binomial experiment? Explain.

4.16 The *Wall Street Journal* (August 20, 1985) reports that many firms are tightening résumé checks of job applicants. The need for this action is supported by a survey of 501 business executives by Ward Howell International, Inc., an executive search firm. According to the *Wall Street Journal*, the survey showed that 17% of the executives said that their new hires misrepresented their job qualifications. Explain why this sampling is or is not a binomial experiment.

4.17 The experiment described in Exercise 1.12 involves measuring the levels of essential vitamins in violent inmates from the California Youth Authority. Does the experiment satisfy the requirements of a binomial experiment? Why or why not?

4.18 A new surgical procedure is said to be successful 80% of the time. Suppose that the operation is performed five times and the results are assumed to be independent of one another. What is the probability that
a. All five operations are successful?
b. Exactly four are successful?
c. Less than two are successful?

4.19 Refer to Exercise 4.18. If less than two operations were successful, how would you feel about the performance of the surgical team?

4.20 Records show that 30% of all patients admitted to a medical clinic fail to pay their bills and that eventually the bills are forgiven. Suppose that $n = 4$ new patients represent a random selection from the large set of prospective patients served by the clinic. Find the probability that
a. All the patients' bills will eventually have to be forgiven.
b. One will have to be forgiven.
c. None will have to be forgiven.

4.21 Consider the medical payment problem in Exercise 4.20 in a more realistic setting. Thirty percent of all patients admitted to a medical clinic fail to pay their bills and the bills are eventually forgiven. If the clinic treats 2000 different patients over a period of one year, what is the mean (expected) number of bills that would have to be forgiven? If x is the number of forgiven bills in the group of 2000 patients, find the variance and standard deviation of x. What can you say about the probability that x will exceed 700? (*Hint:* Use the values of μ and σ, along with Tchebysheff's Theorem, to answer this question.)

4.22 High cholesterol levels are becoming increasingly prevalent in the American adult population. In fact, approximately one-third of Americans 20 years of age or older are at high risk for coronary disease and are considered "candidates for medical advice and intervention" (*New York Times*, July 7, 1989). If six Americans 20 years of age or older are randomly chosen to undergo cholesterol testing, what is the probability that
a. Exactly three are at high risk for coronary disease?
b. At least one is at high risk for coronary disease?
c. At most two are at high risk for coronary disease?

4.23 Consider a metabolic defect that occurs in approximately one of every 100 births. If four infants are born in a particular hospital on a given day, what is the probability that
a. None has the defect?
b. No more than one has the defect?

4.24 Early in the United States missile development program, government and industry defense officials were proclaiming that our missiles were highly reliable, and probabilities of successful firings in the neighborhood of .999 . . . were quoted. Such statements were made even though many missile firings resulted in failure, including the Navy *Vanguard* missile in the 1950s. If the reliability (probability of a successful launch) is even as high as .9, what is the probability of observing three or more failures in a total of four firings? one failure or more? If you observed two or more failures out of four, what would you think about the high claims of reliability of the missiles produced in the 1950s?

4.25 New excitement about the progress in the treatment of AIDS was generated recently at the Fifth International Conference on AIDS (*Press Enterprise*, June 8, 1989). Dr. Samuel Broder announced that the chances of living 18 months or longer after having been diagnosed with

AIDS have increased to more than 60%, compared to 30% in 1982. Suppose that a hospital admitted 480 patients in 1989 with AIDS.

a. What is the minimum number of these patients you would expect to survive for at least 18 months? Justify your answer.

b. What is the maximum number? Justify your answer.

4.26 A manufacturer is thinking of using a sampling plan with $n = 20$ and $a = 1$ because the cost of sampling only 20 items is small and the probability of accepting good lots, say those containing only 1% defectives, is high.

a. Find the probability of accepting lots containing only 1% defectives.

b. The manufacturer's customers want the plan to accept bad lots, those containing a high fraction of defectives (say 10% or more), with a small probability. Will the manufacturer's sampling plan be acceptable to the customers? Explain.

4.27 A radio and television manufacturer who buys large lots of transistors from an electronics supplier selects $n = 25$ transistors from each lot shipped by the supplier and notes the number of defectives.

a. On the same sheet of graph paper, construct the operating characteristic curves for the sampling plans $n = 25$ with $a = 1$, 2, and 3.

b. Which sampling plan best protects the supplier from having acceptable lots rejected and returned by the manufacturer?

c. Which sampling plan best protects the manufacturer from accepting lots for which the fraction of defectives is exceedingly large?

d. How might the sampling inspector arrive at an acceptance level that compromises between the risk to the producer and the risk to the consumer?

4.28 Refer to Exercise 4.27 and assume that the manufacturer wishes the probability to be at least .90 of his accepting lots containing 1% defective, and the probability to be about .90 of rejecting any lot with 10% or more defective. If the manufacturer's sampling inspector samples $n = 25$ items from the supplier's incoming shipments, what is the acceptance number (a) that more nearly meets these requirements?

4.29 A buyer and a seller agree to use sampling plan ($n = 15$, $a = 0$) or sampling plan ($n = 25$, $a = 1$). Sketch the operating characteristic curves for the two sampling plans. If you were a buyer, which of the two sampling plans would you prefer? Why?

4.30 A survey by the American Hospital Association suggests that a majority of Americans favor "pulling the plug" in cases of hopeless illness (*Orlando Sentinel*, March 15, 1983). The survey involved a random sample of 1800 persons. Of these, seven out of ten were in favor of discontinuing expensive life-support systems for patients with little chance of survival. Suppose, in fact, that the percentage of Americans who favor discontinuing expensive life-support systems for patients with little chance of survival is smaller than the percentage found in the sample, say only 50%. If you were to randomly sample 1800 people, find the expected value and standard deviation of x, the number favoring discontinuing life support. If p were really equal to .5, is it probable that as many as seven of ten in the sample would favor this position? Explain.

4.31 Bank failures are eroding the public's confidence in our banking system. A survey by the American Bankers Association suggests that more than one-third of Americans have less confidence in the United States banking system than they had in prior years (*Orlando Sentinel*, October 22, 1984). Suppose that the proportion of adults in the United States who have less confidence in the banking system is one-third. Furthermore, suppose that customers of your local bank are representative of those throughout the United States, and that you randomly sample 600.

a.   What is the expected number and standard deviation of x, the number in the sample with less confidence in the banking system?

b.   Suppose that the number x of persons with decreased confidence at your bank was equal to 270. Would you believe your initial assumption, that your bank's customers were representative of those throughout the United States? Explain.

 4.32   A report by the American Medical Association's Special Task Force on Professional Liability and Insurance notes that every year, 16 of every 100 doctors are subject to malpractice claims. Suppose that a hospital staff consists of 200 physicians and that the occurrence of malpractice claims for one physician is independent of the occurrence of claims for any others.

a.   What is the expected number of the hospital's staff physicians who will be sued for malpractice in a given year? (Assume that the probability that a physician is sued for malpractice is .16.)

b.   Is it likely that the number sued for malpractice could be as large as 50? Explain.

 ## 4.3   THE POISSON PROBABILITY DISTRIBUTION

Another discrete random variable that has numerous applications in business and economics is the **Poisson random variable**. Its probability distribution provides a good model for data that represent the number of occurrences of a specified event in a given unit of time or space. Here are some examples of experiments for which the random variable x can be modeled by the Poission random variable:

1.   The number of calls received by a switchboard during a given period of time.

2.   The number of bacteria per small volume of fluid.

3.   The number of arrivals at a checkout counter during a given minute.

4.   The number of machine breakdowns during a given day.

5.   The number of traffic accidents at a given intersection during a given time period.

In each example, **x represents the number of events occurring in a period of time or space during which an average of μ such events can be expected to occur**. The only assumptions needed when one uses the Poisson distribution to model experiments such as those described above are that the counts or events occur **randomly and independently** of one another. The formula for the Poisson probability distribution as well as its mean and variance are shown in the display.

THE POISSON PROBABILITY DISTRIBUTION

$$p(x) = \frac{\mu^x e^{-\mu}}{x!} \qquad x = 0, 1, 2, 3, \ldots$$

where

$\mu = E(x) =$ mean of random variable x

$\sigma^2 = \mu =$ variance of random variable x

$e = 2.71828 \ldots$ (e is the base of natural logarithms)

Figure 4.3

Poisson probability
distributions for
$\mu = .5, 1,$ and 4

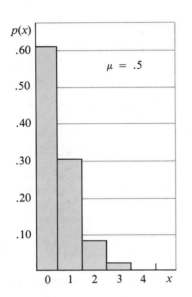

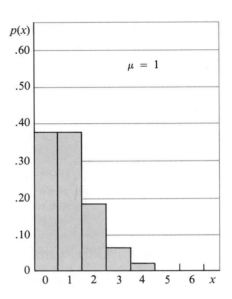

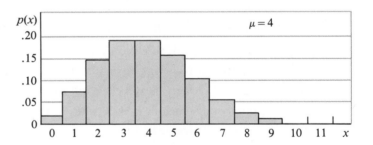

The value of $e^{-\mu}$ can be found by using a calculator or by using Table 2(b) in Appendix III, which provides the values of $e^{-\mu}$ for values of μ between 0 and 10 in increments of .05. Alternatively, the Poisson probabilities can be obtained from Table 2(a) in Appendix III, in which the cumulative probabilities, $P(x \le a) = p(0) + p(1) + \cdots + p(a)$, are given for various values of μ. This table can be used to find Poisson probabilities just as the binomial tables were used in Section 4.2. Graphs of the Poisson probability distribution for $\mu = .5, 1,$ and 4 are shown in Figure 4.3.

EXAMPLE 4.9   The average number of traffic accidents on a certain section of highway is two per week. Assume that the number of accidents follows a Poisson distribution with $\mu = 2$.

a. Find the probability of no accidents on this section of highway during a one-week period.

b. Find the probability of at most three accidents on this section of highway during a two-week period.

Solution   a.   The average number of accidents per week is $\mu = 2$. Therefore, the probability of no accidents on this section of highway during a given week is

$$p(0) = \frac{2^0 e^{-2}}{0!}$$

$$= e^{-2} = .135335$$

b.   During a two-week period the average number of accidents on this section of highway would be $2(2) = 4$. The probability of at most three accidents during a two-week period is

$$P(x \leq 3) = p(0) + p(1) + p(2) + p(3)$$

where

$$p(0) = \frac{4^0 e^{-4}}{0!} \qquad\qquad p(2) = \frac{4^2 e^{-4}}{2!}$$

$$= .018316 \qquad\qquad\quad = .146528$$

$$p(1) = \frac{4^1 e^{-4}}{1!} \qquad\qquad p(3) = \frac{4^3 e^{-4}}{3!}$$

$$= .073264 \qquad\qquad\quad = .195367$$

Therefore,

$$P(x \leq 3) = .018316 + .073264 + .146528 + .195367$$

$$= .433475$$

This value could be read directly from Table 2(a) in Appendix III, indexing $\mu = 4$ and $a = 3$, as $P(x \leq 3) = .433$.

Recall from Section 4.2 that we were able to simplify the calculation of binomial probabilities by using Table 1 in Appendix III. However, binomial tables are seldom available for n greater than 100, and many applications of the binomial experiment with $n = 100$ or more arise in practical situations. Consequently, we need simple, easy-to-compute approximation procedures for calculating binomial probabilities. **The Poisson probability distribution provides good approximations to binomial probabilities when n is large and $\mu = np$ is small, preferably with $np < 7$.** An approximation procedure suitable for larger values of $\mu = np$ will be presented in Chapter 5.

As an illustration of the Poisson approximation procedure, consider the following application. Suppose that a life insurance company insures the lives of 5000 men of age 42. If actuarial studies show the probability of any 42-year-old man dying in a given year to be .001, the exact probability that the company will have to pay $x = 4$ claims during a given year is given by the binomial distribution as

$$P(x = 4) = p(4) = \frac{5000!}{4! 4996!} (.001)^4 (.999)^{4996}$$

for which binomial tables are not available. To compute $P(x = 4)$ without the aid of a computer would be very time-consuming, but the Poisson distribution can be used to provide a good approximation to $P(x = 4)$. Computing $\mu = np = (5000)(.001) = 5$ and substituting into the formula for the Poisson probability distribution, we have

$$p(4) \approx \frac{\mu^4 e^{-\mu}}{4!} = \frac{5^4 e^{-5}}{4!} = \frac{(625)(.006738)}{24} = .175$$

The value of $p(4)$ could also be obtained using Table 2(a) in Appendix III with $\mu = 5$ as

$$p(4) = P(x \leq 4) - P(x \leq 3) = .440 - .265 = .175$$

EXAMPLE 4.10   A manufacturer of power lawn mowers buys one-horsepower, two-cycle engines in lots of 1000 from a supplier. She then equips each of the mowers produced by her plant with one of the engines. Past history shows that the probability of any one engine purchased from the supplier proving unsatisfactory is .001. In a shipment of 1000 engines, what is the probability that none are defective? one is defective? two are? three are? four are?

Solution   This is a binomial experiment with $n = 1000$ and $p = .001$. The expected number of defectives in a shipment of $n = 1000$ engines is $\mu = np = (1000)(.001) = 1$. Since this is a binomial experiment with $np < 7$, the probability of x defective engines in the shipment may be approximated by

$$p(x) = \frac{\mu^x e^{-\mu}}{x!} = \frac{1^x e^{-1}}{x!} = \frac{e^{-1}}{x!}$$

(since $1^x = 1$ for any value of x). Therefore, we have

$$p(0) \approx \frac{e^{-1}}{0!} = \frac{.368}{1} = .368 \qquad p(3) \approx \frac{e^{-1}}{3!} = \frac{.368}{6} = .061$$

$$p(1) \approx \frac{e^{-1}}{1!} = \frac{.368}{1} = .368 \qquad p(4) \approx \frac{e^{-1}}{4!} = \frac{.368}{24} = .015$$

$$p(2) \approx \frac{e^{-1}}{2!} = \frac{.368}{2} = .184$$

$\triangleleft$

The individual and cumulative probabilities for a Poisson distribution with mean μ can be found by using the PDF and the CDF MINITAB commands, followed by the subcommand POISSON μ. The binomial probabilities for $n = 1000$ and $p = .001$, together with the Poisson probabilities for $\mu = 1$, are given in Table 4.3. Comparing the actual binomial probabilities with the corresponding probabilities found by using the Poisson approximation to binomial probabilities, we see that they are quite accurate in this case. Furthermore, we see that the POISSON command also terminates when an individual probability equals 0 within a preassigned level of accuracy.

Table 4.3 MINITAB output of binomial and Poisson probabilities

```
MTB > PDF;                          MTB > PDF;
SUBC > BINOMIAL 1000 .001.          SUBC > POISSON 1.

      BINOMIAL WITH N = 1000  P = 0.001000      POISSON WITH MEAN = 1.000
      K          P(X = K)                       K          P(X = K)
      0          0.3677                          0          0.3679
      1          0.3681                          1          0.3679
      2          0.1840                          2          0.1839
      3          0.0613                          3          0.0613
      4          0.0153                          4          0.0153
      5          0.0030                          5          0.0031
      6          0.0005                          6          0.0005
      7          0.0001                          7          0.0001
      8          0.0000                          8          0.0000
```

EXERCISES Basic Techniques

4.33 Let x be a Poisson random variable with mean $\mu = 2$. Calculate the following probabilities:
  a. $P(x = 0)$ b. $P(x = 1)$ c. $P(x > 1)$ d. $P(x = 5)$

4.34 Let x be a Poisson random variable with mean $\mu = 2.5$. Use Table 2(a) in Appendix III to calculate the following probabilities:
  a. $P(x \geq 5)$ b. $P(x < 6)$ c. $P(x = 2)$ d. $P(1 \leq x \leq 4)$

4.35 Let x be a binomial random variable with $n = 20$ and $p = .1$.
  a. Calculate $P(x \leq 2)$ using Table 1 in Appendix III to obtain the exact binomial probability.
  b. Use the Poisson approximation to calculate $P(x \leq 2)$.
  c. Compare the results of parts (a) and (b). Is the approximation accurate?

4.36 To illustrate how well the Poisson probability distribution approximates the binomial probability distribution, calculate the Poisson approximate values for $p(0)$ and $p(1)$ for a binomial probability distribution with $n = 25$ and $p = .05$. Compare the answers with the exact values obtained from Table 1, Appendix III.

Applications

4.37 The increased number of small commuter planes in major airports has heightened concern over air safety. An eastern airport has recorded a monthly average of five near misses on landings and takeoffs in the past five years.
  a. Find the probability that during a given month there are no near misses on landings and takeoffs at the airport.
  b. Find the probability that during a given month there are five near misses.
  c. Find the probability that there are at least five near misses during a particular month.

4.38 The number x of people entering the intensive care unit at a particular hospital on any one day possesses a Poisson probability distribution with mean equal to five persons per day.
  a. What is the probability that the number of people entering the intensive care unit on a particular day is two? less than or equal to two?
  b. Is it likely that x will exceed ten? Explain.

4.39 Parents who are concerned that their children are "accident-prone" can be reassured, according to a study conducted by the Department of Pediatrics at the University of California, San Francisco (Reprinted by permission of *The Physician and Sportsmedicine*, Vol. 17, No. 9, September, 1989, p. 55. Copyright McGraw-Hill, Inc.). Children who are injured two or more times tend to sustain these injuries during a relatively limited time, usually one year or less. This suggests that the children are experiencing only a "temporary period of heightened vulnerability to injury," perhaps triggered by the biological and psychological changes of adolescence, or stress in the child's environment. If the average number of injuries per year for school-age children is two, what is the probability that
a. A child will sustain two injuries during the year?
b. A child will sustain two or more injuries during the year?
c. A child will sustain at most one injury during the year?

4.40 Refer to Exercise 4.39.
a. Calculate the mean and standard deviation for x, the number of injuries per year sustained by a school-age child.
b. Within what limits would you expect the number of injuries per year to fall?

4.41 If a drop of water is placed on a slide and examined under a microscope, the number x of a particular type of bacteria present has been found to have a Poisson probability distribution. Suppose that the maximum permissible count per water specimen for this type of bacteria is five. If the mean count for your water supply is two and you test a single specimen, is it likely that the count will exceed the maximum permissible count? Explain.

4.42 Corporate drug testing is becoming a more established practice in the United States business community. As an example, MetPath, a commercial drug-testing company, claims that it is performing about 22,000 tests a month for its corporate clients, and that between 5% and 15% of the samples test positive for illegal substances (*Wall Street Journal*, November 29, 1988). On a particular day, suppose that MetPath performs 100 independent drug tests between the hours of 9:00 and 10:00 A.M., and that the probability of a positive test is approximately .05.
a. Let x be the number of positive tests observed. What is the probability distribution for x? What is the mean of this distribution?
b. Use the Poisson distribution to approximate the probability of at most four positive drug tests.

4.4 THE HYPERGEOMETRIC PROBABILITY DISTRIBUTION

Suppose you are selecting a sample of elements from a population and you record whether each element does or does not possess a certain characteristic. Consequently, you are dealing with the "success" or "failure" type of data encountered in the binomial experiment. The ERA survey of Example 4.1 and the sampling for defectives of Example 4.2 are practical illustrations of these sampling situations.

If the number of elements in the population is large relative to the number in the sample (as in Example 4.1), the probability of selecting a success on a single trial is equal to the proportion p of successes in the population. Because the population is large in relation to the sample size, this probability will remain constant (for all practical purposes) from trial to trial, and the number x of successes in the sample will follow a binomial probability distribution. However, **if the number of elements in the population is small in relation to the sample size ($n/N \geq .05$), the** probability of a success for a given trial is dependent on the outcomes of preceding trials. Then the number x of successes follows what is known as a *hypergeometric probability distribution*.

We define the following notation, which is necessary in order to present the formula for the hypergeometric probability distribution.

N = number of elements in population

k = number of elements in population that are successes (that is, number possessing one of two characteristics)

$N - k$ = number of elements in population that are not successes

n = number of elements in sample, selected from N elements in population

x = number of successes in sample

The hypergeometric probability distribution for the random variable x is then as given in the display.

HYPERGEOMETRIC PROBABILITY DISTRIBUTION

$$p(x) = \frac{C_x^k C_{n-x}^{N-k}}{C_n^N}$$

where x can assume integer values $0, 1, 2, \ldots, n$ subject to the restrictions $x \le k$ and $x \ge k + n - N$, and where

$$C_r^n = \frac{n!}{r!(n-r)!}$$

The mean and variance of a hypergeometric random variable are very similar to those for a binomial random variable with a correction for the finite population size.

$$\mu = n\left(\frac{k}{N}\right)$$

$$\sigma^2 = n\left(\frac{k}{N}\right)\left(\frac{N-k}{N}\right)\left(\frac{N-n}{N-1}\right)$$

EXAMPLE 4.11 A case of wine contains 12 bottles, three of which contain spoiled wine. A sample of four bottles is randomly selected from the case.
a. Find the probability distribution for x, the number of bottles of spoiled wine in the sample.
b. What is the mean and variance of x in this case?

Solution For this example $N = 12$, $n = 4$, $k = 3$, and $(N - k) = 9$. Then

$$p(x) = \frac{C_x^3 C_{4-x}^9}{C_4^{12}}$$

a. The possible values for x are 0, 1, 2, and 3. Therefore,

$$p(0) = \frac{C_0^3 C_4^9}{C_4^{12}} = \frac{1(126)}{495} = \frac{126}{495} = \frac{14}{55}$$

$$p(1) = \frac{C_1^3 C_3^9}{C_4^{12}} = \frac{3(84)}{495} = \frac{252}{495} = \frac{28}{55}$$

$$p(2) = \frac{C_2^3 C_2^9}{C_4^{12}} = \frac{3(36)}{495} = \frac{108}{495} = \frac{12}{55}$$

$$p(3) = \frac{C_3^3 C_1^9}{C_4^{12}} = \frac{1(9)}{495} = \frac{9}{495} = \frac{1}{55}$$

b. The mean is given by

$$\mu = 4\left(\frac{3}{12}\right) = 1$$

and the variance is given by

$$\sigma^2 = 4\left(\frac{3}{12}\right)\left(\frac{9}{12}\right)\left(\frac{12-4}{11}\right) = .5455 \qquad \triangleleft$$

EXAMPLE 4.12 A particular industrial product is shipped in lots of 20. Testing to determine whether an item is defective is costly, and hence, the manufacturer samples production rather than using a 100% inspection plan. A sampling plan constructed to minimize the number of defectives shipped to customers calls for sampling five items from each lot and rejecting the lot if more than one defective is observed. (If rejected, each item in the lot is then tested.) If a lot contains four defectives, what is the probability that it will be accepted?

Solution Let x be the number of defectives in the sample. Then $N = 20, k = 4, (N - k) = 16$, and $n = 5$. The lot will be rejected if $x = 2, 3,$ or 4. Then

$$P(\text{accept the lot}) = P(x \leq 1) = p(0) + p(1) = \frac{C_0^4 C_6^{16}}{C_5^{20}} + \frac{C_1^4 C_4^{16}}{C_5^{20}}$$

$$= \frac{\left(\frac{4!}{0!4!}\right)\left(\frac{16!}{5!11!}\right)}{\frac{20!}{5!15!}} + \frac{\left(\frac{4!}{1!3!}\right)\left(\frac{16!}{4!12!}\right)}{\frac{20!}{5!15!}}$$

$$= \frac{91}{323} + \frac{455}{969} = .2817 + .4696 = .7513 \qquad \triangleleft$$

Notice that Example 4.12 is quite similar to Example 4.4. The only difference is that the number x of defectives possesses a hypergeometric probability distribution when the number of elements N in the population is small in relation to the sample size n.

Familiarity with discrete probability distributions and the properties of the experiments that generate them is extremely helpful. Rather than solve the same probability problem over and over again from first principles (as was done in Chapter 3), you need only recognize the type of random variable involved and then substitute into the formula for its probability distribution.

EXERCISES   Basic Techniques

4.43   Evaluate the following probabilities:

a. $\dfrac{C_1^3 C_1^2}{C_2^5}$

b. $\dfrac{C_2^4 C_1^3}{C_3^7}$

c. $\dfrac{C_4^5 C_0^3}{C_4^8}$

4.44   Let x be the number of successes observed in a sample of $n = 5$ items selected from $N = 10$. Suppose that of the $N = 10$ items, six are considered "successes."

a.   Find the probability of observing no successes.

b.   Find the probability of observing at least two successes.

c.   Find the probability of observing exactly two successes.

4.45   Let x be a hypergeometric random variable with $N = 15$, $n = 3$, and $k = 4$.

a.   Calculate $p(0)$, $p(1)$, $p(2)$, and $p(3)$.

b.   Construct the probability histogram for x.

c.   Use the formulas given in Section 4.4 to calculate $\mu = E(x)$ and σ^2.

d.   What proportion of the population of measurements fall within the interval $(\mu \pm 2\sigma)$? within the interval $(\mu \pm 3\sigma)$? Do these results agree with those given by Tchebysheff's Theorem?

4.46   A jar contains two black balls and two white balls. Suppose that the balls have been thoroughly mixed and that two are randomly selected from the jar. Let x be the number of white balls in the selection.

a.   Use the formulas given in this section to calculate the values of $p(x)$, $x = 0, 1, 2$. Compare with the results of Exercise 3.71.

b.   Using the probability distribution from part (a) and the methods of Section 3.8, calculate $\mu = E(x)$ and $\sigma^2 = E(x - \mu)^2$.

c.   Calculate μ and σ^2 using the formulas given in this section. Compare the results of parts (b) and (c).

Applications

4.47   Refer to Exercise 3.76. Use the hypergeometric probability distribution to calculate $p(x)$ for each value of x. Compare your answers with those obtained in Exercise 3.76.

4.48   A company has five applicants for two positions: two women and three men. Suppose that the five applicants are equally qualified and that no preference is given for choosing either sex. Let x equal the number of women chosen to fill the two positions.

a.   Write the formula for $p(x)$, the probability distribution of x.

b.   What are the mean and variance of this distribution?

c.   Construct a probability histogram for x.

4.49   According to the Current Population Survey (CPS) conducted by the Bureau of Labor Statistics, 26% of the workers age 25–64 in the United States labor force are college graduates, up from 21% in 1978 (*Family Economics Review*, Vol. 2, No. 2, U.S. Department of Agriculture, 1989). Suppose that ten in a group of 40 workers in this age category are college graduates. If five workers are hired by a work foreman, what is the probability that exactly one will be a college graduate? at least one?

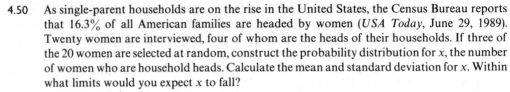

4.50   As single-parent households are on the rise in the United States, the Census Bureau reports that 16.3% of all American families are headed by women (*USA Today*, June 29, 1989). Twenty women are interviewed, four of whom are the heads of their households. If three of the 20 women are selected at random, construct the probability distribution for x, the number of women who are household heads. Calculate the mean and standard deviation for x. Within what limits would you expect x to fall?

4.51   Refer to Exercise 4.50. If three of the 20 women are interviewed, what is the probability that all three of the women are household heads? If this event were to occur, what conclusions might you draw?

TIPS ON PROBLEM SOLVING

1.   If the random variable is the number of occurrences of a specified event in a given unit of *time or space* for which the average number of occurrences per unit time or space is μ, then the random variable has a Poisson distribution.

2.   Suppose that a random sample of n items is selected without replacement from a population of N items in which a proportion p of the items possess a specified property and $q = 1 - p$ do not. The following flowchart is provided to help determine which distribution is appropriate for the situations presented.

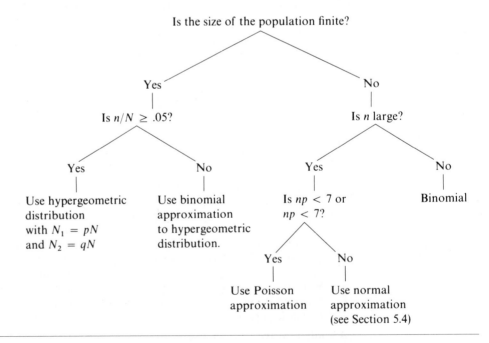

4.5   OTHER DISCRETE DISTRIBUTIONS

In our discussion of probability and random variables in Chapter 3 we often referred to the experiment of tossing a fair die and observing x, the number of dots on the upper face. This is just one example of a more general experiment in which the random variable can assume the values $x = 1, 2, \ldots, N$ with equal probability given by $1/N$. In this case the resulting probability distribution for x is called the **discrete uniform distribution** because the resulting probability histogram has constant or uniform height.

THE DISCRETE UNIFORM DISTRIBUTION

$$p(x) = \frac{1}{N} \qquad \text{for } x = 1, 2, \ldots, N$$

$$\mu = \frac{N + 1}{2}$$

$$\sigma = \sqrt{\frac{N^2 - 1}{12}}$$

In Example 3.26 the probability distribution for the number of dots observed in the throw of a single balanced die is shown to be $p(x) = 1/6$ with $N = 6$. Using the formulas for the mean and standard deviation of x, we find that

$$\mu = \frac{6 + 1}{2} = 3.5 \qquad \text{and} \qquad \sigma = \sqrt{\frac{6^2 - 1}{12}} = 1.7$$

Notice that 100%, or all, of the values of x lie within the interval $\mu \pm 2\sigma$, which is 3.5 ± 3.4, or from .1 to 6.9.

EXAMPLE 4.13   Suppose one chip is drawn at random from a box containing ten chips bearing the numbers 1 through 10. Let x be the number on the chip drawn.
a.   Find the probability distribution for x.
b.   What are the mean and standard deviation of x?
c.   What proportion of the observations lie further than one standard deviation from the mean?

Solution   a.   Since each of the ten outcomes corresponding to $x = 1, 2, \ldots, 10$ are equally likely,

$$p(x) = \frac{1}{10} \qquad x = 1, 2, \ldots, 10$$

b.   In calculating the mean and standard deviation of x, we may use the general formulas for finding expectations, or use the given formulas with $N = 10$. Then

$$\mu = \frac{N + 1}{2} = \frac{10 + 1}{2} = 5.5$$

and

$$\sigma = \sqrt{\frac{N^2 - 1}{12}} = \sqrt{\frac{10^2 - 1}{12}} = 2.9$$

c. To find the proportion of observations lying further than one standard deviation from the mean, we need to evaluate the endpoints of the interval $\mu \pm \sigma$,
which are given by 5.5 ± 2.9, or from 2.6 to 8.4. The values of x outside this
interval are 1, 2, 9, and 10; therefore, 40% of the observations lie further than
one standard deviation from the mean.

Another discrete random variable represents the waiting time until the **first
success** in a series of independent binomial trials, with the probability of success
given by p. Let x be the number of trials until the first success. With S representing a
success and F a failure, the simple events associated with this experiment and their
probabilities follow.

Simple Event, E_i	x	$p(x)$
$E_1: S$	1	p
$E_2: FS$	2	qp
$E_3: FFS$	3	$q^2 p$
$E_4: FFFS$	4	$q^3 p$
$\vdots$	$\vdots$	$\vdots$
$E_i: FFF \quad FS$	i	$q^{i-1} p$

When the trials are independent,

$$P(E_i) = P(FFF \cdots FS) = P(F)P(F) \cdots P(F)P(S)$$

and

$$P(E_i) = q^{i-1} p$$

Because the individual terms of $p(x)$ represent the terms of a geometric progression,
this probability distribution is named the **geometric probability distribution**.

GEOMETRIC PROBABILITY DISTRIBUTION

$$p(x) = q^{x-1} p \qquad x = 1, 2, 3, \ldots$$

$$\mu = \frac{1}{p}$$

$$\sigma = \sqrt{\frac{q}{p^2}}$$

Do the probabilities associated with the geometric distribution sum to 1? Summing,
we have

$$p(x) = p + qp + q^2 p + q^3 p + \cdots$$

The sum of a geometric progression is given by the first term divided by 1 minus the common ratio, provided that the common ratio is less than 1 in absolute value. In this case the first term is p and the common ratio is $q < 1$; therefore, the sum is $p/(1 - q)$, which is equal to 1.

EXAMPLE 4.14   Approximately 40% of the population of the United States have type A blood.
a.   If the order of blood donors entering a blood bank on any given day is random, find the distribution of x, the number of donors entering a blood bank on a given day until the first type A donor is encountered.
b.   What is the mean and standard deviation of the number of donors until the first type A donor is encountered?

Solution   a.   The probability that a random blood donor has type A blood is $p = .4$. Therefore, with $q = 1 - p = .6$,

$$p(x) = (.6)^{x-1}(.4) \qquad x = 1, 2, 3, \ldots$$

b.   The mean and standard deviation of x are

$$\mu = \frac{1}{p} = \frac{1}{.4} = 2.5$$

and

$$\sigma = \sqrt{\frac{q}{p^2}} = \sqrt{\frac{.6}{(.4)^2}} = \sqrt{3.75} = 1.9$$

In examining the formulas for the mean and standard deviation of a geometric random variable, we see that both the mean and standard deviation increase as p gets close to 0, and decrease as p gets close to 1. This is in accordance with reality since, when the occurrence of a success has a very small probability, the waiting time for the first success is long. On the other hand, when the probability of a success is close to 1, the waiting time is short; when $p = 1$, $\mu = 1$ and $\sigma = 0$ because a success is always observed on the first trial.

4.6 SUMMARY

Several useful discrete probability distributions were presented in this chapter: the binomial, the Poisson, the hypergeometric, the uniform, and the geometric distributions. These probability distributions enabled us to calculate the probabilities associated with events that are of interest in the sciences, in business, and in marketing.

The binomial probability distribution allows us to calculate the probability of x successes in a series of n identical independent trials, where the probability of a success in a single trial is equal to p. The binomial experiment is an excellent model for many sampling situations, particularly surveys that result in "yes" or "no" types of data.

The Poisson probability distribution is important because it can be used to approximate certain binomial probabilities when n is large and p is small. Consequently, it can greatly reduce the computations involved in calculating binomial probabilities. In addition, the Poisson probability distribution is important in its own right. It provides an excellent probabilistic model for the number of occurrences of rare events in time or space.

The hypergeometric probability distribution is also related to the binomial probability distribution. It gives the probability of drawing x elements of a particular type from a population where the number N of elements in the population is small in relation to the sample size n. The binomial probability distribution applies to the same situation except that it is appropriate only when N is large in relation to n.

The uniform distribution is useful in modeling situations in which N outcomes have the same probability of occurrence. The geometric distribution is related to the binomial in that both are based on independent trials in which the probability of a success is constant and equal to p. However, the geometric random variable is the number of trials until the first success, whereas the binomial random variable is the number of successes in n trials.

▷ **4.7** MINITAB COMMANDS

```
PDF for values in E [put into E]

    BINOMIAL n = K   p = K
    POISSON  μ = K
```

```
CDF for values in E...E [put into E...E

    BINOMIAL n = K   p = K
    POISSON  μ = K
```

REFERENCES

Feller, W. *An Introduction to Probability Theory and Its Applications*, Vol. 1. 3d ed. New York: Wiley, 1968.

Freund, J. E., and Walpole, R. E. *Mathematical Statistics*. 4th ed. Englewood Cliffs, N. J.: Prentice-Hall, 1987.

Mendenhall, W.; Wackerly, D.; and Scheaffer, R. L. *Mathematical Statistics with Applications*. 4th ed. Boston: PWS-KENT, 1990.

National Bureau of Standards. *Tables of the Binomial Probability Distribution*. Washington, D.C.: Government Printing Office, 1949.

Ryan, T. A.; Joiner, B. L.; and Ryan, B. F. *Minitab Student Handbook*. 2d ed. Boston: Duxbury Press, 1985.

Walpole, R. E. *Introduction to Statistics*. 3d ed. New York: Macmillan, 1982.

Weiss, N. A. *Elementary Statistics*. Reading, Mass.: Addison-Wesley, 1989.

SUPPLEMENTARY EXERCISES

4.52 List the five identifying characteristics of the binomial experiment.

4.53 Under what conditions can the Poisson random variable be used to approximate the probabilities associated with the binomial random variable? What application does the Poisson distribution have other than to estimate certain binomial probabilities?

4.54 Under what conditions would one use the hypergeometric probability distribution in evaluating the probability of x successes in n trials?

4.55 A balanced coin is tossed three times. Let x equal the number of heads observed.
  a. Use the formula for the binomial probability distribution to calculate the probabilities associated with $x = 0, 1, 2$, and 3.
  b. Construct a probability distribution.
  c. Find the expected value and standard deviation of x, using the formulas

$$E(x) = np$$

$$\sigma = \sqrt{npq}$$

  Using the probability distribution in (b), find the fraction of the population measurements lying within one standard deviation of the mean. Repeat for two standard deviations. How do your results agree with Tchebysheff's Theorem and the Empirical Rule?

4.56 Refer to Exercise 4.55. Suppose that the coin is definitely unbalanced and that the probability of a head is equal to $p = .1$. Follow instructions (a), (b), (c), and (d). Note that the probability distribution loses its symmetry and becomes skewed when p is not equal to 1/2.

 4.57 Suppose that the four engines of a commercial aircraft are arranged to operate independently and that the probability of in-flight failure of a single engine is .01. What is the probability that, on a given flight,
  a. No failures are observed?
  b. No more than one failure is observed?

4.58 A buyer and a seller agree to use a sampling plan with sample size $n = 5$ and acceptance number $a = 0$. What is the probability that the buyer will accept a lot having the following fractions defective?
  a. $p = .1$ b. $p = .3$ c. $p = .5$
  d. $p = 0$ e. $p = 1$

  Construct the operating characteristic curve for this plan.

4.59 Repeat Exercise 4.58 for $n = 5, a = 1$.

4.60 Repeat Exercise 4.58 for $n = 10, a = 0$.

4.61 Repeat Exercise 4.58 for $n = 10, a = 1$.

4.62 Graph the operating characteristic curves for the four plans given in Exercises 4.58, 4.59, 4.60, and 4.61 on the same sheet of graph paper. What is the effect of increasing the acceptance number a, when n is held constant? What is the effect of increasing the sample size n, when a is held constant?

4.63 The ten-year survival rate for bladder cancer is approximately 50%. If 20 bladder cancer patients are properly treated for the disease, what is the probability that
  a. At least 1 will survive ten years?

b. At least 10 will survive ten years?

c. At least 15 will survive ten years?

4.64 A city commissioner claims that 80% of all people in the city favor garbage collection by contract to a private concern (in contrast to collection by city employees). To check the theory that the proportion of people in the city favoring private collection is .8, you randomly sample 25 people and find that x, the number of people who support the commissioner's claim, is 22.

a. What is the probability of observing at least 22 who support the commissioner's claim if, in fact, $p = .8$?

b. What is the probability that x is exactly equal to 22?

c. Based on the results of part (a), what would you conclude about the claim that 80% of all people in the city favor private collection? Explain.

4.65 If a person is given the choice of an integer from 0 to 9, is it more likely that the person will choose an integer near the middle of the sequence?

a. If the integers are equally to be chosen, find the probability distribution for x, the number chosen.

b. What is the probability that a person will choose a 4, 5, or 6?

c. What is the probability that a person will not choose a 4, 5, or 6?

4.66 Refer to Exercise 4.65. To answer the question posed in that exercise, 20 persons are asked to select a number from 0 to 9. Eight of them choose a 4, 5, or 6.

a. If the choice of any one number is as likely as any other, what is the probability of observing eight or more choices of the interior numbers 4, 5, or 6?

b. What conclusions would you draw from the results of part (a)?

4.67 "What parents don't know may hurt them—and their children" according to Alan L. Otten. In his column "People Patterns" (*Wall Street Journal*, December 8, 1988) he reported the results of a survey of primary and secondary students and parents across the United States, in which 41% of the children say they have smoked cigarettes, while only 14% of the parents *think* that their children have smoked cigarettes. If a random sample of 25 parents is taken, what is the mean and standard deviation of x, the number of parents who think their children have smoked cigarettes? Within what limits would you expect x to lie?

4.68 Refer to Exercise 4.67. A similar survey is conducted, based on a group of 50 students. Of the 50, 15 say they have smoked cigarettes. If ten of these students are randomly selected to appear on a television talk show involving student drug use, what is the probability that at least three have smoked cigarettes? How many of the ten students would you expect to have smoked cigarettes?

4.69 In early 1985, officials in St. Petersburg, Florida were investigating the cause of death for 12 patients at a local nursing home in the two-week period, November 13–26, 1984 (*New York Times* February 5, 1985). According to the nursing home operator, three to six deaths per month would be "normal." If we take the larger of these two figures to be the mean number of deaths per four-week period, then the mean number of deaths per two-week period, under existing conditions, is three. Based on that mean, is it likely that as many as 12 deaths per two-week period could be from natural causes? Explain.

4.70 An article in the *Orlando Sentinel* (March 28, 1985) notes that football is still the most dangerous sport. Records of the National Center for Sports Injury Research at the University of North Carolina indicate nine deaths in high school and collegiate football in 1983 and 13 in 1984. This represents a decline (probably due to rule changes) from a peak of 36 deaths in 1968. Suppose that the number of football deaths per year possesses a Poisson probability distribution and that the mean number of deaths per year is actually 30.

a.   Would a number x of deaths as small as 13 (the number observed in 1984) be unlikely if, in fact, the mean number of deaths per year were 30?

b.   Can you argue that the rule changes designed to reduce the number of serious football injuries have been effective? Explain.

4.71   A quality-control engineer wishes to study the alternative sampling plans $n = 5, a = 1$ and $n = 25, a = 5$. On the same sheet of graph paper, construct the operating characteristic curves for both plans, making use of acceptance probabilities at $p = .05$, $p = .10$, $p = .20$ $p = .30$, and $p = .40$ in each case.

a.   If you were a seller producing lots with fraction defective ranging from $p = 0$ to $p = .10$, which of the two sampling plans would you prefer?

b.   If you were a buyer wishing to be protected against accepting lots with a fraction defective exceeding $p = .30$, which of the two sampling plans would you prefer?

4.72   Consider a lot acceptance plan with $n = 20, a = 1$. Calculate the probability of accepting lots having a fraction defective of

a. $p = .01$ b. $p = .05$ c. $p = .10$ d. $p = .20$

e.   Sketch the operating characteristic curve for the plan.

4.73   Suppose that early statewide election returns indicate totals of 33,000 votes for candidate A versus 27,000 for candidate B, and that these early returns can be regarded as a random sample selected from the population of all 10,000,000 eligible voters in the state.

a.   If the statewide vote will be split 50–50 (that is, the probability that A will win is .5), find the expected number x of votes for A in the sample of 60,000 early returns.

b.   Find the standard deviation of x.

c.   Is the observed value $x = 33,000$ consistent with the theory in part (a) that the vote will split (that is, $p = .5$), or is x a highly unlikely value?

d.   Do you think that the value $x = 33,000$ is sufficient evidence to indicate that A will win?

4.74   A psychiatrist believes that 80% of all people who visit doctors have problems of a psychosomatic nature. He decides to select 25 patients at random to test his theory.

a.   Assuming that the pyschiatrist's theory is true, what is the expected value of x, the number of the 25 patients who have psychosomatic problems?

b.   What is the variance of x, assuming that the theory is true?

c.   Find $P(x \leq 14)$. (Use tables and assume that the theory is true.)

d.   Based on the probability in part (c), if only 14 of the 25 sampled had psychosomatic problems, what conclusions would you make about the psychiatrist's theory? Explain.

4.75   A particular type of radar installation has a probability of .2 of detecting and tracking an aircraft within a 200-mile radius. Suppose that the radar installations are located along the flight path of an approaching aircraft, and that the installations operate independently.

a.   What is the probability that the third radar installation is the first to detect the approaching aircraft?

b.   What is the probability that the aircraft is detected by a radar installation positioned prior to the fourth installation?

c.   How are parts (a) and (b) related to a geometric distribution with $p = .2$?

4.76   A student government states that 80% of all students favor an increase in student fees to subsidize a new recreational area. A random sample of $n = 25$ students produced 15 in favor of increased fees. What is the probability that 15 or fewer in the sample would favor the issue if student government is correct? Do the data support the student government's assertion, or does it appear that the percentage favoring an increase in fees is less than 80%?

4.77   Most weather forecasters seem to protect themselves very well by attaching probabilities to their forecasts such as, "The probability of rain today is 40%." Then if a particular forecast is incorrect, you are expected to attribute the error to the random behavior of the weather rather than the inaccuracy of the forecaster. To check the accuracy of a particular forecaster, records were checked only for those days when the forecaster predicted rain "with 30% probability." A check of 25 of those days indicated that it rained on 10 of the 25.

a.   If the forecaster is accurate, what is the appropriate value of p, the probability of rain on one of the 25 days?

b.   What are the mean and standard deviation of x, the number of days on which it rained, assuming that the forecaster is accurate?

c.   Calculate the z-score for the observed value, $x = 10$. [*Hint:* Recall from Section 2.9 that z-score $= (x - \mu)/\sigma$.]

d.   Do these data disagree with the forecast of a "30% probability of rain"? Explain.

4.78   A packaging experiment is conducted by placing two different package designs for a breakfast food side by side on a supermarket shelf. The objective of the experiment is to see whether buyers indicate a preference for one of the two package designs. On a given day, twenty-five customers purchased a package from the supermarket. Let x equal the number of buyers who choose the second package design.

a.   If there is no preference for either of the two designs, what is the value of p, the probability that a buyer chooses the second package design?

b.   If there is no preference, use the results of part (a) to calculate the mean and standard deviation of x.

c.   If five of the 25 customers choose the first package design and 20 choose the second design, what would you conclude about the customers' preference for the second package design?

4.79   Although employers are becoming increasingly concerned about the fitness of their employees, few employers encourage their workers by providing personal exercise or fitness programs. In a survey of 200 companies in the high-tech Silicon Valley of California, only one-third of the companies said they offered preventive health-care programs such as weight-control or smoking cessation (*Wall Street Journal*, December 6, 1988). If we assume that 133 of the 200 companies offer such programs, and we randomly select 5 of these companies for a follow-up investigation,

a.   What is the probability that none of the 5 offer preventive health-care programs?

b.   What is the probability that at least two offer preventive health-care programs?

4.80   Refer to Exercise 4.79. Let x be the number of companies offering preventive health-care programs.

a.   Calculate the mean and standard deviation of x.

b.   Within what limits would you expect x to lie?

4.81   The safety record of the Ford Bronco II was called into question when *Consumer Reports* magazine quoted federal statistics concerning Bronco II's fatal accident rate. According to an article in the *Press Enterprise* (May 18, 1989), there were 19 fatal rollovers for every 100,000 Bronco IIs in 1987. Suppose that, in an attempt to prove or disprove this claim, a consumer activist locates the records of 1000 randomly chosen Bronco II owners during that year. Of the 1000 cars, one had been involved in a fatal rollover.

a.   What is the distribution of x, the number of fatal rollovers in a sample of 1000 Bronco IIs?

b.   If the *Consumer Reports* statistics are correct, what is the probability of having one or more fatal rollovers in a sample of 1000 Bronco IIs?

c.   What are the mean and standard deviation of x?

d. Is the observed value of $x = 1$ consistent with the *Consumer Reports* statistics on fatal rollovers? Explain.

4.82 Refer to Exercise 4.81. Use the Poisson approximation to the binomial to approximate the probability found in Exercise 4.81, part (b). Compare the results.

4.83 The November 10, 1985 *Miami Herald* featured a report on the mysterious outbreaks of multiple sclerosis (MS) in Key West. The occurrence of 29 cases, with eight of them being nurses in the same hospital, "challenges the century-old belief that multiple sclerosis is a noncommunicable Snow Belt illness." The rate of occurrence of MS in Key West would be expected to fall in the range of three to ten persons per 100,000. Assume that the number x of cases per 100,000 in Key West under typical conditions can be approximated by a Poisson probability distribution with mean $\mu = 10$. What values of x would lead you to believe that an epidemic of MS might be occurring in Key West? Explain. (*Aside:* The *Miami Herald* gives the rate of MS in nurses at the Florida Keys Hospital as 8000 per 100,000.)

4.84 A manufacturer of videotapes ships them in lots of 1200 tapes per lot. Before shipment, 20 tapes are randomly selected from each lot and tested. If none are defective, the lot is shipped. If one or more is defective, every tape in the lot is tested.
a. What is the probability distribution for x, the number of defective tapes in the sample of 20 tapes?
b. What distribution can be used to approximate probabilities for the random variable x in part (a)?
c. What is the probability that a lot will be shipped if it contains ten defectives? 20 defectives? 30 defectives?

4.85 Use the values calculated in Exercise 4.84 to sketch the operating characteristic curve for this lot acceptance sampling plan.

4.86 An article in the *Orlando Sentinel* (January 29, 1985) reports on a study conducted by the Massachusetts Department of Health on the death rate from cancer for Vietnam veterans. The researchers examined the cause of death for 804 Vietnam veterans and found that nine had died of tumors in muscle or other soft tissue. The expected number of deaths in an equal-size group of non-Vietnam veterans is 1.9.
a. What is the probability distribution for x, the number of veterans in the sample of 804 who died of tumors?
b. What distribution would provide a good approximation to the number x of veterans who died of tumors if the mean number was 1.9 in a sample of 804? Explain.
c. If you observed $x = 9$ deaths in a sample of 804, would you believe that the observation was selected from a binomial population with mean equal to 1.9? Explain.

4.87 Consider sampling $n = 2$ numbers from the set of $N = 5$ numbers given as 1, 2, 5, 7, 12.
a. How many different samples are possible if the order of selection is disregarded?
b. What is the probability of selecting a distinct pair?
c. If x is the sum of the two numbers in the pair, does x have a uniform distribution? Can the mean and variance of x be found by formula? Explain.

4.88 A fair coin is tossed until the first head appears. Let x be the number of tosses required.
a. What is the probability distribution of x?
b. Graph the probability distribution of x.
c. Find the mean and variance of x. Within what limits would you expect x to fall at least 8/9 of the time?
d. If the coin is tossed and the first head occurs on the tenth toss, what conclusions might you draw about the fairness of the coin? Explain.

4.89 According to an article in the *Shands Quarterly* (Vol. 3, no. 3 [Fall 1984]), one out of every 11 women will develop breast cancer during her lifetime. Suppose that you were to select a random sample of 100 women. Let x represent the number in the sample who will develop breast cancer at some time during their lifetime. If the reported breast cancer rate has not changed,

a. Find the expected value of x.

b. Find the standard deviation of x

c. Is it reasonable to expect that as many as 20 or more of the women in the sample might develop breast cancer during their lifetime? Explain.

4.90 Tests by William O'Neil and Eric Topol of the University of Michigan suggest that a new drug, a tissue plasminogen activator (TPA), can unblock clogged arteries and reduce the possibility of second heart attacks by high-risk cardiac patients (*Orlando Sentinel*, December 26, 1985). According to Topol, "TPA is a substance produced by the body that dissolves blood clots in 15 to 20 percent of patients. It can now be mass-produced in the laboratory to help all patients." O'Neil and Topol found that injections of TPA opened blocked arteries in 80% of the 30 patients included in their experiment. Suppose that we want to determine whether the manufactured TPA is effective in unblocking clogged arteries.

a. If the manufactured TPA is ineffective, the proportion p of patients who would have their arteries unblocked (strictly due to the TPA produced in their bodies) is at most .2. If $p = .2$, calculate the mean and standard deviation of x.

b. Are the results of the O'Neill and Topol study likely if in fact $p = .2$? What would you conclude about the effectiveness of the manufactured TPA?

A MYSTERY: CANCERS NEAR A REACTOR

How safe is it to live near a nuclear reactor? Men who lived in a coastal strip that extends 20 miles north from a nuclear reactor in Plymouth, Massachusetts developed some forms of cancer at a rate 50% higher than the statewide rate, according to a study endorsed by the Massachusetts Department of Public Health and reported in the May 21, 1987 edition of the *New York Times*.

The cause of the excess cancers is a mystery, but it was suggested that the cause was linked to the Pilgrim I reactor, which had been shut down for 13 months because of management problems. Boston Edison, the owner of the reactor, acknowledged radiation releases in the mid-1970s that were just above permissible levels. If the reactor was in fact responsible for the excess cancers, the currently acknowledged level of radiation required to cause cancer would have to change. However, confounding the mystery is the fact that women in this same area were seemingly unaffected.

In his report, Dr. Sidney Cobb, an epidemiologist, noted the connection between the radiation releases at the Pilgrim I reactor and 52 cases of hematopoietic cancers. The report indicated that this unexpectedly large number might be attributable to airborne radioactive effluents from Pilgrim I, concentrated along the coast by wind patterns, and not dissipated as assumed by government regulators. How unusual was this number of cancer cases? That is, statistically speaking, is 52 a highly improbable number of cases? If the answer is yes, then either some external factor, possibly radiation, caused this unusually large number, or we have observed a very rare event!

The Poisson probability distribution provides a good approximation to the distributions of variables such as the number of deaths in a region due to a rare disease, the number of accidents in a manufacturing plant per month, or the number of airline crashes per month. Therefore, it is reasonable to assume that the Poisson distribution will provide an appropriate model for the number of cancer cases in this instance.

1. If the 52 reported cases represented a rate 50% higher than the statewide rate, what is a reasonable estimate of μ, the average number of such cancer cases statewide?

2. Based upon your estimate of μ, what is the estimated standard deviation of the number of cancer cases statewide?

3. What is the z-score for the $x = 52$ observed cases of cancer? How would you interpret this z-score in light of the concern about an elevated rate of hematopoietic cancers in this area?

CHAPTER 5

THE NORMAL AND OTHER CONTINUOUS DISTRIBUTIONS

Case Study

If you were the boss, would height play a role in your selection of a successor for your job? Would you purposely choose a successor who was shorter than you? The case study at the end of Chapter 5 examines how the normal curve can be used in a situation involving the height distribution of Chinese males eligible for a very prestigious job.

General Objective

In Chapters 3 and 4 you learned about discrete random variables, their probability distributions, and how these distributions play a role in making inferences about populations based on information contained in a sample. In this chapter you will learn about the uniform, the exponential, and the normal random variables and their distributions. You will learn how to calculate normal probabilities and, under certain conditions, how to use the normal probability distribution to approximate the binomial probability distribution. Then, in Chapter 6 and in the chapters that follow, you will see why the normal probability distribution plays a central role in statistical inference.

Specific Topics

1   Probability distributions for continuous random variables; the uniform and exponential distributions (5.1)
2   The normal probability distribution (5.2)
3   Calculation of areas associated with the normal probability distribution (5.3)
4   The normal approximation to the binomial probability distribution (5.4)

5.1   PROBABILITY DISTRIBUTIONS FOR CONTINUOUS RANDOM VARIABLES

Not all experiments have sample spaces containing a finite or at least countable number of simple events. **Continuous random variables,** such as heights and weights, length of life of a particular product, or experimental laboratory error, can assume the infinitely many values corresponding to points on a line interval. However, we cannot assign a positive probability to each of these infinitely many points and still have the probabilities sum to 1, as for discrete random variables. Therefore, a different approach is used to generate the probability distribution for a continuous random variable. The approach that we adopt uses the concept of a relative frequency histogram, such as that for the 30 grade point averages in Figure 2.1. Recall that the width of the class interval in Figure 2.1 was determined in accordance with the number of measurements involved. If more and more measurements are obtained, we might reduce the width of the class interval. The outline of the histogram would change slightly, for the most part becoming less and less irregular. When the number of measurements becomes very large and the intervals very small, the relative frequency histogram would appear, for all practical purposes, as a smooth curve, as shown in Figure 5.1.

Figure 5.1
A relative frequency histogram for a population

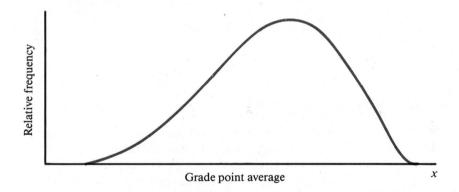

The relative frequency associated with a particular class in the population is the fraction of measurements in the population falling in that interval and also is the probability of drawing a measurement in that class. If the total area under the relative frequency histogram is adjusted to equal 1, areas under the frequency curve correspond to probabilities. In fact, this correspondence was the basis for the application of the Empirical Rule in Chapter 2, which applies when the data are approximately bell-shaped.

Let us construct a model for the probability distribution for a continuous random variable. Assume that the random variable x may take on any value on a real line, as in Figure 5.1. We then distribute 1 unit of probability along the line, much as a person might distribute a handful of sand, each measurement in the population corresponding to a single grain. The probability—grains of sand or measurements—will pile up in certain places, and the result will be the probability distribution shown in Figure 5.2. The depth or density of the probability, which varies with x, may be represented by a mathematical formula $f(x)$, called the *probability distribution*, or the *probability density function*, for the random variable x. The density function $f(x)$, represented graphically in Figure 5.2, provides a mathematical model for the population relative frequency histogram that exists in reality. The total area under the curve $f(x)$ is equal to 1. The area lying above a given interval equals the probability that x will fall in that interval. Thus, the probability that $x_1 < x < x_2$ (x_1 is less than x and x is less than x_2) is equal to the area under the density function between the two points x_1 and x_2. This is the shaded area in Figure 5.2.

Figure 5.2
The probability distribution $f(x)$; $P(x_1 < x < x_2)$ is equal to the shaded area under the curve

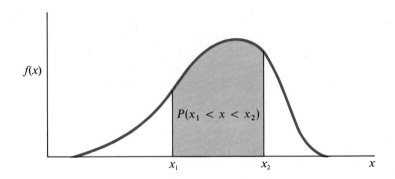

THE UNIFORM DISTRIBUTION

The uniform distribution provides a simple probability model to describe a continuous random variable that can randomly assume any value between two points on a line, say a and b ($a < b$). The uniform probability density function has a rectangular shape over the interval from a to b with height $1/(b - a)$ and an area under the density function equal to 1, as shown in Figure 5.3. Hence, the uniform distribution provides a model for a continuous random variable whose values are

Figure 5.3
The uniform probability distribution on the interval $a \leqslant x \leqslant b$

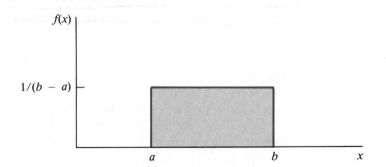

uniformly distributed over an interval. For example, if buses arrive at a given bus stop every 15 minutes, and you arrive at the bus stop at a random time, the time you wait for the next bus to arrive could be described by a uniform distribution over the interval from 0 to 15.

UNIFORM PROBABILITY DENSITY FUNCTION

$$f(x) = \frac{1}{b - a} \quad \text{for} \quad a \le x \le b$$

The mean and standard deviation are

$$\mu = \frac{a + b}{2} \qquad \sigma = \frac{b - a}{\sqrt{12}}$$

The mean of a uniform random variable is equal to the midpoint of the interval, and the standard deviation is directly proportional to the length of the interval.

Since the area under the density function between two points is the probability that x belongs to that interval, the probability that the uniform random variable lies in the interval from c to d when this interval is contained within the interval from a to b is simply the area of the rectangle over the interval from c to d, given by $(d - c)/(b - a)$.*

EXAMPLE 5.1 Suppose that buses arrive at a bus stop every 15 minutes and that the waiting time for the next bus to arrive has a uniform probability distribution on the interval from 0 to 15 minutes.

a. Find the probability that x, a person's waiting time, will exceed 10 minutes.

b. Calculate the mean and standard deviation of x. Graph the probability density function of x, and indicate the location of the mean μ and the endpoints of the intervals $(\mu \pm \sigma)$ and $(\mu \pm 2\sigma)$.

c. Find the containment probabilities for these two intervals. How do they compare with those given by Tchebysheff's Theorem and the Empirical Rule?

Solution a. If a person's waiting time x is uniformly distributed over the interval from 0 to 15, then $P(x > 10)$ is found as the area above the interval from 10 to 15 and is given by

$$P(x > 10) = \frac{15 - 10}{15 - 0} = \frac{5}{15} = \frac{1}{3}$$

Hence, there is a 1/3 chance that a person will wait longer than 10 minutes (and a 2/3 chance of waiting less than 10 minutes).

----

* If you have had a calculus course, you may recall that the area under the curve $f(x) = 1/(b - a)$ between the points c and d is given by

$$\int_c^d \frac{1}{b - a} dx = \frac{d - c}{b - a}$$

b.   To calculate the mean and standard deviation, we replace a and b by 0 and 15, respectively, in the given formulas to find

$$\mu = \frac{a + b}{2} = \frac{0 + 15}{2} = 7.5$$

and

$$\sigma = \frac{b - a}{\sqrt{12}} = \frac{15}{\sqrt{12}} = 4.33$$

Therefore, the interval $(\mu \pm \sigma)$, or (7.5 ± 4.3), has endpoints 3.2 and 11.8. The interval (7.5 ± 8.6) has endpoints -1.1 and 16.1. These interval endpoints and the mean are shown in Figure 5.4.

Figure 5.4
Endpoints and mean for
Example 5.1

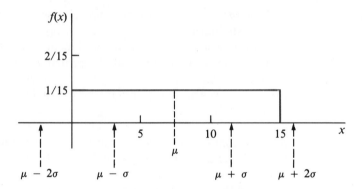

c.   The probability that the waiting time will be between 3.2 and 11.8 minutes is given as

$$P(3.2 < x < 11.8) = (\text{base})(\text{height}) = (11.8 - 3.2)\left(\frac{1}{15}\right)$$

$$= \frac{8.6}{15} = .5733$$

and

$$P(-1.1 < x < 16.1) = (15 - 0)\left(\frac{1}{15}\right) = 1$$

Tchebysheff's Theorem applies to any distribution, and hence, we have exceeded the minimum containment probabilities of 0 and .75. However, the Empirical Rule, which assumes that the distribution is mound-shaped, does not give a reasonable approximation to the actual containment probabilities.

THE EXPONENTIAL DISTRIBUTION

The exponential distribution is another continuous distribution that arises when the random variable represents a waiting time. The time until a machine or one of its components fails, and the waiting time in a service line are examples of random variables that often follow an exponential distribution.

The probability density function of an exponential random variable together with its mean and standard deviation are given in the display.

EXPONENTIAL PROBABILITY DENSITY FUNCTION

$$f(x) = \lambda e^{-\lambda x} \qquad x \geq 0; \quad \lambda > 0$$

The mean and standard deviation are

$$\mu = \frac{1}{\lambda} \qquad \sigma = \frac{1}{\lambda}$$

Curves of the exponential probability density function corresponding to several values of λ are given in Figure 5.5. The shape of an exponential distribution is

Figure 5.5
Exponential probability distributions for $\lambda = .5$, 1, and 2

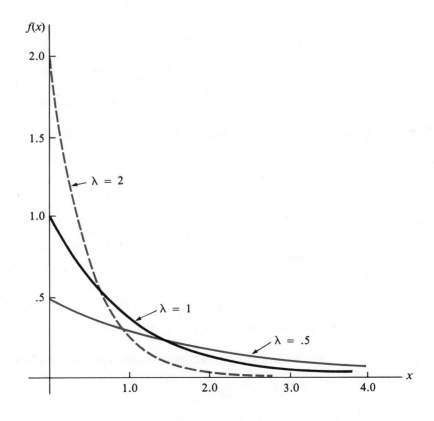

determined by the single parameter λ. In fact, for this distribution the mean and the standard deviation are equal to the common value $1/\lambda$.

When the number of events occurring in a unit time has a Poisson distribution with mean λ, then the waiting time between these events follows an exponential distribution with an average waiting time equal to $1/\lambda$. Therefore, if the number of patients arriving at a hospital emergency room follows a Poisson distribution with an average of $\lambda = 5$ persons per hour, then the time between arrivals follows an exponential distribution with an average waiting time of $\mu = 12$ minutes (that is, $1/\lambda = 1/5 = .2$ hour).

In finding the probabilities associated with an exponential probability distribution, we can use the following relationship.

FINDING RIGHT-TAILED PROBABILITIES
FOR AN EXPONENTIAL RANDOM VARIABLE

$$P(x \geq a) = e^{-\lambda a} \qquad a \geq 0$$

After substituting the appropriate values of λ and a, the value of $e^{-\lambda a}$ can be found from Table 2(b) in Appendix III or by using a calculator that has an exponential function key as one of its features.

EXAMPLE 5.2   Suppose that the time in days between service calls on an office copying machine follows an exponential distribution with $\lambda = .02$.

     a.   What is the probability that the time until the machine again requires service exceeds 60 days?

     b.   What is the probability that the time until the machine again requires service is less than 20 days?

     c.   Find the probability that the time until the machine again requires service is longer than $\mu + 2\sigma$.

Solution   a.   Let x represent the time in days between service calls; then x has an exponential distribution with $\lambda = .02$. Therefore,

$$P(x > 60) = e^{-(\lambda)(60)} = e^{-(.02)(60)} = e^{-1.2} = .301194$$

(Use Table 2(b) or your calculator to evaluate $e^{-1.2}$.)

     b.   Left-tail probabilities are easily found as complements of right-tail probabilities. Therefore,

$$P(x < 20) = 1 - P(x > 20) = 1 - e^{-\lambda(20)} = 1 - e^{-(.02)(20)}$$

$$= 1 - e^{-.4} = 1 - .670320 = .329680$$

     c.   The mean and standard deviation are both equal to

$$\frac{1}{\lambda} = \frac{1}{.02} = 50$$

Therefore,

$$P(x > \mu + 2\sigma) = P(x > 150)$$

$$= e^{-.02(150)} = e^{-3} = .049783$$

Hence, there is only a 5% chance that the machine will not require servicing during the next 150 days or approximately five months.   ◁

How do we choose the model—that is, the probability distribution $f(x)$—appropriate for a given physical situation? Many types of continuous curves are available for modeling, not all of which are mound-shaped as are those shown in Figures 5.1 and 5.2. Fortunately, we will find that many continuous random variables have mound-shaped frequency distributions, often very nearly bell-shaped. A probability model that provides a good approximation to such population distributions is the **normal probability distribution**, which we will study in detail in this chapter.

In actual practice, there will be a difference between the conceptual model $f(x)$ and the relative frequency histogram generated when the experiment is repeated an extremely large number of times. The model $f(x)$ that we adopt can be expected only to *approximate* the population relative frequency curve. This practice can lead to an invalid conclusion if an inappropriate model is chosen out of poor judgment or insufficient knowledge of the phenomenon under study. On the other hand, using a density function $f(x)$ to approximate the population relative frequency distribution for a continuous random variable is a strategy that, when properly applied, has been found to be highly successful in scientific research. The equations, formulas, and various numerical expressions used in all the sciences are simply mathematical models that provide approximations to reality, the goodness of which is evaluated through experimental application. Hence, a model is evaluated in terms of the results of its application—which are decisions or predictions in practical situations. Do the resulting inferences fit in with the body of accumulated evidence? Are the deductions that follow from these inferences verified experimentally? If so, the model has proved its worth.

▷ 5.2 THE NORMAL PROBABILITY DISTRIBUTION

In Section 5.1, we saw that the probabilistic model for the frequency distribution of a continuous random variable involves the selection of a curve, usually smooth, called the **probability distribution**, or **probability density function**.

NORMAL PROBABILITY DENSITY FUNCTION

$$f(x) = \frac{1}{\sigma\sqrt{2\pi}}\, e^{-(x-\mu)^2/2\sigma^2} \qquad -\infty \le x \le \infty$$

The symbols e and π are mathematical constants given approximately by 2.7183 and 3.1416, respectively: μ and σ ($\sigma > 0$) are parameters representing the population mean and standard deviation.

Although these distributions may assume a variety of shapes, a large number of random variables observed in nature possess a frequency distribution that is approximately bell-shaped or, as the statistician would say, is approximately a normal probability distribution.

The normal density function has a total area under its curve equal to 1. The graph of a normal probability distribution with mean μ and standard deviation σ is given in Figure 5.6. From the form of the normal density function and from Figure 5.6, we see that the normal distribution is symmetric about its mean μ. Furthermore, the shape of the distribution is determined by σ, the population standard deviation. Large values of σ reduce the height of the curve and increase the spread; small values of σ increase the height and reduce the spread of the curve.

Figure 5.6
Normal probability
density function

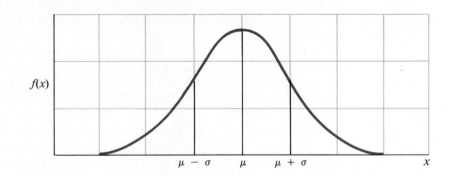

In practice, we seldom encounter variables that range from infinitely small negative values to infinitely large positive values. Nevertheless, many positive random variables such as heights, weights, and times generate a frequency histogram that is well approximated by a normal distribution. The approximation applies because almost all of the values of a normal random variable lie within three standard deviations of the mean, and in these cases, $(\mu \pm 3\sigma)$ almost always encompasses positive values.

5.3 TABULATED AREAS OF THE NORMAL PROBABILITY DISTRIBUTION

In Section 5.1 we explained that the probability that a continuous random variable assumes a value in the interval x_1 to x_2 is the area under the probability density function between the points x_1 and x_2 (see Figure 5.2). The probability model for a continuous random variable differs greatly from the model for a discrete random variable when we consider the probability that x equals some particular value, say a. Since **the area lying over any particular point, say $x = a$, is 0**, it follows from our probability model that the probability that $x = a$ is 0. Thus, the expression $P(x \leq a)$ is the same as $P(x < a)$ because $P(x = a) = 0$. Similarly, $P(x \geq a) =$

$P(x > a)$. This statement is, of course, not true for a discrete random variable because $P(x = a)$ may not equal 0.

To find areas under the normal curve, we first note that the equation for the normal probability distribution (Section 5.2) is dependent on the numerical values of μ and σ and that by supplying various values for these parameters, we could generate an infinitely large number of bell-shaped normal distributions. A separate table of areas for each of these curves is obviously impractical; instead, we would like one table of areas applicable to all. The easiest way to use one table is to work with areas lying within a specified number of standard deviations of the mean, as was done in the case of the Empirical Rule. For instance, we know that approximately .68 of the area will lie within one standard deviation of the mean, .95 within two, and almost all within three. But what fraction of the total area will lie within .7 standard deviation, for instance? Questions of this type can be answered by using Table 3 in Appendix III.

Figure 5.7
Normal distribution: area to the left (or right) of the mean equals .5

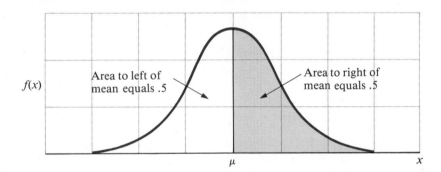

Since the normal curve is symmetrical about the mean, half of the area under the curve lies to the left of the mean and half to the right (see Figure 5.7). Also, because of the symmetry, we can simplify the table of areas by listing the areas between the mean and a specified number z of standard deviations to the right of μ (see Figure 5.8). **An area to the left of the mean can be calculated by using the corresponding and equal area to the right of the mean.** The distance from the mean to a given value of x is $(x - \mu)$. Expressing this distance in units of the standard deviation σ, we obtain

$$z = \frac{x - \mu}{\sigma}$$

Note that there is a one-to-one correspondence between the random variables z and x and, in particular, that $z = 0$ when $x = \mu$. The probability distribution for z is often called the **standardized normal distribution**, because its mean is 0 and its standard deviation is 1. It is shown in Figure 5.8. The area under the standard normal curve between the mean $z = 0$ and a specified value of z, say z_0, is the probability $P(0 \leq z \leq z_0)$. This area is recorded in Table 3 of Appendix III and is shown as the shaded area in Figure 5.8. An abbreviated version of Table 3 in Appendix III is shown here in Table 5.1.

Note that z, correct to the nearest tenth, is recorded in the left-hand column. The second decimal place for z, corresponding to hundredths, is given across the top row. Thus, the area between the mean and $z = .7$ standard deviation to the right, read in the second column of the table opposite $z = .7$, is found to be .2580. Similarly, the area between the mean and $z = 1.0$ is .3413. The area lying within one standard deviation on either side of the mean would be two times .3413, or .6826. The area lying within two standard deviations of the mean, correct to four decimal places, is $2(.4772) = .9544$. These numbers agree with the approximate values, 68% and 95%, used in the Empirical Rule in Chapter 2.

To find the area between the mean and a point $z = .57$ standard deviation to the right of the mean, proceed down the left-hand column to the 0.5 row. Then move across the top row of the table to the .07 column. The intersection of this row-column combination gives the appropriate area, .2157.

Since the normal distribution is continuous, the area under the curve associated with a single point is equal to 0. Keep in mind that this result applies only to

Figure 5.8
Standardized normal
distribution

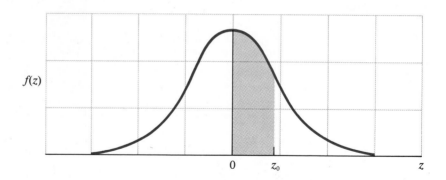

Table 5.1
Abbreviated version of Table 3 in Appendix III

z_0	.00	.01	.02	.03	.04	.05	.06	.07	.08	.09
0.0	.0000	.0040	.0080	.0120	.0160	.0199	.0239	.0279	.0319	.0359
0.1	.0398	.0438	.0478	.0517	.0557	.0596	.0636	.0675	.0714	.0753
0.2	.0793	.0832	.0871	.0910	.0948	.0987	.1026	.1064	.1103	.1141
0.3	.1179	.1217	.1255	.1293	.1331	.1368	.1406	.1443	.1480	.1517
0.4	.1554	.1591	.1628	.1664	.1700	.1736	.1772	.1808	.1844	.1879
0.5	.1915	.1950	.1985	.2019	.2054	.2088	.2123	**.2157**	.2190	.2224
0.6	.2257	:	:	:	:	:	:	:	:	:
0.7	**.2580**									
:	:									
1.0	**.3413**									
:	:									
2.0	**.4772**									

continuous random variables. Later in this chapter we will use the normal probability distribution to approximate the binomial probability distribution. The binomial random variable x is a discrete random variable. Hence, as you know, the probability that x takes some specific value, say $x = 10$, will not necessarily equal 0. Consequently, for discrete random variables $P(x \leq x_0)$ is not the same as $P(x < x_0)$.

Let us now consider some examples.

EXAMPLE 5.3   Find $P(0 \leq z \leq 1.63)$. This probability corresponds to the area between the mean ($z = 0$) and a point $z = 1.63$ standard deviations to the right of the mean (see Figure 5.9).

Figure 5.9
Probability required for
Example 5.3

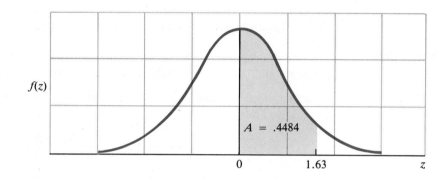

Solution   The area is shaded and indicated by the symbol A in Figure 5.9. Since Table 3 in Appendix III gives areas under the normal curve to the right of the mean, we need only find the tabulated value corresponding to $z = 1.63$. Proceed down the left-hand column of the table to the row corresponding to $z = 1.6$ and across the top of the table to the column marked .03. The intersection of this row and column combination gives the area $A = .4484$. Therefore, $P(0 \leq z \leq 1.63) = .4484$.   ◁

EXAMPLE 5.4   Find $P(-.5 \leq z \leq 1.0)$. This probability corresponds to the area between $z = -.5$ and $z = 1.0$, as shown in Figure 5.10.

Figure 5.10
Area under the normal
curve in Example 5.4

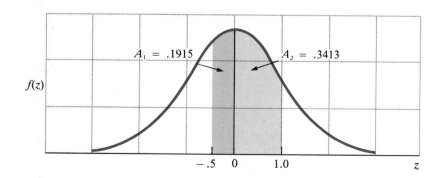

Solution   The area required is equal to the sum of A_1 and A_2 shown in Figure 5.10. From Table 3 in Appendix III we read $A_2 = .3413$. The area A_1 equals the area between $z = 0$ and $z = .5$, or $A_1 = .1915$. Thus, the total area is

$$A = A_1 + A_2 = .1915 + .3413 = .5328$$

That is, $P(-.5 \leq z \leq 1.0) = .5328$.

EXAMPLE 5.5   Find the value of z, say z_0, such that (to four decimal places) .95 of the area is within $\pm z_0$ standard deviations of the mean.

Solution   Half of the area, $(1/2)(.95) = .475$, will lie to the left of the mean and half to the right, because the normal distribution is symmetrical. Thus, we seek the value z_0 corresponding to an area equal to .475. The area .475 falls in the row corresponding to $z = 1.9$ and the .06 column. Hence, $z_0 = 1.96$. Note that this result is very close to the approximate value, $z = 2$, used in the Empirical Rule.

EXAMPLE 5.6   Let x be a normally distributed random variable, with a mean of 10 and a standard deviation of 2. Find the probability that x lies between 11 and 13.6.

Solution   As a first step, we must calculate the values of z corresponding to $x_1 = 11$ and $x_2 = 13.6$. Thus, we have

$$z_1 = \frac{x_1 - \mu}{\sigma} = \frac{11 - 10}{2} = .5 \qquad z_2 = \frac{x_2 - \mu}{\sigma} = \frac{13.6 - 10}{2} = 1.8$$

The desired probability is therefore $P(.5 \leq z \leq 1.8)$ and is the area lying between z_1 and z_2, as shown in Figure 5.11. The area between $z = 0$ and z_1 is $A_1 = .1915$, and the area between $z = 0$ and z_2 is $A_2 = .4641$; these areas are obtained from Table 3. The desired probability is equal to the difference between A_2 and A_1; that is,

$$P(.5 \leq z \leq 1.8) = A_2 - A_1 = .4641 - .1915 = .2726$$

Figure 5.11
Area under the normal
curve in Example 5.6

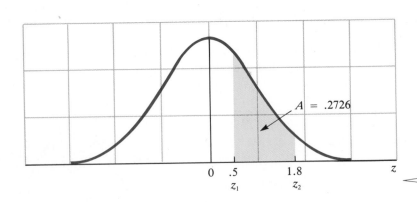

EXAMPLE 5.7   Studies show that gasoline use for compact cars sold in the United States is normally distributed, with a mean use of 25.5 miles per gallon and a standard devia-

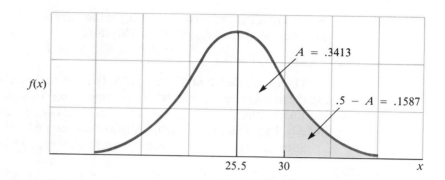

tion of 4.5 miles per gallon. What percentage of compacts obtain 30 or more miles per gallon?

Solution The proportion of compacts obtaining 30 or more miles per gallon is given by the shaded area in Figure 5.12.

We must find the z value corresponding to $x = 30$. Substituting into the formula for z, we obtain

$$z = \frac{x - \mu}{\sigma} = \frac{30 - 25.5}{4.5} = 1.0$$

The area A to the right of the mean, corresponding to $z = 1.0$, is .3413 (from Table 3). Then the proportion of compacts having a miles-per-gallon ratio equal to or greater than 30 is equal to the entire area to the right of the mean, .5, minus the area A:

$$P(x \geq 30) = .5 - P(0 \leq z \leq 1) = .5 - .3413 = .1587$$

The percentage exceeding 30 miles per gallon is

$$100(.1587) = 15.87\%$$

EXAMPLE 5.8 Refer to Example 5.7. In times of scarce energy resources a competitive advantage is given to an automobile manufacturer who can produce a car obtaining substantially better fuel economy than the competitors' cars. If a manufacturer wishes to develop a compact car that outperforms 95% of the current compacts in fuel economy, what must be the gasoline use rate for the new car?

Solution Let x be a normally distributed random variable, with a mean of 25.5 and a standard deviation of 4.5. We want to find the value x_0 such that

$$P(x \leq x_0) = .95$$

As a first step, we find

$$z_0 = \frac{x_0 - \mu}{\sigma} = \frac{x_0 - 25.5}{4.5}$$

and note that our required probability is the same as the area to the left of z_0 for the standardized normal distribution. Therefore,

$$P(z \leq z_0) = .95$$

The area to the left of the mean is .5. The area to the right of the mean between z_0 and the mean is $.95 - .5 = .45$. Thus, from Table 3 we find that z_0 is between 1.64 and 1.65. Notice that the area .45 is exactly halfway between the areas for $z = 1.64$ and $z = 1.65$. Thus, z_0 is exactly halfway between 1.64 and 1.65; that is, $z_0 = 1.645$.
Substituting $z_0 = 1.645$ into the equation for z_0, we have

$$1.645 = \frac{x_0 - 25.5}{4.5}$$

and solving for x_0, we obtain

$$x_0 = (1.645)(4.5) + 25.5 = 32.9$$

The manufacturer's new compact car must therefore obtain a fuel economy of 32.9 miles per gallon to outperform 95% of the compact cars currently available on the U.S. market.

EXERCISES   Basic Techniques

5.1 A random variable x has a uniform probability distribution with density function given as

$$f(x) = 1 \quad \text{for} \quad 0 \leq x \leq 1$$

a. Graph the probability distribution for x.
b. Find the probability that x falls between .5 and .75.
c. Calculate the mean and variance of x.

5.2 A random variable x has an exponential probability distribution with density function given as

$$f(x) = (1/4)e^{-x/4} \quad \text{for} \quad x \geq 0$$

a. Find $P(x = 4)$.   b. Find $P(x \leq 1)$.
c. Find $P(1.5 \leq x \leq 3)$.   d. Find $P(x \geq 1.5)$.

5.3 Using Table 3 in Appendix III, calculate the area under the normal curve between
a. $z = 0$ and $z = 1.6$ b. $z = 0$ and $z = 1.83$
c. $z = 0$ and $z = .90$ d. $z = 0$ and $z = -.90$

5.4 Repeat Exercise 5.3 for
a. $z = -1.4$ and $z = 1.4$ b. $z = -2.0$ and $z = 2.0$ c. $z = -3.0$ and $z = 3.0$

5.5 Repeat Exercise 5.3 for
a. $z = -1.43$ and $z = .68$ b. $z = .58$ and $z = 1.74$
c. $z = -1.55$ and $z = -.44$

5.6 a. Find a z_0 such that $P(z > z_0) = .025$. b. Find a z_0 such that $P(z < z_0) = .9251$.

5.7 a. Find a z_0 such that $P(z > z_0) = .9750$. b. Find a z_0 such that $P(z > z_0) = .3594$.

5.8 Find a z_0 such that $P(-z_0 < z < z_0) = .8262$.

5.9 a. Find a z_0 such that $P(z < z_0) = .9505$. b. Find a z_0 such that $P(z < z_0) = .05$.

5.10 a. Find a z_0 such that $P(-z_0 < z < z_0) = .90$.
      b. Find a z_0 such that $P(-z_0 < z < z_0) = .99$.

5.11 A variable x is normally distributed with mean $\mu = 10$ and standard deviation $\sigma = 2$. Find the probability that
      a. $x > 13.5$ b. $x < 8.2$ c. $9.4 < x < 10.6$

5.12 A variable x is normally distributed with mean $\mu = 1.20$ and standard deviation $\sigma = .15$. Find the probability that
      a. $1.00 < x < 1.10$ b. $x > 1.38$ c. $1.35 < x < 1.50$

5.13 A variable is normally distributed with unknown mean μ and standard deviation $\sigma = 2$. If the probability that x exceeds 7.5 is .8023, find μ.

5.14 A variable is normally distributed with unknown mean μ and standard deviation $\sigma = 1.8$. If the probability that x exceeds 14.4 is .3015, find μ.

5.15 A variable is normally distributed with unknown mean and standard deviation. It is known that the probability that x exceeds 4 is .9772 and the probability that x exceeds 5 is .9332. Find μ and σ.

Applications

5.16 Studies indicate that drinking water supplied by some old lead-lined city piping systems may contain harmful levels of lead. Based on data presented by Karalekas, Ryan, and Taylor ("Control of Lead, Copper, and Iron Pipe Corrosion in Boston," *American Water Works Journal* [February 1983]), it appears that the distribution of lead content readings for individual water specimens has a mean and standard deviation equal (approximately) to .033 mg/l and .10 mg/l, respectively. Explain why you believe that this distribution is or is not normally distributed.

5.17 The sales x of a gasoline distributor have a uniform probability distribution, as shown below. Because of equipment limitations, daily sales will never be less than 10,000 gallons per day and never greater than 50,000 gallons per day. Use the information in the figure.

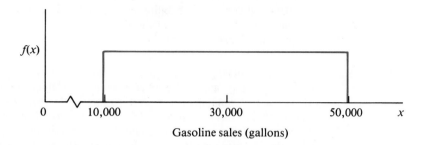

Gasoline sales (gallons)

      a. Find the probability that the distributor sells at least 40,000 gallons per day.
      b. Find the probability that the distributor sells between 20,000 and 30,000 gallons per day.
      c. Find the probability that the distributor sells at most 35,000 gallons per day.
      d. What is the mean and standard deviation of the sales (in gallons)?
      e. What is the probability that the distributor's sales lie in the interval ($\mu \pm \sigma$) gallons?

5.18 Sales of United States-made microchips in Japan has been hampered by long delivery times, according to a Japanese survey conducted by the Ministry of International Trade and

Industry involving 63 major Japanese semiconductor users (*Japan Economic Journal*, February 13, 1988, p. 14). U.S. chips arrive from 6 to 12 months after an order is placed, compared with 45 to 90 days for Japanese chips. Suppose the delivery times for United States-made microchips has an exponential distribution with an average of 6 months.

   a.   What is the probability that delivery of a U.S. shipment takes 12 months or longer?

   b.   What is the probability that delivery of a U.S. shipment takes less than 6 months?

   c.   Use parts (a) and (b) to find the probability that delivery time for a U.S. order is between 6 months and 1 year.

5.19   For a car traveling 30 miles per hour, the distance required to brake to a stop is normally distributed with mean of 50 feet and a standard deviation of 8 feet. Suppose that you are traveling 30 miles per hour in a residential area and a car moves abruptly into your path at a distance of 60 feet.

   a.   If you apply your brakes, what is the probability that you will brake to a stop within 40 feet or less? within 50 feet or less?

   b.   If the only way to avoid a collision is to brake to a stop, what is the probability that you will avoid the collision?

5.20   Suppose that you must establish regulations concerning the maximum number of people who can occupy an elevator. A study of elevator occupancies indicates that if eight people occupy the elevator, the probability distribution of the total weight of the eight people has a mean equal to 1200 pounds and a variance equal to 9800 (pounds)^2. What is the probability that the total weight of eight people exceeds 1300 pounds? 1500 pounds? (Assume that the probability distribution is approximately normal.)

5.21   The discharge of suspended solids from a phosphate mine is normally distributed, with a mean daily discharge of 27 mg/l and a standard deviation of 14 mg/l. What proportion of days will the daily discharge exceed 50 mg/l?

5.22   Philatelists (stamp collectors) often buy stamps at or near retail prices, but when they sell, the price is considerably lower. For example, it may be reasonable to assume that (depending on the mix of a collection, condition, demand, economic conditions, etc.) a collection might be expected to sell at $x\%$ of retail price, where x is normally distributed with a mean equal to 45% and a standard deviation of 4.5%. If a philatelist has a collection to sell that has a retail value of $30,000, what is the probability that the philatelist receives

   a.   more than $15,000 for the collection?

   b.   less than $15,000 for the collection?

   c.   less than $12,000 for the collection?

5.23   T. Sato and associates determined the percentages of iron in bulk specimens of Chilean lumpy iron ore randomly sampled from a 35,325 long ton shipload of ore ("Example of Experiments on Systematic Sampling of Iron Ore," *Reports of Statistical Application Research, Union of Japanese Scientists and Engineers*, 18 [1971]). The mean and standard deviation of a sample of these ore analyses were 62.96% and .61%, respectively. Suppose that we were to use these quantities as approximations to μ and σ, the mean and standard deviation of the population of percentages of iron ore readings that would be obtained if millions of additional ore specimens were analyzed. What proportion of these percentages of iron ore readings would

   a.   exceed 63%?

   b.   be less than 62%?

   c.   lie between 62% and 63%?

5.24   The number of times x an adult human breathes per minute when at rest depends on the age of the human and varies greatly from person to person. Suppose that the probability distribution for x is approximately normal, with the mean equal to 16 and the standard deviation equal to 4. If a person is selected at random and the number x of breaths per minute while at rest is recorded, what is the probability that x will exceed 22?

5.25 One method of arriving at economic forecasts is to use a consensus approach. A forecast is obtained from each of a large number of analysts and the average of these individual forecasts is the consensus forecast. Suppose that the individual 1991 January prime interest rate forecasts of all economic analysts are approximately normally distributed with the mean equal to 10% and with the standard deviation equal to 1.3%. If a single analyst is randomly selected from among this group, what is the probability that the analyst's forecast of the prime interest rate will
  a. exceed 13%? b. be less than 9%?

5.26 The scores on a national achievement test were approximately normally distributed with a mean of 540 and a standard deviation of 110.
  a. If you achieved a score of 680, how far, in standard deviations, did your score depart from the mean?
  b. What percentage of those who took the examination scored higher than you?

5.27 How does the IRS decide on the percentage of income tax returns to audit for each state? Suppose that they did it by randomly selecting 50 values from a normal distribution with a mean equal to 1.55% and a standard deviation equal to .45%. (Computer programs are available for this type of sampling.)
  a. What is the probability that a particular state will have more than 2.5% of its income tax returns audited?
  b. What is the probability that a state will have less than 1% of its income tax returns audited?

5.28 Suppose that the counts on the number of a particular type of bacteria in 1 ml of drinking water tend to be approximately normally distributed with a mean of 85 and a standard deviation of 9. What is the probability that a given 1 ml sample will contain more than 100 bacteria?

5.29 A grain loader can be set to discharge grain in amounts that are normally distributed with mean μ bushels and a standard deviation equal to 25.7 bushels. If a company wishes to use the loader to fill containers that hold 2000 bushels of grain and wants to overfill only one container in 100, at what value of μ should the company set the loader?

5.30 A publisher has discovered that the number of words contained in a new manuscript is normally distributed with a mean equal to 20,000 words in excess of that specified in the author's contract and a standard deviation of 10,000 words. If the publisher wants to be almost certain (say with a probability of .95) that the manuscript will be less than 100,000 words, what number of words should the publisher specify in the contract?

5.31 In 1989, the lifting of Japanese trade restrictions limiting imports of United States beef infused new life into the depressed United States cattle industry, which has seen the U.S. per capita beef consumption drop from 94.2 pounds in 1976 to 72.7 pounds in 1988 (*Time*, March 13, 1989, p. 47). If the individual beef consumption in the United States during 1988 was normally distributed with a standard deviation of 15.7 pounds,
  a. What proportion of the individuals in the United States consumed at least 85 pounds of beef in 1988?
  b. Ninety percent of all Americans consumed more than how many pounds of beef in 1988?

5.32 A stringer of tennis rackets has found that the actual string tension achieved for any individual racket stringing will vary as much as 6 pounds per square inch from the desired tension set on the stringing machine. If the stringer wishes to string at a tension lower than that specified by a customer only 5% of the time, how much above or below the customer's specified tension should the stringer set the stringing machine? (*Note:* Assume that the distribution of string tensions produced by the stringing machine is normally distributed with a mean equal to the tension set on the machine and a standard deviation equal to 2 pounds per square inch.

5.33   Compared with their stay-at-home peers, employed women have higher levels of high-density lipoproteins (HDL), the "good" cholesterol associated with lower heart attack risk. A study of cholesterol levels in 2000 women, age 25 to 64, living in Augsburg, Germany, was conducted by Ursula Haertel, Ulrich Keil, and colleagues at the GSF-Medis Institut in Munich (*Science News*, Vol. 135, June 24, 1989, p. 389). Of these 2000 women, the 48% who worked outside the home had HDL levels that were 2.5 to 3.6 milligrams per deciliter (mg/dl) higher than the HDL levels of their stay-at-home counterparts. If the difference in HDL levels is normally distributed with a mean of 0 (indicating no difference between the two groups of women) and a standard deviation of 1.2 mg/dl, what is the probability of observing a difference in the HDL levels in a single pair of women between 2.5 and 3.6 mg/dl?

5.4 THE NORMAL APPROXIMATION TO THE BINOMIAL PROBABILITY DISTRIBUTION

In Chapter 4 we considered several applications of the binomial probability distribution, all of which required us to calculate the probability that x, the number of successes in n trials, falls in a given region. Most examples involved small values of n because of the lengthy calculations involved in evaluating $p(x)$. However, when n was large and p was small with $np < 7$, the Poisson probability distribution with $\mu = np$ produced satisfactory approximations to binomial probabilities. When n is large and the conditions for using the Poisson approximation are not met, another approximation based on normal curve areas is available.

The binomial probability histogram is symmetrical and bell-shaped for $p = .5$ and is relatively so for values of p not close to 0 or 1 when n is large. In such cases the binomial probability histogram is well approximated by a normal curve with mean $\mu = np$ and variance $\sigma^2 = npq$. Figures 5.13 and 5.14 show the binomial probability

Figure 5.13
The binomial probability distribution for $n = 25$ and $p = .5$ and the approximating normal distribution with $\mu = 12.5$ and $\sigma = 2.5$

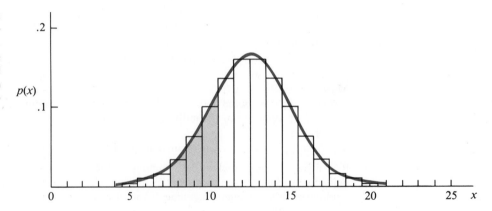

Figure 5.14
The binomial probability
distribution and the
approximating normal
distribution for $n = 25$
and $p = .1$

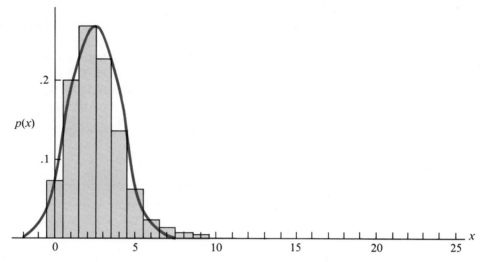

histograms for $n = 25$, $p = .5$, and $p = .1$, respectively, with the approximating normal curves superimposed. The correspondence between the areas for the binomial and normal distributions when $p = .5$ appears to be quite good. It is not so good for $p = .1$, for which the approximating symmetrical normal distribution poorly fits the nonsymmetrical binomial distribution. (However, when $n = 25$ and $p = .1$, $np = 2.5 < 7$, so that the Poisson approximation would be deemed appropriate in this particular case.)

Consider the binomial distribution for x when $n = 25$ and $p = .5$. For this distribution $\mu = np = 25(.5) = 12.5$ and $\sigma = \sqrt{npq} = \sqrt{6.25} = 2.5$. Therefore, we choose an approximating normal distribution with a mean $\mu = 12.5$ and a standard deviation $\sigma = 2.5$. The probability that $x = 8, 9,$ or 10 is equal to the area of the three rectangles lying over $x = 8, 9,$ and 10. This probability can be approximated by the area under the normal curve from $x = 7.5$ to $x = 10.5$, which is the shaded area in Figure 5.13. Therefore,

$$\sum_{x=8}^{10} p(x) \approx P(7.5 < x^* < 10.5)$$

where x^* is the approximating normal random variable. (The symbol $\approx$ means "approximately equal to.") Note that the area under the normal curve from 8 to 10 would not provide a good approximation to the probability that $x = 8, 9,$ or 10 because it excludes half of the probability rectangles corresponding to $x = 8$ and $x = 10$. Thus, we must use the endpoints of the binomial probability rectangles, found by adding or subtracting .5 when calculating the approximating normal probabilities.

How can we tell whether the values of n and p are such that the normal approximation to binomial probabilities will be appropriate? From Section 5.2 (and from the Empirical Rule) we know that approximately 95% of the measurements associated with a normal curve lie within two standard deviations of the mean, and almost all lie within three. The binomial distribution would be nearly symmetrical if the distribution were able to spread out two standard deviations on both sides of the mean. Hence, to determine when the normal approximation is adequate, calculate $\mu = np$ and $\sigma = \sqrt{npq}$. **If the interval ($\mu \pm 2\sigma$) lies within the binomial bounds of 0 to n, the approximation is reasonably good. The approximation will be very good if the interval $\mu \pm 3\sigma$ lies in the interval zero to n.** Note that this criterion is satisfied for the binomial probability distribution in Figure 5.13, but it is not satisfied for the distribution shown in Figure 5.14.

THE NORMAL APPROXIMATION TO THE BINOMIAL PROBABILITY DISTRIBUTION

Approximate the binomial probability distribution using a normal curve with

$$\mu = np$$

$$\sigma = \sqrt{npq}$$

where

$n =$ number of trials

$p =$ probability of success on a single trial

$q = 1 - p$

The approximation will be adequate when n is large and when the interval $\mu \pm 2\sigma$ falls between 0 and n.

EXAMPLE 5.9   To see how well the normal curve can be used to approximate binomial probabilities, refer to the binomial experiment illustrated in Figure 5.13, where $n = 25$ and $p = .5$. Calculate the probability that $x = 8$, 9, or 10, correct to three places, using Table 1 of binomial probabilities in Appendix III. Then calculate the corresponding normal approximation to this probability and compare the results.

Solution   The exact probability P_1 can be calculated by using Table 1 for $n = 10$. Thus, we have

$$P_1 = \sum_{x=8}^{10} p(x) = \sum_{x=0}^{10} p(x) - \sum_{x=0}^{7} p(x) = .212 - .022 = .190$$

As noted earlier in this section, the normal approximation requires the area lying between $x_1 = 7.5$ and $x_2 = 10.5$, where $\mu = 12.5$ and $\sigma = 2.5$. Thus, we have

$$\sum_{x=8}^{10} p(x) \approx P(z_1 \leq z \leq z_2) = P_2$$

where

$$z_1 = \frac{x_1 - \mu}{\sigma} = \frac{7.5 - 12.5}{2.5} = -2.0 \qquad z_2 = \frac{x_2 - \mu}{\sigma} = \frac{10.5 - 12.5}{2.5} = -.8$$

Figure 5.15
Area under the normal
curve for Example 5.9

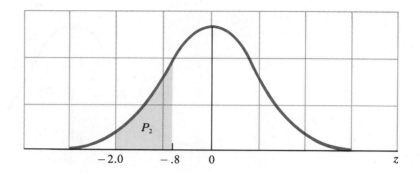

The probability P_2 is shown in Figure 5.15. The area between $z = 0$ and $z = 2.0$ is $A_1 = .4772$. Likewise, the area between $z = 0$ and $z = .8$ is $A_2 = .2881$. From Figure 5.15,

$$P_2 = P(z_1 \leq z \leq z_2) = P(-2.0 \leq z \leq -.8) = A_1 - A_2$$
$$= .4772 - .2881 = .1891$$

The normal approximation is quite close to the binomial probability, .190, obtained from Table 1.

You must be careful not to exclude half of the two extreme probability rectangles when using the normal approximation to the binomial probability distribution. This means that the x values used to calculate z values will always have a 5 in the tenths decimal place. To be certain that you include all the probability rectangles in your approximation, always draw a sketch of the problem.

EXAMPLE 5.10   The reliability of an electrical fuse is the probability that a fuse, chosen at random from production, will function under the conditions for which it has been designed. A random sample of 1000 fuses was tested and $x = 27$ defectives were observed. Calculate the approximate probability of observing 27 or more defectives, assuming that the fuse reliability is .98.

Solution   The probability of observing a defective when a single fuse is tested is $p = .02$, given that the fuse reliability is .98. Then

$$\mu = np = 1000(.02) = 20$$
$$\sigma = \sqrt{npq} = \sqrt{1000(.02)(.98)} = 4.43$$

The probability of 27 or more defective fuses, given $n = 1000$, is

$$P = P(x \geq 27) = p(27) + p(28) + p(29) + \cdots + p(999) + p(1000)$$

The normal approximation to P would be the area under the normal curve to the right of $x = 26.5$. (*Note:* We must use $x = 26.5$ rather than $x = 27$ so as to include the entire probability rectangle associated with $x = 27$.) The z value corresponding

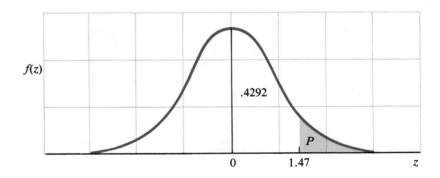

to $x = 26.5$ is

$$z = \frac{x - \mu}{\sigma} = \frac{26.5 - 20}{4.43} = \frac{6.5}{4.43} = 1.47$$

and the area between $z = 0$ and $z = 1.47$ is equal to .4292, as shown in Figure 5.16. Since the total area to the right of the mean is equal to .5,

$$P = P(x \geq 27) \approx P(z \geq 1.47) = .5 - .4292 = .0708 \qquad \triangleleft$$

EXAMPLE 5.11   A producer of soft drinks was fairly certain that her brand of soft drink had a 10% share of the soft drink market. In a market survey involving 2500 consumers of soft drinks, $x = 211$ expressed a preference for her brand. If the 10% figure is correct, find the probability of observing 211 or fewer consumers preferring her brand of soft drink.

Solution   If the producer is correct, then the probability that a consumer prefers this producer's brand of soft drink is $p = .10$. Then

$$\mu = np = 2500(.10) = 250$$
$$\sigma = \sqrt{npq} = \sqrt{2500(.10)(.90)} = \sqrt{225} = 15$$

The probability of observing 211 or fewer preferring her brand is

$$P(x \leq 211) = p(0) + p(1) + \cdots + p(210) + p(211)$$

The normal approximation to this probability would be the area to the left of 211.5 under a normal curve with mean of 250 and a standard deviation of 15.

$$z = \frac{x - \mu}{\sigma} = \frac{211.5 - 250}{15} = -2.57$$

and the area between $z = 0$ and $z = -2.57$ is equal to .4949. Since the area to the left of the mean is equal to .5,

$$P(x \leq 211) \approx .5 - .4949 = .0051$$

The probability of observing a sample value of 211 or fewer when $p = .10$ is so small that we can conclude that one of two things has occurred: either we have observed an unusual sample even though $p = .10$, *or* the sample reflects that the actual value of p is less than .10, and perhaps closer to the value $211/2500 = .08$. ◁

EXERCISES

Basic Techniques

5.34 Let x be a binomial random variable with $n = 25, p = .3$.
a. Use Table 1 in Appendix III to find $P[8 \leq x \leq 10]$.
b. Find μ and σ for the binomial distribution and use the normal approximation to find $P[8 \leq x \leq 10]$. Compare the approximation with the exact value calculated in part (a).

5.35 Find the normal approximation to $P[x \geq 6]$ for a binomial probability distribution with $n = 10, p = .5$.

5.36 Find the normal approximation to $P[x > 6]$ for a binomial probability distribution with $n = 10, p = .5$.

5.37 Find the normal approximation to $P[x > 22]$ for a binomial probability distribution with $n = 100, p = .2$.

5.38 Find the normal approximation to $P[x \geq 22]$ for a binomial probability distribution with $n = 100, p = .2$.

5.39 Let x be a binomial random variable for $n = 25, p = .2$.
a. Use Table 1 in Appendix III to calculate $P(4 \leq x \leq 6)$.
b. Find μ and σ for the binomial probability distribution and use the normal distribution to approximate the probability $P(4 \leq x \leq 6)$. Note that this value is a good approximation to the exact value of $P(4 \leq x \leq 6)$.

5.40 Consider a binomial experiment with $n = 20, p = .4$. Calculate $P(x \geq 10)$ using
a. Table 1 in Appendix III.
b. The normal approximation to the binomial probability distribution.

5.41 Find the normal approximation to $P[355 \leq x \leq 360]$ for a binomial probability distribution with $n = 400, p = .9$.

Applications

5.42 A 1988 Gallup survey, reported in *American Demographics* (September, 1989), indicated that the percent of Americans who do not know the United States population is 56%. Does this hold true for you and your statistics classmates? Assume that it does and that your class contains 50 students. What is the probability that
a. More than 25 do not know the U.S. population?
b. Fewer than 15 do not know the U.S. population?
c. Fewer than 30 *do know* the U.S. population?

5.43 Briggs and King developed the technique of nuclear transplantation in which the nucleus of a cell from one of the later stages of the development of an embryo is transplanted into a zygote (a single-cell fertilized egg) to see if the nucleus can support normal development. If the probability that a single transplant from the early gastrula stage will be successful is .65, what is the probability that more than 70 transplants out of 100 will be successful?

5.44 Americans are becoming more and more health-conscious, limiting cholesterol, cigarettes, and alcohol intake. However, they are still citizens of the "fattest nation on earth" according to a new *Prevention* magazine survey (*Press Enterprise*, May 24, 1989). The survey revealed that 64% of American adults are overweight, compared with 60% in 1984. To check this claim, a random sample of 100 adult Americans is selected, and 51 are found to be overweight.

    a. If the survey is correct, what are the mean and standard deviation of *x*, the number of overweight people in the sample?

    b. What is the probability that 51 or fewer people in the sample are overweight?

    c. Based on the results of part (b), do you have reason to suspect that the percentage of overweight American adults is in fact less than 64%? Explain.

5.45 Data collected over a long period of time show that a particular genetic defect occurs in 1 out of every 1000 children. The records of a medical clinic show $x = 60$ children with the defect in a total of 50,000 examined. If the 50,000 children were a random sample from the population of children represented by past records, what is the probability of observing a value of *x* equal to 60 or more? Would you say that the observation of $x = 60$ children with genetic defects represents a rare event?

5.46 Does the appearance of an athlete or an athletic team on the cover of *Sports Illustrated* increase the athlete's (or the team's) performance? In a study of 271 randomly selected cover subjects from 1954 to 1983, researchers at the University of Southern California found that 58% of the athletes or teams improved their performance after appearing on the cover of *Sports Illustrated* (*USA Today*, July 2, 1984). Suppose that the probability that an athlete's (or team's) improvement is independent of exposure on the cover of *Sports Illustrated* and is, in fact, equal to .5.

    a. What is the probability that as many as 157 (58%) of a sample of 271 would show improvement?

    b. Is the value $x = 157$ (58%) a likely occurrence, assuming that there is no effect due to the athlete's (or team's) appearance on the cover of *Sports Illustrated*? What would you conclude regarding the effect of media coverage for an athlete?

5.47 Airlines and hotels often grant reservations in excess of capacity to minimize losses due to no-shows. Suppose that the records of a motel show that, on the average, 10% of their prospective guests will not claim their reservation. If the motel accepts 215 reservations and there are only 200 rooms in the motel, what is the probability that all guests who arrive to claim a room will receive one?

5.48 Compilation of large masses of data on lung cancer show that approximately 1 of every 40 adults acquire the disease. Workers in a certain occupation are known to work in an air-polluted environment that may cause an increased rate of lung cancer. A random sample of $n = 400$ workers shows 19 with identifiable cases of lung cancer. Do the data provide sufficient evidence to indicate a higher rate of lung cancer for these workers than for the national average?

5.49 "Infant twins are 2 1/2 times more likely to be victims of child abuse than other babies, according to a study of a Kentucky medical center's abuse and neglect registry" (*Jacksonville, Florida Times-Union*, September 20, 1984). Among the 310 cases of child abuse in children age three or younger, 16 were individual twins. This group of 16 included six pairs of twins, both of whom were abused. Based on the proportion of twins in the U.S. population, the expected number of twins in a group of 310 people is 6.2. Does the observation of 16 twins in a group of 310 abused children suggest that the proportion of twins in the population of abused children is higher than the proportion of twins in the general population?

5.50   In Exercise 5.49 we used a normal distribution to approximate a binomial probability for the number x of twins in a sample of 310 cases of child abuse.
a.   Was the normal approximation adequate? Explain.
b.   Would the Poisson probability distribution also provide a good approximation to this binomial probability distribution? Explain. (*Note:* It can be shown, proof omitted, that the Poisson probability distribution is approximately normally distributed for large values of μ. The approximation is satisfactory for $\mu = 9$ and becomes better as μ increases in value.)

5.5   SUMMARY

Data on many continuous random variables observed in nature have relative frequency distributions that are bell-shaped and can be approximated by a normal distribution. In addition, many other random variables, such as the binomial and the Poisson, have distributions that can be approximated by a normal distribution when certain conditions are satisfied. The binomial probability distribution can be approximated by a normal distribution when the n of trials is large and the probability p of success on a single trial is not too close to 0 or 1. The Poisson probability distribution can be approximated by a normal distribution when its mean becomes large. For example, if the mean of the Poisson distribution is as large as 9, the normal approximation to the Poisson distribution is reasonably good.

You might think that this is adequate reason to nominate the normal probability distribution as one of the most important distributions in statistics, but the most compelling reason is explained in Chapter 6. There you will learn that the probability distributions of many sample statistics, sample means, proportions, and so on are approximately normal when certain conditions are satisfied. This result plays an important role in statistical inference and explains why the normal distribution is the most important probability distribution in statistics.

TIPS ON PROBLEM SOLVING

1.   Always sketch a normal curve and locate the probability areas pertinent to the exercise. If you are approximating a binomial probability distribution, sketch in the probability rectangles as well as the normal curve.

2.   Read each exercise carefully to see whether the data come from a binomial experiment or whether they possess a distribution that, by its very nature, is approximately normal. If you are approximating a binomial probability distribution using a normal curve, do not forget to make a half-unit correction, so that you will include the half rectangles at the ends of the interval. If the distribution is not binomial, *do not* make the half-unit corrections. If you make a sketch (as suggested in step 1), you will see why the half-unit correction is or is not needed.

REFERENCES

Devore, J., and Peck, R. *Statistics: The Exploration and Analysis of Data.* St. Paul: West, 1986.

Freund, J. E. *Modern Elementary Statistics.* 6th ed. Englewood Cliffs, N.J.: Prentice-Hall, 1984.

Kendall, M. G., and Stuart, A. *The Advanced Theory of Statistics.* Vol. 2. 4th ed. New York: Hafner, 1979.

Mendenhall, W., and Sincich, T. *Statistics for Engineering and Computer Sciences.* San Francisco: Dellen, 1984.

Ott, L. *An Introduction to Statistical Methods and Data Analysis.* 3d ed. Boston: PWS-KENT, 1988.

Rosner, B. *Fundamentals of Biostatistics.* 3d ed. Boston: PWS-KENT, 1990.

SUPPLEMENTARY EXERCISES

5.51 A random variable x has a uniform probability distribution with density function given as

$$f(x) = 1/2 \qquad \text{for} \qquad 2.5 \le x \le 4.5$$

  a. Find the mean and standard deviation of x.

  b. Graph the probability distribution of x. Calculate the intervals $(\mu \pm k\sigma)$ for $k = 1, 2,$ and 3, and superimpose these intervals on the graph.

  c. Find the probability that x will fall within $(\mu \pm k\sigma)$ for $k = 1, 2, 3$. Do these results agree with those given by Tchebysheff's Theorem? the Empirical Rule?

5.52 A random variable x has an exponential probability distribution with density function given as

$$f(x) = 25e^{-25x} \qquad \text{for} \qquad x \ge 0$$

  a. Find the mean and standard deviation of x.

  b. Graph the probability distribution of x. Calculate the intervals $(\mu \pm k\sigma)$ for $k = 1, 2,$ and 3, and superimpose these intervals on the graph.

  c. Find the probability that x will fall within $(\mu \pm k\sigma)$ for $k = 1, 2, 3$. Do these results agree with those given by Tchebysheff's Theorem? the Empirical Rule?

5.53 Using Table 3 in Appendix III calculate the area under the normal curve between

  a. $z = 0$ and $z = 1.2$ b. $z = 0$ and $z = -.9$

  c. $z = 0$ and $z = 1.46$ d. $z = 0$ and $z = -.42$

5.54 Repeat Exercise 5.53 using

  a. $z = .3$ and $z = 1.56$ b. $z = .2$ and $z = -.2$

5.55 a. Find the probability that z is greater than $-.75$.

  b. Find the probability that z is less than 1.35.

5.56 Find z_0 such that $P(z > z_0) = .5$.

5.57 Find the probability that z lies between $z = -1.48$ and $z = 1.48$.

5.58 Find z_0 such that $P(-z_0 < z < z_0) = .5$.

5.59 The lifespan of oil drilling bits depends on the types of rock and soil that the drill encounters, but it is estimated that the mean length of life is 75 hours. Suppose that an oil exploration company purchases drill bits that have a lifespan that is approximately normally distributed with a mean equal to 75 hours and a standard deviation equal to 12 hours.

a. What proportion of the company's drill bits will fail before 60 hours of use?

b. What proportion will last at least 60 hours?

c. What proportion will have to be replaced after more than 90 hours of use?

5.60 The influx of new ideas into a college or university, introduced primarily by hiring new young faculty, is becoming a matter of concern because of the increasing ages of faculty members; that is, the distribution of faculty ages is shifting upward, due most likely to a shortage of vacant positions and an oversupply of Ph.Ds. Thus faculty members are more reluctant to move and give up a secure position. If the retirement age at most universities is 65, would you expect the distribution of faculty ages to be normal?

5.61 A machine operation produces bearings whose diameters are normally distributed with mean and standard deviation equal to .498 and .002, respectively. If specifications require that the bearing diameter equal .500 inch plus or minus .004 inch, what fraction of the production will be unacceptable?

5.62 A used car dealership has found that the length of time before a major repair is required on the cars it sells is normally distributed with a mean equal to ten months and a standard deviation of three months. If the dealer wants only 5% of the cars to fail before the end of the guarantee period, for how many months should the cars be guaranteed?

5.63 The daily sales (excepting Saturday) at a small restaurant has a probability distribution that is approximately normal, with a mean μ equal to $530 per day and a standard deviation σ equal to $120.

a. What is the probability that the sales will exceed $700 on a given day?

b. The restaurant must have at least $300 in sales per day to break even. What is the probability that on a given day the restaurant will not break even?

5.64 The length of life of a type of automatic washer is approximately normally distributed with mean and standard deviation equal to 3.1 and 1.2 years, respectively. If this type of washer is guaranteed for one year, what fraction of original sales will require replacement?

5.65 Most users of automatic garage door openers activate their openers at distances that are normally distributed with a mean of 30 feet and a standard deviation of 11 feet. In order to minimize interference with other radio-controlled devices, the manufacturer is required to limit the operating distance to 50 feet. What percentage of the time will users attempt to operate the opener outside of its operating limit?

5.66 Consider a binomial experiment with $n = 25$, $p = .4$. Calculate $P(8 \leq x \leq 11)$ using

a. the binomial probabilities given in Table I in Appendix III.

b. the normal approximation to the binomial.

5.67 The average length of time required for a college achievement test was found to equal 70 minutes, with a standard deviation of 12 minutes. When should the test be terminated if we wish to allow sufficient time for 90% of the students to complete the test? (Assume that the time required to complete the test is normally distributed.)

5.68 The length of time required for the periodic maintenance of an automobile will usually have a probability distribution that is mound-shaped and, because some long service times will occur occasionally, is skewed to the right. Suppose that the length of time required to run a 5000-mile check and to service an automobile has a mean equal to 1.4 hours and a standard deviation of .7 hours. Suppose that the service department plans to service 50 automobiles per 8-hour day and that, in order to do so, it must spend no more service time than an average of 1.6 hours per automobile. What proportion of all days will the service department have to work overtime?

5.69 An advertising agency has stated that 20% of all television viewers watch a particular program. In a random sample of 1000 viewers, $x = 184$ viewers were watching the program. Do these data present sufficient evidence to contradict the advertiser's claim?

5.70 A survey by the Justice Department's Bureau of Justice Statistics, reported in the November 6, 1987 edition of the *Gainesville Sun*, shows that there is strong public sympathy for longer prison terms for convicted criminals. The survey, based on interviews with 2000 people concerning their attitudes towards punishment for crimes, reports that 71% of those surveyed wanted jail or prison terms for 26 specific crimes. For those convicted of assault, respondents wanted prison terms ranging from 3.6 to 7.7 years, compared to the average of 2.4 years now served by people convicted of assault. If the standard deviation of the time served by people convicted of assault is 1.5 years, what proportion of those convicted of assault serve between 3.6 and 7.7 years?

5.71 A researcher notes that senior corporation executives are not very accurate forecasters of their own annual earnings. He states that his studies of a large number of company executive forecasts "showed that the average estimate missed the mark by 15%."

   a. Suppose that the distribution of these forecast errors has a mean of 15% and a standard deviation of 10%. Is it likely that the distribution of forecast errors is approximately normal?

     Suppose that the probability is .5 that a corporate executive's forecast error exceeds 15%. If you were to sample the forecasts of 100 corporate executives, what is the probability that more than 60 would be in error by more than 15%?

5.72 A soft drink machine can be regulated to discharge an average of μ ounces per cup. If the ounces of fill are normally distributed with standard deviation equal to .3 ounce, give the setting for μ so that 8-ounce cups will overflow only 1% of the time.

5.73 If you call the Internal Revenue Service with an income tax question, will you get the correct answer? According to an investigation conducted by the General Accounting Office, the IRS is giving incorrect answers to 39% of the questions posed by people who call for help with their taxes (*New York Times*, February 23, 1988). If 65 taxpayers call the IRS with questions,

   a. What is the probability that more than 30 will receive incorrect advice?

   b. What is the probability that fewer than 20 will receive incorrect advice?

5.74 A manufacturing plant uses 3000 electric light bulbs that have a length of life that is normally distributed with mean and standard deviation equal to 500 and 50 hours, respectively. In order to minimize the number of bulbs that burn out during operating hours, all the bulbs are replaced after a given period of operation. How often should the bulbs be replaced if we wish no more than 1% of the bulbs to burn out between replacement periods?

5.75 The admissions office of a small college is asked to accept deposits from a number of qualified prospective freshmen so that, with probability about .95, the size of the freshman class will be less than or equal to 120. Consider that the applicants comprise a random sample from a population of applicants, 80% of whom would actually enter the freshman class if accepted.

   a. How many deposits should the admissions counselor accept?

   b. If applicants in the number determined in part (a) are accepted, what is the probability that the freshman class size will be less than 105?

5.76 An airline finds that 5% of the persons making reservations on a certain flight will not show up for the flight. If the airline sells 160 tickets for a flight with only 155 seats, what is the probability that a seat will be available for every person holding a reservation and planning to fly?

5.77   It is known that 30% of all calls coming into a telephone exchange are long-distance calls. If 200 calls come into the exchange, what is the probability that at least 50 will be long-distance calls?

5.78   Suppose that the random variable x has a binomial distribution corresponding to $n = 20$ and $p = .30$. Use Table 1 in Appendix III to calculate
   a. $P(x = 5)$ b. $P(x \geq 7)$

5.79   Refer to Exercise 5.78. Use the normal approximation to calculate $P(x = 5)$ and $P(x \geq 7)$. Compare with the exact values obtained from Table 1 in Appendix III.

5.80   A purchaser of electric relays buys from two suppliers, A and B. Supplier A supplies 2 of every 3 relays used by the company. If 75 relays are selected at random from those in use by the company, find the probability that at most 48 of these relays come from supplier A. Assume that the company uses a large number of relays.

5.81   Is television dangerous to your diet? Psychologists believe that excessive eating may be associated with emotional states (being upset, bored) and environmental cues (TV, reading, and so on). To test this theory, suppose that we were to randomly select 60 overweight persons and match them by weight and sex in pairs. For a period of two weeks, one of each pair is required to spend evenings reading novels of interest to them. The other member of each pair spends each evening watching television. The calorie count for all snack and drink intake for the evenings is recorded for each person, and we record $x = 19$, the number of pairs for which the TV watchers' calorie intake exceeded the intake of the readers. If there is no difference in the effects of TV and reading on calorie intake, the probability p that the calorie intake of one member of a pair exceeds that of the other member is .5. Do these data provide sufficient evidence to indicate a difference between the effects of TV watching and reading on calorie intake? [*Hint:* Calculate the z-score for the observed value, $x = 19$.]

5.82   In Exercise 5.27 we suggested that the IRS assign auditing rates per state by randomly selecting 50 auditing percentages from a normal distribution with a mean equal to 1.55% and a standard deviation equal to .45%.
   a.   What is the probability that a particular state would have more than 2% of its tax returns audited?
   b.   What is the expected value of x, the number of states that will have more than 2% of their income tax returns audited?
   c.   Is it likely that as many as 15 of the 50 states will have more than 2% of their income tax returns audited?

5.83   There is a difference in sports preferences between men and women, according to a Gallup poll reported in *American Demographics* (September, 1989). Among the ten most popular sports, men include competition-type sports—pool and billiards, basketball, and softball—while women include aerobics, running, hiking, and calisthenics. However, the top recreational activity for men was still the relaxing sport of fishing, with 41% of those surveyed indicating they they had fished during the year. Suppose that 180 randomly selected men are asked whether they had fished in the last year.
   a.   What is the probability that fewer than 50 had fished?
   b.   What is the probability that between 50 and 75 had fished?
   c.   If the 180 men selected for the interview were selected by the marketing department of a sporting goods company based on information obtained from their mailing lists, what would you conclude about the reliability of their survey results?

5.84   Large pieces of equipment, which periodically have worn parts replaced and are subject to periodic maintenance checks and service, will have a length of life that is exponentially distributed. The probability density function for such a piece of equipment with a mean life of 75,000 hours, given by

$$f(x) = \frac{1}{75,000}\, e^{-x/75,000} \qquad x \geq 0$$

is shown in the accompanying figure.

a. What is the probability that one of these pieces of equipment will have an operating life exceeding 100,000 hours?

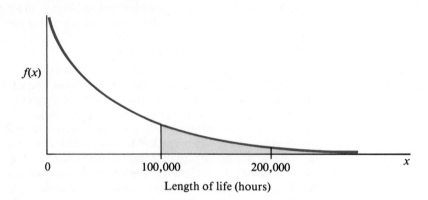

Length of life (hours)

b. What is the probability that one of these pieces of equipment has an operating life of 25,000 hours or less?

c. Between what two values would we find the useful operating lifetimes for 95% of such pieces of equipment if we exclude the largest 2.5% and the smallest 2.5%?

THE LONG AND THE SHORT OF IT

If you were the boss, would height play a role in your selection of a successor for your job? In his July 27, 1981 *Fortune* column, "Keeping Up" (copyright 1981 Times Inc.; all rights reserved), Daniel Seligman discussed his ideas concerning height as a factor in Deng Xiaoping's choice of Hu Yaobang for his replacement as Chairman of the Chinese Communist Party. As Seligman notes, the facts surrounding the case are enough to arouse suspicions when examined in the light of statistics.

Deng, it seems, is only 5 feet tall, a height that is short even in China. Therefore, the choice of Hu Yaobang, who is also 5 feet tall, raised (or lowered) some eyebrows because, as Seligman notes, "the odds against a 'height-blind' decision producing a chairman as short as Deng are about 40 to 1." In other words, if we had the relative frequency distribution of the heights of all Chinese males, only 1 in 41 (i.e., 2.4%) would be 5 feet tall or less. To calculate these odds, Seligman notes in his *Fortune* article that the Chinese equivalent of the U.S. Health Service does not exist and hence that health statistics on the current population of China are difficult to acquire. He says, however, that "it is generally held that a boy's length at birth represents 28.6% of his final height" and that, in prerevolutionary China, the average length of a Chinese boy at birth was 18.9 inches. From this, Seligman deduces that the mean height of mature Chinese males is

$$\frac{18.9}{.286} = 66.08 \text{ inches, or 5 feet 6.08 inches}$$

He then assumes that the distribution of the heights of males in China follows a normal distribution ("as it does in the U.S.") with a mean of 66 inches and a standard deviation equal to 2.7 inches, "a figure that looks about right for that mean."

1. Using Seligman's assumptions, calculate the probability that a single adult Chinese male, chosen at random, will be less than or equal to 5 feet tall, or equivalently, 60 inches tall.

2. Do the results of part 1 agree with Seligman's odds?

3. Comment on the validity of Seligman's assumptions. Are there any basic flaws in his reasoning?

4. Based on the results of parts 1 and 3, do you think that Deng Xiaoping took height into account in selecting his successor?

CHAPTER 6

SAMPLING DISTRIBUTIONS

Case Study

How would you like to try your hand at gambling without the risk of losing? You could do it by simulating the gambling process, making imaginary bets, and observing the results. This technique, called a **Monte Carlo procedure**, is the topic of the case study at the end of this chapter.

General Objective

In the preceding chapters, you were introduced to some useful random variables and their probability distributions. In practical sampling situations, we select a sample of n observations and use these measurements to calculate statistics such as the sample mean and the sample standard deviation. These statistics are used to make inferences about the corresponding parameters in the sampled population.

However, since the value of a statistic depends upon the observed values in the sample, a statistic is itself a random variable that may be either discrete or continuous. The probability distribution of a statistic is called its **sampling distribution**, since it describes the behavior of the statistic in repeated sampling. Our objective then, is to study the sampling distribution of some useful statistics. We shall see that under fairly general conditions, many statistics have sampling distributions that can be approximated by a normal distribution.

Specific Topics

1. Sampling and statistics (6.1)
2. The sampling distribution of a statistic (6.2)
3. The Central Limit Theorem (6.3)
4. The sampling distribution of the sample mean (6.4)

⊳ 6.1 SAMPLING AND STATISTICS

In earlier chapters, we became familiar with random variables and their probability distributions. In fact, we examined several discrete and continuous probability distributions that could serve as widely applicable models for practical situations. These probability distributions allowed us to calculate various descriptive measures, such as a population mean or standard deviation.

How do we apply these probability models in the practice of statistics? Usually, we are able to decide which type of probability distribution might serve as a model in a given situation; however, the values of the parameters that specify the distribution exactly are not available. For example, a pollster could be quite certain that the responses to an "agree/disagree" question on the survey could be modeled using a binomial distribution, without knowing the value of the parameter *p*. Similarly, an agronomist may be willing to assume that the yield per acre of a variety of wheat is normally distributed, without knowing the values of the population mean and standard deviation.

In situations such as these, we rely on the sample to provide information about these unknown population parameters. For example, the sample proportion of those who agree should reflect the actual value of *p* in the population. Similarly, the sample mean and standard deviation of yield per acre should provide information concerning the population mean and standard deviation. In both of these instances, sample information would be used to make inferences about population parameters.

Since the values of sample statistics are unpredictable and vary from sample to sample, they are random variables. Therefore, in order to assess the reliability of an inference based on a statistic computed from sample values, we must be able to determine or approximate the **sampling distribution** of that statistic.

Definition

> The **sampling distribution** of a statistic is the probability distribution for values of the statistic that results when random samples of size *n* are repeatedly drawn from the population.

The sampling distribution of a statistic may be derived mathematically or approximated empirically. Empirical approximations are found by drawing a large number of samples of size *n* from the specified population, calculating the value of the statistic for each sample, and tabulating the results in a relative frequency histogram. When the number of samples is large, the relative frequency histogram should closely approximate the theoretical sampling distribution.

Knowledge of the sampling distributions of statistics allows us to select the best from a group of competing statistics available to estimate a parameter. In addition, the sampling distribution can be used to assess the reliability of a statistic used as an estimator by determining limits within which we would expect the estimator to lie with a specified probability.

6.2 SAMPLING DISTRIBUTIONS

The sample mean, standard deviation, median, and other descriptive measures computed from sample values can be used not only to describe the sample, but also to make inferences in the form of estimates or tests about corresponding population parameters. However, we must know the sampling distribution of the statistic in order to answer questions such as, Does the statistic consistently underestimate or overestimate the value of the parameter? and, Is this statistic less variable than other competitors, and hence more useful as an estimator?

Consider a random sample of size $n = 3$ drawn without replacement from a population of $N = 5$ elements. A simple random sample of size n is selected in such a way that every sample of size n has the same probability of being selected — equal to $1/C_n^N$. Let us begin with a population of $N = 5$ elements whose values are 3, 6, 9, 12, and 15. Since the five elements are distinct, the population probability distribution assigns a probability of $1/5$ to each value of x, so that

$$p(x) = 1/5 \qquad \text{for } x = 3, 6, 9, 12, \text{ and } 15$$

The probability histogram is given in Figure 6.1.

The population mean is found as

$$\mu = \frac{\sum_{i=1}^{N} x_i}{N} = \frac{45}{5} = 9$$

Figure 6.1
Probability histogram for the $N = 5$ population values

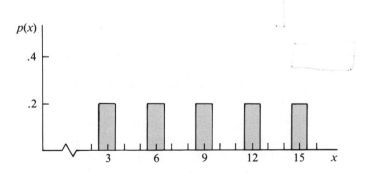

Table 6.1
Values of $\bar{x}$ and m for
simple random sampling
when $n = 3$ and $N = 5$

Sample	Sample values	$\bar{x}$	m
1	3, 6, 9	6	6
2	3, 6, 12	7	6
3	3, 6, 15	8	6
4	3, 9, 12	8	9
5	3, 9, 15	9	9
6	3, 12, 15	10	12
7	6, 9, 12	9	9
8	6, 9, 15	10	9
9	6, 12, 15	11	12
10	9, 12, 15	12	12

Table 6.2
Sampling distributions for
(a) the sample mean and
(b) the sample median

(a)

$\bar{x}$	$p(\bar{x})$
6	.1
7	.1
8	.2
9	.2
10	.2
11	.1
12	.1

(b)

m	$p(m)$
6	.3
9	.4
12	.3

By inspection we see that the population median is $M = 9$ and that $\mu = M$. The number of possible samples is

$$C_3^5 = \frac{5!}{3!2!} = 10$$

and the ten possible samples are given in Table 6.1. For each sample we have calculated the sample mean $\bar{x}$ and the sample median m. Each of the ten samples in Table 6.1 are equally likely. Hence, the values of $\bar{x}$ and m associated with each sample are each assigned probability equal to $1/10$. For example, we will observe a value of $\bar{x} = 6$ only if sample 1 is selected, and this occurs with probability .1. A value of $\bar{x} = 8$ will occur if sample 3 or sample 4 is drawn; therefore, the probability of observing $\bar{x} = 8$ is .2. Continuing in this manner, we can find the sampling distributions for $\bar{x}$ and m given in Table 6.2. The probability histograms of the sampling distributions for $\bar{x}$ and m are given in Figure 6.2.

Since the population of $N = 5$ values is symmetric about the value $x = 9$, both the **population mean** and the **median** equal 9. It would seem reasonable, therefore, to consider using either $\bar{x}$ or m as possible estimators of $M = \mu = 9$. Which

Figure 6.2
Probability histograms for the sampling distributions of the sample mean $\bar{x}$ and the sample median m

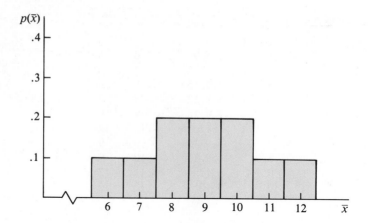

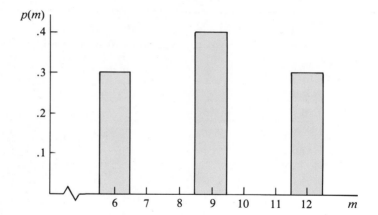

estimator would we choose? From Table 6.2 we see that in using m as an estimator, we would be in error by $9 - 6 = 3$ with probability .3 or by $9 - 12 = -3$ with probability .3. That is, the error in estimation using m would be at least 3 with probability .6. In using $\bar{x}$, however, an error of at least 3 would occur only with probability .2. On these grounds alone, we may wish to use $\bar{x}$ as an estimator in preference to m.

When N is small, sampling distributions can be derived directly. However, when this is not the case, the sampling distribution can be approximated by repeatedly selecting samples of size n and calculating the value of the statistic for each sample. The resulting relative frequency histogram is an approximation to the unknown sampling distribution. Alternatively, **for certain statistics that are sums or means of the sample values, an important theorem that we introduce in the next section will allow us to approximate their sampling distributions when the sample size is large.**

▷ 6.3   THE CENTRAL LIMIT THEOREM

The Central Limit Theorem states that under rather general conditions sums and means of samples of random measurements drawn from a population tend to possess an approximately bell-shaped sampling distribution. The significance of this statement is perhaps best illustrated by an example.

Consider a population of die throws generated by tossing a die infinitely many times, with the resulting **discrete uniform probability distribution** given by Figure 6.3. Draw a sample of $n = 5$ measurements from the population by tossing a die five times and recording each of the five observations, as indicated in Table 6.3. Note that the numbers observed in the first sample are $x = 3, 5, 1, 3, 2$. Calculate the sum of the five measurements as well as the sample mean $\bar{x}$.

For experimental purposes, repeat the sampling procedure 100 times, or preferably an even larger number of times. The results for 100 samples are given in Table 6.3, along with the corresponding values of the sum $\sum\limits_{i=1}^{5} x_i$ and the sample mean $\bar{x}$. Construct a frequency histogram for $\bar{x}\left(\text{or for } \sum\limits_{i=1}^{5} x_i\right)$ for the 100 samples and observe the resulting distribution, shown in Figure 6.4. You will observe an interesting result: Although the values of x in the population ($x = 1, 2, 3, 4, 5, 6$) are equiprobable and hence possess a discrete uniform probability distribution, the distribution of the sample means (or sums) chosen from the population forms a **bell-shaped distribution**. We will make one additional comment without proof. If we should repeat the study outlined here by using larger samples of size $n = 10$, we would find that the distribution of the sample means tends to become more nearly bell-shaped.

The relative frequency distribution (Figure 6.4) provides only a rough approximation to the sampling distribution of the sample mean. An accurate evaluation of the form of the sampling distribution would require an infinitely large number of samples or, at the very least, far more than the 100 samples contained in our experiment. Nevertheless, the relative frequency distribution constructed from the 100 samples provides a good indication of the form of the sampling distribution of

Figure 6.3
Probability distribution for x, the number appearing on a single toss of a die

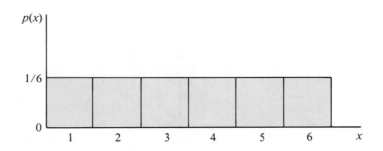

Table 6.3

Sampling from the
population of die throws

Sample number	x, Sample measurements	Σx_i	$\bar{x}$	Sample number	x, Sample measurements	Σx_i	$\bar{x}$
1	3, 5, 1, 3, 2	14	2.8	51	2, 3, 5, 3, 2	15	3.0
2	3, 1, 1, 4, 6	15	3.0	52	1, 1, 1, 2, 4	9	1.8
3	1, 3, 1, 6, 1	12	2.4	53	2, 6, 3, 4, 5	20	4.0
4	4, 5, 3, 3, 2	17	3.4	54	1, 2, 2, 1, 1	7	1.4
5	3, 1, 3, 5, 2	14	2.8	55	2, 4, 4, 6, 2	18	3.6
6	2, 4, 4, 2, 4	16	3.2	56	3, 2, 5, 4, 5	19	3.8
7	4, 2, 5, 5, 3	19	3.8	57	2, 4, 2, 4, 5	17	3.4
8	3, 5, 5, 5, 5	23	4.6	58	5, 5, 4, 3, 2	19	3.8
9	6, 5, 5, 1, 6	23	4.6	59	5, 4, 4, 6, 3	22	4.4
10	5, 1, 6, 1, 6	19	3.8	60	3, 2, 5, 3, 1	14	2.8
11	1, 1, 1, 5, 3	11	2.2	61	2, 1, 4, 1, 3	11	2.2
12	3, 4, 2, 4, 4	17	3.4	62	4, 1, 1, 5, 2	13	2.6
13	2, 6, 1, 5, 4	18	3.6	63	2, 3, 1, 2, 3	11	2.2
14	6, 3, 4, 2, 5	20	4.0	64	2, 3, 3, 2, 6	16	3.2
15	2, 6, 2, 1, 5	16	3.2	65	4, 3, 5, 2, 6	20	4.0
16	1, 5, 1, 2, 5	14	2.8	66	3, 1, 3, 3, 4	14	2.8
17	3, 5, 1, 1, 2	12	2.4	67	4, 6, 1, 3, 6	20	4.0
18	3, 2, 4, 3, 5	17	3.4	68	2, 4, 6, 6, 3	21	4.2
19	5, 1, 6, 3, 1	16	3.2	69	4, 1, 6, 5, 5	21	4.2
20	1, 6, 4, 4, 1	16	3.2	70	6, 6, 6, 4, 5	27	5.4
21	6, 4, 2, 3, 5	20	4.0	71	2, 2, 5, 6, 3	18	3.6
22	1, 3, 5, 4, 1	14	2.8	72	6, 6, 6, 1, 6	25	5.0
23	2, 6, 5, 2, 6	21	4.2	73	4, 4, 4, 3, 1	16	3.2
24	3, 5, 1, 3, 5	17	3.4	74	4, 4, 5, 4, 2	19	3.8
25	5, 2, 4, 4, 3	18	3.6	75	4, 5, 4, 1, 4	18	3.6
26	6, 1, 1, 1, 6	15	3.0	76	5, 3, 2, 3, 4	17	3.4
27	1, 4, 1, 2, 6	14	2.8	77	1, 3, 3, 1, 5	13	2.6
28	3, 1, 2, 1, 5	12	2.4	78	4, 1, 5, 5, 3	18	3.6
29	1, 5, 5, 4, 5	20	4.0	79	4, 5, 6, 5, 4	24	4.8
30	4, 5, 3, 5, 2	19	3.8	80	1, 5, 3, 4, 2	15	3.0
31	4, 1, 6, 1, 1	13	2.6	81	4, 3, 4, 6, 3	20	4.0
32	3, 6, 4, 1, 2	16	3.2	82	5, 4, 2, 1, 6	18	3.6
33	3, 5, 5, 2, 2	17	3.4	83	1, 3, 2, 2, 5	13	2.6
34	1, 1, 5, 6, 3	16	3.2	84	5, 4, 1, 4, 6	20	4.0
35	2, 6, 1, 6, 2	17	3.4	85	2, 4, 2, 5, 5	18	3.6
36	2, 4, 3, 1, 3	13	2.6	86	1, 6, 3, 1, 6	17	3.4
37	1, 5, 1, 5, 2	14	2.8	87	2, 2, 4, 3, 2	13	2.6
38	6, 6, 5, 3, 3	23	4.6	88	4, 4, 5, 4, 4	21	4.2
39	3, 3, 5, 2, 1	14	2.8	89	2, 5, 4, 3, 4	18	3.6
40	2, 6, 6, 6, 5	25	5.0	90	5, 1, 6, 4, 3	19	3.8
41	5, 5, 2, 3, 4	19	3.8	91	5, 2, 5, 6, 3	21	4.2
42	6, 4, 1, 6, 2	19	3.8	92	6, 4, 1, 2, 1	14	2.8
43	2, 5, 3, 1, 4	15	3.0	93	6, 3, 1, 5, 2	17	3.4
44	4, 2, 3, 2, 1	12	2.4	94	1, 3, 6, 4, 2	16	3.2
45	4, 4, 5, 4, 4	21	4.2	95	6, 1, 4, 2, 2	15	3.0
46	5, 4, 5, 5, 4	23	4.6	96	1, 1, 2, 3, 1	8	1.6
47	6, 6, 6, 2, 1	21	4.2	97	6, 2, 5, 1, 6	20	4.0
48	2, 1, 5, 5, 4	17	3.4	98	3, 1, 1, 4, 1	10	2.0
49	6, 4, 3, 1, 5	19	3.8	99	5, 2, 1, 6, 1	15	3.0
50	4, 4, 4, 4, 4	20	4.0	100	2, 4, 3, 4, 6	19	3.8

Figure 6.4
Histogram of sample means for the die-tossing experiment

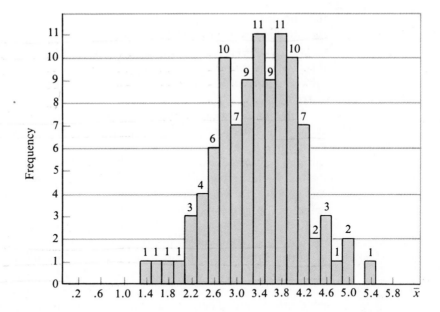

the mean of samples of $n = 5$ observations for the die-tossing experiment and illustrates the basic idea involved in the Central Limit Theorem.

CENTRAL LIMIT THEOREM

If random samples of n observations are drawn from a nonnormal population with finite mean μ and standard deviation σ, then when n is large, the sampling distribution of the sample mean $\bar{x}$ is approximately normally distributed, with mean and standard deviation

$$\mu_{\bar{x}} = \mu \quad \text{and} \quad \sigma_{\bar{x}} = \frac{\sigma}{\sqrt{n}}$$

The approximation will become more accurate as n becomes large.

The Central Limit Theorem can be restated to apply to the **sum of the sample measurements**

$$\sum_{i=1}^{n} x_i$$

which, as n becomes large, would also tend to possess a normal distribution, in repeated sampling, with mean $n\mu$ and standard deviation $\sigma\sqrt{n}$.

It can be shown (proof omitted) that the mean and standard deviation of the sampling distribution of $\bar{x}$ are always related to the mean and standard deviation of the sampled population as well as to the sample size n. The two distributions have the same mean μ, and the standard deviation of the sampling distribution of $\bar{x}$ is equal to the population standard deviation σ divided by $\sqrt{n}$. (It can be shown that this relationship is true regardless of the sample size n.) Consequently, *the spread of the distribution of sample means is considerably less than the spread of the population distribution.*

The significance of the Central Limit Theorem is twofold. First, it explains the rather common occurrence of normally distributed random variables in nature. We might imagine the height of a person as being composed of a number of effects, each random, and associated with such things as the height of the mother, the height of the father, the activity of a particular gland, the environment, and diet. If each of these effects tends to add to the others to yield the measurement of height, then height is the sum of a number of random variables, and according to the Central Limit Theorem, the distribution of heights is approximately normal.

The second and most important contribution of the Central Limit Theorem is in statistical inference. Many estimators that are used to make inferences about population parameters are sums or averages of the sample measurements. When sums or averages are used and the sample size n is sufficiently large, according to the Central Limit Theorem, we expect the estimator to possess (approximately) a normal probability distribution in repeated sampling. We can then use the normal distribution discussed in Chapter 5 to describe the behavior of the inference maker. This aspect of the Central Limit Theorem will be utilized in later chapters dealing with statistical inference.

One disturbing feature of the Central Limit Theorem, and of most approximation procedures, is that we must have some idea of how large the sample size n must be for the approximation to be valid. Unfortunately, there is no clear-cut solution to this problem, since the appropriate value for n will depend on the population probability distribution as well as how the approximation will be used. Although the preceding comment sidesteps the difficulty and suggests that we must rely solely on experience, we may take comfort in the results of the die-tossing experiment discussed earlier in this section. In repeated sampling the distribution of $\bar{x}$, based on a sample of only $n = 5$ measurements, was approximately bell-shaped. This approximation would be even better for larger values of n.

Figure 6.5 gives the results of some other Monte Carlo sampling experiments. We programmed a computer to select random samples of size n, $n = 2, 5, 10$, and 25, from each of three populations, the first having a normal population probability distribution, the second a uniform probability distribution, and the third a negative exponential probability distribution. These population probability distributions are shown in the top row of Figure 6.5. The computer printouts of the approximations to sampling distributions of the sample mean $\bar{x}$ for sample sizes $n = 2$, $n = 5$, $n = 10$, and $n = 25$ are shown in rows 2, 3, 4, and 5 of Figure 6.5.

Figure 6.5 illustrates an important theorem of theoretical statistics. **The sampling distribution of the sample mean is exactly normally distributed (proof**

Figure 6.5

Probability distributions and approximations of the sampling distributions for three populations*

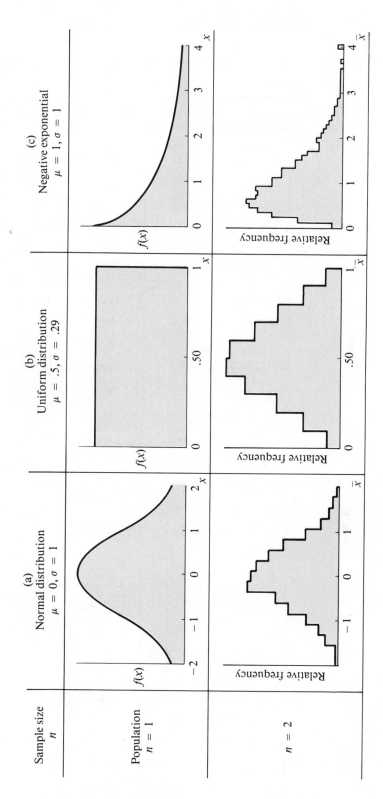

* vertical scale does not remain the same as *n* increases

Figure 6.5
(continued)

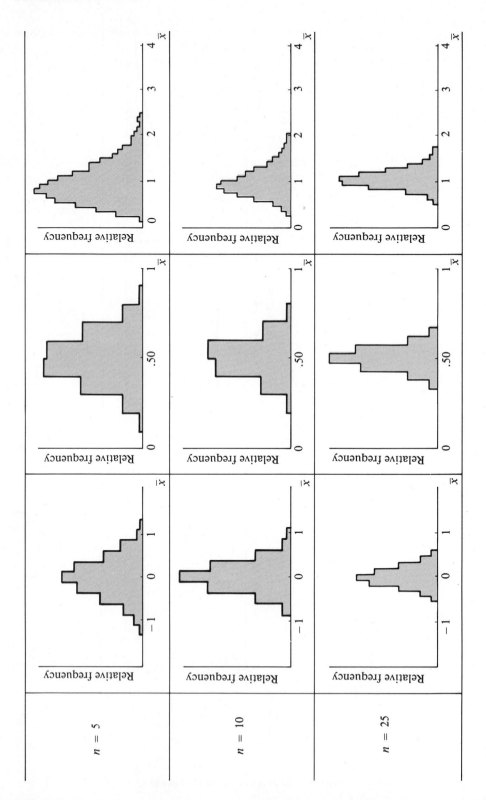

omitted) **regardless of the sample size, when sampling from a population that has a normal distribution.** In contrast, the sampling distributions of $\bar{x}$ for samples selected from populations with uniform and negative exponential probability distributions tend to become more nearly normal as the sample size n increases from $n = 2$ to $n = 25$, more rapidly for the uniform distribution, and more slowly for the highly skewed exponential distribution. But note that the sampling distribution of $\bar{x}$ is normal or approximately normal for sampling from either the uniform or the exponential probability distributions when the sample size is as large as $n = 25$. This suggests that, for many populations, the sampling distribution of $\bar{x}$ will be approximately normal for moderate sample sizes; an exception to this rule is the binomial experiment when either p or $(1 - p)$ is small. The appropriate sample size n will be given for specific applications of the Central Limit Theorem as they are encountered in this chapter and later in the text.

In using the sample mean $\bar{x}$ as a random variable, we must differentiate between the probability density function for a single observation x, which has mean μ and standard deviation σ, and the probability density function for $\bar{x}$, which has mean μ but standard deviation $\sigma/\sqrt{n}$.

6.4 THE SAMPLING DISTRIBUTION OF THE SAMPLE MEAN

Many estimators are available for estimating the population mean, including the median, the trimmed mean, and the midrange (the average of the largest and smallest observation in the set), as well as the sample mean $\bar{x}$. Each generates a sampling distribution in repeated sampling, and depending upon the population and the problem involved, each possesses certain advantages and disadvantages. Some statistics are easier to calculate than others; some statistics produce estimates that are consistently too large or too small; others produce estimates that are highly variable in repeated sampling. The sampling distributions for $\bar{x}$ and m with $n = 3$ for the population of $N = 5$ elements given in Section 6.2 showed that when we use criteria such as these, the sample mean seemed to perform better than the median as an estimator of the population mean μ. In many situations the sample mean $\bar{x}$ has desirable properties as an estimator that are not shared by other competing estimators; therefore, it is more widely used.

THE SAMPLING DISTRIBUTION OF THE SAMPLE MEAN $\bar{x}$

I. If a random sample of n measurements is selected from a population with mean μ and standard deviation σ, the sampling distribution of the sample mean $\bar{x}$ will have a mean

$$\mu_{\bar{x}} = \mu$$

and a standard deviation*

$$\sigma_{\bar{x}} = \frac{\sigma}{\sqrt{n}}$$

2. If the population has a *normal* distribution, the sampling distribution of $\bar{x}$ will be *exactly* normally distributed, *regardless of the sample size, n.*

3. If the population distribution is nonnormal, the sampling distribution of $\bar{x}$ will be, for large samples, approximately normally distributed (by the Central Limit Theorem). Figure 6.5 suggests that the sampling distributions of $\bar{x}$ will be approximately normal for sample sizes as small as $n = 25$ for most populations of measurements.

> The standard deviation of a statistic used as an estimator of a population parameter is often called the **standard error of the estimator**, since it refers to the precision of the estimator. Therefore, the standard deviation of $\bar{x}$, given by $\sigma/\sqrt{n}$, is referred to as the **standard error of the mean**.

EXAMPLE 6.1   Suppose that you select a random sample of $n = 25$ observations from a population with mean $\mu = 8$ and $\sigma = .6$. Find the approximate probability that the sample mean $\bar{x}$ will

a. be less than 7.9
b. exceed 7.9
c. lie within 0.1 of the population mean $\mu = 8$

Solution   a. Regardless of the shape of the population relative frequency distribution, the sampling distribution of $\bar{x}$ will have a mean $\mu_{\bar{x}} = \mu = 8$ and a standard deviation

$$\sigma_{\bar{x}} = \frac{\sigma}{\sqrt{n}} = \frac{.6}{\sqrt{25}} = .12$$

And, for a sample as large as $n = 25$, it is likely (because of the Central Limit Theorem) that the sampling distribution of $\bar{x}$ is approximately normally distribtuted (we will assume that it is). Therefore, the probability that $\bar{x}$ will be less than 7.9 is approximated by the shaded area under the normal sampling

---

* When repeated samples of size n are randomly selected from a *finite* population with N elements whose mean is μ and whose variance is σ^2, the standard deviation of $\bar{x}$ is

$$\sigma_{\bar{x}} = \frac{\sigma}{\sqrt{n}} \sqrt{\frac{N-n}{N-1}}$$

where σ^2 is the population variance. When N is large relative to the sample size n, $\sqrt{(N-n)/(N-1)}$ is approximately equal to 1. Then

$$\sigma_{\bar{x}} = \frac{\sigma}{\sqrt{n}}$$

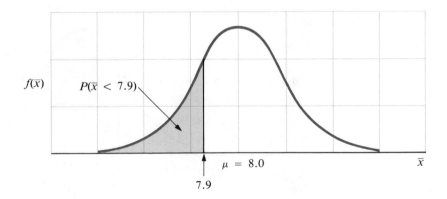

Figure 6.6
The probability that $\bar{x}$ is less than 7.9

distribution in Figure 6.6. To find this area, we need to calculate the value of *z corresponding to* $\bar{x} = 7.9$. The value of $z =$ (variable $-$ expected value)/ (standard deviation of variable) measures the distance from the mean in units of standard deviations, or the distance between $\bar{x} = 7.9$ and $\mu_{\bar{x}} = \mu = 8.0$ expressed in units of

$$\sigma_{\bar{x}} = \frac{\sigma}{\sqrt{n}} = .12$$

Thus,

$$z = \frac{\bar{x} - \mu}{\sigma_{\bar{x}}} = \frac{7.9 - 8.0}{.12} = -.83$$

From Table 3 in Appendix III, we find that the area corresponding to $z = .83$ is .2966. Therefore,

$$P(\bar{x} < 7.9) = .5 - .2967 = .2033$$

[Note that we must use $\sigma_{\bar{x}}$ (not σ) in the formula for z because we are finding an area under the sampling distribution for $\bar{x}$, not under the sampling distribution for x.]

b.   The event that $\bar{x}$ exceeds 7.9 is the complement of the event that $\bar{x}$ is less that 7.9. Thus, the probability that $\bar{x}$ exceeds 7.9 is

$$P(\bar{x} > 7.9) = 1 - P(\bar{x} < 7.9)$$

$$= 1 - .2033$$

$$= .7967$$

c.   The probability that $\bar{x}$ lies within 0.1 of $\mu = 8$ is the shaded area in Figure 6.7. We found in part (a) that the area between $\bar{x} = 7.9$ and $\mu = 8.0$ is .2967. Since the area under the normal curve between $\bar{x} = 8.1$ and $\mu = 8.0$ is identical to the area between $\bar{x} = 7.9$ and $\mu = 8.0$, it follows that

$$P(7.9 < \bar{x} < 8.1) = 2(.2967) = .5934$$

Figure 6.7
The probability that $\bar{x}$ lies
within 0.1 of $\mu = 8$

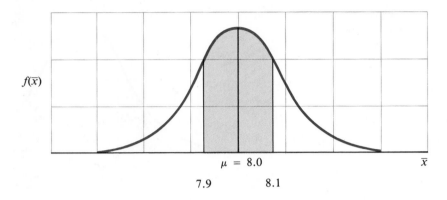

EXAMPLE 6.2   To avoid difficulties with the Federal Trade Commission or state and local consumer protection agencies, a beverage bottler must make reasonably certain that 12-ounce bottles actually contain 12 ounces of beverage. To infer whether a bottling machine is working satisfactorily, one bottler randomly samples 10 bottles per hour and measures the amount of beverage in each bottle. The mean $\bar{x}$ of the 10 fill measurements is used to decide whether to readjust the amount of beverage delivered per bottle by the filling machine. If records show that the amount of fill per bottle is normally distributed with a standard deviation of .2 ounce, and if the bottling machine is set to produce a mean fill per bottle of 12.1 ounces, what is the approximate probability that the sample mean $\bar{x}$ of the 10 test bottles is less than 12 ounces?

Solution   The mean of the sampling distribution of the sample mean $\bar{x}$ is identical to the mean of the population of bottle fills—namely, $\mu = 12.1$ ounces—and the standard deviation (or standard error) of $\bar{x}$ is

$$\sigma_{\bar{x}} = \frac{\sigma}{\sqrt{n}} = \frac{.2}{\sqrt{10}} = .063$$

(*Note:* σ is the standard deviation of the population of bottle fills and n is the number of bottles in the sample.) Since the amount of fill is normally distributed, $\bar{x}$ is also normally distributed. Then the probability distribution of $\bar{x}$ will appear as shown in Figure 6.8.

The probability that $\bar{x}$ will be less than 12 ounces will equal $(.5 - A)$, where A is the area between 12 and the mean $\mu = 12.1$. Expressing this distance in standard deviations, we have

$$z = \frac{\bar{x} - \mu}{\sigma_{\bar{x}}} = \frac{12 - 12.1}{.063} = -1.59$$

Then the area A over the interval $12 < \bar{x} < 12.1$, found in Table 3 of Appendix III, is .4441, and the probability that $\bar{x}$ will be less than 12 ounces is

$$P(\bar{x} < 12) = .5 - A = .5 - .4441 = .0559 \approx .056$$

Figure 6.8
Probability distribution
of $\bar{x}$, the mean of the
$n = 10$ bottle fills in
Example 6.2

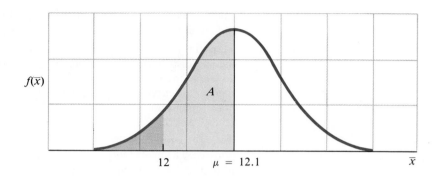

Thus, if the machine is set to deliver an average fill of 12.1 ounces, the mean fill $\bar{x}$ of a sample of 10 bottles will be less than 12 ounces with probability equal to .056. When this danger signal occurs ($\bar{x}$ is less than 12), the bottler takes a larger sample to recheck the setting of the filling machine. ◁

TIPS ON PROBLEM SOLVING

Before attempting to calculate the probability that the statistic $\bar{x}$ falls in some interval, complete the following steps:

1. Calculate the mean and standard deviation of the sampling distribution of $\bar{x}$.
2. Sketch the sampling distribution. Show the location of the mean μ and use the value of $\sigma_{\bar{x}}$ to locate the approximate location of the tails of the distribution.
3. Locate the interval on the sketch from part 2 and shade the area corresponding to the probability that you wish to calculate.
4. Find the z score(s) associated with the value(s) of interest. Use Table 3 in Appendix III to find the probability.
5. When you have obtained your answer, look at your sketch of the sampling distribution and see whether your calculated answer agrees with the shaded area. This provides a very rough check on your calculations.

EXERCISES   Basic Techniques

6.1 Random samples of size n were selected from populations with the means and variances shown below. Find the mean and standard deviation (or standard error) of the sampling distribution of the sample mean for
  a. $n = 25, \mu = 10, \sigma^2 = 9$
  b. $n = 100, \mu = 5, \sigma^2 = 4$
  c. $n = 8, \mu = 120, \sigma^2 = 1$

6.2 Refer to Exercise 6.1.
  a.   If the sampled populations are normal, what is the sampling distribution of $\bar{x}$ for parts (a), (b), and (c)?

    b.   According to the Central Limit Theorem, if the sampled populations are *not* normal, what can be said about the sampling distribution of $\bar{x}$ for parts (a), (b), and (c)?

6.3   Refer to Exercise 6.1, part (b).

    a.   Sketch the sampling distribution for the sample mean and locate the mean and the interval $\mu \pm 2\sigma_{\bar{x}}$ along the $\bar{x}$ axis.

    b.   Shade the area under the curve that corresponds to the probability that $\bar{x}$ lies within .15 unit of the population mean μ.

    c.   Find the probability described in part (b).

6.4   Refer to the die-throwing experiment in Section 6.3 in which x is the number of dots observed when a single die is tossed. The probability distribution for x is given in Figure 6.3, and the relative frequency histogram for $\bar{x}$ is given in Figure 6.4 for 100 random samples of size $n = 5$.

    a.   Verify that the mean and standard deviation of x are $\mu = 3.5$ and $\sigma = 1.71$, respectively.

    b.   Look at the histogram in Figure 6.4. Guess the value of its mean and standard deviation. (*Hint:* The Empirical Rule states that approximately 95% of the measurements associated with a mound-shaped distribution will lie within two standard deviations of the mean.)

    c.   What are the theoretical mean and standard deviation of the sampling distribution of $\bar{x}$? How do these values compare to the guessed values of part (b)?

6.5   Refer to Exercise 6.4. Suppose that the sampling experiment in Section 6.3 was repeated over and over again an infinitely large number of times. Find the mean and standard deviation (or standard error) for the sampling distribution of $\bar{x}$ if each sample consists of

    a. $n = 10$ measurements

    b. $n = 15$ measurements

    c. $n = 25$ measurements

6.6   Refer to Exercises 6.4 and 6.5. What effect does increasing the sample size have on the sampling distribution of $\bar{x}$?

6.7   Exercises 6.4 and 6.5 demonstrate that the standard deviation of the sampling distribution decreases as the sample size increases. To see this relationship more clearly, suppose that a random sample of n observations is selected from a population with standard deviation $\sigma = 1$. Calculate $\sigma_{\bar{x}}$ for $n = 1, 2, 4, 9, 16, 25,$ and 100. Then plot $\sigma_{\bar{x}}$ versus the sample size n and connect the points with a smooth curve. Note the manner in which $\sigma_{\bar{x}}$ decreases as n increases.

6.8   Suppose that a random sample of $n = 5$ observations is selected from a population that is normally distributed with mean equal to 1 and standard deviation equal to .36.

    a.   Give the mean and standard deviation of the sampling distribution of $\bar{x}$.

    b.   Find the probability that $\bar{x}$ exceeds 1.3.

    c.   Find the probability that the sample mean $\bar{x}$ will be less than .5.

    d.   Find the probability that the sample mean will deviate from the population mean $\mu = 1$ by more than .4.

6.9   Suppose that a random sample of $n = 25$ observations is selected from a population that is normally distributed with mean equal to 106 and standard deviation equal to 12.

    a.   Give the mean and the standard deviation of the sampling distribution of the sample mean $\bar{x}$.

    b.   Find the probability that $\bar{x}$ exceeds 110.

    c.   Find the probability that the sample mean will deviate from the population mean $\mu = 106$ by no more than 4.

6.10   A class Monte Carlo experiment: Appendix I presents the computer printout of the relative frequency histogram for the 945 female systolic blood pressures given there. As an experiment, regard this data set as a population. Have each member of the class select a random sample of $n = 4$ observations from this population (using the random-number table, Table 14 in Appendix III) and calculate the sample mean. Construct a relative frequency histogram for the sample means calculated by the class members. This provides an approximation of the sampling distribution of $\bar{x}$.

a.   Compare with the population relative frequency histogram in Appendix I.

b.   Calculate the theoretical mean and standard deviation of the sampling distribution of $\bar{x}$. (*Note:* The mean and standard deviation of the data set are given in Appendix I.) Locate the mean μ and the interval $\mu \pm 2\sigma_{\bar{x}}$ along the horizontal axis of the relative frequency histogram you constructed in part (a). Does μ fall approximately in the center of the histogram? Does the interval $\mu \pm 2\sigma_{\bar{x}}$ include most of the sample means?

c.   Calculate the mean and standard deviation of the sample means used to construct the relative frequency histogram. Are these values close to the values found for μ and $\sigma_{\bar{x}}$ in part (a)?

Applications

6.11   The 1985 SAT scores showed the largest increase in 21 years (*Wall Street Journal*, September 4, 1985). The math and verbal tests, taken by approximately one-third of the nation's high school seniors, showed an increase in the average math score of four points, from 471 to 475. The average of the verbal scores increased five points, from 426 to 431. Why would these very small inceases be viewed as a significant improvement in achievement by educators?

6.12   Explain why the weight of a package of one dozen tomatoes should be approximately normally distributed if the dozen tomatoes represent a random sample.

6.13   Use the Central Limit Theorem to explain why a Poisson random variable—say, the number of a particular type of bacteria in a cubic foot of water—has a distribution that can be approximated by a normal distribution when the mean μ is large. (*Hint:* One cubic foot of water contains 1728 cubic inches of water.)

6.14   Americans are taking better care of themselves, according to a Louis Harris survey conducted for *Prevention* magazine (*Press Enterprise*, May 24, 1989). The results of the survey were used to construct a Prevention Index—a composite score of personal health behaviors, including efforts to limit cholesterol, alcohol consumption, and cigarette smoking, and measures relating to weight and stress levels. A random sample of 1250 American adults scored 65.4 out of a possible 100 points, nearly four points higher than the 1984 average.

a.   Explain why the Prevention Index scores should be approximately normally distributed.

b.   If we assume that the average score for all American adults has not changed since 1984, and is equal to 61.5 with a standard deviation of 12, what is the probability of observing a sample average of 65.4 or larger?

c.   Based on the results of part (b), is it reasonable to assume that the average score for all American adults has not changed since 1984? Explain.

6.15   An important expectation of the recent federal income tax reduction is that consumers will reap a substantial portion of the tax savings. Suppose that estimates of the portion of total tax saved, based on a random sampling of 35 economists, had a mean of 26% and a standard deviation of 12%.

a.   What is the approximate probability that a sample mean, based on a random sample of $n = 35$ economists, will lie within 1% of the mean of the population of the estimates of all economists?

b.   Is it necessarily true that the mean of the population of estimates of all economists is equal to the percent tax savings that will actually be achieved?

6.16   A manufacturer of paper used for packaging requires a minimum strength of 20 pounds per square inch. To check on the quality of the paper, a random sample of ten pieces of paper is selected each hour from the previous hour's production and a strength measurement is recorded for each. The standard deviation σ of the strength measurements, computed by pooling the sum of squares of deviations of many samples, is known to equal 2 pounds per square inch.

a.   What is the approximate probability distribution of the sample mean of $n = 10$ test pieces of paper?

b.   If the mean of the population of strength samples is 21 pounds per square inch, what is the approximate probability that, for a random sample of $n = 10$ test pieces of paper, $\bar{x} < 20$?

c.   What value would you desire for the mean paper strength μ in order that $P(\bar{x} < 20)$ be equal to .001?

6.17   The normal daily human potassium requirement is in the range of 2000 to 6000 mg, with the larger amounts required during hot summer weather. The amount of potassium in food varies, depending on the food. For example, there are approximately 7 mg in a cola drink, 46 mg in a beer, 630 mg in a banana, 300 mg in a carrot, and 440 mg in a glass of orange juice. Suppose that the distribution of potassium in a banana is normally distributed with mean equal to 630 mg and standard deviation equal to 40 mg per banana. Suppose that you eat $n = 3$ bananas per day, and T is the total number of milligrams of potassium you receive from them.

a.   Find the mean and standard deviation of T.

b.   Find the probability that your total daily intake of potassium from the three bananas will exceed 2000 mg.

(*Hint:* Note that T is the sum of three random variables x_1, x_2, and x_3, where x_1 is the amount of potassium in banana number 1, and so on.)

6.18   The total daily sales, x, in the delicatessen section of a local market is the sum of the sales generated by a fixed number of customers who make purchases on a given day.

a.   What kind of probability distribution would you expect the total daily sales to have? Explain.

b.   For this particular market, the average sale per customer in the deli section is $8.50 with $\sigma = \$2.50$. If 30 customers make delicatessen purchases on a given day, give the mean and standard deviation of the probability distribution of the total daily sales, x.

6.5   THE SAMPLING DISTRIBUTION OF THE SAMPLE PROPORTION

Many sampling problems involve consumer preference or opinion polls, which are concerned with estimating the proportion p of people in the population who possess some specified characteristic. These and similar situations provide practical examples of binomial experiments, if the sampling procedure has been conducted in the appropriate manner. If a random sample of n persons is selected from the population and if x of these possess the specified characteristic, then the sample proportion

$$\hat{p} = x/n$$

is used to estimate the population proportion *p*.*

Since each distinct value of *x* results in a distinct value of $\hat{p} = x/n$, the probabilities associated with $\hat{p}$ are equal to the probabilities associated with the corresponding values of *x*. Hence, the sampling distribution of $\hat{p}$ will be the same shape as the binomial probability distribution for *x*. Like the binomial probability distribution, it can be approximated by a normal distribution when the sample size *n* is large. The mean of the sampling distribution of $\hat{p}$ is

$$\mu_{\hat{p}} = p$$

and its standard deviation is

$$\sigma_{\hat{p}} = \sqrt{\frac{pq}{n}}$$

where $q = 1 - p$

PROPERTIES OF THE SAMPLING DISTRIBUTION OF THE SAMPLE PROPORTION $\hat{p}$

1. If a random sample of *n* observations is selected from a binomial population with parameter *p*, the sampling distribution of the sample proportion

   $$\hat{p} = x/n$$

   will have a mean

   $$\mu_{\hat{p}} = p$$

   and a standard deviation

   $$\sigma_{\hat{p}} = \sqrt{\frac{pq}{n}}$$

   where $q = 1 - p$

2. When the sample size *n* is large, the sampling distribution of $\hat{p}$ can be approximated by a normal distribution. The approximation will be adequate if $\mu_{\hat{p}} - 2\sigma_{\hat{p}}$ and $\mu_{\hat{p}} + 2\sigma_{\hat{p}}$ fall in the interval 0 to 1.

EXAMPLE 6.3 The *Wall Street Journal* (March 20, 1985) reports on a survey of 313 children, ages 14 to 22, selected from among the children of the nation's top corporate executives. When asked to identify the best aspect of being one of this privileged group, 55% mentioned material and financial advantages. Describe the sampling distribution of

---

* A hat placed over the symbol of a population parameter is often used to denote a statistic used to estimate the population parameter. For example, the symbol $\hat{p}$ is used to denote the sample proportion.

Figure 6.9
The sampling distribution
of $\hat{p}$ based on a sample
of $n = 313$ children

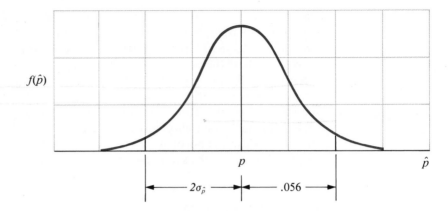

$f(\hat{p})$

the sample proportion $\hat{p}$ of children citing material advantage as the best aspect of their privileged lives.

Solution　We will assume that the 313 children represent a random sample of the children of all top corporate executives and that the true proportion in the population is equal to some unknown value that we will call p. Then the sampling distribution of $\hat{p}$ can be approximated by a normal distribution*, with a mean equal to p (see Figure 6.9) and standard deviation

$$\sigma_{\hat{p}} = \sqrt{\frac{pq}{n}}$$

Examining Figure 6.9, we see that the sampling distribution of $\hat{p}$ centers over its mean p. Even though we do not know the exact value of p (the sample proportion $\hat{p} = .55$ may be larger or smaller than p), we can calculate an approximate value for the standard deviation of the sampling distribution using the sample proportion $\hat{p} = .55$ to approximate the unknown value of p. Thus

$$\sigma_{\hat{p}} = \sqrt{\frac{pq}{n}} \approx \sqrt{\frac{\hat{p}\hat{q}}{n}}$$

$$= \sqrt{\frac{(.55)(.45)}{313}}$$

$$= .028$$

Therefore, we know that approximately 95% of the time $\hat{p}$ will fall within $2\sigma_{\hat{p}} \approx .056$ of the (unknown) value of p.

---

* Checking the conditions that allow the normal approximation to the distribution of $\hat{p}$, we find that $n = 313$ is adequate for values of p near .55, since $\hat{p} \pm 2\sigma_{\hat{p}} = .55 \pm \sqrt{.55(.45)/313} = .55 \pm .056$, or .494 to .606, falls in the interval 0 to 1.

EXAMPLE 6.4 Refer to Example 6.3. Suppose that the population proportion p of children in the population is actually equal to .5. What is the probability of observing a sample proportion as large or larger than the observed value $\hat{p} = .55$?

Figure 6.10
The sampling distribution of $\hat{p}$ for $n = 313$ and $p = .5$

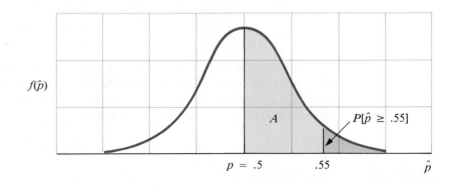

Solution Figure 6.10 shows the sampling distribution of $\hat{p}$ when $p = .5$ with the observed value $\hat{p} = .55$ located on the horizontal axis. From Figure 6.10, we can see that the probability of observing a sample proportion p equal to or larger than .55 can be approximated by the shaded area in the upper tail of a normal distribution with

$$\mu_{\hat{p}} = .5$$

and

$$\sigma_{\hat{p}} = \sqrt{\frac{pq}{n}} = \sqrt{\frac{(.5)(.5)}{313}} = .0283$$

To find this shaded area, we need to know how many standard deviations the observed value $\hat{p} = .55$ lies away from the mean of the sampling distribution $p = .5$. This distance is given by the z value

$$z = \frac{\hat{p} - p}{\sigma_{\hat{p}}} = \frac{.55 - .5}{.0283} = 1.77$$

Table 3 in Appendix III gives the area A corresponding to $z = 1.77$ as

$$A = .4616$$

Therefore, the shaded area in the upper tail of the sampling distribution in Figure 6.10 is

$$P(\hat{p} > .55) = .5 - A = .5 - .4616 = .0384$$

$$\approx .04$$

This tells us that if we were to select a random sample of $n = 313$ observations from a population with proportion p equal to .5, the probability that the sample proportion $\hat{p}$ would be as large or larger than .55 is only .04.

In employing the normal distribution to approximate the binomial probabilities associated with x, we used a correction of $\pm.5$ to improve the approximation. The equivalent correction here would be $\pm(1/2n)$. For example, for $\hat{p}$ the value of z with the correction would be

$$z_1 = \frac{(.55 - .0016) - .5}{\sqrt{\dfrac{(.5)(.5)}{313}}} = 1.71$$

with $A = .4564$ and $(.5 - A) = .0436$. To two-decimal accuracy, this value agrees with our earlier result. When n is large, the effect of using the correction is generally negligible. You should solve problems in this and the remaining chapters *without* the correction factor unless specifically instructed to use it.

EXERCISES Basic Techniques

6.19 Random samples of size n were selected from binomial populations with population parameters p shown below. Find the mean and the standard deviation of the sampling distribution of the sample proportion $\hat{p}$ for
   a. $n = 100, p = .3$
   b. $n = 400, p = .1$
   c. $n = 250, p = .6$

6.20 Sketch each of the sampling distributions in Exercise 6.19. For each, locate the mean p and the interval $p \pm 2\sigma_{\hat{p}}$ along the $\hat{p}$ axis of the graph.

6.21 Refer to the sampling distribution in Exercise 6.19(a).
   a.   Sketch the sampling distribution for the sample proportion and shade the area under the curve that corresponds to the probability that $\hat{p}$ lies within .08 of the population proportion p.
   b.   Find the probability described in part (a).

6.22 Random samples of size $n = 500$ were selected from a binomial population with $p = .1$.
   a.   Is it appropriate to use the normal distribution to approximate the sampling distribution of $\hat{p}$? Check to make sure the necessary conditions are met.

   Using the results of part (a), find the probability that
   b. $\hat{p} > .12$
   c. $\hat{p} < .10$
   d. $\hat{p}$ lies within .02 of p

6.23 Calculate $\sigma_{\hat{p}}$ for $n = 100$ and
   a. $p = .01$ b. $p = .10$ c. $p = .30$ d. $p = .50$
   e. $p = .70$ f. $p = .90$ g. $p = .99$

   Plot $\sigma_{\hat{p}}$ versus p on graph paper and sketch a smooth curve through the points. For what value of p is the standard deviation of the sampling distribution of $\hat{p}$ a maximum? What happens to $\sigma_{\hat{p}}$ when p is near 0 or near 1.0?

6.24 a.   Is the normal approximation to the sampling distribution of $\hat{p}$ appropriate when $n = 400$ and $p = .8$?

b. Use the results of part (a) to find the probability that p is greater than .83.

c. Use the results of part (a) to find the probability that p lies between .76 and .84.

Applications

6.25 Before making the decision to introduce its "new" Coke in 1985, the Coca-Cola Company introduced new Coca-Cola to approximately 40,000 consumers in 30 cities in the United States. With the brands not identified, 55% chose the new Coke over the old (*Fortune*, May 27, 1985). Assume that the 40,000 consumers in the survey represent a random sample of cola drinkers from a population of cola drinkers in the 30 cities.

a. Describe the sampling distribution of $\hat{p}$, the proportion in the sample who favor the new Coke. (*Hint:* Use $\hat{p}$ to approximate p when calculating $\sigma_{\hat{p}}$.)

b. Find the probability that $\hat{p}$ will lie within .005 of the proportion p of cola drinkers in the population who favor the new Coke.

6.26 What are Americans' pet peeves? In the November 6, 1989 edition of the *Wall Street Journal*, David Wessel reported that "Americans don't like anything that wastes their time." According to the *Wall Street Journal's* "American Way of Buying Survey," four out of every ten people surveyed indicated that their pet peeve was staying home for delivery persons or salespeople who don't show. As part of the survey, Peter Hart asked 1034 consumers (about half of the total sample) specifically about service complaints.

a. What is the sampling distribution for $\hat{p}$, the sample proportion surveyed who indicated that their pet peeve was staying home for delivery persons or salespeople who didn't show?

b. What is the probability that the sample value of $\hat{p}$ lies within .02 of the true population value of p?

c. How would the answer to part (b) change if the sample size were doubled?

6.27 Americans continue to buy and use more and more electronic conveniences and devices. In 1987, for example, 22% of American households had computers, 58% had video cassette recorders (VCRs), and 51% had cable television (*Time*, May 22, 1989, p. 36).

a. In a random sample of $n = 1000$ American households, describe the sampling distribution of $\hat{p}$, the sample proportion of households having cable television, if in fact the 51% figure is correct.

b. What is the probability that the sample proportion $\hat{p}$ of households having cable television differs from the true population proportion by at most 2%?

c. If 22% of American households have a computer, what is the probability that the sample proportion based on a random sample of $n = 1000$ lies within .02 of the true population proportion?

d. Why are the answers to parts (b) and (c) different even though both samples consisted of $n = 1000$ households?

6.28 A survey of purchasing agents from 250 industrial companies found that 25% of the buyers reported higher levels of new orders in January 1985 as compared with earlier months (*Wall Street Journal*, February 4, 1985). Assume that the 250 purchasing agents in the sample represent a random sample of company purchasing agents throughout the United States.

a. Describe the sampling distribution of $\hat{p}$, the proportion of buyers in the United States with higher levels of new orders in January. (*Hint:* Use $\hat{p}$ to approximate p when calculating $\sigma_{\hat{p}}$.)

b. What is the probability that $\hat{p}$ will differ from p by more than .01?

▷ **6.6** THE SAMPLING DISTRIBUTION OF THE SUM OF
OR THE DIFFERENCE BETWEEN
TWO INDEPENDENT STATISTICS

A comparison of two populations quite often focuses on the difference between the population means or proportions. We may be interested in the difference between the average production rates for each of two production lines or the difference in the average savings in heating costs for households with and without passive heating devices such as solar collectors. On the other hand, we may be interested in the difference in the proportion of voters favoring two leading mayoral candidates, or we may be interested in the difference in emergence rates for treated and untreated seeds planted in clay soil. Intuitively, the difference between two sample means or proportions would provide the maximum information about the actual difference between two population means or proportions, and this is in fact the case.

Suppose we let the symbols Y_1 and Y_2 represent any two independent statistics (such as $\bar{x}_1$ and $\bar{x}_2$ or $\hat{p}_1$ and $\hat{p}_2$), with means μ_{Y_i} and μ_{Y_2} and variances $\sigma_{Y_1}^2$ and $\sigma_{Y_2}^2$, respectively.

SAMPLING DISTRIBUTION OF THE SUM OF
OR DIFFERENCE BETWEEN
TWO INDEPENDENT STATISTICS

Assume that Y_1 and Y_2 are independent statistics with means μ_{Y_1} and μ_{Y_2} and variances $\sigma_{Y_1}^2$ and $\sigma_{Y_2}^2$, respectively. **Then the sampling distribution of $Y_1 \pm Y_2$ has the following properties:**

1. The mean of the sampling distribution is

$$\mu_{(Y_1 \pm Y_2)} = \mu_{Y_1} \pm \mu_{Y_2}$$

2. The variance of the sampling distribution of $Y_1 \pm Y_2$ is equal to the sum of the variances, that is:

$$\sigma_{(Y_1 \pm Y_2)}^2 = \sigma_{Y_1}^2 + \sigma_{Y_2}^2$$

and

$$\sigma_{(Y_1 \pm Y_2)} = \sqrt{\sigma_{Y_1}^2 + \sigma_{Y_2}^2}$$

3. If Y_1 and Y_2 are exactly (or approximately) normally distributed, then the sampling distribution of $Y_1 \pm Y_2$ is exactly (or approximately) normally distributed.

The properties of the sampling distribution of the difference between two independent statistics can be applied to deduce the properties of the sampling distributions of the difference $(\bar{x}_1 - \bar{x}_2)$ between two sample means and the difference $(\hat{p}_1 - \hat{p}_2)$ between two sample proportions.

PROPERTIES OF THE SAMPLING DISTRIBUTION OF THE DIFFERENCE $(\bar{x}_1 - \bar{x}_2)$ BETWEEN TWO SAMPLE MEANS

When independent random samples of n_1 and n_2 observations have been selected from populations with means μ_1 and μ_2 and variances σ_1^2 and σ_2^2, respectively, the sampling distribution of the difference $(\bar{x}_1 - \bar{x}_2)$ will have the following properties:

1. The mean and the standard deviation of $(\bar{x}_1 - \bar{x}_2)$ will be

$$\mu_{(\bar{x}_1 - \bar{x}_2)} = \mu_1 - \mu_2$$

and

$$\sigma_{(\bar{x}_1 - \bar{x}_2)} = \sqrt{\frac{\sigma_1^2}{n_1} + \frac{\sigma_1^2}{n_2}}$$

2. **If the sampled populations are normally distributed**, then the sampling distribution of $\bar{x}_1 - \bar{x}_2$ is **exactly** normally distributed, regardless of the sample size.

3. **If the sampled populations are not normally distributed**, then the sampling distribution of $\bar{x}_1 - \bar{x}_2$ is **approximately normally distributed when n_1 and n_2 are large, due to the Central Limit Theorem.**

The sampling distribution of the difference between two sample means is shown in Figure 6.11.

Figure 6.11
The sampling distribution of the difference $(\bar{x}_1 - \bar{x}_2)$ between two sample means

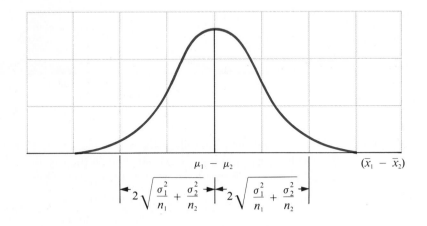

EXAMPLE 6.5   A survey by the National Education Association, reported in the *New York Times* (April 14, 1985), found that the means of teachers' salaries in the 50 states and the District of Columbia ranged from $15,971 in Mississippi to $39,751 in Alaska. The mean salary in New York State was the second largest at $29,000, followed by the District of Columbia at $28,621. If we were to draw a random sample of 40 teachers from the state of New York and 40 from the District of Columbia, what is the

probability that the sample mean salary $\bar{x}_1$ from New York will exceed the sample mean salary $\bar{x}_2$ from the District of Columbia by \$1000 or more? (Assume that the standard deviations for the two population salary distributions are approximately $\sigma_1 = \sigma_2 = \$5000$.)

Solution　From our knowledge of the properties of the sampling distribution of the difference $(\bar{x}_1 - \bar{x}_2)$ in sample means, it follows that

$$\mu_{(\bar{x}_1 - \bar{x}_2)} = \mu_1 - \mu_2 = 29{,}000 - 28{,}621 = 379$$

and

$$\sigma_{(\bar{x}_1 - \bar{x}_2)} = \sqrt{\frac{\sigma_1^2}{n_1} + \frac{\sigma_2^2}{n_2}} = \sqrt{\frac{(5000)^2}{40} + \frac{(5000)^2}{40}}$$

$$= 1118.03$$

We would expect the population salary distributions to have only moderate skewness and, therefore, the sampling distributions of the sample means and of their difference $(\bar{x}_1 - \bar{x}_2)$ to be approximately normal, as shown in Figure 6.12. The probability that $\bar{x}_1$ exceeds $\bar{x}_2$ by \$1000 or more is the shaded area in the figure.

Figure 6.12
The sampling distribution of $(\bar{x}_1 - \bar{x}_2)$ from Example 6.5

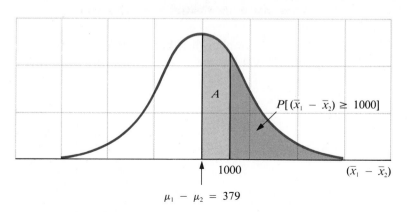

$P[(\bar{x}_1 - \bar{x}_2) \geq 1000]$

1000

$(\bar{x}_1 - \bar{x}_2)$

$\mu_1 - \mu_2 = 379$

To find the probability that $\bar{x}_1$ exceeds $\bar{x}_2$ by \$1000 or more, we must find the z value corresponding to \$1000. This is the distance between 1000 and the mean $\mu_1 - \mu_2 = 379$ expressed in units of $\sigma_{(\bar{x}_1 - \bar{x}_2)}$. Thus,

$$z = \frac{(\bar{x}_1 - \bar{x}_2) - (\mu_1 - \mu_2)}{\sigma_{(\bar{x}_1 - \bar{x}_2)}} = \frac{1000 - 379}{1118.03} = .56$$

The area A (see Figure 6.12) corresponding to $z = .56$ is given in Table 3 in Appendix III as .2123. Therefore,

$$P[(\bar{x}_1 - \bar{x}_2) \geq 1000] = .5 - A = .5 - .2123 = .2877$$

This tells us that the probability that the mean of a random sample of 40 teachers' salaries from New York State exceeds the mean of a sample of 40 teachers' salaries from the District of Columbia is .2877.

Another common statistical problem involves the comparison of two binomial population proportions p_1 and p_2, based on independent random samples of n_1 and n_2 observations, respectively, selected from the two populations. Using the properties of the sampling distribution of the difference in two statistics, it can be shown that the sampling distribution of the difference $(\hat{p}_1 - \hat{p}_2)$ in the sample proportions can be approximated by a normal distribution with the mean and the standard deviation shown in the following display.

PROPERTIES OF THE SAMPLING DISTRIBUTION OF THE DIFFERENCE $(\hat{p}_1 - \hat{p}_2)$ BETWEEN TWO SAMPLE PROPORTIONS

Assume that independent random samples of n_1 and n_2 observations have been selected from binomial populations with parameters p_1 and p_2, respectively. The sampling distribution of the difference between sample proportions $(\hat{p}_1 - \hat{p}_2) = \left(\dfrac{x_1}{n_1} - \dfrac{x_2}{n_2}\right)$ will have the following properties:

1. The mean and the standard deviation of $(\hat{p}_1 - \hat{p}_2)$ will be

$$\mu_{(\hat{p}_1 - \hat{p}_2)} = p_1 - p_2$$

and

$$\sigma_{(\hat{p}_1 - \hat{p}_2)} = \sqrt{\frac{p_1 q_1}{n_1} + \frac{p_2 q_2}{n_2}}$$

2. The sampling distribution of $(\hat{p}_1 - \hat{p}_2)$ can be approximated by a normal distribution when n_1 and n_2 are large due to the Central Limit Theorem.

When we use a normal distribution to approximate binomial probabilities, the interval $(p_1 - p_2) \pm 2\sigma_{(\hat{p}_1 - \hat{p}_2)}$ should be contained within the range of $(\hat{p}_1 - \hat{p}_2)$, which varies from -1 to 1 and not from 0 to 1, as in the case of a single proportion.

EXAMPLE 6.6 A bond proposal for school construction is to be submitted to the voters during the next municipal election. A major portion of the money derived from this bond issue will be used to build new schools in a rapidly developing section of the city, and the remainder will be used for renovating and updating school buildings in the rest of the city. The local newspaper reported that 75% of the residents in the developing section and 60% of the residents in other parts of the city favor passage of the proposed bond issue. Random samples of $n_1 = 50$ residents in the developing section of the city and $n_2 = 100$ residents in other parts of the city are selected, and the residents in the sample are asked whether or not they favor the bond proposal. What is the probability that the difference in magnitude between the sample proportions favoring the bond proposal does not exceed 10%?

Solution If we assume that $p_1 = .75$ and $p_2 = .60$, we can check the conditions that allow the sampling distribution of $\hat{p}_1 - \hat{p}_2$ to be approximated by a normal distribution.

Figure 6.13
The sampling distribution of $(\hat{p}_1 - \hat{p}_2)$ based on sample sizes of $n_1 = 50$ and $n_2 = 100$ for Example 6.6

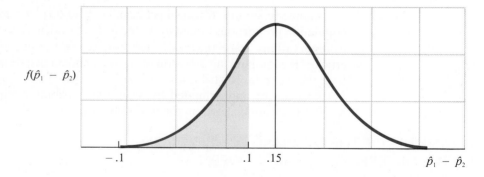

$f(\hat{p}_1 - \hat{p}_2)$

Calculate $(p_1 - p_2) = (.75 - .60) = .15$ and

$$\sigma_{(\hat{p}_1 - \hat{p}_2)} = \sqrt{\frac{p_1 q_1}{n_1} + \frac{p_2 q_2}{n_2}} = \sqrt{\frac{(.75)(.25)}{50} + \frac{(.60)(.40)}{100}} = .0784$$

The interval $(p_1 - p_2) \pm 2\sigma_{(\hat{p}_1 - \hat{p}_2)}$ is calculated as $.15 \pm 2(.0784) = .15 \pm .1568$, or $-.0068$ to $.3068$, which is contained within the range -1 to 1. Hence, the sampling distribution of $(\hat{p}_1 - \hat{p}_2)$ can be approximated by a normal distribution with mean and standard deviation given above.

We wish to find $P(-.1 < \hat{p}_1 - \hat{p}_2 < .1)$, which corresponds to the shaded area in Figure 6.13. With the normal approximation this probability corresponds to $P(z_1 < z < z_2)$, where

$$z_1 = \frac{(-.1) - .15}{.0784} = -3.19 \qquad z_2 = \frac{.1 - .15}{.0784} = -.64$$

The area between 0 and $z_1 = -3.19$ is .5 for all practical purposes, and the area between 0 and $z_2 = -.64$ is .2389. Therefore, the required probability is equal to the difference in these areas, $(.5 - .2389) = .2611$. ◁

TIPS ON PROBLEM SOLVING

1. Begin by reading each exercise carefully to determine the nature of the sampled population or populations. Identify and record the values of population parameters given in the problem, as well as the values of any statistics computed from sample measurements.

2. When approximating the sampling distribution of $\hat{p}$ or $(\hat{p}_1 - \hat{p}_2)$ with a normal distribution, check to see whether the appropriate conditions have been met. For the single sample, $p \pm 2\sigma_{\hat{p}}$ must be contained in the interval 0 to 1, while for the two-sample case, $(p_1 - p_2) \pm 2\sigma_{(\hat{p}_1 - \hat{p}_2)}$ must be contained in the interval -1 to 1. In general, these conditions will be met when the sample sizes are large.

3. Depict the problem by drawing a normal curve and locating the areas under the curve that will be used in solving the problem.

4. The accuracy of your answer will depend on the number of significant digits used in the calculations. For example, in calculating a value of z, use full accuracy for the mean and standard deviation and round your result only after the calculation is completed. This step is especially important in working with $\hat{p}$ and $(\hat{p}_1 - \hat{p}_2)$.

EXERCISES Basic Techniques

6.29 Independent random samples of n_1 and n_2 observations were selected from populations with means μ_1 and μ_2 and variances σ_1^2 and σ_2^2, respectively. Find the mean and standard deviation of the sampling distribution of the difference $(\bar{x}_1 - \bar{x}_2)$ in sample means for

a. $n_1 = 16, \mu_1 = 10, \sigma_1^2 = 4$ and $n_2 = 20, \mu_2 = 20$, and $\sigma_2^2 = 8$

b. $n_1 = 100, \mu_1 = 640, \sigma_1^2 = 1$ and $n_2 = 100, \mu_2 = 642$, and $\sigma_2^2 = 3$

6.30 Refer to Exercise 6.29.

a. If the sampled populations are normal, what is the sampling distribution of $(\bar{x}_1 - \bar{x}_2)$ for parts (a) and (b)?

b. According to the Central Limit Theorem, if the sampled populations are *not* normal, what can be said about the sampling distribution of $(\bar{x}_1 - \bar{x}_2)$ for parts (a) and (b)?

6.31 Refer to Exercise 6.29(b).

a. Sketch the sampling distribution for $(\bar{x}_1 - \bar{x}_2)$ and locate the mean $(\mu_1 - \mu_2)$ on the horizontal axis.

b. Shade the area corresponding to the probability that $(\bar{x}_1 - \bar{x}_2)$ will lie within .3 units of $(\mu_1 - \mu_2)$.

c. Calculate the probability described in part (b).

6.32 Suppose that the means of the populations in Exercise 6.29(b) were unknown. Would it make any difference in the answer to Exercise 6.31(c)? Explain.

6.33 Independent random samples of size $n_1 = n_2 = 50$ were selected from populations with equal means, $\mu_1 = \mu_2$, and with standard deviations $\sigma_1 = 12$ and $\sigma_2 = 15$, respectively. Calculate

a. $P[(\bar{x}_1 - \bar{x}_2) \geq 3]$ b. $P[|\bar{x}_1 - \bar{x}_2| \geq 3]$

6.34 A class Monte Carlo experiment: Turn to the tabulation of the 945 female and 965 male systolic blood pressures in Appendix I. Regard each of these data sets as populations. Have each class member draw a random sample of four observations from each of the two data sets. Calculate the difference $(\bar{x}_1 - \bar{x}_2)$ where $\bar{x}_1$ and $\bar{x}_2$ are the means of the female and male systolic blood pressure samples, respectively. Collect the differences $(\bar{x}_1 - \bar{x}_2)$ for the class and construct a relative frequency histogram for the differences. Note that this provides an approximation to the sampling distribution of $(\bar{x}_1 - \bar{x}_2)$. Calculate the theoretical mean and standard deviation of the sampling distribution of $(\bar{x}_1 - \bar{x}_2)$. (*Note:* The means and standard deviations of the data sets are given in Appendix I.) Locate $(\mu_1 - \mu_2)$ and the interval $(\mu_1 - \mu_2) \pm 2\sigma_{(\bar{x}_1 - \bar{x}_2)}$ along the horizontal axis of your histogram. Does $(\mu_1 - \mu_2)$ fall approximately in the center of the histogram? Does the interval $(\mu_1 - \mu_2) \pm 2\sigma_{(\bar{x}_1 - \bar{x}_2)}$ include most of the differences $(\bar{x}_1 - \bar{x}_2)$ in sample means?

6.35 Independent random samples of n_1 and n_2 were drawn from binomial populations with parameters p_1 and p_2. Find the mean and standard deviation of the sampling distribution of the difference $(\hat{p}_1 - \hat{p}_2)$ in sample proportions for

a. $n_1 = 100, p_1 = .5$ and $n_2 = 300, p_2 = .4$
b. $n_2 = 400, p_1 = .1$ and $n_2 = 400, p_2 = .6$

6.36 a. Sketch the sampling distribution in Exercise 6.35(a), locating the appropriate mean and the interval $(p_1 - p_2) \pm 2\sigma_{(\hat{p}_1 - \hat{p}_2)}$ along the $(\hat{p}_1 - \hat{p}_2)$ axis.
b. Shade the area under the curve corresponding to the probability that the difference $(\hat{p}_1 - \hat{p}_2)$ will differ from $(p_1 - p_2)$ by less than .06.
c. Calculate the probability for part (b).

6.37 Independent random samples of size n_1 and n_2 were drawn from binomial populations with parameters $p_1 = .3$ and $p_2 = .4$, respectively. Find the standard deviation of the sampling distribution of $(\hat{p}_1 - \hat{p}_2)$ if

a. $n_1 = 25, n_2 = 50$ b. $n_1 = 50, n_2 = 50$ c. $n_1 = 500, n_2 = 500$

What is the effect of increasing the sample size on the standard deviation of $(\hat{p}_1 - \hat{p}_2)$?

6.38 Explain how you would conduct a Monte Carlo experiment to obtain an approximation to the sampling distribution for $(\hat{p}_1 - \hat{p}_2)$ if $n_1 = 100, p_1 = .5$ and $n_2 = 300, p_2 = .4$.

Applications

6.39 Analyses of drinking water samples for 100 homes in each of two different sections of a city gave the following means and standard deviations of lead levels (in parts per million).

	Section of city	
	1	2
Sample size	100	100
Mean	34.1	36.0
Standard deviation	5.9	6.0

If the mean lead levels in the drinking water are actually equal for the two sections of the city, what is the probability that the sample means would differ by as much as 1.9 ppm (parts per million)? Assume that σ_1 and σ_2 can be approximated by the sample standard deviations s_1 and s_2.

6.40 In the past decade, the length of time spent traveling has shown a marked increase, due in part to "longer commutes to work, more driving to restaurants, and more travel in pursuit of hobbies and leisure activities—like trips to the video store" (*American Demographics*, September, 1989, p. 10). In particular, the average number of hours per week spent traveling in 1985 was 11 hours for men and 9 hours for women. In an attempt to verify this claim, random samples of 50 men and 50 women were taken in 1985, yielding the following sample information:

	Men	Women
Mean	11.5	8.5
Standard deviation	1.25	1.20

a. If the population means given in *American Demographics* are correct, what is the probability of observing a sample difference as large or larger in absolute value than the one found in these samples? (Use the sample standard deviations to estimate σ_1 and σ_2.)

b. What conclusions can you draw about the accuracy of the magazine's figures based on this sample data?

6.41 An operator of two supermarkets has found that the daily loss due to theft, spoilage, and so on at Store 1 varies with a mean of $1237 per day and a standard deviation of $183 per day. Similar daily losses at Store 2 have a mean of $1485 and standard deviation of $259 per day. What is the probability that the total daily loss from the two stores could exceed $3000 on a given day? Assume that the distributions of daily losses at the two stores are approximately normal.

6.42 An experiment was conducted to test the effect of a new drug on a viral infection. The infection was induced in 100 mice and the mice were randomly split into two groups of 50. The first group, the *control group*, received no treatment for the infection. The second group received the drug. After a 30-day period, the proportions of survivors, $\hat{p}_1$ and $\hat{p}_2$, in the two groups were found to be $\hat{p}_1 = .36$ and $\hat{p}_2 = .60$. If the drug was completely ineffective, what is the probability that the sample proportion $\hat{p}_2$ of survivors who received the drug exceeded the proportion $\hat{p}_1$ for the control group by as much as .24? (*Hint:* Use $\hat{p}_1$ and $\hat{p}_2$ to approximate the values of p_1 and p_2 when calculating $\sigma_{(\hat{p}_1 - \hat{p}_2)}$.)

6.43 A survey of 10,000 people by the Opinion Research Foundation found that approximately 35% of the women and 25% of the men between ages 35 and 40 suffer from allergies (*USA Today*, May 1, 1985). Assume that the 10,000 surveyed were divided evenly between men and women and that the two samples were randomly selected from the population of men and women in the 35–40 age group throughout the United States.
a. What is the sampling distribution of $(\hat{p}_1 - \hat{p}_2)$, the difference in the sample proportions of men and women who suffer from allergies? What is the standard deviation of this distribution?
b. Find the probability that the sample difference $(\hat{p}_1 - \hat{p}_2)$ lies within .01 of the true population difference $(p_1 - p_2)$.
c. If there is actually no difference in the population proportions of men and women who suffer from allergies, and in fact, $p_1 = p_2 = .30$, what is the probability of observing a difference between men and women as large as .10 in absolute value? If the sample proportions did differ by .10, what might you suspect?

6.44 An important source of data for librarians, market research analysts, consultants, and journalists is the *Statistical Abstract of the United States*, compiled yearly by government statisticians. An article in *Time* magazine (May 22, 1989, p. 36) quotes the *Abstract* using a bar graph, which shows 51% of American households having cable television in 1988, up from 20% in 1980.
a. Suppose that 500 randomly selected households were surveyed in 1980 and another 500 in 1988. If the figures given in the *Statistical Abstract* are correct, what is the sampling distribution of the difference in sample proportions, $(\hat{p}_1 - \hat{p}_2)$, from 1980 to 1988?
b. What is the probability of observing a sample difference within .04 of the true difference in population proportions?

6.7 SUMMARY

In a practical sampling situation, we will draw a *single* random sample of n observations from a population, calculate a single value of a sample statistic, and use it to make an inference about a population parameter. But to interpret the statistic, to know how close to the population parameter the computed statistic might be expected to fall, we need to observe the behavior of the statistic in repeated sampling. Thus, if we were to repeat the sampling process over and over again an infinitely large number of times, the distribution of values of the statistic produced by this enormous Monte Carlo experiment would be the sampling (or probability) distribution of the statistic.

This chapter described the properties of the sampling distributions for several useful statistics that we will employ in the following chapters to make inferences about population parameters. First, sample means, proportions, and the differences between a pair of means or proportions have sampling distributions that can be approximated by a normal distribution when the sample sizes are large. Second, these distributions are centered over their respective population parameters. Thus, the mean of the sampling distribution of the sample mean $\bar{x}$ is the population mean μ, the mean of the sampling distribution of the sample proportion $\hat{p}$ is the population proportion p, and so on. Third, the spread of the distributions, measured by their standard deviations, decreases as the sample size increases. As we will see in Chapter 7, this third characteristic is important when we wish to use a sample statistic to estimate its corresponding population parameter. By choosing a larger sample size, we can increase the probability that a sample statistic will fall close to the population parameter.

REFERENCES

Freund, J. E., and Walpole, R. E. *Mathematical Statistics*. 4th ed. Englewood Cliffs, N.J.: Prentice-Hall, 1987.

Hogg, R. V., and Craig, A. T. *Introduction to Mathematical Statistics*. 4th ed. New York: Macmillan, 1986.

Koopmans, L. H. *An Introduction to Contemporary Statistics*. 2d ed. Boston: Duxbury Press, 1987.

Mendenhall, W., and Sincich, T. *Statistics for Engineering and Computer Sciences*. San Francisco: Dellen, 1984.

Mood, A. M., et al. *Introduction to the Theory of Statistics*. 3d ed. New York: McGraw-Hill, 1973.

Ryan, T. A.; Joiner, B. L.; and Ryan, B. F. *Minitab Student Handbook*. 2d ed. Boston: Duxbury Press, 1985.

SAS Institute, Inc. Staff. *SAS User's Guide: Statistics*. Version 5 ed. Cary, N. C.: SAS Institute, Inc., 1985.

SPSS, Inc. Staff. *SPSS-X User's Guide*. 3d ed. New York: McGraw-Hill, 1987.

Weiss, N. A. *Elementary Statistics*. Reading, Mass.: Addison-Wesley, 1989.

SUPPLEMENTARY EXERCISES

[Starred (*) exercises are optional]

6.45   Review the die-tossing experiment in Section 6.3, where we simulated the selection of samples of $n = 5$ observations and obtained an approximation to the sampling distribution for the sample mean. Repeat this experiment, selecting 200 samples of size $n = 3$.

a. Construct the sampling distribution for $\bar{x}$. Note that the sampling distribution of $\bar{x}$ for $n = 3$ does not achieve the bell shape that you observed for $n = 5$ (see Figure 6.4.)

b. The mean and standard deviation of the probability distribution for x, the number of dots that appear when a single die is tossed, are $\mu = 3.5$ and $\sigma = 1.71$. What are the theoretical values of the mean and standard deviation of the sampling distribution of $\bar{x}$ based on samples of $n = 3$?

c. Calculate the mean and standard deviation of the simulated sampling distribution in part (a). Are these values close to the corresponding values obtained for part (b)?

6.46 Refer to the sampling experiment in Exercise 6.45. Calculate the median for each of the 200 samples of size $n = 3$.

a. Use the 200 medians to construct a relative frequency histogram that approximates the sampling distribution of the sample median.

b. Calculate the mean and standard deviation of the sampling distribution in part (a).

c. Compare the mean and standard deviation of this sampling distribution with the mean and standard deviation calculated for the sampling distribution of $\bar{x}$, in Exercise 6.45(b). Which statistic, the sample mean or the sample median, appears to fall closer to μ?

6.47 Independent random samples of $n_1 = 10$ and $n_2 = 8$ observations were randomly selected from populations with means and variances $(\mu_1 = 4, \sigma_1^2 = 6)$ and $(\mu_2 = -3, \sigma_2^2 = 12)$. Find the mean and standard deviations of

a. $\bar{x}_1 - \bar{x}_2$
b. $\bar{x}_1 + \bar{x}_2$

6.48 Refer to Exercise 6.47. Let S_1 and S_2 represent the sum of the observations in samples 1 and 2. Find the mean and standard deviation of the sampling distribution of

a. $S_1 - S_2$
b. $S_1 + S_2$

6.49 If random samples of $n_1 = 400$ and $n_2 = 800$ observations are selected from binomial populations $p_1 = .3$ and $p_2 = .4$, what is the probability that the sample proportions will differ by less than .15?

6.50 Exercise 5.18 described part of a study of water samples taken from the Boston water supply system (P. C. Karalekas, Jr., C. R. Ryan, and F. B. Taylor, "Control of Lead, Copper, and Iron Pipe Corrosion in Boston," *American Water Works Journal* [February 1983]). The researchers determined lead-level readings in drinking water for each of 23 days in 1977. Each daily reading was the average of the lead-level readings for a water specimen collected at each of 40 locations. The mean and standard deviation of the 23 daily lead-level readings were 0.033 and 0.016, respectively. The information given in Exercise 5.18 suggests that the distribution of the individual lead-level readings taken at each of the 40 locations in the piping system is highly skewed toward large values of lead concentration. What can you say about the distribution of the daily lead levels from which the sample of 23 days was selected?

6.51 From each of two normal populations with identical means and with standard deviations of 6.40 and 7.20, independent random samples of 64 observations are drawn. Find the probability that the difference between the means of the samples exceeds .60 in absolute value.

6.52 A survey on the nutritional habits of Americans found that in a sample of 1678 adults who had eaten dinner the night before, 71% had a dinner that was prepared at home, while 4% had a commercially prepared frozen dinner as the main course (*New York Times*, February 24, 1988). Suppose that you drew another random sample of 1678 adults who had eaten dinner last night.

a. If the proportion of adults who eat dinners prepared at home on any given night is .7, what is the probability that your sample percentage of adults who had a dinner that was prepared at home differs by as much as 4% from that obtained in the *New York Times survey*?

b.   If the sample percentages differ by as much as 10%, what might you suspect?

6.53   A finite population consists of the following four elements:

6, 1, 3, 2

a.   How many different samples of size $n = 2$ can be selected from this population if we sample without replacement? (Sampling is said to be *without replacement* if an element cannot be selected twice for the same sample.)

b.   List the possible samples of size $n = 2$.

c.   Compute the sample mean for each of the samples given in part (b).

d.   Find the sampling distribution of $\bar{x}$. Use a probability histogram to graph the sampling distribution of $\bar{x}$.

e.   If all four population values are equally likely, calculate the value of the population mean μ. Do any of the samples listed in part (b) produce a value of $\bar{x}$ exactly equal to μ?

6.54   Refer to Exercise 6.53. Find the sampling distribution for $\bar{x}$ if random samples of size $n = 3$ are selected *without replacement*. Graph the sampling distribution of $\bar{x}$.

6.55   The Central Intelligence Agency (CIA) has compiled some statistics involving the average earnings of citizens in communist and noncommunist countries (*Wall Street Journal*, November 7, 1989). The agency's "Handbook of Economic Statistics 1989" states that the average Soviet citizen earned the equivalent of $8850 last year, compared to $19,970 for the average American citizen.

a.   How would you describe the distributions of earnings in the United States and the Soviet Union?

b.   Under what conditions would the average earnings of American or Soviet citizens based on a random sample of size n be approximately normal?

6.56   The total amount of vegetation held by the earth's forests is important to both ecologists and politicians. Since green plants absorb carbon dioxide, an underestimate of the earth's vegetative mass, or biomass, means that much of the carbon dioxide emitted by human activities (primarily fossil-burning fuels) will not be absorbed, and a climate-altering buildup of carbon dioxide will occur. New studies indicate that the biomass for tropical woodlands, thought to be about 35 kilograms per square meter (kg/m^2), may in fact be too high, and that tropical biomass values vary regionally—from about 5 to 55 kg/m^2 (*Science News*, August 19, 1989, p. 124). Suppose that you measure the tropical biomass in 400 randomly selected square meter plots.

a.   Approximate σ, the standard deviation of the biomass measurements.

b.   What is the probability that your sample average is within two units of the true average tropical biomass?

c.   If your sample average is $\bar{x} = 31.75$, what would you conclude about the overestimation about which the scientists are concerned?

6.57   The safety requirements for hard hats worn by construction workers and others, established by the American National Standards Institute (ANSI), specifies that each of three hats pass the following test (*Wall Street Journal*, November 18, 1977). A hat is mounted on an aluminum head form. An 8-pound steel ball is dropped on the hat from a height of 5 feet, and the resulting force is measured at the bottom of the head form. The force exerted on the head form by each of the three hats must be less than 1000 pounds, and the average of the three must be less than 850 pounds. (The relationship between this test and actual human head damage is unknown.) Suppose that the exerted force is normally distributed and hence a sample mean of three force measurements is normally distributed. If a random sample of three hats is selected from a shipment with a mean equal to 900 and $\sigma = 100$, what is the probability that the sample mean will satisfy the ANSI standard?

6.58 The maximum load (with a generous safety factor) for the elevator in an office building is 2000 pounds. The relative frequency distribution of the weights of all of the men and women using the elevator is mound-shaped (slightly skewed to the heavy weights) with mean μ equal to 150 pounds and standard deviation σ equal to 35 pounds. What is the largest number of people you can allow on the elevator if you want their total weight to exceed the maximum weight with a small probability (say near .01)? (*Hint:* If $x_1, x_2, \ldots, x_n$ are independent observations made on a random variable x, and if x has a mean μ and variance σ^2, then the mean and variance of $\sum_{i=1}^{n} x_i$ are $n\mu$ and $n\sigma^2$, respectively. This result was given in Section 6.3.)

6.59 The number of wiring packages that can be assembled by a company's employees has a normal distribution with a mean equal to 16.4 per hour and a standard deviation of 1.3 per hour.
  a. What is the mean and standard deviation of the number x of packages produced per worker in an 8-hour day?
  b. Would you expect the probability distribution for x to be mound-shaped and approximately normal? Explain.
  c. What is the probability that a worker will produce at least 135 packages per 8-hour day?

6.60 Refer to Exercise 6.59. Suppose the company employs ten assemblers of wiring packages.
  a. Find the mean and standard deviations of the company's daily (8-hour day) production of wiring packages.
  b. What is the probability that the company's daily production would be less than 1280 wiring packages per day?

6.61 A survey of 800 companies reported in *USA Today* (July 6, 1989) indicated that in 1988, 19% of the companies provided health club memberships as a benefit to their executives.
  a. Describe the sampling distribution of $\hat{p}$, the proportion of the sampled companies providing health club memberships. (*Hint:* Use $\hat{p}$ to estimate p in calculating $\sigma_{\hat{p}}$.)
  b. Find the probability that $\hat{p}$ lies within 2% of the true population proportion.

6.62 A new backup test for AIDS antibodies in the blood has been designed by SmithKline Bio-Science Laboratories to confirm results of initial screening procedures currently in use. The maker claims that its test is inconclusive less frequently than the Western blot-confirming test currently in use (*New York Times*, February 26, 1988). According to Dr. John Mills of San Francisco General Hospital, "In tests of more than 2500 blood samples, the new test was inconclusive in 2.5% to 4.2% of the patients, as against 15.6% to 18% inconclusive results for the Western blot assay." Suppose we assume that 2500 blood samples were tested using each of the two methods (SmithKline and Western blot), with inconclusive results in 4.2% and 18% of the samples, respectively.
  a. Describe the sampling distribution of the difference in the inconclusive percentages for the two methods.
  b. What is the approximate probability that the sample difference will lie within 5% of the true difference?
  c. Using the results of part (b), what would you conclude about the reliability of the two backup tests for AIDS antibodies?

6.63 The average per capita fresh apple consumption in the United States during 1970 and 1987 are shown in the following table, along with estimates of the standard deviation of consumption. Samples of $n_1 = n_2 = 100$ people were taken in each of the two years.

1987	1970
$\mu_1 = 20.3$ lb	$\mu_2 = 16.2$ lb
$\sigma_1 \approx 5$ lb	$\sigma_2 \approx 4$ lb

Source: Data based on
information from *Time* magazine,
May 22, 1989, p. 36.

a.   What is the probability that $(\bar{x}_1 - \bar{x}_2)$ exceeds 5 pounds?
b.   What is the probability that $(\bar{x}_1 - \bar{x}_2)$ is less than 0?
c.   If the sample average for 1970 exceeded the sample average for 1987, what conclusions might you draw?

*6.64   If you have access to a computer and a computer program that generates random numbers, you can simulate sampling from a population that has a uniform probability distribution. The numbers produced by a random-number generator are independent of one another, and the probability of observing any one number is the same as the probability of observing any other. Program the computer to generate a large number, say 1000, of samples of $n = 2$ observations and calculate the sample mean for each. Use a computer program to arrange these 1000 sample means in a relative frequency histogram. The resulting histogram will provide a good approximation to the sampling distribution of $\bar{x}$ for samples selected from a population having a uniform relative frequency distribution. It should be similar to the sampling distribution shown in the third column of Figure 6.5 for $n = 2$.

*6.65   Repeat Exercise 6.64 for $n = 5$, 10, and 25. Compare with the corresponding sampling distributions shown in Figure 6.5.

*6.66   Repeat Exercise 6.64 for $n = 100$. Compare with the sampling distributions for $n = 2, 5, 10$, and 25.

*6.67   We stated in Section 6.6 that if Y_1 and Y_2 were two independent statistics (or any random variables) with means and variances $(\mu_{Y_1}, \sigma_{Y_1}^2)$ and $(\mu_{Y_2}, \sigma_{Y_2}^2)$, respectively, then the mean and variance of the sum $(Y_1 + Y_2)$ are

$$\mu_{Y_1 + Y_2} = \mu_{Y_1} + \mu_{Y_2}$$

and

$$\sigma_{Y_1 + Y_2}^2 = \sigma_{Y_1}^2 + \sigma_{Y_2}^2$$

Suppose that Y_3 is a third statistic, independent of Y_1 and Y_2, with mean and variance $(\mu_{Y_3}, \sigma_{Y_3}^2)$. Use the results above to prove that the mean and variance of the sum of the three random variables $S = (Y_1 + Y_2 + Y_3)$ are

$$\mu_S = \mu_{Y_1} + \mu_{Y_2} + \mu_{Y_3}$$
$$\sigma_S^2 = \sigma_{Y_1}^2 + \sigma_{Y2}^2 + \sigma_{Y3}^2$$

(*Note:* The technique used for this proof can also be used to show that the results of Section 6.6 can be extended to the sum or difference of any number of *independent* random variables.)

SAMPLING
THE ROULETTE
AT MONTE CARLO

The technique of simulating a process that contains random elements and repeating the process over and over to see how it behaves is called a **Monte Carlo procedure**. It is widely used in business and other fields to investigate the properties of an operation that is subject to a number of random effects such as weather, human behavior, and so on. For example, you could model the behavior of a manufacturing company's inventory by creating, on paper, daily arrivals and departures of manufactured products from the company's warehouse. Each day a random number of items produced by the company would be received into inventory. Similarly, each day a random number of orders of varying random sizes would be shipped. Based on the input and output of items, you could calculate the inventory, that is, the number of items on hand at the end of each day. The values of the random variables, the number of items produced, the number of orders, and the number of items per order needed for each day's simulation would be obtained from theoretical distributions of observations that closely model the corresponding distributions of the variables that have been observed over time in the manufacturing operation. By repeating the simulation of the supply, the shipping, and the calculation of daily inventory for a large number of days (a sampling of what might really happen), you can observe the behavior of the plant's daily inventory. The Monte Carlo procedure is particularly valuable because it enables the manufacturer to see how the daily inventory would behave when certain changes are made in the supply pattern or in some other aspect of the operation that could be controlled.

In an article entitled "The Road to Monte Carlo" (*Fortune* magazine, April 15, 1985), Daniel Seligman comments on the Monte Carlo method, noting that although the technique is widely used in business schools to study capital budgeting, inventory planning, and cash flow management, no one seems to have used the procedure to study how well we might do if we were to gamble at Monte Carlo.

To follow up on this thought, Seligman programmed his personal computer to simulate the game of roulette. Roulette consists of a wheel whose rim is divided into 38 pockets. Thirty-six of the pockets are numbered 1 to 36 and are alternately colored red and black. The two remaining pockets are colored green and are marked 0 and 00. To play the game, you bet a certain amount of money on one or more pockets. The wheel is spun and turns until it stops. A ball falls into a slot on the wheel to indicate the winning number. If you have money on that number, you win a specified amount. For example, if you were to play the number 20, the payoff is 35 to 1. If the wheel does not stop at that number, you lose your bet. Seligman decided to see how his nightly gains (or losses) would fare if he were to bet $5 on each turn of the wheel and to repeat the process 200 times each night. He did this 365 times, thereby simulating the outcomes of 365 nights at the casino. Not surprisingly, the mean "gain" per $1000 evening for the 365 nights was a *loss* of $55, the average of the winnings retained by the gambling house. The surprise, according to Seligman, was the extreme variability of the nightly "winnings." Seven times out of the 365 evenings, the fictitious gambler lost the $1000 stake, and only once did he win a maximum of $1160. On 141 nights the loss exceeded $250.

1. To evaluate the results of Seligman's Monte Carlo experiment, first find the probability distribution of the gain x on a single $5 bet.

2. Find the expected value and variance of the gain x from part 1.

3. Find the expected value and variance for the evening's gain, the sum of the gains or losses for the 200 bets of $5 each.

4. Use the results of part 2 to evaluate the probability of 7 out of 365 evenings resulting in a loss of the total $1000 stake.

5. Use the results of part 3 to evaluate the probability that the largest evening's winnings were as large as $1160.

CHAPTER 7

▷ ► LARGE-SAMPLE ESTIMATION

Case Study

Do the national polls conducted by the Gallup and Harris organizations, the news media, and others really provide accurate estimates of the percentages of people in the United States who favor various propositions? The case study at the end of this chapter examines the reliability of a poll conducted by the *New York Times* using the theory of large-sample estimation.

General Objective

In previous chapters, we focused attention on the probability distributions of random variables and on the sampling distributions of several statistics that, for large sample sizes, can be approximated by a normal distribution according to the Central Limit Theorem. In this chapter, we present a method for estimating population parameters and illustrate the concept with practical examples. The Central Limit Theorem and the sampling distributions presented in Chapter 6 will play a key role in evaluating the reliability of the estimates.

Specific Topics

1 Types of estimators (7.3)
2 Evaluating the goodness of an estimator (7.4)
3 General formulas for large-sample estimation (7.5)
4 Estimation of a population mean (7.6)
5 Estimating the difference between two means (7.7)
6 Estimating the parameter of a binomial population (7.8)
7 Estimating the difference between two binomial parameters (7.9)
8 Choosing the sample size (7.10)

▷ 7.1 A BRIEF SUMMARY

The preceding six chapters set the stage for the objective of this text, namely, developing an understanding of statistical inference and how it can be applied to the solution of practical problems. In Chapter 1 we stated that statisticians are concerned with making inferences about populations of measurements based on information contained in samples. We showed you how to phrase an inference — that is, how to describe a set of measurements — in Chapter 2. In Chapter 3, we discussed probability, the mechanism for making inferences, and we followed that with a general discussion of probability distributions. Three useful discrete probability distributions were presented in Chapter 4 — the binomial, the Poisson, and the hypergeometric — and three continuous probability distributions, the uniform, exponential, and normal distributions, were presented in Chapter 5.

In Chapter 6 we noted that statistics, computed from the sample measurements, are used to make inferences about population parameters, and we found an important use for the normal probability distribution of Chapter 5. In particular, you learned that some of the most important statistics — sample means and proportions — have sampling distributions that can be approximated by a normal distribution when the sample sizes are large owing to the Central Limit Theorem. These statistics will be used to make inferences about population parameters, and their sampling distributions will provide a means of assessing the reliability of these inferences.

▷ 7.2 INFERENCE: THE OBJECTIVE OF STATISTICS

Inference, specifically decision making and prediction, is centuries old and plays a very important role in our individual lives. Each of us is faced daily with personal decisions and situations that require predictions concerning the future. The government is concerned with predicting short-term and long-term interest rates. The broker would like to forecast the behavior of the stock market. The metallurgist wishes to use the results of an experiment to infer whether a new type of steel is more resistant to temperature changes than another. The consumer wants to know whether detergent A is more effective than detergent B. Hopefully, these inferences will be based on relevant measurements, which are called **observations** or **data.**

In many practical situations the relevant information is abundant, seemingly inconsistent, and, in many respects, overwhelming. As a result, a carefully considered decision or prediction is often little better than an outright guess. You need only refer to the "Market Views" section of the *Wall Street Journal* to observe the diversity of expert opinions concerning future stock market behavior. Similarly, a visual analysis of data by scientists and engineers often yields conflicting opinions regarding conclusions to be drawn from an experiment. Although many individuals tend to feel that their own built-in inference-making equipment is quite good, experience suggests that this may not be the case. It is the job of the mathematical statistician to provide inference-making techniques that are better than subjective guesses.

The objective of statistics is to make inferences about a population based on

information contained in a sample. Since populations are characterized by numerical descriptive measures called **parameters**, statistical inference is concerned with making inferences about population parameters. Typical population parameters are the mean, the standard deviation, the area under the probability distribution above or below some value of the random variable, and the area between two values of the variable. Indeed, all the practical problems mentioned in the first paragraph of this section can be restated in the framework of a population with a specified parameter of interest.

Methods for making inferences about parameters fall into one of two categories. We may **make decisions** concerning the value of the parameter, or we may **estimate** or **predict** the value of the parameter. For example, the circuits in computers and other electronic instruments consist of one or more printed circuit boards (PCBs); computer repairs often consist of simply replacing one or more defective PCBs. In an attempt to determine the proper setting of a plating process applied to one side of a PCB, a production supervisor might *estimate* the thickness of copper plating on printed circuit boards using samples from several days of operation. However, if the supervisor were told by the plant owner that the thickness of the copper plating must not be less than .001 inch in order for the process to be in control, the supervisor might want to *decide* whether or not the average thickness of the copper plating is less than .001 inch. While some statisticians view estimation as a decision-making problem, we will retain the two categories and concentrate separately on estimation and decision-making using tests of hypotheses.

Which method of inference should be used; that is, should the parameter be estimated or should we test a hypothesis concerning its value? The answer to this question is dictated by the practical question posed and is often determined by personal preference. Some people like to test theories concerning parameters; others prefer to express their inference as an estimate. Inasmuch as both estimation and tests of hypotheses are used frequently in scientific literature, we will include both methods in our discussion.

A statistical problem, which involves planning, analysis, and inference making, would be incomplete without reference to a **measure of the goodness** of the inferential procedures. We may define numerous objective methods for making inferences in addition to our own individual procedures based on intuition. Therefore, a measure of goodness must be defined in such a way that one procedure may be compared with another. More than that, we wish to state the goodness of a particular inference in a given practical situation. Thus, to predict that the price of a stock will be $80 next Monday would be insufficient and would stimulate few of us to take action to buy or sell. We would also want to know whether the estimate was correct to within plus or minus $1, $2, or $10. **To summarize, statistical inference contains two elements: (1) the inference and (2) a measure of its goodness.**

7.3 TYPES OF ESTIMATORS

Estimation procedures can be divided into two types: point estimation and interval estimation. Suppose that we wish to estimate the mean weight gain per month of four-month-old golden retriever pups that have been placed on a particular diet. The estimate might be given as a single number—for instance, 3.8 pounds—or we might estimate that the weight gain would fall in an interval such as 2.7 to 4.9 pounds. The first type of estimate is called a **point estimate** because the single number representing the estimate may be associated with a point on a line. The second type, involving two points and defining an interval on a line, is called an **interval estimate**. We will consider each of these methods of estimation.

In order to construct either a point or an interval estimate, we use information from the sample in the form of an estimator. Estimators are functions of sample observations, and hence, by definition, are also **statistics**.

Definition

> An **estimator** is a rule that tells us how to calculate the estimate based on information in the sample and that is generally expressed as a formula.

For example, the sample mean

$$\bar{x} = \frac{\sum_{i=1}^{n} x_i}{n}$$

is an estimator of the population mean μ and explains exactly how the actual numerical value of the estimate can be obtained once the sample values $x_1, x_2, \ldots, x_n$ are known. The sample mean can be used to arrive at a single number to estimate μ, or to construct an interval, two points that are intended to enclose the true value of μ.

Definition

> A **point estimator** of a population parameter is a rule that tells us how to calculate a single number based on sample data. The resulting number is called a **point estimate**.

Definition

> An **interval estimator** of a population parameter is a rule that tells us how to calculate two numbers based on sample data, forming an interval within which the parameter is expected to lie. This pair of numbers is called an **interval estimate** or **confidence interval**.

Both point and interval estimation procedures are developed using the sampling distribution of the best estimator of a specified population parameter. How do we decide which among several estimators of a specified population parameter is best? We will address this question in Section 7.4.

▷ 7.4 EVALUATING THE GOODNESS OF AN ESTIMATOR

The goodness of an estimator is evaluated by observing its behavior in repeated sampling. Let us consider the following analogy. In many respects, point estimation is similar to firing a revolver at a target. The estimator, which generates estimates, is analogous to the revolver; a particular estimate is analogous to the bullet; and the parameter of interest is analogous to the bull's-eye. Drawing a sample from the population and estimating the value of the parameter is equivalent to firing a single shot at the target.

Suppose a man fires a single shot at a target and the shot pierces the bull's-eye. Do we conclude that he is an excellent shot? The answer is no, because not one of us would consent to hold the target while a second shot is fired. On the other hand, if one million shots in succession hit the bull's-eye, we might acquire sufficient confidence in the marksman to hold the target for the next shot, if the compensation were adequate. The point we wish to make is that we cannot evaluate the goodness of an estimation procedure on the basis of a single estimate. Rather, we must observe the results when the estimation procedure is used over and over again, many, many times; then we observe how closely the shots are distributed about the bull's-eye. In fact, since the estimates are numbers, we would evaluate the goodness of the estimator by constructing a frequency distribution of the estimates obtained in repeated sampling and noting how closely the distribution centers about the parameter of interest. This relative frequency distribution would be the sampling distribution of the estimator.

As an illustration, consider the die-tossing experiment used in Chapter 6, where we generated 100 samples of $n = 5$ measurements each and calculated the mean for each sample. Since we know the mean value μ of the number showing on a die toss ($\mu = 3.5$), we can use the results of the die-tossing experiment to see how well the mean of a sample of $n = 5$ measurements estimates μ.

The frequency histogram of the 100 sample means (shown in Figure 7.1) is an approximation to the sampling distribution of the mean $\bar{x}$ of a sample of five observations, one for each die toss. Notice how the estimates group about the population mean $\mu = 3.5$, and that they range from 1.4 to 5.4. Surely this

Figure 7.1
Histogram of sample means for the die-tossing experiment in Section 6.3

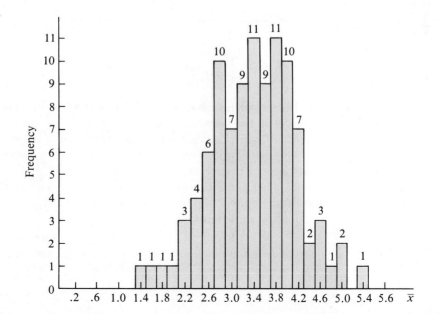

distribution of estimates tells us something about how good a new estimate of μ would be if we were to draw one more sample of $n = 5$ measurements and compute the sample mean $\bar{x}$.

Suppose we let $\hat{\theta}$ denote an estimator of the population parameter θ (μ, σ, or any parameter). What properties should $\hat{\theta}$ have as a desirable estimator? Essentially there are two, and they can be seen by observing the possible types of sampling distributions for $\hat{\theta}$ given in Figures 7.2 and 7.3. First we would like the sampling distribution to be centered over the parameter of interest. *Thus, we would like the mean of the sampling distribution to equal θ.* Such an estimator is said to be **unbiased**.

Definition

If $\hat{\theta}$ is an estimator of a parameter θ and if the mean of the distribution of $\hat{\theta}$ is θ, that is,

$$E(\hat{\theta}) = \theta$$

then $\hat{\theta}$ is said to be **unbiased**. Otherwise, $\hat{\theta}$ is said to be **biased**.

The sampling distributions for an unbiased estimator and a biased estimator are shown in Figure 7.2(a) and (b). Note that the sampling distribution for the biased estimator in Figure 7.2(b) is shifted to the right of θ. This biased estimator is more likely to overestimate θ.

Figure 7.2
Distributions for
unbiased and biased
estimators

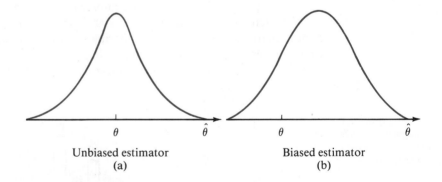

Unbiased estimator
(a)

Biased estimator
(b)

The second desirable property of an estimator is that the spread (measured by the variance) of the sampling distribution of the estimator be as small as possible. This ensures a high probability that an individual estimate will fall close to θ. The sampling distributions for two estimators, one with a small variance* and the other with a larger variance, are shown in Figure 7.3(a) and (b), respectively. Naturally, we would prefer the estimator with the smaller variance, the sampling distribution shown in Figure 7.3(a), because the estimates tend to lie closer to θ than in the distribution shown in Figure 7.3(b).

---

* Statisticians usually use the term *variance of an estimator*, when in fact they mean the variance of the sampling distribution of the estimator. This contractive expression is used almost universally.

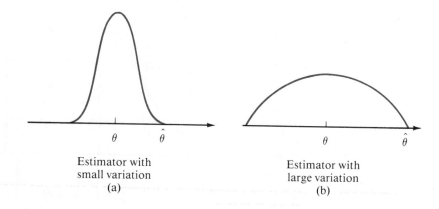

Figure 7.3
Comparison of estimator variability

Estimator with
small variation
(a)

Estimator with
large variation
(b)

In a real-life sampling situation you may know that the sampling distribution of an estimator centers about the parameter that you are attempting to estimate, but you may not know the value of the parameter. All that you have is the estimate computed from the n measurements contained in the sample. How far will your particular estimate be from the estimated parameter? Since the parameter usually lies in the center of the sampling distribution (it is usually the mean of the distribution), the distance between the estimate and the parameter, called the **error of estimation**, is less than or equal to the distance between the center and the tails of the distribution.

| Definition

The distance between an estimate and the estimated parameter is called the **error of estimation**.

We will assume in this chapter that the sample sizes are always large and, therefore, that the estimators we will study have sampling distributions that can be approximated by a normal distribution because of the Central Limit Theorem. Consequently, if we define the difference between a particular estimate $\hat{\theta}$ and the parameter θ it estimates as the **error of estimation**, we would expect the error of estimation to be less than $1.96\sigma_{\hat{\theta}}$, with probability approximately equal to .95 (see Figure 7.4). Thus, most estimates will be less than $1.96\sigma_{\hat{\theta}}$ away from θ, and this quantity provides a practical upper limit or **bound on the error of estimation**. It is possible (with probability .05) that the error of estimation will exceed this bound, but it is very unlikely.

The relative frequency distribution of the 100 values of $\bar{x}$ (Table 6.3), each calculated from a sample $n = 5$ die tosses, illustrates this concept. As noted in Chapter 6, the mean and standard deviation of the population of die tosses is $\mu = 3.5$ and $\sigma = 1.71$. Therefore,

$$1.96\sigma_{\bar{x}} = 1.96 \frac{\sigma}{\sqrt{n}} = 1.96 \frac{1.71}{\sqrt{5}} = 1.50$$

Figure 7.4
Sampling distribution of
$\hat{\theta}$ for large n

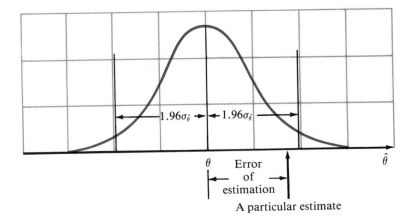

Figure 7.4
Sampling distribution of
$\hat{\theta}$ for large n

Even though this sample size is small and the sampling distribution of $\bar{x}$ is not exactly normally distributed, you can see in Figure 7.5 that all but 5 of the sample means fall within 1.50 of the mean $\mu = 3.5$, that is, in the interval 2.0 to 5.0.

The **goodness of an interval estimator** is analyzed in much the same way as that of a point estimator. Samples of the same size as repeatedly drawn from the population, and the interval estimate is calculated on each occasion. This process will generate a large number of intervals rather than points. **A good interval estimator would successfully enclose the true value of the parameter a large fraction of the time.** Constructing an interval estimate is like attempting to throw a lariat around a fence post. In this case the parameter that you wish to estimate cor-

Figure 7.5
Histogram of sample
means for the die-tossing
experiment in Section 6.3

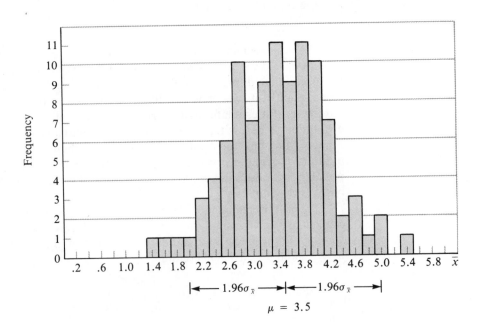

responds to the post and the interval corresponds to the loop formed by the cowboy's lariat. Each time you draw a sample, you construct a confidence interval for a parameter and you hope to "rope it," that is, include it in the interval. But you will not be successful for every sample. The "success rate" is referred to as the **confidence coefficient**.

Definition

> The probability that a confidence interval will enclose the estimated parameter is called the **confidence coefficient**.

To consider a practical example, suppose we want to estimate the mean number of bacteria per cubic centimeter in a polluted stream. If we were to draw ten samples, each containing $n = 30$ observations, and construct a confidence interval for the population mean μ for each sample, the intervals might appear as shown in Figure 7.6. The horizontal line segments represent the ten intervals, and the vertical line represents the location of the true mean number of bacteria per cubic centimeter. Note that the parameter is fixed and that the interval location and width may vary from sample to sample. Thus we speak of "the probability that the interval encloses μ," not "the probability that μ falls in the interval," because μ is fixed. The *interval* is random.

Figure 7.6
Ten confidence intervals for the mean number of bacteria per cubic centimeter (each based on a sample of $n = 30$ observations)

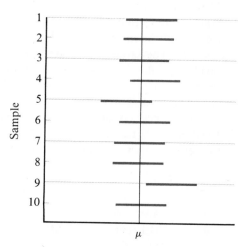

A good confidence interval is one that is as narrow as possible and has a large confidence coefficient, near 1. The narrower the interval, the more exactly we have located the estimated parameter. The larger the confidence coefficient, the more confidence we have that a particular interval encloses the estimated parameter. Remember that the confidence coefficient gives the probability that the interval estimator will produce confidence limits that enclose the estimated parameter. It gives you a measure of the confidence you can place in the confidence limits constructed from the data contained in a sample. In that sense the width of an interval and its associated confidence coefficient measure the goodness of the confidence interval.

What is the effect of larger samples on the width of a confidence interval? Larger samples provide more information to use in forming the interval estimate. Therefore, for a given confidence coefficient, the larger the sample, the narrower will be the resulting confidence interval.

The selection of a "best" estimator—the proper formula to use in calculating the point estimate with bounds on error or the confidence interval estimate—is the task of the theoretical statistician and is beyond the scope of this text. Throughout the remainder of this and succeeding chapters, populations and parameters of interest will be defined and the appropriate estimator indicated along with its mean and standard deviation.

▷ 7.5 ESTIMATION FROM LARGE SAMPLES

The estimation procedures discussed in Sections 7.3 and 7.4 set the stage for the practical estimation problems to be discussed in the remainder of this chapter. A thread of unity runs through all, and, once observed, will simplify your learning process.

In this chapter, we assume that $\hat{\theta}$ is an **unbiased** estimator of a population parameter θ, so that θ is the mean of the sampling distribution of $\hat{\theta}$. The standard deviation of the sampling distribution of the estimator is known and given as $\sigma_{\hat{\theta}}$. **If the standard deviation is not known, we assume that the sample size is large enough to provide a good estimate of any unknown parameters** (for example, σ). In each case, the estimator $\hat{\theta}$ is normally distributed, or approximately so due to the Central Limit Theorem.

Under these conditions, the **point estimator of θ is $\hat{\theta}$**, and the **bound on the error of estimation** is $1.96\sigma_{\hat{\theta}}$. The probability that the error of estimation, the difference between the observed value of $\hat{\theta}$ and θ, is less than the bound $1.96\sigma_{\hat{\theta}}$ is .95.

POINT ESTIMATOR FOR θ

Point Estimator: $\hat{\theta}$
Bound on the Error of Estimation: $1.96\sigma_{\hat{\theta}}$

A large-sample interval estimator, or **large-sample confidence interval**, for a population parameter θ can also be obtained when $\hat{\theta}$ is normally distributed or approximately so due to the Central Limit Theorem. Under the conditions stated above, 95% of the point estimates will lie within $1.96\sigma_{\hat{\theta}}$ of the mean θ. In constructing an interval estimate by measuring $1.96\sigma_{\hat{\theta}}$ on either side of $\hat{\theta}$, 95% of the intervals constructed in this manner will enclose the unknown parameter θ (see Figure 7.7). Similarly, 90% of the intervals constructed by measuring $1.645\sigma_{\hat{\theta}}$ on either side of $\hat{\theta}$ will enclose the unknown parameter θ. In general, for a confidence

Figure 7.7
95% confidence limits
for θ

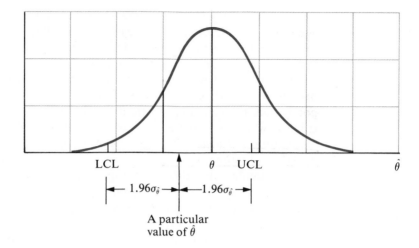

coefficient of $(1 - \alpha)$, a $(1 - \alpha)$ 100% confidence interval estimate for θ is given in the display.

A $(1 - \alpha)$ 100% LARGE-SAMPLE CONFIDENCE INTERVAL FOR θ

$$\hat{\theta} \pm z_{\alpha/2}\sigma_{\hat{\theta}}$$

where $z_{\alpha/2}$ is the z value corresponding to an area $\alpha/2$ in the upper tail of a standard normal distribution.

$\hat{\theta} + z_{\alpha/2}\sigma_{\hat{\theta}}$ is called the **upper confidence limit (UCL)** and $\hat{\theta} - z_{\alpha/2}\sigma_{\hat{\theta}}$ is called the **lower confidence limit (LCL)**.

We will see how these general formulas for point and interval estimation apply for the situations described in the following sections.

▷ 7.6 ESTIMATION OF A POPULATION MEAN

Practical problems very often lead to the estimation of a population mean μ. We may be concerned with the average achievement of college students in a particular university, in the average strength of a new type of steel, in the average number of deaths per capita in a given social class, or in the average demand for a new product. The estimation of μ serves as a very practical application of statistical inference as well as an excellent illustration of the principles of estimation discussed in Section 7.5. Many estimators are available for estimating the population mean μ, including the sample median, the average between the largest and smallest measurements in the sample, and the sample mean $\bar{x}$. Each would have a sampling distribution and, depending on the population and practical problem involved, certain advantages and disadvantages. Although the sample median and the average of the sample extremes are easier to calculate, the sample mean $\bar{x}$ is usually superior in that, for some populations, its variance is a minimum and that, regardless of the population, it is always unbiased.

In Section 6.4 we noted that the sampling distribution of the sample mean had three properties. Regardless of the probability distribution of the sampled population, the sampling distribution of $\bar{x}$ will be approximately normally distributed when the sample size n is reasonably large. The mean of the sampling distribution will always be μ, the mean of the sampled population; therefore, $\bar{x}$ is an unbiased estimator of μ.

The standard deviation of the sampling distribution of $\bar{x}$ is $\sigma_{\bar{x}} = \sigma/\sqrt{n}$. The fact that this standard deviation is proportional to the population standard deviation σ and inversely proportional to the square root of the sample size n is intuitively reasonable. The more variable the population data, measured by σ, the more variable will be $\bar{x}$. On the other hand, more information will be available for estimating μ as n becomes larger. Therefore, the estimates should fall closer to μ, and $\sigma_{\bar{x}}$ should become smaller if you increase the sample size. The sampling distributions for $\bar{x}$ based on random samples of $n = 5$, $n = 20$, and $n = 80$ from a normal distribution are shown in Figure 7.8. Notice how these distributions center about μ and how the spread of the distributions decreases as n increases.

MEAN AND STANDARD DEVIATION OF $\bar{x}$

$$E(\bar{x}) = \mu$$

$$\sigma_{\bar{x}} = \sigma/\sqrt{n}$$

Figure 7.8
Sampling distributions for $\bar{x}$ based on random samples from a normal distribution, $n = 5$, 20, and 80

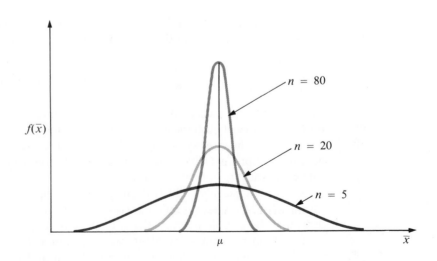

The estimator $\bar{x}$ satisfies all the conditions given in Section 7.5; that is, $\bar{x}$ is an **unbiased estimator** of the population mean μ regardless of the population sampled. **If the sampled population is normal, the $\bar{x}$ has a normal distribution for all sample sizes; furthermore, the distribution of $\bar{x}$ is approximately normal if the sampled population is not normal, but the sample size is large.**

POINT ESTIMATE OF μ

Point estimator: $\bar{x}$

Bound on the Error of Estimation: $1.96\sigma_{\bar{x}} = 1.96\sigma/\sqrt{n}$

If σ is unknown and n is 30 or larger, the sample standard deviation can be used to approximate σ.*

Assumption: $n \geq 30$

EXAMPLE 7.1 Suppose that we wish to estimate the average daily yield of a chemical manufactured in a chemical plant. The daily yield, recorded for $n = 50$ days, produced a mean and standard deviation equal to

$$\bar{x} = 871 \text{ tons}$$

$$s = 21 \text{ tons}$$

Estimate the average daily yield μ.

Solution The estimate of the daily yield is $\bar{x} = 871$ tons. The bound on the error of estimation is

$$1.96\sigma_{\bar{x}} = \frac{1.96\sigma}{\sqrt{n}} = \frac{1.96\sigma}{\sqrt{50}}$$

Although σ is unknown, the sample size is large and we may approximate the value of σ by using s. Thus, the bound on the error of estimation is approximately

$$1.96 \frac{s}{\sqrt{n}} = 1.96 \frac{(21)}{\sqrt{50}} = 5.82$$

We can feel fairly confident that our estimate of 871 tons is within 5.82 tons of the true average yield.

---

* When sampling a normal distribution, the statistic $(\bar{x} - \mu)/(s/\sqrt{n})$ has a t-distribution, which is discussed in Chapter 9. If the sampled population is not normal, but the sample size is *large*, then this statistic is approximately normally distributed.

Similarly, we can construct confidence intervals for μ corresponding to any desired confidence coefficient, say $(1 - \alpha)$, by using the following:

A $(1 - \alpha)$ 100% LARGE-SAMPLE CONFIDENCE INTERVAL FOR A POPULATION MEAN μ

$$\bar{x} \pm z_{\alpha/2} \frac{\sigma}{\sqrt{n}}$$

where $z_{\alpha/2}$ is the z value corresponding to an area $\alpha/2$ in the upper tail of a standard normal z distribution,

n = sample size

σ = standard deviation of the sampled population

If σ is unknown, it can be approximated by the sample standard deviation s when the sample size is large.

Assumption: $n \geq 30$

Figure 7.9
Location of $z_{\alpha/2}$

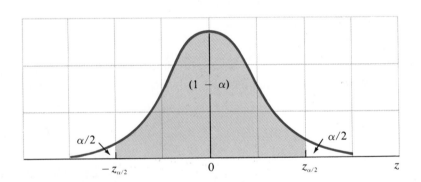

Table 7.1
Confidence limits for μ

Confidence coefficient, $(1 - \alpha)$	α	$z_{\alpha/2}$	LCL	UCL
.90	.10	1.645	$\bar{x} - 1.645 \dfrac{\sigma}{\sqrt{n}}$	$\bar{x} + 1.645 \dfrac{\sigma}{\sqrt{n}}$
.95	.05	1.96	$\bar{x} - 1.96 \dfrac{\sigma}{\sqrt{n}}$	$\bar{x} + 1.96 \dfrac{\sigma}{\sqrt{n}}$
.99	.01	2.58	$\bar{x} - 2.58 \dfrac{\sigma}{\sqrt{n}}$	$\bar{x} + 2.58 \dfrac{\sigma}{\sqrt{n}}$

The normal curve value $z_{\alpha/2}$, which appears in the formula for the confidence interval, is located as shown in Figure 7.9. For example, if you want a confidence coefficient $(1 - \alpha)$ equal to .95, then the tail-end area is $\alpha = .05$, and half of α (.025) is placed in each tail of the distribution. Then $z_{.025}$ is the table z value corresponding to an area of .475 to the right of the mean, or

$$z_{.025} = 1.96$$

Confidence limits corresponding to some of the commonly used confidence coefficients are shown in Table 7.1.

EXAMPLE 7.2   Find a 90% confidence interval for the population mean of Example 7.1. Recall that $\bar{x} = 871$ tons and $s = 21$ tons.

Solution   The 90% confidence limits are

$$\bar{x} \pm 1.645 \frac{\sigma}{\sqrt{n}}$$

Using s to estimate σ, we obtain

$$871 \pm (1.645) \frac{21}{\sqrt{50}} \quad \text{or} \quad 871 \pm 4.89$$

Therefore, we estimate that the average daily yield μ lies in the interval from 866.11 to 875.89 tons. The confidence coefficient .90 implies that in repeated sampling, 90% of the confidence intervals similarly formed would enclose μ.

The confidence interval of Example 7.2 is approximate because we substituted s as an approximation for σ. That is, instead of the confidence coefficient being .95, the value specified in the example, the true value of the coefficient may be .92, .94, or .97. But this discrepancy is of little concern from a practical point of view; as far as our "confidence" is concerned, there is little difference among these confidence coefficients. Most interval estimators employed in statistics yield approximate confidence intervals, because the assumptions upon which they are based are not satisfied exactly. Having made this point, we will not continue to refer to confidence intervals as "approximate." It is of little practical concern as long as the actual confidence coefficient is near the value specified.

Note in Table 7.1 that for a fixed sample size the width of the confidence interval increases as the confidence coefficient increases, a result that is in agreement with our intuition. Certainly, if we wish to be more confident that the interval will enclose μ, we would increase the width of the interval. Since we prefer narrow confidence intervals and large confidence coefficients, we must reach a compromise in choosing the confidence coefficient.

The choice of the confidence coefficient to be used in a given situation is made by the experimenter and depends on the degree of confidence the experimenter wishes to place in the estimate. Most confidence intervals are constructed by using one of the three confidence coefficients shown in Table 7.1. The most popular seem to be 95% confidence intervals. Use of 99% confidence intervals is less common because of the wider interval width that results. Of course, you can always decrease the width by increasing the sample size n.

Note the fine distinction between point estimators and interval estimators. Note also that in placing bounds on the error of a point estimate, for all practical purposes we are constructing an interval estimate when a population mean is being estimated. Although this close relationship exists for most of the parameters estimated in this text, the two methods of estimation are not equivalent. For instance, it is not obvious that the best point estimator falls in the middle of the best interval estimator—in many cases it does not. Furthermore, it is not necessarily true that the best interval estimator is even a function of the best point estimator. Although these problems are of a theoretical nature, they are important and worth mentioning. From a practical point of view, the two methods are closely related, and **the choice between the point and the interval estimator in an actual problem depends on the preference of the experimenter.**

EXERCISES   Basic Techniques

7.1   Explain what is meant by "bound on the error of estimation."

7.2   Give the bound on the error of estimating a population mean μ if
    a. $n = 20, \sigma^2 = 4$ b. $n = 100, \sigma^2 = .9$
    c. $n = 50, \sigma^2 = 12$

7.3   Find and interpret a 95% confidence interval for a population mean μ if
    a. $n = 36, \bar{x} = 13.1, s^2 = 3.42$ b. $n = 64, \bar{x} = 2.73, s^2 = .1047$

7.4   Find a 90% confidence interval for a population mean μ if
    a. $n = 125, \bar{x} = .84, s^2 = .086$ b. $n = 50, \bar{x} = 21.9, s^2 = 3.44$
    c.   Interpret the intervals found in parts (a) and (b).

7.5   Find a $(1 - \alpha)$ 100% confidence interval for a population mean μ if
    a. $\alpha = .01, n = 38, \bar{x} = 34, s^2 = 12$ b. $\alpha = .10, n = 65, \bar{x} = 1049, s^2 = 51$
    c. $\alpha = .05, n = 89, \bar{x} = 66.3, s^2 = 2.48$

7.6   In Exercise 6.4 the mean and standard deviation for the die-toss population were $\mu = 3.5$ and $\sigma = 1.71$, respectively.
    a.   If you were to toss a die $n = 5$ times, what is the approximate probability that the mean $\bar{x}$ of the sample would fall in the interval $2.5 \leq \bar{x} \leq 4.5$?
    b.   If you were to toss a die $n = 10$ times, what is the approximate probability that $\bar{x}$ would fall in the interval $2.5 \leq \bar{x} \leq 4.5$?
    c.   Note in part (b) that $P(2.5 \leq \bar{x} \leq 4.5)$ is large. If you have access to a die, toss it $n = 10$ times and calculate $\bar{x}$. Does $\bar{x}$ fall in the interval $2.5 \leq \bar{x} \leq 4.5$? Suppose that the mean of the population $\mu = 3.5$ were unknown and that you were using your mean to estimate μ. What is your error of estimation?

7.7   A random sample of n measurements is selected from a population with unknown mean μ and known standard deviation $\sigma = 10$. Calculate the width of a 95% confidence interval for μ when
    a. $n = 100$ b. $n = 200$ c. $n = 400$

7.8   Compare the confidence intervals in Exercise 7.7. What is the effect on the width of a confidence interval when
    a.   you double the sample size?
    b.   you quadruple the sample size?

7.9 Refer to Exercise 7.7.
a. Calculate the width of a 90% confidence interval for μ when $n = 100$.
b. Calculate the width of a 99% confidence interval for μ when $n = 100$.
c. Compare the widths of 90, 95, and 99% confidence intervals for μ. What effect does increasing the confidence coefficient have on the width of the confidence interval?

7.10 Class experiment: In our analysis of the blood pressure data in Appendix I, we found that the mean and standard deviation of the 965 systolic male blood pressures given in Appendix I are 118.728 and 14.2343, respectively. Regarding this data set as a population, we will demonstrate the concept of a confidence interval. Each student should select a random sample of $n = 40$ observations from the data set, calculate the sample mean $\bar{x}$ and variance s^2, and then calculate a 95% confidence interval for the population mean $\mu = 118.728$. Note whether the confidence interval encloses μ. Record the confidence intervals for all members of the class. Notice how they shift in a random manner. Theoretically, if there were millions of class members, approximately 95% of the confidence intervals should enclose μ. Calculate the percentage of the confidence intervals constructed by the class that enclose μ. It is unlikely that this percentage will equal 95% (because the number of members in the class is not large enough), but most of the confidence intervals should enclose μ.

Applications

7.11 Geologists are interested in shifts and movements of the earth's surface indicated by fractures (cracks) in the earth's crust. One of the most famous large fractures is the San Andreas fault (moving fracture) in California. A geologist attempting to study the movement of the relative shifts in the earth's crust at a particular location found many fractures in the local rock structure. In an attempt to determine the mean angle of the breaks, she sampled $n = 50$ fractures and found the sample mean and standard deviation to be 39.8 degrees and 17.2 degrees, respectively. Estimate the mean angular direction of the fractures and place a bound on the error of estimation.

7.12 Estimates of the earth's biomass, the total amount of vegetation held by the earth's forests, are important in determining the amount of unabsorbed carbon dioxide that we can expect to remain in the earth's atmosphere (*Science News*, August 19, 1989, p. 124). Suppose that a sample of 75 one-square-meter plots were randomly chosen in North America's boreal (northern) forests, and produced a mean biomass of 4.2 kilograms per square meter (kg/m^2) with a standard deviation of 1.5 kg/m^2. Estimate the average biomass for the boreal forests of North America and place a bound on the error of estimation.

7.13 An increase in the rate of consumer savings is frequently tied to a lack of confidence in the economy and is said to be an indicator of a recessional tendency in the economy. A random sampling of $n = 200$ savings accounts in a local community showed a mean increase in savings account values of 7.2% over the past 12 months with a standard deviation of 5.6%. Estimate the mean percent increase in savings account values over the past 12 months for depositors in the community. Place a bound on your error of estimation.

7.14 The ability to accelerate rapidly is an important attribute for an ice hockey player. G. Wayne Marino investigated some of the variables related to the acceleration and speed of a hockey player from a stopped position ("Selected Mechanical Factors Associated with Acceleration in Ice Skating," *Research Quarterly for Exercise and Sport* 54, [1983]). Sixty-nine hockey players, varsity and intramural, from the University of Illinois were included in the experiment. Each player was required to move as rapidly as possible from a stopped position to cover a distance of 6 meters. The means and standard deviations of some of the variables recorded for each of the 69 skaters are shown on page 254.

	Mean	Standard deviation
Weight (kilograms)	75.270	9.470
Stride length (meters)	1.110	0.205
Stride rate (strides/second)	3.310	0.390
Average acceleration (meters/sec^2)	2.962	0.529
Instantaneous velocity (meters/sec)	5.753	0.892
Time to skate (seconds)	1.953	0.131

a. Give the formula that you would use to construct a 95% confidence interval for one of the population means (e.g., mean time to skate the 6-meter distance).

b. Construct a 95% confidence interval for the mean time to skate. Interpret this interval.

7.15 A study of 392 healthy children living in the area of Tours, France was designed to measure the serum levels of certain fat-soluble vitamins, namely, vitamin A, vitamin E, β-carotene, and cholesterol. Knowledge of the reference levels for children living in an industrial country, with normal food availability and feeding habits, would be important in establishing borderline levels for children living in other less favorable conditions. Results of the study are shown in the table below.

	Mean $\pm$ SD ($n = 392$)	Boys ($n = 207$)	Girls ($n = 185$)
Retinol(μg/dl)	42.5 $\pm$ 12.0	43.0 $\pm$ 13.0	41.8 $\pm$ 10.7
β-carotene(μg/l)	572 $\pm$ 381	588 $\pm$ 406	553 $\pm$ 350
Vitamin E (mg/l)	9.5 $\pm$ 2.5	9.6 $\pm$ 2.7	9.5 $\pm$ 2.2
Cholesterol(g/l)	1.84 $\pm$ 0.42	1.84 $\pm$ 0.47	1.83 $\pm$ 0.38
Vitamin E/cholesterol (mg/g)	5.26 $\pm$ 1.04	5.26 $\pm$ 1.11	5.26 $\pm$ 0.25

Source: Malvy, J.M.D. and colleagues. "Retinol, β-Carotene and a-Tocopherol Status in a French Population of Healthy Children," *International Journal of Vitamin and Nutrition Research*, Vol. 59, 1989, p. 29. Hans Huber Publishers, Toronto, Bern.

a. Describe the sampled population.

b. Find a 99% confidence interval for the mean cholesterol level for the population of boys. Interpret this interval.

c. Find a 99% confidence interval for the mean β-carotene level for the population of girls. Interpret this interval.

d. Can the data collected by the researchers be used to estimate serum levels for all boys and girls in this age category? Why or why not?

7.16 Due to a variation in laboratory techniques, impurities in materials, and other unknown factors, the results of an experiment in a chemistry laboratory will not always yield the same numerical answer. In an electrolysis experiment, a class measured the amount of copper precipitated from a saturated solution of copper sulfate over a 30-minute period. The $n = 30$ students acquired a sample mean and standard deviation equal to .145 and .0051 mole, respectively. Find a 90% confidence interval for the mean amount of copper precipitated from the solution over the period of time.

7.17 Based on repeated measurements of the iodine concentration in a solution, a chemist reports the concentration as 4.614 with an "error margin of .006."

a. How would you interpret the chemist's "error margin"?

b. If the reported concentration is based on a random sample of $n = 30$ measurements with a sample standard deviation $s = .017$, would you agree that the chemist's "error margin" is .006?

7.18   Acid rain, caused by the reaction of certain air pollutants with rainwater, appears to be a growing problem in the northeastern section of the United States. (Acid rain affects the soil and causes corrosion on exposed metal surfaces.) Pure rain falling through clean air registers a pH value of 5.7 (pH is a measure of acidity; 0 is acid, 14 is alkaline). Suppose that water samples from 40 rainfalls are analyzed for pH and that $\bar{x}$ and s are equal to 3.7 and 0.5, respectively. Find a 99% confidence interval for the mean pH in rainfall and interpret the interval. What assumption must be made for the confidence interval to be valid?

7.7 ESTIMATING THE DIFFERENCE BETWEEN TWO MEANS

A problem of equal importance to the estimation of a single population mean is the comparison of two population means. For instance, we may wish to compare the difference in average scores on the Medical College Admission Test (MCAT) for students whose major was biochemistry and students whose major was biology. Or, we might wish to compare the average yield in a chemical plant using raw materials furnished by two suppliers, A and B. Samples of daily yield, one for each of the two suppliers, could be recorded and used to make inferences concerning the difference in mean yield.

For each of these examples there are two populations, the first with mean and variance μ_1 and σ_1^2 and the second with mean and variance μ_2 and σ_2^2. A random sample of n_1 measurements is drawn from population 1, and n_2 from population 2, where the samples are assumed to have been drawn independently of one another. Finally, the estimates of the population parameters are calculated from the sample data using the estimators $\bar{x}_1$, s_1^2, $\bar{x}_2$, and s_2^2.

The point estimator of the difference $(\mu_1 - \mu_2)$ between the population means is $(\bar{x}_1 - \bar{x}_2)$. In Chapter 6 we stated without proof that the sampling distribution of the difference between two normally distributed sample means is normally distributed with a mean and standard deviation given in the following display.

MEAN AND STANDARD DEVIATION OF $(\bar{x}_1 - \bar{x}_2)$

$$E(\bar{x}_1 - \bar{x}_2) = \mu_1 - \mu_2,$$

$$\sigma_{(\bar{x}_1 - \bar{x}_2)} = \sqrt{\frac{\sigma_1^2}{n_1} + \frac{\sigma_2^2}{n_2}}$$

Even if the sampled populations are not normal, the Central Limit Theorem ensures the approximate normality of $\bar{x}_1$ and $\bar{x}_2$ for large sample sizes. Hence, the sampling distribution of $(\bar{x}_1 - \bar{x}_2)$ will also be approximately normal. The sampling distribution of $(\bar{x}_1 - \bar{x}_2)$ is shown in Figure 7.10.

Since $(\mu_1 - \mu_2)$ is the mean of the sampling distribution, it follows that $(\bar{x}_1 - \bar{x}_2)$ is an unbiased estimator of $(\mu_1 - \mu_2)$. Hence, the general formulas of Section 7.5 can be used to construct point and interval estimates.

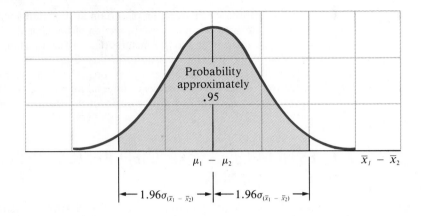

Figure 7.10
The distribution of $(\bar{x}_1 - \bar{x}_2)$ for large samples

POINT ESTIMATION OF $(\mu_1 - \mu_2)$

Estimator: $(\bar{x}_1 - \bar{x}_2)$

Bound on error: $1.96\sigma_{(\bar{x}_1 - \bar{x}_2)} = 1.96\sqrt{\dfrac{\sigma_1^2}{n_1} + \dfrac{\sigma_2^2}{n_2}}$

Note: If σ_1^2 and σ_2^2 are unknown, but both n_1 and n_2 are 30 or more, you can use the sample variances s_1^2 and s_2^2 to estimate σ_1^2 and σ_2^2.

We illustrate this point estimation procedure with an example.

EXAMPLE 7.3   A comparison of the wearing quality of two types of automobile tires was obtained by road-testing samples of $n_1 = n_2 = 100$ tires for each type. The number of miles until wear-out was recorded, where wear-out was defined as a specific amount of tire wear. The test results are given below. Estimate $(\mu_1 - \mu_2)$, the difference in mean miles to wear-out, and place bounds on the error of estimation.

Tire 1	Tire 2
$\bar{x}_1 = 26,400$ miles	$\bar{x}_2 = 25,100$ miles
$s_1^2 = 1,440,000$	$s_2^2 = 1,960,000$

Solution   The point estimate of $(\mu_1 - \mu_2)$ is

$$(\bar{x}_1 - \bar{x}_2) = 26,400 - 25,100 = 1300 \text{ miles}$$

The standard error of $(\bar{x}_1 - \bar{x}_2)$ is

$$\sigma_{(\bar{x}_1 - \bar{x}_2)} = \sqrt{\frac{\sigma_1^2}{n_1} + \frac{\sigma_2^2}{n_2}} \approx \sqrt{\frac{s_1^2}{n_1} + \frac{s_2^2}{n_2}}$$

$$= \sqrt{\frac{1,440,000}{100} + \frac{1,960,000}{100}} = \sqrt{34,000} = 184.4 \text{ miles}$$

where s_1^2 and s_2^2 are used to estimate σ_1^2 and σ_2^2, respectively. We would expect the error of estimation to be less than $(1.96)(184.4) = 361.4$ miles. If the point estimate of the difference in mean miles to wear-out is 1300 miles and the error of estimation is less than 361.4 miles (with a high probability), it seems fairly conclusive that there is a substantial difference in the mean miles to wear-out for the two types of tires. In fact, tire 1 is apparently superior to tire 2 in wearing quality when subjected to the road test. ◁

Following the procedure of Section 7.5, we can construct a confidence interval for $(\mu_1 - \mu_2)$ with confidence coefficient $(1 - \alpha)$.

A $(1 - \alpha)$ 100% CONFIDENCE INTERVAL FOR $(\mu_1 - \mu_2)$

$$(\bar{x}_1 - \bar{x}_2) \pm z_{\alpha/2} \sqrt{\frac{\sigma_1^2}{n_1} + \frac{\sigma_2^2}{n_2}}$$

If σ_1^2 and σ_2^2 are unknown, they can be approximated by the sample variances s_1^2 and s_2^2.

Assumption: n_1 and n_2 are both greater than or equal to 30.

EXAMPLE 7.4   Place a confidence interval on the difference in the mean miles to wear-out for the problem described in Example 7.3. Use a confidence coefficient of .99.

Solution   The confidence interval is

$$(\bar{x}_1 - \bar{x}_2) \pm 2.58 \sqrt{\frac{\sigma_1^2}{n_1} + \frac{\sigma_2^2}{n_2}}$$

Using the results of Example 7.3, we find that the confidence interval is

$$1300 \pm (2.58)(184.4)$$

Therefore, LCL $= 824.2$, UCL $= 1775.8$, and the difference in the mean miles to wear-out is estimated to lie between these two points. Note that the confidence interval is wider than the $\pm 1.96\sigma_{(\bar{x}_1 - \bar{x}_2)}$ bound used in Example 7.3, because we have chosen a larger confidence coefficient. ◁

Examples 7.3 and 7.4 deserve further comment with regard to using sample estimates in place of unknown parameters. The sampling distribution of

$$\frac{(\bar{x}_1 - \bar{x}_2) - (\mu_1 - \mu_2)}{\sqrt{\frac{\sigma_1^2}{n_1} + \frac{\sigma_2^2}{n_2}}}$$

has a standard normal distribution when both sampled populations are normal for all sample sizes, and an approximate standard normal distribution when the sampled populations are not normal, but the sample sizes are large (greater than or equal to 30). When σ_1^2 and σ_2^2 are not known and are estimated by the sample estimates s_1^2 and s_2^2, the resulting statistic will also have an approximate standard normal distribution when the sample sizes are large. The behavior of this statistic when the population variances are unknown and the sample sizes are small will be presented in Section 9.4.

EXERCISES   Basic Techniques

7.19   Independent random samples were selected from two populations 1 and 2. The sample sizes, means, and variances were as follows:

	Population	
	1	2
Sample size	35	49
Sample mean	12.7	7.4
Sample variance	1.38	4.14

Find a bound on the error of estimating the difference in population means ($\mu_1 - \mu_2$).

7.20   Independent random samples were selected from two populations 1 and 2. The sample sizes, means, and variances were as follows:

	Population	
	1	2
Sample size	64	64
Sample mean	2.9	5.1
Sample variance	0.83	1.67

a.   Find a 90% confidence interval for the difference in the population means and interpret your result.
b.   Find a 99% confidence interval for the difference in the population means and interpret your result.

Applications

7.21   A small amount of the trace element selenium, from 50 to 200 micrograms (μg) per day, is considered essential to good health. Suppose that random samples of $n_1 = n_2 = 30$ adults were selected from two regions of the United States, and a day's intake of selenium, from both liquids and solids, was recorded for each person. The mean and standard deviation of the selenium daily intakes for the 30 adults from Region 1 were $\bar{x}_1 = 167.1$ and $s_1 = 24.3$ μg, respectively. The corresponding statistics for the 30 adults from Region 2 were $\bar{x}_2 = 140.9$ and $s_2 = 17.6$. Find a 95% confidence interval for the difference in the mean selenium intake for the two regions. Interpret this interval.

7.22   A study was conducted to compare the mean number of police emergency calls per eight-hour shift in two districts of a large city. Samples of 100 eight-hour shifts were randomly selected from the police records for each of the two regions, and the number of emergency calls was recorded for each shift. The sample statistics are as follows:

	Region	
	1	2
Sample size	100	100
Sample mean	2.4	3.1
Sample variance	1.44	2.64

Find a 90% confidence interval for the difference in the mean number of police emergency calls per shift between the two districts of the city. Interpret the interval.

7.23   One method for solving the electric power shortage employs the construction of floating nuclear power plants located a few miles offshore. Because there is concern about the possibility of ships colliding with the floating (but anchored) plant, an estimate of the density of ship traffic in the area is needed. The number of ships passing within 10 miles of the proposed power-plant location per day, recorded for $n = 60$ days during July and August, has sample mean and variance equal to $\bar{x} = 7.2$ and $s^2 = 8.8$.

a.   Find a 95% confidence interval for the mean number of ships passing within 10 miles of the proposed power-plant location during a one-day time period.

b.   The density of ship traffic was expected to decrease during the winter months. A sample of $n = 90$ daily recordings of ship sightings for December, January, and February gave a mean and variance of $\bar{x} = 4.7$ and $s^2 = 4.9$. Find a 90% confidence interval for the difference in mean density of ship traffic between the summer and winter months.

c.   What is the population associated with your estimate in part (b)? What could be wrong with the sampling procedure in parts (a) and (b)?

7.24   The following statistics are the result of an experiment conducted by P. I. Ward to investigate a theory concerning the moulting behavior of the male *Gammarus pulex*, a small crustacean ("*Gammarus pulex* Control their Moult Timing to Secure Mates," *Animal Behaviour* 32 [1984]). If a male has to moult while paired with a female, he must release her and so loses her. The theory is that the male *Gammarus pulex* is able to postpone the time to moult and thereby reduce the possibility of losing his mate. Ward randomly assigned 100 pairs of males and females to two groups of 50 each. Pairs in the first group were maintained together (Normal); those in the second group were separated (Split). The length of time to moult was recorded for both males and females, and the means, standard deviations, and sample sizes are shown in the table. (The number of crustaceans in each of the four samples is less than 50 because some in each group did not survive until moulting time.) Find a 99% confidence interval for the difference in mean moult time for "Normal" males versus those "Split" from their mates. Interpret the interval.

	Time to moult (days)		
	Mean	*s*	*n*
Males			
Normal	24.8	7.1	34
Split	21.3	8.1	41
Females			
Normal	8.6	4.8	45
Split	11.6	5.6	48

7.25   Refer to the data in Exercise 7.24 and find a 99% confidence interval for the difference in mean moult time for "Normal" females versus those "Split" from their mates. Interpret the interval.

7.26   An experiment was conducted to compare two diets A and B designed for weight reduction. Two groups of 30 overweight dieters each were randomly selected. One group was placed on diet A and the other on diet B and their weight losses were recorded over a 30-day period. The means and standard deviations of the weight loss measurements for the two groups are shown in the accompanying table. Find a 95% confidence interval for the difference in mean weight loss for the two diets. Interpret your confidence interval.

Diet A	Diet B
$\bar{x}_A = 21.3$	$\bar{x}_B = 13.4$
$s_A = 2.6$	$s_B = 1.9$

7.27   In our descriptive analysis of the NIH blood pressure data given in Appendix I we found the means and sample standard deviations of the samples of systolic blood pressures to be as follows:

	Sample size	Mean	Standard deviation
Male	965	118.728	14.2343
Female	945	110.390	12.5753

Find a 99% confidence interval for the difference in population means ($\mu_1 - \mu_2$). Interpret the interval.

7.28   Refer to Exercise 7.15.
a.   Find a 90% confidence interval for the difference between the mean serum cholesterol levels of boys versus girls.
b.   Find a 95% confidence interval for the difference between the mean β-carotene levels of boys versus girls.
c.   Interpret the intervals found in parts (a) and (b).
d.   Do the intervals in parts (a) and (b) contain the value $\mu_1 - \mu_2 = 0$? Why would this be of interest to the researcher?

7.8   ESTIMATING THE PARAMETER OF A BINOMIAL POPULATION

Many surveys have as their objective the estimation of the proportion of people or objects in a large group that possess a particular attribute. Such a survey is a practical example of the binomial experiment discussed in Chapter 4. Estimating the proportion of sales that can be expected in a large number of customer contacts is a practical problem requiring the estimation of a binomial parameter p.

The best point estimator of the binomial parameter p is also the estimator that would be chosen intuitively. That is, the estimator of p, denoted by the symbol $\hat{p}$, is the total number x of successes divided by the total number n of trials. That is,

$$\hat{p} = \frac{x}{n}$$

where x is the number of successes in n trials. [Recall that a "hat" (^) over a parameter is the symbol used to denote the estimator of the parameter.] By "best" we mean that $\hat{p}$ is unbiased and possesses a smaller variance than other possible estimators.

As noted in Chapter 6, the estimator $\hat{p}$ possesses a sampling distribution that can be approximated by a normal distribution because of the Central Limit Theorem. It is an unbiased estimator of the population proportion p, with mean and standard deviation given in the following display.

MEAN AND STANDARD DEVIATION OF $\hat{p}$

$$E(\hat{p}) = p$$

$$\sigma_{\hat{p}} = \sqrt{\frac{pq}{n}}$$

Then from Section 7.5 the point estimation procedure for p is as given in the following display.

POINT ESTIMATOR FOR p

Estimator: $\quad \hat{p} = \dfrac{x}{n}$

Bound on error: $\quad 1.96\sigma_{\hat{p}} = 1.96\sqrt{\dfrac{pq}{n}}$

Estimated bound on error: $\quad 1.96\sqrt{\dfrac{\hat{p}\hat{q}}{n}}$

The corresponding large-sample confidence interval with confidence coefficient $(1 - \alpha)$ is shown below.

A $(1 - \alpha)$ 100% CONFIDENCE INTERVAL FOR p

$$\hat{p} \pm z_{\alpha/2}\sqrt{\frac{\hat{p}\hat{q}}{n}}$$

The sample size will be considered large when we can assume that the distribution of $\hat{p}$ can be approximated by a normal distribution, that is, when $p \pm 2\sigma_{\hat{p}}$ is contained in the interval 0 to 1.

The only difficulty encountered in our procedure will be in calculating $\sigma_{\hat{p}}$, which involves the unknown value of p (and $q = 1 - p$). You will note that we have substituted $\hat{p}$ for the parameter p in the standard deviation $\sqrt{pq/n}$. When n is large, little error will be introduced by this substitution. As a matter of fact, the standard deviation changes only slightly as p changes. This feature can be observed in

Table 7.2
Some calculated values
of $\sqrt{pq}$

p	$\sqrt{pq}$
.5	.50
.4	.49
.3	.46
.2	.40
.1	.30

Table 7.2, where $\sqrt{pq}$ is recorded for several values of p. Note that $\sqrt{pq}$ changes very little as p changes, especially when p is near .5.

EXAMPLE 7.5   A random sample of $n = 100$ voters in a community produced $x = 59$ voters in favor of candidate A. Estimate the fraction of the voting population favoring A and place a bound on the error of estimation.

Solution   The point estimate is

$$\hat{p} = \frac{x}{n} = \frac{59}{100} = .59$$

and the bound on the error of estimation is

$$1.96\sigma_{\hat{p}} = 1.96\sqrt{\frac{pq}{n}} \approx 1.96\sqrt{\frac{(.59)(.41)}{100}} = .096*$$

A 95% confidence interval for p would be

$$\hat{p} \pm 1.96\sqrt{\frac{\hat{p}\hat{q}}{n}}$$

or

$$.59 \pm 1.96(.049).$$

Thus, we would estimate that p lies in the interval .494 to .686 with confidence coefficient .95.   ◁

EXERCISES   **Basic Techniques**

7.29   A random sample of $n = 900$ observations from a binomial population produced $x = 655$ successes. Find a 99% confidence interval for p and interpret the interval.

7.30   A random sample of $n = 300$ observations from a binomial population produced $x = 263$ successes. Find a 90% confidence interval for p and interpret the interval.

---

* Checking the conditions that allow the normal approximation to the sampling distribution of $\hat{p}$, we find that $p \pm 2\sigma_{\hat{p}} \approx .59 \pm 2(.049) = .59 \pm .098$ falls in the interval 0 to 1.

7.31   Suppose that the number of successes observed in $n = 500$ trials of a binomial experiment is 27. Find a 95% confidence interval for p. Why is the confidence interval narrower than the confidence interval in Exercise 7.29?

Applications

7.32   In an article entitled "Doctors Leaving Delivery Room to the Lawyers," the *Orlando Sentinel* (April 6, 1985) provides some statistics on the growing number of obstetricians who have stopped delivering babies. A 1983 survey conducted by the Florida Obstetric and Gynecologic Society focused attention on the rising costs of liability insurance and the increasing number of lawsuits against obstetricians (the second most frequently sued of all medical specialists). The survey found that "25% of the 439 Florida obstetricians who responded had stopped delivering babies and another 30% were considering quitting." How accurate are the estimates derived from this survey? Give bounds on the error of estimation for each estimate.

7.33   It is not uncommon for radio or television stations to air controversial issues during broadcast time and ask for a viewer response as to whether a viewer agrees or disagrees with a given stand on the issue. A poll is conducted by asking those viewers who *agree* to call a 900 telephone number at a charge of $.50 to the calling party, and asking those viewers who *disagree* to call a second 900 telephone number, again at a charge of $.50 per call to the calling party.
  a.   Does this polling technique result in a random sample?
  b.   What can be said about the validity of the results of such a survey? Need we worry about a margin of error in this case?

7.34   Should professional athletes be tested for drugs? Reporting on this question, the *Orlando Sentinel* (October 31, 1985) noted that Donald Fehr, executive director of the Major League Baseball Player's Association, "criticized mandatory testing as being unnecessary."
        Former baseball commissioner Peter Ueberroth and the public did not seem to agree. A *Washington Post*/ABC News survey found that 73% of 1506 people interviewed favored drug tests for professional athletes. Sixty-eight percent said that professional athletes using drugs for the first time should be banned or suspended from professional sports.
  a.   Find a 95% confidence interval for the proportion of the public who favor drug tests for professional athletes.
  b.   Upon what assumptions is your confidence interval in part (a) based?

7.35   Refer to Exercise 7.34 and find a 90% confidence interval for the proportion of the public who favor banning or suspending professional athletes who are found to be using drugs for the first time. What assumptions are required for your confidence interval to be valid?

7.36   Routine chest x-rays, given to many patients on admission to hospitals, are estimated to cost a total of $1.5 billion per year. One study, reported in the *Wall Street Journal* (January 24, 1985), concludes that many of these x-rays may be unnecessary. An examination of chest x-rays ordered for 294 patients at the VA Medical Center in Long Beach, California, found that the x-rays revealed medical problems in 106 cases; of these, however, all but 20 cases had been detected previously. In 19 of these 20, either treatment of the medical problems would have been unaffected by the x-ray information or the problems would have been detected by other diagnostic tests. Using these data, find a 95% confidence interval for the probability p that a chest x-ray for new admissions at the VA hospital in Long Beach will reveal an undetected medical problem. Interpret this interval.

7.37   A *New York Times*/CBS survey of 224 children, ages 9 to 17, conducted shortly after the January 28, 1986 space shuttle disaster, found that two-thirds of the children would like to travel in space (*Orlando Sentinel*, February 3, 1986). Find an approximate bound on the error of estimation for this estimate.

7.38   A survey of the college class of 1992 showed that more incoming college freshmen are smoking, their interest in business careers has declined, and they favor mandatory testing for AIDS and drugs. These findings were based on a survey by the American Council on Education (*New York Times*, January 9, 1989) involving $n = 308,007$ freshmen entering 585 two- and four-year colleges and universities. Specifically, 67% agreed that the best way to control AIDS was through widespread mandatory testing, while 71% agreed that employers should be allowed to test employees or job applicants for drugs.

     a.   Find a 98% confidence interval for the proportion of college freshmen favoring mandatory AIDS testing.

     b.   Find a 99% confidence interval for the proportion of college freshmen favoring drug testing for employees or job applicants. Interpret this interval.

7.39   Refer to Exercise 7.38. The results of this survey were reported to have a sampling error of no more than plus or minus two percentage points. If we interpret this to mean that the bound on error is at most two percentage points, determine whether this claim is true for the two estimates reported in the exercise.

7.40   Exercise 1.5 described a survey conducted by the federal government to measure the extent of tampering with car and light truck emission control systems. In particular, a 14-city survey found that 22% of the 4426 vehicles in the survey showed evidence of deliberate emission control system tampering (*New York Times*, November 4, 1985). Suppose that the 4426 vehicles represent a random sample selected from among all cars and light trucks in the 14 cities.

     a.   How accurate is the estimate—22%—of the percentage of all cars and light trucks in the 14 cities that were manufactured between 1974 and 1984 and that would show evidence of emission control tampering? Find a bound on the error of estimation.

     b.   Will this 22% be a good estimate of the percentage of all cars and light trucks in the United States that were manufactured between 1974 and 1984 and that would show evidence of emission control tampering? Explain.

7.41   A telephone survey conducted by the Florida State University's Policy Sciences Program found that 74% of the 983 responding Florida adults were in favor of raising the drinking age from 19 to 21 (*Gainesville Sun*, April 5, 1985).

     a.   What assumption is necessary for the 74% to be a valid estimate of the percentage of all adult Floridians who favor raising the drinking age from 19 to 21?

     b.   Is it possible that the assumption in part (a) might not be satisfied in a telephone survey?

     c.   Assuming that the assumption in part (a) was satisfied by the pollsters, find a 95% confidence interval for the proportion of all adult Floridians in favor of raising the drinking age from 19 to 21.

▷ **7.9** ESTIMATING THE DIFFERENCE BETWEEN TWO BINOMIAL PARAMETERS

The fourth and final estimation problem considered in this chapter is the estimation of the difference between the parameters of two binomial populations. Assume that the two populations 1 and 2 have parameters p_1 and p_2, respectively. Independent random samples consisting of n_1 and n_2 trials are drawn from their respective populations and the estimates $\hat{p}_1$ and $\hat{p}_2$ are calculated.

The point estimator of $p_1 - p_2$ is the corresponding difference in the estimates $(\hat{p}_1 - \hat{p}_2)$. From Section 6.6, we know that this estimator has a sampling distribution that can be approximated by a normal distribution when the sample sizes are large. The mean and standard deviation of the sampling distribution of $(\hat{p}_1 - \hat{p}_2)$ are shown below.

MEAN AND STANDARD DEVIATION OF $(\hat{p}_1 - \hat{p}_2)$

$$E(\hat{p}_1 - \hat{p}_2) = (p_1 - p_2)$$

$$\sigma_{(\hat{p}_1 - \hat{p}_2)} = \sqrt{\frac{p_1 q_1}{n_1} + \frac{p_2 q_2}{n_2}}$$

Therefore, $(\hat{p}_1 - \hat{p}_2)$ is an unbiased estimator of $(p_1 - p_2)$, and the point estimator of $p_1 - p_2$ is given in the following display.

POINT ESTIMATOR OF $(p_1 - p_2)$

Estimator: $(\hat{p}_1 - \hat{p}_2)$

Bound on error: $1.96\sigma_{(\hat{p}_1 - \hat{p}_2)} = 1.96\sqrt{\frac{p_1 q_1}{n_1} + \frac{p_2 q_2}{n_2}}$

Note: The estimates $\hat{p}_1$ and $\hat{p}_1$ must be substituted for p_1 and p_2 to estimate the bound on the error of estimation.

The $(1 - \alpha)$ 100% confidence interval, appropriate when n_1 and n_2 are large, is shown here.

A $(1 - \alpha)$ 100% LARGE-SAMPLE CONFIDENCE INTERVAL FOR $(p_1 - p_2)$

$$(\hat{p}_1 - \hat{p}_2) \pm z_{\alpha/2}\sqrt{\frac{\hat{p}_1 \hat{q}_1}{n_1} + \frac{\hat{p}_2 \hat{q}_2}{n_2}}$$

Assumption: n_1 and n_2 must be sufficiently large so that the sampling distribution of $(\hat{p}_1 - \hat{p}_2)$ can be approximated by a normal distribution. The interval $(p_1 - p_2) \pm 2\sigma_{(\hat{p}_1 - \hat{p}_2)}$ must be contained in the interval -1 to 1.

EXAMPLE 7.6 A manufacturer of fly sprays wished to compare two new concoctions 1 and 2. Two rooms of equal size, each containing 1000 flies, were used in the experiment. One room was treated with fly spray 1 and the other with an equal amount of fly spray 2. A total of 825 and 760 flies succumbed to sprays 1 and 2, respectively. Estimate the difference in the rate of kill for the two sprays when used in the test environment.

Solution   The point estimate of $(p_1 - p_2)$ is

$$(\hat{p}_1 - \hat{p}_2) = .825 - .760 = .065$$

The bound on the error of estimation is

$$1.96 \sqrt{\frac{p_1 q_1}{n_1} + \frac{p_2 q_2}{n_2}} \approx 1.96 \sqrt{\frac{(.8.25)(.175)}{1000} + \frac{(.76)(.24)}{1000}}$$

$$= .035$$

The corresponding confidence interval, using confidence coefficient .95, is

$$(\hat{p}_1 - \hat{p}_2) \pm 1.96 \sqrt{\frac{\hat{p}_1 \hat{q}_1}{n_1} + \frac{\hat{p}_2 \hat{q}_2}{n_2}}$$

The resulting confidence interval is $.065 \pm .035$.

Hence, we estimate that the difference between the rates of kill $(p_1 - p_2)$ will fall in the interval .030 to .100. We are fairly confident of this estimate because we know that if our sampling procedure were repeated over and over again, each time generating an interval estimate, approximately 95% of the estimates would enclose the quantity $(p_1 - p_2)$.   ◁

EXERCISES   Basic Techniques

7.42   Samples of $n_1 = 500$ and $n_2 = 500$ observations were selected from binomial populations 1 and 2, and $x_1 = 120$ and $x_2 = 147$ were observed. Find a bound on the error of estimating the difference in population proportions $(p_1 - p_2)$.

7.43   Samples of $n_1 = 800$ and $n_2 = 640$ observations were selected from binomial populations 1 and 2, and $x_1 = 337$ and $x_2 = 374$ were observed.
   a.   Find a 90% confidence interval for the difference $(p_1 - p_2)$ in the two population proportions. Interpret the interval.
   b.   What assumptions must you make for the confidence interval to be valid? Are these assumptions met?

7.44   Samples of $n_1 = 1265$ and $n_2 = 1688$ observations were selected from binomial populations 1 and 2, and $x_1 = 849$ and $x_2 = 910$ were observed.
   a.   Find a 99% confidence interval for the difference $(p_1 - p_2)$ in the two population proportions. Interpret the interval.
   b.   What assumptions must you make for the confidence interval to be valid? Are these assumptions met?

Applications

7.45   A study conducted at the University of Massachusetts Medical School in Worcester indicates that heart attack patients given beta blockers have a far lower in-hospital fatality rate than those not given beta blockers. The in-hospital fatality rate for 879 Massachusetts heart attack victims using beta blockers was 9% compared to 27% of 1906 patients who did not use the drug (USA Today, March 12, 1985). Find a 99% confidence interval for the difference between in-hospital fatality rates for those heart attack patients on beta blockers versus those not using the drug. Interpret this interval.

7.46 Federal tax revenues are expected to rise by an average of $74 billion a year through 1993. However, it is also expected that government spending will increase approximately the same amount, leaving the federal deficit above $100 million. In a Roper Organization survey for Citizens for a Sound Economy reported in *Public Opinion* (March/April, 1989), 71% of the Republicans and 75% of the Democrats surveyed indicated that they would prefer to hold down spending rather than raise taxes to reduce the budget deficit.

a. If the number surveyed had included 1500 Republicans and 1500 Democrats, calculate a 99% confidence interval estimate for $(p_1 - p_2)$, the true difference in the proportion of Republicans and Democrats who favor holding down spending to reduce the federal deficit.

b. What is the margin of error associated with this estimate, and how does it compare with the ± 3 percentage points usually associated with such polls?

7.47 In a study of the relationship between law and the use of lethal force, W. B. Waegel classified 459 police shootings in Philadelphia from 1970 to 1978 to see whether new restrictive laws, introduced in 1973, produced a change in the percentage of unjustified uses of lethal force by police (*The Use of Lethal Force by Police: The Effect of Statutory Change, Crime and Delinquency* 30, [January 1984]).

	Total for period 1970–1972	1973	Total for period 1974–1978
Justified	72	30	173
Not justified	11	9	59
Unable to determine	18	7	53
Accidental	10	5	12
Total	111	51	297

Assume that the data in the table represent samples from populations of cases that could have occurred and in which lethal force would have been used, one population corresponding to the legal situation existing prior to 1973 and the other to the legal situation after the new restrictive laws were introduced, the period from 1974–1978. Find a 95% confidence interval for

a. the difference in the proportions of nonjustified uses of lethal force.

b. the difference in the proportions of undecided (unable to be determined) cases.

7.48 A sampling of political candidates—200 randomly chosen from the West and 200 from the East—was classified according to whether the candidate received backing by a national labor union and whether the candidate won. A summary of the data is shown below.

	West	East
Winners backed by union	120	142

Find a 95% confidence interval for the difference between the proportions of union-backed winners in the West versus the East. Interpret this in interval.

7.49 In a study of the relationship between birth order and college success, an investigator found that 126 in a sample of 180 college graduates were first-born or only children. In a sample of 100 nongraduates of comparable age and socioeconomic background, the number of first-born or only children was 54. Estimate the difference between the proportions of first-born or only children in the two populations from which these samples were drawn. Use a 90% confidence interval, and interpret your results.

▷ 7.10 CHOOSING THE SAMPLE SIZE

The design of an experiment is essentially a plan for purchasing a quantity of information. This information, like any other commodity, may be acquired at varying prices depending on the manner in which the data are obtained. Some measurements contain a large amount of information concerning the parameter of interest; others may contain little or none. Since the sole product of research is information, we should try to purchase it at minimum cost.

The **sampling procedure**, or **experimental design**, as it is usually called, affects the quantity of information per measurement. This procedure, along with the sample size n, controls the total amount of relevant information in a sample. With few exceptions, we will be concerned with the simplest sampling situation—random sampling from a relatively large population—and will focus our attention on the selection of the sample size n.

The researcher makes little progress in planning an experiment before encountering the problem of selecting the sample size. Indeed, perhaps one of the questions most frequently asked of the statistician is: **How many measurements should be included in the sample?** Unfortunately, the statistician cannot answer this question without knowing how much information the experimenter wishes to buy. Certainly, the total amount of information in the sample will affect the measure of goodness of the method of inference and must be specified by the experimenter. Referring specifically to estimation, we would like to know how accurate the experimenter wishes the estimate to be. This accuracy may be stated by specifying a bound on the error of estimation.

For instance, suppose we wish to estimate the average daily yield μ of a chemical (Example 7.1) and we wish the error of estimation to be less than 4 tons with a probability of .95. Since approximately 95% of the sample means will lie within $1.96\sigma_{\bar{x}}$ of μ in repeated sampling, we are asking that $1.96\sigma_{\bar{x}}$ equal 4 tons (see Figure 7.11). Then,

$$1.96\sigma_{\bar{x}} = 4 \qquad \text{or} \qquad 1.96\frac{\sigma}{\sqrt{n}} = 4$$

Figure 7.11
Approximate sampling distribution of $\bar{x}$ for large samples

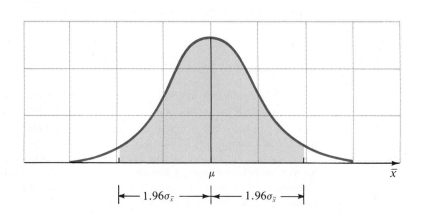

μ

$\bar{x}$

$\longleftarrow 1.96\sigma_{\bar{x}} \longrightarrow \!\!\!\longleftarrow 1.96\sigma_{\bar{x}} \longrightarrow$

Solving for n, we obtain

$$n = \left(\frac{1.96}{4}\right)^2 \sigma^2 \quad \text{or} \quad n = .24\sigma^2$$

You can see that we cannot obtain a numerical value for n unless the population standard deviation σ is known. And certainly, this is exactly what we would expect because the variability of $\bar{x}$ depends on the variability of the population from which the sample was drawn. Lacking an exact value for σ, we would use the best approximation available, such as an estimate s obtained from a previous sample or knowledge of the range in which the measurements will fall. Since the range is approximately equal to 4σ (the Empirical Rule), one-fourth of the range will provide an approximate value for σ. For our example, we would use the results of Example 7.1, which provided a reasonably accurate estimate of σ equal to $s = 21$. Then,

$$n = .24\sigma^2 = .24(21)^2 = 105.9$$

Using a sample of size $n = 106$ or larger, we could be reasonably certain (with probability approximately equal to .95) that our estimate will be less than or equal to $1.96\sigma_{\bar{x}} = 4$ tons of the true average daily yield.

Actually, we should expect the error of estimation to be much less than 4 tons. According to the Empirical Rule, the probability is approximately equal to .68 that the error of estimation is less than $\sigma_{\bar{x}} = 2$ tons. You will note that the probabilities .95 and .68 used in these statements are inexact, since s was substituted for σ. Although this method of choosing the sample size is only approximate for a specified desired accuracy of estimation, it is the best available and is certainly better than selecting the sample size on the basis of our intuition.

The method of choosing the sample size for all the large-sample estimation procedures discussed in preceding sections is identical to that described above.

PROCEDURE FOR CHOOSING THE SAMPLE SIZE

Let θ be the parameter to be estimated and let $\sigma_{\hat{\theta}}$ be the standard deviation of the point estimator. Then proceed as follows:

1. Choose B, the bound on the error of estimation, and a confidence coefficient $(1 - \alpha)$.
2. Solve the following equation for the sample size n:

$$z_{\alpha/2}\sigma_{\hat{\theta}} = B$$

where $z_{\alpha/2}$ is the value of z having area $\alpha/2$ to its right.

> **Note:** For most estimators (all presented in this text), $\sigma_{\hat{\theta}}$ is a function of the sample size n.

We will illustrate with examples.

EXAMPLE 7.7   The reaction of an individual to a stimulus in a psychological experiment can take one of two forms: either an event A will occur or it will not occur. If an experimenter wishes to estimate the probability p that a person will react in favor of A, how many people must be included in the experiment? Assume that he or she will be satisfied if the error of estimation is less than .04 with probability equal to .90. Assume also that he or she expects p to lie somewhere in the neighborhood of .6.

Solution   For this particular example, the bound B on the error of estimation is .04. Since the confidence coefficient is $1 - \alpha = .90$, α must equal .10 and $\alpha/2 = .05$. The z value corresponding to an area equal to .05 in the upper tail of the z distribution is $z_{\alpha/2} = 1.645$.* We then require

$$1.645\sigma_{\hat{p}} = .04$$

or

$$1.645\sqrt{\frac{pq}{n}} = .04$$

Since the variability of $\hat{p}$ is dependent on p, which is unknown, we must use the guessed value of $p = .6$ provided by the experimenter as an approximation. Then,

$$1.645\sqrt{\frac{(.6)(.4)}{n}} = .04$$

or

$$\sqrt{n} = \frac{1.645}{.04}\sqrt{(.6)(.4)}$$

and

$$n = 406$$

Remember, we used an approximate value of p to obtain this value of n. Consequently, the value $n = 406$ is not exact. It is only an approximation to the required sample size. Therefore, it is reasonable to say that a sample of approximately $n = 406$ people will allow you to estimate p with an error of estimation less than .04 with probability equal to .90.   ◁

In the previous example, the value of p was known to be approximately .6. When the value of p is unknown, using $p = .5$ produces the maximum value of $pq = .25$ (see Table 7.2). Hence, the standard deviation of $\hat{p}$ can never be larger than $\sqrt{.25/n}$. **Therefore, using $p = .5$ always produces a value of n satisfying the desired bound.**

---

* *Note:* It is not necessary to carry three decimal places on $z_{\alpha/2}$ for these calculations. The calculated value of n will be very approximate because we are using an approximate value for p in the calculations.

EXAMPLE 7.8   An experimenter wishes to compare the effectiveness of two methods of training industrial employees to perform a certain assembly operation. A number of employees are to be divided into two equal groups, the first receiving training method 1 and the second training method 2. Each employee will perform the assembly operation, and the length of assembly time will be recorded. The measurements for both groups are expected to have a range of approximately 8 minutes. If the estimate of the difference in mean times to assemble is desired to be correct to within 1 minute with probability equal to .95, how many workers must be included in each training group?

Solution   Equating $1.96\sigma_{(\bar{x}_1 - \bar{x}_2)}$ to $B = 1$ minute, we obtain

$$1.96\sqrt{\frac{\sigma_1^2}{n_1} + \frac{\sigma_2^2}{n_2}} = 1$$

Or, since we desire n_1 to equal n_2, we may let $n_1 = n_2 = n$ and obtain the equation

$$1.96\sqrt{\frac{\sigma_1^2}{n} + \frac{\sigma_2^2}{n}} = 1$$

As noted above, the variability of each method of assembly is approximately the same, and hence $\sigma_1^2 = \sigma_2^2 = \sigma^2$. Since the range, equal to 8 minutes, is approximately equal to 4σ, then $4\sigma \approx 8$ and $\sigma \approx 2$. Substituting this value for σ_1 and σ_2 in the above equation, we obtain

$$1.96\sqrt{\frac{(2)^2}{n} + \frac{(2)^2}{n}} = 1$$

$$1.96\sqrt{8/n} = 1$$

$$\sqrt{n} = 1.96\sqrt{8}$$

Solving, we have $n = 31$. Thus, each group should contain approximately $n = 31$ members.   ◁

EXERCISES   Basic Techniques

7.50   Suppose that you wish to estimate a population mean based on a random sample of n observations, and that prior experience suggests that $\sigma = 12.7$. If you wish to estimate μ correct to within 1.6 with probability equal to .95, how many observations should be included in your sample?

7.51   Suppose that you wish to estimate a binomial parameter p correct to within .04 with probability equal to .95. If you suspect that p is equal to some value between .1 and .3 and you want to be certain that your sample is large enough, how large should n be? (*Hint:* When calculating $\sigma_{\hat{p}}$, use the value of p in the interval $.1 < p < .3$ that will give the largest sample size.)

7.52   Independent random samples of $n_1 = n_2 = n$ observations are to be selected from each of two populations 1 and 2. If you wish to estimate the difference between the two population means correct to within .17 with probability equal to .90, how large should n_1 and n_2 be? Assume that you know that $\sigma_1^2 \approx \sigma_2^2 \approx 27.8$.

7.53   Independent random samples of $n_1 = n_2 = n$ observations are to be selected from each of two binomial populations 1 and 2. If you wish to estimate the difference in the two popution binomial parameters correct to within .05 with probability equal to .98, how large should n be? Assume that you have no prior information on the values of p_1 and p_2, but you want to make certain that you have an adequate number of observations in the samples.

Applications

7.54   According to the Sheriff's Office in Marion County, Florida, an automobile salesman who rolled back the mileage on a used car by 60,000 miles increased the car's value by $1375 (*New York Times*, April 14, 1985). Fraud involving the rollback of odometer readings in used cars is prevalent and costly to society. Suppose that it was possible to positively identify an automobile with an altered odometer reading. If you were to accept an appraiser's statement of the amount of the increase in the car's value produced by the odometer alteration, how would you proceed to estimate the mean loss to odometer fraud per used automobile sold?
a.   Explain how you would select your sample.
b.   Explain how you would decide on the number of used automobile purchases to be included in your sample.

7.55   Exercise 7.34 discussed a *Washington Post*/ABC News poll conducted to determine the public's attitudes concerning the use of drugs by professional athletes. Suppose that you were designing a poll of this type.
a.   How would you collect your sample?
b.   If you wanted to estimate the percentage of the population favoring a particular proposition correctly to within 1% with probability equal to .95, approximately how many people would have to be polled? (*Note:* Use a conservative approximation for p, i.e., $p = .5$.)

7.56   If a wildlife service wishes to estimate the mean number of days of hunting per hunter for all hunters licensed in the state during a given season, with a bound on the error of estimation equal to two hunting days, how many hunters must be included in the survey? Assume that data collected in earlier surveys have shown σ to be approximately equal to 10.

7.57   Exercise 7.14 presented statistics from a study of fast starts by ice hockey skaters. The mean and standard deviation of the 69 individual average acceleration measurements over the 6-meter distance were 2.962 and 0.529 meters per second, respectively (G. W. Marino, "Selected Mechanical Factors Associated with Acceleration in Ice Skating." *Research Quarterly for Exercise and Sport* 54, no. 3 [1983]).
a.   Find a 95% confidence interval for this population mean. Interpret the interval.
b.   Suppose that you were dissatisfied with the width of this confidence interval and wanted to cut the interval in half by increasing the sample size. How many skaters (total) would have to be included in the study?

7.58   The mean and standard deviation of the speeds of the sample of 69 skaters at the end of the 6-meter distance in Exercise 7.14 were 5.753 and 0.892 meters per second, respectively (Marino 1983).
a.   Find a 95% confidence interval for the mean velocity at the 6-meter mark. Interpret the interval.
b.   Suppose that you wanted to repeat the experiment and that you wanted to estimate this mean velocity correct to within .1 second with probability .99. How many skaters would have to be included in your sample?

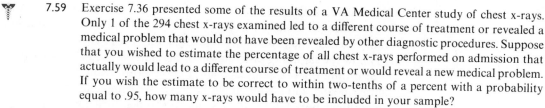

7.59 Exercise 7.36 presented some of the results of a VA Medical Center study of chest x-rays. Only 1 of the 294 chest x-rays examined led to a different course of treatment or revealed a medical problem that would not have been revealed by other diagnostic procedures. Suppose that you wished to estimate the percentage of all chest x-rays performed on admission that actually would lead to a different course of treatment or would reveal a new medical problem. If you wish the estimate to be correct to within two-tenths of a percent with a probability equal to .95, how many x-rays would have to be included in your sample?

7.60 Suppose that you wish to estimate the mean pH of rainfalls in an area that suffers heavy pollution due to the discharge of smoke from a power plant. Assume you know that σ is in the neighborhood of .5 pH and that you wish your estimate to lie within .1 of μ with a probability near .95. Approximately how many rainfalls must be included in your sample (one pH reading per rainfall)? Would it be valid to select all of your water specimens from a single rainfall? Explain.

7.61 Refer to Exercise 7.60. Suppose that you wish to estimate the difference between the mean acidity for rainfalls at two different locations, one in a relatively unpolluted area along the ocean and the other in an area subject to heavy air pollution. If you wish your estimate to be correct to the nearest .1 pH with probability near .90, approximately how many rainfalls (pH values) would have to be included in each sample? (Assume that the variance of the pH measurements is approximately .25 at both locations and that the samples will be of equal size.)

7.62 You want to estimate the difference in grade point average between two groups of college students accurate to within .2 grade point with probability approximately equal to .95. If the standard deviation of the grade point measurements is approximately equal to .6, how many students must be included in each group? (Assume that the groups will be of equal size.)

7.63 Refer to the comparison of the daily adult intake of selenium in two different regions of the United States in Exercise 7.21. Suppose that you wish to estimate the difference in the mean daily intake between the two regions correct to within 5 micrograms with probability equal to .90. If you plan to select an equal number of adults from the two regions (that is, $n_1 = n_2$), how large should n_1 and n_2 be?

▷ 7.11 SUMMARY

This chapter presented the basic concepts of statistical estimation and demonstrated how these concepts can be applied to the solution of some practical problems.

Estimators are rules (usually formulas) that tell how to calculate a parameter estimate based on sample data. Point estimators produce a single number (point) that estimates the value of a population parameter. The properties of a point estimator are contained in its sampling distribution. Thus, we prefer a point estimator that is unbiased (that is, the mean of its sampling distribution is equal to the estimated parameter) and that has a small, preferably a minimum, variance.

The reliability of a point estimator is usually measured by a 1.96 standard deviation (that is, the standard deviation of the sampling distribution of the estimator) bound on the error of estimation. When we use sample data to calculate a particular estimate, the probability that the error of estimation will be less than the bound is approximately .95.

An interval estimator uses the sample data to calculate two points—a confidence interval—that we hope will enclose the estimated parameter. Since we will want to know the probability that the interval will enclose the parameter, we need to know the sampling distribution of the statistic used to calculate the interval.

Four estimators—a sample mean, sample proportion, the difference between two sample means, and the difference between two sample proportions—were used to estimate their population equivalents and to demonstrate the concepts of estimation developed in this chapter. These estimators were chosen for a particular reason. Generally, they are "good" estimators for the respective population parameters in a wide variety of applications. Fortunately, they all have, for large samples, sampling distributions that can be approximated by a normal distribution. This enabled us to use the exact same procedure to construct confidence intervals for the four population parameters μ, p, $(\mu_1 - \mu_2)$, and $(p_1 - p_2)$. Thus, we were able to demonstrate an important role that the Central Limit Theorem (Chapter 6) plays in statistical inference.

TIPS ON PROBLEM SOLVING

In solving the exercises in this chapter, you will be required to answer a practical question of interest to a businessperson, a professional person, a scientist, or a lay person. To find the answers to the questions, you will need to make an inference about one or more population parameters. Consequently, the first step in solving a problem is to decide on the objective of the exercise. What parameters do you wish to make an inference about? Answering the following two questions will help you reach a decision.

1. What *type of data* is involved? This will help you decide which type of parameters you want to make inferences about: binomial proportions (p's) or population means (μ's). Check to see if the data are of the yes-no (two-possibility) variety. If they are, the data are probably binomial, and you will be interested in proportions. If not, the data probably represent measurements on one or more quantitative random variables, and you will be interested in means. To aid you, look for key words such as "proportions," "fractions," and so on, which indicate binomial data. Binomial data often (but not exclusively) evolve from a "sample survey" or an opinion poll.

2. Do you wish to make an inference about a single parameter p or μ or about the *difference between two parameters* $(p_1 - p_2)$ or $(\mu_1 - \mu_2)$? This is an easy question to answer. Check on the number of samples involved. One sample implies an inference about a single parameter; two samples imply a comparison of two parameters. The answers to questions 1 and 2 identify the parameter.

3. After identifying the parameter(s) involved in the exercise, you must identify the exercise objective. It will be either
   a. choosing the sample size required to estimate a parameter with a specified bound on the error of estimation, or
   b. estimating a parameter (or difference between two parameters).

The objective will be very clear if it is (a) because the question will ask for or direct you to find the "sample size." Objective (b) will be clear because the exercise will specifically direct you to estimate a parameter (or the difference between two parameters).

4. Check the conditions required for the sampling distribution of the parameter to be approximated by a normal distribution. For quantitative data, the sample size or sizes must be 30 or more. For binomial data, a large sample size will ensure that $p \pm 2\sigma_{\hat{p}}$ is contained in the interval 0 to 1 $[(p_1 - p_2) \pm 2\sigma_{(\hat{p}_1 - \hat{p}_2)}$ contained in the interval -1 to 1 for the two-sample case].

To summarize these tips, your thought process should follow the decision tree shown in Figure 7.12.

Figure 7.12
Decision tree

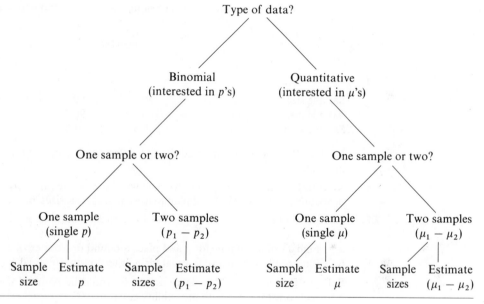

REFERENCES

Freund, J. E., and Walpole, R. E. *Mathematical Statistics*. 4th ed. Englewood Cliffs, N.J.: Prentice Hall, 1987.

Hogg, R. V., and Craig, A. T. *Introduction to Mathematical Statistics*. 4th ed. New York: Macmillan, 1986.

Mendenhall, W.; Wackerly, D.; and Scheaffer, R. L. *Mathematical Statistics with Applications*. 4th ed. Boston: PWS-KENT, 1990.

Mood, A. M., et al. *Introduction to the Theory of Statistics*. 3d ed. New York: McGraw-Hill, 1973.

Neter, J.; Wasserman, W.; and Whitmore, G. A. *Applied Statistics*. 3d ed. Boston: Allyn and Bacon, 1987.

Scheaffer, R. L.; Mendenhall, W.; and Ott, L. *Elementary Survey Sampling*. 4th ed. Boston: PWS-KENT, 1990.

SUPPLEMENTARY EXERCISES

[Starred (*) exercises are optional]

7.64   State the Central Limit Theorem. Of what value is the Central Limit Theorem in statistical inference?

7.65   A random sample of $n = 64$ observations has a mean $\bar{x} = 29.1$ and a standard deviation $s = 3.9$. Give the point estimate for the population mean μ and give a bound on the error of estimation.

7.66   Refer to Exercise 7.65 and find a 90% confidence interval for μ. Interpret this interval.

7.67   Refer to Exercise 7.65. How many observations would be required if you wished to estimate μ with a bound on the error of estimation equal to .5 with probability equal to .95?

7.68   Independent random samples of $n_1 = 50$ and $n_2 = 60$ observations were selected from populations 1 and 2, respectively. The sample sizes and computed sample statistics are shown below.

	Population	
	I	2
Sample size	50	60
Sample mean	100.4	96.2
Sample standard deviation	0.8	1.3

Find a 90% confidence interval for the difference in population means and interpret the interval.

7.69   Refer to Exercise 7.68. Suppose that you wish to estimate $(\mu_1 - \mu_2)$ correct to within .2 with probability equal to .95. If you plan to use equal sample sizes, how large should n_1 and n_2 be?

7.70   A random sample of $n = 500$ observations from a binomial population produced $x = 240$ successes.
   a.   Find a point estimate for p and place a bound on your error of estimation.
   b.   Find a 90% confidence interval for p. Interpret this interval.

7.71   Refer to Exercise 7.70. How large a sample would be required if you wished to estimate p correct to within .025 with probability equal to .90?

7.72   Independent random samples of $n_1 = 40$ and $n_2 = 80$ observations were selected from binomial populations 1 and 2, respectively. The number of successes in the two samples were $x_1 = 17$ and $x_2 = 23$. Find a 90% confidence interval for the difference between the two binomial population proportions. Interpret this interval.

7.73   Refer to Exercise 7.72. Suppose that you wish to estimate $(p_1 - p_2)$ correct to within .06 with probability equal to .90 and that you plan to use equal sample sizes, that is, $n_1 = n_2$. How large should n_1 and n_2 be?

7.74   An experiment was conducted to estimate the effect of smoking on the blood pressure of a group of 25 college-age cigarette smokers. The difference for each participant was obtained by taking the difference in the blood-pressure readings at the time of graduation and again five years later. The sample mean increase, measured in millimeters of mercury, was $\bar{x} = 9.7$. The sample standard deviation was $s = 5.8$. Estimate the mean increase in blood pressure that one would expect for cigarette smokers over the time span indicated by the experiment. Place a bound on the error of estimation. Describe the population associated with the mean that you have estimated.

7.75 Using a confidence coefficient equal to .90, place a confidence interval on the mean increase in blood pressure for Exercise 7.74.

7.76 An unusual survey conducted in Washington and Moscow, reported in the *Tampa Tribune-Times* (January 17, 1988), found Soviets more optimistic but less informed than their American counterparts concerning bilateral issues. The poll, which was conducted by telephone, involved 352 people in Washington, D.C. and 305 nationally. In Moscow, 1000 randomly selected Moscovites were also interviewed by telephone. The news article cited a maximum error for a sample of 1000 as ± 3 percentage points; for 500 as ± 5 percentage points; and for 350, ± 6 percentage points. Do you agree with these error bounds? If not, what are the maximal error rates for samples of size $n = 1000$, $n = 500$, and $n = 350$?

7.77 A study by the General Accounting Office (GAO) of Veterans Administration (VA) hospitals suggests that some of the $8.3 billion spent on the treatment of VA patients in 1984 might have been saved. After evaluating 800 cases at six VA hospitals, the GAO concluded that 244, or 31%, of the patients "did not belong in medical and surgical acute care beds and could have been treated in less costly facilities if available" (*Philadelphia Inquirer*, August 8, 1985). If we can regard the 800 patients as a random sample of all patients treated in VA hospitals in 1984, how accurate is the estimate that 31% could have been treated in less costly facilities? Find a bound on the error of estimation.

7.78 If it is assumed that the heights of men are normally distributed with a standard deviation of 2.5 inches, how large a sample should be taken to be fairly sure (probability .95) that the sample mean does not differ from the true mean (population mean) by more than .50 in absolute value?

7.79 In measuring reaction time, a psychologist estimates that the standard deviation is 0.05 second. How large a sample of measurements must he take in order to be 90% confident that the error of his estimate will not exceed 0.01 second?

7.80 In an experiment 320 out of 400 seeds germinate. Find a confidence interval for the true fraction of germinating seeds; use a confidence coefficient of .98.

7.81 To estimate the proportion of unemployed workers in Panama, an economist selected 400 persons at random from the working class. Of these, 25 were unemployed.
   a. Estimate the true proportion of unemployed workers and place bounds on the error of estimation.
   b. How many persons must be sampled to reduce the bound on error to .02?

7.82 A random sample of 36 cigarettes of a certain brand were tested for nicotine content. The sample gave a mean of 22 and a standard deviation of 4 milligrams. Find a 98% confidence interval for μ, the true mean nicotine content of the brand. Interpret this interval.

7.83 America's public schools are under fire from several directions. Teachers are looking for competitive salaries; parents and children are concerned about quality education; and school administrators are trying to balance quality education and rising educational costs. A Gallup poll conducted for Phi Delta Kappa, a national educational fraternity (*Press Enterprise*, Riverside, Ca., August 25, 1989) found that 60% of the people surveyed supported the controversial idea of allowing parents and students to choose which public school to attend, while nearly two-thirds favored giving school principals broader authority in running their schools in return for making them more responsible for school performance.
   a. If the survey sample size was 3000, find a 95% confidence interval for the true proportion of Americans who support the idea of allowing parents and children to choose which public school to attend.
   b. Estimate the true proportion of Americans who favor broader authority for school principals, and provide a bound on the error of estimation.

7.84   Refer to Exercise 7.83. If the pollster would like the estimates in the education survey to be accurate to within 2 percentage points of the actual percentages, how large a sample should she take? Is the sample size given in Exercise 7.83 large enough to achieve the desired bound?

7.85   In a controlled pollination study involving *Phlox drummondii*, a spring-flowering annual plant common along roadsides and in sandy fields in Central Texas, Karen E. Pittman and Donald A. Levin ("Effects of Parental Identities and Environment on Components of Crossing Success in *Phlox Drummondii*," *American Journal of Botany*, 76(3), p. 409–418, 1989) found that seed survival rates were not affected by water or nutrient deprivation. In the experiment, flowers on plants were counted as males when donating pollen and as females when pollinated by donor pollen in crosses involving plants in the three treatment groups: control, low water, and low nutrient. The following table, which reflects one aspect of the results of the experiment, records the number of seeds surviving to maturity for each of the three groups for both male parents and female parents.

		Male		Female
Treatment	n	Number surviving	n	Number surviving
Control	585	543	632	560
Low water	578	522	510	466
Low nutrient	568	510	589	546

a.   Find the sample estimate of the proportion of surviving seeds and the standard error of the estimate for each of the six treatment-sex combinations.
b.   Find a 99% confidence interval for the difference between survival proportions for low water versus low nutrients for the male category.
c.   Find a 99% confidence interval for the difference between survival proportions for low water versus low nutrients for the female category.
d.   Find a 99% confidence interval for the difference between survival proportions for the male control group compared to the female control group.
e.   Interpret the intervals calculated in parts (a), (b), (c), and (d).

7.86   How large a sample is necessary in order to estimate a binomial parameter p to within .01 of its true value with probability .95? Assume the value of p is approximately .5.

7.87   A dean of men wishes to estimate the average cost of the freshman year at a particular college correct to within $500, with a probability of .95. If a random sample of freshmen is to be selected and asked to keep financial data, how many must be included in the sample? Assume that the dean knows only that the range of expenditure will vary from approximately $4800 to $13,000.

7.88   An experimenter wishes to estimate the difference $(p_1 - p_2)$ for two binomial populations to within .1 of the true difference. Both p_1 and p_2 are expected to assume values between .3 and .7. If samples of equal size are to be selected from the populations to estimate $(p_1 - p_2)$, how large should they be?

7.89   An experimenter fed different rations to two groups of 100 chicks each. Assume that all factors other than rations are the same for both groups. A record of mortality for each group is as follows:

Chicks	Ration A	Ration B
n	100	100
Number died	13	6

Construct a 98% confidence interval for the true difference in mortality rates for the two rations. Interpret this interval.

7.90 You want to estimate the mean hourly yield for a process manufacturing an antibiotic. You observe the process for 100 hourly periods chosen at random, with the results $\bar{x} = 34$ oz/hr and $s = 3$. Estimate the mean hourly yield for the process using a 95% confidence interval.

7.91 A quality control engineer wants to estimate the fraction of defectives in a large lot of light bulbs. From previous experience, he feels that the actual fraction of defectives should be somewhere around .2. How large a sample should he take if he wants to estimate the true fraction to within .01 using a 95% confidence interval?

7.92 Samples of 400 printed-circuit boards were selected from each of two production lines A and B. The number of defectives in each sample was

Line	Number of defectives
A	40
B	80

Estimate the difference between the actual fractions of defectives for the two lines with a confidence coefficient of .90.

7.93 Refer to Exercise 7.92. Suppose that ten samples of $n = 400$ printed circuit boards were tested and a confidence interval was constructed for p for each of the ten samples. What is the probability that exactly one of the intervals will not enclose the true value of p? that at least one interval will not enclose the true value of p?

7.94 It is generally accepted that the medical school experience poses considerable stress for medical students. Moreover, medical students from minorities face additional stress related to perceived or real attitudes about their academic, cultural, and social backgrounds. From 1982 to 1985, Gregory Strayhorn and Henry Frierson conducted a longitudinal study of $n_1 = 50$ black and $n_2 = 392$ white first-year medical students at the University of North Carolina School of Medicine. The following table gives the means and standard deviations for three measures of social well-being reported in their study ("Assessing Correlations between Black and White Students' Perceptions of the Medical School Learning Environment, Their Academic Performances, and Their Well-being," *Academic Medicine,** 64(1989): 468–473).

Aspect measured	Black ($n_1 = 50$)		White ($n_2 = 392$)	
	Mean	Standard deviation	Mean	Standard deviation
Quality of learning environment	112.4	16.4	111.3	13.4
Perception of stress	70.5	11.5	73.5	10.4
Social support	94.0	17.5	85.9	15.2

a. Estimate the true difference in mean scores for the quality of learning environment between blacks and whites with a 90% confidence interval.

b. Does the confidence interval in part (a) enclose the value $(\mu_1 - \mu_2) = 0$? Why might this be of interest to the experimenters?

c. Find a point estimate with bounds on error for the difference in mean scores measuring perception of stress between blacks and whites.

* Formerly called the *Journal of Medical Education.*

HOW RELIABLE IS THAT POLL?

It is almost impossible to read a daily newspaper or listen to the radio or view telecasts without hearing about some opinion poll or economic survey. For many of us comes the inevitable question: How reliable are the percentages derived from these samples of public opinion? Do the national polls conducted by the Gallup and Harris organizations, the news media, and so on really provide accurate estimates of the percentages of people in the United States who favor various propositions?

A report of the results of a poll conducted by the *New York Times* provides a clue to these answers (*New York Times*, February 24, 1988). The object of the poll was to determine eating habits of Americans, focusing especially on the role of women in determining how, when, what, and where their families eat. Nested in the middle of the report is a box entitled "How the Poll was Conducted," shown below.

This *New York Times* poll is based on telephone interviews conducted October 29 through November 5, 1987, with 1870 adults around the United States, excluding Alaska and Hawaii.

The sample of telephone exchanges called was selected by a computer from a complete list of exchanges in the country. The exchanges were chosen so as to ensure that each region of the country was represented in proportion to its population. For each exchange, the telephone numbers were formed by random digits, thus permitting access to both listed and unlisted residential numbers.

Describing the reliability of the poll results, the box states: "In theory, in 19 cases out of 20 the results based on such samples will differ by no more than 2 percentage points in either direction from what would have been obtained by interviewing all adult Americans." In addition, the margin of error for smaller subgroups, which would represent only fractions of the total 1870 adults in the sample, would be larger than 2%. For example, the potential error for married women is ± 4 percentage points, and for married men it is ± 5 percentage points.

1. Verify the bound on the error of estimation given by the pollster.

2. Do the calculations in part 1 confirm the *New York Times* statement that the sample percentages will vary less than 2% from the actual population percentages? What assumptions may have been violated in the *New York Times* sampling procedures?

3. From the potential bounds given for married men and women, can you determine approximately how many people were interviewed in each of these subgroups?

CHAPTER 8

LARGE-SAMPLE TESTS OF HYPOTHESES

Case Study

Will an aspirin a day reduce the risk of heart attack? A very large study of United States physicians showed that a single aspirin tablet taken every other day reduced the risk of heart attack in men by one-half. However, three days later, a British study reported a completely opposite conclusion. How could this be? The case study at the end of this chapter looks at how the studies were conducted, and you will analyze the data using large-sample techniques.

General Objective

In this chapter, the concept of a statistical test of a hypothesis will be formally introduced. The sampling distributions of statistics presented in Chapters 6 and 7 will be used to construct large-sample tests concerning the values of population parameters of interest to the experimenter.

Specific Topics

▷— **8.1** TESTING HYPOTHESES ABOUT POPULATION PARAMETERS

Some practical problems require that we estimate the value of a population parameter; others require that we make decisions concerning the value of the parameter. For example, if a pharmaceutical company were fermenting a vat of antibiotic, it would want to test the potency of samples of the antibiotic and use those samples to estimate the mean potency μ of the antibiotic in the vat. In contrast, suppose that there were no concern that the potency of the antibiotic would be too high; the company's only concern was that the mean potency exceed some government-specified minimum in order that the vat be declared acceptable for sale. In this case, the company would not wish to estimate the mean potency. Rather, it would want to show that the mean potency of the antibiotic in the vat exceeded the minimum specified by the government. Thus, the company would want to decide whether the mean potency exceeded the minimum allowable potency, or whether it did not. The pharmaceutical company's problem illustrates a **statistical test of an hypothesis**.

The reasoning employed in testing an hypothesis bears a striking resemblance to the procedure used in a court trial. In trying a person for theft, the court assumes the accused innocent until proved guilty. The prosecution collects and presents all available evidence in an attempt to contradict the "not guilty" hypothesis and hence to obtain a conviction. However, if the prosecution fails to disprove the "not guilty" hypothesis, this does not prove that the accused is "innocent" but merely that there is not sufficient evidence to conclude that the accused is "guilty."

The statistical problem portrays the potency of the antibiotic as the accused. The hypothesis to be tested, called the **null hypothesis**, is that the potency does not exceed the minimum governmental standard. The evidence in this case is contained in the sample of specimens drawn from the vat. The pharmaceutical company, playing the role of the prosecutor, believes that an **alternative hypothesis** is true—namely, that the potency of the antibiotic does exceed the minimum standard. Hence, the company attempts to use the evidence in the sample to reject the null hypothesis (potency does not exceed minimum standard), thereby supporting the alternative hypothesis (potency exceeds minimum standard). You will recognize this procedure as an essential feature of the scientific method, in which all proposed theories must be compared with reality.

In this chapter we will explain the basic concepts of a test of an hypothesis and demonstrate the concepts with some very useful large-sample statistical tests of the values of a population mean, a population proportion, the difference between a pair of population means, and the difference between two proportions. We will employ the four point estimators discussed in Chapter 7—$\bar{x}, (\bar{x}_1 - \bar{x}_2), \hat{p}$, and $(\hat{p}_1 - \hat{p}_2)$—as test statistics and, in doing so, will obtain a unity in these four statistical tests. All four test statistics will, for large samples, have sampling distributions that are normal or can be approximated by a normal distribution.

▷ 8.2 A STATISTICAL TEST OF AN HYPOTHESIS

The basic reasoning employed in a statistical test of an hypothesis was outlined in Section 8.1 in connection with the test of the potency of an antibiotic. In this section we will attempt to emphasize the specific points involved in a statistical test of an hypothesis.

PARTS OF A STATISTICAL TEST

The objective of a statistical test is to test an hypothesis concerning the values of one or more population parameters. A statistical test involves four elements:

1. Null hypothesis
2. Alternative hypothesis
3. Test statistic
4. Rejection region

Throughout our discussion, the null hypothesis will be denoted by the symbol H_0, and the alternative hypothesis by the symbol H_a.

The specification of these four elements defines a particular test; changing one or more of the parts creates a new test.

The **alternative hypothesis** is the hypothesis that the researcher wishes to support. The **null hypothesis** is a contradiction of the alternative hypothesis; that is, if the null hypothesis is false, the alternative hypothesis must be true. For reasons you will subsequently see, it is easier to show support for the alternative hypothesis by presenting evidence (sample data) that indicates that the null hypothesis is false. Thus, we are building a case in support of the alternative hypothesis by using a method that is analogous to proof by contradiction.

Even though we wish to gain evidence in support of the alternative hypothesis (denoted by the symbol H_a), the null hypothesis (indicated by the symbol H_0) is the hypothesis to be tested. Thus, H_0 will specify hypothesized values for one or more population parameters. For example, we might wish to test the null hypothesis that a population mean is equal to 50, hoping to show, in fact, that the mean exceeds 50. Or we might wish to test the null hypothesis that two population means, say μ_1 and μ_2, are equal, hoping to show, perhaps that μ_1 is larger than μ_2.

The decision to reject or accept the null hypothesis is based on information contained in a sample drawn from the population of interest. The sample values are used to compute a single number, corresponding to a point on a line, which operates as a decision maker. This decision maker is called the **test statistic**. The entire set of values that the test statistic may assume is divided into two sets, or regions. One set, consisting of values that support the alternative hypothesis, is called the **rejection region**. The other, consisting of values that support the null hypothesis, is called the **acceptance region**. If the test statistic computed from a particular sample assumes a value in the rejection region, the null hypothesis is rejected and the alternative hypothesis H_a is accepted. If the test statistic falls in the acceptance region, either the null hypothesis is accepted or the test is judged to be inconclusive. The circumstances leading to this latter decision are explained subsequently.

The decision procedure described above is subject to two types of errors, which are prevalent in a two-choice decision problem.

Definition

> A **type I error** for a statistical test is the error made by rejecting the null hypothesis when it is true. The probability of making a type I error is denoted by the symbol α.
>
> A **type II error** for a statistical test is the error made by accepting (not rejecting) the null hypothesis when it is false and some alternative hypothesis is true. The probability of making a type II error is denoted by the symbol β.

The two possibilities for the null hypothesis—that is, true or false—along with the two decisions the experimenter can make, are indicated in the two-way table, Table 8.1. The occurrences of the type I and type II errors are indicated in the appropriate cells.

Table 8.1
Decision table

Decision	Null hypothesis	
	True	False
Reject H_0	Type I error	Correct decision
Accept H_0	Correct decision	Type II error

The goodness of a statistical test of an hypothesis is measured by the probabilities of making a type I or a type II error, denoted by the symbols α and β, respectively. Since α is the probability that the test statistic falls in the rejection region, assuming H_0 to be true, **an increase in the size of the rejection region increases α** and, at the same time, decreases β for a fixed sample size. Reducing the size of the rejection region decreases α and increases β. If the sample size n is increased, more information is available upon which to base the decision, and for fixed α, β will decrease.

The probability β of making a type II error varies depending upon the true value of the population parameter. For instance, suppose that we wish to test the null hypothesis that the binomial parameter p is equal to $p_0 = .4$. (We will use a subscript 0 to indicate the parameter value specified in the null hypothesis H_0.) Furthermore, suppose that H_0 is false and that p is really equal to an alternative value, say p_a. Which will be more easily detected, a $p_a = .4001$ or a $p_a = 1.0$? Certainly, if p is really equal to 1.0, every single trial will result in a success and the sample results will produce strong evidence to support a rejection of $H_0 : p_0 = .4$. On the other hand, $p_a = .4001$ lies so close to $p_0 = .4$ that it would be extremely difficult to detect p_a without a very large sample. In other words, the probability β of accepting H_0 will vary depending on the difference between the true value of p and the hypothesized value p_0. Ideally, the further p_a lies from p_0, the higher the probability is of rejecting H_0. This probability is measured by $(1 - \beta)$, which is called the **power** of the test.

Definition

> The **power of a statistical test**, given as
>
> $$1 - \beta = P(\text{reject } H_0 \text{ when } H_0 \text{ is false})$$
>
> measures the ability of the test to perform as required.

A graph of the probability of a type II error β as a function of the true value of the parameter is called the operating characteristic curve for the statistical test. Note that the operating characteristic curves for the lot acceptance sampling plans in Chapter 4 were really graphs expressing β as a function of p.

A graph of $(1 - \beta)$, the probability of rejecting H_0 when, in fact, H_0 is false, as a function of the true value of the parameter of interest is called the **power curve** for the statistical test. For constant values of n and α **the power of a test should increase as the distance between the true and hypothesized values of the parameter increases.** An increase in the sample size n will increase the power, $1 - \beta$, for all alternative values of the parameter being tested. Thus, we can create a power curve corresponding to each sample size.

The experimenter should be able to specify values of α and β, measuring the risks of the respective errors he is willing to tolerate, as well as some deviation from the hypothesized value of the parameter he considers of practical importance and wishes to detect. The rejection region for the test will be located in accordance with the specified value of α; the sample size will be chosen large enough to achieve an acceptable value of β for the specified deviation the experimenter wishes to detect. This choice could be made by consulting the power curves, corresponding to various sample sizes, for the chosen test.

In practice, β is often unknown, either because it was never computed before the test was conducted or because it may be extremely difficult to compute for the test. Then rather than accept the null hypothesis when the test statistic falls in the acceptance region, you should withhold judgment. That is, you should not accept the null hypothesis unless you know the risk (measured by β) of making an incorrect decision. Notice that you will never be faced with this "no conclusion" situation when the test statistic falls in the rejection region. Then you can reject the null hypothesis (and accept the alternative hypothesis) because you always know the value of α, the probability of rejecting the null hypothesis when it is true. The fact that β is often unknown explains why we attempt to support the alternative hypothesis by rejecting the null hypothesis. When we reach this decision, the probability α that such a decision is incorrect is known.

In conclusion, keep in mind that **"accepting" a particular hypothesis means deciding in its favor**. Regardless of the outcome of a test, you are never *certain* that the hypothesis you "accept" is true. **There is always a risk of being wrong (measured by α and β).** Consequently, you never "accept" H_0 if β is unknown or its value is unacceptable to you. When this situation occurs, you should withhold judgment and collect more data.

8.3   A LARGE-SAMPLE STATISTICAL TEST

Large-sample tests of hypotheses concerning population means and proportions are similar. The similarity lies in the fact that all of the point estimators discussed in Chapter 7 are unbiased and have sampling distributions that, for large samples, can be approximated by a normal distribution. Therefore, we can use the point estimators as test statistics to test hypotheses about the respective parameters.

In general, we will use the symbol θ to denote one of the four parameters that we estimated in Chapter 7—μ, $(\mu_1 - \mu_2)$, p, or $(p_1 - p_2)$—and $\hat{\theta}$ will denote the corresponding unbiased point estimator (for example, $\hat{\mu} = \bar{x}$, $\hat{p} = x/n$, and so on). Then, when the null hypothesis,

$$H_0 : \theta = \theta_0$$

is true, the sampling distribution of $\hat{\theta}$ is either normal or (for large samples) approximately so with mean θ_0, as shown in Figure 8.1.

Suppose that, from a practical point of view, we are primarily concerned with the rejection of H_0 when θ is greater than θ_0. Then the alternative hypothesis would be $H_a : \theta > \theta_0$ and we will reject the null hypothesis when $\hat{\theta}$ is too large.* Letting $\sigma_{\hat{\theta}}$ represent the standard deviation of the sampling distribution of $\hat{\theta}$, "too large" means too many standard deviations $\sigma_{\hat{\theta}}$ away from θ_0. The rejection region for the test is shown in Figure 8.1. The value of the test statistic that separates the rejection and acceptance regions is called the **critical value** of the test statistic. The critical value of the test statistic for the test indicated in Figure 8.1 is $\theta_0 + C$.

Figure 8.1
Distribution of $\hat{\theta}$ when H_0 is true

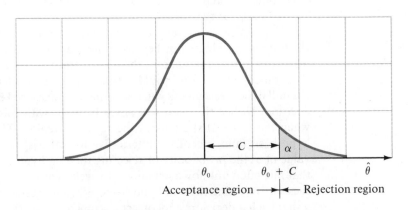

* Note that if the test rejects the null hypothesis $\theta = \theta_0$ in favor of the alternative hypothesis $\theta > \theta_0$, then it will certainly reject a null hypothesis of the form $\theta < \theta_0$, since this is even more contradictory to the alternative hypothesis. For this reason, in this text we state the null hypothesis for a one-tailed test as $\theta = \theta_0$ rather than $\theta \leq \theta_0$.

Definition

> The value of the test statistic that separates the rejection and acceptance regions is called the **critical value** of the test statistic.

The probability of rejecting the null hypothesis, assuming it to be true, would equal the area under the normal curve lying above the rejection region. Thus if we desire $\alpha = .05$, we would reject when $\hat{\theta}$ is more than $1.645\sigma_{\hat{\theta}}$ to the right of θ_0. A test rejecting in one tail of the distribution of the test statistic is called a **one-tailed statistical test**.

Definition

> A **one-tailed statistical test** is one that locates the rejection region in only one tail of the sampling distribution of the test statistic. To detect $\theta > \theta_0$ place the rejection region in the upper tail of the distribution of $\hat{\theta}$. To detect $\theta < \theta_0$ place the rejection region in the lower tail of the distribution of $\hat{\theta}$.

If we wish to detect departures either greater than or less than θ_0, the alternative hypothesis would be

$$H_a : \theta \neq \theta_0$$

that is,

$$\theta > \theta_0$$

or

$$\theta < \theta_0$$

The probability of a type I error α would be equally divided between the two tails of the normal distribution, and we would reject H_0 for values of $\hat{\theta}$ greater than some critical value $\theta_0 + C$ or less than $\theta_0 - C$ (see Figure 8.2). This is called a **two-tailed statistical test**.

Figure 8.2
The rejection region for a two-tailed test

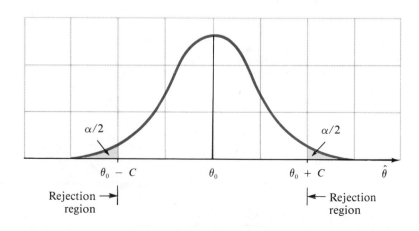

Definition

> A **two-tailed statistical test** is one that locates the rejection region in both tails of the sampling distribution of the test statistic. Two-tailed tests are used to detect either $\theta > \theta_0$ or $\theta < \theta_0$.

The point estimators presented in Chapter 7 satisfy the requirements of the test just described when the sample size is large. The sample size must be large enough so that the point estimator is normally distributed—or approximately normally distributed owing to the Central Limit Theorem—and permits a reasonably good estimate of any unknown parameters.

The mechanics of testing are simplified by using z as a test statistic, as noted in Example 5.11.

LARGE-SAMPLE TEST STATISTIC

$$z = \frac{\hat{\theta} - \theta_0}{\sigma_{\hat{\theta}}}$$

Remember that z is simply the deviation of a normally distributed random variable, $\hat{\theta}$, from θ_0, expressed in units of $\sigma_{\hat{\theta}}$. Thus, for a two-tailed test with $\alpha = .05$, we would reject H_0 when $z > 1.96$ or $z < -1.96$.

The calculation of β for the one-tailed statistical test described above can be facilitated by considering Figure 8.3. When H_0 is false and $\theta = \theta_a$, the test statistic $\hat{\theta}$ will be normally distributed about a mean θ_a, rather than θ_0. The distribution of $\hat{\theta}$, assuming $\theta = \theta_a$, is shown by the black curve. The hypothesized distribution of $\hat{\theta}$, shown by the colored curve, locates the rejection region and the critical value of $\theta_0 + C$. Since β is the probability of accepting H_0, given $\theta = \theta_a$, β would equal the area under the black curve located above the acceptance region. This area, which is shaded, could be calculated using the methods described in Chapter 5.

As we have previously stated, the method of inference used in a given situation will often depend on the preference of the experimenter. Some people wish to express an inference as an estimate; others prefer to test an hypothesis concerning the

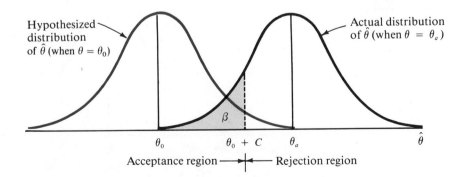

Figure 8.3
Distribution of $\hat{\theta}$ when H_0 is false and $\theta = \theta_a$

Hypothesized distribution of $\hat{\theta}$ (when $\theta = \theta_0$)

Actual distribution of $\hat{\theta}$ (when $\theta = \theta_a$)

β

θ_0 $\theta_0 + C$ θ_a $\hat{\theta}$

Acceptance region ——|—— Rejection region

parameter of interest. The following section will demonstrate the use of the z test in testing an hypothesis concerning a population mean and, at the same time, will illustrate the close relationship between the statistical test and the large-sample confidence intervals discussed in Chapter 7.

8.4 TESTING AN HYPOTHESIS ABOUT A POPULATION MEAN

We can apply the large-sample test in Section 8.3 to test an hypothesis about a population mean. The parameter θ to be tested is μ, the point estimator $\hat{\theta}$ is the sample mean $\bar{x}$, and the standard deviation $\sigma_{\hat{\theta}}$ of the sampling distribution of $\bar{x}$ is $\sigma/\sqrt{n}$. A summary of the test is shown below.

LARGE-SAMPLE STATISTICAL TEST FOR μ

1. Null Hypothesis: $H_0: \mu = \mu_0$.

2. Alternative Hypothesis:

   One-Tailed Test   *Two-Tailed Test*
   $H_a: \mu > \mu_0$ $H_a: \mu \neq \mu_0$
   (or, $H_a: \mu < \mu_0$)

3. Test Statistic: $z = \dfrac{\bar{x} - \mu_0}{\sigma_{\bar{x}}} = \dfrac{\bar{x} - \mu_0}{\sigma/\sqrt{n}}$

   If σ is unknown (which is usually the case), substitute the sample standard deviation s for σ.

4. Rejection Region:

   One-Tailed Test   *Two-Tailed Test*
   $z > z_\alpha$ $z > z_{\alpha/2}$ or $z < -z_{\alpha/2}$
   (or, $z < -z_\alpha$ when the
   alternative hypothesis is
   $H_a: \mu < \mu_0$)

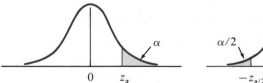
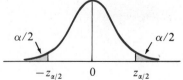

Assumptions: The n observations in the sample were randomly selected from the population and n is large, say $n \geq 30$.

EXAMPLE 8.1   Refer to Example 7.1. Test the hypothesis that the average daily yield of the chemical is $\mu = 880$ tons per day against the alternative that μ is either greater or less than 880 tons per day. The sample (Example 7.1), based on $n = 50$ measurements, yielded $\bar{x} = .871$ and $s = 21$ tons.

Solution   The null and alternative hypotheses are

$$H_0: \mu = 880 \quad \text{and} \quad H_a: \mu \neq 880$$

The point estimate for μ is $\bar{x}$. Therefore, the test statistic is

$$z = \frac{\bar{x} - \mu_0}{\sigma_{\bar{x}}} = \frac{\bar{x} - \mu_0}{\sigma/\sqrt{n}}$$

Using s to approximate σ, we obtain

$$z = \frac{871 - 880}{21/\sqrt{50}} = -3.03$$

For $\alpha = .05$ the rejection region consists of values of $z > 1.96$ and values of $z < -1.96$. Since the calculated value -3.03 of z falls in the rejection region, we reject the hypothesis that $\mu = 880$ tons and conclude that it is less. The probability of rejecting H_0 when H_0 is true is $\alpha = .05$. Hence, we are reasonably confident that our decision is correct. ◁

The statistical test based on a normally distributed test statistic, with a given α, and the $(1 - \alpha)100\%$ confidence interval of Section 7.6 are related. The interval $[\bar{x} \pm (1.96\sigma/\sqrt{n})]$, or approximately (871 ± 5.82) for Example 8.1, is constructed so that in repeated sampling $(1 - \alpha)100\%$ of the intervals will enclose μ. Noting that $\mu = 880$ does not fall in the interval, we would be inclined to reject $\mu = 880$ as a likely value and conclude that the mean daily yield was, indeed, less.

There is another similarity between this test and the confidence interval of Section 7.6. The test is "approximate" because we substituted s, an approximate value, for σ. That is, the probability α of a type I error selected for the test is not .05. It is close to .05, close enough for all practical purposes, but it is not exact. This will be true for most statistical tests because one or more assumptions will not be satisfied exactly.

EXAMPLE 8.2   Refer to Example 8.1. Calculate β and the power of the test, $1 - \beta$, when μ is actually equal to 870 tons.

Solution   The acceptance region for the test of Example 8.1 is located in the interval $(\mu_0 \pm 1.96\sigma_{\bar{x}})$. Substituting numerical values, we obtain

$$880 \pm 1.96 \frac{21}{\sqrt{50}} \quad \text{or} \quad 874.18 \text{ to } 885.82$$

The probability of accepting H_0, given $\mu = 870$, is equal to the area under the frequency distribution for the test statistic $\bar{x}$ in the interval from 874.18 to 885.82.

Figure 8.4
Calculating β in
Example 8.2

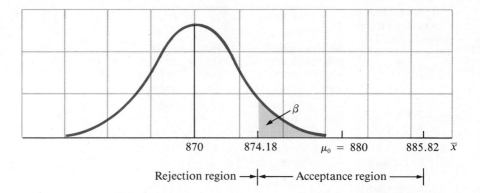

Since $\bar{x}$ is normally distributed with a mean of 870 and $\sigma_{\bar{x}} \approx 21/\sqrt{50} = 2.97$, β is equal to the area under the normal curve located between 874.18 and 885.82 (see Figure 8.4).

Calculating the z values corresponding to 874.18 and 885.82, we obtain

$$z_1 = \frac{\bar{x} - \mu}{\sigma/\sqrt{n}} \approx \frac{874.18 - 870}{21/\sqrt{50}} = 1.41$$

$$z_2 = \frac{\bar{x} - \mu}{\sigma/\sqrt{n}} \approx \frac{885.82 - 870}{21/\sqrt{50}} = 5.33$$

Then

$$\beta = P(\text{accept } H_0 \text{ when } \mu = 870) = P(874.18 < \bar{x} < 885.82 \text{ when } \mu = 870)$$

$$= P(1.41 < z < 5.33)$$

You can see from Figure 8.4 that the area under the normal curve above $\bar{x} = 885.82$ (or $z = 5.33$) is negligible. Therefore,

$$\beta = P(z > 1.41)$$

From Table 3 in Appendix III we find that the area between $z = 0$ and $z = 1.41$ is .4207 and

$$\beta = .5 - .4207 = .0793$$

Hence, the power of the test is

$$1 - \beta = 1 - .0793 = .9207$$

The probability of correctly rejecting H_0, given that μ is really equal to 870, is .9207, or approximately 92 chances in 100.

Values of $(1 - \beta)$ can be calculated for various values of μ_a different from $\mu_0 = 880$ to measure the power of the test. For example, if $\mu_a = 885$,

$$\beta = P(874.18 < \bar{x} < 885.82 \text{ when } \mu = 885)$$

$$= P(-3.64 < z < .28) = .5 + .1103 = .6103$$

Table 8.2
Values of $(1 - \beta)$ for various values of μ_a, Example 8.2

μ_a	$1 - \beta$	μ_a	$1 - \beta$
865	.9990	883	.1726
870	.9207	885	.3897
872	.7673	888	.7673
875	.3897	890	.9207
877	.1726	895	.9990
880	.0500		

Figure 8.5
Power curve for Example 8.2

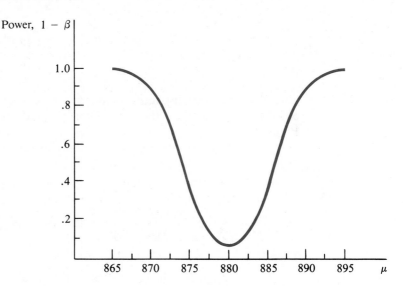

and the power is $(1 - \beta) = .3897$. Table 8.2 shows the power of the test for various values of μ_a, and a power curve is graphed in Figure 8.5. Note that the power of the test increases as the distance between μ_a and μ_0 increases. The result is a U-shaped curve for this two-tailed test.

TIPS ON PROBLEM SOLVING: CALCULATING β

1. Find the critical value or values of $\bar{x}$ used to separate the acceptance and rejection regions.
2. Using one or more values for μ consistent with the alternative hypothesis, H_a, calculate the probability that the sample mean $\bar{x}$ falls in the *acceptance region*. This will produce the value $\beta = P[\text{accept } H_0 \text{ when } \mu = \mu_a]$.

EXERCISES   Basic Techniques

8.1   A random sample of $n = 35$ observations from a population produced a mean $\bar{x} = 2.4$ and standard deviation equal to 0.29. Suppose that your research objective is to show that the population mean μ exceeds 2.3.

a.   Give the null and alternative hypotheses for the test.
b.   Describe the probability of a type I error, α, in terms of the null and alternative hypotheses.
c.   Locate the rejection region for the test if $\alpha = .05$.
d.   Before you conduct the test, glance at the data and use your intuition to decide whether the sample mean $\bar{x} = 2.4$ implies that $\mu > 2.3$. Now conduct the test. Do the data provide sufficient evidence to indicate that $\mu > 2.3$?

8.2   Refer to Exercise 8.1. Suppose that your research objective is to show that the population mean is less than 2.9. Give the null and alternative hypotheses for the test. Is the test one- or two-tailed? Explain.

8.3   Refer to Exercise 8.1. If the research objective is to show that μ differs from 2.9 (i.e., is either greater than or less than 2.9), state the null and alternative hypotheses. Is the test one- or two-tailed?

8.4   Refer to Exercise 8.1. In testing $H_0: \mu = 2.3$ against $H_a: \mu > 2.3$,
a.   Find the critical value of $\bar{x}$ necessary for rejection of H_0.
b.   Calculate $\beta = P[\text{accept } H_0 \text{ when } \mu = 2.4]$. Repeat the calculation for $\mu = 2.3, 2.5,$ and 2.6.
c.   Use the values of β calculated in part (b) to graph the power curve for the test.

Applications

8.5   High airline occupancy rates on scheduled flights are essential to profitability. Suppose that a scheduled flight must average at least 60% occupancy in order to be profitable and that an examination of the occupancy rates for 120 10:00 A.M. flights from Atlanta to Dallas showed a mean occupancy rate per flight of 58% and a standard deviation of 11%.
a.   If μ is the mean occupancy per flight and if the company wishes to determine whether or not this scheduled flight is unprofitable, give the alternative and the null hypotheses for the test.
b.   Does the alternative hypothesis in part (a) imply a one- or a two-tailed test? Explain.
c.   Do the occupancy data for the 120 flights suggest that this scheduled flight is unprofitable? Test using $\alpha = .10$.

8.6   The pH factor is a measure of the acidity or alkalinity of water. A reading of 7.0 is neutral; values in excess of 7.0 indicate alkalinity; those below 7.0 imply acidity. Loren Hill, in *Bassmaster Magazine* (September/October 1980), states that the best chance of catching bass occurs when the pH of the water is in the range of 7.5 to 7.9. Suppose that you suspect that acid rain is lowering the pH of your favorite fishing spot and you wish to determine whether the pH is less than 7.5.
a.   State the alternative and null hypotheses that you would choose for a statistical test.
b.   Would the alternative hypothesis in part (a) imply a one- or a two-tailed test? Explain.
c.   Suppose that a random sample of 30 water specimens gave pH readings with $\bar{x} = 7.3$ and $s = .2$. Just glancing at the data, do you think that the difference $\bar{x} - 7.5 = -2$ is large enough to indicate that the mean pH of the water samples is less than 7.5? (Do not conduct the test.)
d.   Now conduct a statistical test of the null hypothesis in part (a) and state your conclusions. Test using $\alpha = .05$. Compare your statistically based decision with your intuitive decision in part (c).

8.7   A drug manufacturer claimed that the mean potency of one of its antibiotics was 80%. A random sample of $n = 100$ capsules were tested and produced a sample mean of $\bar{x} = 79.7\%$, with a standard deviation $s = .8\%$. Do the data present sufficient evidence to refute the manufacturer's claim? Let $\alpha = .05$.

a.   State the null hypothesis to be tested.

b.   State the alternative hypothesis.

c.   Conduct a statistical test of the null hypothesis and state your conclusion.

8.8   In Exercise 6.14, we described the Prevention Index—a composite index designed to measure fitness based on 21 key health- and safety-promoting activities. A random sample of 1250 American adults (*Press Enterprise*, May 24, 1989) scored 65.4 of a possible 100 points, nearly 4 points higher than the 1984 average. If the mean of the Prevention Index scores in 1984 was 61.5 and the standard deviation of the scores was approximately equal to 12, do the data provide sufficient evidence to indicate that the mean score for all Americans is higher now than it was in 1984? Test using $\alpha = .05$.

8.9   As more and more Americans enter the work force, it is becoming increasingly difficult for the present transportation and communications systems to handle the increasing demand for services during the traditional work hours. Hence, many companies are becoming involved in *flextime*, in which the worker schedules his or her own work hours, or compresses workweeks. A company that was contemplating the installation of a flextime schedule estimated that it needed a minimum mean of 7 hours per day per assembly worker in order to operate effectively. Each of a random sample of 80 of the company's assemblers was asked to submit a tentative flextime schedule. If the mean of the number of hours per day for Monday was 6.7 hours and the standard deviation was 2.7 hours, do the data provide sufficient evidence to indicate that the mean number of hours worked per day on Mondays, for all of the company's assemblers, will be less than 7 hours? Test using $\alpha = .05$.

8.5 TESTING AN HYPOTHESIS ABOUT THE DIFFERENCE BETWEEN TWO POPULATION MEANS

A test of an hypothesis about the difference $(\mu_1 - \mu_2)$ between two population means μ_1 and μ_2, based on independent random samples of n_1 and n_2 observations, respectively, is summarized below.

LARGE-SAMPLE STATISTICAL TEST FOR $(\mu_1 - \mu_2)$

1.   Null Hypothesis: $H_0 : (\mu_1 - \mu_2) = D_0$, where D_0 is some specified difference that you wish to test. For many tests, you will wish to hypothesize that there is no difference between μ_1 and μ_2; that is, $D_0 = 0$.

2.   Alternative Hypothesis:

One-Tailed Test	*Two-Tailed Test*
$H_a : (\mu_1 - \mu_2) > D_0$	$H_a : (\mu_1 - \mu_2) \neq D_0$
(or, $H_a : (\mu_1 - \mu_2) < D_0$)	

3.   Test Statistic: $z = \dfrac{(\bar{x}_1 - \bar{x}_2) - D_0}{\sigma_{(\bar{x}_1 - \bar{x}_2)}} = \dfrac{(\bar{x}_1 - \bar{x}_2) - D_0}{\sqrt{\dfrac{\sigma_1^2}{n_1} + \dfrac{\sigma_2^2}{n_2}}}$.

If σ_1^2 and σ_2^2 are unknown (which is usually the case), substitute the sample variances s_1^2 and s_2^2 for σ_1^2 and σ_2^2, respectively.

4. Rejection Region:

One-Tailed Test

$z > z_\alpha$

(or $z < -z_\alpha$ when the alternative hypothesis is $H_a:(\mu_1 - \mu_2) < D_0$)

Two Tailed-Test

$z > z_{\alpha/2}$ or $z < -z_{\alpha/2}$

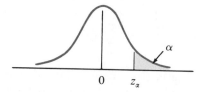 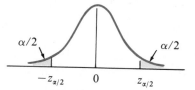

Assumptions: The samples were randomly and independently selected from the two populations and $n_1 \geq 30$ and $n_2 \geq 30$.

EXAMPLE 8.3 A university investigation, conducted to determine whether car ownership affects academic achievement, was based on two random samples of 100 male students, each drawn from the student body. The grade point average for the $n_1 = 100$ nonowners of cars had an average and variance equal to $\bar{x}_1 = 2.70$ and $s_1^2 = .36$, as opposed to a $\bar{x}_2 = 2.54$ and $s_2^2 = .40$ for the $n_2 = 100$ car owners. Do the data present sufficient evidence to indicate a difference in the mean achievement between car owners and nonowners of cars? Test using $\alpha = .10$.

Solution Since we wish to detect a difference, if it exists, between the mean academic achievement for nonowners of cars μ_1 and car owners μ_2, we will wish to test the null hypothesis that there is no difference between the means, against the alternative hypothesis that $(\mu_1 - \mu_2) \neq 0$; that is,

$$H_0:(\mu_1 - \mu_2) = D_0 = 0 \quad \text{and} \quad H_a:(\mu_1 - \mu_2) \neq 0$$

Substituting into the formula for the test statistic, we obtain

$$z = \frac{(\bar{x}_1 - \bar{x}_2) - D_0}{\sqrt{\dfrac{\sigma_1^2}{n_1} + \dfrac{\sigma_2^2}{n_2}}} \approx \frac{2.70 - 2.54}{\sqrt{\dfrac{.36}{100} + \dfrac{.40}{100}}} = 1.84$$

Using a two-tailed test with $\alpha = .10$, we will place $\alpha/2 = .05$ in each tail of the z distribution and reject H_0 if $z > 1.645$ or $z < -1.645$. (See Figure 8.6.) Since $z = 1.84$ exceeds $z_{\alpha/2} = 1.645$, it falls in the rejection region. Therefore, we would

Figure 8.6

Location of the rejection region in Example 8.3

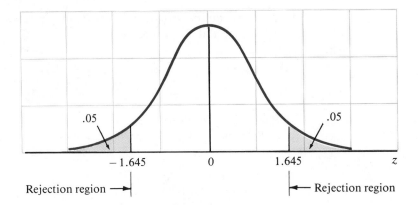

reject the null hypothesis that there is no difference in the average academic achievement of car owners versus nonowners of cars. The chance of rejecting H_0, assuming H_0 is true, is only $\alpha = .10$, and hence we would be inclined to think that we have made a reasonably good decision.

EXERCISES   Basic Techniques

8.10   Independent random samples of $n_1 = 80$ and $n_2 = 80$ were selected from populations 1 and 2, respectively. The population parameters and the sample means and variances are shown below.

	Population	
	1	2
Population mean	μ_1	μ_2
Population variance	σ_1^2	σ_2^2
Sample size	80	80
Sample mean	11.6	9.7
Sample variance	27.9	38.4

a.   If your research objective is to show that μ_1 is larger than μ_2, state the alternative and the null hypotheses that you would choose for a statistical test.

b.   Is the test in part (a) a one- or a two-tailed test?

c.   Give the test statistic that you would use for the test in parts (a) and (b) and give the rejection region for $\alpha = .10$.

d.   Look at the data. Based on your intuition, do you think that the data provide sufficient evidence to indicate that μ_1 is larger than μ_2? [We will employ a statistical test to reach this decision in part (e).]

e.   Conduct the test and draw your conclusions. Do the data present sufficient evidence to indicate that $\mu_1 > \mu_2$?

8.11   Independent random samples of $n_1 = 36$ and $n_2 = 45$ were selected from populations 1 and 2, respectively. The population parameters and the sample means and variances are shown on page 297.

	Population	
	I	2
Population mean	μ_1	μ_2
Population variance	σ_1^2	σ_2^2
Sample size	36	45
Sample mean	1.24	1.31
Sample variance	0.0560	0.0540

a.   If your research objective is to show that μ_1 and μ_2 are different, state the alternative and the null hypotheses that you would choose for a statistical test.

b.   Is the test in part (a) a one- or a two-tailed test?

c.   Give the test statistic that you would use for the test in parts (a) and (b) and give the rejection region for $\alpha = .05$.

d.   Look at the data. Based on your intuition, do you think that the data provide sufficient evidence to indicate that μ_1 differs from μ_2? [We will employ a statistical test to reach this decision in part (e).]

e.   Conduct the test and draw your conclusions. Do the data present sufficient evidence to indicate that $\mu_1 \neq \mu_2$?

8.12   Suppose that you wish to detect a difference between μ_1 and μ_2 (either $\mu_1 > \mu_2$ or $\mu_1 < \mu_2$) and that, instead of running a two-tailed test using $\alpha = .10$, you employ the following test procedure. You wait until you have collected the sample data and have calculated $\bar{x}_1$ and $\bar{x}_2$. If $\bar{x}_1$ is larger than $\bar{x}_2$, you choose the alternative hypothesis $H_a: \mu_1 > \mu_2$ and run a one-tailed test placing $\alpha_1 = .10$ in the upper tail of the z distribution. If, on the other hand, $\bar{x}_2$ is larger than $\bar{x}_1$, you reverse the procedure and run a one-tailed test, placing $\alpha_2 = .10$ in the lower tail of the z distribution. If you use this procedure and if μ_1 actually equals μ_2, what is the probability α that you will conclude that μ_1 is not equal to μ_2 (i.e., what is the probability α that you will incorrectly reject H_0 when H_0 is true)? This exercise demonstrates why statistical tests should be formulated *prior* to observing the data.

Applications

8.13   An experiment was planned to compare the mean time (in days) required to recover from a common cold for persons given a daily dose of 4 milligrams of vitamin C versus those who were not given a vitamin supplement. Suppose that 35 adults were randomly selected for each treatment category and that the mean recovery times and standard deviations for the two groups were as follows:

	Treatment	
	No vitamin supplement	4 mg vitamin C
Sample size	35	35
Sample mean	6.9	5.8
Sample standard deviation	2.9	1.2
Population mean	μ_1	μ_2

a.   Suppose your research objective is to show that the use of vitamin C reduces the mean time required to recover from a common cold and its complications. Give the null and alternative hypotheses for the test. Is this a one- or a two-tailed test?

b.   Conduct the statistical test of the null hypothesis in part (a) and state your conclusion. Test using $\alpha = .05$.

8.14   In Exercise 7.24, we described an experiment conducted by Paul I. Ward to investigate a theory concerning the moulting of the male *Gammarus pulex*, a small crustacean (P. I. Ward, "*Gammarus pulex* Control Their Moult Timing to Secure Mates," *Animal Behaviour* 32 [1984]). If a male has to moult while paired with a female, he must release her and so lose her. The theory is that the male *Gammarus pulex* is able to postpone the time to moult and thereby reduce the possibility of losing his mate. Ward randomly assigned one hundred pairs of males and females to two groups of 50 each. Pairs in the first group were maintained together (Normal); those in the second group were separated (Split). The length of time to moult was recorded for both males and females, and the means, standard deviations, and sample sizes are shown in the table. (The numbers of crustaceans in each of the four samples are less than 50 because some did not survive until moulting time.)

	Time to moult (days)		
	Mean	s	n
Males			
Normal	24.8	7.1	34
Split	21.3	8.1	41
Females			
Normal	8.6	4.8	45
Split	11.6	5.6	48

a.   Do the data present sufficient evidence to indicate that the mean moult time for "Normal" males exceeds the mean time for those "Split" from their mates? State the null and alternative hypotheses that you would use for the test.

b.   Give the rejection region for the test using $\alpha = .05$.

c.   Conduct the test and state your results.

8.15   Refer to the time to moult data for the female *Gammarus pulex* crustaceans in Exercise 8.14. Do the data present sufficient evidence to indicate that the mean moult time for "Normal" females exceeds the mean time for those "Split" from their mates?

8.16   In Exercise 7.94, we presented the results of a longitudinal study of first-year medical students at the University of North Carolina School of Medicine. The data, which is reproduced below, gives the means and standard deviations for three measures of social well-being used in the study (Strayhorn, Gregory, and Frierson, H., "Assessing Correlations between Black and White Students' Perceptions of the Medical School Learning Environment, Their Academic Performances, and Their Well-being," *Academic Medicine,** 64(1989): 468–473).

	Black ($n_1 = 50$)		White ($n_2 = 392$)	
Aspect measured	Mean	Standard deviation	Mean	Standard deviation
Quality of learning environment	112.4	16.4	111.3	13.4
Perception of stress	70.5	11.5	73.5	10.4
Social support	94.0	17.5	85.9	15.2

* Formerly called the *Journal of Medical Education*.

a. Do the data provide sufficient evidence to indicate a difference between mean scores for the quality of learning environment for blacks versus whites? Test using $\alpha = .05$. Interpret the test results.

b. Is there a significant difference between the mean scores measuring perception of stress for blacks versus whites? Test using $\alpha = .05$.

c. Do the data indicate a significant difference between mean scores measuring social support for blacks versus whites? Test using $\alpha = .05$.

d. Discuss the results of parts (a), (b), and (c) in terms of the implications for minority medical students.

8.17 In an article entitled "A Strategy for Big Bucks," Charles Dickey discusses studies of the habits of white-tailed deer that indicate that they live and feed within very limited ranges— approximately 150 to 205 acres (*Field and Stream*, October 1980). To determine whether there was a difference between the ranges of deer located in two different geographical areas, 40 deer were caught, tagged, and fitted with small radio transmitters. Several months later, the deer were tracked and identified, and the distance x from the release point was recorded. The mean and standard deviation of the distances from the release point were as follows:

	Location	
	1	2
Sample size	40	40
Sample mean	2980 ft	3205 ft
Sample standard deviation	1140 ft	963 ft
Population mean	μ_1	μ_2

a. If you have no preconceived reason for believing one population mean to be larger than another, what would you choose for your alternative hypothesis? your null hypothesis?

b. Would your alternative hypothesis in part (a) imply a one- or a two-tailed test? Explain.

c. Do the data provide sufficient evidence to indicate that the mean distances differ for the two geographical locations? Test using $\alpha = .10$.

8.18 In our descriptive analysis of the NIH blood pressure data in Appendix I, we noted that it appears the the distribution of systolic blood pressures for males appears to be shifted more than 8 pressure units above the corresponding distribution for females. Is this difference real or are we just observing a difference that can be explained by the variation within the data? In other words, do the data present sufficient evidence to indicate a difference between mean systolic blood pressures for males versus females in the 15-to-20 age group? The sample sizes, means, and standard deviations for the two data sets (see Appendix I) are shown below. Test the null hypothesis that no difference exists between the means against the two-sided alternative hypothesis that the means differ. Use $\alpha = .01$. Interpret your results.

	Sample size	Mean	Standard deviation
Male	965	118.728	14.2343
Female	945	110.390	12.5753

8.19 *Psychology Today* reports on a study by environmental psychologist Karen Frank and two colleagues, which was conducted to determine whether it is more difficult to make friends in a large city than in a small town. Two groups of graduate students were used for the study: 45 new arrivals at a private university in New York City and 47 at a prestigious university located in a town of 31,000 in an upstate rural area. The number of friends (more than just acquaintances) made by each student was recorded for each student at the end of two months and then again at the end of seven months. The seven-month sample means are shown below. (Reprinted from *Psychology Today.* Copyright ©1981 Ziff-Davis Publishing Company.)

	Location	
	New York City	Upstate rural small town
Sample size	45	47
Sample mean (7 months)	5.1	5.3
Population mean (7 months)	μ_1	μ_2

a. If your theory is that it is more difficult to make friends in a city than in a small town, state the alternative hypothesis that you would use for a statistical test. State your null hypothesis.

b. Suppose that $\sigma_1 = 2.2$ and $\sigma_2 = 2.3$ for the populations of "numbers of friends" made by new graduate students after seven months at a new location. Do the data provide sufficient evidence to indicate that the mean number of friends acquired after seven months differs for the two locations? Test using $\alpha = .05$.

▷ 8.6 TESTING AN HYPOTHESIS ABOUT A POPULATION PROPORTION

The large-sample statistical test for a population proportion p can be summarized as follows:

LARGE-SAMPLE TEST FOR A POPULATION PROPORTION p

1. Null Hypothesis: $H_0:p = p_0$.

2. Alternative Hypothesis:

One-Tailed Test	*Two-Tailed Test*
$H_a:p > p_0$	$H_a:p \neq p_0$
(or, $H_a:p < p_0$)	

3. Test Statistic: $z = \dfrac{\hat{p} - p_0}{\sigma_{\hat{p}}} = \dfrac{\hat{p} - p_0}{\sqrt{\dfrac{p_0 q_0}{n}}}$, with $\hat{p} = \dfrac{x}{n}$

where x is the number of successes in n binomial trials.*

* An equivalent test statistic can be found by multiplying the numerator and denominator of z by n to obtain

$$z = \frac{x - np_0}{\sqrt{np_0 q_0}}$$

4. Rejection Region:

One-Tailed Test

$z > z_\alpha$
(or $z < -z_\alpha$ when the alternative hypothesis is $H_a: p < p_0$)

Two-Tailed Test

$z > z_{\alpha/2}$ or $z < -z_{\alpha/2}$

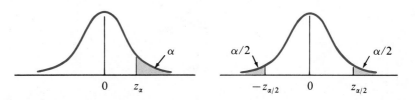

Assumption: The sampling satisfies the assumptions of a binomial experiment (Section 4.2), and n is large enough so that the sampling distribution of $\hat{p}$ can be approximated by a normal distribution. The interval $p \pm 2\sigma_{\hat{p}}$ must be contained in the interval 0 to 1.

EXAMPLE 8.4   Regardless of age, about 20% of American adults participate in fitness activities at least twice a week (*American Demographics*, May 1989, p. 40). However, the fitness activities in which people participate change as they get older, and occasional participants become nonparticipants as they age. In a local survey of $n = 100$ adults randomly selected without regard to age, a total of 27 people indicated that they participated in a fitness activity at least twice a week. Do these data indicate that the local participation rate differs significantly from the 20% figure?

Solution   It is assumed that the sampling procedure satisfies the requirements of a binomial experiment. An answer to the question posed can be determined by testing the hypothesis

$$H_0: p = .2$$

against the alternative

$$H_a: p \neq .2$$

A two-tailed test is used because we wish to detect whether the value of p is either greater or less than .2.

The point estimator of p is $\hat{p} = x/n$, and the test statistic is

$$z = \frac{\hat{p} - p_0}{\sqrt{p_0 q_0 / n}}$$

When H_0 is true, the value of p is $p_0 = .2$, and the sampling distribution of $\hat{p}$ has a mean equal to p_0 and a standard deviation of $\sqrt{p_0 q_0 / n}$. Hence, $\sqrt{\hat{p}\hat{q}/n}$ **is not used to estimate the standard error of $\hat{p}$ in this case, because the test statistic is calculated**

Figure 8.7
Location of the rejection
region in Example 8.4

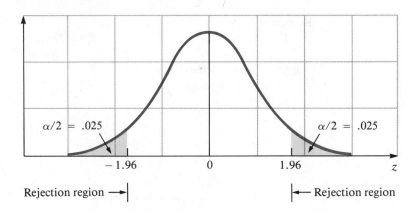

under the assumption that H_0 is true. (When estimating the value of p using the estimator $\hat{p}$, the standard error of $\hat{p}$ is not known and *is estimated* by $\sqrt{\hat{p}\hat{q}/n}$.)

With $\alpha = .05$, we would reject H_0 when $z < -1.96$ or $z > 1.96$ (see Figure 8.7). With $\hat{p} = 27/100 = .27$, the value of the test statistic is

$$z = \frac{\hat{p} - p_0}{\sqrt{\dfrac{p_0 q_0}{n}}} = \frac{.27 - .20}{\sqrt{\dfrac{(.20)(.80)}{100}}} = 1.75$$

The calculated value of the test statistic does not fall in the rejection region, and hence *we do not reject* H_0.

Do we accept H_0? No, not until we have stated alternative values of p different from $p_0 = .2$ that are of *practical significance*. The probability of a type II error should be calculated using these alternative values. If β is sufficiently small, we would accept H_0 with full awareness of the risk of an erroneous decision. ◁

TIPS ON PROBLEM SOLVING

When testing an hypothesis concerning p, use p_0 (not $\hat{p}$) to calculate $\sigma_{\hat{p}}$ in the denominator of the z statistic. The reason for this is that the rejection region is determined by the distribution of $\hat{p}$ when the null hypothesis is true, namely, when $p = p_0$.

Examples 8.1 and 8.4 illustrate an important point. **If the data present sufficient evidence to reject H_0, the probability of an erroneous conclusion α is known in advance because α is used in locating the rejection region. Since α is usually small, we are fairly certain that we have made a correct decision.** On the other hand, if the data present insufficient evidence to reject H_0, the conclusions are not so obvious. Ideally, following the statistical test procedure outlined in Section 8.3, we would have specified a practically significant alternative θ_a in advance and chosen n such that β would be small. Unfortunately, many experiments are not conducted in this ideal manner. Someone chooses a sample size and the experimenter or statistician is left to evaluate the evidence.

The calculation of β is not too difficult for the statistical test procedure outlined in this section but may be extremely difficult, if not beyond the capability of the beginner, in other test situations. **A much simpler procedure is to *not reject* H_0 rather than to accept it, and then to estimate using a confidence interval.** The interval will give you a range of plausible values for θ.

EXERCISES   Basic Techniques

8.20   A random sample of $n = 1000$ observations from a binomial population produced $x = 279$.
   a.   If your research hypothesis is that p is less than .3, what should you choose for your alternative hypothesis? your null hypothesis?
   b.   Does your alternative hypothesis in part (a) imply a one- or a two-tailed statistical test? Explain.
   c.   Do the data provide sufficient evidence to indicate that p is less than .3? Test using $\alpha = .05$.

8.21   A random sample of $n = 1400$ observations from a binomial population produced $x = 529$.
   a.   If your research hypothesis is that p differs from .4, what should you choose for your alternative hypothesis? your null hypothesis?
   b.   Does your alternative hypothesis in part (a) imply a one- or a two-tailed statistical test? Explain.
   c.   Do the data provide sufficient evidence to indicate that p differs from .4? Test using $\alpha = .10$.

8.22   A random sample of 120 observations was selected from a binomial population and 72 successes were observed. Do the data provide sufficient evidence to indicate that p is larger than .5? Test using $\alpha = .05$.

Applications

8.23   Many in Britain believe that to spare the rod is to spoil the child. Writing on this subject, the *New York Times* (August 18, 1985) gives the results of a national poll in Britain conducted in February by Marketing and Opinion Research International for the *Times* of London. Of 604 parents questioned, 63% were in favor of corporal punishment in the schools. Does this provide sufficient evidence to indicate that the majority of the British populace favor corporal punishment in the schools? Test using $\alpha = .01$.

8.24   Voters' reactions to proposed tax increases are often tempered by the ways in which the tax monies are to be used. A report entitled "What Americans Are Saying About Taxes" (*Public Opinion*, March/April, 1989, p. 21) claims that 51% of the voting public view the problem of "providing adequate medical care for all who need it but can't afford it" as one that requires government action even if new taxes are needed. In a random sample of $n = 500$ American voters, 271 agreed that adequate medical care for all requires government action even if new taxes are needed. Use a test of hypothesis to determine if there is sufficient sample evidence to indicate that the true proportion in favor of government action even if taxes are needed differs from $p = .51$.

8.25   The *Wall Street Journal* (August 27, 1985) states that the National Union of Hospital and Health Care Workers won 56 of 80 union representation elections in 1984 compared with an all-health-care union rate of 55%. Do the data provide sufficient evidence to indicate that the National Union of Hospital and Health Care Workers is more successful in winning union representation elections than other unions in the health care industry?

a. State the null and alternative hypothesis that you will use to answer this question.

b. Give the rejection region for the test. Use $\alpha = .05$.

c. What assumptions must be satisfied in order for the test to be valid?

d. Perform the calculations, complete the test, and state your conclusions.

8.26 More than ever before, Americans are working at two jobs, according to a Labor Department survey reported in the *Wall Street Journal* (November 7, 1989). According to the survey, the proportion of employed Americans holding two or more jobs is 6.2% compared to 5.4% in 1985. Assume that the current survey was based on a random sample of 850 employed Americans. If you wish to show that the proportion of Americans holding two or more jobs is greater than the 1985 figure,

a. State the null and alternative hypotheses to be tested.

b. Locate the rejection region for $\alpha = .01$.

c. Conduct the test and state your conclusions.

8.27 Although more than one in ten public and commercial buildings in the United States contain damaged asbestos that could cause cancer, the Environmental Protection Agency (EPA) has declined to order a nationwide cleanup program (*Press Enterprise*, Riverside, Calif., March 1, 1988). An EPA administrator conceded that removal is attractive in concept, but that improperly performed removal could result in high levels of exposure. The survey concluded that 20% of the nation's 3.6 million public or commercial buildings contain asbestos in a form that could be crushed or damaged with simple hand pressure.

a. In a small-scale survey involving 50 randomly selected public and commercial buildings within the county, 15 buildings were found to contain asbestos that could be crushed or damaged by simple hand pressure. Perform a test to determine whether this county rate differs significantly from the postulated national figure of 20%. Use $\alpha = .05$.

b. Find a 95% confidence interval for the proportion of public or commercial buildings in the county that have asbestos that can be damaged by simple hand pressure. Does this interval confirm or contradict the results of part (a)?

8.28 Human skin can be removed from a dead person, stored in a skin bank, and then grafted onto burn victims. Only skin from flat areas, such as the chest, back, and thighs, is acceptable and, because persons with infections, cancer, and so on must be excluded, only approximately 7% of all potential donors are acceptable. The *New York Times* (June 14, 1981) notes that a burn research center at one large city hospital screens approximately 8000 deaths annually and, from these, gets only 80 donors, or 1%. Do the data provide sufficient evidence to indicate that the burn research center is selecting their donors from a population containing a lower percentage of acceptables than found in the population at large (7%)?

8.29 An article in the *Philadelphia Inquirer* (August 15, 1985) states: "The first effective treatment for primary liver cancer, using radioactive antibodies to attack cancer cells, has tripled the average remission time in a group of patients." The article notes that in 104 cases treated by this method since 1979, almost half have gone into remission, compared with the usual rate of only 15%. Suppose we interpret this to mean that 52 patients went into remission. Does this number provide sufficient evidence to indicate that the new treatment is effective in increasing the rate of cases that will go into remission? To answer this question,

a. Give the null and alternative hypotheses used to detect an increase in the rate of remissions.

b. Give the test statistic and the rejection region for the test. Use $\alpha = .01$.

c. Conduct the test and interpret the results.

▷ **8.7 TESTING AN HYPOTHESIS ABOUT THE DIFFERENCE BETWEEN TWO POPULATION PROPORTIONS**

The large-sample statistical test for the difference between two population proportions can be summarized as follows:

A LARGE-SAMPLE STATISTICAL TEST FOR $(p_1 - p_2)$

1. Null Hypothesis: $H_0:(p_1 - p_2) = D_0$ where D_0 is some specified difference that you wish to test. For many tests, you will wish to hypothesize that there is no difference between p_1 and p_2; that is, $D_0 = 0$.

2. Alternative Hypothesis:

   One-Tailed Test
   $H_a:(p_1 - p_2) > D_0$
   (or $H_a:(p_1 - p_2) < D_0$)

   Two-Tailed Test
   $H_a:(p_1 - p_2) \neq D_0$

3. Test Statistic: $z = \dfrac{(\hat{p}_1 - \hat{p}_2) - D_0}{\sigma_{(\hat{p}_1 - \hat{p}_2)}} = \dfrac{(\hat{p}_1 - \hat{p}_2) - D_0}{\sqrt{\dfrac{p_1 q_1}{n_1} + \dfrac{p_2 q_2}{n_2}}}$

   where $\hat{p}_1 = x_1/n_1$ and $\hat{p}_2 = x_2/n_2$. Since p_1 and p_2 are unknown, we will need to approximate their values in order to calculate the standard deviation of $(\hat{p}_1 - \hat{p}_2)$ that appears in the denominator of the z statistic. Approximations are available for two cases.

   Case I: If we hypothesize that p_1 equals p_2, that is,

   $H_0:p_1 = p_2$

   or, equivalently, that

   $(p_1 - p_2) = 0$

   then $p_1 = p_2 = p$ and the best estimate of p is obtained by pooling the data from both samples. Thus, if x_1 and x_2 are the numbers of successes obtained from the two samples, then

   $\hat{p} = \dfrac{x_1 + x_2}{n_1 + n_2}$

   The test statistic would be

   $z = \dfrac{(\hat{p}_1 - \hat{p}_2) - 0}{\sqrt{\dfrac{\hat{p}\hat{q}}{n_1} + \dfrac{\hat{p}\hat{q}}{n_2}}}$

   or

   $z = \dfrac{\hat{p}_1 - \hat{p}_2}{\sqrt{\hat{p}\hat{q}\left(\dfrac{1}{n_1} + \dfrac{1}{n_2}\right)}}$

Case II: On the other hand, if we hypothesize that D_0 is *not* equal to zero; that is,

$$H_0:(p_1 - p_2) = D_0$$

where $D_0 \neq 0$, then the best estimates of p_1 and p_2 are $\hat{p}_1$ and $\hat{p}_2$, respectively. The test statistic would be

$$z = \frac{(\hat{p}_1 - \hat{p}_2) - D_0}{\sqrt{\dfrac{\hat{p}_1 \hat{q}_1}{n_1} + \dfrac{\hat{p}_2 \hat{q}_2}{n_2}}}$$

4.   Rejection Region:

One-Tailed Test

$z > z_\alpha$
(or $z < -z_\alpha$ when the alternative hypothesis is $H_a:(p_1 - p_2) < D_0$)

Two-Tailed Test

$z > z_{\alpha/2}$ or $z < -z_{\alpha/2}$

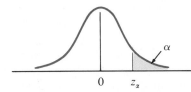

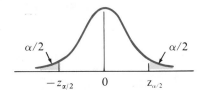

Assumptions: Samples were selected in a random and independent manner from two binomial populations, and n_1 and n_2 are large enough so that the sampling distribution of $(\hat{p}_1 - \hat{p}_2)$ can be approximated by a normal distribution. The interval $(p_1 - p_2) \pm 2\sigma_{(\hat{p}_1 - \hat{p}_2)}$ must be contained in the interval -1 to 1.

EXAMPLE 8.5   The records of a hospital show that 52 men in a sample of 1000 men versus 23 women in a sample of 1000 women were admitted because of heart disease. Do these data present sufficient evidence to indicate a higher rate of heart disease among men admitted to the hospital?

Solution   Assume that the number of patients admitted for heart disease will follow approximately a binomial probability distribution for both men and women with parameters p_1 and p_2, respectively. Then, since we wish to determine whether $p_1 > p_2$, we will test the null hypothesis that $p_1 = p_2$—that is, $H_0:(p_1 - p_2) = 0$—against the alternative hypothesis $H_a:p_1 > p_2$ or, equivalently, that $(p_1 - p_2) > 0$. To conduct this test, we use the z test statistic and approximate the value of $\sigma_{(\hat{p}_1 - \hat{p}_2)}$ using the pooled estimate of p described in Case I on pg. 305. Since H_a implies a one-tailed test, we reject H_0 only for large values of z. Thus, for $\alpha = .05$, we reject H_0 if $z > 1.645$ (see Figure 8.8).

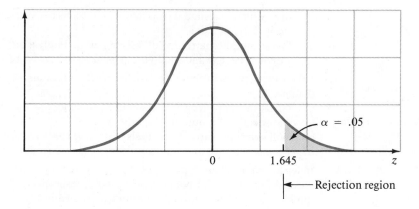

The pooled estimate of p required for $\sigma_{(\hat{p}_1 - \hat{p}_2)}$ is

$$\hat{p} = \frac{x_1 + x_2}{n_1 + n_2} = \frac{52 + 23}{1000 + 1000} = .0375$$

The test statistic is

$$z = \frac{\hat{p}_1 - \hat{p}_2}{\sqrt{\hat{p}\hat{q}\left(\frac{1}{n_1} + \frac{1}{n_2}\right)}} = \frac{.052 - .023}{\sqrt{(.0375)(.9625)\left(\frac{1}{1000} + \frac{1}{1000}\right)}} = 3.41$$

Since the computed value of z falls in the rejection region, we reject the hypothesis that $p_1 = p_2$ and conclude that the data present sufficient evidence to indicate that the percentage of men entering the hospital because of heart disease is higher than that of women. (*Note:* This does not imply that the *incidence* of heart disease is higher in men. Perhaps fewer women enter the hospital when afflicted with the disease!)   ◁

EXERCISES   Basic Techniques

8.30   Independent random samples of $n_1 = 140$ and $n_2 = 140$ observations were randomly selected from binomial populations 1 and 2, respectively. The number of successes in the samples and the population parameters are shown below.

	Population	
	1	2
Sample size	140	140
Number of sucesses	74	81
Binomial parameter	p_1	p_2

a.   Suppose that you have no preconceived theory concerning which parameter, p_1 or p_2, is the larger, and that you only wish to detect a difference between the two parameters if one exists. What should you choose for the alternative hypothesis for a statistical test? for the null hypothesis?

b.   Does your alternative hypothesis in part (a) imply a one- or a two-tailed test?

c.   Conduct the test and state your conclusions. Test using $\alpha = .05$.

8.31   Refer to Exercise 8.30. Suppose that, for practical reasons, you know that p_1 cannot be larger than p_2.

a.   Given this knowledge, what should you choose as the alternative hypothesis for your statistical test? Your null hypothesis?

b.   Will your alternative hypothesis in part (a) imply a one- or a two-tailed test? Explain.

c.   Conduct the test and state your conclusions. Test using $\alpha = .10$.

8.32   Independent random samples of $n_1 = 280$ and $n_2 = 350$ observations were randomly selected from binomial populations 1 and 2, respectively. The number of successes in the samples and the population parameters are shown below.

	Population	
	1	2
Sample size	280	350
Number of successes	132	178
Binomial parameter	p_1	p_2

a.   Suppose that you know that p_1 can never be larger than p_2 and you want to know if p_1 is less than p_2. What should you choose for your null and alternative hypotheses?

b.   Does your alternative hypothesis in part (a) imply a one- or a two-tailed test?

c.   Conduct the test and state your conclusions. Test using $\alpha = .05$.

Applications

8.33   In Exercise 7.47, we presented data collected by William B. Waegel to investigate the relationship between law and the use of lethal force by police. The data, reproduced below, represent the classification of 459 police shootings in Philadelphia over the period 1970 to 1978. The objective of the study was to see whether restrictive laws governing the use of lethal force, introduced in 1973, produced a change (presumably a reduction) in the proportion of cases where lethal force was justifiably used (W.B. Waegel, "The Use of Lethal Force by Police: The Effect of Statutory Change," *Crime and Delinquency* 30, no. 1 [January 1984]).

	Total for period 1970–1972	1973	Total for period 1974–1978
Justified	72	30	173
Not justified	11	9	59
Unable to determine	18	7	53
Accidental	10	5	12
Totals	111	51	297

Assume that the data in the table represent samples from populations of cases that could have occurred and in which lethal force would have been used; one population corresponds to the legal situation existing prior to 1973 and the other population to the situation after the new restrictive laws were introduced.

   a. Do the data provide sufficient evidence to indicate that the proportion of nonjustified shootings for the 1974–1978 period differs from the corresponding proportion for the 1970–1972 period? Test using $\alpha = .05$.

   b. Presumably, the intent of the 1973 change in the law was to reduce the proportion of cases where the use of lethal force was unjustified. Yet, the sample proportion of the total number of shootings over the 1974–1978 period classified as unjustified was larger than for the 1970–1972 period. Can you give a possible explanation for this discrepancy?

8.34 Refer to the data in Exercise 8.33 on the Philadelphia police shootings from 1970 to 1978. Did the new 1973 law placing greater restrictions on the use of lethal force make it more difficult to classify a shooting?

   a. Suppose you want to answer this question by testing for a difference in the proportions of shootings that could not be classified (i.e., they fell in the "Unable to Determine" category). State the null and alternative hypotheses that you would employ for the test.

   b. Conduct the test using $\alpha = .05$.

   c. State the practical conclusions to be derived from the test.

8.35 "Minorities leave their newspaper jobs at a rate three times higher than whites because of what they consider a lack of advancement," concluded a study conducted by the Institute for Journalism Education (*Orlando Sentinel*, May 7, 1985). This conclusion is based on interviews of 175 minority journalists and 125 white journalists. Twenty-three (or 13%) of the minority journalists left journalism compared with 6 (or 5%) of the white journalists.

   a. Do these data provide sufficient evidence to indicate differences in the proportions of minorities and whites leaving the journalism profession? Test using $\alpha = .05$.

   b. If there is evidence of a difference from part (a), does it follow that the difference is due to a perceived lack of advancement opportunities for minorities?

   c. What assumptions must you make for the test in part (a) to be valid? Are these assumptions met?

8.36 "Paralytic shellfish poisoning, PSP, remains a major problem in California," states Shirley Hudgins in her article "New Threats from PSP" (*Sea Grant Today*, May/June 1981). Shellfish containing the denoflagellate *Gonyaulax catenella*, a one-celled organism, are most often found in the cool waters of coastal northern California from May through October, but are also found in decreasing levels as you proceed south along the California coast. The reaction from eating these shellfish varies from a mild case of nausea, stomach cramps, or what appears to be a 24-hour stomach flu, to death within a few hours. Of 400 cases reported between 1927 and 1980, 30 were fatal. In contrast, in 1980 alone there were 100 reported cases and 3 were fatal. There were undoubtedly many more cases of PSP during these time periods, undiagnosed and unreported, but do these data imply a difference in death rate p_1 for reported cases between the 1927–1980 period and the death rate p_2 for the 12 months of 1980?

   a. Suppose you wish only to detect a difference between p_1 and p_2. State the alternative and null hypotheses that you would use for a statistical test.

   b. Based on a visual examination of the data, do you believe that p_1 differs from p_2? [We will conduct the statistical test in part (c).]

   c. Conduct the statistical test using $\alpha = .05$. Do the data present sufficient evidence to indicate a difference in the fatality rates for reported cases between the two time periods?

8.37 Childless women over 50 years of age are more likely to have heart attacks, according to Evelyn Talbot, a University of Pittsburgh researcher, who reported her results at a meeting of the American Heart Association (*Florida Times-Union*, Jacksonville, Fla., November 20, 1986). A study of the medical records of women who died of heart attacks in Allegheny County, Pennsylvania, found that 12 out of 51 heart attack victims were childless. In a similar group of 47 women who survived a heart attack, only two were childless. Do these data present sufficient evidence to conclude at the $\alpha = .05$ level of significance that a difference exists between the fatality rates for heart attacks among childless women versus women of the same age who were not childless? [Hint: Form a 2×2 table to enumerate the number of childless and nonchildless women and their reaction to the heart attack.]

8.8 ANOTHER WAY TO REPORT THE RESULTS OF STATISTICAL TESTS: p-VALUES

The probability α of making a type I error is often called the significance level of the statistical test, a term that originated in the following way. The probability of the observed value of the test statistic, or some value even more contradictory to the null hypothesis, measures, in a sense, the weight of evidence favoring rejection of H_0. Some experimenters report test results as being significant (we would reject H_0) at the 5% significance level but not at the 1% level. This means that we would reject H_0 if α were .05 but not if α were .01.

The smallest value of α for which test results are statistically significant is often called the **p-value**, or the **observed significance level**, for the test. Some statistical computer programs compute p-values for statistical tests correct to four or five decimal places. But if you are using statistical tables to determine a p-value, you will only be able to approximate its value. This is because most statistical tables give the critical values of test statistics only for large differential values of α (for example, .01, .025, .05, .10, and so on). Consequently, the p-value reported by most experimenters is the largest tabulated value of α for which the test remains statistically significant. For example, if a test result is statistically significant for $\alpha = .10$, but not for $\alpha = .05$, then the p-value for the test would be given as p-value $= 10$ or, more precisely, as p-value $< .10$.

Many scientific journals require researchers to report the p-values associated with statistical tests because these values provide a reader with *more information* than simply stating that a null hypothesis is or is not to be rejected for some value of α chosen by the experimenter. In a sense, it allows the reader of published research to evaluate the extent to which the data disagree with the null hypothesis. In particular, it enables each reader to choose his or her own personal value for α and then decide whether or not the data lead to rejection of the null hypothesis.

The procedure for finding the p-value for a test will be illustrated by the following examples.

EXAMPLE 8.6   Find the p-value for the statistical test in Example 8.1. Interpret your results.

Solution   Example 8.1 presents a test of the null hypothesis $H_0: \mu = 880$ against the alternative hypothesis $H_a: \mu \neq 880$. The value of the test statistic, computed from the sample data, was $z = -3.03$. Therefore, the p-value for this two-tailed test is the probability that $z \leq -3.03$ or $z \geq 3.03$ (the shaded areas in Figure 8.9).

From Table 3 in Appendix III, you can see that the tabulated area under the normal curve between $z = 0$ and $z = 3.03$ is .4988 and the area to the right of

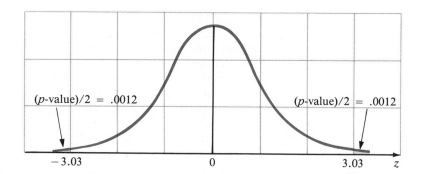

Figure 8.9
Locating the *p*-value for the test in Example 8.1

$(p\text{-value})/2 = .0012$

$(p\text{-value})/2 = .0012$

-3.03 0 3.03 z

$z = 3.03$ is $.5 - .4988 = .0012$. Then, since this was a two-tailed test, the area corresponding to the z values, $z > 3.03$ or $z < -3.03$, is $2(.0012) = .0024$. Consequently, we would report the *p*-value for the test as *p*-value $= .0024$. ◁

EXAMPLE 8.7   Find the *p*-value for the statistical test in Example 8.4. Interpret your results.

Solution   Example 8.4 presented a one-tailed test of the null hypothesis $H_0: p = .10$ against the alternative hypothesis $H_a: p > .10$, and the observed value of the test statistic was $z = 1.41$. Therefore, the *p*-value for the test is the probability of observing a value of the z statistic larger than 1.41. This value is the area under the normal curve to the right of $z = 1.41$ (the shaded area in Figure 8.10).

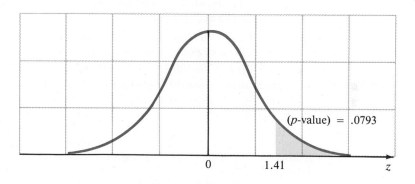

Figure 8.10
Finding the *p*-value for the test in Example 8.4

$(p\text{-value}) = .0793$

0 1.41 z

From Table 3 in Appendix III, the area under the normal curve between $z = 0$ and $z = 1.41$ is .4207. Therefore, the area under the normal curve to the right of $z = 1.41$, the *p*-value for the test, is *p*-value $= .5 - .4207 = .0793$.

Advocating that a researcher report the *p*-value for a test and leave its interpretation to the reader does not violate the traditional statistical test procedure described in the preceding sections. It simply leaves the decision of whether or not to reject the null hypothesis (with the potential for a type I or type II error) to the reader. Thus, it shifts the responsibility for choosing the value of α, and possibly the problem of evaluating the probability β of making a type II error, to the reader. ◁

EXERCISES   Basic Techniques

8.38 Suppose you tested the null hypothesis $H_0: \mu = 94$ against the alternative hypothesis $H_a: \mu < 94$. For a random sample of $n = 52$ observations, $\bar{x} = 92.9$, and $s = 4.1$.
 a. Give the observed significance level for the test. Interpret this value.
 b. If you wish to conduct your test using $\alpha = .05$, what would be your test conclusions?

8.39 Suppose you tested the null hypothesis $H_0: \mu = 94$ against the alternative hypothesis $H_a: \mu \neq 94$. For a random sample of $n = 52$ observations, $\bar{x} = 92.1$, and $s = 4.1$.
 a. Give the observed significance level for the test.
 b. If you wish to conduct your test using $\alpha = .05$, what would be your test conclusions?

8.40 Suppose you tested the null hypothesis $H_0: \mu = 15$ against the alternative hypothesis $H_a: \mu \neq 15$. For a random sample of $n = 38$ observations, $\bar{x} = 15.7$, and $s = 2.4$.
 a. Give the observed significance level for the test.
 b. If you wish to conduct your test using $\alpha = .05$, what would be your test conclusions?

Applications

 8.41 Refer to Exercises 8.14 and 8.15. In a test to detect differences in mean moult time for a species of crustaceans, Paul Ward notes that the p-values to detect differences in the means for both males and females are less than 0.05 (*"Gammarus pulex* Control Their Moult Timing to Secure Mates," *Animal Behaviour* 32 [1984]). Do you agree with Ward's p-values?

 8.42 Give the p-value for the comparison of the mean female and male systolic blood pressures in Exercise 8.18 and interpret it.

 8.43 In Exercise 8.23, we tested to determine whether our data presented sufficient evidence to indicate that the majority of the British favor corporal punishment in their schools. Give the p-value for the test and interpret it.

 8.44 In Exercise 8.33, we conducted a test to determine whether a change in the laws governing the use of lethal force by police produced a change in the proportion of cases that showed an unjustified use of lethal force. Find the p-value for the test and interpret it.

 8.45 A serious problem facing the United States is the reduction in the quality and quantity of our fresh water supply. *Bioscience* (July/August 1981), in "Features and News," reports that the water table is dropping in many areas of the United States. To illustrate, the rate of depletion of a reservoir near Plainview, Texas, from 1936 to 1973, has averaged 76 centimeters (approximately 2.5 feet) per year, and the rate of depletion has increased since 1973. The water level in one well in the Houston area dropped at an average rate of 2.4 meters (8 feet) per year from 1939 to 1973.
 a. Suppose that you wish to determine whether the mean water table in an area has dropped more than 3 feet in a given year. If you plan to sample and to conduct a statistical test, what should you choose for your alternative hypothesis? your null hypothesis?
 b. Given your alternative hypothesis in part (a), will this be a one- or a two-tailed test? Explain.
 c. Suppose that the drop in the water level over a one-year period was measured for 37 wells in the region and that the mean and standard deviation of the drops in water levels were $\bar{x} = 3.2$ feet and $s = 6.4$ inches. Find the observed significance level for the test.
 d. If you wish α to be equal to .05, would you reject the null hypothesis? What conclusions would you derive from the test?

8.46 Does auto safety education increase the use of auto safety devices? Researchers at the University of California at San Francisco gave 78 of a group of 136 expectant couples a special 30-minute lecture on the value of child safety seats (*USA Today*, March 12, 1985). Contacted four to six weeks after the birth of their children, the researchers found that 96% of those who had attended the lecture claimed that they were using the safety seats versus 78% of those who did not attend the lecture.

   a. Is this difference in percentages large enough to imply that the lecture was effective in increasing the percentage of couples using the child safety seats? Test using $\alpha = .05$.

   b. Find the *p*-value for the test, and interpret this value.

8.47 How would you like to live to be 200 years old? For centuries, humankind has sought the key to the mystery of aging. What causes it? How can it be slowed? Recent studies focus on biomarkers, physical or biological changes that occur at a predictable time in a person's life. The theory is that if ways can be found to delay the occurrence of these biomarkers, human life can be extended. A key biomarker, according to scientists, is forced vital capacity (FVC), the volume of air that you can expel after taking a deep breath. A study of 5209 men and women age 30 to 62 showed that FVC declined an average of 3.8 deciliters per decade for men and 3.1 deciliters per decade for women (Tineke, Boddé, "Biomarkers of Aging: Key to a Younger Life?" *Bioscience* 31, no. 8 [1981], 566–567. Copyright © 1981 by the American Institute of Biological Sciences). Suppose that you wished to determine whether a physical fitness program for men and women age 50 to 60 would delay aging and that you measured the FVC for 30 men and 30 women at the beginning and end of the 50-to-60-year age interval and recorded the drop in FVC for each person. The data are shown below.

	Men	Women
Sample size	30	30
Sample average drop in FVC	3.6	2.7
Sample standard deviation	1.1	1.2
Population mean drop in FVC	μ_1	μ_2

   a. Do the data provide sufficient evidence to indicate that the decrease in the mean FVC over the decade for the men on the physical fitness program is less than 3.8 deciliters? Find the observed significance level for the test.

   b. Refer to part (a). If you choose $\alpha = .05$, do the data support the contention that the mean decrease in FVC is less than 3.8 deciliters?

   c. Test to determine whether the FVC drop for women on the physical fitness program was less than 3.1 deciliters for the decade. Find the observed significance level for the test.

   d. Refer to part (c). If you choose $\alpha = .05$, do the data support the contention that the mean decrease in FVC is less than 3.1 deciliters?

8.48 Of those women who are diagnosed to have early-stage breast cancer, one-third eventually die of the disease. Suppose a community public health department instituted a screening program to provide for the early detection of breast cancer and, thus, a consequent increase in the survival rate *p* of those diagnosed to have the disease. A random sample of 200 women was selected from among those who were periodically screened by the program and who eventually were diagnosed to have the disease. Let *x* represent the number of those in the sample who survive the disease.

   a. If we wish to detect whether the community screening program has been effective, state the null hypothesis that should be tested.

b.   State the alternative hypothesis.

c.   If the number of women in the sample of 200 who survive the disease is 164, would you conclude that the community screening program was effective? Test using $\alpha = .05$ and explain the practical conclusions to be derived from your test.

d.   Find the p-value for the test and interpret it.

8.9 SOME COMMENTS ON THE THEORY OF TESTS OF HYPOTHESES

As outlined in Section 8.2, the theory of a statistical test of an hypothesis is indeed a very clear-cut procedure, enabling the experimenter to either reject or accept the null hypothesis with measured risks α and β. Unfortunately, as we noted, the theoretical framework does not suffice for all practical situations.

The crux of the theory requires that we be able to specify a meaningful alternative hypothesis that permits the calculation of the probability β of a type II error for all alternative values of the parameter(s). This indeed can be done for many statistical tests, including the large-sample test discussed in Section 8.3, although the calculation of β for various alternatives and sample sizes may, in some cases, be difficult. On the other hand, in some test situations it is extremely difficult to clearly specify alternatives to H_0 that have practical significance. This may occur when we wish to test an hypothesis concerning the values of a set of parameters, a situation that we will encounter in Chapter 12 in analyzing enumerative data.

The obstacle that we mention does not invalidate the use of statistical tests. Rather, it urges caution in drawing conclusions when insufficient evidence is available to reject the null hypothesis. The difficulty of specifying meaningful alternatives to the null hypothesis, together with the difficulty in calculating and tabulating β for other than the simplest statistical tests, justifies skirting this issue in an introductory text. Hence, we can adopt one of two procedures. We can present the p-value associated with a statistical test and leave the interpretation to you. Or, we can agree to adopt the procedure described in Example 8.4 when tabulated values of β (the operating characteristic curve) are unavailable for the test. When the test statistic falls in the acceptance region, we will *not reject*, rather than *accept*, the null hypothesis. Further conclusions may be made by calculating an interval estimate for the parameter or by consulting one of several published statistical handbooks for tabulated values of β. We will not be too surprised to learn that these tabulations are inaccessible, if not completely unavailable, for some of the more complicated statistical tests.

Finally, we might comment on the choice between a one- or a two-tailed test for a given situation. We emphasize that this choice is dictated by the practical aspects of the problem and will depend on the alternative value of the parameter, say θ, that the experimenter is trying to detect. If we were to sustain a large financial loss if θ were greater than θ_0, but not if it were less, we would concentrate our attention on the detection of values of θ greater than θ_0. Hence, we would reject in the upper tail of the distribution for the test statistics previously discussed. On the other hand, if we are equally interested in detecting values of θ that are either less than or greater than θ_0, we would employ a two-tailed test.

TIPS ON PROBLEM SOLVING

When conducting a statistical test of an hypothesis, it is important to follow the same basic procedure for each problem.

1.   **Determine the type of data** involved (quantitative or binomial) and the number of samples involved (one or two). This will allow you to identify the parameter of interest in the experiment.

2.   **Check the conditions** required for the sampling distribution of the parameter to be approximated by a normal distribution. For quantitative data, the sample size (or sizes) must be 30 or more. For binomial data, a large sample size will ensure that $p \pm 2\sigma_{\hat{p}}$ is contained in the interval 0 to 1 $[(p_1 - p_2) \pm 2\sigma_{(\hat{p}_1 - \hat{p}_2)}$ contained in the interval -1 to 1 for the two-sample case].

3.   **State the null and alternative hypotheses (H_0 and H_a).** The alternative hypothesis is the hypothesis that the researcher wishes to support; the null hypothesis is a contradiction of the alternative hypothesis.

4.   **State the test statistic** to be used in the test of the hypothesis.

5.   **Locate the rejection region** for the test. In this chapter, the rejection region will be found in the tail areas of the standard normal (z) distribution. The exact rejection region will be determined by the desired value of α and the form of the alternative hypothesis (one- or two-tailed).

6.   **Conduct the test**, calculating the observed value of the test statistic based on the sample data.

7.   **Draw conclusions** based on the observed value of the test statistic. If the test statistic falls in the rejection region, the null hypothesis is rejected in favor of the alternative hypothesis. The probability of an incorrect decision is α. However, if the test statistic does not fall in the rejection region, we cannot reject the null hypothesis. There is insufficient evidence to show that the alternative hypothesis is true. Judgment is withheld until more data can be collected.

▷ 8.10   SUMMARY

In this chapter, we have presented the basic concepts of a statistical test of an hypothesis and have demonstrated the procedure with four separate tests. All of the statistical tests described in this chapter are based on the Central Limit Theorem, and hence apply to large samples. When n is large, each of the respective test statistics possesses a sampling distribution that can be approximated by the normal distribution. This result, along with the properties of the normal distribution studied in Chapter 5, permits the calculation of α, β, and p-values for the statistical tests.

In addition to the test statistics presented in this chapter, many other test statistics have sampling distributions that are approximately normal when the sample sizes are large (you will encounter several in Chapter 14). But when the sample sizes are small, the sampling distributions of most test statistics are not normal. One of these nonnormal sampling distributions—the t distribution—will be employed in Chapter 9 to obtain confidence intervals and tests of hypotheses for a single population mean and the difference between two population means.

REFERENCES

Dixon, W. J., and Massey, F. J., Jr. *Introduction to Statistical Analysis.* 4th ed. New York: McGraw-Hill, 1983.

Freund, J. E., and Walpole, R. E. *Mathematical Statistics.* 4th ed. Englewood Cliffs, N.J.: Prentice Hall, 1987.

Hogg. R. V., and Craig, A. T. *Introduction to Mathematical Statistics.* 4th ed. New York: Macmillan, 1986.

Mendenhall, W.; Wackerly, D.; and Scheaffer, R. L. *Mathematical Statistics with Applications.* 4th ed. Boston: PWS-KENT, 1990.

Neter, J.; Wasserman, W.; and Whitmore, G. A. *Applied Statistics.* 3d ed. Boston: Allyn and Bacon, 1987.

Triola, M. F. *Elementary Statistics.* 4th ed. Redwood City, Calif.: Benjamin/Cummings, 1989.

SUPPLEMENTARY EXERCISES

[Starred (*) exercises are optional.]

8.49   Define α and β for a statistical test of an hypothesis.

8.50   What is the observed significance level of a test?

8.51   The daily wages in a particular industry are normally distributed with a mean of $43.20 and a standard deviation of $9.50. Suppose that a company in this industry employing 40 workers pays these workers $41.20 on the average. Based on this sample mean, could these workers be viewed as a random sample from among all workers in the industry?
   a.   Find the observed significance level of the test.
   b.   If you planned to conduct your test using $\alpha = .01$, what would be your test conclusions?

8.52   Refer to Exercise 7.18 and the collection of water samples to estimate the mean acidity (in pH) of rainfalls in the northeastern United States. As noted, the pH for pure rain falling through clean air is approximately 5.7. The sample of $n = 40$ rainfalls produced pH readings with $\bar{x} = 3.7$ and $s = .5$. Do the data provide sufficient evidence to indicate that the mean pH for rainfalls is more acidic ($H_a: \mu < 5.7$ pH) than pure rainwater? Test using $\alpha = .05$. Note that this inference is appropriate only for the area in which the rainwater specimens were collected.

8.53   A manufacturer of automatic washers provides a particular model in one of three colors: A, B, or C. Of the first 1000 washers sold, it is noted that 400 were of color A. Would you conclude that more than one-third of all customers have a preference for color A?
   a.   Find the observed significance level of the test.
   b.   If you planned to conduct your test using $\alpha = .05$, what would be your test conclusions?

8.54 What conditions must be met so that the z test can be used to test an hypothesis concerning a population mean μ?

8.55 In Exercise 7.85, we presented data from a controlled pollination study (Pittman, Karen E., and Levin, Donald A. "Effects of Parental Identities and Environment on Components of Crossing Success in *Phlox Drummondii*," *American Journal of Botany*, 76(3), pp. 409–418, 1989) involving seed survival rates for plants subjected to water or nutrient deprivation. The data, representing the number of seeds surviving to maturity, are reproduced below.

	Male		Female	
Treatment	n	Number surviving	n	Number surviving
Control	585	543	632	560
Low water	578	522	510	466
Low nutrient	568	510	589	546

a. Is there a significant difference between the proportions of surviving seeds for low water versus low nutrients in the male category? Use $\alpha = .01$.

b. Do the data provide sufficient evidence to indicate a difference between survival proportions for low water versus low nutrients for the female category? Use $\alpha = .01$.

c. Look at the data for the other four treatment–sex combinations. Does it appear that there are any significant differences due to low levels of water or nutrients? Explain.

8.56 In Exercise 4.32, we noted that 16 out of every 100 doctors in any given year are subject to malpractice claims (*New York Times*, February 15, 1985). A hospital that has a staff consisting of 300 physicians seems to have an unusually large number of doctors involved in malpractice claims—58 cases in one year. Does this number provide sufficient evidence to indicate that the doctor malpractice claim rate at the hospital differs from the national rate? Test using $\alpha = .05$.

8.57 In a study to assess various effects of using a female model in automobile advertising, each of 100 male subjects was shown photographs of two automobiles matched for price, color, and size, but of different makes. One of the automobiles was shown with a female model to 50 of the subjects (group A), and both automobiles were shown without the model to the other 50 subjects (group B). In group A, the automobile shown with the model was judged as more expensive by 37 subjects, while in group B the same automobile was judged as the more expensive by 23 subjects. Do these results indicate that using a female model influences the perceived expensiveness of an automobile? Use a one-tailed test with $\alpha = .05$.

8.58 Random samples of 200 bolts manufactured by machine A and 200 bolts manufactured by machine B showed 16 and 8 defective bolts, respectively. Do these data present sufficient evidence to suggest a difference in the performance of the machines? Use $\alpha = .05$.

8.59 Suppose that the true fraction p in favor of the death penalty is the same for Democrats as it is for Republicans. Independent random samples are selected, one consisting of 800 Republicans and the other of 800 Democrats. Find the probability that the sample fraction of Republicans favoring the death penalty exceeds that of the Democrats by more than .03 if $p = .5$. What is this probability if $p = .1$?

8.60 In Exercise 6.56, we reported that the biomass for tropical woodlands, thought to be about 35 kilograms per square meter (kg/m^2), may in fact be too high, and that tropical biomass values vary regionally—from about 5 to 55 kg/m^2 (*Science News*, August 19, 1989, p. 124). Suppose that you measure the tropical biomass in 400 randomly selected square-meter plots and obtain $\bar{x} = 31.75$ and $s = 10.5$. Do the data present sufficient evidence to indicate that scientists are overestimating the mean biomass for tropical woodlands, and that the mean is in fact lower than estimated?

a. State the null and alternative hypotheses to be tested.

b. Locate the rejection region for the test with $\alpha = .01$.

c. Conduct the test and state your conclusions.

8.61   A social scientist believes that the fraction p_1 of Republicans in favor of the death penalty is greater than the fraction p_2 of Democrats in favor of the death penalty. She acquired independent random samples of 200 Republicans and 200 Democrats and found 46 Republicans and 34 Democrats favoring the death penalty. Do these data support the social scientist's belief?

a. Find the observed significance level of the test.

b. If you planned to conduct your test using $\alpha = .05$, what would be your test conclusions?

*8.62   Refer to Exercise 8.61. Some thought should have been given to designing a test for which β is tolerably low when p_1 exceeds p_2 by an important amount. For example, find a common sample size n for a test with $\alpha = .05$ and $\beta \leq .20$, when in fact p_1 exceeds p_2 by .1. [*Hint:* The maximum value of $p(1 - p)$ is .25.]

8.63   In comparing the mean weight loss for two diets, the following sample data were obtained.

	Diet I	Diet II
Sample size, n	40	40
Sample mean, $\bar{x}$	10 lb	8 lb
Sample variance, s^2	4.3	5.7

Do the data provide sufficient evidence to indicate that diet I produces a greater mean weight loss than diet II? Use $\alpha = .05$.

8.64   A test of the breaking strengths of two different types of cables was conducted using samples of $n_1 = n_2 = 100$ pieces of each type of cable.

Cable I	Cable II
$\bar{x}_1 = 1925$	$\bar{x}_2 = 1905$
$s_1 = 40$	$s_2 = 30$

Do the data provide sufficient evidence to indicate a difference between the mean breaking strengths of the two cables? Use $\alpha = .10$.

8.65   The braking ability was compared for two types of 1990 automobiles. Random samples of 64 automobiles were tested for each type. The recorded measurement was the distance (in feet) required to stop when the brakes were applied at 40 miles per hour. The computed sample means and variances were

$$\bar{x}_1 = 118 \qquad \bar{x}_2 = 109$$

$$s_1^2 = 102 \qquad s_2^2 = 87$$

Do the data provide sufficient evidence to indicate a difference between the mean stopping distances for the two types of automobiles?

8.66   A fruit grower wishes to test a new spray that a manufacturer claims will *reduce* the loss due to damage by a certain insect. To test the claim, the grower sprays 200 trees with the new spray and 200 other trees with the standard spray. The following data were recorded.

	New spray	Standard spray
Mean yield per tree, $\bar{x}$ (lb)	240	227
Variance, s^2	980	820

a. Do the data provide sufficient evidence to conclude that the new spray is better than the old? (Use $\alpha = .05$.)

b. Construct a 95% confidence interval for the difference between the mean yields for the two sprays.

*8.67 Refer to Example 8.2. Use the procedure described in Example 8.2 to calculate β for several alternative values of μ. (For example, $\mu = 873, 875$, and 877.) Use the three computed values of β along with the value computed in Example 8.2 to construct a power curve for the statistical test.

*8.68 Repeat the procedure described in the preceding exercise for a sample size $n = 25$ (as opposed to $n = 50$ used in Exercise 8.67) and compare the two power curves.

8.69 In Exercise 7.15, we described a study of 392 healthy children living in the area of Tours, France. The table below shows some results of the study, measuring the serum levels of certain fat-soluble vitamins—namely, vitamin A (retinol) and β-carotene—in both boys and girls.

	Boys ($n = 207$)	Girls ($n = 185$)
Retinol(μg/dl)	$\bar{x}_1 = 43.0$	$\bar{x}_2 = 41.8$
	$s_1 = 13.0$	$s_2 = 10.7$
β-carotene(μg/l)	$\bar{x}_1 = 588$	$\bar{x}_2 = 553$
	$s_1 = 406$	$s_2 = 350$

Source: Malvy, J. M. D. and colleagues. "Retinol, β-Carotene and a-Tocopherol Status in a French Population of Healthy Children," *International Journal of Vitamin and Nutrition Research*, Vol. 59, 1989, p. 29.

a. Do the data provide sufficient evidence to indicate a difference between the mean retinol levels for boys versus girls? Use $\alpha = .01$.

b. Do the data indicate a significant difference between the mean β-carotene levels for boys versus girls? Use $\alpha = .01$.

c. Would the results of parts (a) and (b) change if you had used $\alpha = .05$? if $\alpha = .10$?

8.70 A biologist hypothesizes that high concentrations of actinomysin D inhibit RNA synthesis in cells and hence the production of proteins as well. An experiment conducted to test this theory compared the RNA synthesis in cells treated with two concentrations of actinomysin D, .6 and .7 microgram per milliliter, respectively. Cells treated with the lower concentration (.6) of actinomysin D showed that 55 out of 70 developed normally, whereas only 23 out of 70 appeared to develop normally for the higher concentration (.7). Do these data provide sufficient evidence to indicate a difference between the rates of normal RNA synthesis for cells exposed to the two different concentrations of actinomysin D?

a. Find the observed significance level of the test.

b. If you planned to conduct your test using $\alpha = .10$, what would be your test conclusions?

AN ASPIRIN A DAY ...?

On Wednesday, January 27, 1988, the front page of the *New York Times* read "Heart attack risk found to be cut by taking aspirin: Lifesaving effects seen." A very large study of United States physicians showed that a single aspirin tablet taken every other day reduced by one-half the risk of heart attack in men (Greenhouse, Joel B., and Greenhouse, Samuel W., "An Aspirin a Day. . . ?" *Chance: New Directions for Statistics and Computing*, 1(4):24–31). Three days later, a headline in the *Times* read "Value of daily aspirin disputed in British study of heart attacks." How could two seemingly similar studies, both involving doctors as participants, result in such opposite conclusions?

The United States physicians' health study consisted of two randomized clinical trials in one. The first tested the hypothesis that 325 milligrams (mg) of aspirin taken every other day reduces mortality from cardiovascular disease. The second tested whether 50 mg of beta carotene taken on alternate days decreases the incidence of cancer. Using an American Medical Association computer tape, 261,248 male physicians between the ages of 40 and 84 were invited to participate in the trial. Of those responding, 59,285 were willing to participate. After excluding those physicians who had a history of medical disorders, or who were currently taking aspirin or had negative reactions to aspirin, 22,071 physicians were randomized into one of four treatment groups: (1) buffered aspirin and beta carotene; (2) buffered aspirin and a beta carotene placebo; (3) aspirin placebo and beta carotene; and (4) aspirin placebo and beta carotene placebo. Thus, half were assigned to receive aspirin and half received beta carotene.

The study was conducted as a double-blind study, in which neither the participants nor the investigators responsible for following the participants knew to which group a participant belonged. The results of the American study concerning myocardial infarctions (the technical name for heart attacks) are given in the following table:

	American study	
	Aspirin ($n = 11,037$)	Placebo ($n = 11,034$)
Myocardial infarction		
Fatal	5	18
Nonfatal	99	171
Total	104	189

The objective of the British study was to determine whether 500 mg of aspirin taken daily would reduce the incidence of and mortality from cardiovascular disease. In 1978, all male physicians in the United Kingdom were invited to participate. After the usual exclusions, 5139 doctors were randomly allocated to take aspirin, unless some problem developed, and one-third were randomly allocated to *avoid* aspirin. Placebo tablets were not used, and so the study was not blind! The results of the British study are given in the following table:

British study		
	Aspirin ($n = 3429$)	Control ($n = 1710$)
Myocardial infarction		
Fatal	89 (47.3)	47 (49.6)
Nonfatal	80 (42.5)	41 (43.3)
Total	169 (89.8)	88 (92.9)

To account for unequal sample sizes, the British study reported rates per 10,000-subject-years-alive (given in parentheses).

1. Test to see whether the American study does in fact indicate that the rate of heart attacks for physicians taking 325 mg of aspirin every other day was significantly different from the rate for those on the placebo. Is the American claim justified?

2. Repeat the analysis using the data from the British study in which one group took 500 mg of aspirin every day and the control group took none. Based on their data, is the British claim justified?

3. Can you think of some possible explanations as to why the results of these two studies, which were alike in some respects, produced such different conclusions?

CHAPTER 9

▷ ▶ INFERENCE FROM SMALL SAMPLES

Case Study

Is smokeless tobacco harmful to your health? Or is "just a pinch between the cheek and the gum" a safe substitute for cigarettes? The case study at the end of this chapter explores this question based on an experiment conducted at Texas Lutheran College.

General Objective

The basic concepts of statistical estimation and tests of hypotheses were presented in Chapters 7 and 8, along with summaries of the methodologies for some large-sample estimation and test procedures. Large-sample estimation and test procedures for population means and proportions were used to illustrate concepts as well as to give you some useful tools for solving some practical problems. Because all these techniques rely on the Central Limit Theorem to justify the normality of the estimators and test statistics, they apply only when the sample sizes are large. Consequently, the objective of Chapter 9 is to supplement the results of Chapters 7 and 8 by presenting small-sample statistical test and estimation procedures for population means and variances. These techniques differ from those of Chapters 7 and 8 because they require the relative frequency distributions of the sampled populations to be normal or approximately so.

Specific Topics

1   Student's t distribution (9.2)
2   Small-sample inferences concerning a population mean (9.3)
3   Small-sample inferences concerning the difference in two means (9.4)

9.1   INTRODUCTION

Large-sample methods for making inferences concerning population means and the difference between two means were discussed with examples in Chapters 7 and 8. Frequently cost, available time, and other factors limit the size of the sample that can be acquired. In this case large-sample procedures are inappropriate and other tests and estimation procedures must be used. In this chapter we will study several small-sample inferential procedures that are closely related to the large-sample methods presented in Chapters 7 and 8. Specifically, we will consider methods for estimating and testing hypotheses about population means, the difference between two means, a population variance, and a comparison of two population variances. Small-sample tests and confidence intervals for binomial parameters will be omitted from our discussion.*

9.2   STUDENT'S *t* DISTRIBUTION

We introduce our topic by considering the following problem: A very costly experiment has been conducted to evaluate a new process for producing synthetic diamonds. Six diamonds have been generated by the new process with recorded weights .46, .61, .52, .48, .57, and .54 karat.

A study of the process costs indicates that the average weight of the diamonds must be greater than .5 karat in order for the process to be operated at a profitable level. Do the six diamond-weight measurements present sufficient evidence to indicate that the average weight of the diamonds produced by the process is in excess of .5 karat? That is, we wish to test the null hypothesis that $\mu = .5$ against the alternative hypothesis that $\mu > .5$.

According to the Central Limit Theorem,

$$z = \frac{\bar{x} - \mu}{\sigma/\sqrt{n}}$$

---

* A small-sample test for the binomial parameter *p* will be presented in Chapter 14.

has approximately a normal distribution in repeated sampling when n is large. For $\alpha = .05$, we could use a one-tailed statistical test and reject H_0 when $z > 1.645$. This procedure assumes that σ is known or that a good estimate s is available and is based on a reasonably large sample (we have suggested $n \geq 30$). Unfortunately, this latter requirement will not be satisfied for the $n = 6$ diamond-weight measurements. How, then, may we test the hypothesis that $\mu = .5$ against the alternative that $\mu > .5$ when we have a small sample?

The problem that we pose is not new; it is one that received serious attention from statisticians and experimenters at the turn of the century. If a sample standard deviation s were substituted for σ in z, would the resulting quantity have approximately a standardized normal distribution in repeated sampling? More specifically, is the rejection region $z > 1.645$ appropriate; that is, do approximately 5% of the values of the test statistic, computed in repeated sampling, exceed 1.645 when H_0 is true?

The answers to these questions, not unlike many of the problems encountered in the sciences, may be resolved by experimentation. That is, we could draw a small sample, say $n = 6$ measurements, and compute the value of the test statistic. Then we would repeat this process over and over again a very large number of times and construct a frequency distribution for the computed values of the test statistic. The general shape of the distribution and the location of the rejection region would then be evident.

The sampling distribution of the test statistic

$$t = \frac{\bar{x} - \mu}{s/\sqrt{n}}$$

for samples drawn from a normally distributed population was discovered by W. S. Gosset and published in 1908 under the pen name of "Student." He referred to the quantity under study as t, and ever since it has been known as **Student's t**. We omit the complicated mathematical expression for the density function for t, but we will describe some of its characteristics.

In repeated sampling, the distribution of the test statistic

$$t = \frac{\bar{x} - \mu}{s/\sqrt{n}}$$

is, like z, mound-shaped and perfectly symmetrical about $t = 0$. Unlike z, it is much more variable, tailing rapidly out to the right and left, a phenomenon that can be readily explained. The variability of z in repeated sampling is due solely to $\bar{x}$; the other quantities appearing in z (n and σ) are nonrandom. On the other hand, the variability of t is contributed by *two* random quantities, $\bar{x}$ and s, that can be shown to be independent of one another. Thus, when $\bar{x}$ is very large, s may be very small, and vice versa. As a result, t will be more variable than z in repeated sampling (see Figure 9.1). Finally, as we might surmise, the variability of t decreases as n increases because the estimate s of σ is based on more and more information. When n is infinitely large, the t and z distributions are identical. Thus, Gosset discovered that the distribution of t depended on the sample size n.

Figure 9.1
Standard normal *z* and
the *t* distribution based
on *n* = 6 measurements
(5 d.f.)

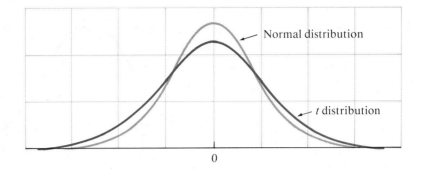

The divisor $(n - 1)$ of the sum of squares of deviations that appears in the formula for s^2 is called the **number of degrees of freedom associated with s^2**. The origin of the term *degrees of freedom* is linked to the statistical theory underlying the probability distribution of s^2 and refers to the number of independent squared deviations available for estimating σ^2. We will not pursue this point further except to say that the test statistic t, based on a sample of n measurements, possesses $(n - 1)$ degrees of freedom (d.f.).

The critical values of t that separate the rejection and acceptance regions for the statistical test are presented in Table 4 in Appendix III. Table 4 is partially reproduced in Table 9.1. The tabulated value t_α records the value of t such that an area α lies to its right, as shown in Figure 9.2. The degrees of freedom d.f. associated with t are shown in the first and last columns of the table, and the values of t_α corresponding to various values of α appear in the top row. Thus, if we wish to find the value of t, such that 5% of the area lies to its right, we would use the column marked $t_{.05}$. The critical value of t for our example, found in the $t_{.05}$ column

Table 9.1
Format of the Student's *t*
table from Table 4 in
Appendix III

d.f.	$t_{.100}$	$t_{.050}$	$t_{.025}$	$t_{.010}$	$t_{.005}$	d.f.
1	3.078	6.314	12.706	31.821	63.657	1
2	1.886	2.920	4.303	6.965	9.925	2
3	1.638	2.353	3.182	4.541	5.841	3
4	1.533	2.132	2.776	3.747	4.604	4
5	1.476	2.015	2.571	3.365	4.032	5
6	1.440	1.943	2.447	3.143	3.707	6
7	1.415	1.895	2.365	2.998	3.499	7
8	1.397	1.860	2.306	2.896	3.355	8
9	1.383	1.833	2.262	2.821	3.250	9
⋮	⋮	⋮	⋮	⋮	⋮	⋮
26	1.315	1.706	2.056	2.479	2.779	26
27	1.314	1.703	2.052	2.473	2.771	27
28	1.313	1.701	2.048	2.467	2.763	28
29	1.311	1.699	2.045	2.462	2.756	29
inf.	1.282	1.645	1.960	2.326	2.576	inf.

Figure 9.2

Tabulated values of Student's t

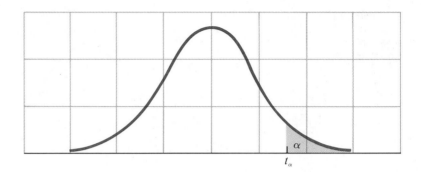

opposite the row corresponding to d.f. $= (n - 1) = (6 - 1) = 5$, is $t = 2.015$ (shaded in Table 9.1). Thus, we would reject $H_0 : \mu = .5$ in favor of $H_a : \mu > .5$ when $t > 2.015$. Since the distribution of t is symmetric about $t = 0$, a left-tailed critical value of t is simply the negative of the corresponding right-tailed value. For example, with 5 d.f., the area to the left of $t = -2.015$ is equal to .05, the area to the right of $t = 2.015$.

Note that the critical value of t is always larger than the corresponding critical value of z for a specified α. For example, where $\alpha = .05$, the critical value of t for $n = 2$ (d.f. $= 1$) is $t = 6.314$, which is very large when compared with the corresponding $z_{.05} = 1.645$. Proceeding down the $t_{.05}$ column, we note that the critical value of t decreases, reflecting the effect of a larger sample size (more degrees of freedom) on the estimation of σ. Finally, when n is infinitely large, the critical value of t equals $z_{.05} = 1.645$.

The reason for choosing $n = 30$ (an arbitrary choice) as the dividing line between large and small samples is apparent. For $n = 30$ (d.f. $= 29$), the critical value of $t_{.05} = 1.699$ is numerically quite close to $z_{.05} = 1.645$. For a two-tailed test based on $n = 30$ measurements and $\alpha = .05$, we would place .025 in each tail of the t distribution and reject $H_0 : \mu = \mu_0$ when $t > 2.045$ or $t < -2.045$. Note that this is very close to $z_{.025} = 1.96$ used in the z test.

Note that the Student's t and corresponding tabulated critical values are based on the assumption that the sampled population has a normal probability distribution. This indeed is a very restrictive assumption because, in many sampling situations, the properties of the population are completely unknown and may be nonnormal. If this restriction were to seriously affect the distribution of the t statistic, the application of the t test would be very limited. Fortunately, this point is of little consequence because **it can be shown that the distribution of the t statistic has nearly the same shape as the theoretical t distribution for populations that are nonnormal but have a mound-shaped probability distribution**. This property of the t statistic and the common occurrence of mound-shaped distributions of data in nature enhance the value of Student's t for use in statistical inference.

Having discussed the origin of Student's t and the tabulated critical values from Table 4 in Appendix III, we now return to the problem of making an inference about the mean diamond weight based on our sample of $n = 6$ measurements. Prior to considering the solution, you may wish to test your built-in inference-making equipment by glancing at the six measurements and arriving at a conclusion concerning the significance of the data.

▷ 9.3 SMALL-SAMPLE INFERENCES CONCERNING A POPULATION MEAN

The statistical test of an hypothesis concerning a population mean may be stated as follows:

SMALL-SAMPLE TEST OF AN HYPOTHESIS CONCERNING A POPULATION MEAN

1. Null Hypothesis: $H_0 : \mu = \mu_0$

2. Alternative Hypothesis:

One-Tailed Test	Two-Tailed Test
$H_a : \mu > \mu_0$	$H_a : \mu \neq \mu_0$
(or, $H_a : \mu < \mu_0$)	

3. Test Statistic: $t = \dfrac{\bar{x} - \mu_0}{s/\sqrt{n}}$

4. Rejection Region:

    One-Tailed Test

    $t > t_\alpha$

    (or, $t < t_\alpha$ when the alternative hypothesis is $H_a : \mu < \mu_0$)

    Two-Tailed Test

    $t > t_{\alpha/2}$ or $t < -t_{\alpha/2}$

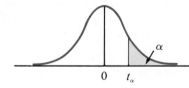 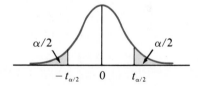

The critical values of t, t_α, and $t_{\alpha/2}$, are based on $(n - 1)$ degrees of freedom. These tabulated values can be found in Table 4 in Appendix III.

> **Assumption:** The sample has been randomly selected from a normally distributed population.

To apply this test to the diamond-weight data, we must first calculate the sample mean $\bar{x}$ and standard deviation s. You can verify that the mean and standard deviation for the six diamond weights are .53 and .0559, respectively.

We wish to test the null hypothesis that the mean diamond weight is .5 against the alternative hypothesis that it is greater than .5. The elements of the test as defined above are

$H_0: \mu = .5$

$H_a: \mu > .5$

Test statistic: $t = \dfrac{\bar{x} - \mu_0}{s/\sqrt{n}} = \dfrac{.53 - .5}{.0559/\sqrt{6}} = 1.32$

Rejection region: The rejection region for $\alpha = .05$ and $(n - 1) = (6 - 1) = 5$ degrees of freedom is $t > 2.015$.

Since the calculated value of the test statistic does not fall in the rejection region, we cannot reject H_0. Thus, the data do not present sufficient evidence to indicate that the mean diamond weight exceeds .5 karat.

The calculation of the probability β of a type II error for the t test is very difficult and is beyond the scope of this text. Therefore, we will avoid this problem and obtain an interval estimate for μ. The large-sample confidence interval for μ, given in Chapter 7, is

$$\bar{x} \pm z_{\alpha/2} \frac{\sigma}{\sqrt{n}}$$

where $z_{\alpha/2} = 1.96$ for a confidence coefficient equal to .95. This result assumes that σ is known and simply involves a measurement of $1.96\sigma_{\bar{x}}$ (or approximately $2\sigma_{\bar{x}}$) on either side of $\bar{x}$ in conformity with the Empirical Rule. When σ is unknown and must be estimated by a small-sample standard deviation s, the large-sample confidence interval will not enclose μ 95% of the time in repeated sampling. To account for the added variability introduced through s, the estimate of σ, the appropriate confidence interval for μ is obtained by replacing the critical value $z_{\alpha/2}$ with the critical value $t_{\alpha/2}$ based on $(n - 1)$ degrees of freedom. The small-sample confidence interval for μ is given in the next display.

SMALL-SAMPLE $(1 - \alpha)$ 100% CONFIDENCE INTERVAL FOR μ

$$\bar{x} \pm t_{\alpha/2} \frac{s}{\sqrt{n}}$$

where $s/\sqrt{n}$ is the estimated standard error of $\bar{x}$ but is often referred to as the **standard error of the mean.**

Assumption: The sampled population is approximately normally distributed.

For our example, a 95% confidence interval for μ is

$$\bar{x} \pm t_{\alpha/2} \frac{s}{\sqrt{n}} \quad \text{or} \quad .53 \pm 2.571 \frac{.0559}{\sqrt{6}} \quad \text{or} \quad .53 \pm .059$$

Therefore, the interval estimate for μ is .471 to .589, with confidence coefficient of .95. If the experimenter wishes to detect a small increase in mean diamond weight in excess of .5 carat, the width of the interval must be reduced by obtaining more diamond weight measurements. Additional measurements will decrease both $1/\sqrt{n}$ and $t_{\alpha/2}$ and thereby decrease the width of the interval. From the standpoint of a statistical test of an hypothesis, an increase in n increases the available information upon which to base a decision and decreases the probability of making a type II error.

EXAMPLE 9.1   A manufacturer of gunpowder has developed a new powder that is designed to produce a muzzle velocity of 3000 feet per second. Eight shells are loaded with the charge and the muzzle velocities measured. The resulting velocities are shown in Table 9.2. Do the data present sufficient evidence to indicate that the average velocity differs from 3000 feet per second?

Table 9.2
Data for Example 9.1

Muzzle velocity (feet per second)	
3005	2995
2925	3005
2935	2935
2965	2905

Solution   Testing the null hypothesis that $\mu = 3000$ feet per second against the alternative that μ is either greater than or less than 3000 feet per second results in a two-tailed statistical test. Thus,

$$H_0: \mu = 3000 \quad \text{and} \quad H_a: \mu \neq 3000$$

Using $\alpha = .05$ and placing .025 in each tail of the t distribution, we find that the critical value of t for $n = 8$ measurements [or $(n - 1) = 7$ d.f.] is $t_{.025} = 2.365$. Hence, we will reject H_0 if $t > 2.365$ or $t < -2.365$. (Recall that the t distribution is symmetrical about $t = 0$.)

The sample mean and standard deviation for the recorded data are

$$\bar{x} = 2958.75 \quad \text{and} \quad s = 39.26$$

Then

$$t = \frac{\bar{x} - \mu_0}{s/\sqrt{n}} = \frac{2958.75 - 3000}{39.26/\sqrt{8}} = -2.97$$

Since the observed value of t falls in the rejection region, we reject H_0 and conclude that the average velocity is less than 3000 feet per second. Furthermore, we are reasonably confident that we have made the correct decision. Using our procedure, we should erroneously reject H_0 only $\alpha = .05$ of the time in repeated applications of the statistical test.   ◁

EXAMPLE 9.2   Find a 95% confidence interval for the mean in Example 9.1.

Solution   Substituting $t_{.020} = 2.365$, $s = 39.26$, and $n = 8$ into the formula

$$\bar{x} \pm t_{\alpha/2} \frac{s}{\sqrt{n}}$$

we obtain

$$2958.75 \pm (2.365) \frac{39.26}{\sqrt{8}}$$

or

$$2958.75 \pm 32.83$$

Thus, we estimate the average muzzle velocity to lie in the interval 2926 to 2992 feet per second. A more accurate interval estimate (a shorter interval) can generally be obtained by increasing the sample size.   ◁

EXAMPLE 9.3   If you planned to report the results of the statistical test in Example 9.1, what p-value would you report?

Solution   The p-value for this test is the probability of observing a value of the t statistic as contradictory to the null hypothesis as the one observed for this set of data, namely, $t = -2.97$. Since this is a two-tailed test, the p-value is the probability that either $t \leq -2.97$ or $t \geq 2.97$.

Unlike the table of areas under the normal curve (Table 3 of Appendix III), the table for t (Table 4 of Appendix III) does not give the areas corresponding to various values of t. Rather, it gives the values of t corresponding to upper-tail areas equal to .10, .05, .025, .010, and .005. Consequently, we can only approximate the upper-tail area that corresponds to the probability that $t > 2.97$. Since the t statistic for this test is based on 7 degrees of freedom, we refer to the d.f. = 7 row of Table 4 and find that 2.97 falls between $t_{.025} = 2.365$ and $t_{.010} = 2.998$. The right-tail area corresponding to the probability that $t > 2.97$ lies between .025 and .010. Since this value represents only half of the required area, the actual p-value lies between 2(.025) = .05 and 2(.010) = .020. Using the tabulated values of t, we could reject H_0 with $\alpha = .05$ but not with $\alpha = .02$. Therefore, the p-value for this test would be reported as p-value = .05. Since most researchers would be using a t table similar or identical to Table 4, the reader of a published research report would realize that for a reported p-value of .05, the exact p-value for the test was probably less than .05 and was between .05 and .02. In fact, some researchers report these results as .02 < p-value < .05, indicating significance at the .05 level but not at the .02 level.   ◁

The MINITAB package contains commands that can be used to implement a small-sample test of a population mean or to produce a small-sample confidence interval for a population mean. The command TTEST requires the user to provide the value of the population mean to be tested, together with the column number in which the data are stored. The subcommand ALTERNATIVE, followed by a 1, −1, or 0, will implement a right-tailed, left-tailed, or two-tailed test procedure. If this subcommand is not used, a two-tailed test is implemented by default.

The results of using TTEST to implement the test in Example 9.1 are given in Table 9.3. In addition to the observed value of $t = -2.97$, the output gives the sample mean $\bar{x} = 2958.7$, the sample standard deviation $s = 39.3$, and the standard error of the mean (SE MEAN), $s/\sqrt{n} = 13.9$. When compared with our results in Example 9.1, the only noticeable difference is in the reported decimal accuracy of the results. In addition, the exact p-value for the test appears in the printout and is given as P VALUE = .021. In using Table 4 to approximate this value, we found that the p-value was between .05 and .02, consistent with the MINITAB result.

Table 9.3
MINITAB printout for
the data in Table 9.2

```
MTB > TTEST 3000 C1

TEST OF MU = 3000.0 VS MU N.E. 3000.0
```

	N	MEAN	STDEV	SE MEAN	T	P VALUE
C1	8	2958.7	39.3	13.9	−2.97	0.021

```
MTB > TINTERVAL C1
```

	N	MEAN	STDEV	SE MEAN	95.0 PERCENT C.I.	
C1	8	2958.7	39.3	13.9	(2925.9,	2991.6)

The MINITAB command TINTERVAL is available for constructing a confidence interval for a population mean. The command requires that the user supply the confidence coefficient and the column location of the data. A 95% confidence interval for μ using the data in Example 9.1 also appears in Table 9.3. Apart from the reported decimal accuracy, the results agree with those found in Example 9.2.

EXERCISES Basic Techniques

9.1 Find the following:
 a. $t_{.05}$ for 5 degrees of freedom (d.f.).   c. $t_{.10}$ for 18 d.f.
 b. $t_{.025}$ for 8 d.f.   d. $t_{.025}$ for 30 d.f.

9.2 Find t_α, given that $P(t > t_\alpha) = \alpha$ for
 a. $\alpha = .10$, 12 d.f.   c. $\alpha = .05$, 16 d.f.
 b. $\alpha = .01$, 25 d.f.   d. $\alpha = .025$, 7 d.f.

9.3 A random sample of $n = 12$ observations from a normal population produced $\bar{x} = 47.1$ and $s^2 = 4.7$.
 a. Test the hypothesis $H_0: \mu = 48$ against $H_a: \mu \neq 48$ using $\alpha = .10$.
 b. What is the observed level of significance for the test in part (a)?
 c. Find a 90% confidence interval for the population mean. Interpret this interval.

9.4   The following $n = 10$ observations resulted when sampling from a normal population.

7.4   7.1   6.5   7.5   7.6   6.3   6.9   7.7   6.5   7.0

a.   Find the mean and the standard deviation of these data.
b.   Find a 99% confidence interval for the population mean μ.
c.   Test $H_0: \mu = 7.5$ versus $H_a: \mu < 7.5$ with $\alpha = .01$.
d.   Do the results of part (b) support your conclusion in part (c)?

Applications

9.5   One family of fish native to tropical Africa can breathe both in air and in water. Thus, the fish can live in water with a relatively low oxygen content by taking a portion of their oxygen from the air. A study by Michael J. Pettit and Thomas L. Beitinger that was conducted to investigate the effects of some variables on oxygen partitioning by reedfish; that is, the proportions of oxygen taken by reedfish from the surrounding air and from the water in which they lived ("Oxygen Acquisition of the Reedfish, *Erpetoichthys calabaricus*," *Journal of Experimental Biology* 114 [1985]). The data shown below represent the oxygen uptake readings from air and water for 11 reedfish in water at 25°C. The first column of the table gives the weights of the fish. The second and third columns give the averages of repeated measurements of oxygen uptake from air and water, respectively, for each fish. The fourth column gives the percentage of total oxygen uptake attributable to air.

Weight (grams)	Oxygen uptake (ml O_2 g^{-1}h^{-1})		Percent*	
	Air	Water	Air	Water
22.1	0.043	0.045	49	51
21.1	0.034	0.066	34	66
13.4	0.026	0.084	24	76
19.5	0.048	0.045	52	48
21.9	0.009	0.056	14	86
19.5	0.028	0.072	28	72
16.5	0.032	0.081	28	72
18.5	0.045	0.051	47	53
18.0	0.061	0.041	60	40
15.4	0.028	0.060	32	68
13.4	0.024	0.113	18	82

Assume that the reedfish included in this experiment represent a random sample of mature reedfish.
a.   Find a 90% confidence interval for the mean weight of mature reedfish. Interpret the interval.
b.   Find a 90% confidence interval for the mean oxygen uptake from the air for reedfish living in an environment similar to that used in this experiment. Interpret your interval.
c.   Find a 90% confidence interval for the mean percentage of oxygen uptake from air in reedfish living in an environment similar to that employed in this experiment. Interpret your interval.

* Percentages shown were calculated by the author based on Pettit and Beitinger's data.

9.6   "Lake Champlain Found to be Polluted by PCBs," reports the *New York Times* (June 16, 1985). PCBs, a group of chemicals used for years as an insulator in some electrical equipment, have been found to cause cancer in laboratory animals and are suspected of having similar effects on humans. Although the federal level of tolerance of PCBs in fish is 2 parts per million, a sampling of 15 American eels in Lake Champlain gave PCB readings ranging from 4.05 to 19.49 parts per million. The mean for the sample was 9.84 parts per million.

   a.   Based on the reported range of the sample measurements, find an approximate value for the standard deviation *s*. Justify your answer.

   b.   Use the sample standard deviation found in part (a) and other pertinent data to construct a 90% confidence interval for the mean PCB concentration in the Lake Champlain American eels.

   c.   Describe the population for which the confidence interval in part (b) applies.

9.7   Scholastic Aptitude Test scores, falling slowly since the inception of the tests, have now begun to rise. Originally, a score of 500 was intended to be "average." The mean scores for 1985 are approximately 431 for the verbal test and 475 for the mathematics test. A sampling of the test scores of 20 seniors at an urban high school produced the following mean scores and standard deviations:

	Verbal	Mathematics
Sample mean	419	455
Sample standard deviation	57	69

   a.   Do the data provide sufficient evidence to indicate that the mean verbal SAT score for the urban high school seniors is less than the 1985 national mean of 431? Test using $\alpha = .05$.

   b.   Find the approximate observed significance level for the test in part (a) and interpret its value.

   c.   Do the data provide sufficient evidence to indicate that the mean mathematics SAT score for the urban high school seniors is less than the 1985 national mean of 475? Test using $\alpha = .05$.

   d.   Find the observed significance level for the test in part (c) and interpret its value.

9.8   Industrial wastes and sewage dumped into our rivers and streams absorb oxygen and thereby reduce the amount of dissolved oxygen available for fish and other forms of aquatic life. One state agency requires a minimum of 5 parts per million of dissolved oxygen in order for the oxygen content to be sufficient to support aquatic life. Six water specimens taken from a river at a specific location during the low-water season (July) gave readings of 4.9, 5.1, 4.9, 5.0, 5.0, and 4.7 parts per million of dissolved oxygen. Do the data provide sufficient evidence to indicate that the dissolved oxygen content is less than 5 parts per million? Test using $\alpha = .05$.

9.9   Studies indicate that some women athletes participating in vigorous training programs are found to possess low levels of estrogen. Other studies indicate that women after menopause show a drop in estrogen levels and an associated reduction in bone mass. In a study to determine whether this same association exists for young (mean age approximately 25 years) women athletes with low estrogen levels, B. L. Drinkwater and colleagues compared the mean bone mass in two groups of 14 women athletes, one group with low estrogen levels (amenorrheic) and one with normal estrogen levels (eumenorrheic) ("Bone Mineral Content of Amenorrheic and Eumenorrheic Athletes," *New England Journal of Medicine* [August 2, 1984]). The two groups were matched in sport, age, weight, and on the frequency and duration of the daily training program. Mineral content and density measurements were taken at two locations S_1 and S_2 in the forearm of each athlete, and a density measurement was also taken in the lumbar vertebrae. The means and standard errors of the means obtained by the authors are shown below. (Recall that the standard error of the mean is the standard deviation of the sampling distribution of the mean.) Find a 95% confidence interval for the mean mineral density in the vertebrae for amenorrheic (low estrogen) women athletes. (The authors note that this mean level of vertebrae bone density is what one would expect to find in a woman aged 51.2 years.)

Site	Amenorrheic	Eumenorrheic	p-value
Radius			
S_1			
Mineral content (g/cm)	0.89 ± 0.03	0.85 ± 0.03	NS
Mineral density (g/cm^2)	0.53 ± 0.02	0.54 ± 0.01	NS
S_2			
Mineral content (g/cm)	0.91 ± 0.02	0.88 ± 0.03	NS
Mineral density (g/cm^2)	0.67 ± 0.02	0.67 ± 0.02	NS
Vertebrae (g/cm^2)	1.12 ± 0.04	1.30 ± 0.03	<0.01

Plus-minus values are means $\pm$ S.E.M. NS denotes "not significant."

9.10   In Exercise 1.9, we presented some data from a study of the infestation of the *T. orientalis* lobster by two types of barnacles, *O. tridens* and *O. lowei*. The carapace lengths of the ten lobsters are reproduced below. Find a 95% confidence interval for the mean carapace length of *T. orientalis* lobsters caught in the seas in the vicinity of Singapore (W. B. Jeffries, H. K. Voris, and C. M. Yang, "Diversity and Distribution of the Pedunculate Barnacle *Octolasmis* Gray, 1825 Epizoic on the Scyllarid Lobster, *Thenus orientalis* (Lund, 1793)," *Crustaceana* 46, no. 3 [1984]).

Lobster field number	A061	A062	A066	A070	A067	A069	A064	A068	A065	A063
Carapace Length (mm)	78	66	65	63	60	60	58	56	52	50

9.11   It is recognized that cigarette smoking has a deleterious effect on lung function. In their study of the effect of cigarette smoking on the carbon monoxide diffusing capacity (DL) of the lung, Ronald J. Knudson, Walter T. Kaltenborn, and Benjamin Burrows (*Am. Rev. Respir. Dis.*, 1989; 140:645–651) found that current smokers had DL readings significantly lower than either exsmokers or nonsmokers. The carbon monoxide diffusing capacities for a random sample of $n = 20$ current smokers follows:

103.768	88.602	73.003	123.086
91.052	92.295	61.675	90.677
84.023	76.014	100.615	88.017
71.210	82.115	89.222	102.754
108.579	73.154	106.755	90.479

Do these data indicate that the mean DL reading for current smokers is significantly lower than 100 DL, the average for nonsmokers? Use $\alpha = .01$.

9.12   Refer to Exercise 9.11 and find a 95% confidence interval estimate for the mean DL reading for current smokers.

9.13   Organic chemists often purify organic compounds by a method known as fractional crystallization. An experimenter wanted to prepare and purify 4.85 grams of aniline. Ten 4.85-gram quantities of aniline were individually prepared and purified to acetanilide. The following dry yields were recorded.

3.85	3.80
3.88	3.85
3.90	3.36
3.62	4.01
3.72	3.82

Estimate the mean number of grams of acetanilide that could be recovered from an initial amount of 4.85 grams of aniline. Use a 95% confidence interval.

9.14   Refer to Exercise 9.13. Approximately how many 4.85-gram specimens of aniline would be required if you wished to estimate the mean number of grams of acetanilide correct to within .06 gram with probability equal to .95?

9.15   As concern grows about toxic chemical emissions in the environment, it becomes important for the public to determine whether emissions data are biased by the inclusion or exclusion of certain contaminants. The data from Exercise 2.53, reproduced below, gives the Environmental Protection Agency's (EPA's) top ten counties for toxic waste in 1987, together with a *USA Today* computer analysis of the same data. Emissions are in millions of pounds.

County	EPA	USA Today
San Bernardino, CA	5200	22.5
Harris, TX	612	78.3
Calhoun, TX	581	579.2
Wayne, MI	512	17.7
Harrison, MS	423	53.0
Brazoria, TX	404	197.9
St. James, LA	349	312.5
Milam, TX	329	329.1
Jefferson, TX	309	198.0
St. Charles, LA	284	209.1

Source: Larry Sanders, *USA Today* database research, "EPA report spurs confusion." Copyright 1989, *USA Today.* Excerpted with permission.

a.   If these ten counties are taken to represent a random sample of all counties receiving toxic waste in 1987, find a 95% confidence interval estimate for the mean amount of toxic waste reported using the EPA's standards.

b.   Find a 99% confidence interval for the mean amount of toxic waste reported using the *USA Today* figures.

c.   What assumptions are required in order to use the estimation procedures in parts (a) and (b)? Is it likely that these assumptions have been satisfied? Explain.

▷ 9.4 SMALL-SAMPLE INFERENCES CONCERNING THE DIFFERENCE BETWEEN TWO MEANS

The physical setting for the problem we consider here is identical to that discussed in Section 7.7. Independent random samples of n_1 and n_2 measurements, respectively, are drawn from two populations, which possess means and variances μ_1, σ_1^2 and μ_2, σ_2^2. Our objective is to make inferences concerning the difference $(\mu_1 - \mu_2)$ between the two population means.

The following small-sample methods for testing hypotheses and placing a confidence interval on the difference between two means are, like the case for a single mean, founded on assumptions regarding the probability distributions of the sampled populations.

> **Assumptions for Small-Sample Inferences Concerning $\mu_1 - \mu_2$**
>
> 1. Random and independent samples are drawn from each of two normal populations.
> 2. The variability of the measurements in the two populations is the same and can be measured by a common variance, σ^2. That is, $\sigma_1^2 = \sigma_2^2 = \sigma^2$.

Although these assumptions may seem to be quite restrictive, they are quite reasonable for many sampling situations.

The point estimator of $(\mu_1 - \mu_2)$ is $(\bar{x}_1 - \bar{x}_2)$, the difference between the sample means. In Section 7.7, we found that in repeated sampling $(\bar{x}_1 - \bar{x}_2)$ was an unbiased estimator of $(\mu_1 - \mu_2)$, with a standard deviation

$$\sigma_{(\bar{x}_1 - \bar{x}_2)} = \sqrt{\frac{\sigma_1^2}{n_1} + \frac{\sigma_2^2}{n_2}}$$

This result was used in placing bounds on the error of estimation, in constructing a large-sample confidence interval, and in testing an hypothesis concerning the difference between two population means. The large-sample test of the hypothesis $H_0: \mu_1 - \mu_2 = D_0$, where D_0 is the hypothesized difference between the means, was based on the z statistic given by

$$z = \frac{(\bar{x}_1 - \bar{x}_2) - D_0}{\sqrt{\dfrac{\sigma_1^2}{n_1} + \dfrac{\sigma_2^2}{n_2}}}$$

When we sample two normal populations with equal variances, and $\sigma_1^2 = \sigma_2^2 = \sigma^2$, then the z statistic can be simplified as follows:

$$z = \frac{(\bar{x}_1 - \bar{x}_2) - D_0}{\sqrt{\dfrac{\sigma^2}{n_1} + \dfrac{\sigma^2}{n_2}}} = \frac{(\bar{x}_1 - \bar{x}_2) - D_0}{\sigma\sqrt{\dfrac{1}{n_1} + \dfrac{1}{n_2}}}$$

For small-sample tests of the hypothesis $H_0: \mu_1 - \mu_2 = D_0$, it seems reasonable to use the test statistic given next.

SMALL-SAMPLE TEST STATISTIC t FOR THE DIFFERENCE BETWEEN TWO MEANS

$$t = \frac{(\bar{x}_1 - \bar{x}_2) - D_0}{s\sqrt{\dfrac{1}{n_1} + \dfrac{1}{n_2}}}$$

That is, we use a sample standard deviation s as an estimator of σ. This test statistic possesses a Student's t distribution in repeated sampling when the stated assumptions are satisfied, a fact that can be proved mathematically or verified by experimental sampling from two normal populations.

The estimate s used in the t statistic could be either s_1 or s_2, the standard deviation for each of the two samples. However, the use of either one alone would be wasteful, since both sample standard deviations are independent estimators of σ. The information from both sample variances can be combined by using a weighted average in which the weights reflect the relative amount of information in s_1^2 and s_2^2. Since s_1^2 is based on $(n_1 - 1)$ degrees of freedom and s_2^2 is based on $(n_2 - 1)$ degrees of freedom, a **pooled estimator of σ^2** using the degrees of freedom as weights is

$$s^2 = \frac{(n_1 - 1)s_1^2 + (n_2 - 1)s_2^2}{(n_1 - 1) + (n_2 - 1)}$$

with $(n_1 - 1) + (n_2 - 1) = n_1 + n_2 - 2$ degrees of freedom.

The pooled estimator s^2 can also be expressed as a sum of the squared deviations in each sample. If x_{1i} and x_{2i} represent the ith observations in samples 1 and 2, then

$$s_1^2 = \frac{\sum_{i=1}^{n_1} (x_{1i} - \bar{x}_1)^2}{(n_1 - 1)} \quad \text{and} \quad s_2^2 = \frac{\sum_{i=1}^{n_2} (x_{2i} - \bar{x}_2)^2}{(n_2 - 1)}$$

Using these expressions, we can write the pooled estimator of σ^2 in either of the following two ways.

POOLED ESTIMATOR OF σ^2

$$s^2 = \frac{\sum_{i=1}^{n_1} (x_{1i} - \bar{x}_1)^2 + \sum_{i=1}^{n_2} (x_{2i} - \bar{x}_2)^2}{(n_1 - 1) + (n_2 - 1)}$$

$$s^2 = \frac{(n_1 - 1)s_1^2 + (n_2 - 1)s_2^2}{(n_1 - 1) + (n_2 - 1)}$$

The first form for calculating s^2 shows that the numerator is the pooled sum of squared deviations from each sample, and the denominator is the pooled sum of the degrees of freedom from each sample. Regardless of the form used for calculation, it can be proved that s^2 is an unbiased estimator of σ^2 and hence is an unbiased estimator of the common population variance. If the pooled estimator s^2 is used to estimate σ^2 and the samples are randomly and independently drawn from normal populations with a common variance, then the statistic

$$t = \frac{(\bar{x}_1 - \bar{x}_2) - (\mu_1 - \mu_2)}{s\sqrt{\dfrac{1}{n_1} + \dfrac{1}{n_2}}}$$

will have a Student's t distribution with $n_1 + n_2 - 2$ degrees of freedom.

The small-sample test for the difference between two means is given in the next display.

TEST OF AN HYPOTHESIS CONCERNING THE DIFFERENCE BETWEEN TWO MEANS

1. Null Hypothesis: $H_0:(\mu_1 - \mu_2) = D_0$, where D_0 is some specified difference that you wish to test. For many tests, you will wish to hypothesize that there is no difference between μ_1 and μ_2; that is, $D_0 = 0$.

2. Alternative Hypothesis:

One-Tailed Test	*Two-Tailed Test*
$H_a:(\mu_1 - \mu_2) > D_0$	$H_a:(\mu_1 - \mu_2) \neq D_0$
(or, $H_a:(\mu_1 - \mu_2) < D_0$)	

3. Test Statistic: $t = \dfrac{(\bar{x}_1 - \bar{x}_2) - D_0}{s\sqrt{\dfrac{1}{n_1} + \dfrac{1}{n_2}}}$

   where $s^2 = \dfrac{\displaystyle\sum_{i=1}^{n_1}(x_{1i} - \bar{x}_1)^2 + \sum_{i=1}^{n_2}(x_{2i} - \bar{x}_2)^2}{n_1 + n_2 - 2}$

4. Rejection Region:

One-Tailed Test	*Two-Tailed Test*
$t > t_\alpha$	$t > t_{\alpha/2}$ or $t < -t_{\alpha/2}$
(or, $t < -t_\alpha$ when the alternative hypothesis is $H_a:(\mu_1 - \mu_2) < D_0$)	

   The critical values of t, t_α, and $t_{\alpha/2}$, will be based on $(n_1 + n_2 - 2)$ degrees of freedom. The tabulated values can be found in Table 4 in Appendix III.

   Assumptions: The samples were randomly and independently selected from normally distributed populations. The variances of the populations σ_1^2 and σ_2^2 are equal.

EXAMPLE 9.4 An assembly operation in a manufacturing plant requires approximately a one-month training period for a new employee to reach maximum efficiency. A new method of training was suggested and a test was conducted to compare the new method with the standard procedure. Two groups of nine new employees were trained for a period of three weeks, one group using the new method and the other following standard training procedure. The length of time in minutes required for each employee to assemble the device was recorded at the end of the three-week period. These measurements appear in Table 9.4. Do the data present sufficient evidence to indicate that the mean time to assemble at the end of a three-week training period is less for the new training procedure?

Table 9.4

Standard procedure	New procedure
32	35
37	31
35	29
28	25
41	34
44	40
35	27
31	32
34	31

Solution Let μ_1 and μ_2 be the mean time to assemble for the standard and the new assembly procedures, respectively. Then, since we seek evidence to support the theory that $\mu_1 > \mu_2$, we will test the null hypothesis $H_0: \mu_1 = \mu_2$ (i.e., $\mu_1 - \mu_2 = 0$) against the alternative hypothesis $H_a: \mu_1 > \mu_2$ (i.e., $\mu_1 - \mu_2 > 0$). To conduct this test, assume that the population distributions of measurements are normal, and that the variability for the two populations of measurements is the same.

The sample means and sums of squared deviations are

$$\bar{x}_1 = 35.22 \quad \text{and} \quad \sum_{i=1}^{9} (x_{1i} - \bar{x}_1)^2 = 195.5556$$

$$\bar{x}_2 = 31.56 \quad \text{and} \quad \sum_{i=1}^{9} (x_{2i} - \bar{x}_2)^2 = 160.2222$$

Then the pooled estimate of the common variance is

$$s^2 = \frac{\sum_{i=1}^{9} (x_{1i} - \bar{x}_1)^2 + \sum_{i=1}^{9} (x_{2i} - \bar{x}_2)^2}{n_1 + n_2 - 2} = \frac{195.5556 + 160.2222}{9 + 9 - 2} = 22.2361$$

and the standard deviation is $s = 4.716$.

The alternative hypothesis $H_a : \mu_1 > \mu_2$, or, equivalently, $\mu_1 - \mu_2 > 0$, implies that we should use a one-tailed statistical test and that the rejection region for the test is located in the upper tail of the t distribution. Referring to Table 4 in Appendix III, we find that the critical value of t for $\alpha = .05$ and $(n_1 + n_2 - 2) = 16$ degrees of freedom is 1.746. Therefore, we will reject the null hypothesis when the calculated value of t is greater than 1.746.

The calculated value of the test statistic is

$$t = \frac{(\bar{x}_1 - \bar{x}_2)}{s\sqrt{\dfrac{1}{n_1} + \dfrac{1}{n_2}}} = \frac{35.22 - 31.56}{4.716\sqrt{\dfrac{1}{9} + \dfrac{1}{9}}} = 1.65$$

Comparing this value with the critical value $t_{.05} = 1.746$, we note that the calculated value does not fall in the rejection region. Therefore, we must conclude that there is insufficient evidence to indicate that the new method of training is superior at the .05 level of significance.

EXAMPLE 9.5   Find the p-value that would be reported for the statistical test of Example 9.4.

Solution   The observed value of t for this one-tailed test was $t = 1.65$. Therefore, the p-value for the test would be the probability that $t > 1.65$. Since we cannot obtain this probability from Table 4 of Appendix III, we would report the p-value for the test as the smallest tabulated value for α that leads to the rejection of H_0. Consulting the row in Table 4 corresponding to 16 degrees of freedom, we see that the observed value, $t = 1.65$, lies between $t_{.10} = 1.337$ and $t_{.05} = 1.746$. Therefore, the probability that $t > 1.65$ is between .05 and .10, and the p-value for this test would be reported as $.05 < p\text{-value} < .10$.

The small-sample confidence interval for $(\mu_1 - \mu_2)$ is based on the same assumptions as the statistical test procedure. This confidence interval, with confidence coefficient $(1 - \alpha)$, is given by the formula in the following display.

SMALL-SAMPLE $(1 - \alpha)$ 100% CONFIDENCE INTERVAL FOR $(\mu_1 - \mu_2)$ BASED ON INDEPENDENT RANDOM SAMPLES

$$(\bar{x}_1 - \bar{x}_2) \pm t_{\alpha/2} s \sqrt{\frac{1}{n_1} + \frac{1}{n_2}}$$

where s is obtained from the pooled estimate of σ^2.

Assumptions: The samples were randomly and independently selected from normally distributed populations. The variances of the populations σ_1^2 and σ_2^2 are equal.

Note the similarity in the procedures for constructing the confidence intervals for a single mean in Section 9.3 and the difference between two means. In both cases, the interval is constructed by using the appropriate point estimator and then adding and subtracting an amount equal to $t_{\alpha/2}$ times the standard error of the point estimator.

EXAMPLE 9.6 Find an interval estimate for $(\mu_1 - \mu_2)$ in Example 9.4 using a confidence coefficient equal to .95.

Solution Substituting into the formula

$$(\bar{x}_1 - \bar{x}_2) \pm t_{\alpha/2} s \sqrt{\frac{1}{n_1} + \frac{1}{n_2}}$$

we find that the interval estimate (or 95% confidence interval) is

$$(35.22 - 31.56) \pm (2.120)(4.716)\sqrt{\frac{1}{9} + \frac{1}{9}} \qquad \text{or} \qquad 3.66 \pm 4.71$$

Thus, we estimate that the difference $(\mu_1 - \mu_2)$ in mean time to assemble falls in the interval from -1.05 to 8.37. Note that the interval width is considerable; it seems advisable to increase the size of the samples and reestimate $(\mu_1 - \mu_2)$ by using this additional information. ◁

To implement a two-sample t procedure with a pooled estimate of variance using the MINITAB package, enter the main command TWOSAMPLE, the columns identifying the two sets of sample values, and a semicolon. The sub-command POOLED, followed by a period, completes the instructions. The calculated value of t, its p-value, and its degrees of freedom are given as output together with a 95% confidence interval estimate of $(\mu_1 - \mu_2)$. Using the subcommand ALTERNATIVE, the user may also specify whether the test is to be left-tailed (-1), right-tailed (1), or two-tailed (0). The test is performed as two-tailed unless another alternative is specified.

The MINITAB output for the TWOSAMPLE command using the data in Table 9.4 appears in Table 9.5. Notice that the entry SE MEAN (standard error of the mean), given for each column, is calculated as $s/\sqrt{n}$. For example, the standard error of the mean for C1 is $4.94/\sqrt{9} = 1.646$, or 1.6. The remaining entries are self-explanatory and can be compared with the results of Examples 9.4, 9.5, and 9.6.

Before concluding our discussion, we should comment on the two assumptions upon which our inferential procedures are based. **Moderate departures from the assumption that the populations possess normal probability distributions do not seriously affect the distribution of the test statistic and the confidence coefficient for the corresponding confidence interval. On the other hand, the population variances should be nearly equal to ensure that the procedures given above are valid.**

If there is reason to believe that the population variances are far from being equal, two changes must be made in the testing and estimation procedures. Since the pooled estimator s^2 is no longer appropriate when the population variances are not equal, the sample variances s_1^2 and s_2^2 are used as estimators of σ_1^2 and σ_2^2. The

Table 9.5
MINITAB output for two
independent samples,
using the data in Table 9.4

```
MTB  >  PRINT  C1  C2
  ROW      C1        C2

    1       32        35
    2       37        31
    3       35        29
    4       28        25
    5       41        34
    6       44        40
    7       35        27
    8       31        32
    9       34        31

MTB   >   TWOSAMPLE  C1  C2;
SUB C  >  POOLED;
SUB C  >  ALTERNATIVE 1.

TWOSAMPLE T FOR C1 VS C2
         N      MEAN      STDEV     SE MEAN
C1  9          35.22       4.94         1.6
C2  9          31.56       4.48         1.5

95 PCT CI FOR  MU C1 — MU C2: (−1.0, 8.4)
TTEST MU C1 = MU C2 (VS GT):   T = 1.65 P = 0.059 DF = 16.0
```

resulting test statistic is

$$\frac{(\bar{x}_1 - \bar{x}_2) - D_0}{\sqrt{\dfrac{s_1^2}{n_1} + \dfrac{s_2^2}{n_2}}}$$

When the sample sizes are *small*, critical values for this statistic are found in Table 4 of Appendix III, using degrees of freedom approximated by the formula

$$\text{d.f.} \approx \frac{\left(\dfrac{s_1^2}{n_1} + \dfrac{s_2^2}{n_2}\right)^2}{\dfrac{\left(\dfrac{s_1^2}{n_1}\right)^2}{(n_1 - 1)} + \dfrac{\left(\dfrac{s_2^2}{n_2}\right)^2}{(n_2 - 1)}}$$

Obviously, this result must be rounded to the nearest integer. In the MINITAB package, the TWOSAMPLE *t* command without the subcommand POOLED implements this procedure.

In Section 9.7 we will present a procedure for testing an hypothesis concerning the equality of two population variances that can be used to determine whether or not the underlying population variances are equal.

If there is reason to believe that the normality assumptions have been violated, you can test for a shift in location of two population distributions using the non-parametric Mann-Whitney *U* test of Chapter 14. This test procedure, which requires fewer assumptions concerning the nature of the population probability distributions, is almost as sensitive in detecting a difference in population means when the conditions necessary for the t test are satisfied. It may be more sensitive when the assumptions are not satisfied.

EXERCISES Basic Techniques

9.16 Give the number of degrees of freedom for s^2, the pooled estimator of σ^2, if
a. $n_1 = 16, n_2 = 8$
b. $n_1 = 10, n_2 = 12$
c. $n_1 = 15, n_2 = 3$

9.17 Calculate s^2, the pooled estimator for σ^2, if
a. $n_1 = 10, n_2 = 4, s_1^2 = 3.4, s_2^2 = 4.9$
b. $n_1 = 12, n_2 = 21, s_1^2 = 18, s_2^2 = 23$

9.18 Two independent random samples of size $n_1 = 4$ and $n_2 = 5$ were selected from each of two normal populations. The data are as follows:

Population	
1	2
12	14
3	7
8	7
5	9
	6

a. Calculate s^2, the pooled estimator of σ^2.
b. Find a 90% confidence interval for $(\mu_1 - \mu_2)$, the difference between the two population means.
c. Test $H_0 : (\mu_1 - \mu_2) = 0$ against $H_a : (\mu_1 - \mu_2) > 0$ for $\alpha = .05$. State your conclusions.

9.19 Independent random samples of $n_1 = 16$ and $n_2 = 13$ observations were selected from two normal populations with equal variances. The sample means and variances are shown below.

	Population	
	1	2
Sample size	16	13
Sample mean	34.6	32.2
Sample variance	4.8	5.9

a. Suppose you wish to detect a difference between the population means. State the null and alternative hypotheses that you would use for the test.
b. Find the rejection region for the test in part (a) for $\alpha = .10$.
c. Find the value of the test statistic.
d. Find the approximate observed significance level for the test.
e. Conduct the test and state your conclusions.

9.20 Refer to Exercise 9.19. Find a 90% confidence interval for $(\mu_1 - \mu_2)$.

Applications

9.21   Jan Lindhe, D.M.D., conducted a study on the effect of an oral antiplaque rinse on plaque buildup on teeth ("Clinical Assessment of Antiplaque Agents," *The Compendium of Continuing Education in Dentistry*, supl. no. 5 [1984]). Fourteen subjects, whose teeth were thoroughly cleaned and polished, were randomly assigned to two groups of seven subjects each. Both groups were assigned to use oral rinses (no brushing) for a two-week period. Group 1 used a rinse that contained an antiplaque agent. Group 2, the control group, received a similar rinse except that, unknown to the subjects, the rinse contained no antiplaque agent. A plaque index x, a measure of plaque buildup, was recorded at 4, 7, and 14 days. The mean and standard deviation for the 14-day plaque measurements are shown below for the two groups.

	Control group	Antiplaque group
Sample size	7	7
Mean	1.26	.78
Standard deviation	.32	.32

a.   State the null and alternative hypotheses that should be used to test the effectiveness of the antiplaque oral rinse.
b.   Do the data provide sufficient evidence to indicate that the oral antiplaque rinse is effective? Test using $\alpha = .05$.
c.   Find the *p*-value for the test.

9.22   Refer to Exercise 9.21. Find a 95% confidence interval for the mean difference between the plaque indexes for the control group and the oral antiplaque rinse group. Interpret the interval.

9.23   Refer to the description of the oxygen partition study in reedfish conducted by Pettit and Beitinger, Exercise 9.5. The table below gives oxygen uptake readings for reedfish exposed to two temperature environments, one group of $n_1 = 11$ reedfish in water at 25°C and the other group of $n_2 = 12$ reedfish in water at 33°C ("Oxygen Acquisition of the Reedfish, *Erpetoichthys calabaricus*," *Journal of Experimental Biology* 114 [1985]).

	Weight (grams)	Oxygen uptake (ml O_2 g^{-1}h^{-1})		Percent	
		Air	Water	Air	Water
25°C	22.1	0.043	0.045	49	51
	21.1	0.034	0.066	34	66
	13.4	0.026	0.084	24	76
	19.5	0.048	0.045	52	48
	21.9	0.009	0.056	14	86
	19.5	0.028	0.072	28	72
	16.5	0.032	0.081	28	72
	18.5	0.045	0.051	47	53
	18.0	0.061	0.041	60	40
	15.4	0.028	0.060	32	68
	13.4	0.024	0.113	18	82

	Weight (grams)	Oxygen uptake (ml O_2 g^{-1} h^{-1})		Percent	
		Air	Water	Air	Water
33°C	22.1	0.020	0.051	28	72
	21.1	0.044	0.036	55	45
	19.5	0.035	0.043	45	55
	21.9	0.051	0.049	51	49
	19.5	0.035	0.051	41	59
	16.5	0.027	0.073	27	73
	18.5	0.043	0.054	44	56
	18.9	0.053	0.057	48	52
	15.4	0.055	0.046	54	46
	21.8	0.075	0.037	67	33
	22.1	0.042	0.049	46	54
	13.2	0.055	0.039	59	41

a. Do the data in the table present sufficient evidence to indicate a difference between the percentages of oxygen uptake by air for reedfish exposed to water at 25°C and those at 33°C? Test using $\alpha = .10$.

b. Find the approximate p-value for the test.

9.24 The data shown below by E.F. Karlin and L. C. Bliss give the percent germination of *Ledum groenlandicum*, a low evergreen heath native to Greenland and North America, for different seed ages ("Germination Ecology of *Ledum groenlandicum* and *Ledum palustre* ssp. *decumbens*," *Arctic and Alpine Research* 15, no. 3 [1983]). Each experimental unit, a set of 50 seeds of a specific age, was placed in a controlled, moist environment, and the percentage x of germinating seeds was recorded. Experimental units were prepared for seeds age 1, 8, 13, and 22 months. The number of experimental units, the sample averages, and standard deviations are given in the table for seed ages of 1 and 22 months.

Seed age (months)	Sample size	$\bar{x}$ (percent)	s (percent)
1	6	58	4
22	3	16	2

a. Suppose you suspect that the percentage of seeds germinating decreases as the age of the seeds increases. If you wished to detect this phenomenon using a test of an hypothesis, what would you choose for your null and alternative hypotheses?

b. Give the rejection region for the test if you wish α to equal .05. Explain how you would complete the test and arrive at a conclusion.

c. What assumptions must be satisfied for your test to be valid?

d. Conduct the test and state your conclusions.

9.25 Historically, the product life cycle (PLC) has been a basis for a variety of marketing tactics and strategies. However, the PLC has gained renewed acceptance in strategic marketing planning in general, and corporate product portfolio analysis in particular. PLC theory suggests that the price of durable goods is affected by forces that tend to drive real prices down. Curry and Reisz* collected information on comparative test studies reported from January 1961 through December 1980 to produce a database containing information on 938 comparative product test studies of 14,000 individual brands or items. In the short data summary that follows, the average and standard deviation of product prices are given in terms of 1980 dollars.

Product	Year	Number of brands	Average price	Standard deviation
FM Antennas	1961	16	$74.59	$10.40
	1973	17	52.62	10.50
Blenders	1965	22	89.37	14.80
	1973	20	62.26	11.10

a. Do the data concerning FM roof antennas support the hypothesis put forward by Curry and Reisz concerning the PLC, namely, that the average price of brands for a fixed product declines over time when expressed in constant dollars? Use $\alpha = .05$.

b. What is the p-value of the test in part (a)?

9.26 Refer to Exercise 9.25. Do the data concerning the price of blenders for the years 1965 and 1973 substantiate the hypothesis that the mean price of brands of blenders is declining over time? What is the p-value of this test?

9.27 In an article in *Science* (September 1985), Ilhan Olmez and Glen E. Gordon state that concentrations of rare earth elements on airborne particles often indicate the source of atmospheric materials, such as air pollutants. One such indicator is the ratio of lanthanum to samarium. Analyses of 14 specimens of Eastern coal produced a mean lanthanum to samarium ratio of 6.1 and a standard error of the mean equal to 0.3. Similar statistics for 22 specimens of Western coal were 9 and 1, respectively. In contrast, the ratio calculated for six specimens of crude oil produced a mean of 12 and a standard error of the mean equal to 16. Explain why a Student's t-test may not be suitable in testing for differences in the ratio of lanthanum to samarium between the two types of coal or between one of the coal types and crude oil. (*Note:* An alternative method for conducting these tests is presented in Section 14.4.)

9.28 Refer to Exercise 9.8, where we measured the dissolved oxygen content in river water to determine whether a stream had sufficient oxygen to support aquatic life. A pollution control inspector suspected that a river community was releasing amounts of semitreated sewage into a river. To check his theory, he drew five randomly selected specimens of river water at a location above the town, and another five below. The dissolved oxygen readings, in parts per million, are as follows:

Above town	4.8	5.2	5.0	4.9	5.1
Below town	5.0	4.7	4.9	4.8	4.9

a. Do the data provide sufficient evidence to indicate that the mean oxygen content below the town is less than the mean oxygen content above? Test using $\alpha = .05$.

* Curry, David J., and Reisz, Peter C., "Prices and Price/Quality Relationships: A Longitudinal Analysis," *Journal of Marketing*, Vol. 52 (January 1988), pp. 36–51. Reprinted with permission from the American Marketing Association.

b. Suppose that you prefer estimation as a method of inference. Estimate the difference in mean dissolved oxygen content between locations above and below the town. Use a 95% confidence interval.

9.29 The table showing the means and standard errors for the study of bone mass in women athletes from Exercise 9.9 is reproduced below (B. L. Drinkwater, et al., "Bone Mineral Content of Amenorrheic and Eumenorrheic Athletes," *New England Journal of Medicine* [August 2, 1984]).

Site	Amenorrheic	Eumenorrheic	p-value
Radius			
S_1			
Mineral content (g/cm)	0.89 ± 0.03	0.85 ± 0.03	NS
Mineral density (g/cm^2)	0.53 ± 0.02	0.54 ± 0.01	NS
S_2			
Mineral content (g/cm)	0.91 ± 0.02	0.88 ± 0.03	NS
Mineral density (g/cm^2)	0.67 ± 0.02	0.67 ± 0.02	NS
Vertebrae (g/cm^2)	1.12 ± 0.04	1.30 ± 0.03	<0.01

Plus-minus values are means ± S.E.M. NS denotes "not significant."

You will note that the authors indicate, by the symbol NS (meaning "not significant," i.e., that there is insufficient evidence to reject the null hypothesis), that there was no evidence of a difference between the means of the amenorrheic and the eumenorrheic groups for measurements taken in the forearm (sites S_1 and S_2). In contrast, they found evidence of a difference in mean density for measurements taken on the vertebrae with a *p*-value of less than 0.01.

a. In testing for differences in mean bone density between amenorrheic and eumenorrheic women at a specific location, should you use a one- or a two-sided test? Explain.

b. Give the null and alternative hypotheses for the test in part (a).

c. Give the rejection region using $\alpha = .05$.

d. Do the data provide sufficient evidence to indicate a difference between the mean bone densities of the amenorrheic versus the eumenorrheic groups at the vertebrae location? Test using $\alpha = .05$.

e. Find the *p*-value for your test. Is it less than the .01 value shown in the table?

9.5 A PAIRED-DIFFERENCE TEST

A manufacturer wishes to compare the wearing qualities of two different types of automobile tires, A and B. For the comparison, a tire of type A and one of type B are randomly assigned and mounted on the rear wheels of each of five automobiles. The automobiles are then operated for a specified number of miles, and the amount of wear is recorded for each tire. These measurements appear in Table 9.6. Do the data present sufficient evidence to indicate a difference in the average wear for the two tire types?

Analyzing the data, we note that the difference between the two sample means is $(\bar{x}_1 - \bar{x}_2) = .48$, a rather small quantity, considering the variability of the data and the small number of measurements involved. At first glance it would seem that

Table 9.6

Automobile	Tire A	Tire B
1	10.6	10.2
2	9.8	9.4
3	12.3	11.8
4	9.7	9.1
5	8.8	8.3
	$\bar{x}_1 = 10.24$	$\bar{x}_2 = 9.76$

there is little evidence to indicate a difference between the population means, a conjecture that we can check by the method outlined in Section 9.4.

The pooled estimate of the common variance σ^2 is

$$s^2 = \frac{\sum\limits_{i=1}^{n_1} (x_{1i} - \bar{x}_1)^2 + \sum\limits_{i=1}^{n_2} (x_{2i} - \bar{x}_2)^2}{n_1 + n_2 - 2} = \frac{6.932 + 7.052}{5 + 5 - 2} = 1.748$$

and

$$s = 1.32$$

The calculated value of t used to test the hypothesis that $\mu_1 = \mu_2$ is

$$t = \frac{\bar{x}_1 - \bar{x}_2}{s\sqrt{\dfrac{1}{n_1} + \dfrac{1}{n_2}}} = \frac{10.24 - 9.76}{1.32\sqrt{\dfrac{1}{5} + \dfrac{1}{5}}} = .57$$

a value that is not nearly large enough to reject the hypothesis that $\mu_1 = \mu_2$.

The corresponding 95% confidence interval is

$$(\bar{x}_1 - \bar{x}_2) \pm t_{\alpha/2} s \sqrt{\frac{1}{n_1} + \frac{1}{n_2}}$$

$$(10.24 - 9.76) \pm (2.306)(1.32)\sqrt{\frac{1}{5} + \frac{1}{5}}$$

or

$$-1.45 \text{ to } 2.41.$$

This interval is quite wide, considering the small difference between the sample means.

A second glance at the data reveals a marked inconsistency with this conclusion. We note that the wear measurement for type A is larger than the corresponding value for type B for *each* of the five automobiles. These differences, recorded as $d = A - B$, are shown in Table 9.7.

Table 9.7
Differences in tire wear, using the data of Table 9.6

Automobile	A	B	$d = A - B$
1	10.6	10.2	.4
2	9.8	9.4	.4
3	12.3	11.8	.5
4	9.7	9.1	.6
5	8.8	8.3	.5
			$\bar{d} = .48$

If there is no difference in mean tire wear for the two tire types, then the probability that tire A shows more wear than tire B is equal to $p = .5$, and the five automobiles correspond to $n = 5$ independent binomial trials. A two-tailed test of the null hypothesis $p = .5$ would include a rejection region consisting of $x = 0$ and $x = 5$ and $\alpha = P(x = 0) + P(x = 5) = 2(1/2)^5 = 1/16 = .0625$. Since five of the differences are positive ($x = 5$), we have evidence to indicate that a difference exists in the mean wear of the two tire types.

You will note that we have used two different statistical tests to test the same hypothesis. Isn't it peculiar that the t test, which uses more information (the actual sample measurements) than the binomial test, fails to supply sufficient evidence for rejection of the hypothesis $\mu_1 = \mu_2$?

There is an explanation for this inconsistency. The t test described in Section 9.4 is not the proper statistical test to be used for our example. The statistical test procedure of Section 9.4 requires that the two samples be *independent and random*. Certainly, the independence requirement was violated by the manner in which the experiment was conducted. The (pair of) measurements, an A and a B tire, for a particular automobile are definitely related. A glance at the data shows that the readings have approximately the same magnitude for a particular automobile, but vary markedly from one automobile to another. This, of course, is exactly what we might expect. Tire wear is largely determined by driver habits, the balance of the wheels, and the road surface. Since each automobile has a different driver, we would expect a large amount of variability in the data from one automobile to another.

The familiarity we have gained with interval estimation has shown us that the width of the large-sample confidence intervals depends on the magnitude of the standard deviation of the point estimator of the parameter. The smaller its value, the better is the estimate and the more likely it is that the test statistic will provide evidence to reject the null hypothesis if it is, in fact, false. Knowledge of this phenomenon was utilized in *designing* the tire wear experiment. The experimenter realized that the wear measurements would vary greatly from auto to auto and that this variability could not be separated from the data if the tires were assigned to the ten wheels in a random manner. (A random assignment of the tires would have implied that the data be analyzed according to the procedure of Section 9.4.) Instead, a comparison of the wear between tire types A and B made on each automobile resulted in the five difference measurements. This design eliminates the effect of the car-to-car variability and yields more information on the mean difference in the wearing quality for the two tire types.

A proper analysis of the data would use the five difference measurements to test the hypothesis that the average difference μ_d is equal to 0 or, equivalently, to test the null hypothesis $H_0: \mu_d = \mu_1 - \mu_2 = 0$ against the alternative hypothesis $H_a \mu_d = (\mu_1 - \mu_2) \neq 0$.

PAIRED-DIFFERENCE TEST FOR $(\mu_1 - \mu_2) = \mu_d$

1. Null Hypothesis: $H_0: \mu_d = 0$
2. Alternative Hypothesis:

    One-Tailed Test   *Two-Tailed Test*

    $H_a: \mu_d > 0$ $H_a: \mu_d \neq 0$

    (or, $H_a: \mu_d < 0$)

3. Test Statistic: $t = \dfrac{\bar{d} - 0}{s_d/\sqrt{n}} = \dfrac{\bar{d}}{s_d/\sqrt{n}}$

    where n = number of paired differences

    $$s_d = \sqrt{\frac{\sum\limits_{i=1}^{n} (d_i - \bar{d})^2}{n-1}}$$

4. Rejection Region:

    One-Tailed Test   *Two-Tailed Test*

    $t > t_\alpha$ $t > t_{\alpha/2}$ or $t < -t_{\alpha/2}$

    (or, $t < -t_\alpha$ when the
    alternative hypothesis
    is $H_a: \mu_d < 0$)

The critical values of t, t_α, and $t_{\alpha/2}$, are based on $(n-1)$ degrees of freedom. These tabulated values are given in Table 4 in Appendix III.

Assumptions: The n paired observations are randomly selected from normally distributed populations.

EXAMPLE 9.7   Do the data in Table 9.6 provide sufficient evidence to indicate a difference in mean wear for tire types A and B? Test using $\alpha = .05$.

Solution   You can verify that the average and standard deviation of the five difference measurements are

$$\bar{d} = .48 \quad \text{and} \quad s_d = .0837$$

Then,

$$H_0: \mu_d = 0 \quad \text{and} \quad H_a: \mu_d \neq 0$$

and

$$t = \frac{\bar{d} - 0}{s_d/\sqrt{n}} = \frac{.48}{.0837/\sqrt{5}} = 12.8$$

The critical value of t for a two-tailed statistical test, $\alpha = .05$ and four degrees of freedom, is 2.776. Certainly, the observed value of $t = 12.8$ is extremely large and highly significant. Hence, we would conclude that there is a difference in the mean amount of wear for tire types A and B. ◁

$(1 - \alpha)$ 100% SMALL-SAMPLE CONFIDENCE INTERVAL FOR $(\mu_1 - \mu_2) = \mu_d$, BASED ON A PAIRED-DIFFERENCE EXPERIMENT

$$\bar{d} \pm t_{\alpha/2} \frac{s_d}{\sqrt{n}}$$

where n = number of paired differences and

$$s_d = \sqrt{\frac{\sum_{i=1}^{n} (d_i - \bar{d})^2}{n - 1}}$$

Assumptions: The n paired observations are randomly selected from normally distributed populations.

EXAMPLE 9.8 Find a 95% confidence interval for $(\mu_1 - \mu_2) = \mu_d$ using the data in Table 9.6.

Solution A 95% confidence interval for the difference between the mean wear would be

$$\bar{d} \pm t_{\alpha/2} \frac{s_d}{\sqrt{n}} = .48 \pm (2.776) \frac{.0837}{\sqrt{5}}$$

or $.48 \pm .10$. ◁

When the units used to compare two or more procedures exhibit marked variability before any experimental procedures are implemented, the effect of this variability can be minimized by comparing the procedures *within* groups of relatively homogeneous units called **blocks**. In this way the effects of the procedures are not masked by the initial variability among the units in the experiment. An experiment conducted in this manner is called a **randomized block design**. In an experiment involving daily sales, blocks may represent days of the week; in an experiment involving product marketing, blocks may represent geographic areas. (Randomized block designs are discussed in more detail in Section 13.6.)

The statistical design of the tire experiment is a simple example of a randomized block design, and the resulting statistical test is often called a *paired-difference test*. You will note that the pairing occurred when the experiment was planned and not after

the data were collected. Comparisons of tire wear were made within relatively homogeneous blocks (automobiles), with the tire types randomly assigned to the two automobile wheels.

The amount of information gained by blocking the tire experiment may be measured by comparing the calculated confidence interval for the unpaired (and incorrect) analysis with the interval obtained for the paired-difference analysis. The confidence interval for $(\mu_1 - \mu_2)$ that might have been calculated, had the tires been randomly assigned to the ten wheels (unpaired), is unknown but probability would have been of the same magnitude as the interval -1.45 to 2.41, which was calculated by analyzing the observed data in an unpaired manner. Pairing the tire types on the automobiles (blocking) and the resulting analysis of the differences produced the interval estimate $.38$ to $.58$. Note the difference in the widths of the intervals, which indicates the very sizable increase in information obtained by blocking in this experiment.

Although blocking proved to be very beneficial in the tire experiment, it may not always be. We observe that the degrees of freedom available for estimating σ^2 are less for the paired than for the corresponding unpaired experiment. If there were actually no differences among the blocks, the reduction in the degrees of freedom would produce a moderate increase in the $t_{\alpha/2}$ employed in the confidence interval and hence would increase the width of the interval. This, of course, did not occur in the tire experiment because the large reduction in the standard error of $\bar{d}$ more than compensated for the loss in degrees of freedom.

Except for notation, the analysis of a paired-difference experiment is the same as that for a single sample presented in Section 9.3. This similarity enables us to use the MINITAB commands TTEST and TINTERVAL to analyze the differences in a paired-difference experiment.

Before concluding, we want to reemphasize a point. Once you have used a paired design for an experiment, you no longer have the option of using the unpaired analysis of Section 9.4. The assumptions upon which that test is based have been violated. Your only alternative is to use the correct method of analysis, the paired-difference test (and associated confidence interval) of this section.

EXERCISES   Basic Techniques

9.30   A paired-difference experiment was conducted using $n = 10$ pairs of observations. Test the null hypothesis $H_0:(\mu_1 - \mu_2) = 0$ against $H_a:(\mu_1 - \mu_2) \neq 0$ for $\alpha = .05$, $\bar{d} = .3$, and $s_d^2 = .16$. Give the approximate p-value for the test.

9.31   Find a 95% confidence interval for $(\mu_1 - \mu_2)$ in Exercise 9.30.

9.32   How many pairs of observations would you need if you wished to estimate $(\mu_1 - \mu_2)$ in Exercise 9.30, correct to within .1 with probability equal to .95?

9.33   A paired-difference experiment consists of $n = 18$ pairs, $\bar{d} = 5.7$, and $s_d^2 = 256$. Suppose that we wish to detect $\mu_d > 0$.
   a.   Give the null and alternative hypotheses for the test.
   b.   Conduct the test and state your conclusions.

9.34   A paired-difference experiment was conducted to compare the means of two populations. The data are shown on page 353.

| | Pairs | | | | |
Population	I	2	3	4	5
1	1.3	1.6	1.1	1.4	1.7
2	1.2	1.5	1.1	1.2	1.8

a. Do the data provide sufficient evidence to indicate that μ_1 differs from μ_2? Test using $\alpha = .05$.
b. Find the approximate observed significance level for the test and interpret its value.
c. Find a 95% confidence interval for $(\mu_1 - \mu_2)$. Compare your interpretation of the confidence interval with your test results in part (a).
d. What assumptions must you make for your inferences to be valid?

Applications

9.35 Persons submitting computing jobs to a computer center are usually required to estimate the amount of computer time required to complete the job. This time is measured in CPUs, the amount of time that a job will occupy a portion of the computer's central processing unit's memory. A computer center decided to perform a comparison of the estimated versus actual CPU times for a particular customer. The corresponding times were available for 11 jobs. The sample data are shown below.

CPU time (minutes)	Job number										
	I	2	3	4	5	6	7	8	9	10	11
Estimated	.50	1.40	.95	.45	.75	1.20	1.60	2.6	1.30	.85	.60
Actual	.46	1.52	.99	.53	.71	1.31	1.49	2.9	1.41	.83	.74

a. Why would you expect these pairs of data to be correlated?
b. Do the data provide sufficient evidence to indicate that, *on the average*, the customer tends to underestimate CPU time required for computing jobs? Test using $\alpha = .10$.
c. Find the observed significance level for the test and interpret its value.
d. Find a 90% confidence interval for the difference in mean estimated CPU time versus mean actual CPU time.

9.36 The data shown below, gathered by Runzheimer and Co., Inc., give the annual operating cost comparisons for automatic and stick-shift versions of four different makes of automobiles (*USA Today*, January 10, 1985). If the data represent a random sampling of the comparative costs of operating automobiles, find a 90% confidence interval for the mean difference in operating costs between automatic and stick-shift versions of the same type of automobile. (Operating costs include oil, maintenance, tires, and fuel.)

| | Annual operating cost | |
	Automatic	Manual
Ford Tempo	$1520	$1440
Plymouth Reliant	1240	1150
AMC Encore S	1380	1150
Chevrolet Cavalier	1150	1100

9.37   An experiment was conducted to compare mean reaction time to two types of traffic signs, prohibitive (No Left Turn) and permissive (Left Turn Only). Ten subjects were included in the experiment. Each subject was presented with 40 traffic signs, 20 prohibitive and 20 permissive, in random order. The mean time to reaction and the number of correct actions were recorded for each subject. The mean reaction times to the 20 prohibitive and 20 permissive traffic signs are shown below for each of the ten subjects.

| | Mean reaction times (ms) for 20 traffic signs | |
Subject	Prohibitive	Permissive
1	824	702
2	866	725
3	841	744
4	770	663
5	829	792
6	764	708
7	857	747
8	831	685
9	846	742
10	759	610

a.   Explain why this is a paired-difference experiment and give reasons why the pairing should be useful in increasing information on the difference between the mean reaction times to prohibitive and permissive traffic signs.

b.   Do the data present sufficient evidence to indicate a difference in mean reaction times to prohibitive and permissive traffic signs? Test using $\alpha = .05$.

c.   Find and interpret the approximate p-value for the test in part (b).

d.   Find a 95% confidence interval for the difference in mean reaction times to prohibitive and permissive traffic signs.

9.38   Two computers are often compared by running a collection of different "benchmark" programs and recording the difference in CPU time required to complete the same program. Six benchmark programs, run on two computers, produced the following CPU times (in minutes):

| | Benchmark program | | | | | |
Computer	1	2	3	4	5	6
1	1.12	1.73	1.04	1.86	1.47	2.10
2	1.15	1.72	1.10	1.87	1.46	2.15

a.   Do the data provide sufficient evidence to indicate a difference in mean CPU times required for the two computers to complete a job? Test using $\alpha = .05$.

b.   Find the approximate observed significance level for the test and interpret its value.

c.   Find a 95% confidence interval for the difference in mean CPU times required for the two computers to complete a job.

9.39   Exercise 9.21 describes a dental experiment conducted to investigate the effectiveness of an oral rinse used to deter the growth of plaque on teeth. Subjects were divided into two groups: one group used a rinse with an antiplaque ingredient, and the control group used a rinse containing inactive ingredients. Suppose that the plaque growth on each person's teeth was measured after using the rinse after 4 hours, and then again after 8 hours. If you wished to estimate the difference in plaque growth from 4 to 8 hours, should you use a confidence interval based on a paired or an unpaired analysis? Explain.

9.40   A study conducted by nutritionists at the University of Toronto has determined that people who nibble all day instead of eating their ordinary-size meals have significantly lower blood cholesterol levels (*Press Enterprise*, October 5, 1989). In the study, seven men ate 2500 calories per day. For two weeks, they ate ordinary-size meals, but for another two weeks, they got the same amount of calories in 17 snacks eaten once an hour. The total cholesterol level, as well as the hazardous low-density lipoprotein cholesterol (LDL), was measured for each subject after each two-week period. Is this a paired or an unpaired experiment? Explain.

9.41   The earth's temperature (which affects seed germination, crop survival in bad weather, and many other aspects of agricultural production) can be measured using either ground-based sensors or infrared-sensing devices mounted in aircraft or space satellites. Ground-based sensoring is tedious, requiring many replications to obtain an accurate estimate of ground temperature. On the other hand, airplane or satellite sensoring of infrared waves appears to introduce a bias in the temperature readings. To determine the bias, readings were obtained at six different locations using both ground- and air-based temperature sensors. The readings, in degrees Celsius, were as follows:

Location	Ground	Air
1	46.9	47.3
2	45.4	48.1
3	36.3	37.9
4	31.0	32.7
5	24.7	26.2

a.   Do the data present sufficient evidence to indicate a bias in the air-based temperature readings? Explain. (Use $\alpha = .05$.)

b.   Estimate the difference in mean temperature between ground- and air-based sensors using a 95% confidence interval.

9.42   Refer to Exercise 9.41. How many paired observations would be required to estimate the difference between mean temperatures for ground- versus air-based sensors correct to within .2°C with probability approximately equal to .95?

9.43   In response to a complaint that a particular tax assessor (A) was biased, an experiment was conducted to compare the assessor named in the complaint with another tax assessor (B) from the same office. Eight properties were selected and each was assessed by both assessors. The assessments (in thousands of dollars) are shown in the table.

Property	Assessor A	Assessor B
1	76.3	75.1
2	88.4	86.8
3	80.2	77.3
4	94.7	90.6
5	68.7	69.1
6	82.8	81.0
7	76.1	75.3
8	79.0	79.1

a.   Do the data provide sufficient evidence to indicate that assessor A tends to give higher assessments than assessor B? Test with $\alpha = .05$.

b.   Estimate the difference in mean assessments for the two assessors.

c.   What assumptions must you make in order for inferences (a) and (b) to be valid?

d.   Suppose that assessor A had been compared with a more stable standard, say the average $\bar{x}$, of the assessments given by four assessors selected from the tax office. Thus, each property would be assessed by A and also by each of the four other assessors, and $x_A - \bar{x}$ would be calculated. If the test in part (a) is valid, could you use the paired-difference t test to test the hypothesis that the bias, the mean difference between A's assessments and the mean of the assessments of the four assessors, is equal to 0? Explain.

9.44   In a study of the well-being of cancer patients, investigators matched each of a group of 104 cancer survivors (those who have lived one or more years past their last cancer treatment) with a control, that is, a person who was considered physically healthy and who had not had cancer (Arthur H. Schmale, et al., "Well-Being of Cancer Survivors." *Psychosomatic Medicine* 45, no. 2 [1983]. Reprinted by permission of Elsevier Publishing Co., Inc. Copyright 1983 by the American Psychosomatic Society, Inc.). Patients and controls were matched according to sex, age, and education. Each person in the study was then administered a questionnaire, "The General Well-Being Measure," which purports to measure a person's well-being. The following table gives the means for 104 cancer survivors (CS) and 104 controls for each of a number of factors associated with well-being and for two composite indices of well-being. Also shown are the t values and associated p-values for testing an hypothesis of "no difference" between the means of the cancer survivors and the controls.

$n_1 = n_2 = 104$

	Means			
	CS	Control	t	2-tailed p
Factors				
Anxiety	12.68	12.39	0.46	0.64
Depression	5.27	5.49	0.74	0.46
Self-control	15.34	16.29	3.30	0.00*
Positive well-being	17.53	16.99	1.09	0.27
Vitality	17.70	17.38	0.68	0.49
General health	13.95	15.20	3.67	0.00*

* Asterisks refer to statistical significance.

$n_1 = n_2 = 104$

	Means			
	CS	Control	t	2-tailed p
Composites				
Mental health index	70.79	71.39	0.41	0.67
General well-being	102.14	103.98	0.88	0.37

a.   Describe the type of experimental design used for this study.

b.   Describe the practical conclusions to be derived from the test results.

c.   Calculate one of the p-values, say the approximate p-value for positive well-being. Does it appear to agree with the p-value given in the author's table?

▷ **9.6** INFERENCES CONCERNING
A POPULATION VARIANCE

We have seen in the preceding sections that an estimate of the population variance σ^2 is fundamental to procedures for making inferences about population means. Moreover, there are many practical situations where σ^2 is the primary objective of an experimental investigation; thus, it may assume a position of far greater importance than that of the population mean.

Scientific measuring instruments must provide unbiased readings with a very small error of measurement. An aircraft altimeter that measured the correct altitude on the *average* would be of little value if the standard deviation of the error of measurement were 5000 feet. Indeed, bias in a measuring instrument can often be corrected, but the precision of the instrument, measured by the standard deviation of the error of measurement, is usually a function of the design of the instrument itself and cannot be controlled.

Machined parts in a manufacturing process must be produced with minimum variability in order to reduce out-of-size and hence defective products. And, in general, it is desirable to maintain a minimum variance in the measurements of the quality characteristics of an industrial product in order to achieve process control and therefore minimize the percentage of poor-quality product.

The sample variance

$$s^2 = \frac{\sum_{i=1}^{n} (x_i - \bar{x})^2}{n - 1}$$

is an unbiased estimator of the population variance σ^2. Thus, the distribution of sample variances generated by repeated sampling will have a probability distribution that commences at $s^2 = 0$ (since s^2 cannot be negative) with a mean equal to σ^2. Unlike the distribution of $\bar{x}$, the sampling distribution of s^2 is nonsymmetrical, the exact form being dependent on the probability distribution of the population.

> **Assumptions:**   For the methodology that follows we will assume that the sample is drawn from a normal population and that s^2 is based on a random sample of n measurements.

Or, using the terminology of Section 9.2, we would say that s^2 has $(n - 1)$ degrees of freedom.

The next and obvious step would be to consider the distribution of s^2 in repeated sampling from a specified normal distribution—one with a specific mean and variance—and to tabulate the critical values of s^2 for some of the commonly used tail areas. If this is done, we will find that the distribution of s^2 is independent of the population mean μ, but has a different distribution for each sample size and each value of σ^2. This task would be quite laborious, but fortunately it may be simplified by *standardizing*, as was done by using z in the normal tables.

The quantity

$$\chi^2 = \frac{(n-1)s^2}{\sigma^2},$$

called a **chi-square variable** by statisticians, admirably suits our purposes. Its distribution in repeated sampling is called, as we might suspect, a **chi-square probability distribution**. The equation of the density function for the chi-square distribution is well known to statisticians who have tabulated critical values corresponding to various tail areas of the distribution. These values are presented in Table 5 in Appendix III.

The shape of the chi-square distribution, like that of the t distribution, will vary with the sample size or, equivalently, with the degrees of freedom associated with s^2. Thus, Table 5 in Appendix III is constructed in exactly the same manner as the t table, with the degrees of freedom shown in the first and last columns. A partial reproduction of Table 5 in Appendix III is shown in Table 9.8. The symbol χ_α^2 indicates that the tabulated χ^2 value is such that an area α lies to its right. (See Figure 9.3.) Stated in probabilistic terms,

$$P(\chi^2 > \chi_\alpha^2) = \alpha$$

Thus, 99% of the area under the χ^2 distribution would lie to the right of $\chi_{.99}^2$. We note that the extreme values of χ^2 must be tabulated for both the lower and upper tail of the distribution because it is nonsymmetrical.

You can check your ability to use the table by verifying the following statements. The probability that χ^2, based on $n = 16$ measurements (d.f. $= 15$), will

Table 9.8 Format of the chi-square table from Table 5 in Appendix III

d.f.	$\chi_{0.995}^2$	$\cdots$	$\chi_{0.950}^2$	$\chi_{0.900}^2$	$\chi_{0.100}^2$	$\chi_{0.050}^2$	$\cdots$	$\chi_{0.005}^2$	d.f.
1	0.0000393		0.0039321	0.0157908	2.70554	3.84146		7.87944	1
2	0.0100251		0.102587	0.210720	4.60517	5.99147		10.5966	2
3	0.0717212		0.351846	0.584375	6.25139	7.81473		12.8381	3
4	0.206990		0.710721	1.063623	7.77944	9.48773		14.8602	4
5	0.411740		1.145476	1.61031	9.23635	11.0705		16.7496	5
6	0.0675727		1.63539	2.20413	10.6446	12.5916		18.5476	6
$\vdots$	$\vdots$		$\vdots$	$\vdots$	$\vdots$	$\vdots$		$\vdots$	$\vdots$
15	4.60094		7.26094	8.54675	22.3072	24.9958		32.8013	15
16	5.14224		7.96164	9.31223	23.5418	26.2962		34.2672	16
17	5.69724		8.67176	10.0852	24.7690	27.5871		35.7185	17
18	6.26481		9.39046	10.8649	25.9894	28.8693		37.1564	18
19	6.84398		10.1170	11.6509	27.2036	30.1435		38.5822	19
$\vdots$	$\vdots$		$\vdots$	$\vdots$	$\vdots$	$\vdots$		$\vdots$	$\vdots$

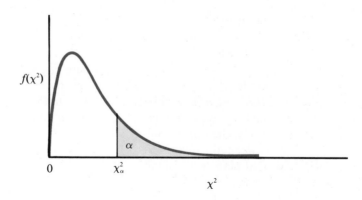

Figure 9.3
A chi-square distribution

exceed 24.9958 is .05. For a sample of $n = 6$ measurements (d.f. $= 5$), 95% of the area under the χ^2 distribution will lie to the right of $\chi^2 = 1.145476$. These values of χ^2 are shaded in Table 9.8. The statistical test of a null hypothesis concerning a population variance

$$H_0: \sigma^2 = \sigma_0^2$$

will employ the test statistic

$$\chi^2 = \frac{(n-1)s^2}{\sigma_0^2}$$

Notice that when H_0 is true, s^2/σ_0^2 should be near 1, so that χ^2 should be close to $(n - 1)$, the degrees of freedom. If σ^2 is really greater than the hypothesized value σ_0^2, the test statistic will tend to be larger than $(n - 1)$ and will probably fall toward the upper tail of the distribution. If $\sigma^2 < \sigma_0^2$, the test statistic will tend to be smaller than $(n - 1)$ and will probably fall toward the lower tail of the chi-square distribution. As in other testing situations, we may use either a one- or a two-tailed statistical test, depending upon the alternative hypothesis under test.

TEST OF AN HYPOTHESIS CONCERNING A POPULATION VARIANCE

1. Null Hypothesis: $H_0: \sigma^2 = \sigma_0^2$
2. Alternative Hypothesis:

   One-Tailed Test   *Two-Tailed Test*
   $H_a: \sigma^2 > \sigma_0^2$ $H_a: \sigma^2 \neq \sigma_0^2$
   (or, $H_a: \sigma^2 < \sigma_0^2$)

3. Test Statistic: $\chi^2 = \frac{(n-1)s^2}{\sigma_0^2}$

4.   Rejection Region:

One-Tailed Test

$\chi^2 > \chi_\alpha^2$

(or, $\chi^2 < \chi_{(1-\alpha)}^2$ when the alternative hypothesis is $H_a: \sigma^2 < \sigma_0^2$), where χ_α^2 and $\chi_{(1-\alpha)}^2$ are, respectively, the upper- and lower-tail values of χ^2 that place α in the tail areas.

Two-Tailed Test

$\chi^2 > \chi_{\alpha/2}^2$ or $\chi^2 < \chi_{(1-\alpha/2)}^2$, where $\chi_{\alpha/2}^2$ and $\chi_{(1-\alpha/2)}^2$ are, respectively, the upper- and lower-tail values of χ^2 that place $\alpha/2$ in the tail areas.

The critical values of χ^2 are based on $(n-1)$ degrees of freedom. These tabulated values are given in Table 5 in Appendix III.

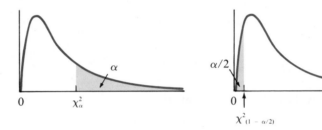

Assumption: The sample has been randomly selected from a normal population.

EXAMPLE 9.9   A cement manufacturer claimed that concrete prepared from his product would have a relatively stable compressive strength and that the strength measured in kilograms per square centimeter would lie within a range of 40 kilograms per square centimeter. A sample of $n = 10$ measurements produced a mean and variance equal to, respectively,

$$\bar{x} = 312$$

$$s^2 = 195$$

Do these data present sufficient evidence to reject the manufacturer's claim?

Solution   As stated, the manufacturer claimed that the range of the strength measurements would equal 40 kilograms per square centimeter. We will suppose that he meant that the measurements would lie within this range 95% of the time and, therefore, that the range would equal approximately 4σ and that $\sigma = 10$. We would then wish to test the null hypothesis

$$H_0: \sigma^2 = (10)^2 = 100$$

against the alternative

$$H_a : \sigma^2 > 100$$

The alternative hypothesis would require a one-tailed statistical test with the entire rejection region located in the upper tail of the χ^2 distribution. The critical value of χ^2 for $\alpha = .05$ and $(n - 1) = 9$ degrees of freedom is $\chi^2 = 16.9190$, which implies that we will reject H_0 if the test statistic exceeds this value.

Calculating, we obtain

$$\chi^2 = \frac{(n - 1)s^2}{\sigma_0^2} = \frac{1755}{100} = 17.55$$

Since the value of the test statistic falls in the rejection region, we conclude that the null hypothesis is false and that the range of concrete-strength measurements will exceed the manufacturer's claim.   ◁

EXAMPLE 9.10   Find the approximate observed significance level for the test in Example 9.9.

Solution   Examining the row corresponding to 9 degrees of freedom in Table 5 in Appendix III, you will see that the observed value of chi-square, $\chi^2 = 17.55$, is larger than the tabulated value, $\chi^2_{.05} = 16.9190$, and less than $\chi^2_{.025} = 19.0228$. Therefore, the observed significance level (p-value) for the test lies between .05 and .025 and we would report the observed significance level for the test as

$$.025 < p\text{-value} < .05$$

This tells us that we would reject the null hypothesis for any value of α equal to or greater than .05.   ◁

A confidence interval for σ^2 with a $(1 - \alpha)$ confidence coefficient can be shown to be

$(1 - \alpha)$ 100% CONFIDENCE INTERVAL FOR σ^2

$$\frac{(n - 1)s^2}{\chi^2_{\alpha/2}} < \sigma^2 < \frac{(n - 1)s^2}{\chi^2_{(1 - \alpha/2)}}$$

where $\chi^2_{\alpha/2}$ and $\chi^2_{(1 - \alpha/2)}$ are the upper and lower χ^2 values, which would locate one-half of α in each tail of the chi-square distribution.

Assumption: The sample has been randomly selected from a normal population.

EXAMPLE 9.11   Find a 90% confidence interval for σ^2 in Example 9.9.

Solution   The first step would be to obtain the tabulated values of $\chi^2_{.95}$ and $\chi^2_{.05}$ corresponding to $(n - 1) = 9$ degrees of freedom.

$$\chi^2_{(1-\alpha/2)} = \chi^2_{.95} = 3.32511$$

$$\chi^2_{\alpha/2} = \chi^2_{.05} = 16.9190$$

Substituting these values and $s^2 = 195$ into the formula for the confidence interval,

$$\frac{(n-1)s^2}{\chi^2_{\alpha/2}} < \sigma^2 < \frac{(n-1)s^2}{\chi^2_{(1-\alpha/2)}}$$

The interval estimate for σ^2 would be

$$\frac{9(195)}{16.9190} < \sigma^2 < \frac{9(195)}{3.32511}$$

or

$$103.73 < \sigma^2 < 527.80.$$

◁

EXAMPLE 9.12   An experimenter was convinced that her measuring equipment had a variability measured by a standard deviation $\sigma = 2$. During an experiment, she recorded the measurements 4.1, 5.2, and 10.2. Do these data disagree with her assumption? Test the hypothesis $H_0: \sigma = 2$ or $\sigma^2 = 4$. Then place a 90% confidence interval on σ^2.

Solution   The calculated sample variance is $s^2 = 10.57$. Since we wish to detect $\sigma^2 > 4$ as well as $\sigma^2 < 4$, we should employ a two-tailed test. When we use $\alpha = .10$ and place .05 in each tail, we will reject H_0 when $\chi^2 > 5.99147$ or $\chi^2 < .102587$.

The calculated value of the test statistic is

$$\chi^2 = \frac{(n-1)s^2}{\sigma_0^2} = \frac{2(10.57)}{4} = 5.29$$

Since the test statistic does not fall in the rejection region, the data do not provide sufficient evidence to reject the null hypothesis $H_0: \sigma^2 = 4$.

The corresponding 90% confidence interval is

$$\frac{(n-1)s^2}{\chi^2_{\alpha/2}} < \sigma^2 < \frac{(n-1)s^2}{\chi^2_{(1-\alpha/2)}}$$

The values of $\chi^2_{(1-\alpha/2)}$ and $\chi^2_{\alpha/2}$ are

$$\chi^2_{(1-\alpha/2)} = \chi^2_{.95} = .102587$$

$$\chi^2_{\alpha/2} = \chi^2_{.05} = 5.99147$$

Substituting these values into the formula for the interval estimate, we obtain

$$\frac{2(10.57)}{5.99147} < \sigma^2 < \frac{2(10.57)}{.102587}$$

or

$$3.53 < \sigma^2 < 206.07$$

Thus, we estimate the population variance to fall in the interval 3.53 to 206.07. This very wide confidence interval indicates how little information on the population variance is obtained in a sample of only three measurements. Consequently, it is not surprising that there was insufficient evidence to reject the null hypothesis $\sigma^2 = 4$. To obtain more information on σ^2, the experimenter needs to increase the sample size.

EXERCISES   Basic Techniques

9.45   A random sample of $n = 25$ observations from a normal population produced a sample variance equal to 21.4. Do these data provide sufficient evidence to indicate that $\sigma^2 > 15$? Test using $\alpha = .05$.

9.46   A random sample of $n = 15$ observations was selected from a normal population. The sample mean and variance were $\bar{x} = 3.91$ and $s^2 = .3214$. Find a 90% confidence interval for the population variance σ^2.

9.47   A random sample of size $n = 7$ from a normal population produced the following measurements:

       1.4   3.6   1.7   2.0   3.3   2.8   2.9

a.   Calculate the sample variance, s^2.
b.   Construct a 95% confidence interval for the population variance, σ^2.
c.   Test $H_0: \sigma^2 = .8$ against $H_a: \sigma^2 \neq .8$ using $\alpha = .05$. State your conclusions.
d.   What is the approximate p-value for the test in part (c)?

Applications

9.48   Refer to the comparison of the estimated versus actual CPU times for computer jobs submitted by the customer in Exercise 9.35.

CPU time (minutes)	Job number										
	I	2	3	4	5	6	7	8	9	10	11
Estimated	.50	1.40	.95	.45	.75	1.20	1.60	2.6	1.30	.85	.60
Actual	.46	1.52	.99	.53	.71	1.31	1.49	2.9	1.41	.83	.74

Find a 90% confidence interval for the variance of the differences between estimated and actual CPU times.

9.49   Refer to Exercise 9.48. Suppose that the computer center expects customers to estimate CPU time correct to within .2 minute most of the time (say 95% of the time).
a.   Do the data provide sufficient evidence to indicate that the customer's error in estimating job CPU time exceeds the computer center's expectation? Test using $\alpha = .10$.
b.   Find the approximate observed significance level for the test and interpret its value.

9.50   A precision instrument is guaranteed to read accurately to within 2 units. A sample of four instrument readings on the same object yielded the measurements 353, 351, 351, and 355. Test the null hypothesis that $\sigma = .7$ against the alternative $\sigma > .7$. Conduct the test at the $\alpha = .05$ level of significance.

9.51   Find a 90% confidence interval for the population variance in Exercise 9.50.

9.52   In order to properly treat patients, drugs prescribed by physicians must have a potency that is accurately defined. Consequently, the distribution of potency values for shipments of a drug must not only have a mean value as specified on the drug's container, but the variation in potency must be small. Otherwise, pharmacists would be distributing drug prescriptions that could be harmfully potent or have a low potency that would be ineffective. A drug manufacturer claims that his drug is marketed with a potency of $5 \pm .1$ milligram per cubic centimeter. A random sample of four containers gave potency readings equal to 4.94, 5.09, 5.03, and 4.90 mg/cc.

   a.   Do the data present sufficient evidence to indicate that the mean potency differs from 5 mg/cc?

   b.   Do the data present sufficient evidence to indicate that the variation in potency differs from the error limits specified by the manufacturer? [*Hint:* It is sometimes difficult to determine exactly what is meant by limits on potency as specified by a manufacturer. Since he implies that the potency values will fall in the interval $5.0 \pm .1$ mg/cc with very high probability—the implication is *always*—let us assume that the range .2, i.e., (4.9 to 5.1) represents 6σ, as suggested by the Empirical Rule. Note that letting the the range equal 6σ rather than 4σ places a stringent interpretation on the manufacturer's claim. We want the potency to fall in the interval $5.0 \pm .1$ with very high probability.]

9.53   Refer to Exercise 9.52. Testing of 60 additional randomly selected containers of the drug gave a sample mean and variance equal to 5.04 and .0063 (for the total of $n = 64$ containers). Using a 95% confidence interval, estimate the variance of the manufacturer's potency measurements.

9.54   A manufacturer of hard safety hats for construction workers is concerned about the mean and and the variation of the forces helmets transmit to wearers when subjected to a standard external force. The manufacturer desires the mean force transmitted by helmets to be 800 pounds (or less), well under the legal 1000-pound limit, and σ to be less than 40. A random sample of $n = 40$ helmets was tested, and the sample mean and variance were found to be equal to 825 pounds and 2350 pounds^2, respectively.

   a.   If $\mu = 800$ and $\sigma = 40$, is it likely that any helmet, subjected to the standard external force, will transmit a force to a wearer in excess of 1000 pounds? Explain.

   b.   Do the data provide sufficient evidence to indicate that when subjected to the standard external force, the mean force transmitted by the helmets exceeds 800 pounds?

   c.   Do the data provide sufficient evidence to indicate that σ exceeds 40?

9.55   A random sample of 15 females was selected from among those listed in Appendix I. The systolic blood pressures for these 15 females are shown below.

84	100	110
110	110	114
112	120	104
112	90	112
130	110	88

   a.   Construct a 95% confidence interval for the population variance, σ^2. Interpret this interval.

   b.   If we consider the 945 female systolic blood pressures to be the population of interest, the variance of this population is given in Appendix I as $\sigma^2 = 158.138$. Does this value fall in the confidence interval constructed in part (a)?

9.56   A manufacturer of industrial light bulbs likes its bulbs to have a mean life that is acceptable to its customers and a variation in life that is relatively small. If some bulbs fail too early in their life, customers become annoyed and shift to competitive products. Large variations above the mean reduce replacement sales, and variation in general disrupts customers' replacement schedules. A sample of 20 bulbs tested produced the following lengths of life (in hours):

2100, 2302, 1951, 2067, 2415, 1883, 2101, 2146, 2278, 2019,

1924, 2183, 2077, 2392, 2286, 2501, 1946, 2161, 2253, 1827

The manufacturer wishes to control the variability in length of life so that σ is less than 150 hours. Do the data provide sufficient evidence to indicate that the manufacturer is achieving this goal? Test using $\alpha = .01$.

9.7 COMPARING TWO POPULATION VARIANCES

The need for statistical methods to compare two population variances is readily apparent from the discussion in Section 9.6. We may frequently wish to compare the precision of one measuring device with that of another, the stability of one manufacturing process with that of another, or even the variability in the grading procedure of one college professor with that of another.

Intuitively, we might compare two population variances σ_1^2 and σ_2^2 using the ratio of the sample variances s_1^2/s_2^2. If s_1^2/s_2^2 is nearly equal to 1, we would find little evidence to indicate that σ_1^2 and σ_2^2 are unequal. On the other hand, a very large or very small value for s_1^2/s_2^2 would provide evidence of a difference in the population variances.

How large or small must s_1^2/s_2^2 be in order for sufficient evidence to exist to reject the null hypothesis

$$H_0: \sigma_1^2 = \sigma_2^2?$$

The answer to this question may be acquired by studying the distribution of s_1^2/s_2^2 in repeated sampling.

When independent random samples are drawn from two normal populations with equal variances, that is, $\sigma_1^2 = \sigma_2^2$, then s_1^2/s_2^2 has a probability distribution in repeated sampling that is known to statisticians as an **F distribution**.

> **Assumptions for s_1^2/s_2^2 to have an F-Distribution**
>
> 1. Random and independent samples are drawn from each of two normal populations.
> 2. The variability of the measurements in the two populations is the same and can be measured by a common variance, σ^2. That is, $\sigma_1^2 = \sigma_2^2 = \sigma^2$.

We need not concern ourselves with the equation of the density function for F except to state that, as we might surmise, it is reasonably complex. For our purposes, it will suffice to accept the fact that the distribution is well known and that critical values have been tabulated. These appear in Tables, 6, 7, 8, 9, and 10 in Appendix III.

The shape of the F distribution is nonsymmetrical and will depend on the number of degrees of freedom associated with s_1^2 and s_2^2. We will represent these quantities as v_1 and v_2, respectively. This fact complicates the tabulation of critical values for the F distribution and necessitates the construction of a table for each value that we may choose for a tail area α. Thus, Tables 6, 7, 8, 9, and 10 in Appendix III present critical values corresponding to $\alpha = .10, .05, .025, .01$, and $.005$, respectively. A typical F probability distribution ($v_1 = 10$ and $v_2 = 10$) is shown in Figure 9.4.

A partial reproduction of Table 7 in Appendix III is shown in Table 9.9. Table 7 records the value $F_{.05}$ such that the probability that F will exceed $F_{.05}$ is $.05$. Another way of saying this is that 5% of the area under the F distribution lies

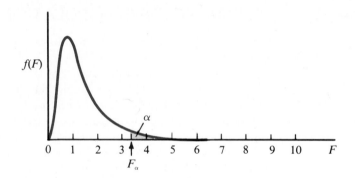

Figure 9.4
An F distribution with $v_1 = 10$ and $v_2 = 10$

Table 9.9   Format of the F table from Table 7 in Appendix III: $\alpha = .05$

v_2(d.f.)	v_1(d.f.) 1	2	3	4	5	6	7	8	9	60	120	∞	v_2(d.f.)
1	161.4	199.5	215.7	224.6	230.2	234.0	236.8	238.9	240.5	252.2	253.3	254.3	1
2	18.51	19.00	19.16	19.25	19.30	19.33	19.35	19.37	19.38	19.48	19.49	19.50	2
3	10.13	9.55	9.28	9.12	9.01	8.94	8.89	8.85	8.81	8.57	8.55	8.53	3
4	7.71	6.94	6.59	6.39	6.26	6.16	6.09	6.04	6.00	5.69	5.66	5.63	4
5	6.61	5.79	5.41	5.19	5.05	4.95	4.88	4.82	4.77	4.43	4.40	4.36	5
6	5.99	5.14	4.76	4.53	4.39	4.28	4.21	4.15	4.10	3.74	3.70	3.67	6
7	5.59	4.74	4.35	4.12	3.97	3.87	3.79	3.73	3.68	3.30	3.27	3.23	7
8	5.32	4.46	4.07	3.84	3.69	3.58	3.50	3.44	3.39	3.01	2.97	2.93	8
9	5.12	4.26	3.86	3.63	3.48	3.37	3.29	3.23	3.18	2.79	2.75	2.71	9
⋮	⋮	⋮	⋮	⋮	⋮	⋮	⋮	⋮	⋮	⋮	⋮	⋮	⋮
15	4.54	3.68	3.29	3.06	2.90	2.79	2.71	2.64	2.59	2.16	2.11	2.07	15
16	4.49	3.63	3.24	3.01	2.85	2.74	2.66	2.59	2.54	2.11	2.06	2.01	16
17	4.45	3.59	3.20	2.96	2.81	2.70	2.61	2.55	2.49	2.06	2.01	1.96	17
18	4.41	3.55	3.16	2.93	2.77	2.66	2.58	2.51	2.46	2.02	1.97	1.92	18
19	4.38	3.52	3.13	2.90	2.74	2.63	2.54	2.48	2.42	1.98	1.93	1.88	19
⋮	⋮	⋮	⋮	⋮	⋮	⋮	⋮	⋮	⋮	⋮	⋮	⋮	⋮

to the right of $F_{.05}$. The degrees of freedom for s_1^2, v_1, are indicated across the top of the table, and the degrees of freedom for s_2^2, v_2, appear in the first column on the left.

Referring to Table 9.9, we note that $F_{.05}$ for sample sizes $n_1 = 7$ and $n_2 = 10$ (that is, $v_1 = 6$, $v_2 = 9$) is 3.37. Likewise, the critical value $F_{.05}$ for sample sizes $n_1 = 9$ and $n_2 = 16$ ($v_1 = 8$, $v_2 = 15$) is 2.64. These values of F are shaded in Table 9.9.

In a similar manner, the critical values for a tail area, $\alpha = .01$, are presented in Table 9 in Appendix III. Thus,

$$P(F > F_{.01}) = .01$$

The statistical test of the null hypothesis

$$H_0 : \sigma_1^2 = \sigma_2^2$$

uses the test statistic

$$F = \frac{s_1^2}{s_2^2}$$

When the alternative hypothesis implies a one-tailed test, that is,

$$H_a : \sigma_1^2 > \sigma_2^2$$

we can use the tables directly. However, when the alternative hypothesis requires a two-tailed test,

$$H_a : \sigma_1^2 \neq \sigma_2^2$$

we note that the rejection region will be divided between the lower and upper tails of the F distribution and that tables of critical values for the lower tail are conspicuously missing. The reason for their absence is explained as follows: We are at liberty to identify either of the two populations as population I. If the population with the larger sample variance is designated as population II, then $s_2^2 > s_1^2$ and we will be concerned with rejection in the lower tail of the F distribution. Since the identification of the populations was arbitrary, we can avoid this difficulty by designating the population with the larger sample variance as population I. In other words, always place the larger sample variance in the numerator of

$$F = \frac{s_1^2}{s_2^2}$$

and designate that population as I. Then, since the area in the right-hand tail will represent only $\alpha/2$, we double this value to obtain the correct value for the probability of a type I error α. Hence, if we use Table 7 in Appendix III for a two-tailed test, the probability of a type I error will be $\alpha = .10$.

TEST OF AN HYPOTHESIS CONCERNING THE EQUALITY OF TWO POPULATION VARIANCES

1. Null Hypothesis: $H_0: \sigma_1^2 = \sigma_2^2$

2. Alternative Hypothesis:

   One-Tailed Test

   $H_a: \sigma_1^2 > \sigma_2^2$

   (or, $H_a: \sigma_2^2 > \sigma_1^2$)

   Two-Tailed Test

   $H_a: \sigma_1^2 \neq \sigma_2^2$

3. Test Statistic:

   One-Tailed Test

   $$F = \frac{s_1^2}{s_2^2}$$

   $$\left(\text{or, } F = \frac{s_2^2}{s_1^2} \text{ for } H_a: \sigma_2^2 > \sigma_1^2 \right)$$

   Two-Tailed Test

   $$F = \frac{s_1^2}{s_2^2}$$

   where s_1^2 is the larger sample variance.

4. Rejection Region:

   One-Tailed Test

   $F > F_\alpha$

   Two-Tailed Test

   $F > F_{\alpha/2}$

   When $F = s_1^2 / s_2^2$, the critical values of F, F_α, and $F_{\alpha/2}$ are based on $v_1 = n_1 - 1$ and $v_2 = n_2 - 1$ degrees of freedom. These tabulated values, for $\alpha = .10, .05, .025, .01,$ and $.005$, can be found in Tables 6, 7, 8, 9, and 10 in Appendix III, respectively.

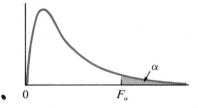 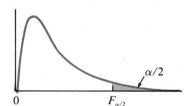

Assumptions: The samples were randomly and independently selected from normally distributed populations.

We will illustrate with examples.

EXAMPLE 9.13   Two samples consisting of 10 and 8 measurements each were observed to have sample variances equal to $s_1^2 = 7.14$ and $s_2^2 = 3.21$, respectively. Do the sample variances present sufficient evidence to indicate that the population variances are unequal?

Solution   Assume that the populations have probability distributions that are reasonably mound-shaped and hence will satisfy, for all practical purposes, the assumption that the populations are normal.

We wish to test the null hypothesis

$$H_0: \sigma_1^2 = \sigma_2^2$$

against the alternative

$$H_a: \sigma_1^2 \neq \sigma_2^2$$

Using Table 7 in Appendix III and doubling the tail area, we will reject H_0 when $F > 3.68$ with $\alpha = .10$.

The calculated value of the test statistic is

$$F = \frac{s_1^2}{s_2^2} = \frac{7.14}{3.21} = 2.22$$

Noting that the test statistic does not fall in the rejection region, we do not reject $H_0: \sigma_1^2 = \sigma_2^2$. Thus, there is insufficient evidence to indicate a difference in the population variances.

The confidence interval, with confidence coefficient $(1 - \alpha)$, for the ratio of two population variances can be shown to equal

A CONFIDENCE INTERVAL FOR σ_1^2 / σ_2^2

$$\frac{s_1^2}{s_2^2} \frac{1}{F_{v_1, v_2}} < \frac{\sigma_1^2}{\sigma_2^2} < \frac{s_1^2}{s_2^2} F_{v_2, v_1}$$

where $v_1 = n_1 - 1$ and $v_2 = n_2 - 1$.

Assumptions: The samples were randomly and independently selected from normally distributed populations.

where F_{v_1, v_2} is the tabulated critical value of F corresponding to v_1 and v_2 degrees of freedom in the numerator and denominator of F, respectively. Similar to the two-tailed test, the α will be double the tabulated value. Thus, F values extracted from Tables 7 and 8 will be appropriate for confidence coefficients equal to .90 and .95, respectively.

EXAMPLE 9.14 Refer to Example 9.13 and find a 90% confidence interval for σ_1^2 / σ_2^2.

Solution The 90% confidence interval for σ_1^2 / σ_2^2 in Example 9.13 is

$$\frac{s_1^2}{s_2^2} \frac{1}{F_{v_1, v_2}} < \frac{\sigma_1^2}{\sigma_2^2} < \frac{s_1^2}{s_2^2} F_{v_2, v_1}$$

where

$$s_1^2 = 7.14 \qquad\qquad s_2^2 = 3.21$$

$$v_1 = (n_1 - 1) = 9 \qquad\qquad v_2 = (n_2 - 1) = 7$$

$$F_{v_1, v_2} = F_{9,7} = 3.68 \qquad\qquad F_{v_2, v_1} = F_{7,9} = 3.29$$

Substituting these values into the formula for the confidence interval, we obtain

$$\left(\frac{7.14}{3.21}\right)\frac{1}{3.68} < \frac{\sigma_1^2}{\sigma_2^2} < \frac{(7.14)(3.29)}{3.21}$$

or

$$.60 < \frac{\sigma_1^2}{\sigma_2^2} < 7.32$$

The calculated interval estimate .60 to 7.32 is observed to include 1.0, the value hypothesized in H_0. This indicates that it is quite possible that $\sigma_1^2 = \sigma_2^2$ and therefore agrees with our test conclusions. Do not reject $H_0 : \sigma_1^2 = \sigma_2^2$.

EXAMPLE 9.15   The variability in the amount of impurities present in a batch of chemical used for a particular process depends on the length of time the process is in operation. A manufacturer using two production lines 1 and 2 has made a slight adjustment to process 2, hoping to reduce the variability as well as the average amount of impurities in the chemical. Samples of $n_1 = 25$ and $n_2 = 25$ measurements from the two batches yield means and variances as follows:

$$\bar{x}_1 = 3.2 \qquad s_1^2 = 1.04$$
$$\bar{x}_2 = 3.0 \qquad s_2^2 = 0.51$$

Do the data present sufficient evidence to indicate that the process variability is less for process 2? Test the null hypothesis $H_0 : \sigma_1^2 = \sigma_2^2$.

Solution   The practical implications of this example are illustrated in Figure 9.5. We believe that the mean levels of impurities in the two production lines are nearly equal (in fact, they may be equal) but that there is a possibility that the variation in the level of impurities is substantially less for line 2. Then distributions of impurity measurements for the two production lines would have nearly the same mean level but they would differ in their variation. A large variance for the level of impurities increases the probability of producing shipments of chemical with an unacceptably high level of impurities. Consequently, we hope to show that the process change in line 2 has made σ_2^2 less than σ_1^2.

Figure 9.5
Distributions of impurity measurements for two production lines

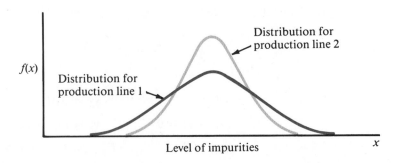

Testing the null hypothesis

$$H_0:\sigma_1^2 = \sigma_2^2$$

against alternative

$$H_a:\sigma_1^2 > \sigma_2^2$$

at an $\alpha' = .05$ significance level, we will reject H_0 when F is greater than $F_{.05} = 1.98$; that is, we will employ a one-tailed statistical test.

We readily observe that the calculated value of the test statistic

$$F = \frac{s_1^2}{s_2^2} = \frac{1.04}{.51} = 2.04$$

falls in the rejection region, and hence we conclude that the variability of process 2 is less than that of process 1.

The 90% confidence interval for the ratio σ_1^2/σ_2^2 is

$$\frac{s_1^2}{s_2^2}\frac{1}{F_{v_1, v_2}} < \frac{\sigma_1^2}{\sigma_2^2} < \frac{s_1^2}{s_2^2} F_{v_2, v_1}$$

$$\frac{1.04}{(.51)(1.98)} < \frac{\sigma_1^2}{\sigma_2^2} < \frac{(1.04)(1.98)}{.51}$$

or

$$1.03 < \frac{\sigma_1^2}{\sigma_2^2} < 4.04$$

Thus, we estimate the reduction in the variance of the amount of impurities to be as large as 4.04 to 1 or as small as 1.03 to 1. The actual reduction would likely be somewhere between these two extremes. This would suggest that the adjustment is quite effective in reducing the variation in the amount of impurities in the chemical.

EXERCISES Basic Techniques

9.57 Independent random samples from two normal populations produced the following variances:

Sample size	Sample variance
16	55.7
20	31.4

a. Do the data provide sufficient evidence to indicate that σ_1^2 differs from σ_2^2? Test using $\alpha = .05$.

b. Find the approximate observed significance level for the test and interpret its value.

9.58 Refer to Exercise 9.57 and find a 95% confidence interval for σ_1^2/σ_2^2.

9.59   Independent random samples from two normal populations produced the following variances:

Sample size	Sample variance
13	18.3
13	7.9

a.   Do the data provide sufficient evidence to indicate that $\sigma_1^2 > \sigma_2^2$? Test using $\alpha = .05$.
b.   Find the approximate observed significance level for the test and interpret its value.

Applications

9.60   Refer to Exercise 9.55. A random sample of $n = 15$ males is now selected from the population in Appendix I. The systolic blood pressures are shown below:

112	120	110
90	120	115
128	130	126
140	106	110
134	90	120

a.   Do the data provide sufficient evidence to indicate that there is a difference in the variability of systolic blood pressures between males and females? Use $\alpha = .01$.
b.   Construct a 95% confidence interval for the ratio of the two population variances.
c.   The ratio of the two population variances, given in Appendix I, is $(14.2343)^2/(12.5753)^2$. Does this value fall within the confidence interval constructed in part (b)? Explain.

9.61   The stability of measurements of the characteristics of a manufactured product is important in maintaining product quality. In fact, it is sometimes better to have small variation in the measured value of some important characteristic of a product and have the process mean be slightly off target than to suffer wide variation with a mean value that perfectly fits requirements. The latter situation may produce a higher percentage of defective products than the former. A manufacturer of light bulbs suspected that one of his production lines was producing bulbs with a high variation in length of life. To test this theory, he compared the lengths of life for $n = 50$ bulbs randomly sampled from the suspect line and $n = 50$ from a line that seemed to be "in control." The sample means and variances for the two samples were as follows:

"Suspect line"	Line "in control"
$\bar{x}_1 = 1{,}520$	$\bar{x}_2 = 1{,}476$
$s_1^2 = 92{,}000$	$s_2^2 = 37{,}000$

a.   Do the data provide sufficient evidence to indicate that bulbs produced by the "suspect line" have a larger variance in length of life than those produced by the line that is assumed to be in control? Use $\alpha = .05$.
b.   Find the approximate observed significance level for the test and interpret its value.

9.62   Use the method of Section 9.7 to obtain a 90% confidence interval for the variance ratio in Exercise 9.61.

9.63 In Exercise 9.23 we conducted a test to detect a difference between the percentages of oxygen uptake by air in reedfish exposed to water at 25°C versus those at 33°C.

a. What assumption had to be made concerning the population variances in order that the test be valid?

b. Do the data provide sufficient evidence to indicate that they violate the assumption in part (a)? Test using $\alpha = .05$.

9.64 A personnel manager, planning to use a Student's t test to compare the mean number of monthly absences for two categories of employees, noticed a possible difficulty. The variation in the number of absences per month seemed to differ for the two groups. As a check, the personnel officer randomly selected five months and counted the number of absences for each group. The data are shown in the table.

Category A	Category B
20	37
14	29
19	51
22	40
25	26

a. Which assumption necessary for use of the t test concerned the personnel officer?

b. Do the data provide sufficient evidence to indicate that the variances differ for the populations of absences for the two employee categories? Test with $\alpha = .10$ and interpret the results of the test.

9.65 A pharmaceutical manufacturer purchases a particular material from two different suppliers. The mean level of impurities in the raw material is approximately the same for both suppliers, but the manufacturer is concerned about the variability of the impurities from shipment to shipment. If the level of impurities tends to vary excessively for one source of supply, it could affect the quality of the pharmaceutical product. To compare the variation in percentage impurities for the two suppliers, the manufacturer selects ten shipments from each of the two suppliers and measures the percentage of impurities in the raw material for each shipment. The sample means and variances are shown in the table.

Supplier A	Supplier B
$\bar{x}_1 = 1.89$	$\bar{x}_2 = 1.85$
$s_1^2 = .273$	$s_2^2 = .094$
$n_1 = 10$	$n_2 = 10$

a. Do the data provide sufficient evidence to indicate a difference in the variability of the shipment impurity levels for the two suppliers? Test using $\alpha = .10$. Based on the results of your test, what recommendation would you make to the pharmaceutical manufacturer?

b. Find a 90% confidence interval for σ_2^2 and interpret your results.

9.66 In Exercise 9.24 we conducted a statistical test to determine whether sufficient evidence exists to indicate that the percentage of seeds germinating is less for seeds in the older (22-month) age group. In order for the test to be valid, the population variances must be approximately equal. Since the sample standard deviations were equal to 4 and 2, respectively, we might wonder whether there is sufficient evidence to indicate that the population variances are unequal. Test using $\alpha = .05$.

9.67 In Exercise 9.27 we raised questions about the validity of the Student's t test for making a comparision of population means. Use the F test to show that the data present sufficient evidence to indicate differences in population variances for the comparison questioned in Exercise 9.27.

$\triangleright$ 9.8 ASSUMPTIONS

As noted earlier, the tests and confidence intervals based on the Student's t, the chi-square, and the F statistic require that the data satisfy specific assumptions in order that the error probabilities (for the tests) and the confidence coefficients (for the confidence intervals) be equal to the values that we have specified. For example, if the assumptions are violated by selecting a sample from a nonnormal population, and the data are used to construct a 95% confidence interval for μ, the actual confidence coefficient might, unbeknownst to us, be equal to .85 instead of .95. The assumptions are summarized next for your convenience.

> Assumptions:
>
> 1. For all tests and confidence intervals described in this chapter, it is assumed that **samples are randomly selected from normally distributed populations**.
>
> 2. When two samples are selected, we assume that they are **selected in an independent manner** except in the case of the paired-difference experiment.
>
> 3. For tests or confidence intervals concerning **the difference between two population means** μ_1 and μ_2 based on independent random samples, we **assume that $\sigma_1^2 = \sigma_2^2$**.

In a practical sampling situation, you never know everything about the probability distribution of the sampled population. If you did, there would be no need for sampling or statistics, Secondly, it is highly unlikely that a population would have exactly the characteristics described above. Consequently, to be useful, the inferential methods described in this chapter must give good inferences when moderate departures from the assumptions are present. For example, if the population has a mound-shaped distribution that is nearly normal, we would like a 95% confidence interval constructed for μ to be one with a confidence coefficient close to .95. Similarly, if we conduct a t test of the null hypothesis $\mu_1 = \mu_2$ based on independent random samples from normal populations where σ_1^2 and σ_2^2 are not exactly equal, we want the probability of incorrectly rejecting the null hypothesis α to be approximately equal to the value we used in locating the rejection region.

A statistical method that is insensitive to departures from the assumptions on which the method is based is said to be robust. The t tests are quite robust to moderate departures from normality. In contrast, the chi-square and F tests are sensitive to departures from normality. The t test for comparing two means is moderately robust to

departures from the assumption $\sigma_1^2 = \sigma_2^2$, when $n_1 = n_2$. However, the test becomes sensitive to departures from this assumption as n_1 becomes large relative to n_2 (or vice versa).

If you are concerned that your data do not satisfy the assumptions prescribed for one of the statistical methods described in this chapter, you may be able to use a nonparametric statistical method to make your inference. These methods, which require few or no assumptions about the nature of the population probability distributions, are particularly useful for testing hypotheses, and some nonparametric methods have been developed for estimating population parameters. Tests of hypotheses concerning the location of a population distribution or a test for the equivalence of two population distributions are presented in Chapter 14. If you can select relatively large samples, you can usually use one of the large-sample estimation or test procedures of Chapters 7 and 8.

$\triangleright$ **9.9   SUMMARY**

It is important to note that the t, χ^2, and F statistics employed in the small-sample statistical methods discussed in the preceding sections are based on the assumption that the sampled populations have a normal probability distribution. This requirement will be adequately satisfied for many types of experimental measurements.

You will observe the very close relationship connecting Student's t and the z statistic and, therefore, the similarity of the methods for testing hypotheses and the construction of confidence intervals. The χ^2 and F statistics employed in making inferences concerning population variances do not, of course, follow this pattern, but the reasoning employed in the construction of the statistical tests and confidence intervals is identical for all the methods we have presented.

TIPS ON PROBLEM SOLVING

To help you decide whether the techniques of this chapter are appropriate for the solution of a problem, ask yourself the following questions:

1. Does the problem imply that an inference should be made about a population mean or the difference between two means? Are the samples small, say $n < 30$? If the answers to both questions are "yes," you may be able to use one of the methods of Sections 9.3, 9.4, or 9.5. If the sample sizes are large, $n \geq 30$, you can use the methods of Chapters 7 and 8. In practice, you would also need to verify that the assumptions underlying each procedure are satisfied. Are the population distributions nearly normal, and have the sampling procedures conformed to those prescribed for the statistical method?

2. When comparing population means, were the observations from the two populations selected in a paired manner? If they were, you must use the paired-difference analysis of Section 9.5. If the samples were selected independently and in a random manner, use the methods of Section 9.4.

3. Is data variation the primary objective of the problem? If it is, you may be required to make an inference about a population variance σ^2 (Section 9.6) or to compare two population variances σ_1^2 and σ_2^2 (Section 9.7).

(*Note:* The tips on problem solving following Section 7.11 will also be helpful in solving problems in this chapter.)

▷ MINITAB COMMANDS

```
TINTERVAL [K% confidence] for C...C
```

```
TTEST [μ = K] on C...C
    ALTERNATIVE = K
```

```
TWOSAMPLE [K% confidence] for C   C
    ALTERNATIVE = K
    POOLED
```

```
TWOT [K% confidence] data in C, groups in C
    ALTERNATIVE = K
    POOLED
```

REFERENCES

Freund, J. E., and Walpole, R. E. *Mathematical Statistics.* 4th ed. Englewood Cliffs, N. J.: Prentice Hall, 1987.

Hogg, R. V., and Craig, A. T., *Introduction to Mathematical Statistics.* 4th ed. New York: Macmillan, 1986.

Koopmans, L. H. *An Introduction to Contemporary Statistics.* 2d ed. Boston: Duxbury Press, 1987.

Mendenhall, W.; Wackerly, D.; and Scheaffer, R. L. *Mathematical Statistics with Applications.* 4th ed. Boston: PWS-KENT, 1990.

Mood, A. M., et al. *Introduction to the Theory of Statistics.* 3d ed. New York: McGraw-Hill, 1973.

Ryan, T. A.; Joiner, B. L.; and Ryan, B. F. *Minitab Student Handbook.* 2d ed. Boston: Duxbury Press, 1985.

SUPPLEMENTARY EXERCISES

9.68   What assumptions are made when Student's t test is employed to test an hypothesis concerning a population mean?

9.69   What assumptions are made about the populations from which independent random samples are obtained when using the t distribution in making small-sample inferences concerning the difference in population means?

9.70   Why use paired observations to estimate the difference between two population means rather than estimation based on independent random samples selected from the two populations? Is a paired experiment always preferable? Explain.

9.71 A manufacturer can tolerate a small amount (.05 milligrams per liter) of impurities in a raw material needed for manufacturing its product. Because the laboratory test for the impurities is subject to experimental error, the manufacturer tests each batch ten times. Assume that the mean value of the experimental error is 0 and hence that the mean value of the ten test readings is an unbiased estimate of the true amount of the impurities in the batch. For a particular batch of the raw material, the mean of the ten test readings is .058 milligrams per liter (mg/l), with a standard deviation of .012 mg/l. Do the data provide sufficient evidence to indicate that the amount of impurities in the batch exceeds .05 mg/l? Find the *p*-value for the test and interpret its value.

9.72 The main stem growth measured for a sample of 17 four-year-old red pine trees produced a mean and standard deviation equal to 11.3 and 3.4 inches, respectively. Find a 90% confidence interval for the mean growth of a population of four-year-old red pine trees subjected to similar environmental conditions.

9.73 The object of a general chemistry experiment is to determine the amount (in milliliters) of sodium (NaOH) solution needed to neutralize 1 gram of a specified acid. This will be an exact amount, but when run in the laboratory, variation will occur as the result of experimental error. Three titrations are made using phenolphthalein as an indicator of the neutrality of the solution (pH equals 7 for a neutral solution). The three volumes of sodium hydroxide required to attain a pH of 7 in each of the three titrations are as follows: 82.10, 75.75, and 75.44 milliliters. Use a 90% confidence interval to estimate the mean number of milliliters required to neutralize 1 gram of the acid.

9.74 Measurements of water intake, obtained from a sample of 17 rats that had been injected with a sodium chloride solution, produced a mean and standard deviation of 31.0 and 6.2 cubic centimeters, respectively. Given that the average water intake for noninjected rats observed over a comparable period of time is 22.0 cubic centimeters, do the data indicate that injected rats drink more water than noninjected rats? Test at the 5% level of significance. Find a 90% confidence interval for the mean water intake for injected rats.

9.75 An experimenter, interested in determining the mean thickness of the cortex of the sea urchin egg, employed an experimental procedure developed by Sakai. The thickness of the cortex was measured for $n = 10$ sea urchin eggs. The following measurements were obtained.

4.5	6.1	3.2	3.9	4.7
5.2	2.6	3.7	4.6	4.1

Estimate the mean thickness of the cortex using a 95% confidence interval.

9.76 A production plant has two extremely complex fabricating systems; one system is twice as old as the other. Both systems are checked, lubricated, and maintained once every two weeks. The number of finished products fabricated daily by each of the systems is recorded for 30 working days. The results are given in the table. Do these data present sufficient evidence to conclude that the variability in daily production warrants increased maintenance of the older fabricating system? Use a 5% level of significance.

New system	Old system
$\bar{x}_1 = 246$	$\bar{x}_2 = 240$
$s_1 = 15.6$	$s_2 = 28.2$

9.77 The data in the table give the diameters and heights of ten fossil specimens of a species of small shellfish, *Rotularia* (*Annelida*) *fallax*, that were unearthed in a mapping expedition near the Antarctic Peninsula (Carlos E. Macellari, "Revision of Serpulids of the Genus *Rotularia* (*Annelida*) at Seymour Island (Antarctic Peninsula) and Their Value in Stratigraphy," *Journal of Paleontology* 58, no. 4 [July 1984]). The first column gives Macellari's identification symbol for the fossil specimen. The second and third columns give the fossil's diameter and height, respectively, in millimeters (mm). The fourth column gives the ratio of diameter to height.

Specimen	Diameter	Height	D/H
OSU 36651	185	78	2.37
OSU 36652	194	65	2.98
OSU 36653	173	77	2.25
OSU 36654	200	76	2.63
OSU 36655	179	72	2.49
OSU 36656	213	76	2.80
OSU 36657	134	75	1.79
OSU 36658	191	77	2.48
OSU 36659	177	69	2.57
OSU 36660	199	65	3.06
$\bar{x}$:	184.5	73	2.54
s:	21.5	5	

Find a 95% confidence interval for:
a. the mean diameter of the specimens
b. the mean ratio of diameter to height
c. the mean height of the specimens

9.78 Refer to Exercise 9.77. Suppose that you wished to estimate the mean diameter of the fossil specimens correct to within 5 mm with probability equal to .95. How many fossils would you have to include in your sample?

9.79 The length of time to recovery was recorded for patients randomly assigned and subjected to two different surgical procedures. The data, recorded in days, are as follows:

Procedure I	Procedure II
$n_1 = 21$	$n_2 = 23$
$\bar{x}_1 = 7.3$	$\bar{x}_2 = 8.9$
$s_1^2 = 1.23$	$s_2^2 = 1.49$

a. Do the data present sufficient evidence to indicate a difference between the mean recovery times for the two surgical procedures? Test using $\alpha = .05$.
b. Find the approximate observed significance level for the test and interpret its value.

9.80 In their study of average price of brands for a fixed product line, Curry and Reisz[*] examined one aspect of the product life cycle (PLC) theory which, in general terms, predicts that because

---

* Curry, David J. and Reisz, Peter C., "Prices and Price/Quality Relationships: A Longitudinal Analysis," *Journal of Marketing*, Vol. 52 (January 1988), pp. 36–51. Reprinted with permission from the American Marketing Association.

of the impact of buyer learning during the PLC, the price variance of a product line declines in magnitude over time.

Product	Year	Number of brands	Average price	Standard deviation
10-Speed Bicycles	1974	16	$209.51	$38.70
	1980	30	206.23	65.50

Do the data given in the accompanying table provide sufficient evidence to contradict the product life cycle theory as it relates to the variances of the prices of 10-speed bicycles? Use $\alpha = .05$. What is the observed significance level of the test?

9.81 A stringer of tennis racquets claims to be able to string racquets to within 3 pounds per square inch of the tension requested by her customers. Laboratory tests of a sample of $n = 10$ of the stringer's racquets produced the following data:

					Racquet					
	1	2	3	4	5	6	7	8	9	10
Requested tension	70	65	65	55	60	68	65	60	65	60
Test tension	68	62	66	54	62	72	64	62	69	60

a. Assume that the stringer's claim to be within 3 pounds per square inch of the requested tension means that three standard deviations of the difference between requested and test tensions is 3 pounds per square inch . Do the data present sufficient evidence to indicate that the stringer's deviations from requested tension exceed the claim? Test using $\alpha = .05$.

b. Do the data present sufficient evidence to indicate bias in the stringing machine, that is, a tendency to string above (or below) the requested tension? Test using $\alpha = .05$.

9.82 The frequency with which the Internal Revenue Service conducts intensive audits of individual tax returns is dependent on several factors, one of which is the region in which the individual lives.* In 1986 the IRS audited only 0.47% of individual returns in Rhode Island, but in contrast examined 2.61% of returns in Nevada. It seems that IRS computers flag returns that trigger statistical alarms for each income level. In recent years the IRS has publicized campaigns to give close scrutiny to "partners and promoters of tax shelters, people who earn tip income, and the self-employed, among others." Hence, the IRS is more likely to audit the return of an individual who lives in an area where the tax paid by individuals is highly variable. Suppose a tax analyst for the IRS wants to compare the variabilities of federal tax collections for individuals in Idaho versus Nevada. If a sample of 121 individual returns from Idaho reveals a standard deviation of tax paid of $s_1 = \$46.22$, while 61 returns from Nevada provided $s_2 = \$77.84$, do these data indicate that a difference exists between the variabilities of federal taxes collected from individuals in Idaho and Nevada? Test by using $\alpha = .02$.

9.83 An experiment was conducted to compare the mean lengths of time required for the bodily absorption of two drugs A and B. Ten people were randomly selected and assigned to each drug treatment. Each person received an oral dosage of the assigned drug, and the length of time (in minutes) for the drug to reach a specified level in the blood was recorded. The means

* Gutfeld, Rose, "Regional Risks: IRS's Audit Rate Shows that All States Aren't Equal," *The Wall Street Journal*, July 22, 1987.

and variances for the two samples are as follows:

Drug A	Drug B
$\bar{x}_1 = 27.2$	$\bar{x}_2 = 33.5$
$s_1^2 = 16.36$	$s_2^2 = 18.92$

a. Do the data provide sufficient evidence to indicate a difference in mean times to absorption for the two drugs? Test using $\alpha = .10$.

b. Find the approximate observed significance level for the test and interpret its value.

9.84 Refer to Exercise 9.83. Find a 95% confidence interval for the difference in mean times to absorption.

9.85 Refer to Exercise 9.84. Suppose that you wished to estimate the difference in mean times to absorption correct to within 1 minute with probability approximately equal to .95.

a. Approximately how large a sample would be required for each drug (assume that the sample sizes are equal)?

b. To conduct the experiment using the sample sizes of part (a) will require a large amount of time and money. Can anything be done to reduce the sample sizes and still achieve the 1-minute bound on the error of estimation?

9.86 Insects hovering in flight expend enormous amounts of energy for their size and weight. The data shown below were taken from a much larger body of data collected by T. M. Casey and colleagues M. L. May and K. R. Morgan ("Flight Energetics of Euglossine Bees in Relation to Morphology and Wing Stroke Frequency," *Journal of Experimental Biology* 116 [1985]). They show the wing stroke frequencies, in hertz, for two different species of bees, $n_1 = 4$ *Euglossa mandibularis* Friese and $n_2 = 6$ *Euglossa imperialis* Cockerell.

Wing stroke frequencies

Euglossa mandibularis Friese	*Euglossa imperialis* Cockerell
235	180
225	169
190	180
188	185
	178
	182

Do these data provide sufficient evidence to indicate a difference in the variances of the populations of wing stroke frequencies corresponding to the two species of bees?

a. Just look at the data. Based on your intuition, do you think that a difference exists between the two population variances?

b. Test to determine whether a difference exists using $\alpha = .05$.

c. Give the approximate p-value for the test in part (b).

d. Explain why a Student's t test would be unsuitable for comparing the mean wing stroke frequencies for the two species of bees.

9.87 Rats, like dogs, mark their territory using urine traces or other secretions. The statistics shown in the table are a portion of those derived from a study by Linda I. A. Birke and Dawn Saddler concerning the marking habits of rats ("Scent Marking Behaviour to Conspecific Odours by the Rat, *Rattus norvegicus*," *Animal Behaviour* 32 [1984]). Ten male rats were each placed in three different environments. The first was odor-free, the second contained female scents, and the third contained male scents. The number of markings made by each rat was recorded. The table gives the mean and sample standard error of the mean (in parentheses) for each of the three environments.

Odor source		
None (controls)	Female	Male
9.0(1.1)	19.4(2.85)	24.2(3.4)

a. Do the data present sufficient evidence to indicate a difference in the mean number of markings for male rats when in the presence of a female scent source as opposed to a male scent source? Test using $\alpha = .05$.

b. Explain why a Student's t test would be unsuitable for comparing the mean number of markings for those exposed to a female scent source versus those exposed to no scent.

9.88 A study in the *American Journal of Botany* examined the differences in a particular plant, *Plantago Major L.*, when grown in full sunlight versus shade conditions (Niklas, Karl J. and Owens, Thomas G., "Physiological and Morphological Modifications of *Plantago Major* (*Plantaginaceae*) in Response to Light Conditions," *Amer. J. Bot.* 76(3): 370–382, 1989). In this study, shaded plants were those receiving direct sunlight for less than 2 hours each day, while full-sun plants were never shaded. A partial summary of the data based on $n_1 = 16$ full-sun plants and $n_2 = 15$ shaded plants is shown below:

	Full-sun		Shade	
	$\bar{x}$	s	$\bar{x}$	s
Leaf area (cm²)	128	43	78.7	41.7
Overlap area (cm²)	46.8	2.21	8.1	1.26
Leaf number	9.75	2.27	6.93	1.49
Thickness (mm)	0.9	0.03	0.5	0.02
Length (cm)	8.70	1.64	8.91	1.23
Width (cm)	5.24	0.98	3.41	0.61

a. What assumptions are required in order to use the small-sample procedures given in this chapter in comparing full-sun versus shade plants? From the summary presented, do you think that any of these assumptions have been violated?

b. Do the data present sufficient evidence to indicate a difference in mean leaf area for full-sun versus shaded plants? Test using $\alpha = .05$.

c. Do the data present sufficient evidence to indicate a difference in mean overlap area for full-sun versus shaded plants? Find the observed level of significance for the test.

9.89 Refer to Exercise 9.88.

a. Find a 95% confidence interval for the difference in mean leaf thickness for full-sun versus shade plants.

b. Find a 95% confidence interval for the difference in mean leaf width for full-sun versus shade plants.

c. Use the confidence intervals from parts (a) and (b) to determine whether there is a significant difference in mean leaf thickness and width for the two populations.

9.90   The table provides statistics on the vertical jumping distance (in centimeters) of elite volleyball players, 8 male members of the U. S. Men's National Team and 14 female members from the Women's University World Games Team (J. Puhl, S. Case, S. Fleck, and
P. V. Handel, "Physical and Physiological Characteristics of Elite Volleyball Players,"
Research Quarterly for Exercise and Sport 53, no. 3 [1982]). Find a 95% confidence interval for the differences in mean jumping distance between men and women. Interpret the
interval.

	Men	Women
Mean	67.0	45.9
Standard deviation	11.5	6.3

9.91   In Exercise 9.90 we gave statistics on the vertical jumping distances for 8 men and 14 women,
all elite volleyball players, all members of U.S. national teams. The data were collected at
the U.S. Olympic Training Center (Puhl, Case, Fleck, and Handel, 1982). Shown below are
similar statistics for the maximum heart rate and maximal oxygen uptake (VO_2 max) measurements for these same athletes.

	Maximum heart rate (beats per minute)		VO_2 max (mg/kg/min)	
	Men	Women	Men	Women
Mean	178	179	56.1	50.6
Standard deviation	6	8	2.2	5.7

a.   Do the data provide sufficient evidence to indicate a difference between the mean
maximum heart rates of men versus women elite volleyball players? Test using $\alpha = .10$.

b.   Find the approximate p-value for the test.

9.92   Scott K. Powers et al., conducted a study to investigate the relationship between several
variables and a competitive runner's finish time in a 10-kilometer race ("Ventilatory
Threshold, Running Economy and Distance Running Performance of Trained Athletes,"
Research Quarterly for Exercise and Sport 54, no. 3 [1983]). The finish time (in minutes) for the
10-kilometer race for each of the nine runners is shown below.

Runner	1	2	3	4	5	6	7	8	9
Finish Time	33.15	33.33	33.50	33.55	33.73	33.86	33.90	34.15	34.90

Assume that these experienced competitive runners' times to complete a 10-kilometer race
represent a random sample of the running times for all competitive runners of 10-kilometer
races. Find a 90% confidence interval for the mean time to complete a 10-kilometer race.
Interpret your interval.

9.93   Four sets of identical twins (pairs A, B, C, and D) were selected at random from a population
of identical twins. One child was selected at random from each pair to form an "experimental
group." These four children were sent to school. The other four children were kept at home as
a control group. At the end of the school year the following IQ scores were obtained.

Pair	Experimental group	Control group
A	110	111
B	125	120
C	139	128
D	142	135

Does this evidence justify the conclusion that lack of school experience has a depressing effect on IQ scores? Use $\alpha = .10$.

9.94 In the study of the physical characteristics of outstanding female junior tennis players, Powers and Walker give statistics on the circumferences of the forearms of ten junior tennis players ("Physiological and Anatomical Characteristics of Outstanding Female Junior Tennis Players," *Research Quarterly for Exercise and Sport* 53, no. 2 [1983]). The circumference was measured on the forearms of both the preferred hand (the one that the player uses most often to grip the racquet) and the nonpreferred hand. The means and standard errors of the means are shown below. If there is any difference in the mean circumference of the forearms in tennis players, the larger circumference should be in the forearm most likely to have the greater buildup in muscle, namely the one for the preferred hand.

	Preferred hand	Nonpreferred hand
Mean	23.88	22.33
Standard error of the mean	0.82	0.41

a. Suppose that you wish to detect this difference, assuming that it exists for female junior tennis players. State the null and alternative hypotheses that you would use for the test.
b. Give the rejection region for $\alpha = .01$.
c. Conduct the test and state your conclusions.
d. What type of experimental design was used for this experiment?
e. Based on your answer to part (d), was your test in parts (a), (b), and (c) the correct procedure for analyzing the data? Explain.

9.95 Seven obese persons were placed on a diet for one month, and the weights, at the beginning and at the end of the month, were recorded. They are shown in the table.

	Weights	
Subjects	Initial	Final
1	310	363
2	295	251
3	287	249
4	305	259
5	270	233
6	323	267
7	277	242
8	299	265

Estimate the mean weight loss for obese persons when placed on the diet for a one-month period. Use a 95% confidence interval and interpret your results. What assumptions must you make in order that your inference be valid?

9.96   A comparison of reaction times for two different stimuli in a psychological word-association experiment produced the following results when applied to a random sample of 16 people.

Stimulus	Reaction time (sec)							
1	1	3	2	1	2	1	3	2
2	4	2	3	3	1	2	3	3

Do the data present sufficient evidence to indicate a difference in mean reaction times for the two stimuli? Test at the $\alpha = .05$ level of significance.

9.97   Refer of Exercise 9.96. Suppose that the word-association experiment had been conducted using eight people as blocks and making a comparison of reaction time within each person; that is, each person would be subjected to both stimuli in a random order. The data for the experiment are as follows:

Person	Reaction time (sec)	
	Stimulus 1	Stimulus 2
1	3	4
2	1	2
3	1	3
4	2	1
5	1	2
6	2	3
7	3	3
8	1	3

Do the data present sufficient evidence to indicate a difference in mean reaction time for the two stimuli? Test at the $\alpha = .05$ level of significance.

9.98   Obtain a 90% confidence interval for $(\mu_1 - \mu_2)$ in Exercise 9.96.

9.99   Obtain a 95% confidence interval for $(\mu_1 - \mu_2)$ in Exercise 9.97.

9.100   Analyze the data in Exercise 9.97 as though the experiment had been conducted in an unpaired manner. Calculate a 95% confidence interval for $(\mu_1 - \mu_2)$ and compare with the answer to Exercise 9.99. Does it appear that blocking increased the amount of information available in the experiment?.

9.101   The following data give readings in foot-pounds of the impact strength of two kinds of packaging material. Determine whether there is evidence of a difference in mean strength between the two kinds of material. Test at the $\alpha = .10$ level of significance.

A	B
1.25	.89
1.16	1.01
1.33	.97
1.15	.95
1.23	.94
1.20	1.02
1.32	.98
1.28	1.00
1.21	.98

A	B
$\sum x_{1i} = 11.13$	$\sum x_{2i} = 8.80$
$\bar{x}_1 = 1.237$	$\bar{x}_2 = .978$
$\sum x_{1i}^2 = 13.7973$	$\sum x_{2i}^2 = 8.6240$

9.102 Would the amount of information extracted from the data in Exercise 9.101 be increased by pairing successive observations and analyzing the differences? Calculate 90% confidence intervals for $(\mu_1 - \mu_2)$ for the two methods of analysis (unpaired and paired) and compare the widths of the intervals.

9.103 When should one employ a paired-difference analysis in making inferences concerning the difference between two means?

9.104 An experiment was conducted to compare the density of cakes prepared from two different cake mixes A and B. Six cake pans received batter A, and six received batter B. Expecting a variation in oven temperature, the experimenter placed an A and a B side by side at six different locations within the oven. The six paired observations are as follows:

Mix	Density (oz/in.³)					
A	.135	.102	.098	.141	.131	.144
B	.129	.120	.112	.152	.135	.163

a. Do the data present sufficient evidence to indicate a difference between the average densities of cakes prepared using the two types of batter? Test at the $\alpha = .05$ level of significance.

b. Place a 95% confidence interval on the difference between the average densities for the two mixes.

9.105 Under what assumptions may the F distribution be used in making inferences about the ratio of population variances?

9.106 A dairy is in the market for a new bottle-filling machine and is considering models A and B manufactured by company X and company Y, respectively. If ruggedness, cost, and convenience are comparable in the two models, the deciding factor is the variability of fills (the model producing fills with the smaller variance being preferred). Let σ_1^2 and σ_2^2 be the fill variances for models A and B, respectively, and consider various tests of the null hypothesis $H_0: \sigma_1^2 = \sigma_2^2$. Obtaining samples of fills from the two machines and using the test statistic s_1^2/s_2^2, one could set up as the rejection region an upper-tail area, a lower-tail area, or a two-tailed area of the F distribution, depending on the point of view. Which type of rejection region would be most favored by

a. the manager of the dairy? Why?

b. a salesman for company X? Why?

c. a salesman for company Y? Why?

9.107 The closing prices of two common stocks were recorded for a period of 15 days. The means and variances are

$$\bar{x}_1 = 40.33 \qquad \bar{x}_2 = 42.54$$

$$s_1^2 = 1.54 \qquad s_2^2 = 2.96$$

a. Do these data present sufficient evidence to indicate a difference between the variabilities of the closing prices of the two stocks for the populations associated with the two samples? Give the p-value for the test and interpret its value.

b. Place a 90% confidence interval on the ratio of the two population variances.

9.108 Refer to Exercise 9.106. Wishing to demonstrate that the variability of fills is less for model A than for model B, a salesman for company X acquired a sample of 30 fills from a machine of model A and a sample of 10 fills from a machine of model B. The sample variances were $s_1^2 = .027$ and $s_2^2 = .065$, respectively. Does this result provide statistical support at the .05 level of significance for the salesman's claim?

9.109   A chemical manufacturer claims that the purity of his product never varies more than 2%. Five batches were tested and given purity readings of 98.2, 97.1, 98.9, 97.7, and 97.9%. Do the data provide sufficient evidence to contradict the manufacturer's claim? (*Hint:* To be generous, let a range of 2% equal 4σ.)

9.110   Refer to Exercise 9.109. Find a 90% confidence interval for σ^2.

9.111   A cannery prints "weight 16 ounces" on its label. The quality control supervisor selects nine cans at random and weighs them. She finds $\bar{x} = 15.7$ and $s = .5$. Do the data present sufficient evidence to indicate that the mean weight is less than that claimed on the label? (Use $\alpha = .05$.)

9.112   A psychologist wishes to verify that a certain drug increases the reaction time to a given stimulus. The following reaction times in tenths of a second were recorded before and after injection of the drug for each of four subjects.

Subject	Reaction time	
	Before	After
1	7	13
2	2	3
3	12	18
4	12	13

Test at the 5% level of significance to determine whether the drug significantly increases reaction time.

9.113   At a time when energy conservation is so important, some scientists think we should give closer scrutiny to the cost (in energy) of producing various forms of food. One recent study compares the mean amount of oil required to produce one acre of different types of crops. For example, suppose that we wish to compare the mean amount of oil required to produce one acre of corn versus one acre of cauliflower. The readings in barrels of oil per acre, based on 20-acre plots, seven for each crop, are shown in the table. Use these data to find a 90% confidence interval for the difference between the mean amounts of oil required to produce these two crops.

Corn	Cauliflower
5.6	15.9
7.1	13.4
4.5	17.6
6.0	16.8
7.9	15.8
4.8	16.3
5.7	17.1

9.114   The effect of alcohol consumption on the body appears to be much greater at high altitudes than at sea level. To test this theory, a scientist randomly selects 12 subjects and randomly divides them into two groups of six each. One group is transported to an altitude of 12,000 feet, where each subject ingests a drink containing 100 cc of alcohol. The second group receives the same drink at sea level. After two hours, the amount of alcohol in the blood (grams per 100 cc) for each subject is measured. The data are shown in the table. Do the data provide sufficient evidence to support the theory that retention of alcohol in the blood is greater at high altitudes? Test with $\alpha = .10$.

Sea level	12,000 feet
.07	.13
.10	.17
.09	.15
.12	.14
.09	.10
.13	.14

9.115 An experiment is conducted to compare two new automobile designs. Twenty people are randomly selected and each person is asked to rate each design on a scale of 1 (poor) to 10 (excellent). The resulting ratings will be used to test the null hypothesis that the mean level of approval is the same for both designs against the alternative hypothesis that one of the automobile designs is preferred. Would these data satisfy the assumptions required for the Student's t test of Section 9.4? Explain.

9.116 The following data were collected on lost-time accidents (the figures given are mean man-hours lost per month over a period of one year) before and after an industrial safety program was put into effect. Data were recorded for six industrial plants. Do the data provide sufficient evidence to indicate whether the safety program was effective in reducing lost-time accidents? (Use $\alpha = .10$.)

	Plant number					
	1	2	3	4	5	6
Before program	38	64	42	70	58	30
After program	31	58	43	65	52	29

9.117 To compare the demand for two different entrées, the manager of a cafeteria recorded the number of purchases for each entrée on seven consecutive days. The data are shown in the table. Do the data provide sufficient evidence to indicate a greater mean demand for one of the entrées?

Day	A	B
Monday	420	391
Tuesday	374	343
Wednesday	434	469
Thursday	395	412
Friday	637	538
Saturday	594	521
Sunday	679	625

 9.118 The EPA limit on the allowable discharge of suspended solids into rivers and streams is 60 milligrams per liter (mg per l) per day. A study of water samples selected from the discharge at a phosphate mine shows that over a long period, the mean daily discharge of suspended solids is 48 mg per l but the day-to-day discharge readings are variable. State inspectors measured the discharge rates of suspended solids for $n = 20$ days and found $s^2 = 39$ (mg/l)^2. Find a 90% confidence interval for σ^2. Interpret your results.

9.119   Exercise 9.44 discussed a study conducted by Dr. Arthur H. Schmale et al., of the well-being of cancer patients who had survived at least one year beyond their last cancer treatment and who appeared to be in good health (Arthur H. Schmale et al., "Well-Being of Cancer Survivors," *Psychosomatic Medicine* 45, no. 2 [1983]. Reprinted by permission of the Elsevier Science Publishing Co., Inc. Copyright 1987 by The American Psychosomatic Society, Inc.). Each of 104 of these cancer survivors was administered a questionnaire called "The General Well-Being Measure." The table below gives the mean scores when the 104 patients are split into groups according to sex and some other sociodemographic variables. The group sample sizes for each of these sociodemographic variables, along with t values and p-values for testing the null hypothesis of "no difference" between a pair of means, are shown in the table.

	Sample size*	Mean	t	2-tailed p
Sex				
Male	44	100.2	1.15	0.25
Female	60	103.5		
Marital status				
Married	72	104.1	2.19	0.03
Other	32	97.7		
Living arrangements				
Not alone	85	103.2	1.94	0.06
Alone	19	95.6		
Change in occupation since cancer diagnosed				
No	88	103.4	2.23	0.03
Yes	16	95.0		

* Sample sizes for the groups were calculated from percentages given in the reference.

a.   Explain the test results.

b.   Using the value of $t = 1.94$ given in the table for comparing the mean well-being scores for patients' living arrangements, calculate the approximate p-value of the test. Does your value appear to agree with the value $p = .06$ given in the table?

9.120   Are Americans putting in more time at work than they once did? John P. Robinson (*American Demographics*, April 1989, p. 68) discusses a survey of men and women who work at least ten hours per week. Respondents kept a record of the number of hours per week spent at paid work, including the commute. The average work week for males, broken down by educational background, follows. (Only the means are reported by Robertson.)

Education	Mean	Standard deviation	Sample size
Grade school	45	10.0	25
High school graduate	42	8.9	25
College graduate	40	8.1	25
Postgraduate	38	9.7	25

a.   Is there a significant difference between mean hours per work week for males with a high school education compared with males having only a grade school education? Use $\alpha = .01$.

b. Can we conclude that a male with postgraduate training works significantly fewer hours per week than a male who is a high school graduate? Use $\alpha = .01$.

9.121 Most working parents in the United States pay large amounts of money for child care; however, they are not always satisfied with the results. In fact, a Louis Harris survey indicated that only 8% of parents of preschoolers believe the child care system is working very well (*American Demographics*, September, 1989, p. 14). Suppose that 12 single mothers and 12 married couples are randomly selected. Their monthly child care costs are recorded below.

Married couple	Single mother
$\bar{x}_1 = \$185$	$\bar{x}_2 = \$211$
$s_1 = 13$	$s_2 = 15$

a. Do the data provide sufficient evidence to indicate that the population variances are different? Use $\alpha = .05$.

b. If the population variances are not significantly different, use the Student's t test to determine whether there is a significant difference between the average monthly child care payments for married couples and single mothers. Determine the approximate p-value for the test.

9.122 American adolescents who espouse traditional values, such as respect for authority, allegiance to a church, or strong belief in a god, may be a minority in the population at large, and as a result, may feel alienated from their peers. In their study of junior and senior high school students in the Midwest, Raymond L. Calabrese and Edgar J. Raymond ("Alienation: Its Impact on Adolescents from Stable Environments," *The Journal of Psychology*, 123(4), 397–404) used the Dean Alienation Scale to assess alienation, isolation, normlessness, and powerlessness. The following table displays means and standard deviations for the measure of isolation for various groups as reported by Calabrese and Raymond. (The sample sizes are not those actually used in their study.)

Group	Mean	Standard deviation	Sample size
Religious affiliation			
Church	68.3	10.1	18
No church	69.8	9.9	12
Religious commitment			
Strong	71.0	13.3	10
Moderate	69.7	12.1	8
None	67.7	9.2	12
Family organization			
Traditional	70.1	9.8	21
Nontraditional	68.9	10.5	9

a. Test for a significant difference in mean isolation scores for those students with a church affiliation and those without a church affiliation. Use $\alpha = .05$.

b. Is there a significant difference in variances for those students with a moderate religious commitment and those with no commitment? Use $\alpha = .05$.

c. Find a 95% confidence interval estimate of the actual difference in mean alienation scores for students from homes having a traditional family organization and those from nontraditional homes.

JUST A PINCH BETWEEN THE CHEEK AND THE GUM

Those interested in controlling and reducing stress should not listen too closely to the advice of the pro-football running backs Walt Garrison and Earl Campbell. If we are to believe a report in the *Orlando Sentinel Star* (November 18, 1981), just a pinch (of "smokeless tobacco," i.e., snuff) between the cheek and the gum can be harmful to your health.

According to the *Sentinel Star*, researchers at Texas Lutheran College found that "oral tobacco increased the work of the heart and drove up blood pressure in humans." These conclusions were based on research reported by William G. Squires, an associate professor of biology, at the American Heart Association's 54th Scientific Sessions. (William G. Squires et al., "Hemodynamics affects of oral tobacco in experimental animals and young adults," *Circulation Supplement* IV

Heart rates and diastolic and systolic blood pressures before and after the use of smokeless tobacco

Subject	Group	HRT1	HRT2	DYS1	DYS2	SYS1	SYS2
1	1	81	105	64	80	127	150
2	1	81	91	79	79	110	125
3	1	68	87	78	81	116	142
4	1	61	86	70	83	124	142
5	1	67	82	72	85	118	134
6	1	74	78	70	71	110	119
7	1	75	87	77	84	125	141
8	1	64	94	70	77	118	130
9	1	70	93	78	74	124	127
10	1	60	90	69	75	116	126
11	2	69	73	72	76	113	126
12	2	65	73	64	72	114	124
13	2	62	71	83	90	136	136
14	2	82	98	78	87	118	147
15	2	71	91	69	80	120	140
16	2	77	87	65	69	112	120
17	2	53	57	74	78	113	120
18	2	69	95	70	79	110	123
19	2	62	90	70	76	117	131
20	2	70	100	70	76	120	130

[October 1981] 186 [Abstract]. By permission of the American Heart Association, Inc., and the author.) The experimental results were based on a study of 20 Texas Lutheran male athletes, ten who used oral tobacco and ten who did not. After obtaining initial electrocardiogram and blood pressure readings, the researchers gave both tobacco users and nonusers "a good healthy man-size dip" (2.5 grams) of tobacco. After 20 minutes, the average heart rate of the tobacco dippers had "increased from 69 to 88 beats per minute" and the "average blood pressure increased from 118 over 72 to 126 over 78." The article further notes that oral tobacco use might pose a health hazard to some persons, particularly those suffering from hypertension.

The data for the experiment, obtained from Professor Squires, are shown in the table on page 390. Each row of the table corresponds to a particular athlete. The first column gives the number assigned to identify the athlete. The second column identifies the athlete as a nonuser (Group 1) or a user (Group 2) of smokeless tobacco. Columns 3 and 4 give the heart rate at the beginning of the experiment (HRT1) and the heart rate after holding the tobacco for a 20-minute period (HRT2). Corresponding before and after readings for the diastolic blood pressures (DYS1 and DYS2) and the systolic blood pressures (SYS1 and SYS2) are shown in columns 5, 6, 7, and 8, respectively.

1. Does the 20-minute use of 2.5 grams of smokeless tobacco increase the mean heart rate for the general population of athletes?

2. Does the 20-minute use of 2.5 grams of smokeless tobacco increase the mean heart rate for the population of athletes who do not use smokeless tobacco?

3. Estimate the mean change in heart rate for nonusers using a 95% confidence interval.

4. Do the data indicate a difference in the mean change in heart rate between the nonuser and user groups?

5. Perform similar analyses (parts 1–4) for the data using diastolic and stystolic blood pressures. Describe the results of your analyses.

CHAPTER 10

LINEAR REGRESSION AND CORRELATION

Case Study

Does it pay to save? Psychologists insist that people are more inclined to save if the amount of government taxation on savings and investments is low. The case study at the end of this chapter investigates this theory using a linear regression analysis.

General Objective

We have presented methods for making inferences about population means when random samples were selected from the populations of interest. The objective of this chapter is to extend this methodology to consider the case in which the mean value of a random variable y is related to another variable, say x. By making simultaneous observations on y and x, we can use information provided by x to estimate the mean value of y and to predict particular values of y for preassigned values of x.

Specific Topics

1   Linear probabilistic models (10.2)
2   Method of least squares (10.3)
3   Calculating s^2 (10.4)
4   Inferences concerning β_1 (10.5)
5   Estimating $E(y)$ for a given value of x (10.6)
6   Predicting y for a particular value of x (10.7)
7   Coefficient of correlation, r (10.8)
8   Assumptions (10.11)

▷ 10.1 INTRODUCTION

An estimation problem of importance to high school seniors, freshmen entering college, their parents, and a university administration concerns the expected academic achievements of a particular student after he or she has enrolled in a university. For example, we might wish to estimate a student's grade point average at the end of the freshman year *before* the student has been accepted or enrolled in the university. At first glance this would seem to be a difficult task.

The statistical approach to this problem is, in many respects, a formalization of the procedure we might follow intuitively. If data were available giving the high school academic grades, psychological and sociological information, and the grades attained at the end of the college freshman year for a large number of students, we might categorize the students into groups having similar characteristics. Highly motivated students who have had a high rank in their high school class, have graduated from a high school with known superior academic standards, and so on, should achieve, on the average, a high grade point average at the end of the college freshman year. On the other hand, students who lack proper motivation or who achieved only moderate success in high school would not be expected, on the average, to do as well. Carrying this line of thought to the ultimate and idealistic extreme, we would expect the grade point average of a student to be a **function** of the many variables that define the psychological and physical characteristics of the individual, as well as those characteristics that define the academic and social environment to which the student will be exposed. Ideally, we would like to have a mathematical equation that would relate a student's grade point average to all relevant variables so that the equation could be used for prediction.

The problem we have defined is of a very general nature. We are interested in a random variable y that is related to a number of independent variables x_1, x_2, $x_3, \ldots$. The variable y for our example would be the student's grade point average, and the independent variables might be

x_1 = rank in high school class

x_2 = score on a mathematics achievement test

x_3 = score on a verbal achievement test

and so on. The ultimate objective is to measure $x_1, x_2, x_3, \ldots, x_k$ for a particular student, substitute these values into the prediction equation, and thereby predict the student's grade point average. To accomplish this objective, we must first locate the related variables $x_1, x_2, x_3, \ldots, x_k$ and obtain a measure of the strength of their relationship to y. Then we must construct a good prediction equation that expresses y as a function of the selected independent predictor variables.

Practical examples of our prediction problem are numerous in business, industry, and the sciences. The stockbroker wishes to predict stock market behavior as a function of a number of "key indices" that are observable and serve as the independent predictor variables $x_1, x_2, x_3, \ldots$. The manager of a manufacturing plant would like to relate yield of a chemical to a number of process variables. The manager would then use the prediction equation to find settings for the controllable process variables that would provide the maximum yield of the chemical. The biologist would like to relate body characteristics to the amounts of various glandular secretions. The political scientist may wish to relate success in a political campaign to the characteristics of a candidate, the nature of the opposition, and various campaign issues, campaign expenditures, and promotional techniques. All these prediction problems involve the same general concept.

In this chapter we will restrict our attention to the simple problem of predicting y as a **linear function** of a **single** variable. The solution for the multivariable problem—for example, predicting a student's grade point average—will be the subject of Chapter 11.

10.2   A SIMPLE LINEAR PROBABILISTIC MODEL

We introduce our topic by considering the problem of predicting a student's final grade in a college freshman calculus course based on his score on a mathematics achievement test administered prior to college entrance. As noted in Section 10.1, we wish to determine whether the achievement test is really worthwhile—that is, whether the achievement test score is actually related to a student's grade in calculus—and to obtain an equation that will be useful for predicting y as a function of x. The evidence, presented in Table 10.1, represents a sample of the achievement test scores and calculus grades for ten college freshmen. We will assume that the ten students constitute a random sample drawn from the population of freshmen who have already entered the university or will do so in the immediate future.

Table 10.1
Mathematics achievement test scores and final calculus grades for college freshmen

Student	Mathematics achievement test score	Final calculus grade
1	39	65
2	43	78
3	21	52
4	64	82
5	57	92
6	47	89
7	28	73
8	75	98
9	34	56
10	52	75

Our initial approach to the analysis of the data of Table 10.1 would be to plot the data as points on a graph, representing a student's calculus grade as y and the corresponding achievement test score as x. The graph, shown in Figure 10.1, is called a **scatter diagram**. You can see that y appears to increase as x increases.

Figure 10.1
Plot of the data in
Table 10.2

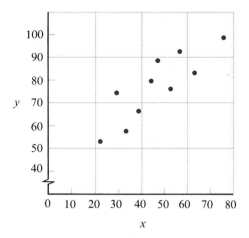

Could this arrangement of the points have occurred due to chance even if x and y were unrelated?

One method of obtaining a prediction equation relating y to x is to place a ruler on the graph and move it about until it seems to pass through the points, thus providing what we might regard as the "best fit" to the data. Indeed, if we were to draw a line through the points, it would appear that our prediction problem had been solved. Certainly, we can now use the graph to predict a student's calculus grade as a function of his score on the mathematics achievement test. Furthermore, we have chosen a **mathematical model** that expresses the supposed functional relation between y and x.

Let us review several facts concerning the graphing of mathematical functions. First, the mathematical **equation of a straight line is**

$$y = \beta_0 + \beta_1 x$$

where β_0 is the **y intercept**, the value of y when $x = 0$, and β_1 is the **slope** of the line, the change in y for a one-unit change in x (see Figure 10.2). Second, the line

Figure 10.2
The y intercept and
slope for a line

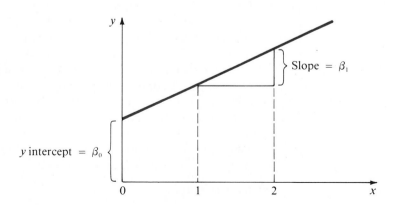

that we graph corresponding to any linear equation is unique. Each equation corresponds to only one line and vice versa. Thus, when we draw a line through the points, we automatically choose a mathematical equation

$$y = \beta_0 + \beta_1 x$$

where β_0 and β_1 have unique numerical values.

EQUATION OF A STRAIGHT LINE

$$y = \beta_0 + \beta_1 x$$

where $\beta_0 = y$ intercept

$\quad = $ value of y when $x = 0$

and $\beta_1 = $ slope of the line

$\quad = $ change in y for a one-unit increase in x

The linear model

$$y = \beta_0 + \beta_1 x$$

is said to be a **deterministic mathematical model** because when a value of x is substituted into the equation, the value of y is determined and no allowance is made for error. Fitting a straight line through a set of points by eye produces a deterministic model. Many other examples of deterministic mathematical models can be found by leafing through the pages of elementary chemistry, physics, or engineering textbooks.

Deterministic models are quite suitable for explaining physical phenomena and predicting when the error of prediction is negligible for practical purposes. Thus, Newton's Law, which expresses the relation between the force F imparted by a moving body with mass m and acceleration a, given by the deterministic model

$$F = ma$$

predicts force with very little error for most practical applications. "Very little" is, of course, a relative concept. An error of .1 inch in forming an I-beam for a bridge is extremely small but would be impossibly large in the manufacture of parts for a wristwatch. Thus, in many physical situations the error of prediction cannot be ignored. Indeed, consistent with our stated philosophy, we would be hesitant to place much confidence in a prediction unaccompanied by a measure of its goodness. For this reason, a visual choice of a line to relate the calculus grade and achievement test score would have limited value.

In contrast to the deterministic model, we might employ a **probabilistic mathematical model**, which is a simple modification of the deterministic model. Rather than saying that y and x are related by the deterministic model

$$y = \beta_0 + \beta_1 x$$

we say that the **mean (or expected) value of y for a given value of x**, denoted by the symbol $E(y|x)$, has a graph that is a straight line. That is, we let

$$E(y|x) = \beta_0 + \beta_1 x$$

For any given value of x, y values will vary in a random manner about the mean $E(y|x)$.

So to summarize, we write the probabilistic model for any particular observed value of y as

$$y = \text{(mean value of } y \text{ for given value of } x) + \text{(random error)}$$
$$\overbrace{E(y|x)}$$
$$= \overbrace{\beta_0 + \beta_1 x} + \epsilon$$

where ϵ is a **random error**, the difference between an observed value of y and the mean value of y for a given value of x. Thus, we assume that for any given value of x the observed value of y varies in a random manner and possesses a probability distribution with a mean value $E(y|x)$. For assistance in conveying this idea, the probability distributions of y about the line of means $E(y|x)$ are shown in Figure 10.3 for three hypothetical values of x—x_1, x_2, and x_3—although, as noted, we imagine a distribution of y values for every value of x.

Figure 10.3
Linear probabilistic
model

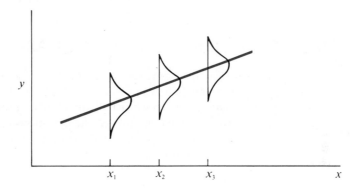

What properties will we assume for the probability distribution of y for a given value of x? Assumptions that seem to fit many practical situations, and upon which the methodology of Sections 10.4, 10.5, 10.6, and 10.7 is based, are given in the display.

> **Assumptions for the Probabilistic Model**   For any given value of x, y possesses a normal distribution, with a mean value given by the equation
>
> $$E(y|x) = \beta_0 + \beta_1 x$$
>
> and with a variance of σ^2. Furthermore, any one value of y is independent of every other value.

These assumptions permit us to construct tests of hypotheses and confidence intervals for β_0, β_1, and $E(y \mid x)$. In addition, we are able to give a prediction interval for y when we use the fitted model (the line that we find to fit a particular set of data) to predict some new value of y for a particular set of x. The practical problems we can solve by using these procedures will become apparent when we have worked some examples.

In the next section we will consider the problem of finding the "best-fitting" line for a given set of data. This line will give us a prediction equation that is also called a **regression line**.

EXERCISES Basic Techniques

10.1 Graph the line corresponding to the equation $y = 2x + 1$ by graphing the points corresponding to $x = 0$, 1, and 2. Give the y intercept and slope for the line.

10.2 Graph the line corresponding to the equation $y = -2x + 1$ by graphing the points corresponding to $x = 0$, 1, and 2. Give the y intercept and slope for the line. How is this line related to the line $y = 2x + 1$ of Exercise 10.1?

10.3 Give the equation and graph for a line with a y intercept equal to 3 and a slope equal to -1.

10.4 Give the equation and graph for a line with a y intercept equal to -3 and a slope equal to 1.

10.5 What is the difference between deterministic and probabilistic mathematical models?

▷ 10.3 THE METHOD OF LEAST SQUARES

The statistical procedure for finding the "best-fitting" straight line for a set of points is, in many respects, a formalization of the procedure employed when we fit a line by eye. For instance, when we visually fit a line to a set of data, we move the ruler until we think that we have minimized the **deviations** of the points from the prospective line. If we denote the predicted value of y obtained from the fitted line as $\hat{y}$, the prediction equation will be

$$\hat{y} = \hat{\beta}_0 + \hat{\beta}_1 x$$

where $\hat{\beta}_0$ and $\hat{\beta}_1$ represent estimates of the parameters β_0 and β_1. This line for the data of Table 10.1 is shown in Figure 10.4. The vertical lines drawn from the prediction line to each point represent the deviations of the points from the predicted value of y. Thus, the deviation of the ith point is

$$y_i - \hat{y}_i$$

where $\hat{y}_i = \hat{\beta}_0 + \hat{\beta}_1 x_i$

Having decided that in some manner or other we will attempt to minimize the deviations of the points in choosing the best-fitting line, we must now define what we mean by "best." That is, we wish to define a criterion for "best fit" that will seem intuitively reasonable, that is objective, and that, under certain conditions, will give the best prediction of y for a given value of x.

We will employ a criterion of goodness that is known as the **principle of least squares**, which may be stated as follows: **Choose as the "best-fitting" line the line that**

Figure 10.4
Linear prediction
equation

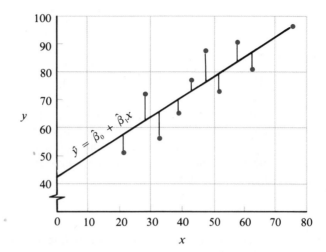

minimizes the sum of squares of the deviations of the observed values of **y from those predicted.*** Expressed mathematically, we wish to choose values for $\hat{\beta}_0$ and $\hat{\beta}_1$ that minimize

$$SSE = \sum_{i=1}^{n} (y_i - \hat{y}_i)^2$$

The symbol **SSE** represents the sum of squares of deviations or, as it is commonly called, the **sum of squares for error**.

Substituting for $\hat{y}_i$ in SSE, we obtain

$$SSE = \sum_{i=1}^{n} [y_i - (\hat{\beta}_0 + \hat{\beta}_1 x_i)]^2$$

The method for finding the numerical values of $\hat{\beta}_0$ and $\hat{\beta}_1$ that minimize SSE uses differential calculus and hence is beyond the scope of this text. We simply state that the least-squares solutions for $\hat{\beta}_0$ and $\hat{\beta}_1$ are given by the following formulas:

LEAST-SQUARES ESTIMATORS OF β_0 AND β_1

$$\hat{\beta}_1 = \frac{S_{xy}}{S_{xx}} \quad \text{and} \quad \hat{\beta}_0 = \bar{y} - \hat{\beta}_1 \bar{x}$$

where

$$S_{xx} = \sum_{i=1}^{n} (x_i - \bar{x})^2 = \sum_{i=1}^{n} x_i^2 - \frac{\left(\sum_{i=1}^{n} x_i\right)^2}{n}$$

* The deviations of the points about the least-squares line satisfy another property. The sum of the deviations equals 0; that is, $\sum_{i=1}^{n} (y_i - \hat{y}_i) = 0$ (proof omitted). Many other lines can be found that satisfy this property.

and

$$S_{xy} = \sum_{i=1}^{n} (x_i - \bar{x})(y_i - \bar{y}) = \sum_{i=1}^{n} x_i y_i - \frac{\left(\sum_{i=1}^{n} x_i\right)\left(\sum_{i=1}^{n} y_i\right)}{n}$$

Note that the sum of squared deviations from the mean for x, S_{xx}, is computed using the shortcut formula presented in Chapter 2. The sum of corrected cross products, S_{xy}, is computed using a similar formula. Once $\hat{\beta}_0$ and $\hat{\beta}_1$ have been computed, substitute their values into the equation of a line to obtain the least-squares prediction equation

$$\hat{y} = \hat{\beta}_0 + \hat{\beta}_1 x$$

EXAMPLE 10.1   Obtain the least-squares prediction line for the data in Table 10.1.

Solution   The calculation of $\hat{\beta}_0$ and $\hat{\beta}_1$ for the data in Table 10.1 is simplified by the use of Table 10.2. Substituting the appropriate sums from Table 10.2 into the least-squares equations, we obtain

$$S_{xx} = \sum_{i=1}^{n} x_i^2 - \frac{\left(\sum_{i=1}^{n} x_i\right)^2}{n} = 23{,}634 - \frac{(460)^2}{10} = 2474$$

$$S_{xy} = \sum_{i=1}^{n} x_i y_i - \frac{\left(\sum_{i=1}^{n} x_i\right)\left(\sum_{i=1}^{n} y_i\right)}{n} = 36{,}854 - \frac{(460)(760)}{10} = 1894$$

$$\bar{y} = \frac{\sum_{i=1}^{n} y_i}{n} = \frac{760}{10} = 76$$

Table 10.2
Calculations for the data in Table 10.1

y_i	x_i	x_i^2	$x_i y_i$	y_i^2
65	39	1521	2535	4225
78	43	1849	3354	6084
52	21	441	1092	2704
82	64	4096	5248	6724
92	57	3249	5244	8464
89	47	2209	4183	7921
73	28	784	2044	5329
98	75	5625	7350	9604
56	34	1156	1904	3136
75	52	2704	3900	5625
Sum 760	460	23,634	36,854	59,816

and

$$\bar{x} = \frac{\sum\limits_{i=1}^{n} x_i}{n} = \frac{460}{10} = 46$$

Hence,

$$\hat{\beta}_1 = \frac{S_{xy}}{S_{xx}} = \frac{1894}{2474} = .7655618 \approx .766$$

and

$$\hat{\beta}_0 = \bar{y} - \hat{\beta}_1 \bar{x} = 76 - (.7655618)(46) = 40.7842 \approx 40.784$$

Then, according to the principle of least squares, the best-fitting straight line relating the calculus grade to the achievement test score is

$$\hat{y} = \hat{\beta}_0 + \hat{\beta}_1 x$$

or

$$\hat{y} = 40.748 + 7.66x$$

The graph of this equation is shown in Figure 10.4. Note that the y intercept, 40.784, is the value of $\hat{y}$ when $x = 0$. The slope of the line, .766, gives the estimated change in y for a one-unit increase in x.

We can now predict y for a given value of x by referring to Figure 10.4 or by substituting into the prediction equation. For example, if a student scored $x = 50$ on the achievement test, the student's predicted calculus grade would be

$$\hat{y} = \hat{\beta}_0 + \hat{\beta}_1 x = 40.784 + (.766)(50) = 79.08$$

How accurate will this prediction be? To answer this question, we need to place a bound on our error of prediction. We will consider this and related problems in succeeding sections.

TIPS ON PROBLEM SOLVING

1. Be careful of rounding errors, which can greatly affect the answer you obtain in calculating S_{xx} and S_{xy}. If you must round a number, it is recommended that you carry six or more significant figures in the calculations. (Note also that in working exercises, rounding errors might cause some slight discrepancies between your answers and the answers given in the back of the text.)

2. Always plot the data points and graph your least-squares line. If the line does not provide a reasonable fit to the data points, you may have made an error in your calculations.

EXERCISES

Basic Techniques

10.6 Given five points whose coordinates are

x	-2	-1	0	1	2
y	1	1	3	5	5

a. Find the least-squares line for the data.

b. As a rough check on the calculations in part (a), plot the five points and graph the line. Does the line appear to provide a good fit to the data points?

10.7 Given the points whose coordinates are

x	1	2	3	4	5	6
y	5.6	4.6	4.5	3.7	3.2	2.7

a. Find the least-squares line for the data.

b. As a rough check on the calculations in part (a), plot the six points and graph the line. Does the line appear to provide a good fit to the data points?

Applications

10.8 An article in the *New York Times* (March 4, 1985) reports that 84% of arriving inmates at state prisons throughout the United States in 1979 were repeat offenders. The percentage of inmates who return to prison within a year of their release seems to be related to age. This percentage is shown below for five different age groups.

Age group interval (years)	Midpoint x of age interval	Percentage y returning within one year
18 to 24	21.5	22
25 to 34	30	12
35 to 44	40	7
45 and older	approx. 55	2

a. Find the least-squares line relating y and x. Use only the three data points corresponding to ages 18 to 44.

b. Plot the three points on graph paper and graph the least-squares line.

c. Plot the data point for the 45 and older group. Note the large deviation between this point and the least-squares line.

d. Can you explain why the relationship between y and x might be approximately linear over the interval 18 to 44 years, but not linear over the interval 18 to 65 years?

10.9 One of the objectives of a study by Scott K. Powers et al., was to investigate the relationship between "running economy" and distance running performance for trained, competitive male runners ("Ventilatory Threshold, Running Economy and Distance Running Performance of Trained Athletes," *Research Quarterly for Exercise and Sport* 54, no. 3 [1983]). Running economy for a runner is defined as the runner's steady-state oxygen consumption for a standardized running speed. The study involved nine trained runners. Each runner competed in two 10-kilometer races, and the best of the two finish times was recorded. In addition, a measure of each runner's "running economy" was recorded. These data, along with the number of years of competitive running experience, are shown below for each runner.

Runner	Years competitive running	Running economy	10 kilometers finish time (min.)
1	9	46.2	33.15
2	13	43.2	33.33
3	5	46.7	33.50
4	7	48.9	33.55
5	12	47.1	33.73
6	6	50.5	33.86
7	4	53.5	33.90
8	5	46.3	34.15
9	3	50.5	34.90

a.   Find the least-squares line relating a runner's running economy x to his finish time y for a 10-kilometer race.

b.   Plot the data points and graph the least-squares line. Does it seem to provide a good fit to the data points?

 10.10 A new and booming area of specialty sales is the gourmet coffee market. Gourmet coffee stores—often small, informal, and quirky—are gaining a bigger and bigger share of the total coffee market, while the conventional coffee market, not including the market for decaffeinated coffee, continues to decline. The following table gives retail sales and forecasts for gourmet coffee (y), in millions of dollars, for the years 1983–1992.

Year	Gourmet coffee
1983	210
1984	270
1985	340
1986	420
1987	500
1988	600
1989	675
1990	750
1991	825
1992	900

Source: Robichaux, Mark, "Boom in Fancy Coffee Pits: Big Marketers, Little Firms," *Wall Street Journal*, November 6, 1989, p. B1.

a.   Fit a least-squares line to the data using $x = $ year $- 1982$.

b.   Plot the data points and graph the least-squares line.

▷ 10.4 CALCULATING s^2, AN ESTIMATOR OF σ^2

Recall that the probabilistic model for y in Section 10.2 assumes that y is related to x by the equation

$$y = \beta_0 + \beta_1 x + \epsilon$$

For a given value of x the y values are independent and possess a normal probability distribution with mean $E(y \mid x)$ and variance σ^2. The variance σ^2 measures the random variation of the y values about the line $E(y \mid x) = \beta_0 + \beta_1 x$. The greater the variation, the larger will be the value of σ^2. Thus, each observed value of y is subject to a random error ϵ that enters into the computations of $\hat{\beta}_0$ and $\hat{\beta}_1$ and introduces error into these estimates. Furthermore, if we use the least-squares line

$$\hat{y} = \hat{\beta}_0 + \hat{\beta}_1 x$$

to predict some future value of y, the random errors affect the error of prediction. Consequently, the variability of the random errors, measured by σ^2, plays an important role when estimating or predicting by using the least-squares line.

The first step toward acquiring a bound on a prediction error requires that we estimate σ^2, the variance of y for a given value of x. For this purpose it seems reasonable to use SSE, the sum of squares of deviations (sum of squares for error) about the predicted line. Indeed, it can be shown that the formula given in the display provides an estimator for σ^2 that is unbiased, based on $(n - 2)$ degrees of freedom.

AN ESTIMATOR FOR σ^2

$$\hat{\sigma}^2 = s^2 = \frac{\text{SSE}}{n - 2}$$

The sum of squares of deviations, SSE, may be calculated directly by using the prediction equation to calculate $\hat{y}$ for each point, then calculating the deviations $(y_i - \hat{y}_i)$, and finally calculating

$$\text{SSE} = \sum_{i=1}^{n} (y_i - \hat{y}_i)^2$$

This procedure tends to be tedious and is rather poor from a computational point of view because the numerous subtractions tend to introduce computational rounding errors. An easier and computationally better procedure is to use the formula given in the display.

FORMULA FOR CALCULATING SSE

$$\text{SSE} = S_{yy} - \hat{\beta}_1 S_{xy} = S_{yy} - \frac{(S_{xy})^2}{S_{xx}}$$

where

$$S_{yy} = \sum_{i=1}^{n} (y_i - \bar{y})^2 = \sum_{i=1}^{n} y_i^2 - \frac{\left(\sum\limits_{i=1}^{n} y_i\right)^2}{n}$$

$$S_{xy} = \sum_{i=1}^{n} x_i y_i - \frac{\left(\sum\limits_{i=1}^{n} x_i\right)\left(\sum\limits_{i=1}^{n} y_i\right)}{n}$$

$$S_{xx} = \sum_{i=1}^{n} x_i^2 - \frac{\left(\sum\limits_{i=1}^{n} x_i\right)^2}{n}$$

Observe that S_{xy} and S_{xx} were used in the calculation of $\hat{\beta}_1$ and hence have already been computed. Furthermore, note that if you must round a number, you should retain six or more significant figures in the calculations to avoid serious rounding errors in the final answer.

EXAMPLE 10.2 Calculate an estimate of σ^2 for the data of Table 10.1.

Solution First, we calculate

$$S_{yy} = \sum_{i=1}^{n} (y_i - \bar{y})^2 = \sum_{i=1}^{n} y_i^2 - \frac{\left(\sum\limits_{i=1}^{n} y_i\right)^2}{n} = 59,816 - \frac{(760)^2}{10} = 2056$$

Substituting this value of S_{yy} and the values of $\hat{\beta}_1$ and S_{xy} calculated in Example 10.1 into the formula for SSE, we obtain

$$\text{SSE} = S_{yy} - \hat{\beta}_1 S_{xy} = 2056 - (.7655618)(1894) = 606.03$$

Then

$$s^2 = \frac{\text{SSE}}{n-2} = \frac{606.03}{8} = 75.754 \qquad \triangleleft$$

How can you interpret these values of SSE and s^2? Refer to Figure 10.4 and note the deviations of the $n = 10$ points from the least-squares line (shown as the vertical line segments between the points and the line). The sum SSE $= 606.03$ is equal to the sum of squares of the numerical values of these deviations. This quantity is then used to calculate $s^2 = 75.754$ and $s = \sqrt{75.754} = 8.70$, estimates of σ^2 and σ.

The practical interpretation that can be given to s ultimately rests on the meaning of σ. Since σ measures the spread of the y values about the line of means $E(y\,|\,x) = \beta_0 + \beta_1 x$ (see Figure 10.3), we would expect (from the Empirical Rule) approximately 95% of the y values to fall within 2σ of that line. Since we do not know σ, $2s$ provides an approximate value for the half width of this interval. Now return to Figure 10.4 and note the location of the data points about the least-squares line. Since we used the $n = 10$ data points to fit the least-squares line, you would not be too surprised to find that most of the points fall within $2s = 2(8.7) = 17.4$ of the line. If you check Figure 10.4, you will see that all ten points fall within $2s$ of the least-squares line. (You will find that, in general, most of the data points used to fit the least-squares line will fall within $2s$ of the line. This estimate provides you with a rough check for your calculated value of s.)

But s will play a much more important role in this chapter than the application described. As mentioned at the beginning of this section, the less the variability of the y values about the line of means (i.e., the smaller the value of σ), the closer the least-squares line will be to the line of means. Consequently, s will play an important role in evaluating the goodness of all of the inferential methods described in this chapter.

TIPS ON PROBLEM SOLVING

1. To reduce rounding error, always carry at least six significant figures when calculating S_{yy} and SSE. You can round when you obtain the answer for SSE if you desire.

2. As a check on your calculated value of s, remember that s measures the spread of the points about the least-squares line. Therefore, you would expect (by the Empirical Rule) most of the points to fall within $2s$ of the least-squares line. For example, if the points appear to fall in a band roughly equal to 4 units in width on the scale of the y variable, and if your calculated value of s is 10, your value of s is too large. You have made an error. For example, perhaps you forgot to divide SSE by $(n - 2)$.

EXERCISES

10.11   Calculate SSE and s^2 for the data in Exercise 10.6.

10.12   Calculate SSE and s^2 for the data in Exercise 10.7.

10.13   Calculate SSE and s^2 for the data in Exercise 10.8.

10.14   Calculate SSE and s^2 for the data in Exercise 10.9.

10.15   Calculate SSE and s^2 for the data in Exercise 10.10.

▷ 10.5   INFERENCES CONCERNING THE SLOPE OF THE LINE, β_1

The initial inference desired in studying the relationship between y and x concerns the existence of the relationship. Does x contribute information for the prediction of y; that is, **do the data present sufficient evidence to indicate that y increases (or decreases) linearly as x increases over the region of observation?** Or, is it quite probable that the points would fall on the graph in a manner similar to that observed in Figure 10.1 when y and x are completely unrelated?

The practical question we pose concerns the value of β_1, which is the average change in y for a one-unit increase in x. Stating that y does not increase (or decrease) linearly as x increases is equivalent to saying that $\beta_1 = 0$. Thus, we would wish to test an hypothesis that $\beta_1 = 0$ against the alternative that $\beta_1 \neq 0$. As we might suspect, the estimator $\hat{\beta}_1$ is extremely useful in constructing a test statistic to test this hypothesis. Therefore, we wish to examine the distribution of estimates $\hat{\beta}_1$ that would be obtained when samples, each containing n points, are repeatedly drawn from the population of interest. If we assume that the random error ϵ is normally distributed, in addition to the previously stated assumptions, it can be shown that both $\hat{\beta}_0$ and $\hat{\beta}_1$ will be normally distributed in repeated sampling and that the expected value and variance of $\hat{\beta}_1$ will be

$$E(\hat{\beta}_1) = \beta_1 \qquad \text{and} \qquad \sigma^2_{\hat{\beta}_1} = \frac{\sigma^2}{S_{xx}}$$

Thus, $\hat{\beta}_1$ is an unbiased estimator of β_1, we know its standard deviation, and hence we can construct a z statistic in the manner described in Section 8.3. Then

$$z = \frac{\hat{\beta}_1 - \beta_1}{\sigma_{\hat{\beta}_1}} = \frac{\hat{\beta}_1 - \beta_1}{\sigma/\sqrt{S_{xx}}}$$

would have a standardized normal distribution in repeated sampling. Since the actual value of σ^2 is unknown, we should obtain the estimated standard deviation of $\hat{\beta}_1$, which is $s/\sqrt{S_{xx}}$. Substituting s for σ in z, we obtain, as in Chapter 9, a test statistic

$$t = \frac{\hat{\beta}_1 - \beta_1}{s/\sqrt{S_{xx}}}$$

which can be shown to follow a Student's t distribution in repeated sampling with $(n - 2)$ degrees of freedom. Note that the number of degrees of freedom associated with s^2 determines the number of degrees of freedom associated with t. Thus, we observe that the test of an hypothesis that β_1 equals some particular numerical value, say β_{10}, is the familiar t test encountered in Chapter 9.

TEST OF AN HYPOTHESIS CONCERNING THE SLOPE OF A LINE

1. Null Hypothesis: $H_0 : \beta_1 = \beta_{10}$
2. Alternative Hypothesis:

   One-Tailed Test   *Two-Tailed Test*

   $H_a : \beta_1 > \beta_{10}$ $H_a : \beta_1 \neq \beta_{10}$

   (or, $\beta_1 < \beta_{10}$)

3. Test Statistic: $t = \dfrac{\hat{\beta}_1 - \beta_{10}}{s/\sqrt{S_{xx}}}$

> When the assumptions given in Section 10.2 (and 10.10) are satisfied, the test statistic will have a Student's t distribution with $(n - 2)$ degrees of freedom.

4.   Rejection Region:

   One-Tailed Test   *Two-Tailed Test*

   $t > t_\alpha$ $t > t_{\alpha/2}$ or $t < -t_{\alpha/2}$

   (or, $t < -t_\alpha$ when the
   alternative hypothesis is
   $H_a : \beta_1 < \beta_{10}$)

The values of t_α and $t_{\alpha/2}$ are given in Table 4 in Appendix III. Use the values of t corresponding to $(n - 2)$ degrees of freedom.

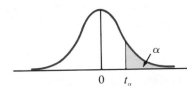 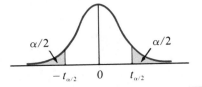

EXAMPLE 10.3   Use the data in Table 10.1 to determine whether the value of $\hat{\beta}_1$ provides sufficient evidence to indicate that β_1 differs from zero; that is, does a linear relationship exist between a freshman's mathematics achievement score x and his or her final calculus grade y?

Solution   We wish to test the null hypothesis

$$H_0 : \beta_1 = 0 \quad \text{against} \quad H_a : \beta_1 \neq 0$$

for the calculus grade-achievement test score data in Table 10.2. The test statistic is

$$t = \frac{\hat{\beta}_1 - 0}{s/\sqrt{S_{xx}}}$$

and, if we choose $\alpha = .05$, we will reject H_0 when $t > 2.306$ or $t < -2.306$. The critical value of t is obtained from the t table using $(n - 2) = 8$ degrees of freedom. Substituting values determined in Examples 10.1 and 10.2 into the test statistic, we obtain

$$t = \frac{\hat{\beta}_1}{s/\sqrt{S_{xx}}} = \frac{.7655618}{8.70/\sqrt{2474}} = 4.38$$

Observing that the test statistic exceeds the critical value of t, we will reject the null hypothesis $\beta_1 = 0$ and conclude that there is evidence to indicate that the calculus final grade is linearly related to the achievement test score.

   In fact, with 8 degrees of freedom, $t = 4.38$ exceeds $t_{.005} = 3.355$. Therefore, the p-value for this test is less than $2(.005) = .01$, indicating that our results are highly significant.

   Once we have decided that x and y are linearly related, we are interested in examining this relationship in detail. If x increases by one unit, what is the estimated change in y and how much confidence can be placed in the estimate? In other words, we require an estimate of the slope β_1. You will not be surprised to observe a continuity in the procedures of Chapters 9 and 10; that is, the confidence interval for β_1, with confidence coefficient $(1 - \alpha)$, can be shown to be

$$\hat{\beta}_1 \pm t_{\alpha/2}(\text{estimated } \sigma_{\hat{\beta}_1})$$

or

A 100$(1 - \alpha)$% CONFIDENCE INTERVAL FOR β_1

$$\hat{\beta}_1 \pm t_{\alpha/2}\frac{s}{\sqrt{S_{xx}}}$$

where $t_{\alpha/2}$ is based on $(n - 2)$ degrees of freedom.

EXAMPLE 10.4   Find a 95% confidence interval for β_1 based on the data in Table 10.2.

Solution   The 95% confidence interval for β_1, using values calculated previously, is

$$\hat{\beta}_1 \pm t_{.025}\frac{s}{\sqrt{S_{xx}}}$$

Substituting, we obtain

$$.766 \pm 2.306\frac{(8.70)}{\sqrt{2474}} \quad \text{or} \quad .766 \pm .403$$

Several points concerning the interpretation of our results deserve particular attention. As we have noted, β_1 is the slope of the assumed line over the region of observation and indicates the linear change in $E(y \mid x)$ for a 1-unit change in x. **Even if we do not reject the null hypothesis that the slope of the line β_1 equals zero, it does not necessarily mean that x and y are unrelated.** In the first place, we must be concerned with the probability of committing a type II error—that is, of accepting the null hypothesis that the slope equals zero when this hypothesis is false. Secondly, it is possible that x and y might be perfectly related in a curvilinear, but not linear, manner. For example, Figure 10.5 depicts a curvilinear relationship between y and x over the domain of $x : a \le x \le f$. We note that a straight line would provide a good

Figure 10.5
Curvilinear relation

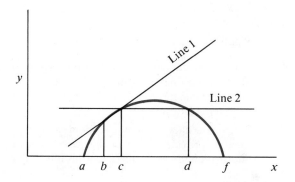

predictor of y if fitted over a small interval in the x domain, say, $b \leq x \leq c$. The resulting line is line 1. On the other hand, if we attempt to fit a line over the region $c \leq x \leq d$, then β_1 equals zero and the best fit to the data is the horizontal line 2. This result would occur even though all the points fell perfectly on the curve and y and x possessed a functional relation as defined in Section 10.2. We must take care in drawing conclusions if we do not find evidence to indicate that β_1 differs from zero. Perhaps we have chosen the wrong type of probabilistic model for the physical situation.

Note that the comments contain a second implication. **If the data provide values of x in an interval $b \leq x \leq c$, then the calculated prediction equation is appropriate only over this region. Extrapolation in predicting y for values of x outside the region $b \leq x \leq c$ for the situation indicated in Figure 10.5 would result in a serious prediction error.**

If the data present sufficient evidence to indicate that β_1 differs from zero, we do not conclude that the true relationship between y and x is linear. Undoubtedly, y is a function of a number of variables, which demonstrate their existence to a greater or lesser degree in terms of the random error ϵ that appears in the model. This random error, of course, is why we have been obliged to use a probabilistic model in the first place. Large errors of prediction imply curvatures in the true relation between y and x, the presence of other important variables that do not appear in the model, or both, as most often is the case. **All we can say is that we have evidence to indicate that y changes as x changes and that we may obtain a better prediction of y by using x and the linear predictor than by simply using $\bar{y}$ and ignoring x. Note that this statement does not imply a causal relationship between x and y. Some third variable may have caused the change in both x and y, producing the relationship that we have observed.**

EXERCISES   Basic Techniques

10.16   Do the data in Exercise 10.6 present sufficient evidence to indicate that y and x are linearly related? (Test the hypothesis that $\beta_1 = 0$; use $\alpha = .05$.)

10.17   Find a 90% confidence interval for the slope of the line in Exercise 10.6. Interpret this interval estimate.

10.18   Do the data in Exercise 10.7 present sufficient evidence to indicate that y and x are linearly related? (Test the hypothesis that $\beta_1 = 0$; use $\alpha = .05$.)

10.19   Find a 95% confidence interval for the slope of the line in Exercise 10.7. Interpret this interval estimate.

Applications

10.20   Refer to the data in Exercise 10.8 relating the percentage y of prison inmates who return to prison within one year to inmate age x. Do the data provide sufficient evidence to indicate that age of the inmates contributes information for the prediction of y over the age interval 18 to 44 years? Test using $\alpha = .05$.

10.21   Refer to the data on running economy and distance running performance in Exercise 10.9. Do the data provide sufficient evidence to indicate that a runner's running economy provides information for the prediction of his finish time y in a 10-kilometer race?

a. Test using $\alpha = .10$.
b. Find the approximate p-value for the test.
c. State the practical consequences of your test results.

10.22 Refer to Exercise 10.10. Suppose that you wish to determine whether the gourmet coffee sales are increasing with time.
a. State the null and alternative hypotheses to be tested.
b. Conduct the test and state your conclusions. Give the observed level of significance for the test.

10.23 Do football coaches get better with age? If Coach Bear Bryant, the legendary football coach for the University of Alabama, is typical of other football coaches, universities may wish to hire old coaches for their teams. In his thirties, Coach Bryant had 59 wins, 23 losses, and 5 ties. In his forties, his record improved to 73 wins, 24 losses, and 8 ties. In his fifties, it was 88, 22, and 3; and in his sixties, his record was 95, 12, and 1. The percentage wins in a given ten-year age interval and the midpoint x of a ten-year age interval are given below for Coach Bryant (*Sports Illustrated* 57, no. 11, 1982).

Midpoint age interval	35	45	55	65
Percent wins	67.8	69.5	77.9	88.0

a. Find the least-squares line to the data.
b. Plot the data points and graph the least-squares line.
c. Calculate s^2 for the simple linear regression analysis.

10.24 The win-loss record for Coach Bryant in Exercise 10.23 is just a sample of what he could have done had he coached and played more games over his long career in coaching. Do the data provide sufficient evidence to indicate that Coach Bryant's probability of winning increased with his age? Test using $\alpha = .05$.

10.25 Our test in Exercise 10.24 is based on the assumption that the variance of the random error ϵ is constant and equal to σ^2 for all values of x.

a. Explain why this assumption is not satisfied for the data in Exercise 10.23. $\left(\text{ } Hint: \text{ In}\right.$

Section 6.5, we learned that $\sigma_p^2 = \dfrac{pq}{n}. \Bigg)$

b. Consider an alternative method of answering the question in Exercise 10.24. Use the data in Exercise 10.23 and the method of Section 8.7 to show that Coach Bryant's probability p_1 of winning in his sixties exceeds the probability p_2 of winning in his thirties.

10.26 A study was conducted to determine the effects of sleep deprivation on subjects' ability to solve simple problems. The amount of sleep deprivation varied over 8, 12, 16, 20, and 24 hours without sleep. A total of ten subjects participated in the study, two at each sleep-deprivation level. After his or her specified sleep-deprivation period, each subject was administered a set of simple addition problems, and the number of errors was recorded. The following results were obtained.

Number of errors, y	8, 6	6, 10	8, 14	14, 12	16, 12
Number of hours without sleep, x	8	12	16	20	24

a. Find the least-squares line appropriate to these data.
b. Plot the points and graph the least-squares line as a check on your calculations.
c. Calculate s^2.

10.27   Do the data in Exercise 10.26 present sufficient evidence to indicate that number of errors is linearly related to number of hours without sleep? (Test using $\alpha = .05$.) Would you expect the relation between y and x to be linear if x were varied over a wider range (say $x = 4$ to $x = 48$)?

10.28   Find a 95% confidence interval for the slope of the line in Exercise 10.27. Give a practical interpretation to this interval estimate.

10.29   Daily readings of air temperatures and soil moisture percentages were collected by William A. Reiners and colleagues Hollinger and Lang at four elevations, 670, 825, 1145, and 1379 meters, in the White Mountains of New Hampshire ("Temperature and Evapotranspiration Gradients of the White Mountains, New Hampshire, U.S.A.," *Arctic and Alpine Research* 16, no. 1 [1984]). Among the analyses of the data are simple linear regressions relating air temperature y (in °C) in a given month to altitude x in meters. For example, the regression analysis for July produced the following statistics: $\hat{\beta}_0 = 21.4$, $\hat{\beta}_1 = -0.0063$, $r^2 = .99$, and the estimate of the standard deviation of $\hat{\beta}_1$ is 0.0007.

a.   Give the least-squares prediction equation relating air temperature y in °C to altitude x in meters for the month of July.

b.   How many degrees of freedom would be associated with SSE and s^2 in this analysis? (Assume that readings were taken each day during the month of July at all four elevations.)

c.   Do the data present sufficient evidence to indicate that altitude x contributes information for predicting air temperature y? Test using $\alpha = .01$.

10.30   Biotin is an important vitamin for poultry; an inadequate supply in the diet can lead to poor growth, poor feed conversion efficiency, and increased mortality. In a 21-day study conducted with male broiler chicks to estimate the availability of biotin from feedstuffs (Blair, R. and R. Miser, "Biotin Bioavailability from Protein Supplements and Cereal Grains for Growing Broiler Chickens," *International Journal of Vitamin Nutrition Research*, Vol. 59, 1989, pp. 55–58. Hans Huber Publishers, Toronto, Bern.), groups of ten-day-old chicks were randomly assigned to seven treatment groups in which a basal diet was supplemented with 0, 50, 100, 200, 250, or 300 μg/kg d-biotin. The following data give the average biotin intake (x) in micrograms per day $(\mu g/d)$ and the average weight gain (y) in grams per day (g/d). A biotin response curve involved only those diets with 0, 50, 100, 150, and 200 μg/kg.

Added biotin	Biotin intake (x)	Weight gain (y)
0	0.14	8.0
50	2.01	17.1
100	6.06	22.3
150	6.34	24.4
200	7.15	26.5

a.   Plot weight gain (y) and biotin intake (x). Do the data appear to be linearly related?

b.   Fit a least-squares line to these data. Plot the fitted line on the graph in part (a). Does the least-squares line seem reasonable?

c.   Find a 95% confidence interval for β_1, the average change in weight for a one-unit increase in biotin intake. Does this confidence interval enclose the value $\beta_1 = 0$? Would you conclude that x and y are linearly related?

▷ ## 10.6   ESTIMATING THE EXPECTED VALUE OF y FOR A GIVEN VALUE OF x

In Chapters 7 and 9 we studied methods for estimating a population mean μ and encountered numerous practical applications of these methods in the examples and exercises. Now we will consider a generalization of this problem.

Estimating the mean value of y for a given value of x—that is, estimating $E(y \mid x)$—can be a very important practical problem. If a corporation's profit y is linearly related to advertising expenditures x, the marketing director may wish to estimate the mean profit for a given expenditure x. Similarly, a research physician might wish to estimate the mean response of a human to a specific drug dosage x, and an educator might wish to know the mean grade expected of calculus students who acquired a mathematics achievement test score of $x = 50$. The least-squares prediction equation can be used to obtain these estimates.

Assume that x and y are linearly related according to the probabilistic model defined in Section 10.2 and therefore that $E(y \mid x) = \beta_0 + \beta_1 x$ represents the expected value of y for a given value of x. **Since the fitted line**

$$\hat{y} = \hat{\beta}_0 + \hat{\beta}_1 x$$

attempts to estimate the line of means $E(y \mid x)$ (that is, we estimate β_0 and β_1), then $\hat{y}$ can be used to estimate the expected value of y as well as to predict some value of y that might be observed in the future. It seems quite reasonable to assume that the errors of estimation and prediction differ for these two cases. Consequently, the two estimation procedures differ. In this section we consider the estimation of the expected value of y for a given value of x.

Observe the two lines in Figure 10.6. The first line represents the line of means

$$E(y \mid x) = \beta_0 + \beta_1 x$$

and the second is the fitted prediction equation

$$\hat{y} = \hat{\beta}_0 + \hat{\beta}_1 x$$

We observe from the figure that the error in estimating the expected value of y when $x = x_p$ is the deviation between the two lines above the point x_p. Also, this error increases as we move to the endpoints of the interval over which x has been

Figure 10.6
Estimating $E(y \mid x)$ when
$x = x_p$

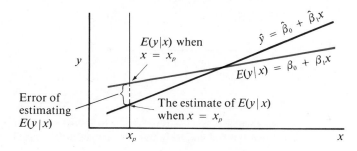

measured. It can be shown that the predicted value

$$\hat{y} = \hat{\beta}_0 + \hat{\beta}_1 x$$

is an unbiased estimator of $E(y \mid x)$—that is, $E(\hat{y}) = \beta_0 + \beta_1 x$—and that $\hat{y}$ is normally distributed, with standard deviation

$$\sigma_{\hat{y}} = \sqrt{\sigma^2 \left[\frac{1}{n} + \frac{(x_p - \bar{x})^2}{S_{xx}} \right]}$$

The corresponding estimated standard deviation of $\hat{y}$ uses s^2 in place of σ^2 in the expression above.

The results outlined above may be used to test an hypothesis about the average or expected value of y for a given value of x, say x_p.* (We may also, of course, test an hypothesis concerning the y intercept β_0, which is the special case where $x_p = 0$.) The null hypothesis is

$$H_0 : E(y \mid x = x_p) = E_0$$

where E_0 is the hypothesized numerical value of $E(y)$ when $x = x_p$. Once again, it can be shown that when H_0 is true, the quantity

$$t = \frac{\hat{y} - E_0}{\text{estimated } \sigma_{\hat{y}}} = \frac{\hat{y} - E_0}{\sqrt{s^2 \left[\frac{1}{n} + \frac{(x_p - \bar{x})^2}{S_{xx}} \right]}}$$

follows a Student's t distribution in repeated sampling with $(n - 2)$ degrees of freedom. The statistical test is conducted in exactly the same manner as the other t test discussed previously.

A TEST CONCERNING THE EXPECTED VALUE OF y WHEN $x = x_p$

1. Null Hypothesis: $H_0 : E(y \mid x = x_p) = E_0$.

2. Alternative Hypothesis:

   One-Tailed Test   *Two-Tailed Test*
   $H_a : E(y \mid x = x_p) > E_0$ $H_a : E(y \mid x = x_p) \neq E_0$
   (or, $E(y \mid x = x_p) < E_0$)

3. Test Statistic:

   $$t = \frac{\hat{y} - E_0}{\sqrt{s^2 \left(\frac{1}{n} + \frac{(x_p - \bar{x})^2}{S_{xx}} \right]}}.$$

* See cautionary comments in Section 10.5. Predicting y for x_p outside the range of x is called **extrapolation**. It is best if x_p lies within the range of the observed values of x.

4. Rejection Region:

    One-Tailed Test   *Two-Tailed Test*

    $t > t_\alpha$ $t > t_{\alpha/2}$ or $t < -t_{\alpha/2}$

    (or, $t < -t_\alpha$ when the
    alternative hypothesis is
    $H_a: E(y \mid x = x_p) < E_0$)

The values of t_α and $t_{\alpha/2}$ are given in Table 4 in Appendix III. Use the value of t corresponding to $(n - 2)$ degrees of freedom.

The corresponding confidence interval, with confidence coefficient $(1 - \alpha)$, for the expected value of y, given $x = x_p$, is

A 100(1 − α)% CONFIDENCE INTERVAL FOR *E*(*y*|*x*) WHEN *x* = *x*ₚ

$$\hat{y} \pm t_{\alpha/2} \sqrt{s^2 \left[\frac{1}{n} + \frac{(x_p - \bar{x})^2}{S_{xx}} \right]}$$

where $t_{\alpha/2}$ is based on $(n - 2)$ degrees of freedom.

EXAMPLE 10.5   Find a 95% confidence interval for the expected value of y, the final calculus grade, given that the mathematics achievement test score is $x = 50$.

Solution   To estimate the mean calculus grade for students whose achievement test score was $x_p = 50$, we would use

$$\hat{y} = \hat{\beta}_0 + \hat{\beta}_1 x_p$$

to calculate $\hat{y}$, the estimate of $E(y \mid x = 50)$. Then, using values calculated previously,

$$\hat{y} = 40.784 + (.766)(50) = 79.08$$

The formula for the 95% confidence interval would be

$$\hat{y} \pm t_{.025} \sqrt{s^2 \left[\frac{1}{n} + \frac{(x_p - \bar{x})^2}{S_{xx}} \right]}$$

Substituting the appropriate quantities into this expression, we find that the 95% confidence interval for the expected (mean) calculus grade, given an achievement test score of 50, is

$$79.08 \pm (2.306) \sqrt{75.754 \left[\frac{1}{10} + \frac{(50 - 46)^2}{2474} \right]}$$

or 79.08 ± 6.55

Thus, we estimate that the mean calculus grade for the population of students acquiring mathematics achievement test scores of $x = 50$ falls in the interval from 72.53 to 85.63.

EXERCISES   Basic Techniques

10.31   Refer to Exercise 10.6. Estimate the expected value of y when $x = 1$ using a 90% confidence interval.

10.32   Refer to Exercise 10.7. Find a 90% confidence interval for the mean value of y when $x = 2$.

10.33   Given the following data:

x	-3	-2	-1	0	1	2	3
y	0	1	3	3	5	7	7

a.   Fit a least-squares line to the data.
b.   Calculate SSE and s^2.
c.   Find a 90% confidence interval for the mean value of y when $x = 1$.

Applications

10.34   Refer to the simple linear regression analysis in Exercise 10.8, relating the recidivism rates of prison inmates to their age. Using the calculations obtained in Exercises 10.8 and 10.13, construct a 90% confidence interval for the average recidivism rate for inmates 30 years of age.

10.35   A marketing research experiment was conducted to study the relationship between the length of time necessary for a buyer to reach a decision and the number of alternative package designs of a product presented. Brand names were eliminated from the packages to reduce the effects of brand preferences. The buyers made their selections using the manufacturer's product descriptions on the packages as the only buying guide. The length of time necessary to reach a decision is recorded for 15 participants in the marketing research study.

Length of decision time, y (sec)	5, 8, 8, 7, 9	7, 9, 8, 9, 10	10, 11, 10, 12, 9
Number of alternatives, x	2	3	4

a.   Find the least-squares line appropriate for these data.
b.   Plot the points and graph the line as a check on your calculations.
c.   Calculate s^2.
d.   Do the data present sufficient evidence to indicate that the length of decision time is linearly related to the number of alternative package designs? (Test at the $\alpha = .05$ level of significance.)
e.   Find the approximate observed significance level for the test and interpret its value.
f.   Estimate the average length of time necessary to reach a decision when three alternatives are presented, using a 95% confidence interval.

10.36   In manufacturing an antibiotic, the yield is a function of time. Data collected show that a process yielded the following pounds of antibiotic for the time periods shown.

Time x (days)	1	2	3	4	5	6
Yield, y	23	31	40	46	52	63

a.   For various reasons, it is convenient to schedule production using a 4-day cycle. Estimate the mean yield for the amount of antibiotic produced over a 4-day period. Use a 95% confidence interval.

b.   In a practical situation, the yield for a zero length of time must be 0. Explain why the least-squares line does not go through the origin. Should it?

10.37   If you try to rent an apartment or buy a house, you find that real estate representatives establish apartment rents and house prices on the basis of square footage of heated floor space. The data in the table give the square footages and sales prices of $n = 12$ houses randomly selected from those sold in a small city.

Square feet x	Price y
1460	$ 88,700
2108	109,300
1743	101,400
1499	91,100
1864	102,400
2391	114,900
1977	105,400
1610	97,000
1530	92,400
1759	98,200
1821	104,300
2216	111,700

a.   Estimate the mean increase in the price for an increase of one square foot for houses sold in the city. Use a 90% confidence interval. Interpret your estimate.

b.   Suppose that you are a real estate salesperson and you desire an estimate of the mean sales price of houses with a total of 2000 square feet of heated space. Use a 95% confidence interval and interpret your estimate.

c.   Calculate the price per square foot for each house and then calculate the sample mean. Why is this estimate of the mean cost per square foot not equal to the answer in part (a)? Should it be? Explain.

10.38   An experiment was conducted to determine the effect of soil applications of various levels of phosphorus on the inorganic phosphorus levels in a particular plant. The following data represent the levels of inorganic phosphorus in micromoles (mM) per gram dry weight of sudan grass roots grown in the greenhouse for 28 days, in the absence of zinc.

Phosphorus applied (x)	Phosphorus in plant (y)
.5 mM	204
	195
	247
	245
.25 mM	159
	127
	95
	144
.10 mM	128
	192
	84
	71

a. Plot the data. Do the data appear to exhibit a linear relationship?
b. Find the least-squares line relating the plant phosphorus levels (y) to the amount of phosphorus applied to the soil (x). Graph the least-squares line as a check on your calculations.
c. Do the data provide sufficient evidence to indicate that the amount of phosphorus present in the plant is linearly related to the amount of phosphorus applied to the soil? Use $\alpha = .05$.
d. Find the approximate p-value for the test in part (c).
e. Estimate the mean amount of phosphorus in the plant if .20 mM of phosphorus is applied to the soil, in the absence of zinc. Use a 90% confidence interval.

10.7 PREDICTING A PARTICULAR VALUE OF y FOR A GIVEN VALUE OF x

Although the expected value of y for a particular value of x is of interest for our example in Table 10.1, we are primarily interested in *using* the prediction equation $\hat{y} = \hat{\beta}_0 + \hat{\beta}_1 x$ based on our observed data to predict the final calculus grade for some prospective student selected from the population of interest. That is, we want to use the prediction equation obtained for the ten measurements in Table 10.2 to predict the final calculus grade for a new student selected from the population. If the student's achievement test score was x_p, we intuitively see that the error of prediction (the deviation between $\hat{y}$ and the actual grade y that the student will obtain) is composed of two elements. Since the student's grade will equal

$$y = \beta_0 + \beta_1 x_p + \epsilon$$

$(y - \hat{y})$ equals the deviation between $\hat{y}$ and the expected value of y, described in Section 10.6 (and shown in Figure 10.6), *plus* the random amount ϵ that represents the deviation of the student's grade from the expected value (see Figure 10.7). **Thus, the variability in the error for predicting a single value of y exceeds the variability for estimating the expected value of y.**

Figure 10.7
Error in predicting a
particular value of y

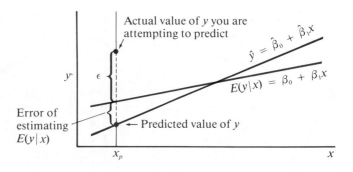

It can be shown that the variance of the error of predicting a particular value of *y* when $x = x_p$, that is, $(y - \hat{y})$, is

$$\sigma^2_{(y-\hat{y})} = \sigma^2_y + \sigma^2_{\hat{y}} = \sigma^2 + \sigma^2\left[\frac{1}{n} + \frac{(x_p - \bar{x})^2}{S_{xx}}\right]$$

$$= \sigma^2\left[1 + \frac{1}{n} + \frac{(x_p - \bar{x})^2}{S_{xx}}\right]$$

When *n* is very large, the second and third terms in the brackets will become small and the variance of the prediction error will approach σ^2. These results may be used to construct the following prediction interval for *y*, given $x = x_p$. The confidence coefficient for the prediction interval is $(1 - \alpha)$.

A $100(1 - \alpha)\%$ PREDICTION INTERVAL FOR *y* WHEN $x = x_p$

$$\hat{y} \pm t_{\alpha/2}\sqrt{s^2\left[1 + \frac{1}{n} + \frac{(x_p - \bar{x})^2}{S_{xx}}\right]}$$

where $t_{\alpha/2}$ is based on $(n - 2)$ degrees of freedom.

EXAMPLE 10.6   Refer to Example 10.1 and predict the final calculus grade for some new student who scored $x = 50$ on the mathematics achievement test.

Solution   The predicted value of *y* would be $\hat{y} = \hat{\beta}_0 + \hat{\beta}_1 x_p$ or

$$\hat{y} = 40.784 + (.766)(50) = 79.08$$

and the 95% prediction interval for the final calculus grade would be

$$79.08 \pm (2.306)\sqrt{75.754\left[1 + \frac{1}{10} + \frac{(50 - 46)^2}{2474}\right]}$$

or 79.08 ± 21.10.

Note that in a practical situation we would probably have the grades and achievement test scores for many more than $n = 10$ students indicated in Table 10.1 and that this would reduce somewhat the width of the prediction interval. In fact, when *n* is large, the prediction interval approaches $\hat{y} \pm z_{\alpha/2}s$ or, for a 95% prediction interval, $\hat{y} \pm 1.96s$.

Again, note the distinction between the confidence interval for $E(y|x)$ discussed in Section 10.6 and the prediction interval presented in this section. $E(y|x)$ is a mean, a parameter of a population of *y* values, and *y* is a random variable that oscillates in a random manner about $E(y|x)$. The mean value of *y* when $x = 50$ is vastly different from some value of *y*, chosen at random from the set of all *y* values for which $x = 50$.

To make this distinction when making inferences, **we always** *estimate* **the value of a parameter and** *predict* **the value of a random variable**. As noted in our earlier discussion and as shown in Figures 10.6 and 10.7, the error of predicting y is different from the error of estimating $E(y \mid x)$. This is evident in the difference in widths of the two prediction and confidence intervals.

A graph of the confidence interval for $E(y \mid x)$ and the prediction interval for a particular value of y for the data in Table 10.2 is shown in Figure 10.8. The plot of the confidence interval is shown by solid lines, and the prediction interval is identified by dashed lines. Note how the widths of the intervals increase as you move to the right or left of $\bar{x} = 46$. In particular, see the confidence interval and prediction interval for $x = 50$ calculated in Examples 10.5 and 10.6.

Figure 10.8
Confidence intervals for $E(y \mid x)$ and prediction intervals for y based on data in Table 10.2

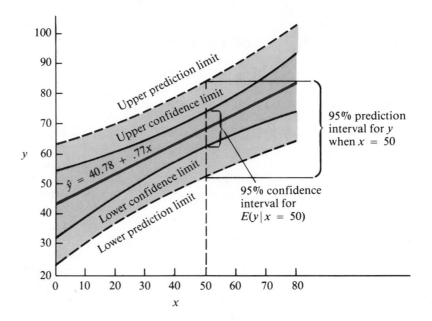

EXERCISES   Basic Techniques

10.39 Refer to Exercise 10.6. Find a 90% prediction interval for some value of y to be observed in the future when $x = 1$.

10.40 Refer to Exercise 10.7. Find a 90% prediction interval for some value of y to be observed in the future when $x = 2$.

10.41 Refer to Exercise 10.33. Find a 90% prediction interval for y when $x = 1$.

Applications

10.42 Refer to Exercise 10.36. Suppose that the production process is operated for a 4-day period at some particular time in the future. Find a 95% prediction interval for the yield. Compare with the confidence interval of the mean yield for $x = 4$ days.

10.43 Refer to Exercise 10.37. Suppose that a house containing 1780 square feet of heated floor space is offered for sale. Give a 90% prediction interval for the price at which the house will sell. Interpret this prediction.

10.44 Refer to Exercise 10.8. If a 30-year-old inmate arrives at a state prison in the United States, predict the probability that he will return to prison within a year of his release. Use the calculations obtained in Exercises 10.8, 10.13, and 10.34.

10.45 Refer to Exercise 10.30. Construct a 99% prediction interval for the mean weight gain in chicks whose biotin intake is 6.00 $\mu g/d$.

10.46 In an effort to increase the production in the semiarid regions of the world, it is necessary that we be able to monitor the amount of moisture in the soil. Two methods are available. One utilizes mathematical formulas, temperatures, rainfalls, and evaporation rates to arrive at a computed value of moisture content. The other relies on direct measurement of the weight of soil before and after evaporation of the soil moisture. A comparison of the theoretically computed moisture content (in percent of dry weight of soil) with the measured moisture content is shown below for nine soil samples randomly selected from a semiarid region.

Computed soil moisture, *x*	10.2	13.6	9.7	7.3	13.1	15.6	9.0	10.4	12.1
Measured soil moisture, *y*	11.8	15.4	9.2	8.0	13.8	16.1	8.4	10.2	10.5

a. Fit a least-squares line to the data.
b. If the computed soil moisture reading at some future time was 10.0, find a 90% prediction interval for the actual measured soil moisture.

10.47 Most sophomore physics students are required to conduct an experiment verifying Hooke's Law. Hooke's Law states that when a force is applied to a body that is long in comparison to its cross-sectional area, the change *y* in its length is proportional to the force *x*; that is,

$$y = \beta_1 x$$

where β_1 is a constant of proportionality. The results of an actual physics student's laboratory experiment are shown in the table. Six lengths of steel wire, .34 millimeters (mm) in diameter and 2 meters (m) long, were used to obtain the six force-length change measurements.

Force *x* (kg)	Change in length *y* (mm)
29.4	4.25
39.2	5.25
49.0	6.50
58.8	7.85
68.6	8.75
78.4	10.00

a. Fit the model $y = \beta_0 + \beta_1 x + \epsilon$ to the data using the method of least squares.
b. Plot the points and graph the line as a check on your calculations.
c. Find a 95% confidence interval for the slope of the line.
d. According to Hooke's Law, the line should pass through the point $(0, 0)$; that is, β_0 should equal 0. Test the hypothesis that $E(y) = 0$ when $x = 0$.

e.   Predict the elongation of a 2-m length of wire when a force of 55 kg is applied. Use a 95% prediction interval.

10.48   In Exercise 10.9, we described an experiment conducted to investigate the relationship between the ventilatory threshold and the running economy on the finish time for experienced runners in a 10-kilometer race (S. K. Powers, et al., "Ventilatory Threshold, Running Economy and Distance Running Performance of Trained Athletes," *Research Quarterly for Exercise and Sport* 54, no. 3 [1983]). The ventilatory threshold x, expressed in milliliters per kilogram per minute, and the finish time y (in minutes) for each of the nine runners are shown below.

Runner	1	2	3	4	5	6	7	8	9
Ventilatory Threshold, x	46.2	44.8	44.2	42.7	41.6	42.7	39.6	38.8	37.0
Finish Time, y	33.15	33.33	33.50	33.55	33.73	33.86	33.90	34.15	34.90

a.   Fit a least-squares line to the data.
b.   Calculate SSE and s^2.
c.   Does the ventilatory threshold x contribute information to predict a runner's finish time? Test using $\alpha = .05$.
d.   Find a 90% prediction interval for the finish time y when x equals 40 ml/kg/m.

10.49   Another simple linear regression analysis presented in the paper by Reiners, Hollinger, and Lang (1984), in Exercise 10.29 involved the relationship between the number y of degree-days (days when the temperature is above freezing) per year and the altitude x in meters ("Temperature and Evapotranspiration Gradients of the White Mountains, New Hampshire, U.S.A.," *Arctic and Alpine Research* 16, no. 1 [1984]). The analysis is based on the recorded number of degree-days for two winters, 1976–77 and 1977–78, at each of the four altitudes. Statistics presented for this analysis are $\hat{\beta}_0 = 3393.93$, $\hat{\beta}_1 = -1.319$, and $r^2 = .99$, and the estimated standard deviation of $\hat{\beta}_1$ is 0.091.
a.   How many degrees of freedom are associated with SSE and s^2 in this analysis?
b.   Do the data provide sufficient evidence to indicate that altitude x provides information for the prediction of degree-days y? Test using $\alpha = .01$.
c.   Give the least-squares prediction equation.
d.   Predict the number of degree-days at an altitude of 825 meters. Do the statistics above enable you to construct a prediction interval for your prediction? Explain.

10.8   A COEFFICIENT OF CORRELATION

Sometimes we wish to obtain an indicator of the strength of the linear relationship between two variables y and x that is independent of their respective scales of measurement. We call this a measure of the **linear correlation between y and x**.

The measure of linear correlation commonly used in statistics is called the **Pearson product moment coefficient of correlation** between y and x. This quantity, denoted by the symbol r, is computed as follows:

PEARSON PRODUCT MOMENT COEFFICIENT OF CORRELATION

$$r = \frac{S_{xy}}{\sqrt{S_{xx}S_{yy}}}$$

We will show you how to compute the Pearson product moment coefficient of correlation for the grade point data in Table 10.1, and then we will explain how it measures the strength of the relationship between y and x.

EXAMPLE 10.7 Calculate the coefficient of correlation for the calculus grade–achievement test score data in Table 10.1.

Solution The coefficient of correlation for the calculus grade–achievement test score data in Table 10.1 can be obtained by using the formula for r and the quantities

$$S_{xy} = 1894$$

$$S_{xx} = 2474$$

and

$$S_{yy} = 2056$$

which were computed previously. Then,

$$r = \frac{S_{xy}}{\sqrt{S_{xx}S_{yy}}} = \frac{1894}{\sqrt{2474(2056)}} = .84 \qquad \lhd$$

A study of the coefficient of correlation r yields rather interesting results and explains the reason for its selection as a measure of linear correlation. We note that the denominators used in calculating r and $\hat{\beta}_1$ will always be positive, since they both involve sums of squares of numbers. Since the numerator used in calculating r is identical to the numerator of the formula for the slope $\hat{\beta}_1$, the coefficient of correlation r will assume exactly the same sign as $\hat{\beta}_1$ and will equal zero when $\hat{\beta}_1 = 0$. **Thus, $r = 0$ implies no linear correlation between y and x. A positive value for r implies that the line slopes upward to the right; a negative value indicates that it slopes downward to the right.**

Figure 10.9 shows four typical scatter diagrams and their associated correlation coefficients. Note that **$r = 0$ implies no linear correlation**, not simply "no correlation." A pronounced curvilinear pattern may exist, as in Figure 10.9(d), but its linear correlation coefficient may equal 0. In general, we can say that r measures the linear association of the two variables y and x. When $r = 1$ or -1, all the points fall on a straight line; when $r = 0$, they are scattered and give no evidence of a *linear* relationship. Any other value of r suggests the degree to which the points tend to be linearly related.

Figure 10.9
Some typical scatter
diagrams with
approximate values of r

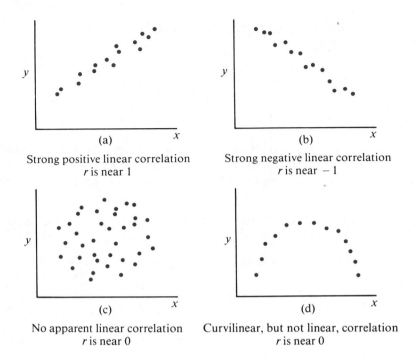

(a)

Strong positive linear correlation
r is near 1

(b)

Strong negative linear correlation
r is near -1

(c)

No apparent linear correlation
r is near 0

(d)

Curvilinear, but not linear, correlation
r is near 0

The interpretation of nonzero values of r can be obtained by comparing the errors of prediction for the prediction equation

$$\hat{y} = \hat{\beta}_0 + \hat{\beta}_1 x$$

with the predictor of y, $\bar{y}$, that would be employed if x were ignored. Figure 10.10(a) and (b) shows the lines $\hat{y} = \hat{\beta}_0 + \hat{\beta}_1 x$ and $\hat{y} = \bar{y}$ fit to the same set of data. Certainly, if x is of any value in predicting y, then SSE, the sum of squares of deviations of y about the linear model, should be less than the sum of squares

Figure 10.10
Two models fit to the
same data

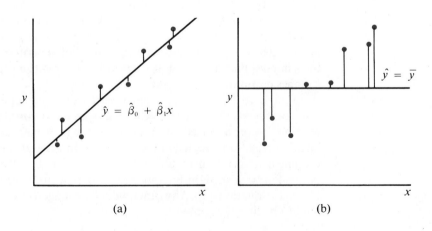

(a)

(b)

of deviations about the predictor $\bar{y}$, which is

$$S_{yy} = \sum_{i=1}^{n} (y_i - \bar{y})^2$$

Indeed, we see that SSE can *never* be larger than

$$S_{yy} = \sum_{i=1}^{n} (y_i - \bar{y})^2$$

because

$$\text{SSE} = S_{yy} - \hat{\beta}_1 S_{xy} = S_{yy} - \left(\frac{S_{xy}}{S_{xx}}\right) S_{xy} = S_{yy} - \frac{(S_{xy})^2}{S_{xx}}$$

Therefore, SSE is equal to S_{yy} minus a nonnegative quantity. Consequently, SSE must always be less than or equal to S_{yy}.

Furthermore, with the aid of a bit of algebraic manipulation, we can show that

$$r^2 = 1 - \frac{\text{SSE}}{S_{yy}} = \frac{S_{yy} - \text{SSE}}{S_{yy}}$$

In other words, r^2 lies in the interval

$$0 \le r^2 \le 1$$

and r will equal $+1$ or -1 only when all the points fall exactly on the fitted line, that is, when SSE equals zero.

Actually, we see that r^2 **is equal to the ratio of the reduction in the sum of squares of deviations obtained by using the linear model to the total sum of squares of deviations about the sample mean $\bar{y}$, which would be the predictor of y if x were ignored**. Thus r^2, called the *coefficient of determination*, would seem to give a more meaningful interpretation of the strength of the relation between y and x than would the correlation coefficient r.

COEFFICIENT OF DETERMINATION

$$r^2 = \frac{S_{yy} - \text{SSE}}{S_{yy}} = \frac{\sum_{i=1}^{n} (y_i - \bar{y})^2 - \text{SSE}}{\sum_{i=1}^{n} (y_i - \bar{y})^2}$$

You will observe that the sample correlation coefficient r is an estimator of a population correlation coefficient ρ (Greek letter rho), which would be obtained if the coefficient of correlation were calculated by using all the points in the population.

A test of the null hypothesis that no correlation exists between y and x, $H_0: \rho = 0$, is exactly equivalent to a test of the hypothesis $H_0: \beta_1 = 0$. The test is summarized in the following display.

TEST OF AN HYPOTHESIS CONCERNING
THE CORRELATION COEFFICIENT ρ

1. Null Hypothesis: $H_0 : \rho = 0$

2. Alternative Hypothesis:

    One-Tailed Test   *Two-Tailed Test*

    $H_a : \rho > 0$ $H_a : \rho \neq 0$

    (or, $\rho < 0$)

3. Test Statistic: $t = \dfrac{r\sqrt{n-2}}{\sqrt{1-r^2}} = \dfrac{\hat{\beta}_1}{s/\sqrt{S_{xx}}}$

> When the assumptions given in Section 10.2 (and 10.10) are satisfied, the test statistic will have a Student's distribution with $(n-2)$ degrees of freedom.

4. Rejection Region:

    One-Tailed Test   *Two-Tailed Test*

    $t > t_\alpha$ $t > t_{\alpha/2}$ or $t < -t_{\alpha/2}$

    (or, $t < t_\alpha$ when the
    alternative hypothesis
    is $H_a : \rho < 0$)

The values of t_α and $t_{\alpha/2}$ are given in Table 4 in Appendix III. Use the values of t corresponding to $(n-2)$ degrees of freedom.

Although r gives a rather nice measure of the goodness of fit of the least-squares line to the fitted data, its use in making inferences concerning ρ may be of dubious practical value in many situations. Ordinarily, we would be interested in testing the null hypothesis that $\rho = 0$ and, since this is algebraically equivalent to testing the hypothesis that $\beta_1 = 0$, we have already considered this problem. If the evidence in the sample suggests that y and x are related, it would seem that we would redirect our attention to the ultimate objective of our data analysis, using the prediction equation to obtain interval estimates for $E(y \mid x)$ and prediction intervals for y. In addition, it seems unlikely that a phenomenon y, observed in the physical sciences, and especially the social sciences, would be a function of a single variable. Thus, the correlation coefficient between a student's predicted grade point average and any one variable would probably be quite small and of questionable value. A larger reduction in SSE could possibly be obtained by constructing a predictor of y based on a set of variables $x_1, x_2, \ldots, x_k$.

One further reminder concerning the interpretation of r is worthwhile. It is not uncommon for researchers in some fields to speak proudly of sample correlation coefficients r in the neighborhood of .5 (and, in some cases, as low as .1) as being indicative of a "relation" between y and x. Certainly, even if these values were *accurate* estimates of ρ, only a very weak relation would be indicated. A value $r = .5$ would imply that the use of x in predicting y reduced the sum of squares of deviations about the prediction line by only $r^2 = .25$, or 25%. A correlation coefficient $r = .1$ would imply only an $r^2 = .01$, or 1%, reduction in the total sum of squares of deviations that could be explained by x.

If the linear coefficients of correlation between y and each of two variables x_1 and x_2 were calculated to be .4 and .5, respectively, it does not follow that a predictor using both variables would account for a $[(.4)^2 + (.5)^2] = .41$, or a 41% reduction in the sum of squares of deviations. Actually, x_1 and x_2 might be highly correlated and therefore contribute virtually the same information for the prediction of y.

Finally, remember that r is a measure of **linear correlation** and that x and y could be perfectly related by some **curvilinear** function when the observed value of r is equal to 0.

EXERCISES   Basic Techniques

10.50 How does the coefficient of correlation measure the strength of the linear relationship between two variables y and x?

10.51 Describe the significance of the algebraic sign and the magnitude of r.

10.52 What value does r assume if all the sample points fall on the same straight line and
a.   the line has positive slope?
b.   the line has negative slope?

10.53 Given the following data:

x	-2	-1	0	1	2
y	2	2	3	4	4

a.   Find the least-squares line for the data.
b.   Plot the data points and graph the least-squares line. Based on your graph, what will be the sign of the sample correlation coefficient?
c.   Calculate r and r^2 and interpret their values.

10.54 For the following data:

x	1	2	3	4	5	6
y	7	5	5	3	2	0

a.   Find the least-squares line.
b.   Sketch the least-squares line on graph paper and plot the six points.
c.   Calculate the sample coefficient of correlation r and interpret.
d.   By what percentage was the sum of squares of deviations reduced by using the least-squares predictor $\hat{y} = \hat{\beta}_0 + \hat{\beta}_1 x$, rather than $\bar{y}$ as a predictor of y?

10.55 Reverse the slope of the line in Exercise 10.54 by reordering the y observations, as follows:

x	1	2	3	4	5	6
y	0	2	3	5	5	7

Repeat the steps of Exercise 10.54. Notice the change in sign of r and the relation between the values of r^2 of Exercise 10.54 and this exercise.

Applications

10.56   Refer to the data relating a runner's "running economy" x and his 10-kilometer race running time in Exercise 10.9.
   a.   Calculate the coefficient of correlation r between y and x and interpret this value.
   b.   Calculate the coefficient of determination and interpret its value.

10.57   In Exercise 1.9, we presented some data from a study of the infestation of the *T. orientalis* lobster by two types of barnacles, *O. tridens* and *O. lowei* (W. B. Jeffries, H. K. Vorhis, and C. M. Yang, "Diversity and Distribution of the Pedunculate Barnacle *Octolasmis* gray, 1825 Epizoic on the Scyllarid Lobster, *Thenus orientalis*, Lund 1793," *Crustaceana* 46, no. 3 [1984]). The data from Exercise 1.9 are reproduced below. Examine the columns of the table that give the numbers of *O. tridens* and *O. lowei* barnacles on each lobster. Does it appear that the barnacles compete for space on the surface of a lobster?

Field number	Carapace length (mm)	Number of barnacles		
		O. tridens	*O. lowei*	Total
AO61	78	645	6	651
AO62	66	320	23	343
AO66	65	401	40	441
AO70	63	364	9	373
AO67	60	327	24	351
AO69	60	73	5	78
AO64	58	20	86	106
AO68	56	221	0	221
AO65	52	3	109	112
AO63	50	5	350	355

   a.   If they do compete, would you expect the number x of *O. tridens* and the number y of *O. lowei* barnacles to be positively or negatively correlated? Explain.
   b.   Calculate the correlation coefficient r between x and y.
   c.   If you wanted to test the theory that the two types of barnacles compete for space by conducting a test of the null hypothesis, "the population correlation coefficient ρ equals 0," what would you use for your alternative hypothesis?
   d.   Conduct the test in part (c) and state your conclusions. Use $\alpha = .05$.

10.58   The first column in the table below gives the final score for each of the top 15 participants in the 1985 ladies' LPGA tournament (*Orlando Sentinel*, July 15, 1985). The second column gives each player's qualifying score, that is, the score in a pretournament play-off that enabled a player to enter the tournament (*New York Times*, July 13, 1985).

	Final score	Qualifying score		Final score	Qualifying score
Kathy Baker	280	142	Ayako Okamoto	290	146
Judy Clark	283	146	Betsy King	290	144
Vicki Alvarez	287	141	Jan Geddes	290	149
Nancy Lopez	288	140	Amy Alcott	292	144
Janet Coles	288	141	Cathy Morse	293	147
Sally Little	289	143	Jan Stephenson	293	145
Penny Pulz	289	149	Pat Bradley	293	149

Is a player's score in a major tournament correlated with the player's qualifying score?

a. If a correlation exists, should it be positive or negative? Explain.

b. Do the data provide sufficient evidence to indicate a correlation of the type given in part (a)? Test using $\alpha = .05$.

10.59 *USA Today* (October 29, 1984) describes a study conducted to investigate the relationship between a student's television viewing time and the student's classroom average. The researcher surveyed 193 fourth, fifth, and sixth grade students and 163 parents to find out how much time the students spent each day viewing TV, and she also recorded each student's average grade. She found that those who indicated that they "always" or "often" watched TV had an average grade of 85. Those who watched it "sometimes" or "seldom" had an average grade of 84. Do you think the results of this study imply little or no correlation between the amount of time a student spends watching TV and a student's average grade in school? Explain.

10.60 In Exercise 7.14 we described a study by G. W. Marino to investigate the variables related to a hockey player's ability to make a fast start from a stopped position. In the experiment, each skater started from a stopped position and attempted to move as rapidly as possible over a 6-meter distance. The correlation coefficient r between a skater's stride rate (number of strides per second) and the length of time to cover the 6-meter distance for the sample of 69 skaters was -0.37.

a. Do the data provide sufficient evidence to indicate a correlation between stride rate and time to cover the distance? Test using $\alpha = .05$.

b. Find the approximate p-value for the test.

c. What are the practical implications of the test in part (a)?

10.61 Refer to Exercise 7.14. Marino calculated the sample correlation coefficient r for the stride rate and the average acceleration rate for the 69 skaters to be .36.

a. Do the data provide sufficient evidence to indicate a correlation between stride rate and average acceleration for the skaters? Test using $\alpha = .05$.

b. Find the approximate p-value for the test in part (a).

10.62 An experiment was conducted in a supermarket to observe the relation between the amount of display space allotted to a brand of coffee (brand A) and its weekly sales. The amount of space allotted to brand A was varied over 3-, 6-, and 9-square-foot displays in a random manner over 12 weeks while the space allotted to competing brands was maintained at a constant 3 square feet for each. The following data were observed:

Weekly sales, y (dollars)	526	421	581	630	412	560	434	443	590	570	346	672
Space allotted, x (ft^2)	6	3	6	9	3	9	6	3	9	6	3	9

a. Find the least-squares line appropriate for the data.

b. Calculate r and r^2. Interpret.

c. Find a 90% confidence interval for the mean weekly sales given that 6 square feet is allotted for display.

d. Use a 90% prediction interval to predict the weekly sales at some time in the future if 6 square feet is allotted for display.

e. By what percentage was the sum of squares of deviations reduced by using the least-squares predictor $\hat{y} = \hat{\beta}_0 + \hat{\beta}_1 x$ rather than $\bar{y}$ as a predictor of y for these data?

f. Would you expect the relation between y and x to be linear if x were varied over a wider range (say $x = 1$ to $x = 30$)?

10.63 Geothermal power is an important source of energy. Since the amount of energy contained in a pound of water is a function of its temperature, you might wonder whether water obtained from deeper wells contains more energy per pound. The data in the table are reproduced from an article on geothermal systems by A. J. Ellis ("Geothermal Systems," *American Scientist* [September–October 1975]).

Location of well	Average (max.) drill hole depth (m)	Average (max.) temperature (°C)
El Tateo, Chile	650	230
Ahuachapan, El Salvador	1000	230
Namafjall, Iceland	1000	250
Larderello (region), Italy	600	200
Matsukawa, Japan	1000	220
Cerro Prieto, Mexico	800	300
Wairakei, New Zealand	800	230
Kizildere, Turkey	700	190
The Geysers, United States	1500	250

a. Find the least-squares line.
b. Sketch the least-squares line on graph paper and plot the nine points.
c. Calculate the sample coefficient of correlation r and interpret it.
d. By what percentage was the sum of squares of deviations reduced by using the least-squares predictor $\hat{y} = \hat{\beta}_0 + \hat{\beta}_1 x$ rather than $\bar{y}$?

10.9 COMPUTER PRINTOUTS FOR A REGRESSION ANALYSIS

Although a simple linear regression problem can be analyzed as we have done in Sections 10.3 through 10.5, these same calculations can be implemented on a computer, using one of several packaged computer programs. Table 10.3 displays the output that resulted when the data in Table 10.1 were analyzed by using the

Table 10.3
MINITAB output for the grade point data of Table 10.1

MTB > REGRESS C1 1 C2

THE REGRESSION EQUATION IS ①
C1 = 40.8 + 0.766 C2

PREDICTOR	COEF ②	STDEV ③	T-RATIO ④	P
CONSTANT	40.784	8.507	4.79	0.000
C2	0.7656	0.1750	4.38	0.002

S = 8.704 ⑤   R-SQ = 70.5% ⑥   R-SQ (ADJ) = 66.8%

ANALYSIS OF VARIANCE

SOURCE	DF	SS	MS	F	P
REGRESSION	1	1450.0	1450.0	19.14	0.002
ERROR	8	606.0	75.8		
TOTAL	9	2056.0			

REGRESS command in the MINITAB package. The values of y and x were stored in columns 1 and 2 of a data array, so that the regression equation found at the top of the printout is given in box #1 as

$$C1 = 40.8 + 0.766\ C2$$

The upper portion of the printout also contains information about the individual parameters in the model. The column labeled PREDICTOR lists the programmer's identification names for the parameters. In this example, the programmer has not specified any particular names; by default, the program labels β_0 as "Constant" and the coefficient of x as "C2," the column in which x was stored. The estimated intercept $\hat{\beta}_0$ and slope $\hat{\beta}_1$ are found in the column labeled COEF (for "coefficient," shaded in box #2) to be $\hat{\beta}_0 = 40.784$ and $\hat{\beta}_1 = .7656$; their rounded values are used in the prediction equation.

The estimated standard deviations of the regression coefficients, $\hat{\beta}_0$ and $\hat{\beta}_1$, are found in the column labeled STDEV. These values are used in testing the null hypothesis

$$H_0\!:\beta_1 = 0 \qquad \text{for} \qquad i = 0 \text{ or } 1$$

and in the construction of confidence intervals for the β_i. For example, in testing $H_0\!:\beta_1 = 0$ versus $H_a\!:\beta_1 \neq 0$, the test statistic is

$$t = \frac{\beta_1}{\text{estimated } \sigma_{\beta_1}} = \frac{.7656}{.1750} = 4.38$$

which, from Example 10.3, has a p-value of less than .01. The value of this t statistic is found in the column labeled T-RATIO in box #4. Similarly, a test of $H_0\!:\beta_0 = 0$ could be performed using the line labeled CONSTANT in this portion of the printout.

In addition to providing the value $s = 8.704$, the sample estimate of σ, in box #5, the printout also provides the value of the coefficient of determination expressed as a percentage (R-SQ $= 70.5\%$) in box #6. That is, 70.5% of the variation in y is accounted for by regression. In the case of linear regression, the sample correlation coefficient, r, is the square root of R-SQ and takes the same sign as β_1.

The lower portion of the printout presents an **analysis of variance** for the data. In regression analysis, an analysis of variance is the technique that partitions the total variation in the response y into one portion associated with random error and another portion associated with the variability accounted for by regression. The ERROR row in this portion of the printout provides the degrees of freedom for error in the DF column, the value of SSE in the SS column, and the value of s^2 in the MS column. The value of the standard deviation s, found in box #5 on the printout, is the square root of s^2. The value found in the TOTAL row and SS column is the total variation in the response y, given as

$$S_{yy} = \sum (y_i - \bar{y})^2 = 2056.0$$

The amount of the total variation that can be accounted for by regression is the difference between the total variation and the error variation, SSE, or SSR $= 1450.0$. Finally, the proportion of this total variation that can be accounted for by regression is calculated as

$$\frac{\text{SSR}}{S_{yy}} = \frac{1450.0}{2056.0} = .705$$

which is, in fact, the coefficient of determination, r^2. The technique called "analysis of variance" will be discussed in detail in Chapter 13.

Table 10.4 contains the SAS output for the same data. Although both computer printouts contain the same basic information, the SAS printout reports results with more decimal accuracy than MINITAB.

Table 10.4
SAS output for the grade point data of Table 10.1

DEP VARIABLE: Y

ANALYSIS OF VARIANCE

SOURCE	DF	SUM OF SQUARES	MEAN SQUARE	F VALUE	PROB > F
MODEL	1	1449.97413	1449.97413	19.141	0.0024
ERROR	8	606.02586904	75.75323363		
C TOTAL	9	2056.00000			

ROOT MSE	8.703633	R-SQUARE	0.7052	
DEP MEAN	76	ADJ R-SQ	0.6684	
C.V.	11.45215			

PARAMETER ESTIMATES

VARIABLE	DF	PARAMETER ESTIMATE	STANDARD ERROR	T FOR HO: PARAMETER = 0	PROB > ITI
INTERCEP	1	40.78415521	8.50686138	4.794	0.0014
X	1	0.76556184	0.17498497	4.375	0.0024

▷ **10.10** ASSUMPTIONS

The assumptions for a regression analysis are given in the accompanying display.

ASSUMPTIONS FOR A REGRESSION ANALYSIS

1. The response y can be represented by the probabilistic model

   $$y = \beta_0 + \beta_1 x + \epsilon$$

2. x is measured without error.

3. ϵ is a random variable such that, for a given value of x,

   $$E(\epsilon) = 0 \quad \text{and} \quad \sigma_\epsilon^2 = \sigma^2$$

   and all pairs, ϵ_i, ϵ_j, are independent in a probabilistic sense.

4. ϵ possesses a normal probability distribution.

At first glance you might fail to understand the significance of the first assumption. Models, deterministic or otherwise, are, as the name implies, only models for real relationships that occur in nature. Consequently, model misspecification is always a possibility. Even if you have obtained a good fit to the data, a large error in prediction is possible if you use the model to predict y for some value of x outside the range of values used to fit the least-squares equation. Of course, this problem will always occur if x is time and you attempt to forecast y at some point in the future. This problem occurs with any model for future time predictions; consequently, you make the forecast but keep the model limitations in mind.

The assumption that the variance of the ϵ's is constant and equal to σ^2 will not be true for all types of data. Furthermore, if x is time, it is possible that y values measured over adjacent time periods will tend to be dependent (an overly large value of y in 1989 might signal a large value of y in 1990). Substantial departure from either of these assumptions will affect the confidence coefficients of interval estimates and significance levels and tests described in this chapter.

Like unequal variances and correlation of the random errors, if the normality assumption (4) is not satisfied, the confidence coefficients and significance levels for interval estimates and tests will not be what we expect them to be. However, modest departures from normality will not seriously disturb these values.

▷ 10.11   A MULTIVARIABLE PREDICTOR

A prediction equation based on a number of variables, $x_1, x_2, \ldots, x_k$, could be obtained by the method of least squares in exactly the same manner as that employed for the simple linear model. For example, we might wish to fit the model

$$y = \beta_0 + \beta_1 x_1 + \beta_2 x_2 + \beta_3 x_3 + \epsilon$$

where y = student grade point average at the end of the freshman year, x_1 = rank in high school class divided by the number in class, x_2 = score on a mathematics achievement test, and x_3 = score on a verbal and written achievement test, to data on the achievement of freshmen college students. (Note that we could add other variables as well as the squares, cubes, and cross products of x_1, x_2, and x_3.)

We would require a random sample of n freshmen selected from the population of interest and would record the values of y, x_1, x_2, and x_3, which, for each student, could be regarded as coordinates of a point in four-dimensional space. Then, ideally, we would like to have a multidimensional "ruler" (in our case, a three-dimensional plane) that we could visually move about among the n points until the deviations of the observed values of y from the predicted values of y would in some sense be a minimum. Although we cannot graph points in four dimensions, you can readily recognize that this device is provided by the method of least squares, which, mathematically, performs the task for us.

The sum of squares of deviations of the observed values of y from the fitted model would be

$$\text{SSE} = \sum_{i=1}^{n} (y_i - \hat{y}_i)^2 = \sum_{i=1}^{n} [y_i - (\hat{\beta}_0 + \hat{\beta}_1 x_{1i} + \hat{\beta}_2 x_{2i} + \hat{\beta}_3 x_{3i})]^2$$

where $\hat{y} = \hat{\beta}_0 + \hat{\beta}_1 x_1 + \hat{\beta}_2 x_2 + \hat{\beta}_3 x_3$ is the fitted model and $\hat{\beta}_0, \hat{\beta}_1, \hat{\beta}_2,$ and $\hat{\beta}_3$ are estimates of the model parameters. We would then use the calculus to find the estimates $\hat{\beta}_0, \hat{\beta}_1, \hat{\beta}_2,$ and $\hat{\beta}_3$ that make SSE a minimum. The estimates, as for the simple linear model, would be obtained as the solution of a set of four simultaneous linear equations known as the least-squares equations.

The reasoning employed in obtaining the least-squares multivariable predictor is identical to the procedure developed for the simple linear model, but the presentation of the least-squares equations and their solutions is beyond the scope of this text. Since most multiple-regression analyses are performed by computers, we introduced you to the computer printout for a simple linear regression analysis in Section 10.9 and will discuss the computer output for a multiple-regression analysis in Chapter 11.

10.12 SUMMARY

Although it was not stressed, you will observe that predicting the value of a random variable y was considered for the most elementary situation in Chapters 7 and 8. Thus, if we possessed no information concerning variables related to y, the sole information available for predicting y would be provided by its probability distribution. If we were to select one value as representative of the population, we would most likely choose μ, or some other measure of central tendency. The estimation of the mean was considered in Chapters 7 and 8.

Chapter 10 was concerned with the problem of predicting y when auxiliary information is available on other variables, say $x_1, x_2, x_3, \ldots, x_k$, which are related to y and hence assist in its prediction. We have concentrated primarily on the problem of predicting y as a linear function of a single variable x, which provides the simplest extension of the prediction problem beyond that considered in Chapters 7 and 8. The more interesting case, where y is a linear function of a set of independent variables, is the subject of Chapter 11.

MINITAB COMMANDS

REGRESS C on **K** pred. **C...C** [store st. res. in **C** [fits in **C**]]

MSE   put into **K**
COEFFICIENTS put into **C**

CORRELATION for **C...C** [put in **M**]

REFERENCES

Draper, N. R., and Smith, H. *Applied Regression Analysis.* 2d ed. New York: Wiley, 1981.

Dunn, O. J., and Clark, V. A. *Applied Statistics: Analysis of Variance and Regression.* 2d ed. New York: Wiley, 1987.

Kleinbaum, D. G.; Kupper, L. L.; and Miller, K. E. *Applied Regression Analysis and Other Multivariable Methods.* Boston: PWS-KENT, 1988.

Mendenhall, W. *An Introduction to Linear Models and the Design and Analysis of Experiments.* Belmont, Calif.: Wadsworth, 1968.

Mendenhall, W., and Sincich, T. *A Second Course in Business Statistics: Regression Analysis.* 2d ed. San Francisco: Dellen, 1986.

Meyers, R. H. *Classical and Modern Regression with Applications.* 2d ed. Boston: PWS-KENT, 1990.

Ryan, T. A.; Joiner, B. L.; and Ryan, B. F. *Minitab Student Handbook.* 2d ed. Boston: Duxbury Press, 1985.

SUPPLEMENTARY EXERCISES

10.64 Graph the line corresponding to the equation $y = -x + 2$ by locating points corresponding to $x = 0, 1,$ and 2.

10.65 Given the linear equation $2x - 3y - 5 = 0$.
a. Give the y intercept and slope for the line.
b. Graph the line corresponding to the equation.

10.66 For what configurations of sample points will s^2 be 0?

10.67 For what parameter is s^2 an unbiased estimator? Explain how this parameter enters into the description of the probabilistic model $y = \beta_0 + \beta_1 x + \epsilon$.

10.68 Given the following data for corresponding values of two variables y and x:

x	2	1.5	1	2.5	2.5	4	5
y	-3	-2	-1	0	1	2	3

a. Find the least-squares line for the data.
b. As a check on the calculations in part (a), plot the seven points and graph the line.
c. Calculate SSE and s^2.
d. Do the data present sufficient evidence to indicate that y and x are linearly related? (Test the hypothesis that $\beta_1 = 0$, using $\alpha = .05$.)
e. Find a 95% confidence interval for the slope of the line, β_1.
f. Obtain a 95% confidence interval for the expected value of y when $x = -1$.
g. Given that $x = 2$, find an interval estimate for a particular value of y. Use a confidence coefficient equal to .90.

10.69 Calculate the coefficient of correlation for the data in Exercise 10.68. By what percentage was the sum of squares of deviations reduced by using the least-squares predictor $\hat{y} = \hat{\beta}_0 + \hat{\beta}_1 x$, rather than $\bar{y}$ as a predictor of y for the data?

10.70 An experiment was conducted to observe the effect of an increase in temperature on the potency of an antibiotic. Three 1-ounce portions of the antibiotic were stored for equal lengths of time at each of the following temperatures: 30°, 50°, 70°, and 90°. The potency readings observed at the temperature of the experimental period were

Potency readings, y	38, 43, 29	32, 26, 33	19, 27, 23	14, 19, 21
Temperature, x	30°	50°	70°	90°

a.   Find the least-squares line appropriate for these data.
b.   Plot the points and graph the line as a check on your calculations.
c.   Calculate s^2.
d.   Estimate the change in potency for a one-unit change in temperature. Use a 90% confidence interval.
e.   Estimate the mean potency corresponding to a temperature of 50°. Use a 90% confidence interval.
f.   Suppose that a batch of the antibiotic were stored at 50° for the same length of time as the experimental period. Predict potency of the batch at the end of the storage period. Use a 90% confidence interval.

10.71   The following data are estimates of the number of reported robberies and arrests in the New York City subway system each month during the first half of 1988.

Month	Robberies	Robbery arrests
January	475	180
February	465	155
March	470	160
April	500	190
May	550	225
June	600	220
July	602	223

Source: Data estimated from graph in the *New York Times*, October 2, 1988.

a.   Without looking at the data, do you think that the correlation between number of robberies and number of robbery arrests should be positive or negative?
b.   Calculate the coefficient of correlation and interpret its value.
c.   Calculate the coefficient of determination and interpret its value.
d.   Do the data provide sufficient evidence to indicate that a positive correlation exists between number of robberies and number of robbery arrests in the New York City subway system? Test using $\alpha = .01$.

10.72   The following table, reproduced from Exercise 2.3, shows the federal government's income through federal income taxes (on a per capita basis) as well as the amount spent per capita in federal aid for each of the 50 states in fiscal year 1986.

State	Taxes	U.S. aid	State	Taxes	U.S. aid
AL	$ 740	$ 434	CO	718	374
AK	3490	1244	CT	1202	471
AZ	975	367	DE	1343	495
AR	770	474	FL	780	278
CA	1144	419	GA	806	448

State	Taxes	U.S. aid	State	Taxes	U.S. aid
HI	1400	446	NM	$ 989	$579
ID	743	434	NY	1278	697
IL	848	434	NC	881	360
IN	810	363	ND	907	638
IA	863	434	OH	843	443
KS	777	359	OK	895	423
KY	863	479	OR	715	497
LA	807	453	PA	898	481
ME	940	573	RI	908	585
MD	1047	439	SC	863	391
MA	1314	528	SD	570	646
MI	1019	476	TN	682	443
MN	1163	501	TX	667	313
MS	731	512	UT	820	485
MO	712	391	VT	923	617
MT	755	723	VA	836	345
NE	700	413	WA	1169	427
NV	1084	434	WV	964	554
NH	472	394	WI	1148	483
NJ	1096	440	WY	1569	929

Source: Data from U.S. Commerce Department, Bureau of the Census, *World Almanac & Book of Facts*, 1989, p. 141.

a. Calculate the correlation between federal per capita taxes and per capita U.S. aid for the 50 states.

b. Is it true that the more a state pays in federal taxes, the more financial aid they will receive? Test using $\alpha = .01$.

 10.73 One of the major causes of the United States trade imbalance is the rising number of import cars sold in the United States. The following data represent the number of import cars sold in the United States (in millions) for the past twenty years.

Year	Year−1969 (x)	Number of import cars (y)	Year	Year−1969 (x)	Number of import cars (y)
1969	0	1.1	1979	10	2.3
1970	1	1.3	1980	11	2.4
1971	2	1.6	1981	12	2.3
1972	3	1.6	1982	13	2.2
1973	4	1.8	1983	14	2.4
1974	5	1.4	1984	15	2.4
1975	6	1.6	1985	16	2.8
1976	7	1.5	1986	17	3.2
1977	8	2.1	1987	18	3.1
1978	9	2.0	1988	19	3.1

Source: Reprinted with permission from *Automotive News: 1989 Market Data Book*, May 31, 1989, p. 33. Copyright Crain Communications, Inc. All rights reserved.

a. Plot the data. Fit a least-squares line to the data and graph the line.

b. Does it appear that there is a significant linear relationship between x and y? Test using $\alpha = .05$.

c. Find a 90% confidence interval for the average yearly change in number of import cars sold.

d. Predict the number of import cars that will be sold in 1989 with a 95% prediction interval. Comment on the problem of predicting a particular value of y outside of the experimental region.

10.74 A comparison of 12 student grade point averages at the end of the college freshman year with corresponding scores on an IQ test produced the following results:

G.P.A., y	2.1	2.2	3.1	2.3	3.4	2.9	2.9	2.7	2.1	1.7	3.3	3.5
IQ score, x	116	129	123	121	131	134	126	122	114	118	132	129

a. Find the least-squares prediction equation appropriate for the data.

b. Graph the points and the least-squares line as a check on your calculations.

c. Calculate s^2.

d. Do the data present sufficient evidence to indicate that x is useful in predicting y? (Test by using $\alpha = .05$.)

e. Calculate the coefficient of correlation for the data.

f. Obtain a 90% confidence interval for the expected grade point average given an IQ score equal to 120.

g. Use the least-squares equation to predict the grade point average of a particular student whose IQ score is equal to 120. Use a 90% prediction interval.

10.75 An experiment was conducted to investigate the effect of a training program on the length of time for a typical male college student to complete the 100-yard dash. Nine students were placed in the program. The reduction y in time to complete the 100-yard dash was measured for three students at the end of two weeks, for three at the end of four weeks, and for three at the end of six weeks of training. The data are shown below.

Reduction in time, y (seconds)	1.6, .8, 1.0	2.1, 1.6, 2.5	3.8, 2.7, 3.1
Length of training, x (weeks)	2	4	6

a. Find the least-squares line for these data.

b. Estimate the mean reduction in running time after four weeks of training. Use a 90% confidence interval.

c. Suppose that only three students had been employed in the experiment and that the reduction in running time was measured for each student at the end of two, four, and six weeks. Would the assumptions required for the confidence interval in part (b) be satisfied? Explain.

10.76 It is generally thought that the temperature on football fields with synthetic turf surfaces is higher than the temperature on a natural grass surface. Jerry D. Ramsey made 81 daily temperature readings of the surface temperature of the synthetic turf at the Texas Tech University football stadium and also at a nearby natural grass surface ("Environmental

Heat from Synthetic and Natural Turf," *Research Quarterly for Exercise and Sport* 53, no. 1 [1982]). Ramsey's computer printout for a simple linear regression analysis is shown below.

a. Give the least-squares prediction equation relating the dry-bulb synthetic surface temperature (DBS), the dependent variable, to the dry-bulb natural surface temperature (DBN).

b. Do Ramsey's data present sufficient evidence to indicate that the dry-bulb natural grass temperature (DBN) provides information for the prediction of the dry-bulb synthetic turf temperature (DBS)? Explain.

Prediction of dry-bulb temperature on synthetic turf — linear regression

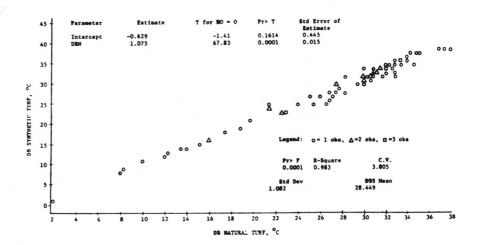

10.77 Suppose that the following data were collected on emphysema patients: the number of years the patient smoked and inhaled (x) and a physician's subjective evaluation of the extent of lung damage (y). The latter variable is measured on a scale of 0 to 100. Measurements taken on ten patients are as follows:

Patient	Years smoking, x	Lung damage, y
1	25	55
2	36	60
3	22	50
4	15	30
5	48	75
6	39	70
7	42	70
8	31	55
9	28	30
10	33	35

a. Calculate the coefficient of correlation r between years smoking (x) and lung damage (y).

b. Calculate the coefficient of determination r^2. Interpret r^2.

c. Fit a least-squares line to the data. Graph the line and plot the data points. Compare with your computed least-squares line and your computed values of r and r^2.

10.78   Some varieties of nematodes, round worms that live in the soil and frequently are so small as to be invisible to the naked eye, feed upon the roots of lawn grasses and other plants. This pest, which is particularly troublesome in warm climates, can be treated by the application of nematicides. Data collected on the percent kill of nematodes for various rates of application (dosages given in pounds per acre of active ingredient) are as follows:

Rate of application, x	2	3	4	5
Percent kill, y	50, 56, 48	63, 69, 71	86, 82, 76	94, 99, 97

   a.   Calculate the coefficient of correlation r between rates of application (x) and percent kill (y).

   b.   Calculate the coefficient of determination r^2 and interpret.

   c.   Fit a least-squares line to the data.

   d.   Suppose that you wish to estimate the mean percent kill for an application of 4 pounds of the nematicide per acre. Do the data satisfy the assumptions that are required for the confidence intervals of Section 10.6?

10.79   If you play tennis, you know that tennis racquets vary in their physical characteristics. The data shown in the accompanying table give measures of bending stiffness and twisting stiffness, as measured by engineering tests, for 12 tennis racquets (R. Gillen, ed., "Equipment Preview 1976: How to Pick the Right Racquet for Your Game," *Tennis USA* [February 1976] 87–91, 99).

Racquet	Bending stiffness x	Twisting stiffness y
Dunlop Maxply Fort	419	227
Garcia 240	407	231
Bancroft Bjorn Borg	363	200
Wilson Jack Kramer	360	211
Davis Classic	257	182
Spalding Smasher III	622	304
Yonex T-7500	424	384
Prince	359	194
Wilson T-4000	346	158
Yamaha YFG-30	556	225
Head Competition II	474	305
Adidas Adistar	441	235

   a.   If a racquet has bending stiffness, is it also likely to have twisting stiffness? Do the data provide evidence that x and y are correlated? (*Hint:* Test $H_0: \rho = 0$.) Find the p-value for the test and interpret its value.

   b.   Calculate the coefficient of determination r^2 and interpret its value.

10.80   In Exercise 9.77, we gave the diameter (mm), height (mm), and the ratio of diameter to height for ten fossil specimens collected east of the Antarctic Peninsula. The table is reproduced on page 441 (Carlos E. Macellari, "Revision of Serpulids of the Genus Rotularia (Annelida) at Seymour Island (Antarctic Peninsula) and Their Value in Stratigraphy," *Journal of Paleontology* 58, no. 4 [July 1984]).

Specimen	Diameter	Height	D/H
OSU 36651	185	78	2.37
OSU 36652	194	65	2.98
OSU 36653	173	77	2.25
OSU 36654	200	76	2.63
OSU 36655	179	72	2.49
OSU 36656	213	76	2.80
OSU 36657	134	75	1.79
OSU 36658	191	77	2.48
OSU 36659	177	69	2.57
OSU 36660	199	65	3.06
Mean:	184.5	73	2.54
s:	21.5	5	

a. Plot these data. Does the relationship between height (x) and diameter (y) appear to be linear?

b. Find the correlation between height and diameter for the ten fossil specimens.

c. Do the data provide sufficient evidence to conclude that the underlying population coefficient of correlation is different from 0? Use $\alpha = .05$.

d. Find the value of r^2 and interpret its value.

 10.81 The table below, reproduced from Exercise 2.53, gives the 1987 toxic chemical emissions for ten counties in the United States as reported by the EPA and *USA Today* (in millions of pounds).

County	EPA	USA Today
San Bernardino, CA	5200	22.5
Harris, TX	612	78.3
Calhoun, TX	581	579.2
Wayne, MI	512	17.7
Harrison, MS	423	53.0
Brazoria, TX	404	197.9
St. James, LA	349	312.5
Milam, TX	329	329.1
Jefferson, TX	309	198.0
St. Charles, LA	284	209.1

Source: Larry Sanders, USA Today database research, "EPA report spurs confusion." Copyright 1989, *USA Today*. Excerpted with permission.

a. Draw a scatterplot of the EPA versus *USA Today* emission amounts. What type of relationship appears to exist?

b. Is there significant evidence of a linear correlation between the reported amounts for the two sources? Test using $\alpha = .01$.

c. Based on the explanation of the data given in Exercise 2.53, of what practical significance is the conclusion drawn in part (b)?

d. What assumptions are necessary in order for the test conducted in part (b) be valid?

10.82   Movement of avocados into the United States from certain areas is prohibited because of the possibility of bringing fruit flies into the country with the avocado shipments. However, certain avocado varieties supposedly are resistant to fruit fly infestation before they soften as a result of ripening. The following data resulted from an experiment in which avocados ranging from 1 to 9 days after harvest were exposed to Mediterranean fruit flies. Penetrability of the avocados was measured on the day of exposure, and the percent of the avocado fruit infested was assessed.

Days after harvest	Penetrability	Percent infected
1	.91	30
2	.81	40
4	.95	45
5	1.04	57
6	1.22	60
7	1.38	75
9	1.77	100

Use the following MINITAB printout of the regression of penetrability (y) on days after harvest (x) to analyze the relationship between these two variables. Explain all pertinent parts of the printout and interpret the results of any tests.

```
MTB > REGRESS C2 1 C1

THE REGRESSION EQUATION IS
C2 = 0.616 + 0.111 C1

PREDICTOR     COEF      STDEV     T-RATIO      P
CONSTANT      0.6159    0.1088    5.66       0.002
C1            0.11085   0.01977   5.61       0.002

S = 0.1353    R-SQ = 86.3%    R-SQ (ADJ) = 83.5%

ANALYSIS OF VARIANCE

SOURCE        DF    SS        MS        F        P
REGRESSION    1     0.57581   0.57581   31.44    0.002
ERROR         5     0.09157   0.01831
TOTAL         6     0.66737
```

10.83   Refer to Exercise 10.82. Suppose that the experimenter wants to examine the relationship between the percent of infected fruit and the number of days after harvest. Does the method of linear regression discussed in this chapter provide an appropriate method of analysis? If not, what assumptions have been violated?

DOES IT PAY TO SAVE?

Does it pay to save? If you had or were to acquire some extra money, should you put it aside for a rainy day (just in case Social Security is not with us in the future)? What is the effect of government taxation on your incentive to put your savings in a bank, a savings and loan association, a money fund, real estate, or some other investment? This question is addressed by Dr. Srully Blotnick, a practicing research psychologist and columnist for *Forbes* magazine. ("Psychology and Investing," *Forbes*, May 25, 1981. Reprinted by permission. Copyright 1981, Forbes, Inc.)

The essence of Dr. Blotnick's article is that we are inclined to save and invest if we are allowed to keep some of the rewards. (It *does* make sense!) But, the greater the amount of government taxation on the return (interest, dividends, capital gains, and so on) from our savings and/or investments, the less we tend to save and invest. (Why should we?) Data suggest that if saving and investment produce low return, people spend their money on entertainment and material goods, and if they have large amounts available for investment, they attempt to place their funds in tax shelters (tax-exempt investments). Thus, according to one theory, heavy taxation decreases capital formation, the life blood needed for investment in research and the modernization of industry, for the revitalization of our economy, and ultimately, for our own individual economic welfare.

To illustrate, it is not uncommon for the combined personal income of a young college-educated husband and wife to equal $25,000 to as much as $50,000 per year. If they have two children, with normal medical, mortgage, and other expenses, and if their joint income totals $50,000 per year, they are required to pay approximately 62% of this income in taxes if they live in Sweden, 26% in the United States, 25% in Italy, 24% in Japan, 20% in Germany, and only 11% in France (prior to the 1981 change to a socialist government). Given the distribution of taxed percentages of earned income for most countries in the Western world, the United States falls near the middle.

In contrast to the taxation on *earned income*, and in comparison with other countries, the United States imposes a tax rate on *invested* income that is among the highest. Suppose that you received the $50,000 annual income as interest, dividends, or capital gains on investment; that is, income from *invested* savings. Dr. Blotnick provides a table that shows, for each of eight countries, the personal savings rate, that is, the percent of total national earned income that is saved, and the approximate percentage of investment income that must be paid in taxes. (The percentage investment income tax liability shown in the table was based on an investment income of $49,000.) If theory is correct, countries that show low taxation of invested income should show a high percentage of savings of earned income and, if the taxation rate is high, the savings rate should be low. The data are shown below.

A comparison of personal savings rates and taxation rates on investment income

Country	Personal savings rate (%)	Investment income tax liability (%)
Italy	23.1	6.4
Japan	21.5	14.4
France	17.2	7.3
West Germany	14.5	11.8
United Kingdom	12.2	32.5
Canada	10.3	30.0
Sweden	9.1	52.7
United States	6.3	33.5

Source: New York Stock Exchange with assistance of Price Waterhouse.

1. Do Dr. Blotnick's data support his theory—that is, does the personal savings rate (on a national basis) decrease as the tax on investment income increases?

2. What assumptions were required in order to answer question 1? Is it likely that any of these assumptions have been violated?

CHAPTER 11

MULTIPLE REGRESSION ANALYSIS

Case Study

If you have ever been responsible for managing a high school or college function, or a recreational facility, or have worked in business or industry, you have probably experienced one of the major problems facing business managers, namely, worker absenteeism. The case study at the end of this chapter uses multiple regression analysis in an attempt to determine variables that are related to worker absenteeism at overseas industrial plants.

General Objective

In Chapter 10 we introduced the concept of simple linear regression and correlation, the ultimate goal being to estimate the mean value of y or to predict a value of y using information contained in a single independent (predictor) variable x. In this chapter, we will extend this concept and relate the mean value of y to one or more independent variables, $x_1, x_2, \ldots, x_k$, in models that are more flexible than the straight-line model of Chapter 10. The process of finding the least-squares prediction equation, testing the adequacy of the model, and conducting tests about and estimating the values of the model parameters is called a multiple regression analysis.

Specific Topics

1 The multiple regression model (11.2)

2 Model formulation for a multiple regression analysis (11.2, Optional 11.7)

3 The least-squares prediction equation (11.3)

4 Estimation and testing for the β parameters (11.3)

▷ 11.1 THE OBJECTIVE OF A MULTIPLE REGRESSION ANALYSIS

The objective of a multiple regression analysis is to relate a response variable y to a set of predictor variables $x_1, x_2, \ldots, x_k$ using a multiple regression model. Ultimately, we want to estimate the mean value of y and/or predict a particular value of y for given values of $x_1, x_2, \ldots, x_k$.

For example, we might want to relate a company's regional sales y of a product to the amount x_1 of the company's television advertising expenditures, to the amount x_2 of newspaper advertising expenditures, and to the number x_3 of sales representatives assigned to the region. We would use data collected on y, x_1, x_2, and x_3 to obtain a mathematical prediction equation relating y to x_1, x_2, and x_3. We would use this prediction equation to predict product sales for the region for specific advertising expenditures (values of x_1 and x_2) and a specific number of sales representatives (x_3). If we are successful in developing a good prediction equation (one that makes accurate sales predictions) we may, by examining the equation, obtain a better understanding of the manner in which these controllable predictor variables x_1, x_2, and x_3 affect the product's sales.

This chapter is intended to be a brief introduction to multiple regression analysis, to help you understand what it is, and to aid you in interpreting the results of a multiple regression computer printout. We do not intend to cover all of the topics usually covered in a discussion of multiple regression analysis, nor do we intend to spend much time discussing the difficult task of formulating the regression model. For additional information on multiple regression, we recommend the references listed at the end of the chapter.

▷ 11.2 THE MULTIPLE REGRESSION MODEL AND ASSOCIATED ASSUMPTIONS

The general linear model for a multiple regression analysis will take the form shown in the following display.

THE GENERAL LINEAR MODEL AND ASSUMPTIONS

$$y = \beta_0 + \beta_1 x_1 + \beta_2 x_2 + \cdots + \beta_k x_k + \epsilon$$

where

1. $y =$ the response variable that you wish to predict.
2. $\beta_0, \beta_1, \beta_2, \ldots, \beta_k$ are constants.
3. $x_1, x_2, \ldots, x_k$ are independent variables that are measured without error.
4. ϵ is a random error that, for any given set of values $x_1, x_2, \ldots, x_k$, is normally distributed with mean 0 and variance equal to σ^2.
5. The random errors, say ϵ_i and ϵ_j, associated with any pair of y values, are independent.

Under these assumptions, it follows that the mean value of y for a given set of values for $x_1, x_2, \ldots, x_k$ is equal to

$$E(y) = \beta_0 + \beta_1 x_1 + \beta_2 x_2 + \cdots + \beta_k x_k$$

The linear model methodology that we use requires that **the β_i's occur in a linear fashion. That is, β_i must be the coefficient of a term that does not involve any unknown parameters.** For example, the term $\beta_2 x^2$ is linear in β_2 since x^2 is a known value. However, the term $\beta_0 e^{\beta_1 x}$ is *not* linear, since the coefficient of β_0, $e^{\beta_1 k}$, contains the unknown parameter β_1.

The variables $x_1, x_2, \ldots, x_k$ that appear in the general linear model need not represent *different* predictor variables. Our assumption requires only that when we observe a value of y, the values of $x_1, x_2, \ldots, x_k$ can be recorded without error. For example, if we wish to relate the mean measure $E(y)$ of worker absenteeism to two predictor variables, say,

$x_1 =$ worker age

$x_2 =$ worker hourly wage rate

we could construct any of a number of models. One possibility is

$$E(y) = \beta_0 + \beta_1 x_1 + \beta_2 x_2$$

This model graphs as a *plane*, a surface in the three-dimensional space defined by y, x_1, and x_2. Or, if we suspect that the response surface relating $E(y)$ to x_1 and x_2 has curvature, we might use the model

$$E(y) = \beta_0 + \beta_1 x_1 + \beta_2 x_2 + \beta_3 x_1 x_2$$

or

$$E(y) = \beta_0 + \beta_1 x_1 + \beta_2 x_2 + \beta_3 x_1 x_2 + \beta_4 x_1^2 + \beta_5 x_2^2$$

In addition to the **first-order terms**, those involving only x_1 or x_2, these models include **second-order terms** such as x_1^2, x_2^2, and the two-variable cross product

$x_1 x_2$.* Thus, all three of these models relate $E(y)$ to only two predictor variables, x_1 and x_2, but the number of terms and interpretations of the models differ.

The process of choosing a good model relating y to a set of independent predictor variables is the most difficult task facing the experimenter. Once the model is chosen, the actual regression analysis is usually accomplished using a packaged computer program. However, if the form of the model is not properly specified, even if the model includes all of the important predictor variables, it still may provide a poor fit to the data.

For example, suppose that $E(y)$ is *perfectly* related to a single predictor variable x_1 by the relationship

$$E(y) = \beta_0 + \beta_1 x_1 + \beta_2 x_1^2$$

If you fit the first-order (straight-line) model

$$E(y) = \beta_0 + \beta_1 x_1$$

to the data, you will obtain the **best-fitting least-squares line**, but it may still provide a poor fit to your data and may be of little value for estimation or prediction (see Figure 11.1a). In contrast, if you fit the second-order model

$$E(y) = \beta_0 + \beta_1 x_1 + \beta_2 x_1^2$$

you obtain a perfect fit to the data (see Figure 11.1b).

Figure 11.1
Two models, each employing one predictor variable

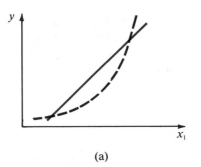
(a)

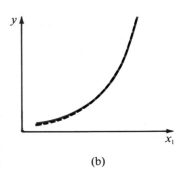
(b)

The lesson is quite clear. Including all of the important predictor variables in the model as first-order terms, that is, $x_1, x_2, \ldots, x_k$, may not (and, most probably, will not) produce a model that provides a good fit to your data. You may have to include second-order terms, such as x_1^2, x_2^2, x_3^2, $x_1 x_2$, and $x_1 x_3$.

This discussion of linear models is purposely brief. First, we want you to understand the importance of model selection as a step that precedes a multiple regression analysis. Secondly, we want to help you understand the logic employed in selecting the model for the multiple regression analysis that follows in Section 11.3.

---

* The order of a term is determined by the sum of the exponents of variables making up that term. Terms involving x_1 or x_2 are first-order. Terms involving x_1^2, x_2^2, or $x_1 x_2$ are second-order.

Although we offer additional comments on model formulation in (Optional) Section 11.7, we have no intention of covering this complex topic in a short introductory chapter on regression analysis. Our aim is to help you understand the output of multiple regression analysis and to understand some of the problems that can be solved using this statistical methodology. For a more thorough discussion of model building, we refer you to Mendenhall and Sincich (1986) listed in the reference section at the end of the chapter.

EXERCISES   Basic Techniques

11.1   Graph the following equations.

    a. $E(y) = 3 + 2x$ b. $E(y) = -1 + x$ c. $E(y) = 4 - \dfrac{x}{2}$

11.2   Graph the following equations (they graph as parabolas).

    a. $E(y) = 2x^2$ b. $E(y) = 1 + 2x^2$

    c. $E(y) = -1 + 2x^2$ d. $E(y) = -2x^2$

    e.   How does the sign of the coefficient of x^2 affect the graph of a parabola?

11.3   Graph the following equation.

    a. $E(y) = 2x^2 - 4x + 2$

    b.   Compare the graph in part (a) with the graph of Exercise 11.2(a). What effect does the inclusion of the first-order term $(-4x)$ have on the graph?

    c.   Suppose that $E(y) = 2x^2 + 4x + 2$. Can you deduce the effect on the parabola of replacing $(-4x)$ by $(+4x)$?

11.4   Suppose that $E(y)$ is related to two predictor variables x_1 and x_2 by the equation

$$E(y) = 3 + x_1 - 2x_2$$

    a.   Graph the relationship between $E(y)$ and x_1 when $x_2 = 2$. Repeat for $x_2 = 1$ and for $x_2 = 0$.

    b.   What relationship do the lines in part (a) have to each other?

11.5   Refer to Exercise 11.4.

    a.   Graph the relationship $E(y)$ and x_2 when $x_1 = 0$. Repeat for $x_1 = 1$ and for $x_1 = 2$.

    b.   What relationship do the lines in part (a) have to each other?

    c.   Suppose that in a practical situation you wanted to model the relationship between $E(y)$ and two predictor variables x_1 and x_2. What would be the implication of using the first-order model $E(y) = \beta_0 + \beta_1 x_1 + \beta_2 x_2$?

11.6   Suppose that $E(y)$ is related to two predictor variables x_1 and x_2 by the equation

$$E(y) = 3 + x_1 - 2x_2 + x_1 x_2$$

    a.   Graph the relationship between $E(y)$ and x_1 when $x_2 = 0$. Repeat for $x_2 = 2$ and for $x_2 = -2$.

    b.   Note that the equation for $E(y)$ is exactly the same as the equation in Exercise 11.4, except that we have added the term $x_1 x_2$. How does the addition of the $x_1 x_2$ term affect the graphs of the three lines?

    c.   What flexibility is added to the first-order model $E(y) = \beta_0 + \beta_1 x_1 + \beta_2 x_2$ by the addition of the term $\beta_3 x_1 x_2$, using the model $E(y) = \beta_0 + \beta_1 x_1 + \beta_2 x_2 + \beta_3 x_1 x_2$?

▷ 11.3   A MULTIPLE REGRESSION ANALYSIS

A multiple regression analysis is performed in much the same manner as a simple linear regression analysis; that is, a multiple regression model, say, $E(y) = \beta_0 + \beta_1 x_1 + \beta_2 x_2 + \cdots + \beta_k x_k$, is fitted to a set of data using the method of least squares, a procedure that finds the prediction equation

$$\hat{y} = \hat{\beta}_0 + \hat{\beta}_1 x_1 + \hat{\beta}_2 x_2 + \cdots + \hat{\beta}_k x_k$$

which minimizes SSE, the sum of squares of deviations of the observed values of y from their predicted values. The major difference between a simple and a multiple regression analysis is that the multiple regression model contains more parameters the computation required for a multiple regression analysis is so complicated that it is usually performed by computer. Almost every computing facility has at least one packaged regression analysis program, such as MINITAB or SAS, that requires only that the user execute the proper commands to activate the program and then submit the problem data. In addition, many statistical packages are available for personal computers (PCs).

Remember, however, that different computing systems using the same program may provide only similar printouts because of differences in the use of computer configurations and printer options. Therefore, for some cases in which we have used a particular computer program in the solution to an example in this text, the results you obtain using the same program may be slightly different in format. In this section, we will present two sets of data, formulate models for each, and present printouts with the results of the appropriate multiple regression analyses. These printouts differ slightly in form and content, but contain the same basic information.

EXAMPLE I I.I   Cucumbers are usually preserved by fermenting them in a low-salt brine (6 to 9% sodium chloride) and then storing them in a high-salt brine until they are used by processors to produce various types of pickles. The high-salt brine is needed to retard softening of the pickles and to prevent freezing when stored outside in northern climates. Data showing the reduction in firmness of pickles stored over time in a high-salt brine are shown in Table 11.1 (R. W. Buescher, J. M. Hudson, J. R. Adams, and D. H. Wallace, "Calcium Makes It Possible to Store Cucumber Pickles in Low-Salt Brine," *Arkansas Farm Research* 30, no. 4 [July/August, 1981]). Choose a linear model to relate y to x.

Table I I.I
Pickle firmness versus storage time

	Weeks in storage at 72°F				
	0	4	14	32	52
Firmness in pounds, y	19.8	16.5	12.8	8.1	7.5

Solution   The first step in choosing a model to describe the data is to plot the data points (see Figure 11.2a). You can see that the firmness y drops rapidly during the first 14 weeks

Figure 11.2
Fitting a second-order
model to firmness–
storage time data

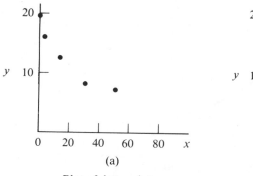

(a)

Plot of data points

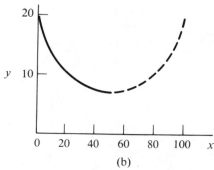

(b)

Graph of a second-order model

of storage, but the decrease becomes more gradual as time of storage increases. Although a number of different models might be used to describe this curvilinear relationship, we will present two.

The first and simplest choice would be the quadratic or second-order model

$$E(y) = \beta_0 + \beta_1 x + \beta_2 x^2$$

This curve graphs as a parabola opening upward, as shown in Figure 11.2(b). The least-squares procedure chooses the parabola that best fits the data by minimizing

$$\text{SSE} = \sum [y_i - (\hat{\beta}_0 + \hat{\beta}_1 x_i + \hat{\beta}_2 x_i^2)]^2$$

The portion of the parabola that passes through our data points is indicated by the solid segment of the curve in Figure 11.2(b), while the portion of the parabola that falls outside of the range of values of x in the data set (values of x larger than 52) is indicated by the dotted curve segment. Since we only predict values of y for values of x within the data set, we would expect the solid portion of the curve in Figure 11.2(b) to provide a satisfactory fit to the data points in Figure 11.2(a).

A second model that might provide a good fit to the data points is the exponential model

$$y = ae^{-bx}$$

where y = measure of firmness, x = storage time, and a and b are parameters of the model. This model graphs as a curve that shows the firmness dropping rapidly over the first few hours of storage time with the decrease in firmness per hour becoming less and less as the storage time increases (see Figure 11.3a). Although not a linear model, we can make it linear by taking the natural logarithm of both sides of the equation

$$\ln y = \ln a - bx$$

Since $\ln y$ (the natural logarithm of the firmness measurement) and x are variables and $\ln a$ and $-b$ are constants, the equation

$$\ln y = \ln a - bx$$

Figure 11.3
Graphs of the
exponential model and
transformed linear
model

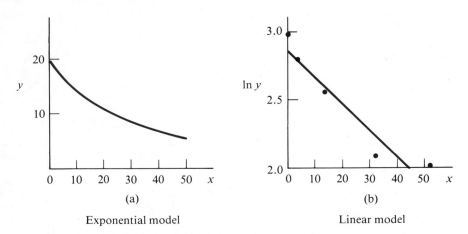

(a)

Exponential model

(b)

Linear model

graphs as a straight line. In our notation,

$y^* = \ln y$ = natural logarithm of the firmness measurement

x = storage time

$\beta_0 = \ln a$

$\beta_1 = -b$

and our linear model is the familiar first-order model of Chapter 10,

$$E(y^*) = \beta_0 + \beta_1 x$$

Figure 11.3(b) shows a plot of the points and a graph of $y^* = \ln y$ versus $x =$ storage time. You can see that a straight line fits the transformed[†] data set reasonably well. ◁

EXAMPLE 11.2   Refer to the pickle-firmness data in Example 11.1. A quadratic model was fitted to the data using the SAS REG multiple regression program, and the resulting output is shown in Table 11.2. Find the least-squares prediction equation for the data and explain the computer output.

Solution   **a.**   The SAS REG output is shown in Table 11.2. The output is similar to the SAS output for simple linear regression shown in Table 10.4, but it contains more entries.

   1.   **Multiple Coefficient of Determination, R^2:**
   In Section 10.9 we showed that the simple linear correlation coefficient r^2 is equal to

$$r^2 = \frac{\text{SSR}}{S_{yy}}$$

[†] Transforming a data set means changing the data according to some rule. In this case, we transformed the data by taking the natural logarithm of each value of y.

Table 11.2   The SAS REG multiple regression analysis printout for Example 11.2

SAS*

DEP VARIABLE: Y   ANALYSIS OF VARIANCE

SOURCE	DF	SUM OF SQUARES	MEAN SQUARE	F VALUE	PROB > F ②
MODEL	2	112.05359	56.02679556	155.975	0.0064
ERROR	2	0.71840887	0.35920444 ⑤		
C TOTAL	4	112.77200			

④ ROOT MSE	0.5993367	R-SQUARE	0.9936 ①
DEP MEAN	12.94	ADJ R-SQ	0.9873
C.V.	4.631659		

PARAMETER ESTIMATES

VARIABLE	DF	③ PARAMETER ESTIMATE	⑥ STANDARD ERROR	⑦ T FOR HO: PARAMETER = 0	⑧ PROB > ITI
INTERCEP	1	19.29681683	0.47114379	40.957	0.0006
X	1	−0.55933158	0.05352823	−10.449	0.0090
XSQ	1	0.006413131	0.001003109	6.393	0.0236

OBS	ID	ACTUAL	PREDICT VALUE	STD ERR PREDICT	LOWER 95% MEAN	UPPER 95% MEAN	RESIDUAL ⑨
1	0	19.8000	19.2968	0.4711	17.2696	21.3240	0.5032
2	4	16.5000	17.1621	0.3594	15.6156	18.7086	−0.6621
3	14	12.8000	12.7231	0.3859	11.0627	14.3836	0.0769
4	32	8.1000	7.9653	0.4791	5.9039	10.0266	0.1347
5	52	7.5000	7.5527	0.5899	5.0145	10.0909	−0.0527

SUM OF RESIDUALS	2.28706E-14
SUM OF SQUARED RESIDUALS	0.7184089
PREDICTED RESID SS (PRESS)	5.808661

* Statistical Analysis System

or

$$r^2 = 1 - \frac{SSE}{S_{yy}}$$

where SSR, SSE, and S_{yy} are found in the **ANALYSIS OF VARIANCE** portion of the printout. In a multiple regression analysis, the total variation is partitioned in exactly the same way, and the quantity

$$R^2 = \frac{SSR}{S_{yy}}$$

called the **multiple coefficient of determination**, gives the proportion of the total variation that is explained by the predictor variables $x_1, x_2, \ldots, x_k$. The remainder of the variation is explained by variables omitted from the model, incorrect model formulation, and experimental error. The multiple coefficient of determination R^2

takes values between 0 and 1. A small value of R^2 means that the model is not providing much information for prediction, while a value near 1 means that the model is providing almost all the information necessary for prediction. Thus, just as r^2 provides a measure of the fit of the simple linear model, R^2 provides a measure of the fit of a more complex regression model.

For this example, the value for R^2 is 0.9936 (box #1). Therefore, 99.4% of the total sum of squares, S_{yy}, is explained by the model.

One must be careful when interpreting the coefficient of determination, R^2. In a regression analysis, R^2 can easily be inflated by simply adding more and more predictor variables to the model. In general, the value of R^2 for a model containing $(k + 1)$ parameters will be larger than one containing k parameters, regardless of whether the new variable is a good predictor of y. **Thus, simply trying to find a regression model where R^2 is as close to 1 as possible is not in general a good practice.**

2. **Testing the Utility of the Model:**
As noted in Section 10.9, the **analysis of variance** portion of the printout presents a partitioning of the total variation, $S_{yy} = \sum(y_i - \bar{y})^2$, into a portion due to error (SSE) and a portion due to regression (SSR). The column labeled SOURCE shows that the TOTAL variation in y, given by S_{yy}, can be decomposed into unexplained ERROR variation and the explained variation due to the MODEL. The DF column gives the degrees of freedom with each source of variation, while the SS column gives the calculated sum of squares for each source of variation. In these two columns, the entries for MODEL and ERROR add to the entry in the TOTAL line. The entries in the MEAN SQUARE column are found by dividing each sum of squares entry by its degrees of freedom. No mean square is calculated for the TOTAL line.

If the model contributes information for the prediction of y, at least one of the model parameters β_1 or β_2 will differ from 0. In testing the hypothesis $H_0 : \beta_1 = \beta_2 = 0$ against the alternative hypothesis H_a: at least one of β_1 or β_2 differs from 0, we use the test statistic

$$F = \frac{\text{MSR}}{\text{MSE}} = \frac{R^2/k}{(1 - R^2)/[n - (k + 1)]}$$

where MSR is the mean square due to regression and MSE is found in the ERROR line of the MEAN SQUARE column. MSR has degrees of freedom $v_1 = k$ equal to the number of parameters in the model *excluding* the intercept, while MSE has degrees of freedom $v_2 = n - (k + 1)$, equal to the number of observations minus the number of parameters in the model *including* the intercept. Since SSR is the variation explained by the model, and SSE is the variation unexplained by the model, a large value of MSR when compared to MSE is cause to reject H_0. Therefore, we reject H_0 for large values of F.

In this example, $F = 155.975$ (box #2) with $v_1 = 2$ and $v_2 = 5 - 3 = 2$ degrees of freedom. Since this value exceeds the tabulated value of $F = 19.00$, with $v_1 = 2$, $v_2 = 2$, and $\alpha = .05$, from Table 7 in Appendix III, we reject H_0 and conclude that at least one of the parameters β_1 or β_2 differs from 0. There is evidence to indicate that the model contributes information for the prediction of y. The observed significance level (*p*-value) for the test, **PROB > F,** shown to the right of the value of the *F* statistic, is 0.0064.

3. **The least-squares prediction equation:**
   The least-squares estimates of the three parameters β_0, β_1, and β_2 in the model

   $$E(y) = \beta_0 + \beta_1 x + \beta_2 x^2$$

   are shown under the heading PARAMETER ESTIMATE (box #3). The column headed VARIABLE displays the names that the programmer assigned to the variables. Thus,

   $$\hat{\beta}_0 = 19.29681683$$

   $$\hat{\beta}_1 = -0.55933158$$

   $$\hat{\beta}_2 = 0.006413131$$

   Rounding these estimates, the prediction equation is

   $$\hat{y} = 19.297 - 0.559x + 0.00641x^2$$

4. **The estimate of σ^2:**
   The computer output gives *s*, the estimate of σ, under the heading ROOT MSE in box #4. The quantity s^2 is defined as

   $$s^2 = \frac{\text{SSE}}{[n - (k + 1)]}$$

   where $k + 1$ is the number of parameters in the multiple regression model. Since SSE and the corresponding degrees of freedom (DF) are given in the ANALYSIS OF VARIANCE portion of the printout, we can find s^2 in the ERROR row and MEAN SQUARE column (box #5) as

   $$s^2 = \frac{\text{SSE}}{\text{DF}} = \text{MSE} = .35920444$$

5. **Estimation of the β_i:**
   The procedure for estimating the regression parameters is identical to the procedure used in Chapter 10 for the simple linear model, except that the computing formulas are much more complicated. However, since the computer output provides the estimated standard deviations of the regression parameters under the heading STANDARD ERROR (box #6), a **100(1 − α)% confidence interval for β_i** is easily calculated as

   $$\hat{\beta}_i \pm t_{\alpha/2}(\text{estimated } \sigma_{\hat{\beta}_i})$$

   where $t_{\alpha/2}$ is based on $n - (k + 1)$, the degrees of freedom for error.
   For example, a 95% confidence interval for β_1 is

   $$-0.559 \pm 4.303(0.0535)$$

   or

   $$-0.559 \pm .230$$

where $t_{.025}$ is based on $v = n - $ (number of parameters in the model) $= n - (k + 1) = 5 - 3 = 2$ degrees of freedom.

6. **Tests of Significance for the β_i:**

A test of an hypothesis that a particular parameter, say β_i, equals 0 can be conducted by using a t statistic:

$$t = \frac{\hat{\beta}_i}{(\text{estimated } \sigma_{\hat{\beta}_i})}$$

with $v = n - (k + 1)$, the degrees of freedom for error. The procedure is similar to the procedure used in Chapter 10, except that the estimated standard deviation for $\hat{\beta}_i$ is more complex. The computed value of the t statistic for each of the model parameters is shown in box #7 under the heading T FOR HO: PARAMETER = 0, and the observed significance levels or **p-values for a two-tailed test** are shown in box #8 under the column heading PROB > |T|.

For example, the value of the t statistic for the test of $H_0: \beta_2 = 0$ is $t = 6.393$. The number of degrees of freedom for testing is $n - (k + 1) = 5 - 3 = 2$ and, with $\alpha = .05$, the two-tailed rejection region is $|t| > 4.303$. Since the computed value of t falls in the rejection region, there is evidence to indicate that β_2 differs from 0. The observed significance level, p-value $= .0236$, would lead to the same conclusion.

The predicted value of y for each of the five observed values of x is shown in box #9 under PREDICT VALUE. The corresponding values of x are shown under the column headed ID and the observed values of y are shown in the column headed ACTUAL. The 95% lower and upper confidence limits on $E(y)$ for these five values of x are given in the columns headed LOWER 95% MEAN and UPPER 95% MEAN. The RESIDUAL, shown in the last column, gives the deviation between the observed and predicted values. For example, when $x = 0$, the observed value of y is 19.8000, the predicted value is 19.2968, and the residual is $(19.8000 - 19.2968) = 0.5032$. The 95% confidence interval for the mean when $x = 0$ is 17.2696 to 21.3240.

b. A graph of the second-order prediction curve is shown in Figure 11.4 along with the plotted data points. You can see that the prediction curve provides a good fit to the data.

Figure 11.4
A graph of the least-squares prediction equation for a second-order model

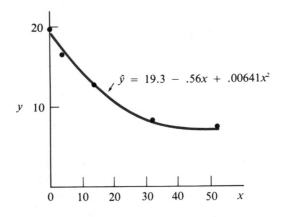

EXAMPLE 11.3   Fit the exponential model in Example 11.1 to the pickle firmness–storage time data using the SAS REG multiple regression computer package. Interpret the output.

Solution   As explained in Example 11.1, the exponential model

$$y = ae^{-bx}$$

is not linear, but we can fit the first-order model

$$E(y^*) = \beta_0 + \beta_1 x$$

to the natural logarithms of the firmness measurements, that is,

$$y^* = \ln y$$

$$x = \text{storage time}$$

$$\beta_0 = \ln a$$

$$\beta_1 = -b$$

The transformed data are shown in Table 11.3.

Table 11.3
$y^* = \ln y$ versus storage time x

Storage time x	0	4	14	32	52
$\ln y$	2.98568	2.80336	2.54945	2.09186	2.01490

Table 11.4 shows the SAS REG computer printout of the simple linear regression analysis for fitting a least-squares line to the natural logarithms of the firmness readings given in Table 11.3. You can see from the printout that

$$\hat{\beta}_0 = 2.87597185 \quad \text{and} \quad \hat{\beta}_1 = -0.01896671$$

Therefore, the least-squares prediction equation is

$$\hat{y}^* = 2.87597 - 0.01897x.$$

A plot of the data points and a graph of the least-squares line are shown in Figure 11.5.

Examining the computer printout, you can see that

$$\text{SSE} = 0.06283237 \text{ with d.f.} = 3$$

$$\text{MSE} = 0.02094412$$

and

$$s = \text{ROOT MSE} = 0.1447209$$

The model contributes information for the prediction of ($F = 31.934$ with p-value $= 0.0110$) but the value of R^2, $R^2 = 0.9141$, is not nearly as large as the value ($R^2 = 0.9936$) for the second-order model. Although both models contribute information for the prediction of y, the y are fitted on two different scales, and are therefore not directly comparable.

Table 11.4   The SAS REG computer printout for fitting the exponential model in Example 11.1

STATISTICAL ANALYSIS SYSTEM
ANALYSIS OF VARIANCE

DEP VARIABLE: LOGY

SOURCE	DF	SUM OF SQUARES	MEAN SQUARE	F VALUE	PROB > F ②
MODEL	1	0.66882153	0.66882153	31.934	0.0110
ERROR	3	0.06283237	0.02094412 ⑤		
C TOTAL	4	0.73165390			

④ ROOT MSE	0.1447209	R-SQUARE	0.9141 ①	
DEP MEAN	2.489051	ADJ R-SQ	0.8855	
C.V.	5.814299			

PARAMETER ESTIMATES

VARIABLE	DF	③ PARAMETER ESTIMATE	⑥ STANDARD ERROR	⑦ T FOR H0: PARAMETER = 0	⑧ PROB > ITI
INTERCEP	1	2.87597185	0.09421741	30.525	0.0001
X	1	−0.01896671	0.003356356	−5.651	0.0110

OBS	ID	ACTUAL	PREDICT VALUE	STD ERR PREDICT	LOWER 95% MEAN	UPPER 95% MEAN	RESIDUAL ⑨
1	0	2.9857	2.8760	0.0942	2.5761	3.1758	0.1097
2	4	2.8034	2.8001	0.0850	2.5297	3.0705	.0032554
3	14	2.5494	2.6104	0.0682	2.3934	2.8275	−0.0610
4	32	2.0919	2.2690	0.0755	2.0287	2.5094	−0.1772
5	52	2.0149	1.8897	0.1242	1.4943	2.2851	0.1252

SUM OF RESIDUALS	1.77636E-15
SUM OF SQUARED RESIDUALS	0.06283237
PREDICTED RESID SS (PRESS)	0.3284966

Figure 11.5
A graph of the least-squares prediction equation for the linear model $\ln y = \ln a - bx$

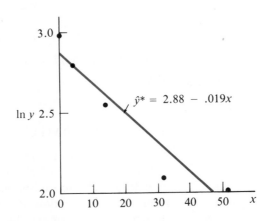

$\hat{y}^* = 2.88 - .019x$

The independent predictor variables that contribute information for the prediction of y can be of two types, **quantitative** and **qualitative**. As you will see, the way in which an independent variable is entered into a regression model depends on its type.

Definition

> A **quantitative variable** is a variable that can take values corresponding to the points on the real line. Variables that are not quantitative are said to be **qualitative variables**.

Interest rates, agricultural yield in tons per acre, number of customers, and number of incorrect bookkeeping entries are four examples of quantitative variables. In contrast, the sex of a respondent is a qualitative variable that often accounts for significantly different response patterns. Geographic location is another qualitative variable that explains differences in responses among local, state, or national regions.

We have investigated multiple regression models in which the independent or predictor variables were quantitative variables. However, qualitative variables may contain as much or possibly more information concerning a response than other associated quantitative variables. For example, the amount of money in a savings account per household would vary by family income, number in the household, occupation, state, or section of the country among a possible list of predictor variables. Notice that some of these variables are quantitative, and others are qualitative.

Qualitative independent variables are entered into a model by using **dummy variables**, the number of dummy variables always being one less than the number of levels (categories) associated with the independent variable.

Definition

> A **dummy variable** is an independent variable used to include or exclude the effect of a qualitative factor.

The effect of dummy variables is to produce several regression models simultaneously. For example, to include the effect of three different locations in a multiple regression model, two dummy variables can be defined as

$$x_1 = \begin{cases} 1 & \text{if Location 2} \\ 0 & \text{otherwise} \end{cases}$$

$$x_2 = \begin{cases} 1 & \text{if Location 3} \\ 0 & \text{otherwise} \end{cases}$$

The first few terms of this multiple regression model are given as

$$E(y) = \beta_0 + \beta_1 x_1 + \beta_2 x_2 \\ + \{\text{terms associated with other predictor variables}\} + \epsilon$$

Substituting values of x_1 and x_2 into the multiple regression equation produces models for the three different locations:

Location 1: $x_1 = x_2 = 0$

$$E(y) = \beta_0 + \{\text{terms associated with other predictor variables}\} + \epsilon$$

Location 2: $x_1 = 1, x_2 = 0$

$$E(y) = \beta_0 + \beta_1 + \{\text{terms associated with other predictor variables}\} + \epsilon$$

Location 3: $x_1 = 0, x_2 = 1$

$$E(y) = \beta_0 + \beta_2 + \{\text{terms associated with other predictor variables}\} + \epsilon$$

EXAMPLE 11.4   A study was conducted to examine the relationship between university salary y, the number of years of experience of the faculty member, and the sex of the faculty member. If we expect a straight-line relationship between mean salary and years of experience for both males and females, write the model relating mean salary to the two predictor variables

1. Years of experience (quantitative)
2. Sex of the professor (qualitative)

Solution   Since we might suspect the mean salary lines for females and males to be different, we want to construct a model for mean salary $E(y)$ that will account for this difference.

A straight-line relationship between $E(y)$ and years of experience x_1 would imply the model

$$E(y) = \beta_0 + \beta_1 x_1$$

which graphs as a straight line. The qualitative variable, sex, can only assume two "values," male and female. Therefore, we can enter the predictor variable "sex" into the model using a dummy (or indicator) variable x_2 as

$$E(y) = \beta_0 + \beta_1 x_1 + \beta_2 x_2$$

$$x_2 = \begin{cases} 1 & \text{if male} \\ 0 & \text{if female} \end{cases}$$

If the faculty member is male, then $x_2 = 1$ and

$$E(y) = (\beta_0 + \beta_2) + \beta_1 x_1$$

If the faculty member is female, $x_2 = 0$ and

$$E(y) = \beta_0 + \beta_1 x_1$$

The two lines have the same slopes, but different intercepts, and graph as two parallel lines.

The fact that we wish to allow the slopes of the two lines to differ means that we think that the two predictor variables **interact**; that is, the change in $E(y)$ corresponding to a change in x_1 depends on whether the professor is male or female. To allow for this **interaction** (difference in slopes), we introduce the interaction term $x_1 x_2$ into the model. The complete model that would characterize the graph

Figure 11.6
Hypothetical relationship between mean salary $E(y)$, years of experience (x_1), and sex (x_2)

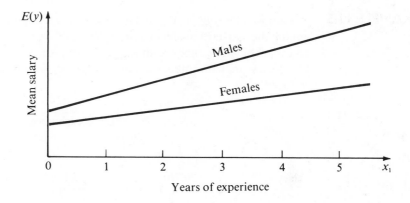

in Figure 11.6 is

$$E(y) = \beta_0 + \beta_1 x_1 + \underset{\substack{\uparrow \\ \text{dummy} \\ \text{variable} \\ \text{for sex}}}{\beta_2 x_2} + \underset{\substack{\uparrow \\ \text{interaction}}}{\beta_3 x_1 x_2},$$

where the $\beta_1 x_1$ term is labeled *years of experience*.

where

$$x_1 = \text{years of experience}$$

$$x_2 = \begin{cases} 1 & \text{if a male} \\ 0 & \text{if a female} \end{cases}$$

The interpretation of the model parameters can be seen by assigning values to the **dummy variable** x_2. If the faculty member is female, $x_2 = 0$ and

$$E(y) = \beta_0 + \beta_1 x_1 + \beta_2(0) + \beta_3 x_1(0)$$

$$= \beta_0 + \beta_1 x_1$$

Similarly, if the faculty member is male, $x_2 = 1$ and

$$E(y) = \beta_0 + \beta_1 x_1 + \beta_2(1) + \beta_3 x_1(1)$$

$$= \underbrace{(\beta_0 + \beta_2)}_{y \text{ intercept}} + \underbrace{(\beta_1 + \beta_3)}_{\text{slope}} x_1$$

The y-intercept of the female line is β_0, while the y-intercept of the male line is $(\beta_0 + \beta_2)$. Hence, β_2 is the difference in the y-intercepts for the two lines. Similarly, since the slope of the male line is $(\beta_1 + \beta_3)$, and the slope of the female line is β_1, it follows that $(\beta_1 + \beta_3) - \beta_1 = \beta_3$ must be the difference in the slopes of the two lines.

EXAMPLE 11.5   Random samples of six female and six male assistant professors were selected from among the assistant professors in a College of Arts and Sciences. The data on salary and years of experience are shown in Table 11.5. Note that both samples contained two professors with three years of experience, but no male professor had two years of experience.

Table 11.5
Salary versus years of experience

| | Years of experience, x_1 | | | | |
	1	2	3	4	5
Salary y, Males	20,710		23,160 23,210	24,140	25,760 25,590
Salary y, Females	19,510	20,440	21,340 21,760	22,750	23,200

a. Explain the output of the MINITAB multiple regression computer printout.
b. Graph the predicted salary lines.

Solution   a.   The MINITAB multiple regression computer printout for the data in Table 11.5 is shown in Table 11.6. The explanation of this printout is similar to the explanation of the SAS REG computer printout given in Example 11.2. The relevant portions of the printout are shaded and numbered.

Table 11.6
MINITAB multiple regression analysis computer printout for Example 11.5

```
THE REGRESSION EQUATION IS
Y = 18593 + 969 X1 + 867 X2 + 260 X1*X2

PREDICTOR      COEF ③    STDEV ⑥    T-RATIO ④      P ⑤
CONSTANT     18593.0      207.9        89.41      0.000
X1             969.00      63.67        15.22      0.000
X2             866.7      305.3         2.84      0.022
X1*X2          260.13      87.06         2.99      0.017

S = 201.3 ⑧      R-SQ = 99.2% ①      R-SQ (ADJ) = 98.9%

ANALYSIS OF VARIANCE

SOURCE         DF          SS          MS        ② F        P
REGRESSION      3     42108776     14036259     346.24     0.000
ERROR           8       324315       40539 ⑦
TOTAL          11     42433092 ⑨

SOURCE         DF        SEQ SS
X1              1      33294036
X2              1       8452797
X1*X2           1        361944
```

1. The value of the multiple coefficient of determination,

$$R^2 = \frac{\text{SSR}}{S_{yy}} = 1 - \frac{\text{SSE}}{S_{yy}}$$

provides a measure of how well the model fits the data. From box #1, 99.2% of the sum of squares of deviations of the y values about $\bar{y}$ is explained by all of the terms in the model.

2.   If the model contributes information for the prediction of y, at least one of the model parameters, β_1, β_2, or β_3, will differ from 0. Consequently, we wish to test the null hypothesis $H_0: \beta_1 = \beta_2 = \beta_3 = 0$ against the alternative hypothesis H_a: at least one of the parameters β_1, β_2, or β_3 differs from 0. The test statistic is $F = 346.24$, shown in box #2. Since this value exceeds the tabulated value for F, with $v_1 = 3$, $v_2 = 8$, and $\alpha = .05$, $F = 4.07$ (Table 7 in Appendix III), we reject H_0 and conclude that at least one of the parameters β_1, β_2, or β_3 differs from 0. There is evidence to indicate that the model contributes information for the prediction of y. The observed significance level (p-value) for the test, **P**, shown to the right of the value of the F statistic, is less than .001.

3.   The estimates of the four model parameters are shown under the heading COEF. Thus, $\hat{\beta}_0 = 18593.0$, $\hat{\beta}_1 = 969.00$, $\hat{\beta}_2 = 866.7$, and $\hat{\beta}_3 = 260.13$ and, rounding these estimates, we obtain the prediction equation $\hat{y} = 18593 + 969x_1 + 867x_2 + 260x_1x_2$.

4.   Tests of hypotheses concerning each of the individual model parameters can be conducted using Student's t tests. The test statistic is

$$t = \frac{\text{parameter estimate}}{\text{estimated standard deviation of the estimator}}$$

The computed value of the t statistic for each of the model parameters is shown in box #4 under the column T-RATIO. For example, the value of the t statistic for the test $H_0: \beta_1 = 0$ is 15.22. The number of degrees of freedom for SSE, s^2, and therefore, t, will always be based on $v = n -$ (number of parameters in the model) degrees of freedom. For our example, this will be $v = 12 - 4 = 8$ degrees of freedom. For $\alpha = .05$, the tabulated value of t, $t_{.025}$, for a two-tailed test is given in Table 4 in Appendix III as 2.306. Since the computed t value corresponding to β_1 exceeds this value, there is evidence to indicate that β_1 differs from 0.

5.   The observed significance levels (p-values) for the t tests described in (4) are shown in box #5 under the column headed P. These values are calculated for two-tailed tests. **If you wish to conduct a one-tailed test, the p-value is half of the value shown in the printout.** For example, the observed significance level for a test of the hypothesis $H_0: \beta_2 = 0$, $H_a: \beta_2 \neq 0$, is .0218 for this two-tailed test. If the alternative hypothesis is $H_a: \beta_2 > 0$, that is, you wish to conduct a one-tailed test of H_0, then the p-value is equal to $(.0218)/2 = .0109$. If you were to test H_0 using $\alpha = .05$, these small p-values would imply rejection of H_0.

6.   The estimated standard deviations of the estimators are shown in box #6 under the heading STDEV. For example,

$$s_{\hat{\beta}_1} = 63.67$$

This quantity is useful in calculating $100(1 - \alpha)\%$ confidence intervals for the model parameters. For example, a $100(1 - \alpha)\%$ confidence interval for β_1 is equal to

$$\hat{\beta}_1 \pm t_{\alpha/2}(\text{estimated } \sigma_{\hat{\beta}_1})$$

Therefore, a 95% confidence interval for β_1 is

$$969.0 \pm (2.306)(63.67)$$

or

$$969.0 \pm 146.82$$

where $t_{\alpha/2}$ is based on $v = n - $ (number of parameters in the model) $= 12 - 4 = 8$ degrees of freedom. Confidence intervals for the other model parameters are constructed in the same way.

7. The values of SSE $= 324315$ and $s^2 = 40539$ are shown in box #7 in the row corresponding to ERROR and under the columns headed, respectively, SUM OF SQUARES and MEAN SQUARE.

8. The standard deviation $s = 201.3$ is shown in box #8, labeled S.

9. The total sum of squares of deviations S_{yy} of the y values about the mean is shown in box #9 to be 42433092. You can verify that the tabulated value of R^2 is equal to $1 - \text{SSE}/S_{yy}$.

b. A graph of the two salary lines is shown in Figure 11.7. Note that the salary line corresponding to the male faculty members appears to be rising at a more rapid rate than the salary line for females. Is this real or could it be due to chance? The following example will answer this question.

Figure 11.7
A graph of the faculty
salary prediction lines

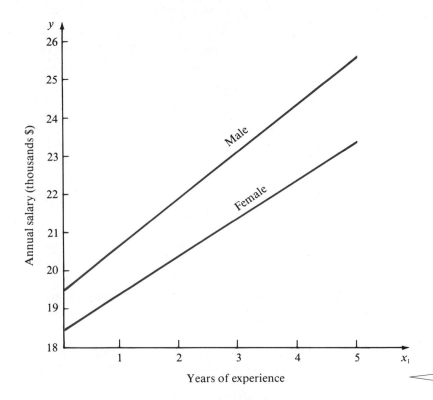

EXAMPLE 11.6 Refer to Example 11.5. Do the data provide sufficient evidence to indicate that the annual rate of increase in male junior faculty salaries exceeds the annual rate of increase for women faculty salaries? That is, do the data provide sufficient evidence to indicate that the slope of the men's faculty salary line exceeds the slope of the women's faculty salary line?

Solution Since β_3 measures the difference in slopes, the slopes of the two lines will be identical if $\beta_3 = 0$. Therefore, we wish to test

$$H_0: \beta_3 = 0$$

against the alternative hypothesis

$$H_a: \beta_3 > 0$$

which implies that the slope of the male faculty salary line is greater than the slope of the female faculty salary line.

The calculated value of t corresponding to β_3, $t = 2.99$, is shown in box #4 of the computer printout. Since we wish to detect values of $\beta_3 > 0$, we will conduct a one-tailed test and reject H_0 if $t > t_\alpha$. The tabulated t value for $\alpha = .05$ and $v = 8$ degrees of freedom is $t_{.05} = 1.860$. The calculated value of t exceeds this value, and thus there is evidence to indicate that the annual rate of increase in men's faculty salaries exceeds the corresponding annual rate of increase in faculty salaries for women.*

Some computer program multiple regression packages (SAS is one of them) can be instructed to print confidence intervals for the mean value of y and/or prediction intervals for y for specific values of the independent variables. You can also instruct some programs to print a test statistic for testing the hypothesis that each parameter in a subset of the model parameters, say β_2 and β_3, simultaneously equals 0, that is, $H_0: \beta_2 = \beta_3 = 0$. This latter topic is addressed in Optional Exercise 11.19.

▷ 11.4 A COMPARISON OF COMPUTER PRINTOUTS

Several popular statistical program packages contain multiple regression programs. Three of these, MINITAB, SAS, and SPSS-X, are referenced at the end of this chapter. Some programs can only be used on specific computers; others are more versatile. If you plan to conduct a multiple regression analysis, you will want to determine the statistical program packages available at your computer center so that you may become familiar with their output.

· We presented and explained SAS and MINITAB multiple regression computer outputs in Section 11.3. Most other computer outputs are very similar. They vary somewhat in the format in which the output is presented and, in some cases, the type of output that can be requested. In this section we will compare the output of three

---

* If we wish to determine whether the data provide sufficient evidence to indicate that the male faculty members start at higher salaries, we would test $H_0: \beta_2 = 0$ against the alternative hypothesis $H_a: \beta_2 > 0$.

Table 11.7 The SAS GLM, MINITAB, and SPSS-X printouts for Example 11.5

(a) SAS GLM PRINTOUT
DEPENDENT VARIABLE: Y

SQUARE	DF	SUM OF SQUARES	MEAN SQUARE	② F VALUE	PR > F	① R-SQUARE	C.V.
MODEL	3	42108777.02898556	14036259.00966185	346.24	0.0001	0.992357	0.8897
ERROR	8	324314.63768142	40539.32971018 ⑦		ROOT MSE		Y MEAN
CORRECTED TOTAL	11	42433091.66666698 ⑨			201.34380971 ⑧		22630.83333333

SOURCE	DF	TYPE I SS	F VALUE	PR > F	DF	TYPE III SS	F VALUE	PR > F
X1	1	33294036.23595509	821.28	0.0001	1	9389610.00000008	231.62	0.0001
X2	1	8452796.51598297	208.51	0.001	1	326808.74399183	8.06	0.0218
X1*X2	1	361944.27704750	8.93	0.0174	1	361944.27704750	8.93	0.0174

PARAMETER	③ ESTIMATE	④ T FOR HO: PARAMETER = 0	⑤ PR > ITI	⑥ STD ERROR OF ESTIMATE
INTERCEPT	18593.00000000	89.41	0.0001	207.94699250
X1	969.00000000	15.22	0.0001	63.67050315
X2	866.71014493	2.84	0.0218	305.25678646
X1*X2	260.13043478	2.99	0.0174	87.05798112

(b) MINITAB PRINTOUT

THE REGRESSION EQUATION IS
Y = 18593 + 969 X1 + 867 X2 + 260 X1*X2

PREDICTOR	COEF ③	STDEV ⑥	T-RATIO ④	P ⑤
CONSTANT	18593.0	207.9	89.41	0.000
X1	969.00	63.67	15.22	0.000
X2	866.7	305.3	2.84	0.022
X1*X2	260.13	87.06	2.99	0.017

S = 201.3 ⑧ R-SQ = 99.2% ① R-SQ (ADJ) = 98.9%

ANALYSIS OF VARIANCE

SOURCE	DF	SS	MS	F	P ②
REGRESSION	3	42108776	14036259	346.24	0.000
ERROR	8	324315	40539 ⑦		
TOTAL	11	42433092 ⑨			

SOURCE	DF	SEQ SS
X1	1	33294036
X2	1	8452797
X1*X2	1	361944

MTB >

(c) SPSS-X PRINTOUT
******************************* MULTIPLE REGRESSION ******************************* VARIABLE LIST 1
REGRESSION LIST 1

DEPARTMENT VARIABLE..   Y   SALARY

VARIABLE(S) ENTERED ON STEP NUMBER 1..   X3
X1
X2

		ANALYSIS OF VARIANCE	DF	SUM OF SQUARES	MEAN SQUARE
MULTIPLE R	0.99617				
R SQUARE	0.99236 ①	REGRESSION	3.	42108777.02899	14036259.00966
ADJUSTED R SQUARE	0.98949	RESIDUAL	3.	324314.63768	40539.32971 ⑦
STANDARD ERROR	201.34381 ⑧				

F = 346.23806   SIGNIF F = .0001 ②

-------------- VARIABLES IN THE EQUATION -------------- VARIABLES NOT IN THE EQUATION --------------

VARIABLE	③ B	⑥ SE B	BETA	④ T	⑤ SIG T
X1X2					
	260.1304				
X1	969.0000	87.05798	0.27739	2.99	.0174
X2	866.7101	63.67050	0.70168	15.22	.0001
(CONSTANT)	18593.00	305.25679	0.23045	2.84	.0218

multiple regression computer packages, SAS GLM, MINITAB, and SPSS-X, for the data of Example 11.5. The three computer printouts for this set of data are shown in Table 11.7.

Corresponding output is shaded and marked with the same number in each printout. For example, parameter estimates on the MINITAB printout appear under the column heading COEF, and on the SPSS-X printout they appear under the column heading B.

Both the MINITAB and the SPSS-X printouts give a value of R^2 that is adjusted for the number of degrees of freedom associated with SSE. It appears as R-SQUARED ADJUSTED FOR DF on the MINITAB printout and ADJUSTED R SQUARE on the SPSS-X printout. We have not utilized this quantity in our analysis of the data in Section 11.3, and we will therefore omit it from our discussion.

The computer printouts for the MINITAB, SAS, SPSS-X, and other multiple regression computer program packages are very similar. Once you can read one, it is likely that you will be able to read the output of the computer programs available at your computer center. If you have difficulty understanding the output, consult the package instruction manual.

EXERCISES   Basic Techniques

11.7   Suppose that you were to fit the model

$$E(y) = \beta_0 + \beta_1 x_1 + \beta_2 x_2 + \beta_3 x_3$$

to 15 data points and found R^2 equal to .94.
a.   Interpret the value of R^2.
b.   Do the data provide sufficient evidence to indicate that the model contributes information for the prediction of y? Test using $\alpha = .05$.

11.8   The computer output for the multiple regression analysis in Exercise 11.7 provides the following information:

$$\hat{\beta}_0 = 1.04 \qquad \hat{\beta}_1 = 1.29 \qquad \hat{\beta}_2 = 2.72 \qquad \hat{\beta}_3 = .41$$

$$s_{\hat{\beta}_1} = .42 \qquad s_{\hat{\beta}_2} = .65 \qquad s_{\hat{\beta}_3} = .17$$

a.   Give the least-squares prediction equation.
b.   Which, if any, of the independent variables, x_1, x_2, and x_3 contribute information for the prediction of y? Test using $\alpha = .05$.
c.   On the same sheet of graph paper, graph y versus x_1 when $x_2 = 1$ and $x_3 = 0$ and when $x_2 = 1$ and $x_3 = .5$. What relationship do the two lines have to each other?
d.   What is the practical interpretation of the parameter β_1?
e.   Find a 90% confidence interval for β_1 and interpret it.

11.9   When will an F statistic equal the square of a t statistic?

11.10   The least-squares equation relating the dependent variable y to two independent variables x_1 and x_2 is given as

$$\hat{y} = 3.47 + 0.2x_1 - 1.64x_2 + 0.63x_1 x_2$$

a.   Write the least-squares equation as a function of x_1 when $x_2 = 1$, 2, and 3.
b.   Graph the three equations found in part (a) on the same sheet of graph paper.

c. What type of curves are represented by the three equations? Does there appear to be an interaction between x_1 and x_2?

11.11 A dependent variable y is recorded along with two independent variables x_1 and x_2 in the accompanying table.

Observation	y	x_1	x_2	Observation	y	x_1	x_2
1	50	0	0	9	15	1	3
2	40	0	1	10	10	1	4
3	30	0	2	11	40	2	0
4	25	0	3	12	30	2	1
5	30	0	4	13	10	2	2
6	35	1	0	14	15	2	3
7	30	1	1	15	23	2	4
8	20	1	2				

a. Plot the dependent variable y as a function of x_1 when $x_2 = 0$, 2, and 4. What type of relationship appears to exist between y and x_1?

b. Plot y as a function of x_2 when $x_1 = 0$, 1, and 2. What type of relationship appears to exist between y and x_2?

c. A MINITAB regression program was used to fit the second-order model.

$$y = \beta_0 + \beta_1 x_1 + \beta_2 x_2 + \beta_3 x_1^2 + \beta_4 x_2^2 + \beta_5 x_1 x_2 + \epsilon$$

to the data set. The printout is shown below. What is the least-squares equation for the model given above?

```
MTB > REGRESS C1 5 C2-C6

THE REGRESSION EQUATION IS
C1 = 51.7 - 20.9 C2 - 15.6 C3 + 7.30 C4 + 2.40 C5 + 0.30 C6

PREDICTOR      COEF      STDEV      T-RATIO        P
CONSTANT      51.676     3.761       13.74      0.000
C2           -20.900     5.809       -3.60      0.006
C3           -15.552     3.232       -4.81      0.000
C4             7.300     2.598        2.81      0.020
C5            2.4048     0.7319       3.29      0.009
C6            0.300      1.061        0.28      0.784

S = 4.743     R-SQ = 89.4%      R-SQ (ADJ) = 83.4%

ANALYSIS OF VARIANCE

SOURCE         DF         SS         MS        F         P
REGRESSION      5      1699.25     339.85     15.11     0.000
ERROR           9       202.49      22.50
TOTAL          14      1901.73

SOURCE         DF      SEQ SS
C2              1      324.90
C3              1      952.03
C4              1      177.63
C5              1      242.88
C6              1        1.80
```

d. Does the second-order model contribute information for the prediction of y? Test using $\alpha = .05$.

e. Interpret the value of R^2 given in the printout.

11.12 A publisher of college textbooks conducted a study to relate profit per text to cost of sales over a six-year period in which its sales force (and sales costs) were growing rapidly. The following inflation-adjusted data (in thousands of dollars) were collected.

Profit per text, y	16.5	22.4	24.9	28.8	31.5	35.8
Sales cost per text, x	5.0	5.6	6.1	6.8	7.4	8.6

Expecting profit per book to rise and then plateau, the publisher fitted the model $E(y) = \beta_0 + \beta_1 x + \beta_2 x^2$ to the data.

a. What sign would you expect the actual value of β_2 to assume? The SAS computer printout is shown below. Find the value of $\hat{\beta}_2$ on the printout and see if the sign agrees with your answer.

SAS output for
Exercise 11.12

DEPENDENT VARIABLE: Y

SOURCE	DF	SUM OF SQUARES	MEAN SQUARE	F VALUE	PR > F	R-SQUARE	C.V.
MODEL	2	234.95514252	117.47757126	332.53	0.0003	0.995509	2.2303
ERROR	3	1.05985748	0.35328583		ROOT MSE		Y MEAN
CORRECTED TOTAL	5	236.01500000			0.59437852		26.65000000

SOURCE	DF	TYPE I SS	F VALUE	PR > F	DF	TYPE III SS	F VALUE	PR > F
X	1	227.81864814	644.86	0.0001	1	15.20325554	43.03	0.0072
X*X	1	7.13649437	20.20	0.0206	1	7.13649437	20.20	0.0206

PARAMETER	ESTIMATE	T FOR H0: PARAMETER = 0	PR > ITI	STD ERROR OF ESTIMATE
INTERCEPT	−44.19249551	−5.33	0.0129	8.28688218
X	16.33386317	6.56	0.0072	2.48991042
X*X	−0.81976920	−4.49	0.0206	0.18239471

OBSERVATION	OBSERVED VALUE	PREDICTED VALUE	RESIDUAL	LOWER 95% CL FOR MEAN	UPPER 95% CL FOR MEAN
1	16.50000000	16.98259044	−0.48259044	15.33066719	18.63451369
2	22.40000000	21.56917626	0.83082374	20.56714686	22.57120566
3	24.90000000	24.94045805	−0.04045805	23.93483233	25.94608377
4	28.80000000	28.97164643	−0.17164643	27.81528749	30.12800537
5	31.50000000	31.78753079	−0.28753079	30.65188830	32.92317327
6	35.80000000	35.64859804	0.15140196	33.81460946	37.48258661
7*		27.34236657		26.23203462	28.45269852

*OBSERVATION WAS NOT USED IN THIS ANALYSIS

SUM OF RESIDUALS	−0.00000000
SUM OF SQUARED RESIDUALS	1.05985748
SUM OF SQUARED RESIDUALS − ERROR SS	−0.00000000
PRESS STATISTIC	12.11621807
FIRST ORDER AUTOCORRELATION	−0.39797399
DURBIN-WATSON D	2.55457960

b. Find SSE and s^2 on the printout.

c. How many degrees of freedom do SSE and s^2 have? Show that $s^2 = \text{SSE}/(\text{degrees of freedom})$.

d. Do the data provide sufficient evidence to indicate that the model contributes information for the prediction of y? Test using $\alpha = .05$.

e. Find the observed significance level for the test in part (d) and interpret its value.

f. Do the data provide sufficient evidence to indicate curvature in the relationship between $E(y)$ and x (i.e., evidence to indicate that β_2 differs from 0)? Test using $\alpha = .05$.

g. Find the observed significance level for the test in part (f) and interpret its value.

h. Find the prediction equation and graph the relationship between $\hat{y}$ and x.

i. Find R^2 on the printout and interpret its value.

j. Use the prediction equation to estimate the mean profit per text when the sales cost per text is $6500. (Express the sales cost in thousands of dollars before substituting into the prediction equation.) We instructed the SAS program to print this confidence interval. The confidence interval when $x = 6.5$, 26.23203462 to 28.45269852, is shown at observation 7 in the printout.

11.13 Use the values of SSE and S_{yy} in the printout in Exercise 11.12 to calculate R^2. Compare your calculated value with the value of R^2 given in the printout.

11.14 Refer to Example 11.5. The t value for testing the hypothesis $H_0: \beta_1 = 0$ is

$$ t = \frac{\hat{\beta}_1 - 0}{s_{\hat{\beta}_1}} = \frac{\hat{\beta}_1}{s_{\hat{\beta}_1}} $$

where $\hat{\beta}_1$ is the estimate of β_1 and $s_{\hat{\beta}_1}$ is the estimated standard deviation (or standard error) of $\hat{\beta}_1$. Both of these quantities are shown on the SAS, MINITAB, and SPSS-X printouts in Table 11.7. Examine the printouts and verify that the corresponding printed values for $\hat{\beta}_1$ and $s_{\hat{\beta}_1}$ are identical except for rounding errors. Calculate $t = \hat{\beta}_1/s_{\hat{\beta}_1}$. Verify that your calculated value of t is equal to the value printed on the SAS and the MINITAB printouts, that is, $t = 15.22$.

11.15 Refer to Exercise 10.30. Groups of ten-day-old chicks were randomly assigned to seven treatment groups in which a basal diet was supplemented with 0, 50, 100, 200, 250, or 300 μg/kg d-biotin. The following data give the average biotin intake (x) in micrograms per day (μg/d) and the average weight gain (y) in grams per day (g/d).

Added biotin	Biotin intake (x)	Weight gain (y)
0	0.14	8.0
50	2.01	17.1
100	6.06	22.3
150	6.34	24.4
200	7.15	26.5
250	9.65	23.4
300	12.50	23.3

Source: Blair, R. and R. Miser, "Biotin Bioavailability from Protein Supplements and Cereal Grains for Growing Broiler Chickens," *International Journal of Vitamin Nutrition Research*, Vol. 59, 1989, pp. 55–58. Hans Huber Publishers, Toronto, Bern.

In the MINITAB printout that follows, the second-order polynomial model

$$E(y) = \beta_0 + \beta_1 x + \beta_2 x^2$$

was fitted to the data. Use the printout to answer the following questions.

MTB > REGRESS C3 2 C1 C2

THE REGRESSION EQUATION IS
Y = 8.59 + 3.82 X − 0.217 X∗X

PREDICTOR	COEF	STDEV	T-RATIO	P
CONSTANT	8.585	1.641	5.23	0.006
X	3.8208	0.5683	6.72	0.003
X∗X	−0.21663	0.04390	−4.93	0.008

S = 1.833   R-SQ = 94.4%   R-SQ (ADJ) = 91.5%

ANALYSIS OF VARIANCE

SOURCE	DF	SS	MS	F	P
REGRESSION	2	224.75	112.37	33.44	0.003
ERROR	4	13.44	3.36		
TOTAL	6	238.19			

SOURCE	DF	SEQ SS
X	1	142.92
X∗X	1	81.83

MTB >

---

a.   What is the fitted least-squares line?
b.   Find R^2 and interpret its value.
c.   Do the data provide sufficient evidence to conclude that the model contributes significant information for predicting y?
d.   Find the results of the test of $H_0: \beta_2 = 0$. Is there sufficient evidence to indicate that the quadratic model provides a better fit to the data than a simple linear model does?

11.16   The waiting time y that elapses between the time a computing job is submitted to a large computer and the time at which the job is initiated (computing commences) is a function of many variables, including the priority assigned to the job, the number and sizes of the jobs already on the computer, the size of the job being submitted, and so on. A study was initiated to investigate the relationship between waiting time y (in hours) for a job and x_1, the estimated CPU time (in seconds) for the job, and x_2, the CPU utilization factor. The estimated CPU time x_1 is an estimate of the amount of time that a job will occupy a portion of the computer's central processing unit's memory. The CPU utilization factor x_2 is the percentage of the memory bank of the central processing unit that is occupied at the time the job is submitted. We would expect the waiting time y to increase as the size of the job x_1 increases and as the CPU utilization factor x_2 increases. To conduct the study, 15 jobs of varying sizes were submitted to the computer at randomly assigned times throughout the day. The job waiting time y, estimated CPU time x_1, and CPU utilization factor x_2 were recorded for each job. The data* are shown below.

	Job														
	1	2	3	4	5	6	7	8	9	10	11	12	13	14	15
x_1	2.0	9.3	5.6	3.7	12.4	18.1	13.5	26.6	34.2	38.8	56.1	60.3	4.4	2.6	20.9
x_2	45	80	23	25	67	30	55	21	79	40	22	37	50	66	42
y	0.001	1.140	0.030	0.001	0.780	0.300	0.600	0.200	2.240	0.440	0.001	0.320	0.160	0.290	0.490

A second-order model, $E(y) = \beta_0 + \beta_1 x_1 + \beta_2 x_2 + \beta_3 x_1 x_2 + \beta_4 x_1^2 + \beta_5 x_2^2$, was selected to model mean waiting time $E(y)$. The SAS multiple regression analysis for the data is shown in the following printout.

SAS output for Exercise 11.16

DEPENDENT VARIABLE: Y

SOURCE	DF	SUM OF SQUARES	MEAN SQUARE	F VALUE	PR > F	R-SQUARE	C.V.
MODEL	5	4.74848837	0.94969767	159.23	0.0001	0.988822	16.5655
ERROR	9	0.05367803	0.00596423		ROOT MSE		Y MEAN
CORRECTED TOTAL	14	4.80216640			0.07722840		0.46620000

SOURCE	DF	TYPE I SS	F VALUE	PR > F	DF	TYPE III SS	F VALUE	PR > F
X1	1	0.08032055	13.47	0.0052	1	0.00448637	0.75	0.4083
X2	1	3.21553051	539.14	0.0001	1	0.13905895	23.32	0.0009
X1*X2	1	0.96272106	161.42	0.0001	1	0.48988534	82.14	0.0001
X1*X1	1	0.22455767	37.65	0.0002	1	0.16181969	27.13	0.0006
X2*X2	1	0.26535859	44.49	0.0001	1	0.26535859	44.49	0.0001

PARAMETER	ESTIMATE	T FOR H0: PARAMETER = 0	PR > ITI	STD ERROR OF ESTIMATE
INTERCEPT	0.43806816	2.71	0.0239	0.16147813
X1	0.00526474	0.87	0.4083	0.00607025
X2	-0.03017242	-4.83	0.0009	0.00624867
X1*X2	0.00068770	9.06	0.0001	0.00007588
X1*X1	-0.00037952	-5.21	0.0006	0.00007286
X2*X2	0.00040726	6.67	0.0001	0.00006106

a. Find the values of SSE and s^2.
b. Find the prediction equation.
c. Find R^2 and interpret its value.
d. Do the data provide sufficient evidence to indicate that the model contributes information for the prediction of y? Test using $\alpha = .10$.

* Waiting time data frequently violate the assumptions required for significance tests and confidence intervals in a regression analysis. The probability distribution for waiting times is often skewed, and its variance increases as the mean waiting time increases. Methods are available for coping with this problem (*see* Mendenhall and Sinich 1986), but we will ignore it for the purposes of this introductory discussion.

e.   Find the observed significance level for the test in part (d) and interpret its value.

f.   Note that the observed significance level for the test $H_0: \beta_1 = 0$ is large (p-value = .4083). Does this mean that there is little evidence to indicate that x_1 contributes information for the prediction of y?

g.   Predict the waiting time when the estimated CPU time for a job is 30 seconds and the CPU utilization factor is 55%.

h.   Graph the estimated mean waiting time as a function of estimated CPU time x_1 for CPU utilization, $x_2 = 30\%$. Also graph the curves for $x_2 = 50\%$ and $x_2 = 70\%$. Observe the behavior of the estimated mean waiting time as x_1 and x_2 increase in value.

11.5   STEPWISE REGRESSION ANALYSIS

Suppose that we wished to construct a model to predict the grade point average of a college freshman at the end of the freshman year based on information available at the end of the student's senior year in high school. Since data for high school seniors are available for many different variables, such as grades in courses, achievement test scores, and so on, which of this large number of independent variables should be included in our model? One way to answer this question is to use a computer program package that performs a stepwise regression analysis.

Suppose that we have data available on y and a number of possible independent variables, $x_1, x_2, \ldots, x_k$. The first step in a stepwise regression analysis is to fit the model

$$E(y) = \beta_0 + \beta_1 x$$

for each of the k independent variables. The best fitting of these k prediction equations is chosen (for most programs) as the one with the largest t value for β_1. We will denote this independent variable as x_1.

The second step in a stepwise regression analysis is to fit all possible two-variable models. Most computer programs retain x_1 and fit only models of the type

$$E(y) = \beta_0 + \beta_1 x_1 + \beta_2 x_2$$

where x_2 is one of the $(k - 1)$ remaining variables. For these programs, the variable to be denoted as x_2 and retained in the model is again the one with the largest t value for β_2. Other programs try all possible two-variable models because it is conceivable that the best two-variable model will not include x_1, the variable entered in Step 1. These programs usually choose as the best two-variable model the one with the largest value of R^2.

The third and succeeding steps in a stepwise regression analysis are identical to the second step. The third step fits the model

$$E(y) = \beta_0 + \beta_1 x_1 + \beta_2 x_2 + \beta_3 x_3$$

The variable x_3 chosen to be retained in the model is either the one for which the t value for β_3 is the largest or the value for which R^2 is the largest (depending on the computer program). The procedure stops when the t value for the entering value is less than some predetermined value that would imply statistical significance (or when R^2 exceeds some predetermined value).

A stepwise regression analysis is an easy way to locate some variables that contribute information for the prediction of y, but it is not a foolproof method for finding a good model. Stepwise regression procedures almost always fit first-order models composed of the variables entered into the computer, that is, models of the type

$$E(y) = \beta_0 + \beta_1 x_1 + \beta_2 x_2 + \cdots + \beta_k x_k$$

If you want to introduce second-order terms to account for curvature in the response surface, you will have to enter them into the computer as though they were new variables. In other words, if you had three independent variables $x_1, x_2,$ and x_3, you could enter these variables into the computer along with

$$x_4 = x_1^2, \quad x_5 = x_2^2, \quad x_6 = x_3^2, \quad x_7 = x_1 x_2, \quad \text{and so on.}$$

If the number of independent variables is large, the total number of first-order and second-order terms may be unmanageable. One alternative, when this situation occurs, is to perform a stepwise regression analysis on the original set of independent variables, thereby acquiring a small set of information-contributing variables. If the model requires improvement, a second stepwise regression analysis could be performed using these variables as well as their squares and cross products. This procedure is unlikely to yield the best second-order model to fit the data, but it is a reasonable alternative to the massive task involved in fitting all possible second-order models.

The computer printout for a stepwise regression analysis is a sequence of individual regression analyses, one for each step in the procedure. Most will be easy to read. They will differ, depending on the computer program that you use, but the printout for each step will be similar to the printout for a standard multiple regression analysis.

$\triangleright$ 11.6 MISINTERPRETATIONS IN REGRESSION ANALYSES

There are several common misinterpretations of the output of a regression analysis, the first of which involves the importance of model selection. **If a model does not fit a set of data, it does not mean that the variables included in the model contribute little or no information for the prediction of y.** The variables may be very important contributors of information, but you may not have entered the variables into the model in an appropriate way. For example, a second-order model in the variables might provide a very good fit to the data when a first-order model appears to be completely useless in describing the response variable y.

Secondly, **you must be careful not to deduce that a causal relationship exists between a response y and a variable x.** Just because a variable x contributes information for the prediction of y, it does not imply that changes in x cause changes in y. It is possible to *design* an experiment to detect causal relationships. For example, if you randomly assign experimental units to each of two levels of a variable x, say $x = 5$ and $x = 10$, and the data show that the mean value of y is larger when $x = 10$, then you could say that the change in the level of x caused a change in the mean value of y. But in most regression analyses, where the experiments are not designed, there is no guarantee that an important predictor variable, say x_1, caused y to change. It is quite possible that some variable that is not even in the model caused **both** y and x_1 to change.

A third common misinterpretation concerns the magnitude of the regression coefficients. **Neither the size of a regression coefficient nor its t value indicates the importance of the associated variable as an information contributor.** For example, suppose that you wish to predict a college student's calculus grade y as a function of the student's high school average mathematics grade x_1 and the student's score x_2 on a college mathematics replacement test. A regression analysis, using the first-order model $E(y) = \beta_0 + \beta_1 x_1 + \beta_2 x_2$, would likely show that both x_1 and x_2 contribute information for the prediction of y. However, it is conceivable that the t value associated with one of the regression coefficients would not be statistically significant. This is because much of the information contained in x_1 is the same information contained in x_2. When this occurs, the one-variable model

$$E(y) = \beta_0 + \beta_1 x_1 \qquad (\text{or } E(y) = \beta_0 + \beta_2 x_2)$$

may be almost as useful in predicting y as the two-variable model

$$E(y) = \beta_0 + \beta_1 x_1 + \beta_2 x_2$$

When two or more of the independent variables are related and, therefore, contribute overlapping information for the prediction of y, we say that **multicollinearity** exists in the independent variables; that is, they are correlated. Since multicollinearity exists in most regression analyses, the individual terms in the model should be viewed as information contributors. The primary decision to be made is whether one or more of the terms contribute sufficient information for the prediction of y and whether they should be retained in the model.

▷ 11.7 SOME COMMENTS ON MODEL FORMULATION (OPTIONAL)

The variables encountered in a regression analysis can be one of two types, **quantitative** or **qualitative**. For example, the age of a person is a quantitative variable because its values express the quantity or amount of something, in this case, age. In contrast, the nationality of a person is a variable that varies from person to person, but the values of the variable cannot be quantified. They can only be classified. To use the methods we present, the response variable y must always be (according to the assumptions of Section 11.2) a quantitative variable. In contrast, predictor variables may be either quantitative or qualitative. For example, suppose that we wish to predict the annual income of a person. Both the age and the nationality of the person could be important predictor variables, and there are probably many others.

When two or more quantitative independent variables appear in a regression model, the resulting response function produces a graph called a response surface in three or more dimensions. These graphs become difficult to produce when three or more independent variables are included in the model. A model involving quantitative variables is said to be a first-order model if each independent variable appears in the model with power one. The model

$$E(y) = \beta_0 + \beta_1 x_1 + \beta_2 x_2 + \cdots + \beta_k x_k$$

is a **first-order model** involving k independent variables, since the model is linear in each x. The graph of a first-order model is a response plane, which means that the surface is "flat" but has some directional tilt with respect to its axes. Second-order linear models in k quantitative predictor variables include all the terms in a first-order model, all cross-product terms such as $x_1 x_2, x_1 x_3, x_2 x_3, \ldots, x_{k-1} x_k$, and all pure quadratic terms $x_1^2, x_2^2, \ldots, x_k^2$. A **second-order model** with two predictor variables is given as

$$E(y) = \beta_0 + \beta_1 x_1 + \beta_2 x_2 + \beta_3 x_1^2 + \beta_4 x_1 x_2 + \beta_5 x_2^2$$

The quadratic terms x_1^2 and x_2^2 allow for curvature, while the cross-product or interaction term $x_1 x_2$ allows for warping or twisting of the response surface.

Definition

> Two predictor variables are said to **interact** if the change in $E(y)$ corresponding to a change in one predictor variable depends upon the value of the other variable.

To understand the concept of interaction, consider the first-order (planar) model

$$E(y) = \beta_0 + \beta_1 x_1 + \beta_2 x_2$$

If you were to graph $E(y)$ as a function of x_1, *with x_2 held constant*, you would obtain a straight line. Repeating this process for other values of x_2, you would obtain a set of parallel lines (with slopes β_1 but intercepts depending on the particular value of x_2) that might appear as shown in Figure 11.8. (Graphs of $E(y)$ versus x_2 for various values of x_1 would also produce a set of parallel lines but with common slope β_2.)

Figure 11.8 shows that, regardless of the value of x_2, a change in x_1, say from $x_1 = a$ to $x_1 = b$, will always produce the same change Δ in $E(y)$. When this situation occurs, that is, when the change in $E(y)$ for a change in one variable does not depend on the value of the second variable, we say that the variables **do not interact**.

In contrast, consider the second-order (interaction) model

$$E(y) = \beta_0 + \beta_1 x_1 + \beta_2 x_2 + \beta_3 x_1 x_2$$

If you were to again graph $E(y)$ versus x_1 for various values of x_2, you would obtain a set of lines, but they would not be parallel (see Figure 11.9). In Figure 11.9 you can see that $E(y)$ rises very slowly as x_1 increases when $x_2 = 1$, more rapidly when $x_2 = 2$, and even more rapidly when $x_2 = 3$. Therefore, the effect on $E(y)$ of a

Figure 11.8
No interaction between
x_1 and x_2

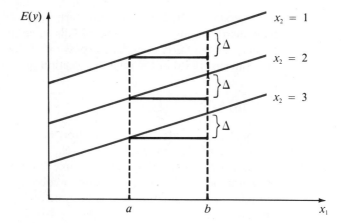

Figure 11.9
Interaction between
x_1 and x_2

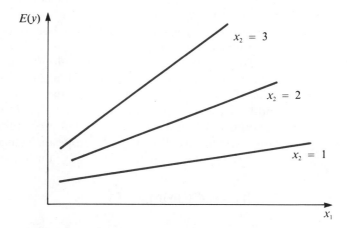

change in x_1 **depends** on the value of x_2. When this situation occurs, we say that the predictor variables **interact**. If you attempt to fit the noninteractive first-order model to data that are graphed as shown in Figure 11.9, you obtain a very poor fit to your data. **The warning is clear. You may need interaction terms in your model to obtain a good fit to a set of data.**

In contrast to quantitative predictor variables, qualitative predictor variables are entered into a model using **dummy** (or **indicator**) variables. **Models with qualitative predictor variables that can assume any number, say k, of values may be constructed in a similar manner using $(k - 1)$ terms involving dummy variables.** These terms can be added to models containing other predictor variables, quantitative or qualitative, and you can also include terms involving the cross products (interaction terms) of the dummy variables with other variables that appear in the model.

In this optional section we discussed two important aspects of model formulation: the concept of predictor variable interaction (and how to cope with it) and the method for introducing qualitative predictor variables into a model. Clearly, this is not enough to make you proficient in model formulation, but it will help you to understand why and how some terms are included in regression models, and it may help you to avoid some pitfalls encountered in model construction. An elementary but fairly complete discussion of this topic can be found in Mendenhall and Sincich (1986) in the reference section of this chapter.

▷ **11.8** SUMMARY

Chapter 11 provides an introduction to an extremely useful area of statistical methodology, one that enables us to relate the mean value of a response variable y to a set of predictor variables. The multiple regression model expresses $E(y)$ as the sum of a number of terms, each term involving the product of a single unknown parameter and a function of one or more of the predictor variables. Values of y are observed when the predictor variables assume specified values, and these data are used to estimate the unknown parameters in the model using the method of least squares. The procedure for estimating the unknown parameters of the model, testing its utility, or testing hypotheses about sets of the model parameters is known as a multiple regression analysis.

The practical utility of a fitted multiple regression equation lies in its use as a prediction equation—to estimate the mean value of y or to predict a particular value of y for a given set of values of the predictor variables. Scientists, engineers, and business managers all face the same problem: forecasting some quantitative response based on the values of a set of predictor variables that describe current or future conditions. When the appropriate assumptions are satisfied, a multiple regression analysis can be used to solve this important problem.

▷ MINITAB COMMANDS

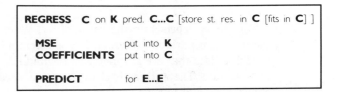

REGRESS C on **K** pred. **C...C** [store st. res. in **C** [fits in **C**]]	
MSE	put into **K**
COEFFICIENTS	put into **C**
PREDICT	for **E...E**

REFERENCES

Constas, K., and Vichas, R. P. "An Interpretive Policy-Making Model of Absenteeism with Reference to the Marginal Worker in Overseas Plants." *Journal of Management Science* Vol. 17, No. 2 (1980).

Draper, N. R., and Smith, H. *Applied Regression Analysis.* 2d ed. New York: Wiley, 1981.

Dunn, O. J., and Clark, V. A. *Applied Statistics: Analysis of Variance and Regression.* 2d ed. New York: Wiley, 1987.

Kleinbaum, D. G.; Kupper, L. L.; and Miller, K. E. *Applied Regression Analysis and Other Multivariable Methods.* Boston: PWS-KENT, 1988.

Mendenhall, W., and Sincich, T. *A Second Course in Business Statistics: Regression Analysis.* 2d ed. San Francisco: Dellen, 1986.

Meyers, R. H. *Classical and Modern Regression with Applications.* 2d. ed. Boston: PWS-KENT, 1990.

Ryan, T. A.; Joiner, B. L.; and Ryan, B. F. *Minitab Student Handbook.* 2d ed. Boston: Duxbury Press, 1985.

SAS Institute, Inc. Staff. *SAS User's Guide: Statistics.* Version 5 ed. Cary, N. C.: SAS Institute, Inc., 1985.

SPSS, Inc. Staff. *SPSS-X User's Guide.* 3d ed. New York: McGraw-Hill, 1987.

SUPPLEMENTARY EXERCISES

[Starred (*) exercises are optional]

11.17 Utility companies, which must plan the operation and expansion of electricity generation, are vitally interested in predicting customer demand over both short and long periods of time. A short-term study was conducted to investigate the effect of mean monthly daily temperature x_1 and cost per kilowatt x_2 on the mean daily consumption (kwh) per household. The company expected the demand for electricity to rise in cold weather (due to heating), fall when the weather was moderate, and rise again when the temperature rose and there was a need for air conditioning. They expected demand to decrease as the cost per kilowatt-hour increased, reflecting greater attention to conservation. Data were available for two years, a period in which the cost per kilowatt-hour, x_2, increased due to the increasing costs of fuel. The company fitted the model

$$E(y) = \beta_0 + \beta_1 x_1 + \beta_2 x_1^2 + \beta_3 x_2 + \beta_4 x_1 x_2 + \beta_5 x_1^2 x_2$$

to the data shown below.

Price per kwh x_2		Mean daily consumption (kwh) per household											
8 ¢	Mean Daily Temperature, (°F)x_1	31	34	39	42	47	56	62	66	68	71	75	78
	Mean Daily Consumption, y	55	49	46	47	40	43	41	46	44	51	62	73
10 ¢	Mean Daily Temperature, x_1	32	36	39	42	48	56	62	66	68	72	75	79
	Mean Daily Consumption, y	50	44	42	42	38	40	39	44	40	44	50	55

Before we analyze the data, let us examine the logic employed in formulating the model.

a. Examine the relationship between $E(y)$ and temperature x_1 for a fixed price per kilowatt-hour (kwh) x_2. Substitute a value for x_2, say 10¢, into the equation for $E(y)$. What type of curve will the model relating $E(y)$ to x_1 represent?

b. If the company's theory regarding the relationship between mean daily consumption $E(y)$ and temperature x_1 is correct, what should be the sign (positive or negative) of the coefficient of x_1^2?

c. Examine the relationship between $E(y)$ and price per kwh x_2, when temperature x_1 is held constant. Substitute a value for x_1 into the equation for $E(y)$, say $x_1 = 50°F$. What type of curve will the model relating $E(y)$ to x_2 represent?

d. Refer to (c). If the company's theory regarding the relationship between mean daily consumption $E(y)$ and price per kwh x_2 is correct, what should be the sign of the coefficient of x_2?

e. What effect do the terms $\beta_4 x_1 x_2$ and $\beta_5 x_1^2 x_2$ have on the curves relating $E(y)$ to x_1 for various values of price per kwh?

11.18 The SAS multiple regression computer printout for the data in Exercise 11.17 is shown below.

a. Find SSE and s^2 on the printout.

b. How many degrees of freedom do SSE and s^2 have? Show that $s^2 = $ SSE/(degrees of freedom).

c. Do the data provide sufficient evidence to indicate that the model contributes information for the prediction of mean daily kwh consumption per household? Test using $\alpha = .10$.

d. Use Tables 6, 7, 8, 9, and 10 in Appendix III to find the approximate observed significance level for the test in part (c).

e. Find the total SS, S_{yy}, on the printout. Find the value of R^2 and interpret its value. Show that $R^2 = 1 - $ SSE/S_{yy}.

SAS output for Exercise 11.18

DEPENDENT VARIABLE: Y

SOURCE	DF	SUM OF SQUARES	MEAN SQUARE	F VALUE	PR > F	R-SQUARE	C.V.
MODEL	5	1346.44751546	269.28950309	31.85	0.0001	0.898455	6.2029
ERROR	18	152.17748454	8.45430470		ROOT MSE		Y MEAN
CORRECTED TOTAL	23	1498.62500000			2.90762871		46.87500000

SOURCE	DF	TYPE I SS	F VALUE	PR > F	DF	TYPE III SS	F VALUE	PR > F
X1	1	140.71101071	16.64	0.0007	1	104.43456772	12.35	0.0025
X1*X1	1	892.77914943	105.60	0.0001	1	125.58263370	14.85	0.0012
X2	1	192.44266992	22.76	0.0002	1	46.78392523	5.53	0.0302
X1*X2	1	57.83521130	6.84	0.0175	1	50.03623552	5.92	0.0256
X1*X1*X2	1	62.67947420	7.41	0.0140	1	62.67947420	7.41	0.0140

PARAMETER	ESTIMATE	T FOR H0: PARAMETER = 0	PR > |T|	STD ERROR OF ESTIMATE
INTERCEPT	325.60644505	3.92	0.0010	83.06412820
X1	−11.38255606	−3.51	0.0025	3.23859476
X1*X1	0.11319735	3.85	0.0012	0.02944828
X2	−21.69920900	−2.35	0.0302	9.22432344
X1*X2	0.87302921	2.43	0.0256	0.35886030
X1*X1*X2	−0.00886945	−2.72	0.0140	0.00325742

f. Use the prediction equation to predict mean daily consumption when the mean daily temperature is $60°F$ and the price per kwh is $8¢$.

g. If you have access to a computer, use it to perform a multiple regression analysis for the data in Exercise 11.17. Compare your computer output with the SAS output on page 480.

11.19 Refer to Exercises 11.17 and 11.18.

a. Graph the curve depicting $\hat{y}$ as a function of temperature x_1 when the cost per kwh is $x_2 = 8¢$. Construct a similar graph for the case when $x_2 = 10¢$ per kwh. Does it appear that the consumption curves differ?

b. If cost per kwh is unimportant in predicting usage, then we do not need the terms involving x_2 in the model. Therefore, the null hypothesis H_0: "x_2 does not contribute information for the prediction of y" is equivalent to the hypothesis $H_0: \beta_3 = \beta_4 = \beta_5 = 0$ (if $\beta_3 = \beta_4 = \beta_5 = 0$, the terms involving x_2 disappear from the model). This hypothesis is tested by computing SSE_1, the SSE that you would obtain if $\beta_3 = \beta_4 = \beta_5 = 0$, that is, using the Reduced Model,

Model 1 (Reduced Model):

$$E(y) = \beta_0 + \beta_1 x_1 + \beta_2 x_1^2$$

and then computing SSE_2, the SSE for the Complete Model,

Model 2 (Complete Model):

$$E(y) = \beta_0 + \beta_1 x_1 + \beta_2 x_1^2 + \beta_3 x_2 + \beta_4 x_1 x_2 + \beta_5 x_1^2 x_2$$

If the terms involving β_3, β_4, and β_5 contribute information for the prediction of y, then SSE_2 should be much less than SSE_1. Therefore, the test of $H_0: \beta_3 = \beta_4 = \beta_5$ is conducted as follows:

Null Hypothesis: $H_0: \beta_3 = \beta_4 = \beta_5 = 0$.

Alternative Hypothesis: H_a: at least one of the parameters β_3, β_4, or β_5 differs from 0.

Test Statistic: $F = \dfrac{\dfrac{SSE_1 - SSE_2}{v_1}}{\dfrac{SSE_2}{v_2}} = \dfrac{\dfrac{SSE_1 - SSE_2}{v_1}}{s^2}$

where v_1 = number of parameters involved in H_0 (in this exercise, $v_1 = 3$),

v_2 = number of degrees of freedom for SSE_2 and s^2 (in this exercise, $v_2 = 18$),

s^2 = estimate of σ^2 obtained for the complete model.

Rejection Region: Reject H_0 if $F > F_\alpha$ where F_α is the tabulated value of F (Tables 6, 7, 8, 9, and 10 in Appendix III), based on v_1 and v_2 degrees of freedom.

The SAS multiple regression computer printout, obtained by fitting the Reduced Model to the data, is shown on page 482. The following steps enable you to conduct a test to determine whether the data provide sufficient evidence to indicate that price per kwh x_2 contributes information for the prediction of y. Test using $\alpha = .05$.

i. Find SSE_1 and SSE_2.

ii. Give the values for v_1 and v_2.

iii. Calculate the value of F.

iv. Find the tabulated value of $F_{.05}$.

v. State your conclusions.

SAS output for Exercise 11.19(b)

DEPENDENT VARIABLE: Y

SOURCE	DF	SUM OF SQUARES	MEAN SQUARE	F VALUE	PR > F	R-SQUARE	C.V.
MODEL	2	1033.49016004	516.74908002	23.33	0.0001	0.689626	10.0401
ERROR	21	465.13432996	22.14927809		ROOT MSE		Y MEAN
CORRECTED TOTAL	23	1498.62500000			4.70630196		48.87500000

SOURCE	DF	TYPE I SS	F VALUE	PR > F	DF	TYPE III SS	F VALUE	PR > F
X1	1	140.71101071	6.35	0.0199	1	810.33211666	36.59	0.0001
X1*X1	1	892.77914933	40.31	0.0001	1	892.77914933	40.31	0.0001

PARAMETER	ESTIMATE	T FOR H0: PARAMETER = 0	PR > |T|	STD ERROR OF ESTIMATE
INTERCEPT	130.00929147	8.74	0.0001	14.87580615
X1	−3.30171859	−6.05	0.0001	0.57893460
X1*X1	0.03337133	6.35	0.0001	0.00525631

11.20   A department store conducted an experiment to investigate the effects of advertising expenditures on the weekly sales for its men's wear, children's wear, and women's wear departments. Five weeks were randomly selected for each department, and an advertising budget x_1 (hundreds of dollars) was assigned for each. The weekly sales (thousands of dollars) are shown in the table for each of the fifteen one-week sales periods.

Department	Advertising expenditures (hundreds of dollars)				
	1	2	3	4	5
Men's Wear A	5.2	5.9	7.7	7.9	9.4
Children's Wear B	8.2	9.0	9.1	10.5	10.5
Women's Wear C	10.0	10.3	12.1	12.7	13.6

If we expect weekly sales $E(y)$ to be linearly related to advertising expenditure x_1, and if we expect the slopes of the lines corresponding to the three departments to differ, then an appropriate model for $E(y)$ is

$$E(y) = \beta_0 + \underbrace{\beta_1 x_1}_{\substack{\text{Advertising} \\ \text{expenditure}}} + \underbrace{\beta_2 x_2 + \beta_3 x_3}_{\substack{\text{Dummy variables} \\ \text{used to introduce} \\ \text{the qualitative} \\ \text{variable "Department"} \\ \text{into the model}}} + \underbrace{\beta_4 x_1 x_2 + \beta_5 x_1 x_3}_{\substack{\text{Interaction terms} \\ \text{that introduce} \\ \text{differences in} \\ \text{slopes}}}$$

where

$x_1 =$ advertising expenditure

$x_2 = \begin{cases} 1 & \text{if children's wear department B} \\ 0 & \text{if not} \end{cases}$

$$x_3 = \begin{cases} 1 & \text{if women's wear department C} \\ 0 & \text{if not} \end{cases}$$

a. Find the equation of the line relating $E(y)$ to advertising expenditure x_1 for the men's wear department A. (*Hint:* According to the coding employed for the dummy variables, the model represents mean sales $E(y)$ for the men's wear department A when $x_2 = x_3 = 0$. Substitute $x_2 = x_3 = 0$ into the equation for $E(y)$ to find the equation of this line.)

b. Find the equation of the line relating $E(y)$ to x_1 for the children's wear department B. (*Hint:* According to the coding, the model represents $E(y)$ for the children's wear department when $x_2 = 1$ and $x_3 = 0$.)

c. Find the equation of the line relating $E(y)$ to x_1 for the women's wear department C.

d. Find the difference between the intercepts of the $E(y)$ lines corresponding to the children's wear B and men's wear A departments.

e. Find the difference in slopes between $E(y)$ lines corresponding to the women's wear C and men's wear A departments.

f. Refer to part (e). Suppose that you wish to test the null hypothesis that the slopes of the lines corresponding to the three departments are equal. Express this as a test of an hypothesis about one or more of the model parameters.

11.21 If you have access to a computer and a multiple regression computer program package, perform a multiple regression analysis for the data in Exercise 11.20.

a. Verify that SSE = 1.2190, that $s^2 = .1354$, and that $s = .368$.

b. How many degrees of freedom will SSE and s^2 have?

c. Verify that the total $SS = S_{yy} = 76.0493$.

d. Calculate R^2 and interpret its value.

e. Verify that the parameter estimates (values rounded) are approximately equal to

$$\hat{\beta}_0 \approx 4.10 \qquad \hat{\beta}_1 \approx 1.04 \qquad \hat{\beta}_2 \approx 3.53$$

$$\hat{\beta}_3 \approx 4.76 \qquad \hat{\beta}_4 \approx -.43 \qquad \hat{\beta}_5 \approx -.08$$

f. Find the prediction equation and graph the three department sales lines.

g. Examine the graphs in part (f). Do the slopes of the lines corresponding to the children's wear B and the men's wear A departments appear to differ? Test the null hypothesis that the slopes do not differ (i.e., $H_0: \beta_4 = 0$) against the alternative hypothesis that the slopes do differ (i.e., $H_a: \beta_4 \neq 0$). Test using $\alpha = .05$. (*Note:* Verify on your computer printout that the estimated standard deviation of $\hat{\beta}_4$ is $s_{\hat{\beta}_4} \approx .165$.)

h. Find a 95% confidence interval for the difference between the slopes of lines B and A.

i. Do the data provide sufficient evidence to indicate a difference in the slopes between lines C and A? Test using $\alpha = .05$. (*Hint:* $s_{\hat{\beta}_5} \approx .165$.)

11.22 As noted in Example 11.1, "Cucumber pickles usually are preserved by fermentation followed by storage in brine until it is convenient for processors to manufacture various finished products. The primary function of high salt content concentration levels in brine is to retard softening of the pickles and to prevent freezing." Recent studies by R. W. Buescher and colleagues Hudson, Adams, and Wallace suggest that pickle firmness can be maintained for longer periods by adding small amounts of calcium chloride to the brine solution ("Calcium Makes It Possible to Store Cucumber Pickles in Low Salt Brine," *Arkansas Farm Research* 30, no. 4 [July/August 1981]). To compare this brine with the standard salt (sodium chloride) solution, pickles were stored in the two types of brine for 4, 14, 32, and 52 weeks. Pickle firmness y was measured at the start of the experiment and at each of four time periods. The data are shown on the next page.

	Weeks in storage at 72°F				
Brine	0	4	14	32	52
	Firmness in pounds				
Standard	19.8	16.5	12.8	8.1	7.5
Low Salt with Calcium Chloride	19.8	22.0	19.6	20.0	20.0

a. Plot the ten data points and sketch the curve that relates hardness to time for each of the two types of brine.

b. Explain why the following model might be a good choice for characterizing the relationship between mean firmness $E(y)$ and the two predictor variables, time (weeks in storage) and "brine type."

$$E(y) = \beta_0 + \beta_1 x_1 + \beta_2 x_1^2 + \beta_3 x_2 + \beta_4 x_1 x_2 + \beta_5 x_1^2 x_2$$

x_1 = time in weeks

x_2 = 1 if standard brine

x_2 = 0 if not

The qualitative variable brine type is entered into the model using the dummy (indicator) variable x_2.

c. The SAS computer printout for the regression analysis is shown below. Does the model contribute information for the prediction of y? Test using $\alpha = .05$.

SAS output for Exercise 11.22

DEPENDENT VARIABLE: Y

SOURCE	DF	SUM OF SQUARES	MEAN SQUARE	F VALUE	PR > F	R-SQUARE	C.V.
MODEL	5	247.22102336	49.44420467	48.86	0.0011	0.983890	6.0565
ERROR	4	4.04797664	1.01199416		ROOT MSE		Y MEAN
CORRECTED TOTAL	9	251.26900000			1.00597920		16.61000000

SOURCE	DF	TYPE I SS	F VALUE	PR > F	DF	TYPE III SS	F VALUE	PR > F
X1	1	54.85657530	54.21	0.0018	1	0.22880167	0.23	0.6592
X1·X1	1	8.66901488	8.57	0.0429	1	0.11034618	0.11	0.7578
X2	1	134.68900000	133.09	0.0003	1	1.62363682	1.60	0.2740
X1·X2	1	42.88308477	42.37	0.0029	1	16.72909295	16.53	0.0153
X1·X1·X2	1	6.12334841	6.05	0.0697	1	6.12334841	6.05	0.0697

PARAMETER	ESTIMATE	T FOR H0: PARAMETER = 0	PR > |T|	STD ERROR OF ESTIMATE
INTERCEPT	20.71339998	26.19	0.0001	0.79080905
X1	−0.04272103	−0.48	0.6592	0.08984648
X1·X1	0.00055598	0.33	0.7578	0.00168371
X2	−1.41658315	−1.27	0.2740	1.11837288
X1·X2	−0.51661055	−4.07	0.0153	0.12706211
X1·X1·X2	0.00585716	2.46	0.0697	0.00238112

d. Find the observed significance level for the test in part (c) and interpret its value.

e. On the same piece of graph paper, graph the predicted value of y as a function of time of storage x_1 for: (1) the standard brine, and (2) the brine containing calcium chloride.

11.23 Refer to Exercise 11.22. If there is no difference in mean response for the two types of brine, then the variable x_2 need not be included in the model. A test of this null hypothesis is equivalent to a test of the hypothesis that the terms involving x_2 contribute no information for the prediction of y, that is, $H_0: \beta_3 = \beta_4 = \beta_5 = 0$. The SAS computer printout appropriate for fitting the model

$$E(y) = \beta_0 + \beta_1 x_1 + \beta_2 x_1^2$$

to the data is shown below. Use the method of Exercise 11.19 to test the null hypothesis that there is no difference in mean firmness $E(y)$ between the two brine types. Test using $\alpha = .05$. State your conclusions.

SAS regression analysis: reduced model

DEPENDENT VARIABLE: Y

SOURCE	DF	SUM OF SQUARES	MEAN SQUARE	F VALUE	PR > F	R-SQUARE	C.V.
MODEL	2	63.52559018	31.76279509	1.18	0.3606	0.252819	31.1791
ERROR	7	187.74340982	26.82048712		ROOT MSE		Y MEAN
CORRECTED TOTAL	9	251.26900000			5.17884998		16.61000000

SOURCE	DF	TYPE I SS	F VALUE	PR > F	DF	TYPE III SS	F VALUE	PR > F
X	1	54.85657530	2.05	0.1958	1	22.72033479	0.85	0.3880
XI*XI	1	8.66901488	0.32	0.5874	1	8.66901488	0.32	0.5874

PARAMETER	ESTIMATE	T FOR HO: PARAMETER = 0	PR > |T|	STD ERROR OF ESTIMATE
INTERCEPT	20.00510841	6.95	0.0002	2.87873018
XI	−0.30102630	−0.92	0.3380	0.32706224
XI*XI	0.00348455	0.57	0.5874	0.00612908

11.24 In step 2 of our explanation of the SAS multiple regression computer printout in Example 11.2, we noted that the F statistic employed to test the utility of the complete model is related to the multiple coefficient of determination R^2 by the formula

$$F = \frac{R^2/k}{(1 - R^2)/[n - (k + 1)]}$$

Use the value of R^2 given in the computer printout in Table 11.2, $R^2 = 0.9936$, to calculate the value of the F statistic. Check your answer with the correct answer shown in the computer printout in Table 11.2.

11.25 Nancy Kay Butts conducted a study to investigate the relationship between the physical characteristics of female cross-country runners and their completion time y for a 3000-meter race ("Physiological Profiles of High School Female Cross Country Runners," *Research Quarterly for Exercise and Sport* 53, no. 1 [1982]). Measurements were taken on 14 physical characteristics for each of 127 runners—age, height, weight, percent fat, length of time on a treadmill, VO_2 max, maximal ventilation volume (VE max), and so on—along with the runner's completion time y for a 3000-meter race. Butts employed a stepwise regression analysis to develop a prediction equation. Variables entering the first-order stepwise regression model, in order of entry, were

$x_1 =$ minutes on treadmill

$x_2 = VO_2$ max

$x_3 = $ VE max

The multiple correlation coefficient R values associated with these one-, two-, and three-variable models were 0.58, 0.66, and 0.67, respectively.

a. How many degrees of freedom would be associated with SSE for this three-variable model?

b. Does it appear that the three-variable stepwise regression model

$$E(y) = \beta_0 + \beta_1 x_1 + \beta_2 x_2 + \beta_3 x_3$$

provides a good fit to the data?

c. Suppose that you want to determine whether the complete three-variable model contributes information for the prediction of completion time y. Calculate the value of the F statistic required for the test. (*Hint:* See item 2 of the explanation of the SAS multiple regression computer printout in Example 11.2.)

d. Conduct the test in part (c) and state your conclusions. Use $\alpha = .05$.

 11.26 Because dolphins (and other large marine mammals) are considered to be the top predators in the marine food chain, the heavy metal concentrations in striped dolphins were measured as part of a marine pollution study. The concentration of mercury, the heavy metal reported in this study, is expected to differ for males and females, since the mercury in a female is apparently transferred to her offspring during gestation and nursing. This study involved 28 males between the ages of 0.21 and 39.5 years, and 17 females between the ages of 0.80 and 34.5 years. For the data that follow,

$x_1 = $ age of the dolphin in years

$x_2 = \begin{cases} 0 & \text{if female} \\ 1 & \text{if male} \end{cases}$

$y = $ mercury concentration ($\mu g/g$) in the liver

y	x_1	x_2	y	x_1	x_2
1.70	0.21	1	481.00	22.50	1
1.72	0.33	1	485.00	24.50	1
8.80	2.00	1	221.00	24.50	1
5.90	2.20	1	406.00	25.50	1
101.00	8.50	1	252.00	26.50	1
85.40	11.50	1	329.00	26.50	1
118.00	11.50	1	316.00	26.50	1
183.00	13.50	1	445.00	26.50	1
168.00	16.50	1	278.00	27.50	1
218.00	16.50	1	286.00	28.50	1
180.00	17.50	1	315.00	29.50	1
264.00	20.50	1	241.00	31.50	1

y	x_1	x_2	y	x_1	x_2
397.00	31.50	1	142.00	17.50	0
209.00	36.50	1	180.00	17.50	0
314.00	37.50	1	174.00	18.50	0
318.00	39.50	1	247.00	19.50	0
2.50	0.80	0	223.00	21.50	0
9.35	1.58	0	167.00	21.50	0
4.01	1.75	0	157.00	25.50	0
29.80	5.50	0	177.00	25.50	0
45.30	7.50	0	475.00	32.50	0
101.00	8.05	0	342.00	34.50	0
135.00	11.50	0			

```
MTB > REGRESS CI 4 C2−C5

THE REGRESSION EQUATION IS
Y = − 17.7 + 10.7 XI − 28.7 X2 + 12.7 XI∗X2 − 0.371 XISQ∗X2

PREDICTOR      COEF      STDEV      T-RATIO      P
CONSTANT      −17.65      32.02      −0.55      0.585
XI             10.737      1.690      6.35      0.000
X2            −28.69      48.84      −0.59      0.560
XI∗X2          12.722      4.440      2.87      0.007
XISQ∗X2       −0.3707     0.1053     −3.52      0.001

S = 71.29     R-SQ = 76.5%     R-SQ (ADJ) = 74.1%

ANALYSIS OF VARIANCE

SOURCE         DF       SS        MS        F         P
REGRESSION      4     661313    165328    32.53    0.000
ERROR          40     203285     5082
TOTAL          44     864598

SOURCE         DF     SEQ SS
XI              I     585782
X2              I      11185
XI∗X2           I       1371
XISQ∗X2         I      62974

UNUSUAL OBSERVATIONS
OBS.     XI      Y       FIT     STDEV. FIT    RESIDUAL    ST. RESID
 13     22.5   481.0    293.8       18.1         187.2       2.71R
 14     24.5   485.0    305.9       17.2         179.1       2.59R
 28     39.5   318.0    302.0       41.8          16.0       0.28X
 44     32.5   475.0    331.3       32.9         143.7       2.27R

R DENOTES AN OBS. WITH A LARGE ST. RESID.
X DENOTES AN OBS. WHOSE X VALUE GIVES IT LARGE INFLUENCE.
```

Use the **MINITAB** printout to answer the following questions.
a.   What is the model that has been fitted?
b.   What is the fitted prediction equation?

c. Find R^2 and interpret its value.

d. Do the data provide sufficient evidence for us to conclude that the model contributes significant information for predicting y?

e. What is the prediction equation for predicting the mercury concentration in a female dolphin as a function of her age?

f. What is the prediction equation for predicting the mercury concentration in a male dolphin as a function of his age?

g. Does the quadratic term in the prediction equation for males contribute significantly to the prediction of mercury concentration in a male dolphin?

h. Are there any other important conclusions that you feel were not considered regarding the fitted prediction equation?

PREDICTING WORKER ABSENTEEISM

One of the major problems facing business managers, among others, is **worker absenteeism**. Can you imagine being responsible for the planning and production of college homecoming activities when the participants, seemingly at random, miss the organizational and functional meetings? Or can you imagine coaching a professional football team when the players, seemingly at random, are absent from games or practice sessions? Or can you imagine managing an automobile assembly line when various members of the assembly team are absent on any given day? As worker absenteeism increases, the productivity of an operating group, an office force, a manufacturing plant, and so on, decreases, and the quality of the product usually decreases also. Consequently, business managers seek to identify the causes of worker absenteeism so that they can take action to control or reduce it.

K. Constas and R. P. Vichas employed a multiple regression analysis in an attempt to identify variables related to worker absenteeism at two different types of overseas industrial plants ("An Interpretative Policy-Making Model of Absenteeism with Reference to the Marginal Worker in Overseas Plants," *Journal of Management Studies* 17, no. 2 [1980]). They confined their attention to independent variables that characterized a worker's "life space," that is, variables that personally characterized the worker, as opposed to the work space or working conditions to which the worker was exposed.

The measure of worker absenteeism was the number of days the worker was absent from his or her job during a one-year period. The predictor variables and their type—qualitative or quantitative—and the symbols that represent them were:

1. Sex (qualitative)
   Coding: $x_1 = 0$ if male
   $\quad\quad\quad x_1 = 1$ if female

2. Marital Status (qualitative)
   Coding: $x_2 = 0$ if married or divorced
   $\quad\quad\quad x_2 = 1$ if single

3. Age of Employee (quantitative)
   Coding: $x_3 =$ age of the employee

4. Number of Dependents (quantitative)
   Coding: $x_4 =$ number of dependents

5. Automobile Ownership (qualitative)
   Coding: $x_5 = 0$ if employee owns a vehicle
   $x_5 = 1$ if not

   The data for the Constas-Vichas study consisted of observations on 50 workers randomly selected from among the work force of an electronics assembly plant and observations on 120 workers at a petrochemical plant, both located in southern Puerto Rico. The authors fitted separate first-order regression models

   $$E(y) = \beta_0 + \beta_1 x_1 + \beta_2 x_2 + \cdots + \beta_5 x_5$$

Table 1
Results of two multiple regression analyses

	Plant A	Plant B
Sample Size, n	50	120
Coefficient of determination, R^2	0.452	0.282
F	2.266	1.969

Table 2
Results of two multiple regression analyses

Plant A		Variable				
		Sex	Marital status	Age of employee	Number of dependents	Automobile ownership
Parameter	β_0	β_1	β_2	β_3	β_4	β_5
Parameter estimate	12.388	5.113	4.756	−0.315	0.675	1.906
Estimated standard deviation of estimate		3.139	2.856	0.152	0.729	2.209
Computed value of t		1.63	1.67	−2.07*	0.93	0.86

Plant B		Variable				
		Sex	Marital status	Age of employee	Number of dependents	Automobile ownership
Parameter	β_0	β_1	β_2	β_3	β_4	β_5
Parameter estimate	13.180	−13.101	3.043	−0.064	−1.218	10.714
Estimated standard deviation of estimate		8.914	5.831	0.178	0.933	4.150
Computed value of t		− 1.47	0.52	−0.36	−1.31	2.58*

* Stars attached to t values indicate that they are statistically significant for $\alpha = .05$.

to each of the two sets of data. The sample sizes, their computed values of R^2, and F values for tests of the complete models are shown in Table 1. The parameter estimates, their estimated standard deviations (standard errors), and the computed values of the t statistics are presented in Table 2.

1. Discuss each of the five variables selected for the model. Why do you think the variable was selected, and how do you think it will affect the response variable y?

2. Does the model contribute information for the prediction of worker absenteeism y for plants A and B?

3. Discuss the relationship between absenteeism y and the worker's age, x_3. If the variable x_3 is entered into the model as a first-order term, is the model likely to provide a good fit to the data?

4. Discuss the fit of the models as measured by the values of R^2.

CHAPTER 12

▷ ━ ANALYSIS OF
ENUMERATIVE DATA

Case Study

What do you think about drinking wine from a can? The marketer says cans are lighter in weight, more compact, and less susceptible to damage. But will the consumer buy them? The case study at the end of this chapter discusses methods for analyzing a consumer preference poll concerning canned versus bottled wine.

General Objective

Many types of surveys and experiments result in qualitative rather than quantitative response variables, so that the responses can be classified but not quantified. Data from these experiments consist of the count or number of observations falling in each of the response categories included in the experiment. This chapter is concerned with methods of analysis for count (or enumerative) data.

Specific Topics

1. The multinomial experiment (12.1)
2. A test of specific cell probabilities (12.2)
3. Contingency tables (12.3)
4. A test for homogeneity of multinomial populations (12.4)
5. Other applications (12.8)
6. Assumptions (12.9)

▷ 12.1   A DESCRIPTION OF THE EXPERIMENT

Many experiments, particularly in the social sciences, result in enumerative (or count) data. For instance, the classification of people into five income brackets results in an enumeration or count corresponding to each of the five income classes. Or we might be interested in studying the reaction of a mouse to a particular stimulus in a psychological experiment. If the mouse were to react in one of three ways when the stimulus was applied, and if a large number of mice were subjected to the stimulus, the experiment would yield three counts, indicating the number of mice falling in each of the reaction classes. Similarly, a traffic study might require a count and classification of the type of motor vehicles using a section of highway. An industrial process manufactures items that fall into one of three quality classes: acceptable, seconds, and rejects. A student of the arts might classify paintings in one of k categories according to style and period in order to study trends in style over time. We might wish to classify ideas in a philosophical study or style in the field of literature. The results of an advertising campaign would yield count data indicating a classification of consumer reaction. Indeed, many observations in the physical sciences are not amenable to measurement on a continuous scale and hence result in enumerative or classificatory data.

The illustrations in the preceding paragraph exhibit, to a reasonable degree of approximation, the following characteristics that define a **multinomial experiment**:

THE CHARACTERISTICS OF A MULTINOMIAL EXPERIMENT

1.   The experiment consists of n identical trials.

2.   The outcome of each trial falls into one of k classes or cells.

3.   The probability that the outcome of a single trial will fall in a particular cell, say cell i, is $p_i(i = 1, 2, \ldots, k)$ and remains the same from trial to trial. Note that $0 \leq p_i \leq 1$ for all i, and

$$p_1 + p_2 + p_3 + \cdots + p_k = 1$$

4.   The trials are independent.

5.   The experimenter is interested in $n_1, n_2, n_3, \ldots, n_k$, where $n_i(i = 1, 2, \ldots, k)$ is equal to the number of trials in which the outcome falls in cell i. Note that $n_1 + n_2 + n_3 + \cdots + n_k = n$.

A multinomial experiment is analogous to tossing n balls at k boxes, where each ball must fall into one of the boxes. The boxes are arranged so that the probability that a ball will fall into a box varies from box to box but remains the same for a particular box in repeated tosses. Finally, the balls are tossed in such a way that the trials are independent. At the conclusion of the experiment, we observe n_1 balls in the first box, n_2 in the second, . . . , and n_k in the kth. The total number of balls is equal to

$$\sum_{i=1}^{k} n_i = n$$

Note the similarity between the binomial and multinomial experiments and, in particular, note that the binomial experiment represents the special case of the multinomial experiment when $k = 2$. The two cell probabilities p and q of the binomial experiment are replaced by the k cell probabilities, $p_1, p_2, \ldots, p_k$ of the multinomial experiment. The objective of this chapter is to make inferences about the cell probabilities $p_1, p_2, \ldots, p_k$. The inferences will be expressed in terms of a statistical test of an hypothesis concerning their specific numerical values or their relationship to one another.

If we were to proceed as in Chapter 4, we would derive the probability of the observed sample $(n_1, n_2, \ldots, n_k)$ for use in calculating the probability of the type I and type II errors associated with a statistical test. Fortunately, we have been relieved of this chore by the British statistician Karl Pearson, who proposed a very useful test statistic for testing hypotheses concerning $p_1, p_2, \ldots, p_k$ and gave its approximate sampling distribution.

▷ 12.2 THE CHI-SQUARE TEST

Suppose that $n = 100$ balls were tossed at the cells (boxes) and that we knew that p_1 was equal to .1. How many balls would be expected to fall into the first cell? Referring to Chapter 4 and utilizing knowledge of the binomial experiment, we would calculate

$$E(n_1) = np_1 = 100(.1) = 10$$

In like manner, the expected number falling into the remaining cells can be calculated using the formula

$$E(n_i) = np_i, \qquad i = 1, 2, \ldots, k$$

Now suppose that we hypothesize values for $p_1, p_2, \ldots, p_k$ and calculate the expected value for each cell. Certainly, if our hypothesis is true, the cell counts n_i should not deviate greatly from their expected values $np_i (i = 1, 2, \ldots, k)$. Hence, it would seem intuitively reasonable to use a test statistic involving the k deviations,

$$(n_i - np_i), \qquad i = 1, 2, \ldots, k$$

In 1900, Karl Pearson proposed the following test statistic, which is a function of the squares of the deviations of the observed counts from their expected values, weighted by the reciprocals of their expected values:

CHI-SQUARE TEST STATISTIC

$$
X^2 = \sum_{i=1}^{k} \frac{[n_i - E(n_i)]^2}{E(n_i)}
$$
$$
= \sum_{i=1}^{k} \frac{[n_i - np_i]^2}{np_i}
$$

Although the mathematical proof is beyond the scope of this text, it can be shown that when n is large, X^2 has, approximately, a chi-square probability distribution in repeated sampling. Experience has shown that the cell counts n_i should not be too small in order that the chi-square distribution provide an adequate approximation to the distribution of X^2. As a rule of thumb, **we will require that all expected cell counts equal or exceed 5**, although Cochran (see the references) has noted that this value can be as low as 1 for some situations.

Recall from Section 9.6 that the chi-square probability distribution is used for testing an hypothesis concerning a population variance σ^2 and that the shape of the chi-square distribution varies according to the number of degrees of freedom associated with s^2. We discussed the use of Table 5 in Appendix III, which presents the critical values of χ^2 corresponding to various right-hand tail areas of the distribution. Therefore, we must know which chi-square distribution to use in approximating the distribution of X^2. That is, we must determine the number of degrees of freedom associated with the appropriate chi-square distribution. Furthermore, we must determine whether the rejection region for the test is one-tailed or two-tailed.

The latter problem is easily solved. Since large deviations of the observed cell counts from those expected would tend to contradict the null hypothesis concerning the cell probabilities $p_1, p_2, \ldots, p_k$, we reject the null hypothesis when X^2 is large and employ a one-tailed statistical test using the upper-tail values of χ^2 to locate the rejection region.

The determination of the appropriate number of degrees of freedom for the test can be rather difficult and therefore will be specified for the practical applications described in the following sections. In addition, we will state the principle involved (which is fundamental to the mathematical proof of the approximation) so that you may understand why the number of degrees of freedom changes with various applications. The principle states that **the appropriate number of degrees of freedom equals the number of cells k less one degree of freedom for each independent linear restriction placed on the observed cell counts**. For example, one linear restriction is always present because the sum of the cell counts must equal n; that is,

$$n_1 + n_2 + n_3 + \cdots + n_k = n$$

Other restrictions will be introduced for some applications because of the necessity for estimating unknown parameters required in the calculation of the expected cell frequencies or because of the method in which the sample is collected. These restrictions will become apparent as we consider various practical examples.

▷ 12.3 A TEST OF AN HYPOTHESIS CONCERNING SPECIFIED CELL PROBABILITIES

The simplest hypothesis concerning the cell probabilities is one that specifies numerical values for each cell. For example, suppose that a psychologist performs the following experiment. One or more rats are attracted to the end of a ramp that divides, leading to three different doors. The objective of the experiment is to determine whether the rats have or acquire a preference for one of the three doors. We wish to test the hypothesis that the rats have no preference in the choice of a door, so that over the long run, we would expect the rats to choose each door approximately one-third of the time. Therefore,

$$H_0 : p_1 = p_2 = p_3 = \frac{1}{3}$$

versus

$$H_a : \text{at least one } p_i \text{ is different from } \frac{1}{3}$$

where p_i is the probability that a rat will choose door i, $i = 1, 2,$ or 3.

Suppose that the rat was sent down the ramp $n = 90$ times and that the three observed cell frequencies were $n_1 = 23$, $n_2 = 36$, and $n_3 = 31$. The expected cell frequencies are the same for each cell; namely, $E(n_i) = np_i = 90(1/3) = 30$.

The observed and expected cell frequencies are presented in Table 12.1. Noting the discrepancy between the observed and expected cell frequencies, we would wonder whether the data present sufficient evidence to warrant rejection of the hypothesis of no preference.

Table 12.1
Observed and expected cell counts for the rat experiment

	Door		
	1	2	3
Observed cell frequency	$n_1 = 23$	$n_2 = 36$	$n_3 = 31$
Expected cell frequency	30	30	30

The chi-square test statistic for this example has $(k - 1) = 2$ degrees of freedom since the only linear restriction on the cell frequencies is that

$$n_1 + n_2 + \cdots n_k = n$$

or, for our example,

$$n_1 + n_2 + n_3 = 90$$

Therefore, if we choose $\alpha = .05$, we would reject the null hypothesis when $X^2 > 5.99147$ (see Table 5 in Appendix III).

Substituting into the formula for X^2, we obtain

$$X^2 = \sum_{i=1}^{k} \frac{[n_i - E(n_1)]^2}{E(n_i)} = \sum_{i=1}^{k} \frac{[n_i - np_i]^2}{np_i}$$

$$= \frac{(23 - 30)^2}{30} + \frac{(36 - 30)^2}{30} + \frac{(31 - 30)2}{30} = 2.87$$

Since X^2 is less than the tabulated critical value of χ^2, 5.991, the null hypothesis is not rejected and we conclude that the data do not present sufficient evidence to indicate that the rats have a preference for a particular door.

EXERCISES   Basic Techniques

12.1   List the characteristics of a multinomial experiment.

12.2   Give the value of χ_α^2 for
   a. $\alpha = .05$, d.f. $= 3$ c. $\alpha = .10$, d.f. $= 15$
   b. $\alpha = .01$, d.f. $= 8$ d. $\alpha = .10$, d.f. $= 11$

12.3   Give the rejection region for a chi-square test of specified cell probabilities if the experiment involves k cells, where
   a. $k = 7, \alpha = .10$ c. $k = 14, \alpha = .05$
   b. $k = .10, \alpha = .01$ d. $k = 3, \alpha = .05$

12.4   Suppose that a response can fall in one of $k = 5$ categories with probabilities $p_1, p_2, \ldots, p_5$, respectively, and that $n = 300$ responses produced the following category counts:

Category	1	2	3	4	5
Observed count	47	63	74	51	65

   a.   If you were to test $H_0 : p_1 = p_2 = \cdots = p_5$ using the chi-squre test, how many degrees of freedom would the test statistic have?
   b.   If $\alpha = .05$, find the rejection region for the test.
   c.   What is your alternative hypothesis?
   d.   Conduct the test in part (a) using $\alpha = .05$. State your conclusions.
   e.   Find the approximate observed significance level for the test and interpret its value.

12.5   Suppose that a response can fall in one of $k = 3$ categories with probabilities p_1, p_2, and p_3, respectively, and that $n = 300$ responses produced the following category counts:

Category	1	2	3
Observed count	130	98	72

   a.   If you were to test $H_0 : p_1 = p_2 = p_3$ using the chi-square test, how many degrees of freedom would the test statistic have?
   b.   If $\alpha = .05$, find the rejection region for the test.
   c.   What is your alternative hypothesis?
   d.   Conduct the test in part (a) using $\alpha = .05$. State your conclusions.
   e.   Find the approximate observed significance level for the test and interpret its value.

Applications

12.6   A city expressway utilizing four lanes in each direction was studied to see whether drivers preferred to drive on the inside lanes. A total of 1000 automobiles were observed during the heavy early-morning traffic, and their respective lanes were recorded. The results were as

follows:

Lane	1	2	3	4
Observed count	294	276	238	192

Do the data present sufficient evidence to indicate that some lanes are preferred over others? (Test the hypothesis that $p_1 = p_2 = p_3 = p_4 = 1/4$, using $\alpha = .05$.)

12.7 The Mendelian theory states that the number of peas of a certain type falling into the classifications round and yellow, wrinkled and yellow, round and green, and wrinkled and green should be in the ratio $9:3:3:1$. Suppose that 100 such peas revealed 56, 19, 17, and 8 in the respective classes. Do these data disagree with the Mendelian theory? Use $\alpha = .05$.

12.8 In theory, it seems likely that the frequency of assaults and fights in non-air-conditioned jails and prisons would increase in weather that was especially humid, hot, or cold. Randy Atlas classified 1329 prison assaults by month of occurrence for four Florida state prisons ("Violence in Prison," *Environment and Behavior* 16, no. 3 [1984] © 1984 Sage Publications, Inc., with permission). The table below gives the number of assaults per month for one of these, the non-air-conditioned maximum security Florida State Prison at Raiford. Use a chi-square test to test the null hypothesis that the frequency of occurrence of assaults at the Florida State Prison is the same for any one month as it is for any other.

Month	Jan	Feb	Mar	Apr	May	June	July	Aug	Sept	Oct	Nov	Dec
Frequency of assault	30	30	33	42	25	28	41	32	48	28	28	24

a. State the null hypothesis in terms of cell probabilities.
b. State the alternative hypothesis.
c. Give the rejection region for the test for $\alpha = .05$.
d. Conduct the test and state your conclusions.

12.9 Medical statistics show that deaths due to four major diseases—call them A, B, C, and D—account for 15, 21, 18, and 14%, respectively, of all nonaccidental deaths. A study of the causes of 308 nonaccidental deaths at a hospital gave the following counts of patients dying of diseases A, B, C, and D.

Disease	Number of deaths
A	43
B	76
C	85
D	21
Other	83
	308

Do these data provide sufficient evidence to indicate that the proportions of people dying of diseases A, B, C, and D at this hospital differ from the proportions accumulated for the population at large?

12.10   Research has suggested a link between the prevalence of schizophrenia and birth during particular months of the year in which viral infections are prevalent. Suppose you are working on a similar problem and you suspect a linkage between a disease observed in later life and month of birth. You have records of 400 cases of the disease and you classify them according to month of birth. The data appear in the table. Do the data present sufficient evidence to indicate that the proportion of cases of the disease per month varies from month to month? Test with $\alpha = .10$.

Month	Jan	Feb	Mar	Apr	May	June	July	Aug	Sept	Oct	Nov	Dec
Number of births	38	31	42	46	28	31	24	29	33	36	27	35

12.11   Portable personal computers, sometimes called "laptops," represent a fast-growing segment of the PC market. According to Market Intelligence Research Company, the use of portable "laptop" computers can be classified in the following user segments ("Laptop's Three Musts for Success in Sales," *Sales and Marketing Management*, February 1988, p. 91):

Business–professional	69%
Government	21%
Education	7%
Home	3%

A sample of $n = 150$ laptop computer owners were surveyed, and the user segments were tabulated as follows:

Business–professional	102
Government	32
Education	`12
Home	4

Do the data provide sufficient evidence to indicate that the figures given by Market Research Company are not accurate? Calculate the approximate *p*-value for the test, and use it to make your decision.

▷ 12.4   CONTINGENCY TABLES

A problem frequently encountered in the analysis of count data concerns the independence of two methods of classifying observed events. For example, suppose we wish to classify defects found on furniture produced in a manufacturing plant—first, according to the type of defect and, second, according to the production shift during which the piece of furniture was produced. If the proportions of the various types of defects are constant from shift to shift, then classification by defects is independent of the classification by production shift. On the other hand, if the proportions of the various defects vary from shift to shift, then the classification by defects is **contingent** upon the shift classification, and the classifications are dependent. In investigating whether one method of classification is contingent upon another, we display the data by using a cross-classification in an array called a **contingency table**.

A total of $n = 309$ furniture defects were recorded and the defects were classified as being one of four types: A, B, C, or D. At the same time, each piece of furniture was identified according to the production shift in which it was manufactured. These counts are presented as a contingency table in Table 12.2. (*Note:* Numbers in parentheses are the expected cell frequencies.)

Table 12.2
Contingency table

| Shift | Type of defect | | | | |
	A	B	C	D	Total
1	15(22.51)	21(20.99)	45(38.94)	13(11.56)	94
2	26(22.99)	31(21.44)	34(39.77)	5(11.81)	96
3	33(28.50)	17(26.57)	49(49.29)	20(14.63)	119
Total	74	69	128	38	309

Let p_A be the unconditional probability that a defect will be of type A. Similarly, define $p_B, p_C,$ and p_D as the probabilities of observing the three other types of defects. Then these probabilities, which we will call the **column probabilities** of Table 12.2, will satisfy the requirement

$$p_A + p_B + p_C + p_D = 1$$

In like manner, let p_i ($i = 1, 2,$ or 3) equal the **row probability** that a defect will have occurred on shift i, where

$$p_1 + p_2 + p_3 = 1$$

Then, if the two classifications are independent of each other, a cell probability will equal the product of its respective row and column probabilities in accordance with the multiplicative law of probability.

For example, the probability that a particular defect will occur on shift 1 and be of type A is $(p_1)(p_A)$. Thus, we observe that the numerical values of the cell probabilities are unspecified in the problem under consideration. **The null hypothesis specifies only that each cell probability will equal the product of its respective row and column probabilities and therefore imply independence of the two classifications.** The alternative hypothesis is that this equality does not hold for at least one cell.

The analysis of the data obtained from a contingency table differs from the problem discussed in Section 12.3 because we must **estimate** the row and column probabilities so that we can estimate the expected cell frequencies.

If proper estimates of the cell probabilities are obtained, the estimated expected cell frequencies can be substituted for the $E(n_i)$ in X^2, and X^2 will continue to have a distribution in repeated sampling that is approximated by the chi-square probability distribution. The proof of this statement, as well as a discussion of the methods for obtaining the estimates, is beyond the scope of this text. Fortunately, the procedures for obtaining the estimates, known as the **method of maximum likelihood and the method of minimum chi-square**, yield estimates that are intuitively obvious for our relatively simple applications.

It can be shown that the maximum likelihood estimator of a column probability will equal the column total divided by $n = 309$. If we denote the total of column j as c_j, then

$$\hat{p}_A = \frac{c_1}{n} = \frac{74}{309} \qquad \hat{p}_C = \frac{c_3}{n} = \frac{128}{309}$$

$$\hat{p}_B = \frac{c_2}{n} = \frac{69}{309} \qquad \hat{p}_D = \frac{c_4}{n} = \frac{38}{309}$$

Likewise, the row probabilities p_1, p_2, and p_3 can be estimated using the row totals r_1, r_2, and r_3:

$$\hat{p}_1 = \frac{r_1}{n} = \frac{94}{309} \qquad \hat{p}_2 = \frac{r_2}{n} = \frac{96}{309} \qquad \hat{p}_3 = \frac{r_3}{n} = \frac{119}{309}$$

Denote the observed frequency of the cell in row i and column j of the contingency table by n_{ij}. Then the estimated expected value of n_{11} will be

$$\hat{E}(n_{11}) = n[\hat{p}_1 \hat{p}_A] = n\left(\frac{r_1}{n}\right)\left(\frac{c_1}{n}\right)$$

$$= \frac{r_1 c_1}{n}$$

where $(\hat{p}_1 \hat{p}_A)$ is the estimated cell probability. Likewise, we can find the estimated expected value for any other cell, say $\hat{E}(n_{23})$:

$$\hat{E}(n_{23}) = n[\hat{p}_2 \hat{p}_c] = n\left(\frac{r_2}{n}\right)\left(\frac{c_3}{n}\right) = \frac{r_2 c_3}{n}$$

Thus, we see that the estimated expected value of the observed cell frequency n_{ij} for a contingency table is equal to the product of its respective row and column totals divided by the total frequency; that is,

ESTIMATED EXPECTED CELL FREQUENCY

$$\hat{E}(n_{ij}) = \frac{r_i c_j}{n}$$

where

r_i = total for row i

c_j = total for column j

The estimated expected cell frequencies for our example are shown in parentheses in Table 12.2.

We can now use the expected and observed cell frequencies shown in Table 12.2 to calculate the value of the test statistic

$$X^2 = \sum_{j=1}^{4} \sum_{i=1}^{3} \frac{[n_{ij} - \hat{E}(n_{ij})]^2}{\hat{E}(n_{ij})}$$

$$= \frac{(15 - 22.51)^2}{22.51} + \frac{(26 - 22.99)^2}{22.99} + \cdots + \frac{(20 - 14.63)^2}{14.63}$$

$$= 19.18$$

The next step is to determine the appropriate number of degrees of freedom associated with the test statistic. We give the rule here, which we will later attempt to justify. **The degrees of freedom associated with a contingency table having r rows and c columns will always equal $(r - 1)(c - 1)$.** Thus, for our example, we will compare X^2 with the critical value of χ^2 with $(r - 1)(c - 1) = (3 - 1)(4 - 1) = 6$ degrees of freedom.

You will recall that the number of degrees of freedom associated with the X^2 statistic equals the number of cells (in this case, $k = rc$) less one degree of freedom for each independent linear restriction placed on the observed cell frequencies. The total number of cells for the data of Table 12.2 is $k = 12$. From this we subtract one degree of freedom because the sum of the observed cell frequencies must equal n; that is,

$$n_{11} + n_{12} + \cdots + n_{34} = 309$$

In addition, we used the cell frequencies to estimate three of the four column probabilities. Note that the estimate of the fourth column probability will be determined once we have estimated p_A, p_B, and p_C because

$$p_A + p_B + p_C + p_D = 1$$

Thus, we lose $(c - 1) = 3$ degrees of freedom for estimating the column probabilities.

Finally, we used the cell frequencies to estimate $(r - 1) = 2$ row probabilities and, therefore, we lose $(r - 1) = 2$ additional degrees of freedom. The total number of degrees of freedom remaining will be

$$\text{d.f.} = 12 - 1 - 3 - 2 = 6$$

In general, we see that the total number of degrees of freedom associated with an $r \times c$ contingency table is

$$\text{d.f.} = rc - 1 - (c - 1) - (r - 1)$$

$$= rc - c - r + 1 = (r - 1)(c - 1)$$

Therefore, if we use $\alpha = .05$, we will reject the null hypothesis that the two classifications are independent if $X^2 > 12.5916$. Since the value of the test statistic, $X^2 = 19.18$, exceeds the critical value of χ^2, we reject the null hypothesis. The data present sufficient evidence to indicate that the proportion of the various types of defects varies from shift to shift. A study of the production operations for the three shifts would likely reveal the cause or causes of the observed discrepancies.

EXAMPLE 12.1 A survey was conducted to evaluate the effectiveness of a new flu vaccine that had been administered in a small community. The vaccine was provided free of charge in a two-shot sequence over a period of two weeks. Some people received the two-shot sequence, some appeared only for the first shot, and others received neither.

A survey of 1000 local inhabitants the following spring provided the information shown in Table 12.3. Do the data present sufficient evidence to indicate that the vaccine was successful in reducing the number of flu cases in the community?

Table 12.3
Data tabulation for
Example 12.1

	No vaccine	One shot	Two shots	Total
Flu	24(14.4)	9(5.0)	13(26.6)	46
No flu	289(298.6)	100(104.0)	565(551.4)	954
Total	313	109	578	1000

Solution The question concerning the success of the vaccine in reducing flu cases can be restated in terms of whether the data provide sufficient evidence to indicate a dependence between the vaccine classification and the occurrence or nonoccurrence of flu. Therefore, we analyze the data as a contingency table.

The estimated expected cell frequencies may be calculated using the appropriate row and column totals,

$$\hat{E}(n_{ij}) = \frac{r_i c_j}{n}$$

Thus,

$$\hat{E}(n_{11}) = \frac{r_1 c_1}{n} = \frac{46(313)}{1000} = 14.4$$

$$\hat{E}(n_{12}) = \frac{r_1 c_2}{n} = \frac{46(109)}{1000} = 5.0$$

$$\vdots$$

$$\hat{E}(n_{23}) = \frac{r_2 c_3}{n} = \frac{954(578)}{1000} = 551.4$$

These values are shown in parentheses in Table 12.3.

The value of the test statistic X^2 can now be computed and compared with the critical value of χ^2 having $(r-1)(c-1) = (1)(2) = 2$ degrees of freedom. Then, for

$\alpha = .05$, we will reject the null hypothesis when $X^2 > 5.99147$. Substituting into the formula for X^2, we obtain

$$X^2 = \frac{(24 - 14.4)^2}{14.4} + \frac{(289 - 298.6)^2}{298.6} + \cdots + \frac{(565 - 551.4)^2}{551.4}$$

$$= 17.35$$

Since the observed value of X^2 falls in the rejection region, we reject the null hypothesis of independence of the two methods of classification. Based on our sample evidence, we believe there is a dependence between the vaccine classification and the occurrence or nonoccurrence of the flu.

Has the vaccine been successful? A comparison of the percentage incidence of flu for each of the three categories would suggest that those receiving the two-shot sequence were less susceptible to the disease. Further analysis of the data could be obtained by deleting one of the three categories—the second column, for example—to compare the effect of the vaccine with that of no vaccine. This could be done by using a 2 × 2 contingency table or treating the two categories as two binomial populations and using the methods of Section 8.7. Or, we might wish to analyze the data by comparing the results of the two-shot vaccine sequence with those of the combined no vaccine–one shot group; that is, we would combine the first two columns of the 2 × 3 table into one.   ◁

EXERCISES   Basic Techniques

12.12   Calculate the value and give the number of degrees of freedom for X^2 for the following contingency tables.

a.

	Columns			
Rows	1	2	3	4
1	120	70	55	16
2	79	108	95	43
3	31	49	81	140

b.

	Columns		
Rows	1	2	3
1	35	16	84
2	120	92	206

12.13   Suppose that a consumer survey summarizes the responses of $n = 307$ people in a contingency table that contains three rows and five columns. How many degrees of freedom will be associated with the chi-square test statistic?

12.14   A survey of 400 respondents produced the following cell counts in a 2 × 3 contingency table.

	Columns			
Rows	1	2	3	Total
1	37	34	93	164
2	66	57	113	236
Total	103	91	206	400

a. If you wish to test the null hypothesis of "independence"—that the probability that a response falls in any one row is independent of the column it will fall in—and you plan to use a chi-square test, how many degrees of freedom will be associated with the χ^2 statistic?

b. Find the value of the test statistic.

c. Find the rejection region for $\alpha = .10$.

d. Conduct the test and state your conclusions.

e. Find the approximate observed significance level for the test and interpret its value.

Applications

12.15 A recent study conducted by J. E. Brush and colleagues suggests that the initial electrocardiogram of a suspected heart attack victim can be used to predict in-hospital complications of an acute nature ("Use of the Initial Electrocardiogram to Predict In-Hospital Complications of Acute Myocardial Infarction," *New England Journal of Medicine* [May 2, 1985]). The study included 469 patients with suspected myocardial infarction. Each of these patients was categorized in the table below according to whether their initial electrocardiogram was positive or negative and whether the person suffered life-threatening complications of an acute myocardial infarction subsequently in the hospital. The tabled values give the number of persons in each category.

Electrocardiogram results	Subsequent in-hospital life-threatening complications		
	No	Yes	Total
Negative	166	1	167
Positive	260	42	302
	426	43	469

Do the data present sufficient evidence to indicate that the probability of having life-threatening acute myocardial infarctions is dependent on the electrocardiogram results? To answer this question,

a. State the null hypothesis to be tested and the alternative hypothesis.

b. Give the formula for the test statistic and substitute the numerical values into it.

c. Give the rejection region for the test, using $\alpha = .05$.

d. State your conclusions for each of the two possible outcomes of the test.

e. What assumptions are necessary in order that this test be valid? Are they satisfied for these data?

f. Complete the necessary calculations, conduct the test, and state your conclusion.

12.16   In Exercise 8.33, we presented data giving the classifications of 459 police shootings in Philadelphia over the period 1970–78. The purpose of the study, conducted by William B. Waegel ("The Use of Lethal Force by Police: The Effect of Statutory Change," *Crime and Delinquency* 30, no. 1 [January 1984]), was to determine whether a 1973 change in the law (that placed greater restrictions on when a police officer could use lethal force) affected the proportion of unjustified uses of lethal force. The table below shows how the 459 shootings over the 1970–78 period were classified.

	Total for period 1970–1972	1973	Total for period 1974–1978
Justified	72	30	173
Not justified	11	9	59
Unable to determine	18	7	53
Accidental	10	5	12
Total	111	51	297

a.   Do the data provide sufficient evidence to indicate that the distribution of shootings according to whether they are justified, not justified, and so on differed among the three time periods? Test using OC = .05.

b.   Give the approximate *p*-value for the test.

12.17   The data shown below are the results of a study by J. Farwell and J. T. Flannery to determine whether the parents, siblings, and offspring of children with central nervous system tumors have a higher incidence of cancer than the general public ("Cancer in Relatives of Children with Central-Nervous-System Neoplasms," *New England Journal of Medicine* 311, no. 12 [September 20, 1984]). Among a group of 643 children with this type of cancer, 73 had at least one relative with cancer. In contrast, of 360 healthy children (controls), 35 had at least one relative with cancer.

	At least one relative with cancer	No relatives with cancer	Total
Children with central nervous system tumors	73	570	643
Controls	35	325	360
Total	108	895	1003

a.   Do the data present sufficient evidence to indicate a dependence between the occurrence of central nervous system tumors in children and the incidence of cancer in their parents, siblings, and children? Test using the chi-square test with $\alpha = .05$.

b.   Let p_1 represent the proportion of children with a central nervous system tumor who had at least one relative with cancer, and let p_2 represent the corresponding proportion for the children in the control group. State the null and alternative hypotheses employed in the test in part (a).

c.   Suppose that you only wanted to detect p_2 larger than p_1. State the null and alternative hypotheses for this test.

d.   How would you conduct the test in part (c)?

12.18 In 1987, the number of fatalities resulting from automobile accidents decreased substantially, with the decrease being attributed to more stringent seat belt laws and a crackdown on drivers under the influence of alcohol or drugs. Is there a difference in seat belt usage between men and women? Suppose that a survey was conducted in which the interviewee was asked how often he or she wore a seat belt. The results of the survey are shown below.

	Seat belt usage			
	Always	Most of the time	Sometimes	Never
Male	37	60	54	64
Female	39	58	49	39

Source: Adapted from *Statistical Abstract of the U.S.,* 107th Edition, 1987, p. 591.

a. Do the data provide sufficient evidence to indicate that there is a difference in seat belt usage between men and women? Test using $\alpha = .05$.

b. Find the approximate p-value of the test in part (a).

12.19 According to one theory, people with Type A behavior—hard-driving, impatient, competitive people—are much more prone to heart problems than those classified as low-key, Type B, individuals. In a study conducted by Robert B. Case, et al. ("Type A Behavior and Survival after Acute Myocardial Infarction," *New England Journal of Medicine* 312, no. 12 [March 21, 1985]), 516 heart attack patients were administered the Jenkins Activity Survey (JAS), a questionnaire that purports to provide an approximate measure of individual Type A behavior. The scores ranged from a low of -21 (Type B behavior) to a high of 23 (Type A behavior). The higher the score, the greater the level of Type A behavior in an individual. Each of the 516 patients was categorized according to whether the patient's JAS score was less than -5 (Type B), -5 to 5 (neutral), or greater than 5 (Type A). The number of patients in each of the three classes, along with the number who died during a three-year follow-up period, are shown in the table.

	JAS score		
	Less than -5	-5 to 5	Greater than 5
Total number of patients	180	171	165
Number dying in 3-year follow-up	21	17	11

The authors used the chi-square test to test for a dependence between the follow-up mortality rate and the level of Type A behavior as measured by the JAS score.

a. Compute the value of X^2.

b. Using Table 5 in Appendix III, would you conclude that the p-value for the test is larger than 0.10?

c. Based on the data in the table, what do you think the authors concluded about the dependence between the mortality rate of heart attack patients and their JAS scores?

12.20 Has the rise over the last 20 years in the number of children involved in organized sports (Little League baseball, organized football, and soccer teams) produced a decrease in the numbers involved in self-directed physical games? To study this problem, D. A. Kleiber and

G. C. Roberts interviewed 78 male and 77 female fourth- and fifth-grade students from public grammar schools in the Champaign-Urbana area of Illinois ("The Relationship Between Game and Sport Involvement in Later Childhood: A Preliminary Investigation," *Research Quarterly for Exercise and Sport* 54, no. 2 [1983]). Each student was questioned to determine whether they actively participated in one or more organized sports and the frequency of involvement in self-directed games. The following table shows the numbers of girls falling into three levels of participation in self-directed games for those participating in organized sports and for those who were nonparticipants.

Frequency of participation in self-directed physical games	Participants	Nonparticipants
Less than once a week	4	16
More than once a week but less than daily	15	20
Every day	7	15
Total	26	51

a. Do the data provide sufficient evidence to indicate that the frequencies in the three levels of involvement in self-directed games differ depending on whether the girls were or were not involved in organized sports? Test using $\alpha = .05$.

b. Find the approximate *p*-value for the test.

c. Check your answers with the results published in the paper by Kleiber and Roberts. They show $X^2 = 3.02$, d.f. $= 2$, *p*-value $= .22$. Do you agree?

12.5 TABLES WITH FIXED ROW OR COLUMN TOTALS: TESTS OF HOMOGENEITY

In the previous section we described the analysis of an $r \times c$ contingency table, using examples that for all practical purposes fit the multinomial experiment described in Section 12.1. While methods of collecting data in many surveys obviously satisfy the requirements of a multinomial experiment, other methods do not. For example, we may not wish to randomly sample the population described in Example 12.1 because we might find that, owing to chance, one category contains a very small number of people, or is completely missing. To avoid such outcomes, we may decide beforehand to interview a specified number of people in each column category, thereby fixing the column totals in advance.

Suppose that we decide to fix each column total at 300 people. With this sampling plan we would not have a multinomial experiment with $r \times c = 2 \times 3 = 6$ cells; instead we would have three binomial experiments. The first consists of $n_1 = 300$ people who received no vaccine, with the probability of an individual getting the flu given by $P(\text{flu}) = p_1$. The second consists of $n_2 = 300$ people who received one shot, with $P(\text{flu}) = p_2$. The third consists of $n_3 = 300$ people who received both shots, with $P(\text{flu}) = p_3$. By fixing the column totals at 300, we are assured that the unconditional probability that an individual in the study belongs to group 1, 2, or 3 is $300/900 = 1/3$.

Consider the joint multinomial cell probabilities p_{ij}, for $i = 1, 2$ and $j = 1, 2, 3$ when the vaccine has no effect in preventing the flu. The probability that an individual gets the flu is the same for the three groups and $p_1 = p_2 = p_3 = p$. Therefore, the unconditional probability of a person getting the flu is p, and the joint probability that a person in group j gets the flu is

$$p_{1j} = \left(\frac{1}{3}\right)p \quad \text{for} \quad j = 1, 2, 3$$

Similarly, the probability of a person in group j *not* getting the flu is

$$p_{2j} = \left(\frac{1}{3}\right)q \quad \text{for} \quad j = 1, 2, 3$$

If, on the other hand, the vaccine is effective, then p_1, p_2, and p_3 will not be equal to p. Consequently,

$$p_{1j} \neq \left(\frac{1}{3}\right)p \quad \text{and} \quad p_{2j} \neq \left(\frac{1}{3}\right)q$$

However, we see that this is the same as asking whether row and column classifications are independent. Thus, the test statistic is calculated as though you were testing for independence using Pearson's chi-square with $(r - 1)(c - 1)$ degrees of freedom.

As you can see, when the column totals are fixed, testing for independence comes down to a test of equality of three binomial p's. Tests of this sort are called **tests of homogeneity** of several binomial populations. If there were three or more row classifications with fixed column totals, the test would be one of **homogeneity** of several multinomial populations.

EXAMPLE 12.2   A survey of voter sentiment was conducted in four midcity political wards to compare the fraction of voters favoring candidate A. Random samples of 200 voters were polled in each of the four wards with results as shown in Table 12.4. Do the data present sufficient evidence to indicate that the fractions of voters favoring candidate A differ in the four wards?

Table 12.4
Data tabulation for
Example 12.2

	\multicolumn{4}{c}{Ward}			
	1	2	3	4	Total
Favor A	76(59)	53(59)	59(59)	48(59)	236
Do not favor A	124(141)	147(141)	141(141)	152(141)	564
Total	200	200	200	200	800

Solution   You will observe that the test of an hypothesis concerning the equivalence of the parameters of the four binomial populations corresponding to the four wards is identical to the test of an hypothesis implying independence of the row and column classifications. If we denote the fraction of voters favoring A as p and hypothesize that p is the same for all four wards, we imply that the first- and second-row probabilities are equal to p and $(1 - p)$, respectively. The probability that a member of the sample of $n = 800$ voters falls in a particular ward will equal one-fourth, since this was fixed in advance. Then the cell probabilities for the table would be obtained as usual.

Notice that

$$\hat{p}_1 = .38 \qquad \hat{p}_3 = .30$$

$$\hat{p}_2 = .27 \qquad \hat{p}_4 = .24$$

while the pooled estimate of the unconditional probability of favoring A is

$$\hat{p} = \frac{236}{800} = .28$$

The estimated expected cell frequencies, calculated using the row and column totals, appear in parentheses in Table 12.4. We see that

$$X^2 = \sum_{j=1}^{4} \sum_{i=1}^{2} \frac{[n_{ij} - \hat{E}(n_{ij})]^2}{\hat{E}(n_{ij})}$$

$$= \frac{(76 - 59)^2}{59} + \frac{(124 - 141)^2}{141} + \cdots + \frac{(152 - 141)^2}{141}$$

$$= 10.72$$

The critical value of χ^2 for $\alpha = .05$ and $(r - 1)(c - 1) = (1)(3) = 3$ degrees of freedom is 7.81473. Since X^2 exceeds this critical value, we reject the null hypothesis and conclude that the fraction of voters favoring candidate A is not the same for all four wards.

EXERCISES   Basic Techniques

12.21   Random samples of 200 observations were selected from each of three populations, and each observation was classified according to whether it fell into one of three mutually exclusive categories. The cell counts are shown below.

Population	Category 1	2	3	Total
1	108	52	40	200
2	87	51	62	200
3	112	39	49	200

Do the data provide sufficient evidence to indicate that the proportions of observations in the three categories depend on the population from which they were drawn?

a.   Give the value of X^2 for the test.

b.   Give the rejection region for the test for $\alpha = .10$.

c.   State your conclusions.

d.   Find the approximate p-value for the test and interpret its value.

12.22   Suppose that you wish to test the null hypothesis that three binomial parameters p_A, p_B, and p_C are equal, against the alternative hypothesis that at least two of the parameters differ. Independent random samples of 100 observations were selected from each of the populations. The data are shown below.

	Population			
	A	B	C	Total
Number of successes	24	19	33	76
Number of failures	76	81	67	224
Total	100	100	100	300

a.   Find the value of X^2, the test statistic.

b.   Give the rejection region for the test for $\alpha = .05$.

c.   State your test conclusions.

d.   Find the observed significance level for the test and interpret its value.

Applications

12.23   In Exercise 12.8, we described a study by Randy Atlas of the relationship between environmental conditions and the frequency of prison assaults ("Violence in Prison," *Environment and Behavior* 16, no. 3 [1984], © 1984 Sage Publications, Inc., with permission). In Florida, the months of June, July, August, and September are usually hot and humid. October, November, April, and May are the most temperate, and the remaining winter months consist of a mixture of temperate and cold days. The following table gives the number of assaults for these three periods of the year for two types of prisons, the air-conditioned minimum security Dade Correctional Institution (DCI) and the non-air-conditioned maximum security Union Correctional Institution (UCI). The data enable us to compare the incidence of assaults in a prison environment that is reasonably pleasant throughout the year with one subject to high and unpleasant levels of temperature and humidity. Do the data present sufficient evidence to indicate that the frequencies of assaults in the three seasonal time periods differ for the two prisons? Test using $\alpha = .05$ and state your conclusions.

	Period of year			
Prison	June–Sept	Oct, Nov, Apr, May	Dec, Mar	Total
DCI	32	20	37	89
UCI	241	216	191	648

12.24   A study to determine the effectiveness of a drug (serum) for arthritis resulted in the comparison of two groups, each consisting of 200 arthritic patients. One group was inoculated with the serum; the other received a placebo (an inoculation that appears to contain serum but actually is nonactive). After a period of time, each person in the study was asked to state whether his arthritic condition had improved. The following results were observed.

	Treated	Untreated
Improved	117	74
Not improved	83	126

Do these data present sufficient evidence to indicate that the serum was effective in improving the condition of arthritic patients?

a.   Test by means of the X^2 statistic. Use $\alpha = .05$.

b.   Test by use of the z test in Section 8.7.

 12.25   A particular poultry disease is thought to be noncommunicable. To test this theory, 30,000 chickens were randomly partitioned into three groups of 10,000. One group had no contact with diseased chickens, one had moderate contact, and the third had heavy contact. After a six-month period, the following data were collected on the number of diseased chickens in each group of 10,000. Do the data provide sufficient evidence to indicate a dependence between the amount of contact between diseased and nondiseased fowl and the incidence of the disease? Use $\alpha = .05$.

	No contact	Moderate contact	Heavy contact
Number of diseased chickens	87	89	124
Number of nondiseased chickens	9,913	9,911	9,876
Total	10,000	10,000	10,000

 12.26   A study of the purchase decisions for three stock portfolio managers A, B, and C was conducted to compare the rates of stock purchases that resulted in profits over a time period that was less than or equal to one year. One hundred randomly selected purchases obtained for each of the managers gave the following results:

	Manager		
	A	B	C
Purchases that resulted in a profit	63	71	55
Purchases that resulted in no profit	37	29	45
Total	100	100	100

Do the data provide evidence of differences among the rates of successful purchases for the three managers? Use $\alpha = .05$.

 12.27   H. W. Menard has conducted research involving manganese nodules, a mineral-rich concoction found abundantly on the deep-sea floor ("Time, Chance and the Origin of Manganese Nodules," *American Scientist* [September–October 1976]). In one portion of his report, Menard provides data relating the magnetic age of the earth's crust to the "probability of finding manganese nodules." The data shown in the table give the number of samples of the earth's core and the percentage of those that contain manganese nodules for each of a set of magnetic-crust ages. Do the data provide sufficient evidence to indicate that the probability of finding manganese nodules in the deep-sea earth's crust is dependent on the magnetic-age classification? Test with $\alpha = .05$.

Age	Number of samples	Percentage with nodules
Miocene–recent	389	5.9
Oligocene	140	17.9
Eocene	214	16.4
Paleocene	84	21.4
Late Cretaceous	247	21.1
Early and Middle Cretaceous	1120	14.2
Jurassic	99	11.0

12.28   In a survey of 500 voters in each of four regions of the United States as they left voting booths during the 1988 presidential election in which the Bush/Quayle ticket won, the following numbers of voters indicated their party and the party of the presidential candidate of their choice.

	Region			
Voted for	East	Midwest	South	West
Republican	275	228	202	248
Democrat	225	272	298	252
Totals	500	500	500	500

Data adapted from: Survey by ABC News/*Washington Post*, reported in *Public Opinion*, January/February, 1989, p. 26

Do these samples provide sufficient information to conclude that the proportion of voters favoring the 1988 Republican ticket varied significantly among the four regions of the country? Use $\alpha = .01$.

12.6   A COMPUTER ANALYSIS FOR AN $r \times c$ CONTINGENCY TABLE

Most statistical packages include a program for analyzing data contained in an $r \times c$ contingency table. To illustrate, the SAS, SPSS-X, and MINITAB computer printouts for the analysis of the data in Example 12. 1 are shown in Table 12.5. The computed value of X^2 and its observed significance level are boxed on the computer printouts.

At the top of the SAS printout, Table 12.5(a), is the chi-square table showing the observed and expected values in their respective cells. The computed value of the test statistic, $X^2 = 17.313$, is shown directly below the table. The number of degrees of freedom for X^2, DF $= 2$, and the observed significance level, PROB $= 0.0002$, are shown to the right of X^2. The last four lines of the printout, PHI, CONTINGENCY

Table 12.5   SAS, SPSS-X, and MINITAB computer printouts for the chi-square analysis of the data in Example 12.1

(a) SAS

TABLE OF FLU BY SHOTS

FLU   SHOTS

FREQUENCY EXPECTED	NONE	ONE	TWO	TOTAL
NO	289	100	565	954
	298.6	104.0	551.4	
YES	24	9	13	46
	14.4	5.0	26.6	
TOTAL	313	109	578	1000

STATISTICS FOR 2-WAY TABLES

CHI-SQUARE	17.313	DF = 2	PROB = 0.0002
PHI	0.132		
CONTINGENCY COEFFICIENT	0.130		
CRAMER'S V	0.132		
LIKELIHOOD RATIO CHISQUARE	17.252	DF = 2	PROB = 0.0002

(b) SPSS-X

*** CROSSTABULATION OF ***
FLU BY SHOTS

*** PAGE 1 OF 1

SHOTS

COUNT ROW PCT COL PCT TOT PCT	NONE	ONE	TWO	ROW TOTAL
FLU				
NO	289	100	565	954
	30.3	10.5	59.2	95.4
	92.3	91.7	97.8	
	28.9	10.0	56.5	
YES	24	9	13	46
	52.2	19.6	28.3	4.6
	7.7	8.3	2.2	
	2.4	0.9	1.3	
COLUMN TOTAL	313	109	578	1000
	31.3	10.9	57.8	100.0

CHI-SQUARE	D.F.	SIGNIFICANCE	MIN E.F.	CELLS WITH E.F. < 5
17.31296	2	0.0002	5.0	NONE

NUMBER OF MISSING OBSERVATIONS = 0

(c) MINITAB

EXPECTED COUNTS ARE PRINTED BELOW OBSERVED COUNTS

	ZERO	ONE	TWO	TOTAL
1	24	9	13	46
	14.40	5.01	26.59	
2	289	100	565	954
	298.60	103.99	551.41	
TOTAL	313	109	578	1000

CHISQ =   6.404 + 3.169 + 6.944 +
           0.309 + 0.153 + 0.335 = 17.313
DF = 2

COEFFICIENT, CRAMER'S V, and LIKELIHOOD RATIO CHISQUARE, are not pertinent to our analysis.

Four numbers are shown in each cell of the 2×3 table for the SPSS-X printout in Table 12.5(b). These are, top to bottom, the cell frequency, and the percentages of the row total, the column total, and the grand total (n) that this cell frequency represents. For example, the cell in the first column, had a cell frequency of 289. This frequency represented 30.3% of the row total, 954, 92.3% of the column total, 313, and 28.9% of the total of $n = 1000$ persons included in the experiment. The value, $X^2 = 17.31296$, its degrees of freedom, d.f. = 2, and observed significance level, $p = .0002$, are shown directly below the table. The quantities printed beneath the value of X^2 are irrelevant to our discussion.

The MINITAB printout, Table 12.5(c), shows the same 2×3 contingency table except that row 1 corresponds to "Yes" and row 2 to "No." The numbers shown in the cells are the observed and expected cell counts, respectively. The computed value, $X^2 = 17.313$, is shown below the table along with its degrees of freedom, d.f. = 2.

The information shown on the printouts differs slightly from one printout to another, as do some of the computed numbers (due to rounding), but the basic information is the same. All three show the calculated value of X^2 and its degrees of freedom. These two quantities, along with the table of the critical values of chi-square, Table 5 in Appendix III, are all that we need to test in order to detect dependence between the two qualitative variables represented in the contingency table. The value of the observed significance level, given in the SAS and the SPSS-X printouts, eliminates the need for the chi-square table and gives us a measure of the weight of evidence favoring rejection of the null hypothesis.

▷ 12.7 OTHER APPLICATIONS

The applications of the chi-square test in analyzing enumerative data, which we have described, represent only a few of the interesting classification problems that can be approximated by one or more multinomial experiments and for which our method of analysis is appropriate. By and large, these applications are complicated to a greater or lesser degree because the numerical values of the cell probabilities are unspecified and hence require the estimation of one or more population parameters. Then, as in Sections 12.4, 12.5, and 12.7, we can estimate the expected cell frequencies and use X^2 as a test statistic. Although we omit the mechanics of the statistical tests, several additional applications of the chi-square test are worth mentioning.

GOODNESS-OF-FIT TESTS

Goodness-of-fit tests are used to determine whether observed cell counts are consistent with expected cell counts calculated under the hypothesis that a specified probability model is true. This type of problem was presented in Section 12.3.

As another example, suppose that we wish to test an hypothesis stating that a population has a normal probability distribution. The cells of a sample frequency histogram (for example, Figure 2.1) would correspond to the k cells of the multinomial experiment, and the observed cell frequencies would be the number of measurements falling in each cell of the histogram. Given the hypothesized normal probability distribution for the population, we could use the areas under the normal curve to calculate the theoretical cell probabilities and hence the expected cell frequencies. The difficulty arises when μ and σ are unspecified for the normal population and these parameters must be estimated to obtain the estimated cell probabilities. This difficulty can, of course, be surmounted.

TIME-DEPENDENT MULTINOMIALS

A second and interesting application of our methodology is its use in the investigation of the rate of change of a multinomial (or binomial) population as a function of time. For example, we might study the decision-making ability of a human (or any animal) as he or she is subjected to an educational program and tested over time. If, for instance, he or she is tested at prescribed intervals of time and the test is of the yes or no type yielding a number of correct answers x that would follow a binomial probability distribution, we would be interested in the behavior of the probability of a correct response p as a function of time. If the number of correct responses was recorded for c time periods, the data would fall in a $2 \times c$ table similar to that in Example 12.2. We would then be interested in testing the hypothesis that p is equal to a constant (that is, that no learning has occurred), and we would then proceed to more interesting hypotheses to determine whether the data present sufficient evidence to indicate a gradual (say, linear) change over time as opposed to an abrupt change at some point in time. The procedures we have described could be extended to decisions involving more than two alternatives.

Our learning example is common to business, to industry, and to many other fields, including the social sciences. For example, we might wish to study the rate of consumer acceptance of a new product for various types of advertising campaigns as a function of the length of time that the campaign has been in effect. Or, we might wish to study the trend in the lot fraction defective in a manufacturing process as a function of time. Both of these examples, as well as many others, require a study of the behavior of a binomial (or multinomial) process as a function of time.

MULTIDIMENSIONAL CONTINGENCY TABLES

The construction of a two-way table to investigate dependency between two classifications can be extended to three or more classifications. For example, if we wish to test the mutual independence of three classifications, we would employ a three-dimensional "table" or rectangular parallelepiped. The reasoning and methodology associated with the analysis of both the two- and three-way tables are identical, although the analysis of the three-way table is a bit more complex.

One area of research, concerned with **log-linear models**, assumes that $\ln p_{ij}$ is a linear function of row and/or column parameters or a linear function of regressor variables. The interested reader is referred to the text by Bishop, Fienberg, and Holland listed in the references at the end of this chapter.

The examples that we have just described are intended to suggest the relatively broad application of the chi-square analysis of enumerative data, a fact that should be borne in mind by the experimenter concerned with this type of data. The statistical test employing X^2 as a test statistic requires care in the determination of the appropriate estimates and the number of degrees of freedom for X^2, which, for some of these problems, may be rather complex.

▷ 12.8   ASSUMPTIONS

The following assumptions must be satisfied if X^2 is to have approximately a chi-square distribution and, consequently, if the tests described in this chapter are to be valid.

> **Assumptions**
> 1. The cell counts $n_1, n_2, \ldots, n_k$ satisfy the conditions of a multinomial experiment (or a set of multinomial experiments created by restrictions on row or column totals).
> 2. The expected values of all cell counts should equal or exceed 5.

Assumption 1 must be satisfied. The chi-square goodness-of-fit tests, of which these tests are special cases, compare observed frequencies with expected frequencies and apply only to data generated by a multinomial experiment.

The larger the sample size n, the closer the chi-square distribution will approximate the distribution of X^2. We have stated in assumption 2 that n must be large enough so that all the expected cell frequencies will be equal to 5 or more. This is a safe figure. Actually, the expected cell frequencies can be smaller for some tests. For information on the minimum expected cell frequencies for specific goodness-of-fit tests, see the paper by Cochran listed in the references.

▷ **12.9** SUMMARY

The preceding material concerned a test of an hypothesis regarding the cell probabilities associated with one or more multinomial experiments. When the number of observations n is large, the test statistic X^2 can be shown to have, approximately, a chi-square probability distribution in repeated sampling, the number of degrees of freedom being dependent on the particular application. In general, we assume that n is large and that the minimum expected cell frequency is equal to or greater than five.

Several words of caution concerning the use of the X^2 statistic as a method of analyzing enumerative data are in order. The determination of the correct number of degrees of freedom associated with the X^2 statistic is very important in locating the rejection region. If the number is incorrectly specified, erroneous conclusions might result. Also, note that nonrejection of the null hypothesis does not imply that it should be accepted. We would have difficulty in stating a meaningful alternative hypothesis for many practical applications and, therefore, we would lack knowledge of the probability of making a type II error. For example, we hypothesize that the two classifications of a contingency table are independent. A specific alternative would have to specify some measure of dependence, which may or may not have practical significance to the experimenter. Finally, if parameters are missing and the expected cell frequencies must be estimated, the estimators of missing parameters should have certain properties in order that the test be valid.

▷ MINITAB COMMANDS

CHISQUARE test on table stored in **C ...C**

REFERENCES

Bishop, Y. M. M.; Fienberg, S. E.; and Holland, P. W. *Discrete Multivariate Analysis: Theory and Practice*. Cambridge, Mass.: Massachusetts Institute of Technology Press, 1975.

Cochran, W. G. "The χ^2 Test of Goodness of Fit," *Annotated Mathematical Statistics 23* (1952): 315–345.

Hogg, R. V., and Craig, A. T. *Introduction to Mathematical Statistics*. 4th ed. New York: Macmillan, 1986.

Kendall, M. G., and Stuart, A. *The Advanced Theory of Statistics. Vol. 2*. 4th ed. New York: Hafner, 1979.

Koopmans, L. H. *An Introduction to Contemporary Statistics*. 2d ed. Boston: Duxbury Press, 1987.

Ryan, T. A.; Joiner, B. L.; and Ryan, B. F. *Minitab Student Handbook*. 2d ed. Boston: Duxbury Press, 1985.

SAS Institute, Inc. Staff. *SAS User's Guide: Statistics*. Version 5 ed. Cary, N.C.: SAS Institute, Inc., 1985.

SPSS, Inc, Staff. *SPSS-X User's Guide*. 3d ed. New York: McGraw-Hill, 1987.

SUPPLEMENTARY EXERCISES

[Starred (*) exercises are optional.]

12.29 A manufacturer of floor polish conducted a consumer preference experiment to see whether a new floor polish A was superior to those produced by four of his competitors. A sample of 100 housekeepers viewed five patches of flooring that had received the five polishes, and each indicated the patch that he or she considered superior in appearance. The lighting, background, and so on were approximately the same for all five patches. The results of the survey are as follows:

Polish	A	B	C	D	E
Frequency	27	17	15	22	19

Do these data present sufficient evidence to indicate a preference for one or more of the polished patches of floor over the others? If one were to reject the hypothesis of "no preference" for this experiment, would this imply that polish A is superior to the others? Can you suggest a better method of conducting the experiment?

12.30 A survey was conducted to investigate interest of middle-aged adults in physical fitness programs in Rhode Island, Colorado, California, and Florida. The objective of the investigation was to determine whether adult participation in physical fitness programs varies from one region of the United States to another. A random sample of people were interviewed in each state and the following data were recorded.

	Rhode Island	Colorado	California	Florida
Participate	46	63	108	121
Do not participate	149	178	192	179

Do the data indicate a difference in adult participation in physical fitness programs from one state to another?

12.31 An analysis of accident data was made to determine the distribution of numbers of fatal accidents for automobiles of three sizes. The data for 346 accidents are as follows.

	Size of auto		
	Small	Medium	Large
Fatal	67	26	16
Not fatal	128	63	46

Do the data indicate that the frequency of fatal accidents is dependent on the size of automobiles?

12.32 An experiment was conducted to investigate the effect of general hospital experience on the attitudes of physicians toward lower-class people. A random sample of 50 physicians who had just completed four weeks of service in a general hospital were categorized according to their concern for lower-class people before and after their general hospital experience. The data are shown on the next page. Do the data provide sufficient evidence to indicate a change in "concern" due to the general hospital experience?

Concern before experience in a general hospital	After experience in a general hospital		Total
	High	Low	
Low	27	5	32
High	9	9	18

12.33 In a study of alcohol advertising and adolescent drinking, Lieberman and Orlandi found that of $n = 1747$ young adolescents in the study, 63% recalled the specific brand of alcohol being advertised, and 89% of 1108 adolescents perceived the ads as depicting young adults drinking in social situations, partying, and having fun.[†] The following were the results after $n_1 = 100$ adolescents and $n_2 = 100$ adults were shown these same advertisements, and were asked to identify the ages of the people in the advertisements.

Perceived age	Adolescents	Adults
Young adults	78	60
Teens/kids	10	22
Mixed ages	7	15
Older adults	5	3

Do the data present sufficient evidence to indicate that there is a difference in the perceived age(s) of people specified in alcohol advertisements between adolescents and adults? Test using $\alpha = .05$.

12.34 If you find yourself forgetting names and appointments, two tablespoons daily of the common health food staple lecithin may improve your memory, according to Florence Safford and Barry Baumel, who presented the results of their study to the Gerontological Society of America annual conference in 1988 (*Gainesville Sun*, February 9, 1989). Their experiment involved 61 people age 50 to 80 years and excluded any Alzheimer's patients. Of the 41 subjects receiving lecithin, 37 reported a significant decrease in memory lapse. Among the twenty subjects given only a placebo, 12 reported more memory lapses. The data summary follows:

Groups	Memory lapses		Total
	Decrease	No decrease	
Lecithin	37	4	41
Placebo	8	12	20

Do the data present sufficient evidence to indicate that a decrease in memory lapses depends on whether a subject has or has not been on a daily regimen of lecithin? Use $\alpha = .05$.

 12.35 In a study of changing evaluations of floodplain hazards, Robert J. Payne and John J. Pigram describe the attitudes of people at risk of flood hazards in the Hunter River Valley in Australia ("Changing Evaluations of Flood Plain Hazard: The Hunter Valley, Australia," *Environment and Behavior* 13, no 4 [1981] 461–480. © 1981 Sage Publications, Inc., reprinted with per-

[†] Lieberman, L. R. and M. A. Orlandi, "Alcohol Advertising and Adolescent Drinking," *Alcohol Health and Research World*, Vol. 12, no. 1 (Fall 1987), p. 30. Sponsored by the National Institute for Alcohol Abuse and Alcoholism.

mission). In part of the study, each respondent was asked to classify the damage (major or minor) that the respondent expected would incur if a flood were to occur. The respondent was also asked whether his/her preparation for a flood would result in a low, moderate, or high cost. The numbers of responses in the six cost-damage categories are shown below.

	Damage		
Cost	Major	Minor	Total
Low	43	16	59
Moderate	10	28	38
High	4	9	13
Total	57	53	110

a.   Do the proportions of people taking the three levels of preparation prior to a flood differ depending on whether the respondents perceive the prospective flood damage as major or minor? Test using $\alpha = .05$.

b.   Find the approximate observed significance level for the test and interpret its values.

 12.36   A group of 306 people were interviewed to determine their opinion concerning a particular current American foreign-policy issue. At the same time their political affiliation was recorded. The data are as follows:

	Approve of policy	Do not approve of policy	No opinion
Republicans	114	53	17
Democrats	87	27	8

Do the data present sufficient evidence to indicate a dependence between party affiliation and the opinion expressed for the sampled population?

12.37   A survey was conducted to determine student, faculty, and administration attitudes on a new university parking policy. The distribution of those favoring or opposing the policy is shown below.

	Student	Faculty	Administration
Favor	252	107	43
Oppose	139	81	40

Do the data provide sufficient evidence to indicate that attitudes regarding the parking policy are independent of student, faculty, or administration status?

*12.38   The chi-square test used in Exercise 12.24 is equivalent to the two-tailed z test of Section 8.7, provided α is the same for the two tests. Show algebraically that the chi-square test statistic X^2 is the square of the test statistic z for the equivalent test.

*12.39   It is often not clear whether all properties of a binomial experiment are actually met in a given application. A "goodness-of-fit" test is desirable for such cases. Suppose that an experiment consisting of four trials was repeated 100 times. The number of repetitions on which a given number of successes was obtained is recorded in the following table.

Possible results (number of successes)	0	1	2	3	4
Number of times obtained	11	17	42	21	9

Estimate p (assuming that the experiment was binomial), obtain estimates of the expected cell frequencies, and test for goodness of fit. To determine the appropriate number of degrees of freedom for X^2, note that p was estimated by a linear combination of the observed frequencies.

12.40 A problem that sometimes occurs during surgical operations is the occurrence of infections during blood transfusions. An experiment was conducted to determine whether the injection of antibodies reduced the probability of infection. An examination of the records of 138 patients produced the data shown in the accompanying table. Do the data provide sufficient evidence to indicate that injections of antibodies affect the likelihood of transfusional infections? Test by using $\alpha = .10$.

	Infection	No infection
Antibody	4	78
No antibody	11	45

12.41 By tradition, U.S. labor unions have been content to leave the management of the company to the managers and corporate executives. But in Europe, worker participation in management decision making is an accepted idea and one that is continually spreading. To study the effect of worker satisfaction with worker participation in managerial decision making, 100 workers were interviewed in each of two separate West German manufacturing plants. One plant had active worker participation in managerial decision making; the other did not. Each selected worker was asked whether he or she generally approved of the managerial decisions made within the firm. The results of the interviews are shown in the table.

	Participative decision making	No participative decision making
Generally approve of the firm's decisions	73	51
Do not approve of the firm's decisions	27	49

a. Do the data provide sufficient evidence to indicate that approval or disapproval of management's decisions depends on whether workers participate in decision making? Test by using the X^2 test statistic. Use $\alpha = .05$.

b. Do these data support the hypothesis that workers in a firm with participative decision making more generally approve of the firm's managerial decisions than those employed by firms without participative decision making? Test by using the z test presented in Section 8.7. This problem requires a one-tailed test. Why?

12.42 An occupant-traffic study was conducted to aid in the remodeling of an office building that contains three entrances. The choice of entrance was recorded for a sample of 200 persons entering the building. Do the data shown in the table indicate that there is a difference in preference for the three entrances? Find a 90% confidence interval for the proportion of persons favoring entrance 1.

	Entrance		
	I	2	3
Number entering	83	61	56

12.43 Most high school required reading lists have changed little over the last 25 years, in spite of conservative critics' allegations of watered-down curricula and a retreat from the classics. Arthur N. Applebee, the author of a survey by the Center for the Learning and Teaching of Literature at the State University of New York at Albany[†], indicated that only one of the ten most frequently assigned titles was written by a woman—*To Kill a Mockingbird* by Harper Lee—and none by minorities. This survey indicated that 69% of 322 public schools, 67% of 80 catholic schools, and 47% of private schools included *To Kill a Mockingbird* on their required reading lists. Do these data provide sufficient evidence to indicate that the proportion of schools that include Harper Lee's book as required reading varies according to school classification?

	Public	Catholic	Private
Required	222	54	40
Not required	100	26	46

12.44 Refer to Exercise 12.43. A survey by the Center for Learning and Teaching of Literature at the State University of New York at Albany revealed that 84% of 322 public schools, 63% of catholic schools, and 66% of private schools included *Romeo and Juliet* by William Shakespeare among the ten most frequently assigned titles. A summary of this information follows.

	Schools		
Required	Public	Catholic	Private
Yes	270	50	57
No	52	30	29
Totals	322	80	86

Is this sufficient information to conclude that the proportion of schools requiring the reading of *Romeo and Juliet* varies significantly among the three categories of schools? Use $\alpha = .05$.

12.45 Refer to Exercise 12.43. The novel *Huckleberry Finn* by Mark Twain was also included among the ten most frequently assigned works in 70% of the 322 public schools, 76% of the 80 catholic schools, and 56% of the 86 private schools in the survey. Do these data, summarized in the accompanying table, indicate that the proportion of schools with *Huckleberry*

[†] Reported in the *Press Enterprise*, May 21, 1989, Riverside, Calif. *Source:* Center for Learning and Teaching of Literature, State University of New York at Albany.

Finn on their required reading list varies significantly according to the classification of school? Use $\alpha = .05$.

	Schools		
	Public	Catholic	Private
Required	225	61	48

12.46   The following table shows the categorization of 204 men awaiting bypass heart surgery according to the relative degree of each man's coronary artery obstruction and according to his perceived level of discomfort due to angina pectoris (C. David Jenkins, et al., "Correlates of Angina Pectoris among Men Awaiting Coronary By-Pass Surgery," *Psychosomatic Medicine* 45, no. 2 [1983]. Reprinted by permission of Elsevier Publishing Co., Inc. Copyright 1983 by The American Psychosomatic Society, Inc.). Do the data present sufficient evidence to indicate that the level of angina is dependent on the level of coronary artery obstruction? The authors report the *p*-value for a chi-square test to be $p = 0.01$.

a.   Compute the value of X^2 for the data.
b.   Find the *p*-value for the test and compare with the author's value $p = 0.01$.
c.   What conclusions would you reach based on your analysis?

	Arteries obstructed 75% or more			
Level of angina	0 or 1	2	3 to 6	Total
None	3	21	20	44
Mild	2	12	9	23
Moderate	26	20	31	77
Moderate/Severe	13	10	18	41
Severe	7	5	7	19
Total	51	68	85	204

12.47   A recent complaint on the part of blacks and minorities has resulted in at least one lawsuit. According to the complaint, culturally biased IQ tests result in many black students being assigned to EMR (educable mentally retarded) classes. Typical of data in support of this contention would be the following: A school district has an enrollment of 6000 elementary and junior high school students of which 1640 are black and 4360 are white. A total of 280 students are enrolled in EMR classes, and of these 163 are black. The data are given in the table. Does it appear that student assignment to EMR classes is dependent on race? Test with $\alpha = .01$.

School class assignment	Race		Total
	Black	White	
EMR	163	117	280
Non-EMR	1477	4243	5720
Total	1640	4360	6000

12.48 According to a 1987 Sports Participation Survey by the National Sporting Goods Association reported in *American Demographics*, May 1989, p. 41, the fitness activities in which people participate change as they get older. The following table gives the results of a survey of 500 males and 500 females in which individuals who were frequent participants in fitness activities were classified by sex and type of exercise.

			Type of activity			
Sex	Walking	Cycling	Aerobics	Running	Calesthenics	Swimming
Male	60	85	28	113	79	179
Female	106	81	138	55	89	90

Do these data provide sufficient evidence to indicate that the type of fitness activity participation varies by sex? Use $\alpha = .05$.

12.49 In a public opinion poll that summarized American opinions on federal taxes prior to the 1988 election, respondents were asked to choose one of five responses to a question regarding the problem of halting the smuggling of drugs into the United States. The data summary follows:

Not much of a problem	2%
A problem, but no government action required	6%
A problem that requires government action only if no new taxes are needed	24%
A problem that requires government action even if new taxes are needed	66%
Don't know	2%

Source: Survey by the Gallup Organization for the Times Mirror Company, May 13–22, 1988 (*Public Opinion*, March/April 1989).

In a similar survey conducted in 1990, $n = 300$ respondents were asked the same questions, with 4, 12, 81, 201, and 2 falling in the five opinion categories given above. Does this data provide sufficient evidence to indicate that opinions on this subject have changed since the 1988 survey was conducted? Use $\alpha = .01$.

12.50 In Exercise 12.20, we examined the relationship between the involvement of fourth- and fifth-grade girls in self-directed physical games and their involvement in organized sports. Similar data for the 78 boys involved in the study are shown on page 526. (D. D. Kleiber and G. C. Roberts, "The Relationship Between Game and Sport Involvement in Later Childhood: A Preliminary Investigation," *Research Quarterly for Exercise and Sport* 54, no. 2 [1983]).

a. Do the data provide sufficient evidence to indicate that the frequencies in the three levels of involvement in self-directed games differ depending on whether the boys were or were not involved in organized sports? Test using $\alpha = .05$.

b. Find the approximate *p*-value for the test.

c. Check your answers with the results published in the paper by Kleiber and Roberts. They show $X^2 = 1.19$, d.f. $= 2$, $p = .55$. Do you agree?

Frequency of participation in self-directed physical games	Participants	Nonparticipants
Less than once a week	19	12
More than once a week but less than daily	23	8
Every day	11	5
Total	53	25

NO WINE BEFORE ITS TIME

In an article entitled, "They Will Can No Wine before Its Time," *Washington Post* staff writer Nicholas D. Kristof surveys industry response to a novel method for packaging wine—canning! (*Washington Post*, August 23, 1981.)

Both the Geyser Creek Winery (of California) and Wine Spectrum, a subsidiary of Coca Cola, envision great sales prospects for this new method of marketing wine. Packaged in six-packs, similar to beer, the cans are more compact, lighter in weight, and less susceptible to damage than bottles. For example, a case of wine in bottles weights 48 pounds, while an equivalent amount packed in aluminum cans weighs 22 pounds and occupies 30% less space. This feature is particularly appealing to the airlines, where weight and space play an important role in a flight's profit equation.

But what about the concept and the taste? Will it be appealing to consumers? "*Trés, trés* tacky," is the reported response of Jean Jacques Moreau, a wine connoisseur and producer of fine wines in the Chablis region of France. Moreau is further quoted as saying, "*C'est ridicule, ça ne va pas. Ça n'a pas de sense d'acheter du bon vin en autre chose qu'uane bouteille de verre. Ça ne marcherait pas en Europe,*" which Kristof translates to mean, in short, "Canned wine is gauche."

But what does the consumer say? Reynolds Metals Company, which makes the aluminum cans, claims that the cans are coated on the inside to prevent the wine from acquiring a metallic taste. Reynolds claims to have conducted a blind taste test, one in which the participants were given two glasses of the same wine, one obtained from a can and the other from a bottle. Fifty percent of the participants preferred the wine from the can, 39% from the bottle, and 11% could detect no difference between the two.

We would like to use the Reynolds Metals Company taste-test data to determine whether consumers can detect a difference in taste between canned and bottled wine. Although the author of the article did not report the total number n of tasters, we will assume, for the sake of example, that $n = 100$.

1. State the null and alternative hypotheses of interest to the experimenter.
2. Let p_1 be the probability that a taster prefers canned wine; let p_2 be the probability that a taster prefers bottled wine; and let p_3 be the probability that a taster has no

preference. If $p_1 = p_2$, it can be shown (proof omitted) that the best estimates* of p_1, p_2, and p_3 are

$$\hat{p}_1 = \hat{p}_2 = \frac{1}{2}\left(\frac{n_1 + n_2}{n}\right) \quad \text{and} \quad \hat{p}_3 = 1 - \hat{p}_1 - \hat{p}_2$$

Use these estimates to obtain the estimated expected cell counts.

3. Test the null hypothesis using the chi-square statistic. How many degrees of freedom
4. are associated with the chi-square statistic?

What conclusions can you draw from your analysis?

---

* These are known in statistics as maximum likelihood estimates. They represent the values of p_1, p_2, and p_3 that give the highest probability of observing this particular set of sample data. Maximum likelihood estimates often agree with our intuition. Thus, if we assume that $p_1 = p_2$, it seems reasonable to find the estimate of p_1 by letting it equal $1/2$ of the estimated probability for the *combined* two cells; that is, $1/2$ of $(n_1 + n_2)/n$.

CHAPTER 13

▷ ► # EXPERIMENTAL DESIGN AND THE ANALYSIS OF VARIANCE

Case Study

Are offensive linemen more at risk for coronary heart disease (CHD) than football players who play other positions? A comparison of various CHD risk factors for football players in different playing positions provides the basis for the case study at the end of this chapter.

General Objective

In this chapter we will demonstrate to you how certain factors affect the quantity of information contained in a sample. Methods for comparing two population means, based on two independent random samples, and on a paired-difference experiment, were presented in Chapter 9. In this chapter we extend these analyses to the comparison of three or more means using a technique called the analysis of variance. We will explain the logic of an analysis of variance and give the analysis for two designs that are generalizations of the unpaired and paired experiments in Chapter 9.

Specific Topics

1 The design of an experiment (13.1)
2 The analysis of variance (13.2)
3 The completely randomized design (13.3)
4 The analysis of variance and estimation for the completely randomized design (13.4, 13.5)

▷ 13.1 THE DESIGN OF AN EXPERIMENT

An **experiment** is the process by which an observation (or measurement) is obtained. In designing an experiment, the statistician attempts to gather a maximum amount of information for the least cost. The information contained in a sample can be measured by the width (or half-width) of a confidence interval for the parameter of interest constructed using the sample information. For example, a 95% confidence interval estimate for the mean of a normal population is given by

$$\bar{x} \pm 1.96 \frac{\sigma}{\sqrt{n}}$$

or from $\text{LCL} = \bar{x} - 1.96 \frac{\sigma}{\sqrt{n}}$ to $\text{UCL} = \bar{x} + 1.96 \frac{\sigma}{\sqrt{n}}$. The width of the confidence interval is given by

$$w = 2\left(1.96 \frac{\sigma}{\sqrt{n}}\right)$$

The widths of almost all of the commonly employed confidence intervals depend upon the population variation and the sample size. **The length of the interval increases as σ, the population standard deviation, increases, and decreases as the sample size n increases.** Basically, these two factors, variation and sample size, affect the quantity of information in the sample. In terms of the audio engineer, variation corresponds to noise in a transmission, and the sample size corresponds to the volume of the signal transmitted. The greater the noise, the less information in the sample. The louder the signal, the more information in the sample and the more likely the signal will penetrate the noise and be received. Obviously, the design of an experiment focuses on ways to control or reduce the variation in the data. At the same time, we decide on the size of the samples to be included in the experiment, thereby buying a maximum amount of information for a minimum cost.

In earlier chapters, we described data sets and/or used summary statistics to make inferences about the populations sampled in the form of estimates or decisions concerning values of population parameters. In designing an experiment, an individual or object on which an observation is made is called an **experimental unit**.

Definition   The objects on which measurements are made are called **experimental units**.

Treatments are those procedures applied to the experimental units whose effects on the units are to be estimated and compared. A treatment might be the baking temperature for an angel food cake, the manufacturer of an automobile tire, or the various amounts of nitrogen, phosphorus, and potassium in a fertilizer. Each of these treatments involves one or more **independent experimental variables** called **factors**, whose values are controlled and varied by the experimenter.

Definition   The intensity setting of a factor is called a **level**.

If the baking temperatures for the angel food cake were $350°$, $400°$, and $425°$ Fahrenheit, then the factor "temperature" had three levels.

Definition   A **treatment** is a specific combination of factor levels.

In designing an experiment, we must first specify the factor(s) to be included in the experiment as well as the number of levels of each factor; that is, we must specify the number of treatments to be included. Secondly, we need to decide on the number of experimental units to be included for each treatment, based on predetermined accuracy in estimation; and finally, we must decide how the treatments will be assigned or allocated to the experimental units.

Experimental designs are often referred to as **volume-increasing** or **noise-reducing** designs. A volume-increasing design is one which minimizes the length of a confidence interval for a fixed number of observations.

The next two examples illustrate how the allocation of a fixed number of observations can be used to minimize the length of a confidence interval.

EXAMPLE 13.1   In estimating the difference between two population means, $(\mu_1 - \mu_2)$, compare the standard deviation of $(\bar{x}_1 - \bar{x}_2)$ when $\sigma_1 = 1$, $\sigma_2 = 2$, and

a. $n_1 = n_2 = 6$
b. $n_1 = 4, n_2 = 8$

Solution   The standard deviation of the difference between two sample means, $(\bar{x}_1 - \bar{x}_2)$, is given as

$$\sigma_{(\bar{x}_1 - \bar{x}_2)} = \sqrt{\frac{\sigma_1^2}{n_1} + \frac{\sigma_2^2}{n_2}}$$

a.   When $n_1 = n_2 = 6$,

$$\sigma_{(\bar{x}_1 - \bar{x}_2)} = \sqrt{\frac{1}{6} + \frac{4}{6}} = \sqrt{\frac{5}{6}} = \sqrt{.83}$$

b. When $n_1 = 4$ and $n_2 = 8$,

$$\sigma_{(\bar{x}_1 - \bar{x}_2)} = \sqrt{\frac{1}{4} + \frac{4}{8}} = \sqrt{\frac{6}{8}} = \sqrt{.75}$$

The standard deviation of $(\bar{x}_1 - \bar{x}_2)$ is smaller for the sample size allocation in part (b). In general, $\sigma_{(\bar{x}_1 - \bar{x}_2)}$ is a minimum when $n_1/n_2 = \sigma_1/\sigma_2$, that is, when the ratio of the sample sizes is the same as the ratio of the standard deviations. This implies that equal allocation of sample sizes is best only when $\sigma_1 = \sigma_2$.   ◁

EXAMPLE 13.2   Evaluate the standard deviation of the sample slope $\hat{\beta}_1$ in the linear regression problem, given by

$$\sigma_{\hat{\beta}_1} = \frac{\sigma}{\sqrt{\sum(x_i - \bar{x})^2}}$$

for the two allocations of $n = 5$ data points given by
a. $x_1 = 1, x_2 = 2, x_3 = 3, x_4 = 4$, and $x_5 = 5$
b. $x_1 = x_2 = 1, x_3 = 3$, and $x_4 = x_5 = 5$

Solution   For both parts (a) and (b), $\bar{x} = 3$.
a. $\sum(x_i - \bar{x})^2 = (-2)^2 + (-1)^2 + 0^2 + (1)^2 + (2)^2 = 10$ so that

$$\sigma_{\hat{\beta}_1} = \frac{\sigma}{\sqrt{10}}$$

b. $\sum(x_i - \bar{x})^2 = (-2)^2 + (-2)^2 + 0^2 + (2)^2 + (2)^2 = 16$ so that

$$\sigma_{\hat{\beta}_1} = \frac{\sigma}{\sqrt{16}}$$

Therefore, the value of $\sigma_{\hat{\beta}_1}$ in part (a) is larger than it is in part (b). In general, the standard deviation of the slope is minimized when half of the observations are run at the lower end and the other half at the upper end of the region in x where the model $E(y) = \beta_0 + \beta_1 x$ is to be fitted. Most statisticians would also suggest running one or two observations near the middle of the region, to detect departures from a simple linear model.   ◁

Noise-reducing designs are those for which the method of assigning treatments to experimental units tends to reduce the standard deviation of the difference between two means in much the same way as the paired difference design of Chapter 9. The generalization of the paired difference design to three or more treatments is called the **randomized block design**. This design will be presented in Section 13.6.

EXERCISES   Basic Techniques

13.1   How can one measure the quantity of information in a sample pertinent to a specific population parameter?

13.2   Give the two factors that affect the quantity of information in an experiment.

13.3   What is a factor? a treatment? an experimental unit?

13.4   Suppose that one wishes to compare the means for two populations. If $\sigma_1^2 = \sigma_2^2$, what is the optimal allocation of n_1 and n_2, $(n_1 + n_2 = n)$, for a fixed sample size n? Does this operation employ the principle of noise reduction or signal amplification?

13.5   Refer to Exercise 13.4. Suppose that $\sigma_1^2 = 9$, $\sigma_2^2 = 25$, and $n = 90$. What allocation of $n = 90$ to the two samples will result in the maximum amount of information on $(\mu_1 - \mu_2)$?

13.6   Refer to Exercise 13.5. Suppose that we allocate $n_1 = n_2$ observations to each sample. How large must n_1 and n_2 be in order to obtain the same amount of information as that implied by the solution to Exercise 13.5?

13.7   Suppose that you wish to estimate the slope of a line using the method of least squares, that you can afford only 12 experimental points, and that the assumptions of Section 10.2 are satisfied. How can you select the locations of the points (the settings of the independent variable x) so as to minimize the variance of $\hat{\beta}_1$? Note that this selection of settings accomplishes an increase in "volume" without increasing the sample size. Essentially, it shifts information within the design so that it focuses on the slope β_1.

Applications

13.8   Suppose that one wishes to study the effect of the stimulant digitalis on the blood pressure of rats over a dosage range of $x = 2$ to $x = 5$ units. The response is expected to be linear over the region. Six rats are available for the experiment, and each rat can receive only one dose. What dosages of digitalis should be employed in the experiment and how many rats should be run at each dosage to maximize the quantity of information on β_1, the slope of the regression line? Which aspect of design—noise reduction or signal amplification—is implied in this experiment?

13.9   Refer to Exercise 13.8. Consider two methods for selecting the dosages. Method 1 assigns three rats to the dosage $x = 2$ and three rats to $x = 5$. Method 2 equally spaces the dosages between $x = 2$ and $x = 5$, $(x = 2, 2.6, 3.2, 3.8, 4.4, 5.0)$. Suppose that σ is known and that the relationship between $E(y)$ and x is truly linear (see Chapter 10). How much larger will the confidence interval be for the slope β_1 for method 2 in comparison with method 1? Approximately how many observations would be required to obtain the same confidence interval as obtained by the optimal assignment of method 1?

13.10   Refer to Exercise 13.8. Why might it be advisable to assign one or two points at $x = 3.5$?

13.2   THE ANALYSIS OF VARIANCE

The methodology for the analysis of experiments involving several independent variables can be explained in terms of the multiple linear regression model of Chapter 11. However, we will attempt an intuitive discussion using a procedure known as the analysis of variance. Further information on this subject can be found in Mendenhall (1968), which is listed in the references.

The analysis of variance is a procedure that attempts to analyze the variation of a response and to assign portions of this variation to each of a set of independent variables. The reasoning is that the response variable varies only because of variation in a set of unknown independent variables. Since the experimenter will rarely, if ever, include all the variables affecting the response in his experiment, random variation in the response is observed even though all independent variables considered important are held constant. The objective of an analysis of variance is to locate important independent variables in a study and to determine how they interact and affect the response.

The rationale underlying the analysis of variance can be indicated best with a symbolic discussion. The variability of a set of n measurements is proportional to the sum of squares of deviations, $S_{xx} = \sum_{i=1}^{n} (\bar{x}_i - \bar{x})^2$, and this quantity is used to calculate the sample variance. **An analysis of variance partitions S_{xx}, called the total sum of squares of deviations, into parts, each of which is attributed to one of the controlled independent variables in the experiment, plus a remainder that is associated with random error.** This may be shown diagrammatically as indicated in Figure 13.1 for three independent variables.

Figure 13.1
Partitioning of the total sum of squares of deviations

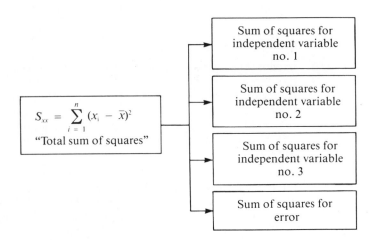

For the cases that we consider, and when the response is unrelated to the independent variables, it can be shown that each of the pieces of the total sum of squares of deviations, divided by an appropriate constant, provides an independent and unbiased estimator of σ^2, the variance of the experimental error. When a variable is highly related to the response, its portion (called the "sum of squares" for the variable) of variability will be inflated. This condition can be detected by comparing the estimate of σ^2 for a particular independent variable with that obtained from SSE using an F test (similar to the one encountered in Section 9.7). If the estimate for the independent variable is significantly larger than the estimate calculated from SSE, the F test will reject the hypothesis of "no effect for the independent variable" and thereby produce evidence to indicate a relation to the response.

The logic behind an analysis of variance can be illustrated by considering a familiar example, the comparison of two population means for an unpaired experiment, which was analyzed in Chapter 9 using a Student's t statistic. We begin with a graphic and intuitive explanation of the procedure.

Figure 13.2

Graphic portrayal of the deviations of the x values about their means

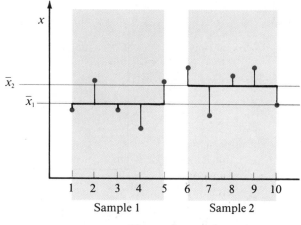

Suppose that we have selected random samples of five observations each from two populations 1 and 2 and that the x values are plotted as shown in Figure 13.2. Note that the $n_1 = 5$ observations from population 1 lie to the left, and the $n_2 = 5$ observations from population 2 lie to the right. The sample means $\bar{x}_1$ and $\bar{x}_2$ are shown as horizontal lines in the figure, and the deviations of the x values about their respective means are the vertical line segments. Now examine Figure 13.2. Do you think the data provide sufficient evidence to indicate a difference between the two population means? Before we explain, let us look at another figure.

The same two sets of five points are plotted in Figure 13.3 except that the relative distance between the two sets is greater in (b) than in (a), and even greater in (c). Therefore, the distance between $\bar{x}_1$ and $\bar{x}_2$ increases as you move from Figure 13.3(a) to (b) and then to (c), but the relative variation within each set is held constant.

Now view the three plots—(a), (b), and (c)—and decide which of the three situations provides the greatest evidence to indicate a difference between μ_1 and μ_2. We think you will choose Figure 13.3(c) because that plot shows the greatest difference between sample means in comparison with the variation of the points about their respective sample means. This latter variation was held constant for all three plots.

The population means may actually differ for Figure 13.3(a), but this fact would not be apparent, because the variation of the points about their respective sample means is too large in comparison with the difference between $\bar{x}_1$ and $\bar{x}_2$. Figure 13.4 shows the same difference between sample means as in Figure 13.3(a), but the variation within samples has been reduced. Now it appears that a difference does exist between μ_1 and μ_2.

Now let us leave our intuitive discussion and consider the two-sample comparison of means for sample sizes n_1 and n_2. In particular, we will want to see how the total sum of squares of deviations can be partitioned into portions corresponding to the difference between the sample means and the other to the variation

Figure 13.3
Three fictitious sets of measurements, $n_1 = n_2 = 5$ (the relative positions of points within each set are held constant)

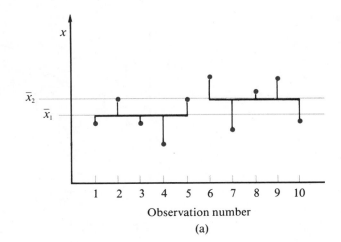

(a)

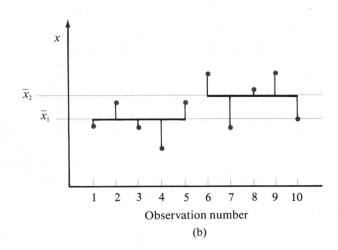

(b)

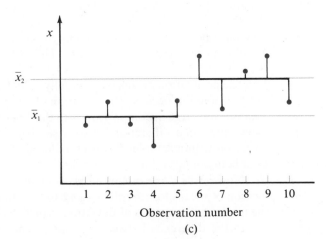

(c)

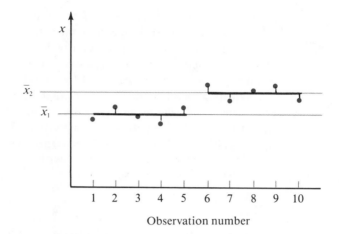

Figure 13.4
Small amount of within-sample variation in comparison with the difference between sample means

Observation number

within the two samples. The total sum of squared deviations of all $(n_1 + n_2)$ x-values about the general mean is

$$\text{Total SS} = \sum_{i=1}^{2} \sum_{j=1}^{n_i} (x_{ij} - \bar{x})^2$$

where $\bar{x}$ is the average of all $(n_1 + n_2)$ observations. It can be shown that

$$\text{Total SS} = \sum_{i=1}^{2} \sum_{j=1}^{n_i} (x_{ij} - \bar{x})^2 = \underbrace{\sum_{i=1}^{2} n_i (\bar{x}_i - \bar{x})^2}_{\text{SST}} + \underbrace{\sum_{i=1}^{2} \sum_{j=1}^{n_i} (x_{ij} - \bar{x}_i)^2}_{\text{SSE}}$$

where $\bar{x}_i$ is the average of the observations in the ith sample, $i = 1, 2$. The first quantity to the right of the equal sign is called the **sum of squares for treatments** and denoted by SST. It can be shown that for two samples

$$\text{SST} = \frac{n_1 n_2}{n_1 + n_2} (\bar{x}_1 - \bar{x}_2)^2$$

Thus SST, which increases as the difference between $\bar{x}_1$ and $\bar{x}_2$ increases, measures the variation between the sample means. The second quantity to the right of the equal sign is the pooled sum of squares of deviations computed in the t test of Section 9.4. It represents the variation of the measurements about their respective sample means. This within-sample variation is usually attributed to experimental error and is consequently called the **sum of squares for error** and denoted by SSE.

The quantities SST and SSE measure the two kinds of variation that we viewed in the graphic representation of Figure 13.3. The greater the variation between means (the larger SST) in comparison with the variation within samples (SSE), the greater the weight of evidence to indicate that $\mu_1 \neq \mu_2$. As in Chapter 9, when there are two samples and $\sigma_1^2 = \sigma_2^2 = \sigma^2$,

$$s^2 = \text{MSE} = \frac{\text{SSE}}{n_1 + n_2 - 2}$$

provides an unbiased estimator of σ^2. In the context of analysis of variance, s^2 is usually denoted by MSE, meaning **mean square for error**. Also, when the null hypothesis is true and $\mu_1 = \mu_2$, SST divided by an appropriate number of degrees of freedom yields a second unbiased estimator of σ^2 denoted by MST. For this example, the number of degrees of freedom for MST is equal to 1.

When the null hypothesis is true (that is, $\mu_1 = \mu_2$), MSE (the mean square for error) and MST (the mean square for treatments) estimate the same quantity and should be "roughly" of the same magnitude. When the null hypothesis is false and $\mu_1 \neq \mu_2$, MST will almost always be larger than MSE.

The preceding discussion, along with a review of the variance ratio given in Section 9.7, suggests the use of

$$\frac{\text{MST}}{\text{MSE}}$$

as a test statistic to test the hypothesis $\mu_1 = \mu_2$ against the alternative $\mu_1 \neq \mu_2$. Indeed, when both populations are normally distributed, it can be shown that MST and MSE are independent in a probabilistic sense and that they can be used in a test statistic, given by $F = \text{MST}/\text{MSE}$.

TEST STATISTIC FOR THE NULL HYPOTHESIS $H_0: \mu_1 = \mu_2$

$$F = \frac{\text{MST}}{\text{MSE}}$$

The test statistic F follows the F probability distribution of Section 9.7. Disagreement with the null hypothesis is indicated by a large value of F, and hence the rejection region for a given α consists of values of $F > F_\alpha$. Thus, **the analysis of variance test results in a one-tailed F test**. The degrees of freedom for F will be those associated with MST and MSE, which we denote as v_1 and v_2, respectively. Although we have not indicated how we determine v_1 and v_2, in general, for the two-sample experiment described earlier, $v_1 = 1$ and $v_2 = (n_1 + n_2 - 2)$.

EXAMPLE 13.3   The coded values for the measure of elasticity in plastic, prepared by two different processes, are given below for samples of six drawn randomly from each of the two processes.

Process A	Process B	
6.1	9.1	
7.1	8.2	
7.8	8.6	
6.9	6.9	
7.6	7.5	
8.2	7.9	
Total 43.7	48.2	91.9
Mean 7.2833	8.0333	7.6583

Solution   Although the Student's t could be used as the test statistic for this example, we will use the analysis of variance F test, since it is more general and can be used to compare more than two means.

The two desired sums of squares of deviations are

$$SST = \sum_{i=1}^{2} n_i(\bar{x}_i - \bar{x})^2$$

$$= 6(7.2833 - 7.6583)^2 + 6(8.0333 - 7.6583)^2$$

$$= .84375 + .84375 = 1.6875$$

$$SSE = \sum_{i=1}^{2} \sum_{j=1}^{6} (x_{ij} - \bar{x}_i)^2 = \sum_{j=1}^{6} (x_{1j} - \bar{x}_1)^2 + \sum_{j=1}^{6} (x_{2j} - \bar{x}_2)^2 = 5.8617$$

(You can verify that SSE is the pooled sum of squares of the deviations for the two samples discussed in Section 9.4. Also, note that Total SS = SST + SSE.) The mean squares for treatment and error are

$$MST = \frac{SST}{1} = 1.6875$$

$$MSE = \frac{SSE}{n_1 + n_2 - 2} = \frac{5.8617}{10} = .5862$$

To test the null hypothesis $\mu_1 = \mu_2$, we compute the test statistic

$$F = \frac{MST}{MSE} = \frac{1.6875}{.5862} = 2.88$$

The critical value of the F statistic for $\alpha = .05$ is 4.96 (see Table 7 in Appendix III). Although the mean square for treatments is almost three times as large as the mean square for error, it is not large enough to reject the null hypothesis. Consequently, there is not sufficient evidence to indicate a difference between μ_1 and μ_2. ◁

The purpose of the preceding example was to illustrate the computations involved in a simple analysis of variance. The F test for comparing two means is equivalent to a Student's t test, because an F statistic with one degree of freedom in the numerator is equal to t^2. Had the t test been used for Example 13.3, we would have found $t = -1.6967$, which satisfies the relationship $t^2 = (-1.6967)^2 = 2.88 = F$. This relationship also holds for the critical values. You can verify that the square of $t_{.025} = 2.228$ (used for the two-tailed test with $\alpha = .05$ and $\nu = 10$ degrees of freedom) is equal to $F_{.05} = 4.96$. Since each value of F corresponds to two values of t, one positive and one negative, the F test with one degree of freedom in the numerator always corresponds to a two-tailed t test.

Of what value is the Total SS? The answer is that it provides an easy way to compute SSE. Since the Total SS partitions into SST and SSE, that is,

Total SS = SST + SSE   and   SSE = Total SS − SST

Both the Total SS and SST are easy to compute. Hence, we can easily find SSE by substituting into the expression above. For Example 13.3 we have

$$\text{Total SS} = \sum_{i=1}^{2} \sum_{j=1}^{6} (x_{ij} - \bar{x})^2 = \sum_{i=1}^{2} \sum_{j=1}^{6} x_{ij}^2 - \frac{\left(\sum_{i=1}^{2} \sum_{j=1}^{6} x_{ij} \right)^2}{12}$$

$$= (\text{sum of squares of all } x \text{ values}) - \frac{(\text{total of all } x \text{ values})^2}{12}$$

$$= 711.35 - \frac{(91.9)^2}{12}$$

$$= 7.5492$$

Then

$$\text{SSE} = \text{Total SS} - \text{SST} = 7.5492 - 1.6875 = 5.8617$$

This is exactly the same value obtained by the tedious computation and pooling of the sums of squares of deviations from the individual samples.

13.3 THE COMPLETELY RANDOMIZED DESIGN

An analysis of variance to detect a difference in a set of more than two population means is a simple generalization of the analysis of variance of Section 13.2. The random selection of independent samples from k populations is known as a **completely randomized experimental design**. The analysis of variance for a completely randomized design is based on the following assumptions.

> **Assumptions for the Analysis of Variance of a Completely Randomized Design**
>
> 1. Independent random samples have been drawn from k normal populations with means $\mu_1, \mu_2, \ldots, \mu_k$.
> 2. The variability of the measurements in each of the sampled populations is equal to σ^2.

To be completely general, we allow the sample sizes to be unequal and we let n_i, $i = 1, 2, \ldots, k$, be the number in the sample drawn from the ith population. The total number of observations in the experiment is $n = n_1 + n_2 + \cdots + n_k$.

Let x_{ij} denote the measured response on the jth experimental unit in the ith sample and **let T_i and $\bar{T}_i$ represent the total and the mean, respectively, for the observations in the ith sample**. (The modification in the symbols for sample totals and averages will simplify the computing formulas for the sums of squares.) Then, as in the analysis of variance involving two means,

$$\text{Total SS} = \text{SST} + \text{SSE}$$

where

$$\text{Total SS} = \sum_{i=1}^{k} \sum_{j=1}^{n_i} (x_{ij} - \bar{x})^2 = \sum_{i=1}^{k} \sum_{j=1}^{n_i} x_{ij}^2 - \text{CM}$$

$$= (\text{sum of squares of all } x \text{ values}) - \text{CM}$$

$$\text{CM} = \frac{(\text{total of all observations})^2}{n} = \frac{\left(\sum_{i=1}^{k} \sum_{j=1}^{n_i} x_{ij}\right)^2}{n} = n\bar{x}^2$$

(the term CM denotes "correction for the mean"),

$$\text{SST} = \sum_{i=1}^{k} n_i(\bar{T}_i - \bar{x})^2 = \sum_{i=1}^{k} \frac{T_i^2}{n_i} - \text{CM}$$

$$= \left\{ \begin{array}{l} \text{sum of squares of treatment totals, with each square divided} \\ \text{by the number of observations in that particular total} \end{array} \right\} - \text{CM}$$

$$\text{SSE} = \text{Total SS} - \text{SST}$$

Although the easy way to compute SSE is by subtraction, as shown above, it is interesting to note that SSE is the pooled sum of squares for all k samples and is equal to

$$\text{SSE} = \sum_{i=1}^{k} \sum_{j=1}^{n_i} (x_{ij} - \bar{T}_i)^2$$

The unbiased estimator of σ^2, based on $(n_1 + n_2 + \cdots + n_k - k)$ degrees of freedom, is

$$s^2 = \text{MSE} = \frac{\text{SSE}}{n_1 + n_2 + \cdots + n_k - k}$$

The mean square for treatments possesses $(k - 1)$ degrees of freedom—that is, one less than the number of means—and is given by

$$\text{MST} = \frac{\text{SST}}{k - 1}$$

To test the null hypothesis

$$H_0: \mu_1 = \mu_2 = \cdots = \mu_k$$

against the alternative that at least one of the equalities does not hold, MST is compared with MSE by using the F statistic, based on $v_1 = k - 1$ and

$$v_2 = \sum_{i=1}^{k} n_i - k = n - k$$

degrees of freedom. The null hypothesis will be rejected if

$$F = \frac{MST}{MSE} > F_{\alpha}$$

where F_{α} is the critical value of F, based on $(k - 1)$ and $(n - k)$ degrees of freedom, for a probability α of a type I error.

Intuitively, **the greater the difference between the observed treatment means $\bar{T}_1$, $\bar{T}_2, \ldots, \bar{T}_k$, the greater is the evidence to indicate a difference between their corresponding population means**. We can see from the formula for SST that SST $= 0$ when all the observed treatment means are identical because then $\bar{T}_1 = \bar{T}_2 = \cdots = \bar{T}_k = \bar{x}$, and the deviations appearing in SST, $(\bar{T}_i - \bar{x})$, $i = 1, 2, \ldots, k$, equal zero. As the treatment means get farther apart, the deviations $(\bar{T}_i - \bar{x})$ increase in absolute value and SST increases in magnitude. Consequently, **the larger the value of SST, the greater is the weight of evidence favoring a rejection of the null hypothesis**. This same line of reasoning applies to the F tests employed in the analysis of variance for all designed experiments.

The test is summarized in the display.

F TEST FOR COMPARING k POPULATION MEANS

1. Null Hypothesis: $H_0: \mu_1 = \mu_2 = \cdots = \mu_k$.

2. Alternative Hypothesis: H_a: One or more pairs of population means differ.

3. Test Statistic: $F = MST/MSE$, where F is based on $v_1 = (k - 1)$ and $v_2 = (n - k)$ degrees of freedom.

4. Rejection Region: Reject if $F > F_{\alpha}$, where F_{α} lies in the upper tail of the F distribution (with $v_1 = k - 1$ and $v_2 = n - k$) and satisfies the expression $P(F > F_{\alpha}) = \alpha$.

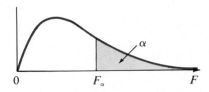

The assumptions underlying the analysis-of-variance F tests deserve particular attention. The samples are assumed to have been randomly selected from the k populations in an independent manner. The populations are assumed to be normally distributed with equal variances σ^2 and means $\mu_1, \mu_2, \ldots, \mu_k$. Moderate departures from these assumptions will not seriously affect the properties of the test when the sample sizes are equal, that is, when $n_1 = n_2 = \cdots = n_k$. When the sample sizes are unequal and one or more of the parametric assumptions are violated, the Kruskal-Wallis test, a nonparametric procedure discussed in Section 14.6, should be used.

EXAMPLE 13.4   Groups of students were randomly assigned to be taught by four different teaching techniques. They were tested at the end of a specified period of time. Because of dropouts in the experimental groups (sickness, transfers, and so on), the number of students varied from group to group. Do the following data present sufficient evidence to indicate a difference in the mean achievement for students taught using the four teaching techniques?

	Techniques		
1	2	3	4
65	75	59	94
87	69	78	89
73	83	67	80
79	81	62	88
81	72	83	
69	79	76	
	90		
T_i 454	549	425	351
$\bar{T}_i$ 75.67	78.43	70.83	87.75

Solution   We must compute the following quantities:

$$CM = \frac{\left(\sum_{i=1}^{4} \sum_{j=1}^{n_i} x_{ij}\right)^2}{n} = \frac{(\text{total of all observations})^2}{n}$$

$$= \frac{(1779)^2}{23} = 137{,}601.8$$

$$\text{Total SS} = \sum_{i=1}^{4} \sum_{j=1}^{n_i} x_{ij}^2 - CM = (\text{sum of squares of all } x \text{ values}) - CM$$

$$= (65)^2 + (87)^2 + (73)^2 + \cdots + (88)^2 - CM$$

$$= 139{,}511 - 137{,}601.8 = 1909.2$$

$$SST = \sum_{i=1}^{4} \frac{T_i^2}{n_i} - CM$$

$$= \left\{ \begin{array}{l} \text{sum of squares of treatment totals, with each square} \\ \text{divided by the number of observations in that} \\ \text{particular total} \end{array} \right\} - CM$$

$$= \frac{(454)^2}{6} + \frac{(549)^2}{7} + \frac{(425)^2}{6} + \frac{(351)^2}{4} - CM$$

$$= 138{,}314.4 - 137{,}601.8 = 712.6$$

$$SSE = \text{Total SS} - SST = 1196.6$$

The mean squares for treatment and error are

$$MST = \frac{SST}{k-1} = \frac{712.6}{3} = 237.5$$

$$MSE = \frac{SSE}{n_1 + n_2 + \cdots + n_k - k} = \frac{SSE}{n-k} = \frac{1196.6}{19} = 63.0$$

The test statistic for testing the hypothesis $\mu_1 = \mu_2 = \mu_3 = \mu_4$ is

$$F = \frac{MST}{MSE} = \frac{237.5}{63.0} = 3.77$$

where

$$v_1 = (k-1) = 3 \quad \text{and} \quad v_2 = \sum_{i=1}^{k} n_i - 4 = 19$$

The critical value of F for $\alpha = .05$ is $F_{.05} = 3.13$. Since the computed value of F, 3.77, exceeds $F_{.05} = 3.13$, we reject the null hypothesis and conclude that the evidence is sufficient to indicate a difference in mean achievement for the four training programs. Since the observed value of $F = 3.77$ lies between $F_{.05} = 3.13$ and $F_{.025} = 3.90$, the observed significance level of the test lies between 0.25 and .05, so that

$$.025 < p\text{-value} < .05$$

You may feel that the above conclusion could have been made on the basis of visual observation of the treatment means. However, it is not difficult to construct a set of data that will lead the "visual" decision maker to erroneous results.

13.4   AN ANALYSIS-OF-VARIANCE TABLE FOR A COMPLETELY RANDOMIZED DESIGN

The calculations of the analysis of variance are usually displayed in an analysis-of-variance (ANOVA or AOV) table. The table for the design of Section 13.3 involving k treatment means is shown in Table 13.1. Column 1 shows the sources of variation corresponding to each sum of squares of deviations; column 2 gives the respective degrees of freedom; columns 3 and 4 give the corresponding sums of squares and

Table 13.1
ANOVA table for a comparison of means, completely randomized design

Source	d.f.	SS	MS	F
Treatments	$k-1$	SST	$MST = SST/(k-1)$	MST/MSE
Error	$n-k$	SSE	$MSE = SSE/(n-k)$	
Total	$n-1$	Total SS		

Table 13.2
ANOVA table for
Example 13.4

Source	d.f.	SS	MS	F
Treatments	3	712.6	237.5	3.77
Error	19	1,196.6	63.0	
Total	22	1,909.2		

mean squares, respectively. A calculated value of F, comparing MST and MSE, is usually shown in column 5. Note that the degrees of freedom and sums of squares add to their respective totals.

The ANOVA table for Example 13.4, shown in Table 13.2, gives a compact presentation of the appropriate computed quantities for the analysis of variance. Computer packages for analysis of variance are readily available through your computer center (or personal microcomputer). The MINITAB command ONEWAY for an analysis of variance of the data for Example 13.4 produced the output that appears in Table 13.3. The values to be analyzed were stored in column 1, and the corresponding treatment group designations (1, 2, 3, or 4) were stored in column 2. The analysis of variance table, which appears as the first section of the printout, is identical to that in Table 13.1, which was found by direct calculations. Notice that the source of variation due to treatments is identified by row C2, which contains the treatment designations. The second section of the printout provides the sample size, the sample mean, and the standard deviation for each treatment group (identified as level 1, 2, 3, or 4). The pooled standard deviation is equal to $s = \sqrt{\text{MSE}} = 7.936$. The program also provides a graphical display locating the sample mean as well as the 95% lower and upper confidence limits for each treatment mean using the pooled standard deviation.

Table 13.3 MINITAB printout for Example 13.4

```
MTB > ONEWAY C1 C2

ANALYSIS OF VARIANCE ON C1
SOURCE    DF       SS        MS       F
C2         3      712.6     237.5    3.77
ERROR     19     1196.6      63.0
TOTAL     22     1909.2

                                    INDIVIDUAL 95 PCT CIS FOR MEAN
                                    BASED ON POOLED STDEV
LEVEL     N     MEAN    STDEV    ------+---------+---------+---------+
  1       6    75.667   8.165          ( ------*----- )
  2       7    78.429   7.115            ( -----*------ )
  3       6    70.833   9.579    ( ------*------ )
  4       4    87.750   5.795                  ( --------*------- )
                                    ------+---------+---------+---------+
POOLED STDEV =    7.936             70        80        90       100
```

Table 13.4   SAS printout for Example 13.4

ANALYSIS OF VARIANCE PROCEDURE

DEPENDENT VARIABLE X

SOURCE	DF	SUM OF SQUARES	MEAN SQUARE	F VALUE	PR > F	R-SQUARE	C.V.
MODEL	3	712.58643892	237.52881297	3.77	0.0280	0.373235	10.2602
ERROR	19	1196.63095238	62.98057644				
CORRECTED					ROOT MSE		X MEAN
TOTAL	22	1909.21739130			7.93603027		77.34782608

SOURCE	DF	ANOVA SS	F VALUE	PR > F
TRTMENTS	3	712.58643892	3.77	0.0280

The printout from an equivalent SAS* program for implementing a one-way analysis of variance for these same data is given in Table 13.4. You will note that the analysis of variance is broken down into two parts. The upper part of the table partitions that total sum of squares into two sources, MODEL and ERROR. The numbers appearing in the ERROR row of Table 13.4 are identical (except for rounding) to the numbers appearing in the "Error" row of the analysis of variance given in Table 13.2. The MODEL row of the printout in Table 13.4 corresponds to all effects other than error. In this case there is only one other source of variation, namely, treatments. Consequently, the row identified as MODEL gives the values corresponding to treatment, as well as the calculated value of F. These values, apart from rounding, agree with those in the "Treatments" row of Table 13.2.

The lower part of the table breaks down the MODEL source of variation into its components. In this particular case, there is only one source, TRTMENTS. Consequently, this line repeats the value of SST (712.58 . . .), gives the computed F value (3.77) for a test of the null hypothesis "no difference between treatment means," and gives the probability of observing a value of F as large as or larger than 3.77, given that the null hypothesis is true. This probability, 0.0280, is the significance level (p-value) for the test. With p-value $= 0.0280$, we could reject H_0 with $\alpha = .05$ but not with $\alpha = .01$.

The value of $s = \sqrt{\text{MSE}} = 7.936 \ldots$, given at the right in the top part of the printout, can be used to construct confidence intervals.

▷ 13.5   ESTIMATION FOR THE COMPLETELY RANDOMIZED DESIGN

Confidence intervals for a single treatment mean and the difference between a pair of treatment means are very similar to those given in Chapter 9. The confidence interval for the mean of treatment i or the difference between treatments i and j are given in the display.

* SAS is the registered trademark of SAS Institute, Inc., Cary, NC, USA.

COMPLETELY RANDOMIZED DESIGN

A $100(1 - \alpha)\%$ Confidence Interval for a Single Treatment Mean and the Difference Between Two Treatment Means

A Single Treatment Mean:

$$\bar{T}_i \pm t_{\alpha/2} s / \sqrt{n_i}$$

The Difference Between Two Treatment Means:

$$(\bar{T}_i - \bar{T}_j) \pm t_{\alpha/2} s \sqrt{\frac{1}{n_i} + \frac{1}{n_j}}$$

where

$$s = \sqrt{s^2} = \sqrt{\text{MSE}} = \sqrt{\frac{\text{SSE}}{n - k}}, \, n = n_1 + n_2 + \cdots + n_k$$

and $t_{\alpha/2}$ is based upon $(n - k)$ degrees of freedom.

The confidence intervals are appropriate for single-treatment means or a comparison of a pair of means **selected prior to observation of the data**. The stated confidence coefficients are based on random sampling. If you were to look at the data and then compare the largest and smallest sample means, the assumption of randomness would be violated. Certainly the difference between the largest and smallest sample means is expected to be larger than for a pair selected at random.

EXAMPLE 13.5   Find a 95% confidence interval for the mean score for teaching technique 1 in Example 13.4.

Solution   The 95% confidence interval for the mean score is

$$\bar{T}_1 \pm \frac{t_{.025} s}{\sqrt{n}} \quad \text{or} \quad 75.67 \pm \frac{(2.093)(7.94)}{\sqrt{6}} \quad \text{or} \quad 75.67 \pm 6.78$$

Thus, we infer that the interval 75.67 ± 6.78 or 68.89 to 82.45 encloses the mean score for students subjected to teaching technique 1.   ◁

EXAMPLE 13.6   Find a 95% confidence interval for the difference in mean score for teaching techniques 1 and 4 in Example 13.4.

Solution   The 95% confidence interval for $(\mu_1 - \mu_4)$ is

$$(\bar{T}_1 - \bar{T}_4) \pm t_{\alpha/2} s \sqrt{\frac{1}{n_1} + \frac{1}{n_4}},$$

$$(75.67 - 87.75) \pm (2.093)(7.94) \sqrt{\frac{1}{6} + \frac{1}{4}}, \quad \text{or} \quad -12.08 \pm 10.73.$$

Thus, we estimate that the interval -12.08 ± 10.73 or -22.81 to -1.35 encloses the difference in mean scores for teaching techniques 1 and 4. Because all points in the interval are negative, we infer that μ_4 is larger than μ_1. Also, note that the variability in scores within students is rather large. Consequently, the sample sizes should be increased if the experimenter wishes to reduce the width of the confidence interval. ◁

TIPS ON PROBLEM SOLVING

The following suggestions apply to all the analyses of variance in this chapter.

1. When calculating sums of squares, be certain to carry at least six significant figures before performing subtractions.

2. Remember, sums of squares can never be negative. If you obtain a negative sum of squares, you have made a mistake in arithmetic.

3. Always check your analysis-of-variance table to make certain that the degrees of freedom sum to the total degrees of freedom $n - 1$ and that the sums of squares sum to the Total SS.

EXERCISES   Basic Techniques

13.11   The data shown below are observations collected using a completely randomized design.

	Sample	
1	2	3
3	4	2
2	3	0
4	5	2
3	2	1
2	5	

a. Calculate CM and Total SS.
b. Calculate SST and MST.
c. Calculate SSE and MSE.
d. Construct an ANOVA table for the data.
e. State the null and alternative hypotheses for an analysis-of-variance F test.
f. Give the rejection region for the test using $\alpha = .05$.
g. Conduct the test and state your conclusions.

13.12   Refer to Exercise 13.11. Do the data provide sufficient evidence to indicate a difference between μ_2 and μ_3? Test using the t test of Section 9.4 with $\alpha = .05$.

13.13   Refer to Exercise 13.11.
a. Find a 90% confidence interval for μ_1.
b. Find a 90% confidence interval for the difference $(\mu_1 - \mu_3)$.

13.14 A portion of the ANOVA table for a completely randomized design is shown below.

Source	d.f.	SS	MS	F
Treatments	4	26.3		
Error		52.8		
Total	29			

a. How many independent random samples were selected in this experiment?
b. Does the ANOVA table provide the information necessary to determine the sample sizes?
c. How many observations were involved in the complete design?
d. Fill in the blanks in the ANOVA table.
e. Give the rejection region for the analysis-of-variance F test.
f. Do the data provide sufficient evidence to indicate a difference among at least two of the population means? Test using $\alpha = .05$.

Applications

13.15 Analyze the data of Exercise 9.96 by the procedure outlined in Section 13.2. Determine whether there is evidence of a difference in mean reaction times for the two stimuli. Test at the $\alpha = .05$ level of significance. If you have not worked Exercise 9.96, analyze the same data using Student's t test (as described in Exercise 9.96). Note that both methods lead to the same conclusion and that the computed values of F and t, for the two methods, are related. That is, $F = t^2$ (this will hold true only for the comparison of *two* population means).

13.16 A clinical psychologist wished to compare three methods for reducing hostility levels in university students. A certain psychological test (HLT) was used to measure the degree of hostility. High scores on this test were taken to indicate great hostility. Eleven students obtaining high and nearly equal scores were used in the experiment. Five were selected at random from among the eleven problem cases and treated by method A. Three were taken at random from the remaining six students and treated by method B. The other three students were treated by method C. All treatments continued throughout a semester. Each student was given the HLT test again at the end of the semester, with the following results:

Method	Scores on the HLT tests				
A	73	83	76	68	80
B	54	74	71		
C	79	95	87		

a. Perform an analysis of variance for this experiment.
b. Do the data provide sufficient evidence to indicate a difference in mean student response to the three methods after treatment?

13.17 Refer to Exercise 13.16. Let μ_A and μ_B, respectively, denote the mean scores at the end of the semester for the populations of extremely hostile students who are treated throughout that semester by method A and method B.

a. Find a 95% confidence interval for μ_A.
b. Find a 95% confidence interval for μ_B.
c. Find a 95% confidence interval for $\mu_A - \mu_B$.
d. Is it correct to claim that the confidence intervals found in parts (a), (b), and (c) are jointly valid?

13.18 Three groups of fourth graders were randomly selected and assigned, one group each, to three different physical exercise programs to determine whether the programs were effective in increasing the children's abilities to throw an object. Twenty-eight students were involved in the experiment—ten in a control group (no exercise) and nine each assigned to two different exercise regimes, each of which lasted four weeks. The velocity at which a child was able to throw a test ball was measured before and after the four weeks of exercise, and the gain (or loss) in velocity y (in feet per second) was recorded. A table of mean gains for the three groups and a partially completed analysis-of-variance table for the data are as shown.

Sample means

Control	Exercise regime A	Exercise regime B
−1.34	.32	3.69

ANOVA table

Source	d.f.	SS	MS
Exercise regimes	—	64.31	—
Error	—	—	—
Total	—	402.33	

a. Fill in the missing numbers in the analysis-of-variance table.
b. Do the data provide sufficient evidence to indicate a difference in population means for the three groups? Explain the implications of the test results.
c. Find a 95% confidence interval for the difference in mean gain between children in the control group versus those on exercise regime B.
d. Find a 95% confidence interval for the mean gain for children on exercise regime B.

13.19 G. J. Bacher and F. I. Norman present data on the mercury concentration in the wing muscles of ten species of Australian waterfowl ("Mercury Concentrations in Ten Species of Australian Waterfowl [Family *Anatidae*]," *Australian Wildlife Research* 11 [1984]). The data below give the sample sizes, means, and standard deviations for mercury concentration readings (in micrograms per gram) taken on waterfowl from three of these species.

Species	Sample size	Mean	Standard deviation
Australian Shelduck	6	0.09	0.06
Australian Shoveler	3	0.10	0.05
Blue-billed Duck	18	0.12	0.05

a. Calculate the sample total for each of the three samples.
b. Calculate CM.
c. Calculate SST.
d. Calculate the sum of squares of deviations of the observations about the mean for each of the samples.
e. Calculate SSE. (*Note:* SSE is the pooled sum of squares of deviations for the three samples, that is, the sum of the three quantities obtained in part d.)
f. Construct an ANOVA table for your analysis of variance.

g. Do the data present sufficient evidence to indicate differences in the mean concentration of mercury among the three species of waterfowl? Test using $\alpha = .05$.

13.20 An experiment was conducted to compare the effectiveness of three training programs A, B, and C in training assemblers of a piece of electronic equipment. Fifteen employees were randomly assigned, five each, to the three programs. After completion of the courses, each person was required to assemble four pieces of the equipment, and the average length of time required to complete the assembly was recorded. Due to resignation from the company, only four employees completed program A, and only three completed B. The data are shown in the accompanying table. An SAS computer printout of the analysis of variance for the data is also shown below. Use the information in the printout to answer questions (a) through (d).

Training program	Average assembly time (min)				
A	59	64	57	62	
B	52	58	54		
C	58	65	71	63	64

a. Do the data provide sufficient evidence to indicate a difference in mean assembly time for people trained by the three programs? Give the p-value for the test and interpret its value.
b. Find a 90% confidence interval for the difference in mean assembly time between persons trained by programs A and B.
c. Find a 90% confidence interval for the mean assembly time for persons trained in program A.
d. Do you think the data will satisfy (approximately) the assumption that they have been selected from normal populations? Why?

STATISTICAL ANALYSIS SYSTEM
ANALYSIS OF VARIANCE PROCEDURE

DEPENDENT VARIABLE: X

SOURCE	DF	SUM OF SQUARES	MEAN SQUARE	F VALUE	PR > F	R-SQUARE	C.V.
MODEL	2	170.45000000	85.22500000	5.70	0.0251	0.559005	6.3802
ERROR	9	134.46666667	14.94074074		ROOT MSE		Y MEAN
CORRECTED TOTAL	11	304.91666667			3.86532544		60.58333333

SOURCE	DF	ANOVA SS	F VALUE	PR > F
TRTMENTS	2	170.45000000	5.70	0.0251

13.21 In Exercise 9.24 we presented some of the data from a study by E. F. Karlin and L. C. Bliss (1983) of the germination rate of seeds as a function of seed age ("Germination Ecology of *Ledum groenlandicum* and *Ledum palustre* ssp. *decumbens*," *Arctic and Alpine Research* 15, no. 3 [1983]). Each experimental unit, a set of 50 seeds of a specific age, was placed in a controlled moist environment and the percentage y of germinating seeds was recorded. Experimental units were prepared for seeds aged 1, 8, 13, and 22 months. The number of experimental units, the sample average, and the standard deviation are shown on page 552 for each of the four seed ages.

Seed age (months)	Sample size	$\bar{x}$ (percent)	s (percent)
1	6	58	4
8	7	47	4
13	3	16	3
22	3	16	2

a. Calculate the sample total for each of the four samples.
b. Calculate CM.
c. Calculate SST.
d. Calculate the sum of squares of deviations of the observations about the sample mean for each of the four samples.
e. Calculate SSE. (*Note:* SSE is equal to the sum of the four quantities calculated in part d.)
f. Construct an ANOVA table for your analysis of variance.
g. Do the data present sufficient evidence to indicate differences in the mean percentage of seeds germinating among the four age groups? Test using $\alpha = .05$.

 13.22 An ecological study was conducted to compare rate of growth of vegetation at four swampy undeveloped sites and to determine the cause of any differences that might be observed. Part of the study involved the measurement of leaf lengths of a particular plant species at a preselected date in May. Six plants were randomly selected at each of the four sites to be used in the comparison. The following data represent the mean leaf length per plant, in centimeters, for a random sample of ten leaves per plant.

Location	Mean leaf length (cm)					
1	5.7	6.3	6.1	6.0	5.8	6.2
2	6.2	5.3	5.7	6.0	5.2	5.5
3	5.4	5.0	6.0	5.6	4.9	5.2
4	3.7	3.2	3.9	4.0	3.5	3.6

a. You will recall that the test and estimation procedures for an analysis of variance require that the observations be selected from normally distributed (at least, roughly so) populations. Why might you feel reasonably confident that your data satisfy this assumption?
b. Do the data provide sufficient evidence to indicate a difference in mean leaf length among the four locations?
c. Suppose, prior to seeing our data, that we decided to compare the mean leaf length of locations 1 and 4. Test the null hypothesis that $\mu_1 = \mu_4$ against the alternative that $\mu_1 \neq \mu_4$.
d. Refer to part (c). Find a 90% confidence interval for $(\mu_1 - \mu_4)$.
e. Rather than use an analysis-of-variance F test, it would seem simpler to examine one's data, select the two locations that have the smallest and largest sample mean lengths, and then compare these two means using a Student's t test. If there is evidence to indicate a difference in these means, there is clearly evidence of a difference among the four. (If you were to use this logic, there would be no need for the analysis-of-variance F test.) Explain why this procedure is invalid.

13.23 In Exercise 9.121 we examined the differences in monthly child-care expenses for married couples and single mothers. In a similar study, the differences between various ethnic groups was also noted. A random sample of 12 sets of parents were chosen from each of three different ethnic groups, and the monthly child-care expenses were recorded.

	White	Black	Hispanic
$\bar{x}$	$181	$258	$187
s	11	14	9

   a. Use the values of $\bar{x}$ to calculate CM, the treatment totals, and SST.
   b. Use the values of s to calculate SSE as a pooled sum of squares.
   c. Construct the ANOVA table for the experiment.
   d. Do the data provide sufficient evidence to indicate that there is a difference in average monthly child-care expenses among ethnic groups? What is the approximate p-value associated with the test?
   e. Assuming you were particularly interested in differences between blacks and Hispanics, construct a 95% confidence interval for the difference between these two population means.

13.24 Water samples were taken at four different locations in a river to determine whether the quantity of dissolved oxygen, a measure of water pollution, varied from one location to another. Locations 1 and 2 were selected above an industrial plant, one near the shore and the other in midstream; location 3 was adjacent to the industrial water discharge for the plant; and location 4 was slightly downriver in midstream. Five water specimens were randomly selected at each location, but one specimen, corresponding to location 4, was lost in the laboratory. The data are shown below (the greater the pollution, the lower the dissolved oxygen readings).

Location	Mean dissolved oxygen content				
1	5.9	6.1	6.3	6.1	6.0
2	6.3	6.6	6.4	6.4	6.5
3	4.8	4.3	5.0	4.7	5.1
4	6.0	6.2	6.1	5.8	

   a. Do the data provide sufficient evidence to indicate a difference in mean dissolved oxygen content for the four locations?
   b. Compare the mean dissolved oxygen content in midstream above the plant with the mean content adjacent to the plant (location 2 versus location 3). Use a 95% confidence interval.

13.6 THE RANDOMIZED BLOCK DESIGN

The completely randomized design discussed in Sections 13.3 and 13.4 is the appropriate design to use when the experimental material or experimental units are relatively homogeneous. When the experimental material is not homogeneous, we may be able to find groups of homogeneous units, called **blocks**, within which the means associated with the treatments under investigation may be compared. With such a design the comparisons among the treatments are made within homogeneous blocks of material; therefore, the effects of the blocks as well as the block-to-block variability are removed from the error variation.

Definition

> A **randomized block design** consists of b blocks, each containing k experimental units. The k treatments are randomly assigned to the units in each block, and each treatment appears once in every block.

For example, in assessing the effects of three package designs on the amount of product sales, we might decide to use a completely randomized design and select 12 supermarkets, four of which would be assigned to display and promote one of the three package designs. However, the difference in sales could be due to more than just differences among the packaging designs. Supermarkets with a large volume of business would be expected to have large overall sales of the product, but supermarkets with a small volume of business would be expected to have a smaller volume of product sales. Hence, an alternative design that uses the blocking principle is the randomized block design in which only four supermarkets are used but all three packaging designs are displayed and available for sale in each of the four stores. In this design for which $b = 4$ supermarkets (blocks) and $k = 3$ package designs (treatments), the store-to-store variability has been eliminated from uncontrolled error through the choice of design. The resulting randomized block design would appear symbolically as shown in Figure 13.5.

Figure 13.5
Randomized block design for supermarket experiment

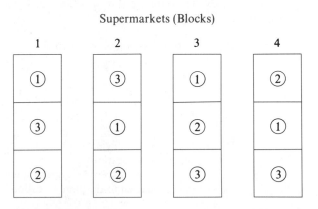

The word *randomized* in the name of the design implies that the treatments are randomly assigned within the block. For our experiment the position within the block pertains to the position in the sequence when assigning a particular package design to a given supermarket over time. For instance, the experiment may be conducted over a period of three weeks, with each store being randomly assigned a different package design to sell each week. The purpose of the randomization (that is, the position in the block) is to eliminate bias due to time. That is, in case the demand for the product in any one week is higher or lower than in another week, it gives every package an equal chance of being displayed during that week.

Blocks may represent time, location, or experimental material. If three treatments are to be compared and there is a suspected trend in the mean response over time, a substantial part of the time variation may be removed by blocking. All three treatments would be randomly applied to experimental units in one small block of time. This procedure would be repeated in succeeding blocks of time until the required amount of data is collected.

As we have seen, a comparison of the sale of competitive or differently designed products in supermarkets should be made within supermarkets, thus using the supermarkets as blocks and removing store-to-store variability. An experiment designed to test and compare subject response to a set of stimuli uses the subjects as blocks to remove subject-to-subject variability. That is, each person is subjected to the complete set of stimuli, with the stimuli spaced in time (to reduce the possibility of a residual effect from the previous stimulus) and assigned in random order. Experiments in medicine often utilize children within a single family as blocks, applying all the treatments, one each, to children within a family. Because of heredity, children within a family are more homogeneous than those in different families. This type of blocking removes the family-to-family variation, just as the stimulus-response experiment is designed to remove the subject-to-subject variation and the market preference experiment is designed to remove the store-to-store variation.

13.7   THE ANALYSIS OF VARIANCE FOR A RANDOMIZED BLOCK DESIGN

The randomized block design implies the presence of two qualitative independent variables: blocks and treatments. Consequently, the total sum of squares of deviations of the response measurements about their mean can be partitioned into three parts: the sums of squares for blocks, treatments, and error.

We denote the total and mean of all observations in block i as B_i and $\bar{B}_i$, respectively. Similarly, we let T_j and $\bar{T}_j$ represent the total and the mean for all observations receiving treatment j. Then, for a randomized block design involving b blocks and k treatments,

$$\text{Total SS} = \text{SSB} + \text{SST} + \text{SSE}$$

where

$$\text{Total SS} = \sum_{i=1}^{b} \sum_{j=1}^{k} (x_{ij} - \bar{x})^2 = \sum_{i=1}^{b} \sum_{j=1}^{k} x_{ij}^2 - \text{CM}$$

$$= (\text{sum of squares of all } x \text{ values}) - \text{CM}$$

$$\text{SSB} = k \sum_{i=1}^{b} (\bar{B}_i - \bar{x})^2 = \frac{\sum_{i=1}^{b} B_i^2}{k} - \text{CM}$$

$$= \frac{\text{sum of squares of all block totals}}{\text{number of observations in a single total}} - \text{CM}$$

$$\text{SST} = b \sum_{j=}^{k} (\bar{T}_j - \bar{x})^2 = \frac{\sum_{j=1}^{k} T_j^2}{b} - \text{CM}$$

$$= \frac{\text{sum of squares of all treatment totals}}{\text{number of observations in a single total}} - \text{CM}$$

where

$$\bar{x} = (\text{average of all } n = bk \text{ observations}) = \frac{\sum_{i=1}^{b} \sum_{j=1}^{k} x_{ij}}{n}$$

and

$$\text{CM} = \frac{(\text{total of all observations})^2}{n} = \frac{\left(\sum_{i=1}^{b} \sum_{j=1}^{k} x_{ij}\right)^2}{n}, \text{ with } n = bk$$

The analysis of variance for the randomized block design is presented in Table 13.5. The degrees of freedom associated with each sum of squares is shown in the second column. Mean squares are calculated by dividing the sums of squares by their respective degrees of freedom.

Table 13.5
ANOVA table for a randomized block design

Source	d.f.	SS	MS	F
Blocks	$b - 1$	SSB	$\text{SSB}/(b - 1)$	MSB/MSE
Treatments	$k - 1$	SST	$\text{SST}/(k - 1)$	MST/MSE
Error	$(b - 1)(k - 1)$	SSE	$\text{SSE}/(b - 1)(k - 1)$	
Total	$n - 1$	Total SS		

To test the null hypothesis "there is no difference in treatment means," we use the F statistic

$$F = \frac{\text{MST}}{\text{MSE}}$$

and reject H_0 if $F > F_\alpha$ based on $v_1 = (k - 1)$ and $v_2 = (b - 1)(k - 1)$ degrees of freedom.

Blocking not only reduces the experimental error; it also provides an opportunity to see whether evidence exists to indicate a difference in the mean response for blocks. Under the null hypothesis that there is no difference in mean response for blocks, MSB provides an unbiased estimator for σ^2 based on $(b - 1)$ degrees of freedom. Where a real difference exists in the block means, MSB will tend to be inflated in comparison with MSE. Therefore,

$$F = \frac{\text{MSB}}{\text{MSE}}$$

can be used as a test statistic to test for differences among block means. As in the test for treatments, the rejection region for the test will be

$$F > F_\alpha$$

based on $v_1 = b - 1$ and $v_2 = (b - 1)(k - 1)$ degrees of freedom.

F TEST FOR COMPARING k TREATMENTS, USING A RANDOMIZED BLOCK DESIGN

1. Null Hypothesis: H_0: The population treatment means are equal.
2. Alternative Hypothesis: H_a: One or more pairs of population treatment means differ.
3. Test Statistic: $F = \text{MST}/\text{MSE}$, where F is based on $v_1 = (k - 1)$ and $v_2 = (b - 1)(k - 1)$ degrees of freedom.
4. Rejection Region: Reject if $F > F_\alpha$.

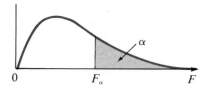

Assumptions:

1. The populations corresponding to the treatment–block combinations are normally distributed with common variances equal to σ^2.
2. The k treatments are randomly assigned, one treatment to each of the k experimental units within a block.

EXAMPLE 13.7   A stimulus-response experiment involving three treatments was laid out in a randomized block design using four subjects. The response was the length of time to reaction measured in seconds. The data (treatment identification numbers are circled) are as follows:

Subject

1	2	3	4
① 1.7	③ 2.1	① .1	② 2.2
③ 2.3	① 1.5	② 2.3	① .6
② 3.4	② 2.6	③ .8	③ 1.6

Do the data present sufficient evidence to indicate a difference in the mean response for stimuli (treatments)? for subjects?

Solution The treatment and block totals are as follows: $T_1 = 3.9$, $T_2 = 10.5$, $T_3 = 6.8$, $B_1 = 7.4$, $B_2 = 6.2$, $B_3 = 3.2$, $B_4 = 4.4$.

The sums of squares for the analysis of variance are shown individually below, and jointly in the analysis-of-variance table. Thus,

$$CM = \frac{(\text{Total})^2}{n} = \frac{(21.2)^2}{12} = 37.4533$$

$$\text{Total SS} = \sum_{i=1}^{4} \sum_{j=1}^{3} (x_{ij} - \bar{x})^2 = \sum_{i=1}^{4} \sum_{j=1}^{3} x_{ij}^2 - CM$$

$$= (\text{sum of squares of all } x \text{ values}) - CM$$

$$= (1.7)^2 + (2.3)^2 + \cdots + (1.6)^2 - CM$$

$$= 46.8600 - 37.4533 = 9.4067$$

$$SSB = \frac{\sum_{i=1}^{4} B_i^2}{3} - CM$$

$$= \frac{\text{sum of squares of all block totals}}{\text{number of observations in a single total}} - CM$$

$$= \frac{(7.4)^2 + (6.2)^2 + (3.2)^2 + (4.4)^2}{3} - CM$$

$$= 40.9333 - 37.4533 = 3.4800$$

$$SST = \frac{\sum_{j=1}^{3} T_j^2}{4} - CM$$

$$= \frac{\text{sum of squares of all treatment totals}}{\text{number of observations in a single total}} - CM$$

$$= \frac{(3.9)^2 + (10.5)^2 + (6.8)^2}{4} - CM$$

$$= 42.9250 - 37.4533 = 5.4717$$

$$SSE = \text{Total SS} - SSB - SST$$

$$= 9.4067 - 3.4800 - 5.4717 = .4550$$

The analysis-of-variance table for Example 13.7 is shown in Table 13.6.

We use the ratio of mean-square treatments to mean-square error to test the hypothesis of no difference in the expected response for treatments. Thus,

$$F = \frac{MST}{MSE} = \frac{2.74}{.076} = 36.05$$

Table 13.6
ANOVA table for
Example 13.7

Source	d.f.	SS	MS	F
Blocks	3	3.4800	1.16	15.26
Treatments	2	5.4717	2.74	36.05
Error	6	.4550	.076	
Total	11	9.4067		

The critical value of the F statistic ($\alpha = .05$) for $v_1 = 2$ and $v_2 = 6$ degrees of freedom is $F_{.05} = 5.14$ (Table 7 of Appendix III). Since the computed value of F exceeds the critical value, there is sufficient evidence to reject the null hypothesis and conclude that a real difference does exist in the expected response for the four stimuli.

A similar test may be conducted for the null hypothesis that no difference exists in the mean response for subjects. Rejection of this hypothesis would imply that subject-to-subject variability does exist, and that blocking is desirable. The computed value of F based on $v_1 = 3$ and $v_2 = 6$ degrees of freedom is

$$F = \frac{MSB}{MSE} = \frac{1.16}{.076} = 15.26$$

Since this value of F exceeds the corresponding tabulated critical value, $F_{.05} = 4.76$, we reject the null hypothesis and conclude that a real difference exists in the expected response in the group of subjects.

Table 13.7
MINITAB TWOWAY
printout for Example 13.7

```
MTB > NAME C2    'TRTS'  C3        'BLOCKS'
MTB > PRINT C1 - C3

ROW      C1      TRTS      BLOCKS

 1      1.7       1          1
 2      3.4       2          1
 3      2.3       3          1
 4      1.5       1          2
 5      2.6       2          2
 6      2.1       3          2
 7       .1       1          3
 8      2.3       2          3
 9       .8       3          3
10       .6       1          4
11      2.2       2          4
12      1.6       3          4

MTB > TWOWAY  C1  C2  C3

ANALYSIS OF VARIANCE ON C1

SOURCE           DF         SS         MS
TRTS              2      5.4717     2.7358
BLOCKS            3      3.4800     1.1600
ERROR             6       .4550      .0758
TOTAL            11      9.4067

MTB >
```

Table 13.8 SAS printout for Example 13.7

ANALYSIS OF VARIANCE PROCEDURE

DEPENDENT VARIABLE X

SOURCE	DF	SUM OF SQUARES	MEAN SQUARE	F VALUE	PR > F	R-SQUARE	C.V.
MODEL	5	8.95166667	1.79033333	23.61	0.0007	0.951630	15.5875
ERROR	6	.45500000	.07583333				
CORRECTED TOTAL	11	9.40666667			ROOT MSE .27537853		X MEAN 1.7666667

SOURCE	DF	ANOVA SS	F VALUE	PR > F
BLOCKS	3	3.48000000	15.30	0.0032
TRTMENTS	2	5.47166667	36.08	0.0005

A randomized block design corresponds to a two-way classification of the observations by treatments and blocks. The TWOWAY analysis of variance command in the MINITAB package was used to analyze the data in Example 13.7. For this program, each observation must be identified according to the block in which it is located and the treatment applied. The values to be analyzed were stored in column 1; the corresponding treatment designation (1, 2, or 3) was stored in column 2; and the corresponding block designation (1, 2, 3, or 4) was stored in column 3. For ease in reading, columns 2 and 3 were labeled as TRTS and BLOCKS, respectively. A listing of the data and the resulting MINITAB output is given in Table 13.7. Table 13.8 provides the same results using the SAS analysis-of-variance procedure. ◁

13.8 ESTIMATION FOR THE RANDOMIZED BLOCK DESIGN

The confidence interval for the difference between a pair of means is exactly the same as for the completely randomized design in Section 13.5.

RANDOMIZED BLOCK DESIGN

A 100(1 − α)% Confidence Interval for the Difference Between Two Treatment Means

$$(\bar{T}_i - \bar{T}_j) \pm t_{\alpha/2} s \sqrt{\frac{2}{b}}$$

where $n_i = n_j = b$, the number of observations contained in a treatment mean, and $t_{\alpha/2}$ is based on $(b - 1)(k - 1)$ degrees of freedom.

The difference between the confidence intervals for the completely randomized and the randomized block designs is that s, appearing in the expression above, will probably be smaller for the randomized block design.

Similarly, one may construct a $100(1 - \alpha)\%$ confidence interval for the difference between a pair of block means. Each block contains k observations corresponding to the k treatments. Therefore, the confidence interval is

RANDOMIZED BLOCK DESIGN

A $100(1 - \alpha)\%$ Confidence Interval for the Difference Between Two Block Means

$$(\bar{B}_i - \bar{B}_j) \pm t_{\alpha/2} s \sqrt{\frac{2}{k}}$$

where $t_{\alpha/2}$ is based on $(b - 1)(k - 1)$ degrees of freedom.

EXAMPLE 13.8 Construct a 95% confidence interval for the difference between treatments 1 and 2 in Example 13.7.

Solution The confidence interval for the difference in mean response for a pair of treatments is

$$(\bar{T}_i - \bar{T}_j) \pm t_{\alpha/2} s \sqrt{\frac{2}{b}}$$

where, for our example, $t_{.025}$ is based on 6 degrees of freedom. Then $t_{.025} = 2.447$ (Table 4 of Appendix III) and $s = \sqrt{\text{MSE}} = \sqrt{.076} = .28$. Substituting these values along with the values of $\bar{T}_1$ and $\bar{T}_2$, we have

$$(.98 - 2.63) \pm (2.447)(.28)\sqrt{\frac{2}{4}} \quad \text{or} \quad -1.65 + .48$$

Thus, we infer that the interval $-1.65 \pm .48$ or -2.13 to -1.17 encloses the difference in mean reaction times for stimuli 1 and 2. Because all points in the interval are negative, we conclude that the mean time to react for stimulus 2 is larger than that for stimulus 1.

TIPS ON PROBLEM SOLVING

Be careful of this point: Unless the blocks have been randomly selected from a population of blocks, you cannot obtain a confidence interval for a single treatment mean. This limitation occurs because the sample treatment mean is biased by the positive and negative effects that the blocks have on the response. Further discussion of this topic can be found in the references.

EXERCISES Basic Techniques

13.25 The data shown on page 562 are observations collected from an experiment that compared four treatments A, B, C, and D within each of three blocks, using a randomized block design.

	Treatment				
Block	A	B	C	D	Total
1	6	10	8	9	33
2	4	9	5	7	25
3	12	15	14	14	55
Total	22	34	27	30	113

a. Calculate CM and Total SS.
b. Calculate SST and MST.
c. Calculate SSB and MSB.
d. Calculate SSE and MSE.
e. Construct an ANOVA table for the data.
f. Do the data present sufficient evidence to indicate differences among the treatment means? Test using $\alpha = .05$.
g. Do the data present sufficient evidence to indicate differences among the block means? Test using $\alpha = .05$.
h. Does it appear that the use of a randomized block design for this experiment was justified? Explain.

13.26 Refer to Exercise 13.25. Find a 90% confidence interval for the difference $(\mu_A - \mu_B)$.

13.27 The data shown below are observations collected from an experiment that compared three treatments A, B, and C within each of five blocks, using a randomized block design.

	Block					
Treatment	1	2	3	4	5	Total
A	2.1	2.6	1.9	3.2	2.7	12.5
B	3.4	3.8	3.6	4.1	3.9	18.8
C	3.0	3.6	3.2	3.9	3.9	17.6
Total	8.5	10.0	8.7	11.2	10.5	48.9

a. Calculate CM and Total SS.
b. Calculate SST and MST.
c. Calculate SSB and MSB.
d. Calculate SSE and MSE.
e. Construct an ANOVA table for the data.
f. Do the data present sufficient evidence to indicate differences among the treatment means? Test using $\alpha = .05$.
g. Do the data present sufficient evidence to indicate differences among the block means? Test using $\alpha = .05$.
h. Does it appear that the use of a randomized block design for this experiment was justified? Explain.
i. Find a 99% confidence interval for the difference $(\mu_C - \mu_A)$.

13.28   The partially completed ANOVA table for a randomized block design is shown below.

Source	d.f.	SS	MS	F
Treatments	4	14.2		
Blocks		18.9		
Error	24			
Total	34	41.9		

a.   How many blocks were involved in the design?
b.   How many observations are in each treatment total?
c.   How many observations are in each block total?
d.   Fill in the blanks in the ANOVA table.
e.   Do the data present sufficient evidence to indicate differences among the treatment means? Test using $\alpha = .10$.
f.   Do the data present sufficient evidence to indicate differences among the block means? Test using $\alpha = .10$.

Applications

13.29   A study was conducted to compare automobile gasoline mileage for three brands of gasoline A, B, and C. Four automobiles, all of the same make and model, were employed in the experiment and each gasoline brand was tested in each automobile. Using each brand within the same automobile has the effect of eliminating (blocking out) automobile-to-automobile variability. The data, in miles per gallon, are as follows:

Gasoline brand	Automobile			
	1	2	3	4
A	15.7	17.0	17.3	16.1
B	17.2	18.1	17.9	17.7
C	16.1	17.5	16.8	17.8

a.   Do the data provide sufficient evidence to indicate a difference in mean mileage per gallon for the three gasolines?
b.   Is there evidence of a difference in mean mileage for the four automobiles?
c.   Suppose that *prior to looking at the data*, we had decided to compare the mean mileage per gallon for gasoline brands A and B. Find a 90% confidence interval for this difference.

13.30   An experiment was conducted to compare the effect of four different chemicals A, B, C, and D in producing water resistance in textiles. A strip of material, randomly selected from a bolt, was cut into four pieces and the pieces were randomly assigned to receive one of the four chemicals A, B, C, or D. This process was replicated three times, thus producing a randomized block design. The design, with moisture-resistance measurements, is as shown (low readings indicate low moisture penetration). An SAS computer printout of the analysis of variance for the data is also presented. Use the information in the printout to answer the following questions.

a.   Do the data provide sufficient evidence to indicate a difference in the mean moisture penetration for fabric treated with the four chemicals?

b.   Do the data provide evidence to indicate that blocking increased the amount of information in the experiment?

c.   Find a 95% confidence interval for the difference in mean moisture penetration for fabrics treated by chemicals A and D. Interpret the interval.

Blocks (bolt samples)

1	2	3
C 9.9	D 13.4	B 12.7
A 10.1	B 12.9	D 12.9
B 11.4	A 12.2	C 11.4
D 12.1	C 12.3	A 11.9

STATISTICAL ANALYSIS SYSTEM
ANALYSIS OF VARIANCE PROCEDURE

DEPENDENT VARIABLE: X

SOURCE	DF	SUM OF SQUARES	MEAN SQUARE	F VALUE	PR > F	R-SQUARE	C.V.
MODEL	5	12.37166667	2.47433333	27.75	0.0004	0.958549	2.5023
ERROR	6	0.53500000	0.08916667		ROOT MSE		Y MEAN
CORRECTED TOTAL	11	12.90666667			0.29860788		11.93333333

SOURCE	DF	ANOVA SS	F VALUE	PR > F
BLOCKS	2	7.17166667	40.21	0.0003
TRTMENTS	3	5.20000000	19.44	0.0017

 13.31   An experiment was conducted to compare the glare characteristics of four types of automobile rearview mirrors. Forty drivers were randomly selected to participate in the experiment. Each driver was exposed to the glare produced by a headlight located 30 feet behind the rear window of the experimental automobile. The driver then rated the glare produced by the rearview mirror on a scale of 1 (low) to 10 (high). Each of the four mirrors was tested by each driver; the mirrors were assigned to a driver in random order. An analysis of variance of the data produced the following ANOVA table.

Source	d.f.	SS	MS	F
Mirrors		46.98		
Drivers			8.42	
Error				
Total		638.61		

a. Fill in the blanks in the ANOVA table.
b. Do the data present sufficient evidence to indicate differences in the mean glare ratings of the four rearview mirrors? Test using $\alpha = .10$.
c. Do the data present sufficient evidence to indicate that the level of glare perceived by the drivers varied from driver to driver? Test using $\alpha = .10$.

13.32 An experiment was conducted to determine the effect of three methods of soil preparation on the first-year growth of slash pine seedlings. Four locations (state forest lands) were selected and each location was divided into three plots. Since it was felt that soil fertility within a location was more homogeneous than between locations, a randomized block design was employed using locations as blocks. The methods of soil preparation were A (no preparation), B (light fertilization), and C (burning). Each soil preparation was randomly applied to a plot within each location. On each plot the same number of seedlings were planted and the observation recorded was the average first-year growth of the seedlings on each plot.

Soil preparation	Location			
	1	2	3	4
A	11	13	16	10
B	15	17	20	12
C	10	15	13	10

a. Conduct an analysis of variance. Do the data provide sufficient evidence to indicate a difference in the mean growth for the three soil preparations?
b. Is there evidence to indicate a difference in mean rates of growth for the four locations?
c. Use a 90% confidence interval to estimate the difference in mean growth for methods A and B.

13.33 A study was conducted to compare the effect of three levels of digitalis on the level of calcium in the heart muscle of dogs. A description of the actual experimental procedure is omitted, but it is sufficient to note that the general level of calcium uptake varies from one animal to another so that comparison of digitalis levels (treatments) had to be blocked on heart muscles. That is, the tissue for a heart muscle was regarded as a block, and comparisons of the three treatments were made within a given muscle. The calcium uptakes for the three levels of digitalis A, B, and C were compared based on the heart muscle of four dogs. The results are as

follows:

	Dogs		
1	2	3	4
A	C	B	A
1342	1698	1296	1150
B	B	A	C
1608	1387	1029	1579
C	A	C	B
1881	1140	1549	1319

a. Calculate the sums of squares for this experiment and construct an analysis-of-variance table.
b. How many degrees of freedom are associated with SSE?
c. Do the data present sufficient evidence to indicate a difference in the mean uptake of calcium for the three levels of digitalis?
d. Do the data indicate a difference in the mean uptake in calcium for the four heart muscles?
e. Give the standard deviation of the difference between the mean calcium uptake for two levels of digitalis.
f. Find a 95% confidence interval for the difference in mean response between treatments A and B.

13.34 An experiment was conducted to investigate the toxic effects of three chemicals A, B, and C on the skin of rats. One-inch squares of skin were treated with the chemicals and then scored from 0 to 10 depending on the degree of irritation. Three adjacent one-inch squares were marked on the backs of eight rats and each of the three chemicals was applied to each rat. Thus, the experiment was blocked on rats to eliminate the variation in skin sensitivity from rat to rat. The data are as follows:

	Rats						
1	2	3	4	5	6	7	8
B	A	A	C	B	C	C	B
5	9	6	6	8	5	5	7
A	C	B	B	C	A	B	A
6	4	9	8	8	5	7	6
C	B	C	A	A	B	A	C
3	9	3	5	7	7	6	7

a. Do the data provide sufficient evidence to indicate a difference in the toxic effect of the three chemicals?
b. Estimate the difference in mean score for chemicals A and B using a 95% confidence interval.

$ 13.35 A building contractor employs three construction engineers A, B, and C to estimate and bid on jobs. To determine whether one tends to be a more conservative (or liberal) estimator than the others, the contractor selects four projected construction jobs and has each estimator independently estimate the cost (dollars per square foot) of each job. The data are shown in the table.

Estimator (treatments)	Construction job (blocks)				Total
	1	2	3	4	
A	35.10	34.50	29.25	31.60	130.45
B	37.45	34.60	33.10	34.40	139.55
C	36.30	35.10	32.45	32.90	136.75
Total	108.85	104.20	94.80	98.90	406.75

a. Do the data provide sufficient evidence to indicate a difference in the mean building costs estimated by the three estimators? Test by using $\alpha = .05$.

b. Find a 90% confidence interval for the difference in the mean of the estimates produced by estimators A and B. Interpret the interval.

c. Do the data support the contention that the mean estimate of the cost per square foot varies from job to job?

13.36 In Exercise 9.21, we described an experiment conducted by J. Lindhe, D.M.D., (1984) to investigate the effectiveness of an oral antiplaque rinse on the buildup of plaque on teeth ("Clinical Assessment of Antiplaque Agents," *The Compendium of Continuing Education in Dentistry*, suppl no. 5 [1984]). Fourteen subjects were randomly divided into two groups of seven subjects each. One group of seven used an oral rinse (no brushing) that contained an active antiplaque agent. The second group, the **control** group, used a similar oral rinse but, unknown to these subjects, the rinse contained no active antiplaque agent. A plaque index, a measure of plaque growth, was recorded after 4, 7, and 14 days. Suppose that we wish to compare plaque index means after 4, 7, and 14 days for the seven subjects who used the oral antiplaque rinse.

a. If you wish to compare the means, should you use an analysis of variance for a completely randomized design or a randomized block design? Justify your answer.

b. Lindhe's article gives the following means and standard deviations of the plaque index measurements. Does this table provide you with the information you need to perform the analysis in part (a)? Explain.

	At end of		
	4 days	7 days	14 days
Sample Size	7	7	7
Mean	.42	.67	.78
Standard Deviation	·30	.38	.32

▷ 13.9   SOME CAUTIONARY COMMENTS ON BLOCKING

Two points need to be emphasized when using block designs. The first is that a randomized block design should not be used when treatments and blocks correspond to two **experimental** factors. In situations such as this, the effect of one factor usually depends upon the level of the other, and the factors are said to **interact** rather than act independently. The randomized block design requires that the block and treatment factors **do not interact** in order to provide an unbiased estimate of the error variation. When there is interaction between the treatment and block factors, the analysis of variance could lead to erroneous conclusions regarding the relationship between the treatments and the response. Information on designs for multi-factor experiments can be found in any standard text on the design of experiments (see the references).

The second point is that **blocking is not always beneficial**. A randomized block design produces a gain in information if the variation between blocks is greater than the variation within blocks. Then blocking removes this larger source of variation from SSE, and s^2 assumes a smaller value (as does the population variance σ^2). However, at the same time, you lose information because blocking reduces the number of degrees of freedom associated with SSE and s^2. To see this, suppose that we replicate each of k treatments b times so that $n = bk$, first using a completely randomized design, and then using a randomized block design. The difference in the degrees of freedom for error are $(bk - k) - (b - 1)(k - 1) = b - 1$. Consequently, if blocking is to be beneficial, **the gain in information due to the elimination of block variation must outweigh the loss due to the reduction in the number of degrees of freedom for error**. However, unless blocking leaves only a small number of degrees of freedom for error, the loss of $(b - 1)$ degrees of freedom for error causes only a small reduction in the information in the experiment. Consequently, if you have reason to suspect that the experimental units are not homogeneous, and can group the units into blocks, it usually pays to use the randomized block design.

▷ 13.10   SELECTING THE SAMPLE SIZE

Selecting the sample size for the completely randomized or the randomized block design is an extension of the procedure of Section 7.10. We confine our attention to the case of equal sample sizes, $n_1 = n_2 = \cdots = n_k$, for the k treatments of the completely randomized design. The number of observations per treatment is equal to b for the randomized block design. Thus, the problem is to select n_i or b for these two designs so as to purchase a specified quantity of information.

The selection of sample size follows a similar procedure for both designs; hence, we will outline a general method. **First the experimenter must decide on the parameter (or parameters) of major interest.** Usually, she will wish to compare a pair of treatment means. **Secondly, she must specify a bound on the error of estimation that** she is willing to tolerate. Once determined, she need only select n_i (the number of observations in a treatment mean) or, correspondingly, b (the number of blocks in the randomized block design) that will reduce the half-width of the confidence interval for the parameter so that it is less than or equal to the specified bound on the error of estimation. It should be emphasized that the sample-size solution will *always* be an approximation, since σ is unknown and s is unknown until the sample is acquired. The best available value will be used for s in order to produce an approximate solution. We will illustrate the procedure with an example.

EXAMPLE 13.9   A completely randomized design is to be conducted to compare five teaching techniques on classes of equal size. Estimation of the difference in mean response to an achievement test is desired correct to within five test-score points, with probability equal to .95. It is expected that the test scores for a given teaching technique will have a range approximately equal to 40. Find the approximate number of observations required for each sample in order to acquire the specified information.

Solution   The confidence interval for the difference between a pair of treatment means is

$$(\bar{T}_i - \bar{T}_j) \pm t_{\alpha/2} s \sqrt{\frac{1}{n_i} + \frac{1}{n_j}}$$

Therefore, we will wish to select n_i and n_j so that

$$t_{\alpha/2} s \sqrt{\frac{1}{n_i} + \frac{1}{n_j}} \leq 5$$

The value of σ is unknown and s is a random variable. However, an approximate solution for $n_i = n_j$ can be obtained by guessing s to be roughly equal to one-fourth of the range. Thus, $s \approx (40)/4 = 10$. The value of $t_{\alpha/2}$ will be based on $(n_1 + n_2 + \cdots + n_5 - 5)$ degrees of freedom, and for even moderate values of n_i, $t_{.025}$ will be approximately equal to 2. Then

$$t_{.025} s \sqrt{\frac{1}{n_i} + \frac{1}{n_j}} \approx 2(10) \sqrt{\frac{2}{n_i}} \leq 5$$

or, solving for n_i,

$$n_i \approx 32, \qquad i = 1, 2, \ldots, 5 \qquad \qquad \triangleleft$$

EXAMPLE 13.10   An experiment is to be conducted to compare the fire-retardant effect of three different chemicals on various fabrics. The effect of the chemicals was expected to vary substantially from fabric to fabric. Therefore, all three chemicals were to be tested on each fabric, thereby blocking out fabric-to-fabric variability.

The standard deviation of the experimental error was unknown, but prior experimentation involving several applications of a given chemical on the same fabric suggested a range of response measurements of about five units.

Find a value of b such that the error of estimating the difference between a pair of treatment means is less than one unit, with probability equal to .95.

Solution   A very approximate value for s would be one-fourth the range, or $s \approx 1.25$. Then we wish to select b so that

$$t_{.025} s \sqrt{\frac{1}{b} + \frac{1}{b}} = t_{.025} s \sqrt{\frac{2}{b}} \leq 1$$

Since $t_{.025}$ will depend on the degrees of freedom associated with s^2, which will be $(b - 1)(k - 1)$, we will guess $t_{.025} \approx 2$. Then

$$2(1.25) \sqrt{\frac{2}{b}} \leq 1$$

or

$$b \approx 13, \qquad i = 1, 2, 3$$

Thus, approximately 13 fabrics will be required to obtain the required information.

The degrees of freedom associated with s^2 will be 24 based on this solution. Therefore, the guessed value of t would seem to be adequate for this approximate solution.   ◁

The sample-size solutions for Examples 13.9 and 13.10 are very approximate, and are intended to provide only a rough estimate of approximate size and consequent cost of the experiment. The experimenter will obtain information on σ as the data are being collected and can recalculate a better approximation to n as he or she proceeds.

EXERCISES   Basic Techniques

13.37   Suppose you wish to compare three treatment means using a completely randomized design. Past experience suggests that the observations within a sample have a range equal to .8. If you wish to estimate the difference between a pair of population means correct to within .1 with probability equal to .95, how many observations should be included in each sample (assume that the sample sizes are equal)?

13.38   Refer to Exercise 13.37. Suppose you wish to estimate each individual population mean correct to within .05 with probability equal to .95. How many observations should be included in each sample?

13.39   A preliminary randomized block design, conducted to compare four treatment means, produced an estimate s, based on 18 degrees of freedom, equal to .24. Suppose that you wish to conduct a new experiment that utilizes a randomized block design and that you wish to estimate the difference between a pair of treatment means correct to within .1 with probability equal to .90.
a.   How many blocks would have to be included in the design?
b.   What is the total number of observations for the experiment?

Applications

13.40   Refer to Exercise 13.22. How many plants would have to be selected and analyzed for each location if we wished to estimate the difference in mean leaf length for two locations with an error of less than .3 and a probability approximately equal to .95?

13.41   Refer to Exercise 13.34. Approximately how many rats would be required in order to estimate the difference in mean scores for two chemicals correct to within one unit?

13.42   Refer to Exercise 13.33. Approximately how many replications would be required for each treatment (observations per treatment) in order that the error of estimating the difference in mean response for a pair of treatments be less than 20 (with probability equal to .95)? Assume that additional observations would be made within a randomized block design.

13.43   A manufacturer wished to compare the mean daily production for four production lines A, B, C, and D that manufacture the same product. Since daily production is affected by the quality of raw materials and labor associated with supply to the lines, the decision was made to make the comparisons **within** each day (thereby eliminating day-to-day variations associated with supply).
   a.   What type of experimental design was the manufacturer using to collect his data?
   b.   Suppose that the mean daily yield per line is in the neighborhood of 1000 units and that an earlier analysis of the differences in daily production, recorded each day over a long period of time, suggests that the variance of $(x_1 - x_2)$, the difference in daily yields for production lines A and B, is equal to 5000. For how many days must the daily production be recorded for the production lines to estimate the difference in mean production between any pair of lines correct to within 25 units?

13.44   Refer to Exercise 13.30. Suppose you wish to estimate the difference in mean moisture absorption between fabrics treated by two different chemicals correct to within .2 units, with probability equal to .95. Approximately how many bolts of fabric (blocks) should be used for the experiment?

13.45   An experiment is to be conducted to compare the mean amount of effluents discharged by three chemical plants. Suppose that the variance of the effluent measurements at all three plants is approximately equal to .03. If you wish to estimate the mean amount of effluents (pounds/gallon) at a plant correct to within .05 lb/gal, with probability near .95, approximately how many effluent specimens would have to be randomly selected at a single point?

13.11   ASSUMPTIONS OF ANALYSIS OF VARIANCE

The assumptions about the probability distribution of the random response x that result in a valid analysis of variance are given in the following display.

   Assumptions

   1.   For any treatment or block combination or combination of factor levels, x is normally distributed with variance equal to σ^2.

   2.   The observations have been selected independently so that any pair of x values, x_i and x_j, are independent in a probabilistic sense for all i and $j(i \neq j)$.

In a practical situation you can never be certain that the assumptions are satisfied, but you will often have a fairly good idea whether the assumptions are reasonable for your data. To illustrate, the inferential methods of Chapters 7 through 11 are not seriously affected by moderate departures from the assumptions of normality, but you would want the probability distribution of x to be at least mound-shaped. So if x is a discrete random variable that can assume only three values, say $x = 10, 11, 12$, then it is *unreasonable* to assume that the probability distribution of x is approximately normal.

The assumption of a constant variance for x for the various experimental conditions should be approximately satisfied, although violation of this assumption is not too serious if the sample sizes for the various experimental conditions are equal. However, suppose the response is binomial—say, the proportion p of people who favor a particular type of investment. We know that the variance of a proportion (Section 6.5) is

$$\sigma_{\hat{p}}^2 = \frac{pq}{n} \qquad \text{where} \qquad q = 1 - p$$

and therefore that the variance is dependent on the expected (or mean) value of $\hat{p}$, namely, p. Then the variance of $\hat{p}$ will change from one experimental setting to another, and the assumptions of the analysis of variance will have been violated.

A similar situation occurs when the response measurements are Poisson data (say, the number of industrial accidents per month in a manufacturing plant). If the response possesses a Poisson probability distribution, the variance of the response equals the mean (Section 4.3). Consequently, Poisson response data also violate the analysis of variance assumptions.

Many kinds of data are not measurable and hence are unsuitable for an analysis of variance. For example, many responses cannot be measured but can be ranked. Product preference studies yield data of this type. You know you like product A better than B, and B better than C, but you have difficulty assigning an exact value to the strength of your preferences.

What do you do when the assumptions of an analysis of variance are not satisfied? For example, suppose the variances of the responses for various experimental conditions are not equivalent. This situation can sometimes be remedied by transforming the response measurements. That is, instead of using the original response measurements, we might use their square roots, logarithms, or some other function of the response x. Transformations that tend to stabilize the variance of the response have been found to make the probability distributions of the transformed responses more nearly normal. See the texts listed in the references for discussions of these topics.

When nothing can be done to satisfy (even approximately) the assumptions of the analysis of variance, or if the data are rankings, you should use nonparametric testing and estimation procedures. These procedures, which rely on the comparative magnitudes of measurements (often ranks), are almost as powerful in detecting treatment differences as the tests presented in this chapter. When the assumptions are not satisfied, the nonparametric procedures may be more powerful. Nonparametric tests—what they are and how to use them—are presented in Chapter 14.

▷ 13.12 SUMMARY

The completely randomized and the randomized block designs are procedures that specify the manner in which the treatments are assigned to the units to be used in an experiment. The **completely randomized design** is used to investigate the effect of one independent variable corresponding to treatments when the experimental units are relatively homogeneous. The **randomized block design**, which involves two independent variables corresponding to treatments and blocks, is used when the experimental units are not homogeneous but can be grouped into blocks such that the variability within blocks is less than the variability between blocks. The randomized block design is used to control the effect of the block variable and remove its effect from experimental error.

The **analysis of variance** is a technique that partitions the total sum of squares of deviations of the observations about their mean into portions associated with the independent variables in the experiment and a portion associated with error. The mean squares corresponding to the independent variables are tested against the mean square for error by using the F statistic. A significantly large value of F indicates that the mean square under test is larger than would be expected if the corresponding independent variable had no effect on the response of interest.

This chapter represents a brief introduction to the design of experiments and the widely used technique for analyzing designed experiments called the analysis of variance. Designs are available for experiments involving several design variables as well as several treatment factors. **Design variables** correspond to factors whose effect we wish to control and hence remove from experimental error; **treatment factors** are the experimental variables whose effects we wish to investigate. A properly designed experiment can be analyzed by using an analysis of variance. Experiments in which the levels of one or more variables are measured (as opposed to being set at preselected values) can be analyzed by using *multiple regression analysis*, which was the subject of Chapter 11.

▷ MINITAB COMMANDS

AOVONEWAY on **C...C**

ONEWAY aov, data **C,** levels in **C** [put resids in **C** [fits in **C**]]

TWOWAY aov, data in **C,** levels in **C C** [put resids in **C** [fits in **C**]]
  ADDITIVE model
  MEAN for factors **C** [and **C**]

REFERENCES

Cochran, W. G.; and Cox, G. M. *Experimental Designs.* 2d ed. New York: Wiley, 1957.

Dixon, W. J., et al. *BMDP Statistical Software Manual* Vol. 1. Berkeley: University of California, 1988.

Dunn, O. J.; and Clark, V. A. *Applied Statistics: Analysis of Variance and Regression.* 2d ed. New York: Wiley, 1987.

Hicks, C. R. *Fundamental Concepts in the Design of Experiments.* 3d ed. New York: Holt, Rinehart, and Winston, 1982.

Mendenhall, W. *An Introduction to Linear Models and the Design and Analysis of Experiments.* Belmont, Calif.: Wadsworth, 1968.

Montgomery, D. C. *Design and Analysis of Experiments.* New York: Wiley, 1984.

Ryan, T. A.; Joiner, B. L.; and Ryan, B. F. *Minitab Student Handbook.* 2d ed. Boston: Duxbury Press, 1985.

SAS Institute, Inc. Staff. *SAS User's Guide: Statistics.* Version 5 ed. Cary, N. C.: SAS Institute, Inc., 1985.

Snedecor, G. W.; and Cochran, W. G. *Statistical Methods.* 8th ed. Ames: Iowa State University Press, 1989.

SPSS, Inc. Staff. *SPSS-X User's Guide.* 3d ed. New York: McGraw-Hill, 1987.

SUPPLEMENTARY EXERCISES

13.46 State the assumptions underlying the analysis of variance of a completely randomized design.

13.47 Refer to Example 13.4. Calculate SSE by pooling the sums of squares of deviations within each of the four samples, and compare with the value obtained by subtraction. Note that this is an extension of the pooling procedure used in the two-sample case discussed in Section 13.2.

13.48 A study was initiated to investigate the effect of two drugs, administered simultaneously, in reducing human blood pressure. It was decided to utilize three levels of each drug and to include all nine combinations in the experiment. Nine patients with high blood pressure were selected for the experiment, and one was randomly assigned to each of the nine drug combinations. The response observed was the drop in blood pressure over a fixed interval of time.

a. Is this a randomized block design?

b. Suppose that two patients were assigned to each of the nine drug combinations. What type of experimental design is this?

13.49 Refer to Exercise 13.48. Suppose that prior experimentation suggests that $\sigma = 20$.

a. How many replications would be required to estimate any treatment (drug combination) mean correct to within ± 10?

b. How many degrees of freedom will be available for estimating σ^2 when using the number of replications determined in part (a)?

c. Give the approximate half-widths of a confidence interval for the difference in mean response for two treatments when using the number of replications determined in part (a).

13.50 A completely randomized design was conducted to compare the effect of five different stimuli on reaction time. Twenty-seven people were employed in the experiment, which was conducted using a completely randomized design. Regardless of the results of the analysis of variance, it was desired to compare stimuli A and D. The results of the experiment were as

follows:

Stimulus	Reaction time (sec)							Total	Mean
A	.8	.6	.6	.5				2.5	.625
B	.7	.8	.5	.5	.6	.9	.7	4.7	.671
C	1.2	1.0	.9	1.2	1.3	.8		6.4	1.067
D	1.0	.9	.9	1.1	.7			4.6	.920
E	.6	.4	.4	.7	.3			2.4	.48

a. Conduct an analysis of variance, and test for a difference in mean reaction time due to the five stimuli.

b. Compare stimuli A and D to see if there is a difference in mean reaction time.

13.51 The experiment in Exercise 13.50 might have been more effectively conducted using a randomized block design with people as blocks, since we would expect mean reaction time to vary from one person to another. Hence, four people were used in a new experiment and each person was subjected to each of the five stimuli in a random order. The reaction times (in seconds) were as follows:

Subject	Stimulus				
	A	B	C	D	E
1	.7	.8	1.0	1.0	.5
2	.6	.6	1.1	1.0	.6
3	.9	1.0	1.2	1.1	.6
4	.6	.8	.9	1.0	.4

Conduct an analysis of variance, and test for differences in treatments (stimuli).

13.52 An experiment was conducted to compare the price of a loaf of bread (a particular brand) at four city locations. Four stores were randomly sampled in locations 1, 2, and 3, but only two were selected from location 4 (only two carried the brand). Note that a completely randomized design was employed. Conduct an analysis of variance for the data.

Location	Prices (cents)			
1	139	143	145	141
2	138	141	144	143
3	134	139	135	138
4	149	150		

a. Do the data provide sufficient evidence to indicate a difference in mean price of the bread in stores located in the four areas of the city?

b. Suppose that prior to seeing the data, we wished to compare the mean prices between locations 1 and 4. Estimate the difference in means using a 95% confidence interval.

13.53 An experiment was conducted to examine the effect of age on heart rate when a person is subjected to a specific amount of exercise. Ten male subjects were randomly selected from four age groups: 10–19, 20–39, 40–59, and 60–69. Each subject walked on a treadmill at a fixed grade for a period of 12 minutes, and the increase in heart rate, the difference before and after exercise, was recorded (in beats per minute). These data are shown in the table.

	Age			
	10–19	20–39	40–59	60–69
	29	24	37	28
	33	27	25	29
	26	33	22	34
	27	31	33	36
	39	21	28	21
	35	28	26	20
	33	24	30	25
	29	34	34	24
	36	21	27	33
	22	32	33	32
Total	309	275	295	282

a. Do the data provide sufficient evidence to indicate a difference in mean increase in heart rate among the four age groups? Test by using $\alpha = .05$.

b. Find a 90% confidence interval for the difference in mean increase in heart rate between age groups 10–19 and 60–69.

c. Find a 90% confidence interval for the mean increase in heart rate for the age group 20–39.

d. Approximately how many people would you need in each group if you wanted to be able to estimate a group mean correct to within two beats per minute with probability equal to .95?

13.54 Fifteen subjects with relatively equal reaction times and physical attributes were randomly grouped in fives. The first five were to react to a light stimulus, the second five to a sound stimulus, and the third five to a stimulus of both sound and light. The results were as follows:

Stimulus	Reaction time (sec)				
Light	.36	.41	.45	.42	.31
Sound	.49	.42	.35	.48	.40
Both	.29	.37	.42	.41	.47

Is there sufficient evidence to indicate a difference in mean reaction times for the three stimuli? Use $\alpha = .05$.

13.55 A company wished to study the differences among four sales-training programs on the sales abilities of their sales personnel. Thirty-two people were randomly divided into four groups of equal size, and the groups were then subjected to the different sales-training programs. Because there were some dropouts during the training programs, due to illness, vacations, and so on, the number of trainees completing the programs varied from group to group. At the end of the training programs, each salesperson was randomly assigned a sales area from a group of sales areas that were judged to have equivalent sales potentials. The numbers of sales made by each of the four groups of salespeople during the first week after completing the training program are listed in the table.

a. Do the data present sufficient evidence to indicate a difference in the mean achievement for the four training programs?

b. Find a 90% confidence interval for the difference in the mean of sales that would be expected for persons subjected to training programs 1 and 4. Interpret the interval.

Training program

	1	2	3	4
	78	99	74	81
	84	86	87	63
	86	90	80	71
	92	93	83	65
	69	94	78	86
	73	85		79
		97		73
		91		70
Total	482	735	402	588

c. Find a 90% confidence interval for the mean number of sales by persons subjected to training program 2.

13.56 In Exercise 9.87 we presented data on the mean number of rat markings when exposed to three different odor sources (L. I. A. Birke and D. Saddler, "Scent Marking Behavior to Conspecific Odours by the Rat, *Rattus norvegicus*," *Animal Behavior* Vol. 32 [1984]). Ten male rats were exposed to each of the odor environments—no odor, a female odor, and a male odor. The means and standard errors for the three samples are shown below.

Odor source

None (controls)	Female	Male
9.0(1.1)	19.4(2.85)	24.2(3.4)

a. Calculate each of the three treatment totals.
b. Calculate CM.
c. Calculate SST.
d. Calculate the sample variance for each of the three samples.
e. Calculate the sum of squares of deviations of the x values about their mean for each of the three samples.
f. Calculate SSE. (*Hint:* SSE is equal to the pooled sum of squares of deviations of the x values about their mean, that is, the sum of the three quantities calculated in part e.)
g. Construct an analysis-of-variance table for the data.
h. Do the data present sufficient evidence to indicate differences in the mean number of markings for the three odor sources? Test using $\alpha = .05$.
i. The data suggest that one of the assumptions required for validity of an analysis of variance may not be satisfied. Which assumption is it? Justify your answer. (*Note:* An alternative method of analysis is described in Section 14.6.)

13.57 The whitefly, which causes defoliation of shrubs and trees and a reduction in salable crop yields, has emerged as a pest in Southern California. In a study to determine factors affecting the life cycle of the whitefly, an experiment was conducted in which whiteflies were placed on two different types of plants at three different temperatures. The observation of interest was the total number of eggs laid by caged females under one of the six possible treatment combinations. Each treatment combination was run using five cages.

	Plant type					
	Cotton			Cucumber		
Temperature	70°	77°	82°	70°	77°	82°
	37	34	46	50	59	43
	21	54	32	53	53	62
	36	40	41	25	31	71
	43	42	36	37	69	49
	31	16	38	48	51	59

a. Do the data provide sufficient evidence to indicate that the mean number of eggs laid varies with differences in plant type and temperature?

b. Plot the treatment means for cotton as a function of temperature. Plot the treatment means for cucumber as a function of temperature. Comment on the similarity or disparity of these two plots.

c. Find the mean number of eggs laid on cotton and cucumber based on 15 observations each. Calculate a 95% confidence interval for the difference in the underlying population means. (*Hint:* The standard error is calculated as usual, but with $n_1 = 15$ and $n_2 = 15$.)

 13.58 Four chemical plants, producing the same product and owned by the same company, discharge effluents into streams in the vicinity of their locations. To check on the extent of the pollution created by the effluents and to determine if this varies from plant to plant, the company collected random samples of liquid waste, five specimens for each of the four plants. The data are shown in the accompanying table. An SAS computer printout of the analysis of variance for the data is also shown. Use the information in the printout to answer questions (a) through (c).

Plant	Polluting effluents (lb/gal of waste)				
A	1.65	1.72	1.50	1.37	1.60
B	1.70	1.85	1.46	2.05	1.80
C	1.40	1.75	1.38	1.65	1.55
D	2.10	1.95	1.65	1.88	2.00

STATISTICAL ANALYSIS SYSTEM
ANALYSIS OF VARIANCE PROCEDURE

DEPENDENT VARIABLE: X

SOURCE	DF	SUM OF SQUARES	MEAN SQUARE	F VALUE	PR > F	R-SQUARE	C.V.
MODEL	3	0.46489500	0.15496500	5.20	0.0107	0.493679	10.1515
ERROR	16	0.47680000	0.02980000		ROOT MSE		Y MEAN
CORRECTED TOTAL	19	0.94169500			0.17262677		1.70050000

SOURCE	DF	ANOVA SS	F VALUE	PR > F
TRTMENTS	3	0.46489500	5.20	0.0107

a. Do the data provide sufficient evidence to indicate a difference in the mean amount of effluents discharged by the four plants?

b. If the maximum mean discharge of effluents is 1.5 lb/gal, do the data provide sufficient evidence to indicate that the limit is exceeded at plant A?

c. Estimate the difference in the mean discharge of effluents between plants A and D, using a 95% confidence interval.

13.59 The yields of wheat in bushels per acre were compared for five different varieties A, B, C, D, and E at six different locations. Each variety was randomly assigned to a plot at each location. The results of the experiment are shown in the accompanying table. An SAS computer printout for the analysis of variance for the data is also shown below. Use the information in the printout to answer questions (a) through (c).

Varieties	Location					
	1	2	3	4	5	6
A	35.3	31.0	32.7	36.8	37.2	33.1
B	30.7	32.2	31.4	31.7	35.0	32.7
C	38.2	33.4	33.6	37.1	37.3	38.2
D	34.9	36.1	35.2	38.3	40.2	36.0
E	32.4	28.9	29.2	30.7	33.9	32.1

STATISTICAL ANALYSIS SYSTEM
ANALYSIS OF VARIANCE PROCEDURE

DEPENDENT VARIABLE: X

SOURCE	DF	SUM OF SQUARES	MEAN SQUARE	F VALUE	PR > F	R-SQUARE	C.V.
MODEL	9	210.81166667	23.42351852	12.22	0.0001	0.846152	4.0499
ERROR	20	38.33000000	1.91650000		ROOT MSE		Y MEAN
CORRECTED TOTAL	29	249.14166667			1.38437712		34.18333333

SOURCE	DF	ANOVA SS	F VALUE	PR > F
BLOCKS	5	68.14166667	7.11	0.0006
TRTMENTS	4	142.67000000	18.61	0.0001

a. Do the data provide sufficient evidence to indicate a difference in the mean yield for the five varieties of wheat?

b. Do the data provide sufficient evidence to indicate that blocking on "location" increased the amount of information in the experiment?

c. Find a 95% confidence interval for the difference in mean yields for varieties C and E.

13.60 Reference levels of retinol, β-carotene, and α-tocopherol (fat-soluble vitamins) serum levels are necessary to assess borderline vitamin deficiency in children. J. M. D. Malvy, et al.[†] undertook a pilot study to determine the blood fat-soluble levels in a population of children 1–16 years of age living in the countryside of Tours, France. The average vitamin E serum levels (mg/l) for six age groups of children follow.

Age	Sample size*	Mean	Standard deviation
1–3	10	4.45	3.2
3–6	13	8.0	2.5
6–9	11	9.7	2.7
9–12	10	10.0	2.0
12–14	8	9.7	2.5
14–16	9	10.0	2.1

* The sample sizes are not those given in Malvy et al.

a. Do these data provide sufficient evidence to conclude that the mean vitamin E serum level varies with age?

b. Plot the mean vitamin E level against the midpoint of the age interval for each group. Describe any relationship that is apparent from this plot of means.

c. The point estimate of the difference in mean vitamin E levels for the 6–9 and the 12–14 age groups is 0. Use a 95% confidence interval to provide a range of possible values for the true mean difference in vitamin E levels for these two groups.

[†] Malvy, J. M. D., M. S. Mourey, C. Carlier, P. Caces, L. Dostolova, B. Montagnon, and O. Amedee-Manesme. "Retinol, β-Carotene and α-Tocopherol Status in a French Population of Healthy Children." *International Journal of Vitamin Nutrition Research*, 59 (1989), 29–34. Hans Huber Publishers, Toronto, Bern.

CORONARY HEART DISEASE AND COLLEGE FOOTBALL

In an article entitled, "Coronary Heart Disease: Risk Profiles of College Football Players," Mindy Millard-Stafford* and associates found, as expected, that linemen were somewhat taller and heavier than players of other positions. However, they also found that offensive linemen had the highest mean value for total cholesterol (TC) and the lowest mean value for high-density lipoprotein cholesterol (HDL-C) of all positions, independent of a player's race. More importantly, they indicated that although *average* blood pressures and blood lipids appear normal, individual football players—especially offensive linemen—may have an increased risk profile for coronary heart disease. A summary of their data concerning total cholesterol (TC) is given in the following table.

Position	Sample size	Mean	Standard deviation
Offensive linemen	17	206.1	34.2
Defensive linemen	15	181.2	42.2
Offensive backs	18	167.8	37.7
Defensive backs	19	196.0	36.2
Receivers	13	185.8	14.5
Linebackers	13	193.7	62.0

1. What type of experimental design has been used in this experiment?

2. Analyze these data using the appropriate analysis-of-variance technique. What conclusions can you draw regarding the differences in mean total cholesterol levels among the various positions?

3. Construct a confidence interval estimate for the difference in mean total cholesterol between offensive and defensive linemen. Interpret your results.

4. Construct a confidence interval estimate for the difference in mean total cholesterol between offensive backs and receivers. Are your results surprising? Why or why not?

* Millard-Stafford, Mindy, Linda B. Rosskopf, and Phillip B. Sparling, "Coronary Heart Disease: Risk Profiles of College Football Players." Reprinted by permission of *The Physician and Sportsmedicine*, 17(9), September 1989. Copyright McGraw-Hill, Inc.

CHAPTER 14

▷ ▶ NONPARAMETRIC STATISTICS

Case Study

Does the national rate of savings tend to decrease as the investment-income tax increases? The case study of Chapter 10 is reexamined in light of the fact that the assumptions required for a linear regression analysis may not have been satisfied. A nonparametric analog of the simple correlation coefficient is used in the case study at the end of this chapter.

General Objective

In Chapters 7, 8, and 9 we presented statistical techniques for comparing two populations by comparing their respective population parameters (usually their respective population means). The techniques in Chapters 7, 8, and 9 are applicable to data measured on a continuum, and the techniques in Chapter 9 are applicable to data having normal distributions. The purpose of this chapter is to present several statistical tests for comparing populations for the many types of data that do not satisfy the assumptions specified in Chapters 7, 8, and 9.

Specific Topics

1   Parametric versus nonparametric tests (14.1)
2   The Sign Test (14.2)
3   The Mann-Whitney U Test (14.4)
4   The Wilcoxon Signed-Rank Test (14.5)
5   The Kruskal-Wallis H Test (14.6)
6   The Friedman F_r Test (14.7)
7   The rank correlation coefficient (14.8)

▷ 14.1 INTRODUCTION

Some experiments yield response measurements that defy quantification. That is, they generate response measurements that can be ordered (ranked), but the location of the response on a scale of measurement is arbitrary. For example, the data may only admit pairwise directional comparisons, that is, whether one observation is larger than another or vice versa. Or, we may be able to rank all observations in a data set but may not know the exact values of the measurements. To illustrate, suppose a judge is employed to evaluate and rank the sales abilities of four salespeople, the edibility and taste characteristics of five brands of cornflakes, or the relative appeal of five new automobile designs. Clearly it is impossible to give an exact measure of sales competence, palatability of food, or design appeal, but it is usually possible to rank the salespeople, brands, and so on according to which one we think is best, second best, etc. Thus, the response measurements here differ markedly from those presented in preceding chapters. Although experiments that produce this type of data occur in almost all fields of study, they are particularly evident in social science research and in studies of consumer preference. The data that they produce can be analyzed using **nonparametric statistical methods**.

Nonparametric statistical procedures are also useful in making inferences in situations where serious doubt exists about the assumptions that underlie standard methodology. For example, the t test for comparing a pair of means, Section 9.4, is based on the assumption that both populations are normally distributed with equal variances. The experimenter will never know whether these assumptions hold in a practical situation, but will often be reasonably certain that departures from the assumptions will be small enough so that the properties of the statistical procedure will be undisturbed. That is, α and β will be approximately what the experimenter thinks they are. On the other hand, it is not uncommon for the experimenter to seriously question the assumptions and wonder whether he or she is using a valid statistical procedure. This difficulty may be circumvented by using a nonparametric statistical test and thereby avoiding reliance on a very uncertain set of assumptions.

Research has shown that nonparametric statistical tests are almost as capable of detecting differences among populations as the parametric methods of preceding chapters when normality and other assumptions are satisfied. They may be, and often are, more powerful in detecting population differences when the assumptions are not satisfied. For this reason, many statisticians advocate the use of nonparametric statistical procedures in preference to their parametric counterparts.

In this chapter we will discuss only the most common situations in which nonparametric methods are used. For a more complete treatment of the subject, see the texts listed in the references.

▷ **14.2** THE SIGN TEST FOR COMPARING TWO POPULATIONS

Without emphasizing the point, we employed a nonparametric statistical test as an alternative procedure for determining whether evidence existed to indicate a difference in the mean wear for the two types of tires in the paired-difference experiment in Section 9.5. Each pair of responses was compared and X (the number of times A exceeded B) was used as the test statistic. This nonparametric test is known as the **sign test** because x is the number of positive (or negative) signs associated with the differences. **The implied null hypothesis is that the two population distributions are identical**, and the resulting technique is completely independent of the form of the distribution of differences. Thus, regardless of the distribution of differences, **the probability that A exceeds B for a given pair will be $p = .5$ when the null hypothesis is true** (that is, when the distributions for A and B are identical). Then x will have a binomial probability distribution and a rejection region for x can be obtained using the binomial probability distribution of Chapter 4.

THE SIGN TEST

Null hypothesis H_0: The two population distributions are identical and $P(A$ exceeds B for a given pair$) = p = .5$.

Alternative hypothesis H_a: The two population distributions are not identical and $p \neq .5$. Or H_a: The population of A measurements is shifted to the right of the population of B measurements and $p > .5$.*

Test statistic: For n, the number of pairs in which no ties were reported, use x, the number of times that $(A - B)$ was positive, or equivalently, the number of times A exceeded (or was preferred to) B.

Rejection region: For the two-tailed test $H_a : p \neq .5$ and a given value of α, reject H_0 if $x \leq x_L$ or $x \geq x_U$, where x_L and x_U are the lower- and upper-tailed values of a binomial distribution, with $p = .5$ satisfying $P(x \leq x_L) \leq \alpha/2$ and $P(x \geq x_U) \leq \alpha/2$. For the one-tailed test of $H_a : p > .5$ and a given value of α, reject H_0 if $x \geq x_U$, where x_U is the upper-tailed value of a binomial distribution, with $p = .5$ satisfying $P(x \geq x_U) \leq \alpha$.*

The following example will help you understand how the sign test is constructed and how it is used in a practical situation.

EXAMPLE 14.1   The number of defective electrical fuses proceeding from each of two production lines A and B was recorded daily for a period of ten days, with the following results:

---

* For the one-tailed test when the alternative hypothesis is $H_a : p < .5$, simply interchange the letters A and B.

| | Production lines | |
Day	A	B
1	170	201
2	164	179
3	140	159
4	184	195
5	174	177
6	142	170
7	191	183
8	169	179
9	161	170
10	200	212

Assume that both production lines produced the same daily output. Compare the number x_A of defectives produced each day by production line A with the number x_B produced by production line B and let x equal the number of days when x_A exceeded x_B. Do the data present sufficient evidence to indicate that one production line tends to produce more defectives than the other or, equivalently, that $P(x_A > x_B) \neq 1/2$? State the null hypothesis to be tested and use x as a test statistic.

Solution   Note that this is a paired-difference experiment. Let x be the number of times that the observation for production line A exceeds that for B in a given day. Under the null hypothesis that the two distributions of defectives are identical, the probability p that A exceeds B for a given pair is $p = .5$. Or, equivalently, we may wish to test the null hypothesis that the binomial parameter p equals .5 against the alternative hypothesis $p \neq .5$.

Very large or very small values of x are most contradictory to the null hypothesis. Therefore, the rejection region for this two-tailed test will be located by including the most extreme values of x that at the same time provide an α that is feasible for the test.

Suppose we want α to be about .05 or .10. We begin the selection of the rejection region by including $x = 0$ and $x = 10$ and calculating the α associated with this region using $p(x)$ (the probability distribution for the binomial random variable, Chapter 4). Thus, with $n = 10$, $p = .5$,

$$\alpha = p(0) + p(10) = C_0^{10}(.5)^{10} + C_{10}^{10}(.5)^{10} = .002$$

Since this value of α is too small, the rejection region is expanded by including the next pair of x values that are most contradictory to the null hypothesis, namely, $x = 1$ and $x = 9$. The value of α for this rejection region ($x = 0, 1, 9, 10$) is obtained from Table 1 in Appendix III.

$$\alpha = p(0) + p(1) + p(9) + p(10) = .022$$

This α is also small, so we again expand the region. This time we include $x = 0, 1, 2, 8, 9, 10$. You can verify that the corresponding value of α is .11. At this point,

Figure 14.1
Rejection region for
Example 14.1

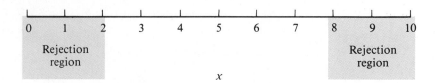

we have a choice. We can either choose α equal to .022 or .11, depending on the risk we are willing to take of making a type I error. For this example, we choose $\alpha = .11$ and, consequently, employ $x = 0, 1, 2, 8, 9, 10$ as the rejection region for the test. (See Figure 14.1.)

From the data, we observe that $x = 1$ and therefore we reject the null hypothesis. Thus, we conclude that sufficient evidence exists to indicate that the population distributions of number of defects are not identical. A quick glance at the data suggests that the mean number of defects for production line B tends to exceed that for A. The probability of rejecting the null hypothesis when true is only $\alpha = .11$, and therefore we are reasonably confident of our conclusion.

The experimenter in this example is using the test procedure as a rough tool for detecting faulty production lines. The rather large value of α is not likely to be disturbing because additional data can easily be collected if the experimenter is concerned about making a type I error in reaching a conclusion.   ◁

One problem that may occur when you are conducting a sign test is that the observations associated with one or more pairs may be equal and therefore result in ties. When this situation occurs, delete the tied pairs and reduce n, the total number of pairs.

You will also encounter situations where n, the number of pairs, is large. Then the values of α associated with the sign test can be obtained by using the normal approximation to the binomial probability distribution discussed in Section 5.3. You can verify (by comparison of exact probabilities with their approximations) that these approximations will be quite adequate for n as small as 10. This is due to the symmetry of the binomial probability distribution for $p = .5$.

For $n \geq 25$, you can conduct the sign test by using the z statistic,

$$z = \frac{x - np}{\sqrt{npq}} = \frac{x - .5n}{.5\sqrt{n}}$$

as the test statistic. In using z, you are testing the null hypothesis $p = .5$ against the alternative $p \neq .5$ for a two-tailed test or against the alternative $p > .5$ (or $p < .5$) for a one-tailed test. The tests would utilize the familiar rejection regions of Chapter 8.

SIGN TEST FOR LARGE SAMPLES: $n \geq 25$

1. Null Hypothesis: $H_0: p = .5$ (one treatment is not preferred to a second treatment, and vice versa).

2. Alternative Hypothesis: $H_a: p \neq .5$, for a two-tailed test. (*Note:* We use the two-tailed test as an example. Many analyses might require a one-tailed test.)

3. Test Statistic: $z = \dfrac{x - .5n}{.5\sqrt{n}}$.

4. Rejection Region: Reject H_0 if $z \geq z_{\alpha/2}$ or if $z \leq -z_{\alpha/2}$, where $z_{\alpha/2}$ is the z value from Table 3, Appendix III, corresponding to an area of $\alpha/2$ in the upper tail of the normal distribution.

EXAMPLE 14.2 A production superintendent claims that there is no difference between the employee accident rates for the day versus the evening shifts in a large manufacturing plant. The number of accidents per day is recorded for both the day and evening shifts for $n = 100$ days. It is found that the number of accidents per day for the evening shift x_E exceeded the corresponding number of accidents on the day shift x_D on 63 of the 100 days. Do these results provide sufficient evidence to indicate that more accidents tend to occur on one shift than on the other, or equivalently, that $P(x_E > x_D) \neq 1/2$?

Solution This study is a paired-difference experiment, with $n = 100$ pairs of observations corresponding to the 100 days. To test the null hypothesis that the two distributions of accidents are identical, we use the test statistic

$$z = \frac{x - .5n}{.5\sqrt{n}}$$

where x represents the number of days in which the number of accidents on the evening shift exceeded the number of accidents on the day shift. Then for $\alpha = .05$ we reject the null hypothesis if $z \geq 1.96$ or $z \leq -1.96$. Substituting into the formula for z, we have

$$z = \frac{x - .5n}{.5\sqrt{n}} = \frac{63 - (.5)(100)}{.5\sqrt{100}} = \frac{13}{5} = 2.60$$

Since the calculated value of z exceeds $z_{\alpha/2} = 1.96$, we reject the null hypothesis. The data provide sufficient evidence to indicate a difference between the accident rate distributions for the day versus evening shifts.

Note that the data in Example 14.1 are the result of a paired-difference experiment. Suppose that the paired differences were normally distributed with variance σ_d^2. Would the sign test detect a shift in location of the two populations as effectively as the Student's t test? Intuitively, we would suspect the answer to be no, and this is correct, because the Student's t test utilizes comparatively more information. In addition to giving the sign of the difference, the t test uses the magnitudes of the observations to obtain a more accurate value of the sample means and variances. Thus, one might say that the sign test is not as "efficient" as the Student's t test, but this statement is meaningful only if the populations conform to the assumption stated above; that is, the differences in paired observations are normally distributed with variance σ_d^2. The sign test might be more efficient when these assumptions are not satisfied.

EXERCISES   Basic Techniques

14.1 Suppose that you wish to use the sign test to test $H_0:p = .5$ against $H_a:p > .5$ for a paired-difference experiment with $n = 25$ pairs.
   a. State the practical situation that would dictate the alternative hypothesis given above.
   b. Use Table 1 in Appendix III to find values of α ($\alpha < .15$) available for the test.

14.2 Repeat the instructions of Exercise 14.1 for $H_a:p \neq .5$.

14.3 Repeat the instructions of Exercises 14.1 and 14.2 for $n = 10$, 15, and 20.

14.4 A paired-difference experiment was conducted to compare two populations. The data are shown below. Use a sign test to determine whether the population distributions are different.

				Pairs			
Population	1	2	3	4	5	6	7
1	8.9	8.1	9.3	7.7	10.4	8.3	7.4
2	8.8	7.4	9.0	7.8	9.9	8.1	6.9

   a. State the null and alternative hypotheses for the test.
   b. Determine an appropriate rejection region with $\alpha \approx .01$.
   c. Calculate the observed value of the test statistic.
   d. Do the data present sufficient evidence to indicate that populations 1 and 2 are different?

Applications

14.5 In Exercise 9.43 we compared property evaluations of two tax assessors A and B. Their assessments for eight properties are shown in the table.

	Assessor	
Property	A	B
1	76.3	75.1
2	88.4	86.8
3	80.2	77.3
4	94.7	90.6
5	68.7	69.1
6	82.8	81.0
7	76.1	75.3
8	79.0	79.1

   a. Use the sign test to determine whether the data present sufficient evidence to indicate that one of the assessors tends to be consistently more conservative than the other; that is, $P(x_A > x_B) \neq 1/2$. Test by using a value of α near .05. Find the p-value for the test and interpret its value.

b. Exercise 9.43 uses the t statistic to test the null hypothesis that there is no difference in the mean level of property assessments between assessors A and B. Check the answer (in the answer section) for Exercise 9.43 and compare it with your answer to part (a). Do the test results agree? Explain why the answers are (or are not) consistent.

14.6 Two gourmets rated 22 meals on a scale of 1 to 10. The data are shown in the table. Do the data provide sufficient evidence to indicate that one of the gourmets tends to give higher ratings than the other? Test by using the sign test with a value of α near .05.

Meal	A	B	Meal	A	B
1	6	8	12	8	5
2	4	5	13	4	2
3	7	4	14	3	3
4	8	7	15	6	8
5	2	3	16	9	10
6	7	4	17	9	8
7	9	9	18	4	6
8	7	8	19	4	3
9	2	5	20	5	4
10	4	3	21	3	2
11	6	9	22	5	3

a. Use the binomial tables in Appendix III to find the exact rejection region for the test.

b. Use the large-sample z statistic. (*Note:* Although the large-sample approximation is suggested for $n \geq 25$, it will work fairly well for values of n as small as 15.)

c. Compare the results of parts (a) and (b).

14.7 A report in the August *American Journal of Public Health**—the first to follow blood lead levels in law-abiding handgun hobbyists using indoor firing ranges—documents a significant risk of lead poisoning. Lead exposure measurements were made on 17 members of a law enforcement trainee class before, during, and after a three-month period of firearm instruction at a state-owned indoor firing range. No trainee had elevated blood lead levels before the training, but 15 of the 17 ended their training with blood lead levels deemed "elevated" by the Occupational Safety and Health Administration (OSHA). If the use of an indoor firing range has no positive effect on blood lead levels, then p, the probability that a person's blood lead level increases, is less than or equal to .5. If, however, use of the indoor firing range causes an increase in a person's blood lead levels, then $p > .5$. Use the sign test to determine whether using an indoor firing range has the effect of increasing a person's blood lead level with $\alpha = .05$. (*Hint:* The normal approximation to binomial probabilities will be fairly accurate for $n = 17$.)

14.8 Clinical data concerning the effectiveness of two drugs in treating a particular disease were collected from ten hospitals. The numbers of patients treated with the drugs varied from one

* Reported in *Science News*, August 19, 1989, Vol. 136, p. 126.

hospital to another. The data, in percent recovery, are shown below:

	Drug A			Drug B		
Hospital	Number in group	Number recovered	Percent recovered	Number in group	Number recovered	Percent recovered
1	84	63	75.0	96	82	85.4
2	63	44	69.8	83	69	83.1
3	56	48	85.7	91	73	80.2
4	77	57	74.0	47	35	74.5
5	29	20	69.0	60	42	70.0
6	48	40	83.3	27	22	81.5
7	61	42	68.9	69	52	75.4
8	45	35	77.8	72	57	79.2
9	79	57	72.2	89	76	85.4
10	62	48	77.4	46	37	80.4

Do the data present sufficient evidence to indicate a higher recovery rate for one of the two drugs?

a. Test using the sign test. Choose your rejection region so that α is near .10.

b. Why might it be inappropriate to use a Student's t test in analyzing the data?

14.3   A COMPARISON OF STATISTICAL TESTS

We have introduced two different statistical tests to test an hypothesis based on the same set of data. Which test, if either, is better? One way to answer this question would be to hold the sample size and α constant for both procedures and compare β, the probability of a type II error. Actually, statisticians prefer a comparison of the **power of a test** where

$$\text{power} = 1 - \beta$$

Since β is the probability of failing to reject the null hypothesis when it is false, the power of the test is the probability of rejecting the null hypothesis when it is false and some specified alternative is true. It is the probability that the test will do what it was designed to do, that is, detect a departure from the null hypothesis when a departure exists.

Probably the most common method of comparing two test procedures is in terms of the relative efficiency of a pair of tests. **Relative efficiency** is the ratio of the sample sizes for the two test procedures required to achieve the same α and β for a given alternative to the null hypothesis.

In some situations, you may not be too concerned whether you are using the most powerful test. For example, you might choose to use the sign test over a more powerful competitor because of its ease of application. Thus, you might view tests as microscopes that are utilized to detect departures from an hypothesized theory. One need not know the exact power of a microscope to use it in a biological investigation, and the same applies to statistical tests. If the test procedure detects a departure from the null hypothesis, we are delighted. If not, we can reanalyze the data by using a more powerful test, or we can increase the power of the microscope (test) by increasing the sample size.

14.4 THE MANN-WHITNEY *U* TEST: INDEPENDENT RANDOM SAMPLES

The *t* test for comparing two population means, Section 9.4, is a test to detect differences in the location of two normal population frequency distributions. The Mann-Whitney *U* test is a nonparametric alternative to this test. It is used when we have doubts that the assumptions of normality and/or equal variances required for the Student's *t* test are satisfied. The Mann-Whitney *U* test (and all of the tests that follow in this chapter) is based on an analysis of the ranks of the sample observations.

Assume that you have independent random samples of sizes n_1 and n_2 from two populations—say 1 and 2. The first step in finding the Mann-Whitney *U* statistic is to rank all $(n_1 + n_2)$ observations in order of magnitude, assigning a 1 to the smallest observation, a 2 to the second smallest, and so on. **Ties in the observations can be handled by averaging the ranks that would have been assigned to the tied observations and assigning this average to each.** Then calculate the sums of the ranks, T_1 and T_2, for the two samples.

For example, two samples of four observations each ($n_1 = n_2 = 4$) are shown in Table 14.1(a). The ranks of the $n_1 + n_2 = 8$ observations are shown in parentheses. Notice that the similar samples each contain some of the larger and some of the smaller observations. There is little evidence to indicate a difference in the two population distributions, and the rank sums, $T_1 = T_2 = 18$, are equal. In contrast, Table 14.1(b) shows different samples, in which all of the small observations are in one sample and all of the large observations are in the other. The rank sums $T_1 = 10$ and $T_2 = 26$ differ by the maximum amount and provide the greatest amount of evidence to indicate a difference in location for the two population distributions.

The formula for the Mann-Whitney *U* statistic can be given in terms of T_1 or T_2; one value of *U* will be larger than the other, but the sum of the two *U* values will

Table 14.1
Hypothetical rankings of eight observations

	(a) Similar samples		(b) Different samples	
	1	2	1	2
	28 (6)	25 (4)	18 (2)	27 (5)
	16 (1)	18 (2)	21 (3)	35 (8)
	35 (8)	27 (5)	16 (1)	28 (6)
	21 (3)	30 (7)	25 (4)	30 (7)
Rank Sum	$T_1 = 18$	$T_2 = 18$	$T_1 = 10$	$T_2 = 26$

always equal $n_1 n_2$. Since it is easier to construct a table of probabilities of U for only one tail of the U distribution, we will agree always to use the smaller value of U as a test statistic. The formulas for the two values of U, which we will denote as U_1 and U_2, are given in the display.

FORMULAS FOR THE MANN-WHITNEY U STATISTIC

$$U_1 = n_1 n_2 + \frac{n_1(n_1 + 1)}{2} - T_1$$

$$U_2 = n_1 n_2 + \frac{n_2(n_2 + 1)}{2} - T_2$$

where

n_1 = number of observations in sample 1

n_2 = number of observations in sample 2

T_1 and T_2 are the rank sums of samples 1 and 2, respectively. (*Note:* It can be shown that $U_1 + U_2 = n_1 n_2$ and $T_1 + T_2 = \dfrac{n(n + 1)}{2}$, where $n = n_1 + n_2$.)

As you can see from the formulas for U_1 and U_2, U_1 will be small when T_1 is large, a situation that likely will occur when the population 1 distribution of measurements is shifted to the right of the population 2 distribution of measurements. Consequently, to conduct a one-tailed test to detect a shift in distribution 1 to the right of distribution 2, you will reject the null hypothesis of "no difference in the population distributions" if U_1 less than some specified value U_0; that is, you will reject H_0 for small values of U_1. Similarly, to conduct a one-tailed test to detect a shift of distribution 2 to the right of distribution 1, you would reject H_0 if U_2 is less than some specified value, say U_0. Consequently, the rejection region for the Mann-Whitney U test would appear as shown in Figure 14.2.

Figure 14.2
Rejection region for a
Mann-Whitney U test

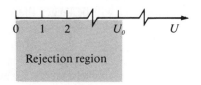

Table 11 in Appendix III gives the probability that an observed value of U will be less than some specified value, say U_0. This is the value of α for a one-tailed test. To conduct a two-tailed test—that is, to detect a shift in the population distributions for the measurements in either direction—we use U, the smaller of U_1 or U_2, as the test statistic and reject H_0 for $U < U_0$ (see Figure 14.2). The value of α for the two-tailed test will be double the tabulated value given in Table 11 in Appendix III.

Table 14.2

An abbreviated version of Table 11 in Appendix III; $P(U \leqslant U_0)$ for $n_2 = 5$

U_0	n_1				
	1	2	3	4	5
0	.1667	.0476	.0179	.0079	.0040
1	.3333	.0952	.0357	.0159	.0079
2	.5000	.1905	.0714	.0317	.0159
3		.2857	.1250	.0556	.0278
4		.4286	.1964	.0952	.0476
5		.5714	.2857	.1429	.0754
⋮			⋮	⋮	⋮

To see how to locate the rejection region for the Mann-Whitney U test, suppose that $n_1 = 4$ and $n_2 = 5$. Then you would consult the table corresponding to $n_2 = 5$ in Table 11 of Appendix III. The first few lines of Table 11 for $n_2 = 5$ are shown in Table 14.2. Note that the table is constructed assuming that $n_1 \leq n_2$. Therefore, we identify the population with the smaller sample size as population 1.

Across the top of Table 14.2 you see values of n_1. Values of U_0 are shown down the left side of the table. The entries give the probability that U will assume a small value, namely, the probability that $U \leq U_0$. Since, for this example, $n_1 = 4$, we move across the top of the table to $n_1 = 4$. Move to the second row of the table corresponding to $U_0 = 2$ for $n_1 = 4$. The probability that U will be less than or equal to 2 is found as .0317. Similarly, moving across the row for $U_0 = 3$, you see that the probability that U is less than or equal to 3 is .0556. (This value is shaded in Table 14.2.) So, if you want to conduct a one-tailed Mann-Whitney U test with $n_1 = 4$ and $n_2 = 5$ and would like α to be near .05, you would reject the null hypothesis of equality of population relative frequency distributions when $U \leq 3$. The probability of a type I error for the test would be $\alpha = .0556$. If you use this same rejection region for a two-tailed test, that is, $U \leq 3$, α would be double the tabulated value, or $\alpha = 2(.0556) = .1112$.

Table 11 in Appendix III can also be used to find the observed significance level for a test. For example, if $n_1 = 5$, $n_2 = 5$, and $U = 4$, then from Table 14.2, the p-value for a one-tailed test would be

$$P\{U \leq 4\} = .0476$$

If the test were two-tailed, the p-value would be

$$2(.0476) \qquad \text{or} \qquad .0952$$

When applying the test to a set of data, you may find that some of the observations are of equal value. **Recall that ties in the observations can be handled by averaging the ranks that would have been assigned to the tied observations and assigning this average to each.** Thus, if three observations were tied and were due to receive ranks 3, 4, and 5, we would assign the rank of 4 to all three. The next observation in the sequence would receive the rank of 6, and ranks 3 and 5 would not appear. Similarly, if two observations were tied for ranks 3 and 4, each would receive a rank of 3.5, and ranks 3 and 4 would not appear.

THE MANN-WHITNEY U TEST

1. Null Hypothesis: H_0: The population relative frequency distributions 1 and 2 are identical.

2. Alternative Hypothesis: H_a: The two population relative frequency distributions are shifted with respect to their relative locations (a two-tailed test). Or H_a: The population relative frequency distribution 1 is shifted to the right of the relative frequency distribution for population 2 (a one-tailed test).*

3. Test Statistic: For a two-tailed test, use U, the smaller of

$$U_1 = n_1 n_2 + \frac{n_1(n_1 + 1)}{2} - T_1$$

and

$$U_2 = n_1 n_2 + \frac{n_2(n_2 + 1)}{2} - T_2$$

where T_1 and T_2 are the rank sums for samples 1 and 2, respectively. For a one-tailed test, use U_1.

4. Rejection Region:
   a. For the two-tailed test and a given value of α, reject H_0 if $U \leq U_0$, where $P(U \leq U_0) = \alpha/2$. [*Note:* U_0 is the value such that $P(U \leq U_0)$ is equal to half of α.]
   b. For a one-tailed test and a given value of α, reject H_0 if $U_1 \leq U_0$, where $P(U_1 \leq U_0) = \alpha$.

Ties are assigned a rank equal to the average of the ranks that the observations would have received if they had been slightly different in value.

EXAMPLE 14.3   The bacteria counts per unit volume are shown below for two types of cultures A and B. Four observations were made for each culture.

Culture A	Culture B
27 (3)	32 (7)
31 (6)	29 (5)
26 (2)	35 (8)
25 (1)	28 (4)

Do the data present sufficient evidence to indicate a difference in the population distributions of bacteria counts? Test using a value of α near .05.

---

* For the sake of convenience, we will describe the one-tailed test as one designed to detect a shift in distribution 1 measurements to the right of distribution 2 measurements. To detect a shift in distribution 2 to the right of distribution 1, just interchange the numbers 1 and 2 in the discussion.

Solution   We wish to test the null hypothesis

H_0: The population distributions of bacteria counts are identical

against the alternative hypothesis

H_a: The population distributions differ in location

This alternative hypothesis implies a two-tailed test. Since Table 11, Appendix III, gives values of $P(U \leq U_0)$ for specified sample sizes and values of U_0, we must double the tabulated value to find α. Suppose that we desire a value of α near .05. Checking Table 11, Appendix III, for $n_1 = n_2 = 4$, we find $P(U \leq 1) = .0286$. Using $U \leq 1$ as the rejection region, α will equal $2(.0286) = .0572$ or, rounding to three decimal places, $\alpha = .057$. The ranks for the $n_1 + n_2 = 8$ bacteria counts are shown in parentheses next to each data value, and the rank sums are $T_1 = 12$ and $T_2 = 24$. Then

$$U_1 = n_1 n_2 + \frac{n_1(n_1 + 1)}{2} - T_1$$

$$= (4)(4) + \frac{4(4 + 1)}{2} - 12$$

$$= 14$$

and

$$U_2 = n_1 n_2 + \frac{n_2(n_2 + 1)}{2} - T_2$$

$$= (4)(4) + \frac{4(4 + 1)}{2} - 24$$

$$= 2$$

The smaller of U_1 and U_2 is the test statistic, $U = 2$, which does not fall in the rejection region. Hence, there is not sufficient evidence to show a shift in locations of the population distributions of bacteria counts for cultures 1 and 2. ◁

The Mann-Whitney test can be implemented by using the MANN-WHITNEY command in the MINITAB package. As input, the program uses the values from the first and second samples, which are stored in two separate columns. The MINITAB output for the Mann-Whitney test using the data in Example 14.3 is given in Table 14.3. Data summaries appear in lines 1 and 2; point and interval estimates for the difference in population medians are given in lines 3 and 4. The value $W = 12.0$ in line 5 is the sum of the ranks for the values stored in column 1—and in general, for the column number given first in the program command. The value of W can be used to calculate U_A and U_B if desired. The significance level (.1124) of the test is given in line 6. Since the significance level of .1124 is greater than $\alpha = .05$, consistent with our earlier results, we cannot conclude that the under-lying population distributions are different.

Table 14.3
MINITAB output for the
data of Example 14.3

```
MTB > PRINT C1 C2
  ROW   C1   C2

     1    27   32
     2    31   29
     3    26   35
     4    25   28


MTB > MANNWHITNEY C1 C2

MANN-WHITNEY CONFIDENCE INTERVAL AND TEST

C1            N = 4    MEDIAN =       26.500
C2            N = 4    MEDIAN =       30.500
POINT ESTIMATE FOR ETA1-ETA2 IS    −3.4993
97.0   PCT  C.I.  FOR ETA1−ETA2 IS (   −10.0,    3.0)
W =       12.0
TEST OF ETA1 = ETA2   VS.   ETA1 N.E. ETA2 IS SIGNIFICANT AT 0.1124

CANNOT REJECT AT ALPHA = 0.05
```

EXAMPLE 14.4   An experiment was conducted to compare the strength of two types of kraft papers, one a standard kraft paper of a specified weight and the other the same standard kraft paper treated with a chemical substance. Ten pieces of each type of paper, randomly selected from production, produced the strength measurements shown in the accompanying table. Test the hypothesis of "no difference in the distributions of strengths for the two types of paper" against the alternative hypothesis that the treated paper tends to be of greater strength (i.e., its distribution of strength measurements is shifted to the right of the corresponding distribution for the untreated paper).

Solution   The ranks are shown in parentheses alongside the $n_1 + n_2 = 10 + 10 = 20$ strength measurements, and the rank sums T_1 and T_2 are shown below the columns. Since we wish to detect a shift in the distribution 1 measurements to the right of the distribution 2 measurements, we will conduct a one-tailed test and reject the null hypothesis of "no difference in population strength distributions" when T_2 is excessively large, or equivalently when U_2 is small.

	Standard 1	Treated 2
	1.21 (2)	1.49 (15)
	1.43 (12)	1.37 (7.5)
	1.35 (6)	1.67 (20)
	1.51 (17)	1.50 (16)
	1.39 (9)	1.31 (5)
	1.17 (1)	1.29 (3.5)
	1.48 (14)	1.52 (18)
	1.42 (11)	1.37 (7.5)
	1.29 (3.5)	1.44 (13)
	1.40 (10)	1.53 (19)
Rank sums	$T_1 = 85.8$	$T_2 = 124.5$

Suppose we choose a value of α near .05. Then we can find U_0 by consulting the portion of Table 11, Appendix III, corresponding to $n_2 = 10$. The probability, $P(U \leq U_0)$, nearest .05 is .0526 and corresponds to $U_0 = 28$. Hence, we will reject if $U_2 \leq 28$.

Calculating U_2, we have

$$U_2 = n_1 n_2 + \frac{n_2(n_2 + 1)}{2} - T_2$$

$$= (10)(10) + \frac{(10)(11)}{2} - 124.5$$

$$= 30.5$$

As you can see, U_2 is not less than $U_0 = 28$. Therefore, we cannot reject the null hypothesis. At the $\alpha = .05$ level of significance, there is not sufficient evidence to indicate that the treated kraft paper is stronger than the standard. ◁

A simplified large-sample test ($n_1 \geq 10$ and $n_2 \geq 10$) can be obtained by using the familiar z statistic of Chapter 8. When the population distributions are identical, it can be shown that the U statistic has expected value and variance,

$$E(U) = \frac{n_1 n_2}{2} \quad \text{and} \quad \sigma_U^2 = \frac{n_1 n_2(n_1 + n_2 + 1)}{12}$$

The distribution of

$$z = \frac{U - E(U)}{\sigma_U}$$

tends to normality with mean 0 and variance equal to 1 as n_1 and n_2 become large. **This approximation will be adequate when n_1 and n_2 are both greater than or equal to 10.** Thus, for a two-tailed test with $\alpha = .05$, we would reject the null hypothesis if $|z| \geq 1.96$.

Observe that by using the z statistic, you will reach the same conclusion as when using the exact U test for Example 14.4. Thus,

$$z = \frac{30.5 - [(10)(10)/2]}{\sqrt{[(10)(10)(10 + 10 + 1)]/12}} = \frac{30.5 - 50}{\sqrt{2100/12}} = -1.47$$

For a one-tailed test with $\alpha = .05$ located in the lower tail of the z distribution, we will reject the null hypothesis if $z < -1.645$. Since $z = -1.47$ does not fall in the rejection region, we reach the same conclusion as the exact U test of Example 14.4.

THE MANN-WHITNEY *U* TEST FOR LARGE SAMPLES:
$n_1 \geq 10$ and $n_2 \geq 10$

1. Null Hypothesis: H_0: The population relative frequency distributions are identical.

2. Alternative Hypothesis: H_a: The two population relative frequency distributions differ in location (a two-tailed test). Or H_a: The population 1 relative frequency

distribution is shifted to the right (or left) of the relative frequency distribution for population 2 (a one-tailed test).

3. Test Statistic: $z = \dfrac{U - (n_1 n_2 / 2)}{\sqrt{n_1 n_2 (n_1 + n_2 + 1)/12}}$

    Let $U = U_1$.

4. Rejection Region: Reject H_0 if $z > z_{\alpha/2}$ or $z < -z_{\alpha/2}$ for a two-tailed test. For a one-tailed test, place all of α in one tail of the z distribution. To detect a shift in distribution 1 to the right of distribution 2, let $U = U_1$ and reject H_0 when $z < -z_\alpha$. To detect a shift in the opposite direction, let $U = U_1$ and reject H_0 when $z > z_\alpha$. Tabulated values of z are given in Table 3, Appendix III.

EXERCISES   Basic Techniques

14.9   Suppose you wish to detect a shift in distribution 1 to the right of distribution 2 based on sample sizes $n_1 = 6, n_2 = 8$.
    a.   Should you use U_1 or U_2 for your test statistic?
    b.   Give the rejection region for the test if you wish α to be close to but less than .10.
    c.   Give the value of α for the test.

14.10   Suppose that the alternative hypothesis for Exercise 14.9 is that distribution 1 is shifted either to the left or to the right of distribution 2.
    a.   Should you use U_1 or U_2 for your test statistic?
    b.   Give the rejection region for the test if you wish α to be close to but less than .10.
    c.   Give the value of α for the test.

14.11   Observations from two random and independent samples, drawn from populations 1 and 2, are reproduced below.

Sample 1	Sample 2
1	4
3	7
2	6
3	8
5	6

Use the Mann-Whitney U test to determine whether population 1 is shifted to the left of population 2.
    a.   State the null and alternative hypotheses to be tested.
    b.   Find the rejection region for $\alpha \approx .05$.
    c.   Rank the combined sample from smallest to largest. Calculate T_1, T_2, U_1, and U_2.
    d.   What is the observed value of the test statistic?
    e.   Do the data provide sufficient evidence to indicate that population 1 is shifted to the left of population 2?

14.12   Independent random samples of size $n_1 = 20$ and $n_2 = 25$ are drawn from nonnormal populations A and B. The combined sample is ranked and $T_A = 252$, $T_B = 783$. Use the large-sample approximation to the Mann-Whitney U test to determine whether there is a difference in the two population distributions. Calculate the observed significance level, or p-value for the test.

14.13   Suppose you wish to detect a shift in direction 1 to the right of distribution 2 based on sample sizes $n_1 = 12$ and $n_2 = 14$. If $T_1 = 193$, what do you conclude? Use $\alpha = .05$.

Applications

14.14   In Exercise 9.86 we presented data on the wing stroke frequencies of two species of Euglossine bees (T. M. Casey, M. L. May, and K. R. Morgan, "Flight Energetics of Euglossine Bees in Relation to Morphology and Wing Stroke Frequency," *Journal of Experimental Biology* 116 [1985]). The wing frequencies for a sample of $n_1 = 4$ *Euglossa mandibularis Friese* and $n_2 = 6$ *Euglossa imperialis Cockerell* are shown in the table.

Wing stroke frequencies

Euglossa mandibularis Friese	Euglossa imperialis Cockerell
235	180
225	169
190	180
188	185
	178
	182

Do the data present sufficient evidence to indicate that the distributions of wing stroke frequencies differ in location for the two species?
a.   Test using the Mann-Whitney U test with α less than but close to .10.
b.   Find the approximate p-value for the test.

14.15   The observations below represent the dissolved oxygen content in water. The higher the dissolved oxygen, the greater the ability of a river, lake, or stream to support aquatic life. In this experiment, a pollution-control inspector suspected that a river community was releasing amounts of semitreated sewage into a river. To check this theory, five randomly selected specimens of river water were selected at a location above the town, and another five below. The dissolved oxygen readings, in parts per million, are as follows:

Above town	4.8	5.2	5.0	4.9	5.1
Below town	5.0	4.7	4.9	4.8	4.9

a.   Use a one-tailed Mann-Whitney U test with $\alpha = .05$ to answer the question.
b.   Use a Student's t test (with $\alpha = .05$) to analyze the data. Compare the conclusions reached in parts (a) and (b).

14.16   In an investigation of visual scanning behavior of deaf children, measurements of eye-movement rate were taken on nine deaf and nine hearing children. From the data given, does it appear that the distributions of eye-movement rates for deaf children (A) and hearing children (B) differ?

	Deaf children (A)	Hearing children (B)
	(15) 2.75	.89 (1)
	(11) 2.14	1.43 (7)
	(18) 3.23	1.06 (4)
	(10) 2.07	1.01 (3)
	(14) 2.49	.94 (2)
	(12) 2.18	1.79 (8)
	(17) 3.16	1.12 (5.5)
	(16) 2.93	2.01 (9)
	(13) 2.20	1.12 (5.5)
Rank sum	126	45

14.17　The following table depicts the life in months of service before failure of the color television picture tube for eight television sets manufactured by firm A and ten sets manufactured by firm B.

Firm	Life of picture tube (months)									
A	32	25	40	31	35	29	37	39		
B	41	39	36	47	45	34	48	44	43	33

Use the Mann-Whitney U test to analyze the data, and test to see if the life in months of service before failure of the picture tube differs for the picture tubes produced by the two manufacturers. (Use $\alpha = .10$.)

14.18　A comparison of the weights of turtles caught in two different lakes was conducted to compare the effects of the two lake environments on turtle growth. All the turtles were of the same age and were tagged before being released in the lakes. The weight measurements for $n_1 = 10$ tagged turtles caught in lake 1 and $n_2 = 8$ caught in lake 2 are as follows:

Lake	Weight (oz.)									
1	14.1	15.2	13.9	14.5	14.7	13.8	14.0	16.1	12.7	15.3
2	12.2	13.0	14.1	13.6	12.4	11.9	12.5	13.8		

Do the data provide sufficient evidence to indicate a difference in the distributions of weights for the tagged turtles exposed to the two lake environments? Use a Mann-Whitney U test with $\alpha = .05$ to answer the question.

14.19　Cancer treatment by means of chemicals—chemotherapy–utilizes chemicals that kill both cancerous and normal cells. In some instances the toxicity of the cancer drug, that is, its effect on normal cells, can be reduced by the simultaneous injection of a second drug. A study was conducted to determine whether a particular drug injection was beneficial in reducing the harmful effects of a chemotherapy treatment on the survival time for rats. Two randomly selected groups of 12 rats were used in an experiment in which both groups, call them A and B, received the toxic drug in a dosage large enough to cause death, but in addition, group B received the antitoxin, which was to reduce the toxic effect of the chemotherapy on normal cells. The test was terminated at the end of 20 days, or 480 hours. The lengths of survival time for the two groups of rats, to the nearest four hours, are shown in the table. Do the data provide sufficient evidence to indicate that rats receiving the antitoxin tend to survive longer after chemotherapy than those not receiving the toxin? Use the Mann-Whitney U test with a value of α near .05.

Chemotherapy only A	Chemotherapy plus drug B
84	140
128	184
168	368
92	96
184	480
92	188
76	480
104	244
72	440
180	380
144	480
120	196

14.5   THE WILCOXON SIGNED-RANK TEST FOR A PAIRED EXPERIMENT

A signed-rank test proposed by F. Wilcoxon can be used to analyze the paired-difference experiment of Section 9.5 by considering the paired differences of the two treatments 1 and 2. Under the null hypothesis of "no differences in the distributions for 1 and 2," you would expect (on the average) half of the differences in pairs to be negative and half to be positive; that is, the expected number of negative differences between pairs would be $n/2$ (where n is the number of pairs). Furthermore, it would follow that positive and negative differences of equal absolute magnitude should occur with equal probability. If one were to order the differences according to their absolute values and rank them from smallest to largest, the expected rank sums for the negative and positive differences would be equal. Sizable differences in the sums of the ranks assigned to the positive and negative differences would provide evidence to indicate a shift in location between the distributions of responses for the two treatments 1 and 2.

CALCULATION OF THE TEST STATISTIC FOR THE WILCOXON SIGNED-RANK TEST

1. Calculate the differences $(x_A - x_B)$ for each of the n pairs. Differences equal to 0 are eliminated and the number of pairs, n, is reduced accordingly.

2. Rank the **absolute values** of the differences, assigning 1 to the smallest, 2 to the second smallest, and so on. Tied observations are assigned the average of the ranks that would have been assigned with no ties.

3. Calculate the **rank sum** for the **negative** differences and label this value T^-. Similarly, calculate T^+, the **rank sum** for the **positive** differences.

For a **two-tailed test**, we use the **smaller of these two quantities T as a test statistic** to test the null hypothesis that the two population relative frequency histograms are identical. The smaller the value of T, the greater will be the weight of evidence favoring rejection of the null hypothesis. **Therefore, we will reject the null hypothesis if T is less than or equal to some value, say T_0.**

To detect the **one-sided alternative**, that **distribution 1 is shifted to the right of distribution 2, use the rank sum T^-** of the negative differences and reject the null hypothesis for small values of T^-, say $T^- \leq T_0$. If we wish to detect a **shift of distribution 2 to the right of distribution 1, we use the rank sum T^+** of the positive differences as a test statistic and reject for small values of T^+, say $T^+ \leq T_0$.

The probability that T is less than or equal to some value T_0 has been calculated for a combination of sample sizes and values of T_0. These probabilities, given in Table 12 in Appendix III, can be used to find the rejection region for the T test.

An abbreviated version of Table 12, Appendix III, is shown in Table 14.4. Across the top of the table you see the number of differences (the number of pairs) n. Values of α (denoted by the symbol P in Wilcoxon's table) for a one-tailed test appear in the first column of the table. The second column gives values of α (P in Wilcoxon's notation) for a two-tailed test. Table entries are the critical values of T. You will recall that the critical value of a test statistic is the value that locates the boundary of the rejection region.

For example, suppose you have $n = 7$ pairs and you are conducting a two-tailed test of the null hypothesis that the two population relative frequency distributions are identical. Checking the $n = 7$ column of Table 14.4 and using the

Table 14.4
An abbreviated version of Table 12 in Appendix III; critical values of T

One-sided	Two-sided	$n = 5$	$n = 6$	$n = 7$	$n = 8$	$n = 9$	$n = 10$
$P = .05$	$P = .10$	1	2	4	6	8	11
$P = .025$	$P = .05$		1	2	4	6	8
$P = .01$	$P = .02$			0	2	3	5
$P = .005$	$P = .01$				0	2	3

One-sided	Two-sided	$n = 11$	$n = 12$	$n = 13$	$n = 14$	$n = 15$	$n = 16$
$P = .05$	$P = .10$	14	17	21	26	30	36
$P = .025$	$P = .05$	11	14	17	21	25	30
$P = .01$	$P = .02$	7	10	13	16	20	24
$P = .005$	$P = .01$	5	7	10	13	16	19

One-sided	Two-sided	$n = 17$
$P = .05$	$P = .10$	41
$P = .025$	$P = .05$	35
$P = .01$	$P = .02$	28
$P = .005$	$P = .01$	23

second row (corresponding to $P = \alpha = .05$ for a two-tailed test), you see the entry 2 (shaded in Table 14.4). This value is T_0, the critical value of T. As noted earlier, the smaller the value of T, the greater will be the evidence to reject the null hypothesis. Therefore, you will reject the null hypothesis for all values of T less than or equal to 2. The rejection region for the Wilcoxon signed-rank test for a paired experiment is always of the form: reject H_0 if $T \leq T_0$, where T_0 is the critical value of T. The rejection region is shown symbolically in Figure 14.3.

Figure 14.3
Rejection region for the Wilcoxon signed-rank test for a paired experiment (reject H_0 if $T \leq T_0$)

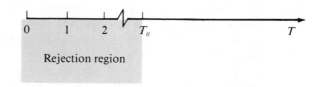

WILCOXON SIGNED-RANK TEST FOR A PAIRED EXPERIMENT

1. Null Hypothesis: H_0: The two population relative frequency distributions are identical.

2. Alternative Hypothesis: H_a: The two population relative frequency distributions differ in location (a two-tailed test). Or H_a: The population 1 relative frequency distribution is shifted to the right of the relative frequency distribution for population 2 (a one-tailed test).

3. Test Statistic:
   a.   For a two-tailed test, use T, the smaller of the rank sum for positive and the rank sum for negative differences.
   b.   For a one-tailed test (to detect the alternative hypothesis described above), use the rank sum T^- of the negative differences.

4. Rejection Region:
   a.   For a two-tailed test, reject H_0 if $T \leq T_0$, where T_0 is the critical value given in Table 12 in Appendix III.
   b.   For a one-tailed test (to detect the alternative hypothesis described above), use the rank sum T^- of the negative differences. Reject H_0 if $T^- \leq T_0$.*

   $\left(\text{Note: It can be shown that } T^+ + T^- = \dfrac{n(n + 1)}{2} \right)$

EXAMPLE 14.5   Test an hypothesis of no difference in population distributions of cake density for the paired-difference experiment in Exercise 9.104.

---

* To detect a shift of distribution 2 to the right of the distribution 1, use the rank sum T^+ of the positive differences as the test statistic and reject H_0 if $T^+ \leq T_0$.

Solution   The original data and differences in density for the six pairs of cakes are as follows:

	Density (oz/in.³)					
x_1	.135	.102	.098	.141	.131	.144
x_2	.129	.120	.112	.152	.135	.163
Difference $(x_1 - x_2)$	.006	−.018	−.014	−.011	−.004	−.019
Rank	2	5	4	3	1	6

As with other nonparametric tests, the null hypothesis to be tested is that the two population frequency distributions of cake densities are identical. The alternative hypothesis, which implies a two-tailed test, is that the distributions are different.

Because the amount of data is small, we will conduct our test using $\alpha = .10$. From Table 12, the critical value of T for a two-tailed test, $\alpha = .10$, is $T_0 = 2$. Hence, we will reject H_0 if $T \leq 2$.

The differences $x_1 - x_2$ are calculated and ranked according to their absolute value in the table. The sum of the positive ranks is $T^+ = 2$, and the rank sum of the negative ranks is $T^- = 19$. The test statistic is the smaller of these two rank sums, or $T = 2$. Since $T = 2$ falls in the rejection region, we reject H_0 and conclude that the two population frequency distributions of cake densities differ.   ◁

Although Table 12 in Appendix III is applicable for values of n (number of data pairs) as large as $n = 50$, it is worth noting that T^+, **like the Mann-Whitney U, will be approximately normally distributed when the null hypothesis is true and n is large (say 25 or more). This enables us to construct a large-sample z test,** where

$$E(T) = \frac{n(n + 1)}{4}$$

$$\sigma_T^2 = \frac{n(n + 1)(2n + 1)}{24}$$

Then the z statistic

$$z = \frac{T^+ - E(T^+)}{\sigma_{T^+}} = \frac{T^+ - \dfrac{n(n + 1)}{4}}{\sqrt{\dfrac{n(n + 1)(2n + 1)}{24}}}$$

can be used as a test statistic. Thus, for a two-tailed test and $\alpha = .05$, we would reject the hypothesis of "identical population distributions" when $|z| \geq 1.96$.

A LARGE-SAMPLE WILCOXON SIGNED-RANK TEST FOR A PAIRED EXPERIMENT: $n \geq 25$

1.   Null Hypothesis: H_0: The population relative frequency distributions 1 and 2 are identical.

2. Alternative Hypothesis: H_a: The two population relative frequency distributions differ in location (a two-tailed test). Or H_a: The population 1 relative frequency distribution is shifted to the right (or left) of the relative frequency distribution for population 2 (a one-tailed test).

3. Test Statistic: $z = \dfrac{T^+ - [n(n + 1)/4]}{\sqrt{[n(n + 1)(2n + 1)]/24}}$

4. Rejection Region: Reject H_0 if $z > z_{\alpha/2}$ or $z < -z_{\alpha/2}$ for a two-tailed test. For a one-tailed test, place all of α in one tail of the z distribution. To detect a shift in distribution 1 to the right of distribution 2, reject H_0 when $z > z_\alpha$. To detect a shift in the opposite direction, reject H_0 if $z < -z_\alpha$. Tabulated values of z are given in Table 3 in Appendix III.

The MINITAB package includes a command WTEST for implementing the Wilcoxon signed-rank test for a paired experiment. The initial input and output for this program using the data in Example 14.5 is given in Table 14.5. The differences of the paired values are set in column 3. The command WTEST 0 C3 specifies that the Wilcoxon signed-rank procedure will be used to test for a zero median using the values in column 3. Notice that none of the six differences was equal to 0, so all six were used in calculating the value of the test statistic given as 2.0. Since the p-value of .093 exceeds $\alpha = .05$, there is not sufficient evidence to conclude that the underlying population distributions differ significantly.

Table 14.5
MINITAB output for the data of Example 14.5

```
MTB > PRINT C3
C3
  0.0060000    -0.0180000    -0.0140000    -0.0110000    -0.0040000    -0.0190000

MTB > WTEST 0 C3

TEST OF MEDIAN = 0 VERSUS MEDIAN N.E. 0
```

	N	N FOR TEST	WILCOXON STATISTIC	P-VALUE	ESTIMATED MEDIAN
C3	6	6	2.0	0.093	-0.01100

EXERCISES Basic Techniques

14.20 Suppose you wish to detect a difference in location between two population distributions based on a paired-difference experiment consisting of $n = 30$ pairs.
  a. Give the null and alternative hypotheses for the Wilcoxon signed-rank test.
  b. Give the test statistic.
  c. Give the rejection region for the test for $\alpha = .05$.
  d. If $T^+ = 249$, what are your conclusions? (*Note:* $T^+ + T^- = n(n + 1)/2$.)

14.21 Refer to Exercise 14.20. Suppose you wish to detect only a shift in distribution 1 to the right of distribution 2.
  a. Give the null and alternative hypotheses for the Wilcoxon signed-rank test.

b.   Give the test statistic.

c.   Give the rejection region for the test for $\alpha = .05$.

d.   If $T^+ = 249$, what are your conclusions? (*Note:* $T^+ + T^- = n(n + 1)/2$.)

14.22   Refer to Exercise 14.20. Conduct the test using the large-sample z test. Compare your results with the nonparametric test results in Exercise 14.20 (d).

14.23   Refer to Exercise 14.21. Conduct the test using the large-sample z test. Compare your results with the nonparametric test results in Exercise 14.21 (d).

14.24   Refer to Exercise 14.4. The data reproduced below are from a paired-difference experiment with $n = 7$ pairs of observations.

				Pairs			
Population	I	2	3	4	5	6	7
1	8.9	8.1	9.3	7.7	10.4	8.3	7.4
2	8.8	7.4	9.0	7.8	9.9	8.1	6.9

a.   Use Wilcoxon's signed-rank test to determine whether there is a significant difference between the two populations.

b.   Compare the results of part (a) with the result obtained in Exercise 14.4. Are the results the same? Explain.

Applications

14.25   In Exercise 14.5 we used the sign test to determine whether the data provided sufficient evidence to indicate a shift in the distributions of property assessments for assessors A and B.

a.   Use the Wilcoxon signed-rank test for a paired experiment to test the null hypothesis that there is no difference in the distributions of property assessments between assessors A and B. Test by using a value of α near .05.

b.   Compare the conclusion of the test in part (a) with the conclusions derived from the t test in Exercise 9.43, and the sign test in Exercise 14.5. Explain why these test conclusions are (or are not) consistent.

14.26   The number of machine breakdowns per month was recorded for nine months on two identical machines used to make wire rope. The data are shown in the table.

	Machines	
Month	A	B
1	3	7
2	14	12
3	7	9
4	10	15
5	9	12
6	6	6
7	13	12
8	6	5
9	7	13

a. Do the data provide sufficient evidence to indicate a difference in the monthly break-down rates for the two machines? Test by using a value of α near .05.

b. Can you think of a reason why the breakdown rates for the two machines might vary from month to month?

14.27 Refer to the comparison of gourmet meal ratings in Exercise 14.6, and use the Wilcoxon signed-rank test to determine whether the data provide sufficient evidence to indicate a difference in the ratings of the two gourmets. Test by using a value of α near .05. Compare the results of this test with the results of the sign test in Exercise 14.6. Are the test conclusions consistent?

14.28 Two methods for controlling traffic, A and B, were employed at each of $n = 12$ intersections for a period of one week, and the numbers of accidents occurring during this time period were recorded. The order of use (which method would be employed for the first week) was selected in a random manner.

Intersection	Method A	Method B	Intersection	Method A	Method B
1	5	4	7	2	3
2	6	4	8	4	1
3	8	9	9	7	9
4	3	2	10	5	2
5	6	3	11	6	5
6	1	0	12	1	1

Do the data provide sufficient evidence to indicate a shift in the distributions of accident rates for traffic control methods A and B?

a. Analyze using a sign test.

b. Analyze using the Wilcoxon signed-rank test for a paired experiment.

14.29 One explanation for the bright coloration of male birds versus the dull coloration of females and chicks is the predator-deflection theory, which suggests that males have developed a bright color to attract predators to themselves and away from their nesting females and young chicks. To test this theory, G. S. Butcher placed a stuffed and mounted sharp-shinned hawk (*Accipiter striatus*) a distance of 12 meters from each of 12 northern orioles' nests ("The Predator-Deflection Hypothesis for Sexual Colour Dimorphism: A Test on the Northern Oriole," *Animal Behaviour* 32 [1984]). For each nest, he recorded the total number of hits and swoops on the hawk by both the male and female oriole. He also noted which of the pair made the first attack on the hawk and which was more aggressive. The data are shown in the accompanying table.

Do the data present sufficient evidence to indicate a difference in the distributions of numbers of hits and swoops for male versus female members of the nest?

a. Test using Wilcoxon's signed-rank test with $\alpha = .05$. State your conclusions.

b. Give the approximate p-value for the test and interpret it.

Presentation number*	Total number of hits and swoops		More aggressive sex (M or F)†	Sex seen first
	M	F		
1	4	18	F	F
2	11	47	F	F
3	16	33	F	F
4	14	0	M	F
5	27	2	M	F
6	71	59	M	M
7	76	5	M	M
8	61	0	M	M
9	0	25	F	F
10	6	21	F	M
11	35	50	F	F
12	3	59	F	F

* For each of the 12 trials, the mounted *Accipiter* was placed within 12 meters of the nest for a 10-minute period.

† M = male and F = female of the pair who fed offspring in the nest closest to the mounted *Accipiter*.

14.30 Refer to Exercise 14.29. Do the data present sufficient evidence to indicate that the male member of a nest is more aggressive than the female? Explain.

14.31 Eight subjects were asked to perform a simple puzzle-assembly task under normal conditions and under stressful conditions. During the stressful time, a mild shock was delivered to subjects 3 minutes after the start of the experiment and every 30 seconds thereafter until the task was completed. Blood pressure readings were taken under both conditions. The following data represent the highest readings during the experiment.

Subject	Normal	Stressful
1	126	130
2	117	118
3	115	125
4	118	120
5	118	121
6	128	125
7	125	130
8	120	120

Do the data present sufficient evidence to indicate higher blood pressure readings under stressful conditions? Analyze the data using the Wilcoxon signed-rank test for a paired experiment.

THE KRUSKAL-WALLIS *H* TEST FOR COMPLETELY RANDOMIZED DESIGNS

Just as the Mann-Whitney *U* test is the nonparametric alternative to the Student's *t* test for a comparison of population means, the Kruskal-Wallis *H* test is the nonparametric alternative to the analysis-of-variance *F* test for a completely randomized design. It is used to detect differences in location among more than two population distributions based on independent random sampling.

The procedure for conducting the Kruskal-Wallis *H* test is similar to that used for the Mann-Whitney *U* test. Suppose that we are comparing *k* populations based on independent random samples n_1 from population 1, n_2 from population 2, . . . , and n_k from population *k* where

$$n_1 + n_2 + \cdots + n_k = n$$

The first step is to rank all *n* observations from the smallest (rank 1) to the largest (rank *n*). **Tied observations are assigned a rank equal to the average of the ranks they would have received if they had been nearly equal but not tied.** We then calculate the rank sums $T_1, T_2, \ldots, T_k$ for the *k* samples and calculate the test statistic

$$H = \frac{12}{n(n+1)} \sum_{i=1}^{k} \frac{T_i^2}{n_i} - 3(n+1)$$

The greater the differences in location among the *k* population distributions, the larger will be the value of the *H* statistic. Thus, we reject the null hypothesis that the *k* population distributions are identical for large values of *H*.

How large is large? It can be shown (proof omitted) that when the sample sizes are moderate to large, say each sample size is equal to 5 or larger, and when H_0 is true, the *H* statistic will have approximately a chi-square distribution with $(k-1)$ degrees of freedom. Therefore, for a given value of α, we reject H_0 when the *H* statistic exceeds χ_α^2 (see Figure 14.4).

Figure 14.4
Approximate distribution of the *H* statistic when H_0 is true

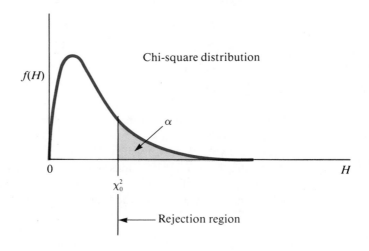

EXAMPLE 14.6   The data given in Example 13.4 were collected using a completely randomized design. They represent the achievement test scores for four different groups of students, each group taught by a different teaching technique. The objective of the experiment is to test the hypothesis of "no difference in the population distributions of achievement test scores" against the alternative that they differ in location—that is, at least one of the distributions is shifted above the others. Conduct the test using the Kruskal-Wallis H test with $\alpha = .05$.

Techniques (ranks in parentheses)			
1	2	3	4
65 (3)	75 (9)	59 (1)	94 (23)
87 (19)	69 (5.5)	78 (11)	89 (21)
73 (8)	83 (17.5)	67 (4)	80 (14)
79 (12.5)	81 (15.5)	62 (2)	88 (20)
81 (15.5)	72 (7)	83 (17.5)	
69 (5.5)	79 (12.5)	76 (10)	
	90 (22)		
Rank sum $T_1 = 63.5$	$T_2 = 89$	$T_3 = 45.5$	$T_4 = 78$

Solution   The first step is to rank the $n = 23$ observations from the smallest (rank 1) to the largest (rank 23). These ranks are shown in parentheses in the table. Note how the ties are handled. For example, two observations at 69 were tied for rank 5. Therefore, they were assigned the average 5.5 of the two ranks (5 and 6) that they would have occupied if they had been slightly different. The rank sums T_1, T_2, T_3, and T_4 for the four samples are shown in the bottom row of the table. Substituting rank sums and sample sizes into the formula for the H statistic, we obtain

$$H = \frac{12}{n(n+1)} \sum_{i=1}^{k} \frac{T_i^2}{n_i} - 3(n+1)$$

$$= \frac{12}{23(24)} \left[\frac{(63.5)^2}{6} + \frac{(89)^2}{7} + \frac{(45.5)^2}{6} + \frac{(78)^2}{4} \right] - 3(24)$$

$$= 79.775102 - 72$$

$$= 7.775102$$

The rejection region for the H statistic for $\alpha = .05$ includes values of $H \geq \chi^2_{.05}$ where $\chi^2_{.05}$ is based on $k - 1 = 4 - 1 = 3$ degrees of freedom. This value of χ^2, given in Table 5 in Appendix III, is $\chi^2_{.05} = 7.81473$. The observed value of the H statistic, $H = 7.775102$, does not fall in the rejection region for the test. Therefore, there is insufficient evidence to indicate differences in the distributions of achievement test scores for the four teaching techniques.   ◁

EXAMPLE 14.7   Compare the results of the analysis-of-variance *F* test and the Kruskal-Wallis *H* test for testing for differences in the distributions of achievement test scores for the four teaching techniques.

Solution   Examining Example 13.2, you can see that the *F* test detected differences among the population means (for $\alpha = .05$). The *H* test did not detect a shift in population distributions. Although these conclusions seem to be far apart, the test results are very close. The *p*-value corresponding to $F = 3.77$, $v_1 = 3$, and $v_2 = 10$, is slightly less than .05 in contrast to the *p*-value for $H = 7.78$, $v = 3$, which is slightly larger than .05. A person viewing the *p*-values for the two tests would see little difference in the results of the *F* and *H* tests. However, if we adhere to our choice of $\alpha = .05$, we could not reject H_0 using the *H* test.   ◁

THE KRUSKAL-WALLIS *H* TEST FOR COMPARING MORE THAN TWO POPULATIONS: COMPLETELY RANDOMIZED DESIGN (INDEPENDENT RANDOM SAMPLES)

1. **Null Hypothesis:**   The *k* population distributions are identical.

2. **Alternative Hypothesis:**   At least two of the *k* population distributions differ in location.

3. **Test Statistic:**

$$H = \frac{12}{n(n + 1)} \sum_{i=1}^{k} \frac{T_i^2}{n_i} - 3(n + 1)$$

   where n_i = sample size for population *i*
      T_i = rank sum for population *i*
      n = total number of observations
         $= n_1 + n_2 + \cdots + n_k$

4. **Rejection Region for a Given α:**

   $$H > \chi_\alpha^2$$

   Assumptions:

   1. All sample sizes are greater than or equal to 5.
   2. Ties assume the average of the ranks that they would have occupied if they had not been tied.

Table 14.6 contains the output generated using the MINITAB command KRUSKAL-WALLIS for the data in Example 14.6. The data was stored in C1, and the treatment subscripts in C2.

Table 14.6
MINITAB printout for
the data in Example 14.6

```
MTB > KRUSKAL-WALLIS C1 C2

    LEVEL    NOBS     MEDIAN     AVE. RANK    Z VALUE
      1        6       76.00        10.6        −0.60
      2        7       79.00        12.7         0.33
      3        6       71.50         7.6        −1.86
      4        4       88.50        19.5         2.43
    OVERALL   23                    12.0

  H = 7.78   D.F. = 3   P = 0.051
  H = 7.79   D.F. = 3   P = 0.051 (ADJ. FOR TIES)

  ⊛ NOTE ⊛ ONE OR MORE SMALL SAMPLES
  MTB >
```

EXERCISES   Basic Techniques

14.32   Three treatments were compared using a completely randomized design. The data are shown below.

Treatment		
1	2	3
26	27	25
29	31	24
23	30	27
24	28	22
28	29	24
26	32	20
	30	21
	33	

Do the data provide sufficient evidence to indicate a difference in location for at least two of the population distributions? Test using the Kruskal-Wallis H statistic with $\alpha = .05$.

14.33   Four treatments were compared using a completely randomized design. The data are shown below.

Treatment			
1	2	3	4
124	147	141	117
167	121	144	128
135	136	139	102
160	114	162	119
159	129	155	128
144	117	150	123
133	109		

Do the data provide sufficient evidence to indicate a difference in location for at least two of the population distributions? Test using the Kruskal-Wallis H statistic with $\alpha = .05$.

Applications

 14.34 Exercise 13.22 presents data on the rate of growth of vegetation at four swampy under-developed sites. Six plants were randomly selected at each of the four sites to be used in the comparison. The following data represent the mean leaf length per plant, in centimeters, for a random sample of ten leaves per plant.

Location	Mean leaf length (cm)					
1	5.7	6.3	6.1	6.0	5.8	6.2
2	6.2	5.3	5.7	6.0	5.2	5.5
3	5.4	5.0	6.0	5.6	4.9	5.2
4	3.7	3.2	3.9	4.0	3.5	3.6

  a. Do the data present sufficient evidence to indicate differences in location for at least two of the distributions of mean leaf length corresponding to the four locations? Test using the Kruskal-Wallis *H* test with $\alpha = .05$.
  b. Find the approximate *p*-value for the test.
  c. We analyzed this same set of data in Exercise 13.22 using an analysis of variance. Find the *p*-value for the *F* test used to compare the four location means in Exercise 13.22.
  d. Compare the *p*-values in parts (b) and (c), and explain the implications of the comparison.

14.35 Exercise 13.53 presents data on the heart rate for samples of ten men randomly selected from each of four age groups. Each subject walked a treadmill at a fixed grade for a period of 12 minutes, and the increase in heart rate (the difference before and after exercise) was recorded in beats per minute. The data are shown in the table.

	Age		
10–19	20–39	40–59	60–69
29	24	37	28
33	27	25	29
26	33	22	34
27	31	33	36
39	21	28	21
35	28	26	20
33	24	30	25
29	34	34	24
36	21	27	33
22	32	33	32
Total 309	275	295	282

  a. Do the data present sufficient evidence to indicate differences in location for at least two of the four age groups? Test using the Kruskal-Wallis *H* test with $\alpha = .10$.
  b. Find the approximate *p*-value for the test in part (a).
  c. Since the *F* test in Exercise 13.53 and the *H* test in part (a) are both tests to detect differences in location of the four heart-rate populations, how do the test results compare? Compare the *p*-values for the two tests and explain the implications of the comparison.

 14.36 A sampling of the acidity of rain for ten randomly selected rainfalls was recorded at three different locations in the United States, the Northeast, the Middle Atlantic region, and the

Southeast. The pH readings for these thirty rainfalls are shown below. (*Note:* pH readings range from 0 to 14; 0 is acid, 14 is alkaline. Pure water falling through clean air has a pH reading of 5.7.)

Northeast	Middle Atlantic	Southeast
4.45	4.60	4.55
4.02	4.27	4.31
4.13	4.31	4.84
3.51	3.88	4.67
4.42	4.49	4.28
3.89	4.22	4.95
4.18	4.54	4.72
3.95	4.76	4.63
4.07	4.36	4.36
4.29	4.21	4.47

a. Do the data present sufficient evidence to indicate differences in the levels of acidity in rainfalls for the three different locations? Test using the Kruskal-Wallis H test with $\alpha = .05$.

b. Find the approximate p-value for the test in part (a), and interpret it.

▷ 14.7 THE FRIEDMAN F_r TEST FOR RANDOMIZED BLOCK DESIGNS

The Friedman F_r test, proposed by Nobel prize–winning economist Milton Friedman, is a nonparametric test for comparing the distributions of measurements for k treatments laid out in b blocks using a randomized block design. The procedure for conducting the test is very similar to that used for the Kruskal-Wallis H test. The first step in the procedure is to rank the k treatment observations within each block. Ties are treated in the usual way; that is, they receive an average of the ranks occupied by the tied observations. The rank sums $T_1, T_2, \ldots, T_k$ are then obtained and the test statistic

$$F_r = \frac{12}{bk(k+1)} \sum_{i=1}^{k} T_i^2 - 3b(k+1)$$

is calculated.

The value of the F_r statistic will be at a minimum when the rank sums are equal—that is, $T_1 = T_2 = \cdots = T_k$—and will increase in value as the differences among the rank sums increase. When either the number k of treatments or the number b of blocks is larger than 5, the sampling distribution of F_r can be approximated by a chi-square distribution with $(k - 1)$ degrees of freedom. Therefore, like the Kruskal-Wallis H test, the rejection region for the F_r test consists of values of F_r for which

$$F_r > \chi_\alpha^2$$

We will illustrate the test using the reaction time data in Exercise 13.50.

EXAMPLE 14.8 Suppose that you wished to compare the reaction times of people subjected to one of six different stimuli. A reaction time measurement is obtained by subjecting a person to a stimulus and then measuring the length of time until the person presents some specified reaction. The objective of the experiment is to determine whether differences exist in the magnitudes of the reaction times for the stimuli employed in the experiment.

In order to eliminate the person-to-person variation in reaction time, four persons (subjects) were employed in the experiment and each person's reaction time (in seconds) was measured for each of the six stimuli. The data are reproduced below (ranks of the observations within each block are shown in parentheses). Use the Friedman F_r test to determine whether the data present sufficient evidence to indicate differences in the locations of the distributions of reaction times for the six stimuli. Test using $\alpha = .05$.

Subject	A	B	C	D	E	F
1	.6 (2.5)	.9 (6)	.8 (5)	.7 (4)	.5 (1)	.6 (2.5)
2	.7 (3.5)	1.1 (6)	.7 (3.5)	.8 (5)	.5 (1.5)	.5 (1.5)
3	.9 (3)	1.3 (6)	1.0 (4.5)	1.0 (4.5)	.7 (1)	.8 (2)
4	.5 (2)	.7 (5)	.8 (6)	.6 (3.5)	.4 (1)	.6 (3.5)
Rank Sum	$T_1 = 11$	$T_2 = 23$	$T_3 = 19$	$T_4 = 17$	$T_5 = 4.5$	$T_6 = 9.5$

Solution We wish to test

H_0: The distributions of reaction times for the six stimuli are identical.

against the alternative hypothesis

H_a: At least two of the distributions of reaction times for the six stimuli differ in location.

The table shows the ranks (in parentheses) of the observations within each block and the rank sums for each of the six stimuli (the treatments). The value of the F_r statistic for these data is

$$F_r = \frac{12}{bk(k+1)} \sum_{i=1}^{k} T_i^2 - 3b(k+1)$$

$$= \frac{12}{(4)(6)(7)} [(11)^2 + (23)^2 + (19)^2 + \cdots + (9.5)^2] - 3(4)(7)$$

$$= 100.75 - 84 = 16.75$$

Since the number $k = 6$ of treatments exceeds 5, the sampling distribution of F_r can be approximated by a chi-square distribution with $k - 1 = 6 - 1 = 5$ degrees of freedom. Therefore, for $\alpha = .05$, we reject H_0 if

$$F_r > \chi^2_{.05} \quad \text{where} \quad \chi^2_{.05} = 11.0705$$

This rejection region is shown in Figure 14.5.

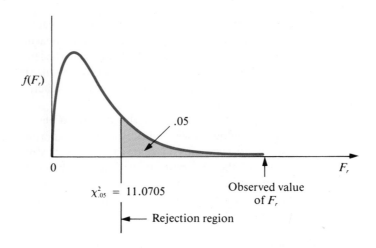

Figure 14.5
Rejection region for
Example 14.8

$f(F_r)$

.05

0   F_r

$\chi^2_{.05} = 11.0705$ Observed value
of F_r

← Rejection region

Since the observed value $F_r = 16.75$ exceeds $\chi^2_{.05} = 11.0705$, it falls in the rejection region. We therefore reject H_0 and conclude that at least two of the distributions in reaction times differ in location. ◁

EXAMPLE 14.9   Find the approximate p-value for the test in Example 14.8.

Solution   Consulting Table 5 in Appendix III with 5 degrees of freedom, we see that the observed value of $F_r = 16.75$ exceeds the table value $\chi^2_{.005} = 16.7496$. Hence, the p-value is very close to, but slightly less than, .005. ◁

THE FRIEDMAN F_r TEST FOR A RANDOMIZED BLOCK DESIGN

1.   Null Hypothesis: The k population distributions are identical.
2.   Alternative Hypothesis: At least two of the k population distributions differ in location.
3.   Test Statistic:

$$F_r = \frac{12}{bk(k+1)} \sum_{i=1}^{k} T_i^2 - 3b(k+1)$$

    where b = number of blocks
              k = number of treatments
              T_i = rank sum for treatment i, $i = 1, 2, \ldots, k$

4.   Rejection Region:

     $F_r > \chi^2_\alpha$,   where χ^2_α is based on $(k-1)$ degrees of freedom.

     Assumption: Either the number k of treatments or the number b of blocks is larger than 5.

Table 14.7
MINITAB printout for
the data in Example 14.8

MTB > FRIEDMAN C1 C2 C3

FRIEDMAN TEST OF C1 BY C2 BLOCKED BY C3

S = 16.75   D.F. = 5   P = 0.005
S = 17.37   D.F. = 5   P = 0.004 (ADJUSTED FOR TIES)

		EST.	SUM OF
C2	N	MEDIAN	RANKS
1	4	0.6500	11.0
2	4	1.0000	23.0
3	4	0.8000	19.0
4	4	0.7500	17.0
5	4	0.5000	4.5
6	4	0.6000	9.5

GRAND MEDIAN = 0.7167

MTB >

Table 14.7 contains the output generated using the MINITAB command FRIEDMAN for the data in Example 14.8. The data were stored in C1, the treatment subscripts in C2, and the block subscripts in C3.

EXERCISES   Basic Techniques

14.37   A randomized block design is employed to compare three treatments in six blocks. The data are shown below.

	Treatment		
Block	1	2	3
1	3.2	3.1	2.4
2	2.8	3.0	1.7
3	4.5	5.0	3.9
4	2.5	2.7	2.6
5	3.7	4.1	3.5
6	2.4	2.4	2.0

a.   Use the Friedman F_r test to detect differences in location among the three treatment distributions. Test using $\alpha = .05$.
b.   Find the approximate p-value for the test.
c.   Perform an analysis of variance and give the ANOVA table for the analysis.
d.   Give the value of the F statistic for testing the equality of the three treatment means.
e.   Give the approximate p-value for the F statistic in part (d).
f.   Compare the p-values for the tests in parts (a) and (d) and explain the practical implications of the comparison.

14.38   A randomized block design is employed to compare four treatments in eight blocks. The data are shown on page 618.

| | Treatment | | | |
Block	1	2	3	4
1	89	81	84	85
2	93	86	86	88
3	91	85	87	86
4	85	79	80	82
5	90	84	85	85
6	86	78	83	84
7	87	80	83	82
8	93	86	88	90

a. Use the Friedman F_r test to detect differences in location among the four treatment distributions. Test using $\alpha = .05$.

b. Find the approximate p-value for the test.

c. Perform an analysis of variance and give the ANOVA table for the analysis.

d. Give the value of the F statistic for testing the equality of the four treatment means.

e. Give the approximate p-value for the F statistic in part (d).

f. Compare the p-values for the tests in parts (a) and (d) and explain the practical implications of the comparison.

Applications

14.39 In a comparison of the prices of items at five supermarkets, six items were randomly selected for the comparison, and the price of each was recorded for each of the five supermarkets. The objective of the study was to see whether the data indicated differences in the level of prices among the five supermarkets. The data are shown below.

| | Supermarket | | | | |
Item	Kash n' Karry	Publix	Winn-Dixie	Albertsons	Food 4 Less
Celery	.33	.34	.69	.59	.58
Colgate Toothpaste	1.28	1.49	1.44	1.37	1.28
Campbell's Beef Soup	1.05	1.19	1.23	1.19	1.10
Crushed Pineapple	.83	.95	.95	.87	.84
Mueller's Spaghetti	.68	.79	.83	.69	.69
Heinz Ketchup	1.41	1.69	1.79	1.65	1.49

a. Does the distribution of the prices of items differ in location from one supermarket to another? Test using the Friedman F_r test with $\alpha = .05$.

b. Find the approximate p-value for the test and interpret it.

14.40 Exercise 13.34 describes an experiment conducted to compare the effects of three toxic chemicals A, B, and C on the skin of rats. One-inch squares of skin were treated with the chemicals and then scored from 0 to 10 depending on the degree of irritation. Three adjacent one-inch squares were marked on the backs of eight rats, and each of the three chemicals was applied to each rat. Thus, the experiment was blocked on rats to eliminate the variation in skin sensitivity from rat to rat. The data are as follows:

Rats

I	2	3	4	5	6	7	8
B	A	A	C	B	C	C	B
5	9	6	6	8	5	5	7
A	C	B	B	C	A	B	A
6	4	9	8	8	5	7	6
C	B	C	A	A	B	A	C
3	9	3	5	7	7	6	7

a. Do the data provide sufficient evidence to indicate a difference in the toxic effect of the three chemicals? Test using the Friedman F_r test with $\alpha = .05$.
b. Find the approximate p-value for the test and interpret it.
c. Compare your conclusion in part (a) with the conclusion reached using the analysis of variance in Exercise 13.34. Do they agree? Explain.

14.8 RANK CORRELATION COEFFICIENT

In the preceding sections we have used ranks to indicate the relative magnitude of observations in nonparametric tests for comparison of treatments. We will now employ the same technique in testing for a relation between two ranked variables. Two common rank-correlation coefficients are the **Spearman r_s** and the **Kendall τ**. We will present the Spearman r_s because its computation is identical to that for the sample correlation coefficient r of Chapter 1. Kendall's rank correlation coefficient is discussed in detail in Kendall and Stuart (1979).

Suppose that eight elementary science teachers have been ranked by a judge according to their teaching ability, and that all have taken a "national teachers' examination." The data are as follows:

Teacher	Judge's rank	Examination score
1	7	44
2	4	72
3	2	69
4	6	70
5	1	93
6	3	82
7	8	67
8	5	80

Do the data suggest an agreement between the judge's ranking and the examination score? That is, is there a correlation between ranks and test scores?

The two variables of interest are rank and test score. The former is already in rank form, and the test scores can be ranked similarly, as shown below. **The ranks for tied observations are obtained by averaging the ranks that the tied observations would have had if no ties had been observed.**

Teacher	Judge's rank (x_i)	Test rank (y_i)
1	7	1
2	4	5
3	2	3
4	6	4
5	1	8
6	3	7
7	8	2
8	5	6

The Spearman rank correlation coefficient r_s is calculated by using the ranks as the paired measurements on the two variables x and y in the formula for r, Chapter 10.

SPEARMAN'S RANK CORRELATION COEFFICIENT

$$r_s = \frac{S_{xy}}{\sqrt{S_{xx}S_{yy}}}$$

where x_i and y_i represent the ranks of the ith pair of observations and

$$S_{xy} = \sum_{i=1}^{n}(x_i - \bar{x})(y_i - \bar{y}) = \sum_{i=1}^{n}x_iy_i - \frac{\left(\sum_{i=1}^{n}x_i\right)\left(\sum_{i=1}^{n}y_i\right)}{n}$$

$$S_{xx} = \sum_{i=1}^{n}(x_i - \bar{x})^2 = \sum_{i=1}^{n}x_i^2 - \frac{\left(\sum_{i=1}^{n}x_i\right)^2}{n}$$

$$S_{yy} = \sum_{i=1}^{n}(y_i - \bar{y})^2 = \sum_{i=1}^{n}y_i^2 - \frac{\left(\sum_{i=1}^{n}y_i\right)^2}{n}$$

When there are no ties in either the x observations or the y observations, the above expression for r_s algebraically reduces to the simpler expression

$$r_s = 1 - \frac{6\sum_{i=1}^{n}d_i^2}{n(n^2 - 1)}, \qquad \text{where } d_i = x_i - y_i$$

If the number of ties is small in comparison with the number of data pairs, little error will result in using this shortcut formula.

EXAMPLE 14.10   Calculate r_s for the teacher-judge test-score data.

Solution   The differences and squares of differences between the two rankings are as follows:

Teacher	x_i	y_i	d_i	d_i^2
1	7	1	6	36
2	4	5	-1	1
3	2	3	-1	1
4	6	4	2	4
5	1	8	-7	49
6	3	7	-4	16
7	8	2	6	36
8	5	6	-1	1
Total				144

Substituting into the formula for r_s,

$$r_s = 1 - \frac{6 \sum_{i=1}^{n} d_i^2}{n(n^2 - 1)} = 1 - \frac{6(144)}{8(64 - 1)}$$

$$= -.714 \qquad\qquad \triangleleft$$

The Spearman rank correlation coefficient can be employed as a test statistic to test an hypothesis of "no association" between two populations. We assume that the n pairs of observations (x_i, y_i) have been randomly selected and, therefore, "no association between the populations" implies a random assignment of the n ranks within each sample. Each random assignment (for the two samples) represents a simple event associated with the experiment, and a value of r_s can be calculated for each. Thus, it is possible to calculate the probability that r_s assumes a large absolute value due solely to chance and thereby suggests an association between populations when none exists.

The rejection region for a two-tailed test is shown in Figure 14.6. If the alternative hypothesis is that the correlation between x and y is negative, you would reject H_0 for negative values of r_s that are close to -1 (in the lower tail of Figure 14.6). Similarly, if the alternative hypothesis is that the correlation between x and y is positive, you would reject H_0 for large positive values of r_s (in the upper tail of Figure 14.6).

Figure 14.6
Rejection region for a two-tailed test of the null hypothesis of "no association," using Spearman's rank correlation test

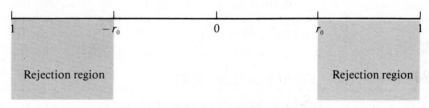

r_s: Spearman's rank correlation coefficient

Table 14.8

An abbreviated version of Table 13 in Appendix III; for Spearman's rank correlation test

n	$\alpha = .05$	$\alpha = 0.25$	$\alpha = .01$	$\alpha = .005$
5	0.900	—	—	—
6	0.829	0.886	0.943	—
7	0.714	0.786	0.893	—
8	0.643	0.738	0.833	0.881
9	0.600	0.683	0.783	0.833
10	0.564	0.648	0.745	0.794
11	0.523	0.623	0.736	0.818
12	0.497	0.591	0.703	0.780
13	0.475	0.566	0.673	0.745
14	0.457	0.545		
15	0.441	0.525		
16	0.425			
17	0.412			
18	0.399	$\vdots$	$\vdots$	$\vdots$
19	0.388			
20	0.377			
$\vdots$	$\vdots$			

The critical values of r_s are given in Table 13 in Appendix III. An abbreviated version of Table 13 is shown here in Table 14.8.

Across the top of Table 14.8 (and Table 13 in Appendix III) are recorded values of α that you might wish to use for a one-tailed test of the null hypothesis of "no association" between x and y. The number of rank pairs n appears at the left side of the table. The table entries give the critical value r_0 for a one-tailed test. Thus, $P(r_s \geq r_0) = \alpha$.

For example, suppose you have $n = 8$ rank pairs and the alternative hypothesis is that the correlation between the ranks is positive. Then you would want to reject the null hypothesis of "no association" only for large positive values of r_s and would use a one-tailed test. Referring to Table 14.8 and using the row corresponding to $n = 8$ and the column for $\alpha = .05$, you read $r_0 = 0.643$. Therefore, you would reject H_0 for all values of r_s greater than or equal to 0.643.

The test is conducted in exactly the same manner if you wish to test only the alternative hypothesis that the ranks are negatively correlated. The only difference is that you would reject the null hypothesis if $r_s \leq -0.643$. That is, you just use the negative of the tabulated value of r_0 to get the lower-tail critical value.

To conduct a two-tailed test, you reject the null hypothesis if $r_s \geq r_0$ or $r_s \leq -r_0$. The value of α for the test will be double the value shown at the top of the table. For example, if $n = 8$ and you choose the 0.025 column, you will reject H_0 if $r_s \geq .738$ or $r_s \leq -.738$. The α value for the test will be $2(.025) = .05$.

SPEARMAN'S RANK CORRELATION TEST

1. Null Hypothesis: H_0: There is no association between the rank pairs.

2. Alternative Hypothesis: H_a: There is an association between the rank pairs (a two-

tailed test). Or H_a: The correlation between the rank pairs is positive or negative (a one-tailed test).

3. Test Statistic: $r_s = \dfrac{S_{xy}}{\sqrt{S_{xx}S_{yy}}}$

    where x_i and y_i represent the ranks of the ith pair of observations.

4. Rejection Region: For a two-tailed test, reject H_0 if $r_s \geq r_0$ or $r_s \leq -r_0$, where r_0 is given in Table 13 in Appendix III. Double the tabulated probability to obtain the value of α for the two-tailed test. For a one-tailed test, reject H_0 if $r_s \geq r_0$ (for an upper-tailed test) or $r_s \leq -r_0$ (for a lower-tailed test). The α value for a one-tailed test is the value shown in Table 13, Appendix III.

EXAMPLE 14.11   Test an hypothesis of "no association" between the populations for Example 14.10.

Solution   The critical value of r_s for a one-tailed test with $\alpha = .05$ and $n = 8$ is .643. Let us assume that a correlation between judge's rank and the teachers' test scores could not possibly be positive. (Low rank means good teaching and should be associated with a high test score if the judge and test measure teaching ability.) The alternative hypothesis is that the population rank correlation coefficient ρ_s is less than 0, and we are concerned with a one-tailed statistical test. Thus, α for the test is the tabulated value for .05, and we will reject the null hypothesis if $r_s \leq -.643$.

The calculated value of the test statistic, $r_s = -.714$, is less than the critical value for $\alpha = .05$. Hence, the null hypothesis would be rejected at the $\alpha = .05$ level of significance. It appears that some agreement does exist between the judge's rankings and the test scores. However, it should be noted that this agreement could exist when *neither* provides an adequate yardstick for measuring teaching ability. For example, the association could exist if both the judge and those who constructed the teacher's examination had a completely erroneous, but similar, concept of the characteristics of good teaching.

EXERCISES   Basic Techniques

14.41   Give the rejection region for a test to detect positive rank correlation if the number of pairs of ranks is 16 and
a. $\alpha = .05$ b. $\alpha = .01$

14.42   Give the rejection region for a test to detect negative rank correlation if the number of pairs of ranks is 12 and
a. $\alpha = .05$ b. $\alpha = .01$

14.43   Give the rejection region for a test to detect rank correlation if the number of pairs of ranks is 25 and
a. $\alpha = .05$ b. $\alpha = .01$

14.44   The following paired observations were obtained on two variables x and y.

x	1.2	.8	2.1	3.5	2.7	1.5
y	1.0	1.3	.1	$-.8$	$-.2$	.6

a.   Calculate the Spearman's rank correlation coefficient r_s.

b.   Do the data present sufficient evidence to indicate a correlation between x and y? Test using $\alpha = .05$.

Applications

14.45   A political scientist wished to examine the relationship of the voter image of a conservative political candidate and the distance between the residences of the voter and the candidate. Each of 12 voters rated the candidate on a scale of 1 to 20. The data are as follows:

Voter	Rating	Distance
1	12	75
2	7	165
3	5	300
4	19	15
5	17	180
6	12	240
7	9	120
8	18	60
9	3	230
10	8	200
11	15	130
12	4	130

a.   Calculate the Spearman rank correlation coefficient r_s.

b.   Do these data provide sufficient evidence to indicate a negative correlation between rating and distance?

14.46   Is the number of years of competitive running experience related to a runner's distance running performance? The data on nine runners, obtained from the study by Scott Powers, et al., Exercise 10.9, is reproduced below ("Ventilatory Threshold, Running Economy and Distance Running Performance of Trained Athletes," *Research Quarterly for Exercise and Sport* 54, no. 3 [1983]).

Runner	Years of competitive running	10 kilometers finish time (min)
1	9	33.15
2	13	33.33
3	5	33.50
4	7	33.55
5	12	33.73
6	6	33.86
7	4	33.90
8	5	34.15
9	3	34.90

a.   Calculate the rank correlation coefficient between years of competitive running x and a runner's finish time y in the 10-kilometer race.

b.   Do the data provide sufficient evidence to indicate a rank correlation between y and x? Test using $\alpha = .10$.

14.47   In addition to engineering tests, *Tennis Industry Magazine* in Florida, a leading tennis business magazine, obtained a subjective evaluation of racket vibration, stiffness, and torque (twist) by 29 male and female tennis professionals. A summary of the subjective ratings of the professionals for each racket is shown in the table. (R. Gillen, ed., "Equipment Preview 1976: How to Pick the Right Racquet for Your Game," *Tennis USA* [February 1976] 87–91, 99). Calculate r_s for the following pairs of variables.

Racket	Vibration center (sweet-spot) hits	Vibration off-center hits	Flex	Torque
Wilson T-4000	55	71	46	71
Davis Classic	59	71	43	65
Yamaha YFG-30	55	61	59	65
P.D.P. Fiberstaff	46	55	74	54
Aldila Cannon	50	61	66	53
Dunlop-Fort	56	65	60	60
Garcia "Pro" Royal	59	68	51	66
Head Comp. II	54	66	63	59
Yonex T-7500	51	53	69	58
F. Willys-Devastator	76	75	73	73
Wilson Kramer	58	66	55	65
Garcia 240	60	65	70	63

a.   Flex and torque.
b.   Vibration from sweet-spot hits and vibration from off-center hits.
c.   Torque and vibration from off-center hits.
Test to determine whether the data provide sufficient evidence to indicate a correlation between each of the pairs of variables in parts (a), (b), and (c). Test using a value of α near .05.

14.48   A school principal suspected that a teacher's attitude toward a first grader depended on his original judgment of the child's ability. The principal also suspected that much of that judgment was based on the first grader's IQ score, which was usually known to the teacher. After three weeks of teaching, a teacher was asked to rank the nine children in his class from 1 (highest) to 9 (lowest) as to his opinion of their ability. Calculate r_s for the following teacher–IQ ranks:

Teacher	1	2	3	4	5	6	7	8	9
IQ	3	1	2	4	5	7	9	6	8

14.49   Refer to Exercise 14.48. Do the data provide sufficient evidence to indicate a positive correlation between the teacher's ranks and the ranks of the IQs? Use $\alpha = .05$.

14.50   Two art critics each ranked ten paintings by contemporary (but anonymous) artists in accordance with their appeal to the respective critics. The ratings are shown in the table. Do the critics seem to agree on their ratings of contemporary art? That is, do the data provide sufficient evidence to indicate a positive correlation between critics A and B? Test by using a value of α near .05.

Paintings	Critic A	Critic B
1	6	5
2	4	6
3	9	10
4	1	2
5	2	3
6	7	8
7	3	1
8	8	7
9	5	4
10	10	9

14.51 An experiment was conducted to study the relationship between the ratings of a tobacco leaf grader and the moisture content of the corresponding tobacco leaves. Twelve leaves were rated by the grader on a scale of 1 to 10, and corresponding readings of moisture content were made. The data are as follows:

Leaf	Grader's rating	Moisture content
1	9	.22
2	6	.16
3	7	.17
4	7	.14
5	5	.12
6	8	.19
7	2	.10
8	6	.12
9	1	.05
10	10	.20
11	9	.16
12	3	.09

Calculate r_s. Do the data provide sufficient evidence to indicate an association between the grader's ratings and the moisture content of the leaves?

14.52 Is the national poverty rate related to the per capita expenditure (adjusted for inflation) on poverty programs? The data shown below were contained in an article in the *Boston Herald* (June 24, 1985) that described a study by Professors Lowell Galloway and Richard Vedder of Ohio State University. Based on this and other data, the article stated that Galloway and Vedder contended that public aid increases rather than decreases the number of people on the poverty rolls.

a. If the poverty rate is correlated with the per capita expenditures on federal public aid to reduce poverty, should the correlation be positive or negative?

b. Calculate the rank correlation coefficient between the annual poverty rate and the per capita expenditure (in 1980 dollars) on federal public aid to reduce poverty.

c. Test for the rank correlation described in part (a) using $\alpha = .05$.

POVERTY-WELFARE TRENDS 1953–83				
Year	Poverty Rate %	Poverty Number (Millions)	Fed. Pub. Aid Real Per Capita (Constant 1980 Dollars)	National Income Real Per Capita (Constant 1980 Dollars)
Steady progress against poverty				
1953	25.8%	41.3	$26.31	$5833
1958	22.5%	39.3	29.88	5985
1963	19.5%	36.4	42.75	6971
1968	12.8%	25.4	76.74	8551
Progress slowed				
1973	11.1%	23.0	158.25	9506
Progress reversed				
1978	11.4%	24.5	227.19	9986
1983	15.2%	35.3	197.54	9331

Sources: Census Bureau; Commerce Department.

14.53 Refer to Exercise 14.52.
   a. Calculate the rank correlation coefficient between the annual number of persons on the poverty rolls and the annual per capita expenditure on federal public aid to reduce poverty.
   b. Do the data provide sufficient evidence to indicate a correlation (either positive or negative) between the annual number of persons on the poverty rolls and the annual per capita expenditure on federal public aid to reduce poverty? Test using $\alpha = .10$.
   c. What are the practical conclusions to be derived from the test in part (b)? Do the results of the test support the theory that increased federal aid increases rather than decreases the number of people on poverty rolls? Explain.

▷ **14.9 SUMMARY**

The nonparametric statistical tests presented in the preceding pages represent only a few of the many nonparametric statistical methods of inference available. A much larger collection of nonparametric test procedures, along with worked examples, are given by Siegel and Castellan (1988) and Conover (1980) (see references).

   We have indicated that nonparametric statistical procedures are particularly useful when the experimental observations can be ordered but cannot be measured on a quantitative scale. Parametric statistical procedures usually cannot be applied to this type of data; hence, all inferential procedures must be based on nonparametric methods.

   A second application of nonparametric statistical methods is in testing hypotheses associated with populations of quantitative data when uncertainty exists concerning the satisfaction of assumptions about the form of the population distributions. Just how useful are nonparametric methods for this situation?

In this chapter we presented a number of useful nonparametric methods along with illustrations of their applications. The Mann-Whitney U test can be used to compare the locations of two population frequency distributions when the observations can be ranked according to their relative magnitudes and when the samples have been randomly and independently selected from the two populations. The Kruskal-Wallis H test provides similar methodology for comparing the locations of three or more population frequency distributions. The simplest nonparametric test—the sign test—provides a rapid procedure for comparing the locations of two population distributions when the observations have been independently selected in matched pairs. If the differences between pairs can be ranked according to their relative magnitudes, you can use the Wilcoxon signed-rank test for comparing the two populations. This latter test utilizes more sample information than the sign test and consequently is more likely to detect a difference in location if a difference exists. The Friedman F_r test enables us to extend this comparison to more than two population distributions when the data have been collected in matched sets, that is, according to a randomized block design. Finally, we presented a nonparametric method—Spearman's rank correlation test—for testing the correlation between the ranks of two variables when the observations associated with each variable can be ranked according to their relative magnitudes.

▷ MINITAB COMMANDS

> **FRIEDMAN** data in **C**, treat. in **C**, block in **C** [put resids in **C** [fits in **C**]]

> **KRUSKAL-WALLIS** test for data in **C**, levels in **C**

> **MANN-WHITNEY** [confidence = **K**] on **C C**
> **ALTERNATIVE = K**

> **WTEST** [of center = **K**] on **C...C**
> **ALTERNATIVE = K**

REFERENCES

Conover, W. J. *Practical Nonparametric Statistics.* 2d ed. New York: Wiley, 1980.

Daniel, W. W. *Applied Nonparametric Statistics.* 2d ed. Boston: PWS–KENT, 1990.

Friedman, M. "The Use of Ranks to Avoid the Assumption of Normality Implicit in the Analysis of Variance." *Journal of the American Statistical Association* 32(1937).

Hollander, M., and Wolfe, D. A. *Nonparametric Statistical Methods.* New York: Wiley, 1973.

Kruskal, W. H. "A Nonparametric Test for the Several Sample Problem." *Annals of Mathematical Statistics* 23(1952).

Mann, H. B., and Whitney, D. R. "On a Test of Whether One of Two Random Variables is Stochastically Larger than the Other." *Annals of Mathematical Statistics* 18(1947): 50–60.

Ryan, T. A.; Joiner, B. L.; and Ryan, B. F. *Minitab Student Handbook*. 2d ed. Boston: Duxbury Press, 1985.

Seigel, S., and Castellan, N. J. *Nonparametric Statistics*. 2d ed. New York: McGraw–Hill, 1988.

Wilcoxon, F. "Individual Comparisons by Ranking Methods." *Biometrics* 1(1945): 80–83.

SUPPLEMENTARY EXERCISES

14.54 A psychological experiment was conducted to compare the lengths of response time (in seconds) for two different stimuli. In order to remove natural person-to-person variability in the responses, both stimuli were applied to each of nine subjects, thus permitting an analysis of the difference between stimuli *within* each person.

Subject	Stimulus 1	Stimulus 2
1	9.4	10.3
2	7.8	8.9
3	5.6	4.1
4	12.1	14.7
5	6.9	8.7
6	4.2	7.1
7	8.8	11.3
8	7.7	5.2
9	6.4	7.8

a. Use the sign test to determine whether sufficient evidence exists to indicate a difference in mean response for the two stimuli. Use a rejection region for which $\alpha \le .05$.

b. Test the hypothesis of no difference in mean response using Student's t test.

14.55 Refer to Exercise 14.54. Test the hypothesis that no difference exists in the distributions of responses for the two stimuli, using the Wilcoxon signed-rank test. Use a rejection region for which α is as near as possible to the α achieved in Exercise 14.54(a).

14.56 To compare two junior high schools A and B in academic effectiveness, an experiment was designed requiring the use of ten sets of identical twins, each twin having just completed the sixth grade. In each case, the twins in the same set had obtained their schooling in the same classrooms at each grade level. One child was selected at random from each pair of twins and assigned to school A. The remaining children were sent to school B. Near the end of the ninth grade, a certain achievement test was given to each child in the experiment. The results are shown in the following table.

School	Twin pair									
	1	2	3	4	5	6	7	8	9	10
A	67	80	65	70	86	50	63	81	86	60
B	39	75	69	55	74	52	56	72	89	47

a. Test (using the sign test) the hypothesis that the two schools are the same in academic effectiveness, as measured by scores on the achievement test, against the alternative that the schools are not equally effective. Use a level of significance as near as possible to $\alpha = .05$.

b.   Suppose it was known that junior high school A had a superior faculty and better learning facilities. Test the hypothesis of equal academic effectiveness against the alternative that school A is superior. Use a level of significance as near as possible to $\alpha = .05$.

14.57   Refer to Exercise 14.56. What answers are obtained if Wilcoxon's signed-rank test is used in analyzing the data? Compare with the answers to Exercise 14.56.

14.58   The coded values for a measure of brightness in paper (light reflectivity), prepared by two different processes, are given below for samples of nine observations drawn randomly from each of the two processes:

Process	Brightness								
A	6.1	9.2	8.7	8.9	7.6	7.1	9.5	8.3	9.0
B	9.1	8.2	8.6	6.9	7.5	7.9	8.3	7.8	8.9

Do the data present sufficient evidence ($\alpha = .10$) to indicate a difference in the populations of brightness measurements for the two processes?
a.   Use the Mann-Whitney U test.
b.   Use the Student's t test.

14.59   If (as in the case of measurements produced by two well-calibrated measuring instruments) the means of two populations are equal, it is possible to use the Mann-Whitney U statistic for testing hypotheses concerning the population variances as follows:

1.   Rank the combined sample.

2.   Number the ranked observations "from the outside in"; that is, number the smallest observation 1; the largest, 2; the next-to-smallest, 3; the next-to-largest, 4; and so on. This final sequence of numbers induces an ordering on the symbols A (population A items) and B (population B items). If $\sigma_A^2 > \sigma_B^2$, one would expect to find a preponderance of A's near the first of the sequences, and thus a relatively small "sum of ranks" for the A observations.

a.   Given the following measurements produced by well-calibrated precision instruments A and B, test at near the $\alpha = .05$ level to determine whether the more expensive instrument B is more precise than A. (Note that this would simply a one-tailed test.) Use the Mann-Whitney U test.

Instrument A	Instrument B
1060.21	1060.24
1060.34	1060.28
1060.27	1060.32
1060.36	1060.30
1060.40	

b.   Test, using the F statistic of Section 9.7.

14.60   An experiment was conducted to compare the tenderness of meat cuts subjected to two different meat tenderizers A and B. To reduce the effect of extraneous variables, the data were paired by the specific meat cut, by applying the tenderizers to two cuts taken from the same steer, by cooking paired cuts together, and by using a single judge for each pair. After cooking, each cut was rated by a judge on a scale of 1 to 10, with 10 corresponding to the most tender condition of the cooked meat. The data, shown for a single judge, are given below. Do the data provide sufficient evidence to indicate that one of the two tenderizers tends to receive higher ratings than the other? Use a value of α near .05. [*Note:* $(1/2)^8 = .003906$.] Would a Student's t test be appropriate for analyzing these data? Explain.

Cut	Tenderizer	
---	A	B
Shoulder roast	5	7
Chuck roast	6	5
Rib steak	8	9
Brisket	4	5
Club steak	9	9
Round steak	3	5
Rump roast	7	6
Sirloin steak	8	8
Sirloin tip steak	8	9
T-bone steak	9	10

14.61 A large corporation selects college graduates for employment, using both interviews and a psychological achievement test. Interviews conducted at the home office of the company were far more expensive than the tests that could be conducted on campus. Consequently, the personnel office was interested in determining whether the test scores were correlated with interview ratings and whether tests could be substituted for interviews. The idea was not to eliminate interviews but to reduce their number. To determine whether correlation was present, ten prospects were ranked during interviews, and tested. The paired scores are as follows:

Subject	Interview rank	Test score
1	8	74
2	5	81
3	10	66
4	3	83
5	6	66
6	1	94
7	4	96
8	7	70
9	9	61
10	2	86

Calculate the Spearman rank correlation coefficient r_s. Rank 1 is assigned to the candidate judged to be the best.

14.62 Refer to Exercise 16.61. Do the data present sufficient evidence to indicate that the correlation between interview rankings and test scores is less than 0? If this evidence does exist, can we say that tests could be used to reduce the number of interviews?

14.63 A comparison of reaction times for two different stimuli in a psychological word-association experiment produced the following results when applied to a random sample of

16 people:

Stimulus	Reaction time (sec)							
1	1	3	2	1	2	1	3	2
2	4	2	3	3	1	2	3	3

Do the data present sufficient evidence to indicate a difference in mean reaction time for the two stimuli? Use the Mann-Whitney U statistic and test using $\alpha = .05$. (*Note:* This test was conducted using Student's t in the exercises for Chapter 9. Compare your results.)

14.64 The following table gives the scores of a group of 15 students in mathematics and art.

Student	Math	Art		Student	Math	Art
1	22	53		8	60	71
2	37	68		9	62	55
3	36	42		10	65	74
4	38	49		11	66	68
5	42	51		12	56	64
6	58	65		13	66	67
7	58	51		14	67	73
				15	62	65

Use Wilcoxon's signed-rank test to determine if the median scores for these students differ significantly for the two subjects.

14.65 Refer to Exercise 14.64. Compute Spearman's rank correlation coefficient for these data and test $H_0: \rho_s = 0$ at the 10% level of significance.

14.66 Exercise 13.59 presents an analysis of variance of the yields of five different varieties of wheat, observed on one plot each at each of six different locations. The data from this randomized block design are shown below.

| Varieties | Location | | | | | |
	1	2	3	4	5	6
A	35.3	31.0	32.7	36.8	37.2	33.1
B	30.7	32.2	31.4	31.7	35.0	32.7
C	38.2	33.4	33.6	37.1	37.3	38.2
D	34.9	36.1	35.2	38.3	40.2	36.0
E	32.4	28.9	29.2	30.7	33.9	32.1

a. Use the Friedman F_r test to determine whether the data provide sufficient evidence to indicate a difference in the levels of yield for the five different varieties of wheat. Test using $\alpha = .05$.

b. Exercise 13.59 presents a computer printout of the analysis of variance for comparing the mean yield for the five varieties of wheat. How do the results of the analysis-of-variance F test compare with the Friedman F_r test in part (a)? Explain.

14.67 In Exercise 13.55 we compared the number of sales per trainee after completing one of four different sales training programs. Six trainees completed training program 1, eight completed 2, and so on. The number of sales per trainee is shown on page 633.

		Training program		
	I	2	3	4
	78	99	74	81
	84	86	87	63
	86	90	80	71
	92	93	83	65
	69	94	78	86
	73	85		79
		97		73
		91		70
Total	482	735	402	588

a. Do the data present sufficient evidence to indicate that the location of the distribution of number of sales per trainee differs from one training program to another? Test using the Kruskal-Wallis H test with $\alpha = .05$.

b. How do the test results in part (a) compare with the results of the analysis-of-variance F test in Exercise 13.55?

14.68 In Exercise 13.58 we performed an analysis of variance to compare the mean level of effluents in water at four different industrial plants. Five samples of the liquid waste (lb/gal) were taken at the output of each of the four industrial plants. The data are shown below.

Plant	Polluting effluents (lb/gal of waste)				
A	1.65	1.72	1.50	1.37	1.60
B	1.70	1.85	1.46	2.05	1.80
C	1.40	1.75	1.38	1.65	1.55
D	2.10	1.95	1.65	1.88	2.00

a. Do the data present sufficient evidence to indicate a difference in the level of pollutants for the four different industrial plants? Test using the Kruskal-Wallis H test with $\alpha = .05$.

b. Find the approximate p-value for the test and interpret its value.

c. Compare the test results in part (a) with the analysis-of-variance test in Exercise 13.58. Do the results agree? Explain.

14.69 In comparing the 1987 and 1988 median prices of a single-family home in 59 metropolitan statistical areas as defined by the U.S. Office of Budget and Management (*Statistical Abstract of the U.S.*, 1989, p. 799), 41 areas showed an increase in median price, 16 showed a decrease, and 2 showed no change. Do these data contain sufficient evidence to conclude that the median price of a single-family dwelling has increased significantly from 1987 to 1988? Use $\alpha = .05$. What is the observed significance level of this test?

14.70 Scientists have shown that a newly developed vaccine can shield rhesus monkeys from infection by a virus closely related to the AIDS-causing human immunodeficiency virus (HIV). In their work, Ronald C. Resrosiers and his colleagues of the New England Regional Primate Research Center gave each of the $n = 6$ rhesus monkeys five inoculations with the simian immunodeficiency virus (SIV) vaccine. One week after the last vaccination, each monkey received an injection of live SIV. Two of the six vaccinated monkeys showed no evidence of SIV infection for as long as a year and a half after the SIV injection (*Proceedings* *of the National Academy of Sciences*, Vol. 86, No. 16: reported in *Science News*, Vol. 136, August 19, 1989, p. 116). Scientists were able to isolate the SIV virus from the other four vaccinated monkeys, although these animals showed no sign of the disease. Does this information contain sufficient evidence to indicate that the vaccine is effective in protecting monkeys from SIV? Use $\alpha = .10$.

14.71   In the traditional American Western, you could immediately recognize the bad guys by their black hats. Mark Frank and Thomas Gilovich ("The Dark Side of Self- and Social Perception: Black Uniforms and Aggression in Professional Sports," *Journal of Personality and Social Psychology* (1988), Vol 54, No. 1, 74–85. Copyright 1988 by the American Psychological Association. Reprinted by permission.) ran an experiment to determine whether the color of a person's clothing has a significant effect on others' perception of his or her behavior. As part of this experiment, they had 25 subjects rate the uniforms of all National Football League (NFL) teams on several seven-point scales. Five of the 28 teams have uniforms that were considered "black"—namely, the Pittsburgh Steelers, the New Orleans Saints, the Los Angeles Raiders, the Cincinnati Bengals, and Chicago Bears. The resulting "malevolence indexes" are given in the following table.

Malevolence ratings of the uniforms of professional football teams

Football team	Rating	Football team	Rating
LA RAIDERS	5.10	Minnesota	3.90
PITTSBURGH	5.00	Atlanta	3.87
CINCINNATI	4.97	San Francisco	3.83
NEW ORLEANS	4.83	Indianapolis	3.83
CHICAGO	4.68	Seattle	3.82
Kansas City	4.58	Denver	3.80
Washington	4.40	Tampa Bay	3.77
St. Louis	4.27	New England	3.60
NY Jets	4.12	Buffalo	3.53
LA Rams	4.10	Detroit	3.38
Cleveland	4.05	NY Giants	3.27
San Diego	4.05	Dallas	3.15
Green Bay	4.00	Houston	2.88
Philadelphia	3.97	Miami	2.80

Note: Teams in boldface capitals are those with black uniforms. The malevolence ratings represent the average rating of three semantic differential scales: good/bad, timid/aggressive, and nice/mean.

Use the large-sample Mann–Whitney U test to determine whether there is sufficient information to conclude that "black" uniforms receive higher malevolence scores. Use $\alpha = .05$.

DOES IT PAY TO SAVE?
A SECOND LOOK

You may recall the case study of Chapter 10 that discussed Dr. Srully Blotnick's theory concerning the relationship between the percentage of total national income saved y and the approximate percentage of investment income x that must be paid in taxes. We asked whether the national savings rate y was negatively correlated with the investment income tax rate x. Then we warned that it was quite possible that the data did not satisfy the assumptions required for the test for linear correlation in Chapter 10. The data from the case study of Chapter 10 are reproduced below.

A comparison of personal savings rates and taxation rates on investment income

Country	y Personal savings rate (%)	x Investment income tax liability (%)
Italy	23.1	6.4
Japan	21.5	14.4
France	17.2	7.3
West Germany	14.5	11.8
United Kingdom	12.2	32.5
Canada	10.3	30.0
Sweden	9.1	52.7
United States	6.3	33.5

Source: New York Stock Exchange with assistance of Price Waterhouse.

1. Use a nonparametric analysis to test for negative correlation between the national savings rate and the investment income tax rate.

2. Do the results agree with those obtained in the case study of Chapter 10?

NATIONAL INSTITUTES OF HEALTH BLOOD PRESSURE (DIASTOLIC AND SYSTOLIC) DATA SET FOR 965 MALES AND 945 FEMALES AGES 15 THROUGH 20

NIH data: 965 males, ages 15 through 20

OBS	AGE	SYSBP	DIASBP	OBS	AGE	SYSBP	DIASBP	OBS	AGE	SYSBP	DIASBP
1	16	116	74	51	15	110	74	101	19	170	52
2	16	110	82	52	18	144	60	102	15	114	70
3	18	110	68	53	15	100	70	103	19	106	72
4	16	110	74	54	19	150	84	104	16	116	74
5	17	116	72	55	16	112	80	105	15	138	72
6	15	116	68	56	15	116	70	106	15	120	68
7	17	100	54	57	15	134	74	107	17	120	78
8	18	110	66	58	15	130	80	108	15	134	78
9	19	112	48	59	15	120	90	109	16	140	88
10	16	98	42	60	18	120	70	110	19	134	86
11	19	116	76	61	16	114	74	111	20	116	74
12	18	116	50	62	17	108	90	112	18	130	58
13	19	124	70	63	16	120	84	113	16	106	54
14	20	120	66	64	19	125	86	114	15	118	72
15	16	98	64	65	15	120	80	115	18	132	80
16	20	106	70	66	18	118	80	116	16	104	76
17	19	110	70	67	16	126	70	117	16	140	66
18	19	118	78	68	17	124	78	118	20	168	106
19	18	136	84	69	17	120	80	119	16	108	60
20	16	124	74	70	16	128	78	120	17	118	78
21	16	140	100	71	16	114	70	121	17	115	70
22	17	100	80	72	15	140	60	122	15	128	52
23	18	124	76	73	20	110	86	123	15	80	60
24	17	120	78	74	20	126	70	124	19	125	74
25	17	130	74	75	20	130	80	125	15	115	80
26	18	150	110	76	20	166	90	126	19	130	90
27	16	98	60	77	15	112	66	127	15	115	70
28	19	130	80	78	19	108	64	128	15	110	50
29	16	110	64	79	20	124	76	129	18	110	65
30	20	114	70	80	17	122	84	130	16	120	62
31	19	120	84	81	19	122	80	131	16	104	60
32	16	108	68	82	19	116	82	132	16	105	65
33	15	102	66	83	20	122	76	133	17	130	88
34	16	118	68	84	17	104	74	134	16	105	70
35	15	110	·	85	18	128	76	135	16	124	60
36	16	86	64	86	20	104	72	136	20	122	78
37	17	116	74	87	16	132	80	137	15	116	66
38	16	140	56	88	18	110	70	138	18	144	78
39	16	118	46	89	18	105	75	139	15	120	60
40	18	100	70	90	19	110	85	140	17	138	86
41	15	112	76	91	15	128	68	141	15	128	74
42	15	114	70	92	15	135	75	142	15	112	68
43	17	116	62	93	15	128	76	143	17	110	72
44	17	108	84	94	19	108	70	144	19	108	76
45	15	112	66	95	16	90	60	145	18	120	78
46	20	134	76	96	16	110	60	146	16	124	78
47	20	126	76	97	17	118	65	147	17	100	56
48	20	210	110	98	19	120	80	148	17	110	74
49	20	102	78	99	18	110	85	149	18	120	78
50	17	114	88	100	20	115	45	150	17	110	80

NIH data: 965 males, ages 15 through 20 (continued)

OBS	AGE	SYSBP	DIASBP	OBS	AGE	SYSBP	DIASBP	OBS	AGE	SYSBP	DIASBP
151	17	112	66	201	15	108	56	251	17	134	84
152	18	116	76	202	15	110	72	252	15	112	76
153	19	132	78	203	17	116	80	253	18	110	74
154	20	118	76	204	19	90	72	254	18	102	66
155	20	146	70	205	15	102	78	255	19	106	60
156	20	134	80	206	15	106	68	256	18	102	66
157	20	144	70	207	15	96	56	257	17	126	66
158	15	114	78	208	16	110	84	258	15	130	90
159	15	120	72	209	20	118	86	259	16	118	78
160	17	136	90	210	18	112	62	260	16	122	60
161	20	130	90	211	15	102	70	261	15	106	70
162	16	130	88	212	17	120	64	262	17	130	80
163	16	122	74	213	18	110	70	263	18	116	76
164	16	134	84	214	18	128	88	264	15	110	60
165	19	140	84	215	16	128	80	265	15	118	80
166	16	104	76	216	16	100	70	266	19	136	68
167	17	124	70	217	19	126	74	267	18	114	84
168	19	102	86	218	17	124	84	268	18	106	64
169	15	122	56	219	15	120	70	269	15	114	76
170	17	118	50	220	16	124	72	270	20	100	70
171	15	116	84	221	17	130	96	271	20	106	74
172	19	126	72	222	16	136	66	272	20	114	74
173	17	114	78	223	16	130	80	273	20	108	72
174	19	128	78	224	18	130	94	274	19	130	88
175	18	120	70	225	19	142	70	275	16	150	94
176	17	106	70	226	17	114	72	276	19	120	70
177	19	132	96	227	15	106	72	277	19	126	86
178	20	134	80	228	15	98	64	278	17	120	76
179	20	128	92	229	15	98	60	279	15	112	70
180	16	140	60	230	15	130	74	280	18	132	78
181	19	130	68	231	16	112	70	281	18	144	86
182	19	134	84	232	16	106	66	282	18	138	94
183	19	120	80	233	15	118	78	283	16	120	74
184	19	118	84	234	19	116	80	284	17	130	70
185	17	122	74	235	20	112	86	285	20	124	74
186	17	124	70	236	20	150	70	286	20	156	90
187	19	142	90	237	15	120	70	287	20	114	76
188	18	140	100	238	19	130	90	288	17	144	80
189	15	146	90	239	19	140	90	289	15	114	46
190	18	120	80	240	15	120	80	290	18	102	70
191	19	128	94	241	17	116	78	291	15	100	70
192	16	126	86	242	17	116	80	292	18	110	80
193	17	120	70	243	17	120	78	293	15	114	72
194	15	108	76	244	16	154	100	294	19	110	66
195	18	130	94	245	19	128	76	295	18	126	88
196	17	120	74	246	15	124	84	296	16	118	86
197	18	140	90	247	20	120	78	297	19	130	96
198	16	108	76	248	20	120	96	298	16	120	80
199	17	116	58	249	18	122	80	299	18	104	66
200	17	110	60	250	17	94	56	300	19	110	88

NIH data: 965 males, ages 15 through 20 (continued)

OBS	AGE	SYSBP	DIASBP	OBS	AGE	SYSBP	DIASBP	OBS	AGE	SYSBP	DIASBP
301	16	154	66	351	18	112	68	401	16	90	50
302	18	96	64	352	16	125	92	402	16	140	70
303	17	112	66	353	19	108	70	403	16	120	80
304	15	136	82	354	19	118	58	404	15	140	76
305	17	108	78	355	20	130	78	405	19	120	74
306	17	128	80	356	20	130	78	406	17	132	74
307	18	136	86	357	20	145	95	407	16	142	86
308	15	114	76	358	18	140	90	408	15	·	80
309	16	134	66	359	18	134	72	409	20	150	70
310	16	110	76	360	17	110	64	410	20	110	60
311	19	146	66	361	16	90	70	411	20	130	86
312	16	118	78	362	17	140	100	412	20	90	58
313	16	114	74	363	16	120	82	413	17	130	80
314	17	122	78	364	19	104	80	414	17	130	86
315	18	126	84	365	19	132	88	415	15	106	84
316	16	112	72	366	18	130	80	416	19	126	80
317	15	115	65	367	20	100	70	417	19	106	76
318	19	130	70	368	16	126	82	418	16	100	40
319	18	108	82	369	17	116	70	419	18	110	60
320	15	80	50	370	16	106	68	420	16	114	68
321	15	112	58	371	19	134	98	421	17	112	80
322	19	115	78	372	19	122	76	422	18	94	56
323	16	108	58	373	20	118	74	423	18	108	64
324	17	115	92	374	16	96	60	424	20	112	76
325	16	110	60	375	16	126	88	425	17	100	60
326	18	108	70	376	16	130	86	426	18	120	70
327	15	100	60	377	18	110	60	427	15	140	80
328	20	120	82	378	18	108	80	428	18	122	76
329	16	118	66	379	18	140	86	429	15	120	70
330	19	104	72	380	20	136	62	430	19	120	70
331	19	142	94	381	20	104	70	431	17	118	86
332	16	126	88	382	20	126	68	432	18	130	80
333	18	104	84	383	15	110	94	433	19	130	80
334	18	106	72	384	15	90	60	434	16	120	70
335	16	114	76	385	19	90	60	435	15	118	70
336	18	114	76	386	15	140	70	436	15	110	60
337	16	100	70	387	15	120	80	437	18	110	70
338	18	120	76	388	19	108	70	438	17	110	70
339	17	130	·	389	19	130	80	439	18	120	70
340	16	108	80	390	19	160	100	440	19	120	70
341	19	120	80	391	15	90	60	441	19	136	70
342	16	130	80	392	15	114	74	442	19	120	80
343	20	140	80	393	15	90	60	443	17	120	70
344	20	116	80	394	18	118	60	444	18	104	76
345	17	128	80	395	16	100	68	445	16	110	70
346	15	128	80	396	17	100	70	446	17	110	70
347	16	140	100	397	16	120	74	447	17	112	70
348	18	148	90	398	17	100	70	448	18	110	82
349	18	115	90	399	16	120	70	449	19	134	68
350	17	140	100	400	20	145	90	450	20	162	86

NIH data: 965 males, ages 15 through 20 (continued)

OBS	AGE	SYSBP	DIASBP	OBS	AGE	SYSBP	DIASBP	OBS	AGE	SYSBP	DIASBP
451	20	120	70	501	18	112	95	551	16	110	65
452	20	110	60	502	19	98	65	552	16	120	75
453	18	140	80	503	18	110	70	553	17	105	70
454	15	118	64	504	17	140	55	554	15	112	78
455	19	150	78	505	19	150	102	555	19	110	72
456	16	124	92	506	20	110	80	556	20	100	74
457	18	112	80	507	20	120	60	557	17	120	86
458	16	110	70	508	20	128	75	558	17	104	68
459	17	122	72	509	20	125	90	559	17	150	90
460	15	110	64	510	15	130	80	560	16	120	90
461	19	98	66	511	15	100	60	561	15	130	80
462	17	112	65	512	19	110	70	562	17	130	100
463	15	104	60	513	17	100	60	563	20	120	82
464	20	105	55	514	19	120	70	564	17	132	68
465	19	130	90	515	15	110	70	565	17	132	84
466	17	100	60	516	15	120	70	566	17	110	70
467	15	104	74	517	16	120	80	567	16	130	68
468	19	124	70	518	17	110	70	568	18	110	88
469	19	128	88	519	15	120	70	569	19	110	76
470	15	112	78	520	19	120	70	570	15	122	70
471	15	100	60	521	20	120	80	571	15	122	72
472	20	130	68	522	20	120	70	572	17	120	70
473	20	122	86	523	20	120	70	573	19	128	70
474	17	134	58	524	20	130	80	574	18	110	76
475	17	136	80	525	19	110	54	575	17	130	80
476	17	118	66	526	18	130	76	576	15	130	90
477	17	142	80	527	18	90	56	577	18	120	80
478	15	116	80	528	17	110	86	578	16	142	84
479	16	120	78	529	18	108	68	579	19	120	76
480	16	130	64	530	19	100	74	580	15	110	70
481	18	116	70	531	17	108	64	581	18	126	70
482	19	142	86	532	17	114	82	582	18	120	70
483	20	130	74	533	17	120	80	583	18	108	70
484	19	130	92	534	16	154	90	584	15	120	80
485	15	114	72	535	18	100	70	585	17	106	80
486	20	106	64	536	20	96	50	586	20	140	80
487	16	120	70	537	20	100	50	587	20	118	70
488	18	115	80	538	20	106	70	588	20	120	70
489	19	120	90	539	17	118	70	589	20	130	78
490	19	102	60	540	18	110	70	590	17	116	56
491	18	110	65	541	16	112	70	591	16	136	76
492	19	115	82	542	16	115	65	592	15	116	76
493	17	124	80	543	15	108	62	593	17	126	66
494	16	110	80	544	16	106	70	594	19	106	82
495	19	112	78	545	16	122	82	595	19	118	82
496	19	110	65	546	19	112	54	596	15	106	68
497	18	112	65	547	17	114	70	597	15	120	64
498	19	120	80	548	19	134	76	598	20	116	82
499	19	126	90	549	16	112	70	599	15	100	70
500	19	112	78	550	18	122	70	600	15	120	68

NIH data: 965 males, ages 15 through 20 (continued)

OBS	AGE	SYSBP	DIASBP	OBS	AGE	SYSBP	DIASBP	OBS	AGE	SYSBP	DIASBP
601	18	112	80	651	18	134	88	701	15	110	70
602	19	120	90	652	19	138	66	702	19	94	66
603	17	120	80	653	16	106	64	703	16	106	80
604	15	106	60	654	15	116	60	704	19	96	70
605	17	130	80	655	18	120	66	705	18	110	76
606	18	116	74	656	15	140	80	706	18	90	60
607	15	110	68	657	17	140	70	707	16	110	74
608	19	130	80	658	16	126	90	708	18	116	74
609	17	136	70	659	15	122	76	709	18	118	78
610	15	124	70	660	17	140	70	710	16	100	56
611	18	128	80	661	16	110	76	711	16	126	68
612	17	128	58	662	16	140	92	712	18	118	90
613	17	108	70	663	19	102	70	713	19	122	58
614	17	110	60	664	16	112	60	714	19	136	106
615	18	110	65	665	17	110	72	715	17	124	84
616	17	115	82	666	17	120	90	716	15	116	66
617	17	110	70	667	18	124	86	717	19	106	76
618	18	130	68	668	20	134	78	718	17	104	70
619	17	128	60	669	20	108	64	719	20	150	74
620	16	105	65	670	19	126	68	720	20	154	88
621	17	130	70	671	16	124	86	721	20	142	72
622	15	115	70	672	19	148	90	722	20	122	84
623	17	100	80	673	18	162	66	723	20	124	84
624	15	100	68	674	16	136	90	724	15	90	60
625	18	118	80	675	17	118	76	725	15	90	60
626	15	130	60	676	15	110	70	726	15	120	70
627	18	130	68	677	19	108	68	727	18	140	80
628	18	115	68	678	16	118	74	728	18	100	70
629	17	110	60	679	17	126	80	729	17	100	60
630	17	92	70	680	19	116	70	730	16	120	70
631	17	112	82	681	19	126	64	731	18	120	70
632	19	110	80	682	19	128	70	732	17	90	60
633	17	84	54	683	18	110	74	733	19	100	60
634	16	120	70	684	18	122	64	734	17	120	75
635	15	104	60	685	18	126	78	735	15	120	70
636	18	112	88	686	19	154	90	736	20	120	70
637	16	112	80	687	18	116	70	737	20	120	80
638	19	120	58	688	16	116	58	738	19	130	92
639	19	110	70	689	18	124	70	739	15	120	82
640	16	108	68	690	15	114	74	740	18	128	70
641	15	112	88	691	20	110	64	741	15	96	80
642	15	120	70	692	18	110	70	742	16	124	78
643	15	110	80	693	16	120	70	743	15	104	62
644	20	112	68	694	16	120	70	744	18	138	82
645	20	115	90	695	16	100	70	745	17	116	78
646	16	130	70	696	16	120	70	746	16	120	78
647	18	104	72	697	16	100	70	747	16	124	84
648	18	110	60	698	15	120	70	748	16	120	82
649	17	102	56	699	20	120	70	749	19	122	86
650	16	170	100	700	18	110	70	750	16	140	78

NIH data: 965 males, ages 15 through 20 (continued)

OBS	AGE	SYSBP	DIASBP	OBS	AGE	SYSBP	DIASBP	OBS	AGE	SYSBP	DIASBP
751	17	138	74	801	20	120	70	851	18	104	60
752	16	108	70	802	15	110	60	852	16	110	70
753	18	140	64	803	18	130	92	853	17	120	70
754	18	110	86	804	17	70	.	854	16	120	80
755	18	120	74	805	16	110	80	855	15	100	60
756	16	140	70	806	17	120	80	856	15	120	70
757	16	150	80	807	19	110	78	857	15	120	70
758	15	80	68	808	16	100	80	858	16	140	90
759	18	130	88	809	20	130	64	859	17	120	80
760	18	130	70	810	17	120	70	860	19	120	70
761	18	110	80	811	18	104	66	861	15	90	60
762	18	120	80	812	17	118	72	862	19	120	70
763	15	110	70	813	17	106	80	863	17	120	70
764	20	110	90	814	18	140	80	864	18	130	80
765	15	110	80	815	19	122	70	865	19	130	80
766	16	108	80	816	19	126	80	866	17	100	70
767	16	110	70	817	16	130	80	867	15	120	70
768	16	120	50	818	17	130	58	868	16	120	70
769	17	90	60	819	16	118	80	869	16	120	70
770	18	118	98	820	16	124	76	870	15	120	60
771	18	100	58	821	15	116	60	871	15	120	70
772	16	105	80	822	17	110	80	872	20	150	90
773	17	100	68	823	20	120	88	873	20	120	70
774	15	122	80	824	16	130	76	874	20	130	90
775	20	110	70	825	16	110	78	875	20	120	70
776	15	120	78	826	17	90	60	876	20	100	60
777	16	104	48	827	15	120	80	877	20	130	80
778	18	106	64	828	15	98	60	878	20	170	100
779	17	120	76	829	15	130	90	879	20	150	100
780	17	116	60	830	17	92	.	880	19	110	70
781	18	138	82	831	15	100	60	881	15	138	90
782	15	132	68	832	16	100	60	882	17	120	80
783	15	134	80	833	15	100	60	883	17	110	76
784	18	116	66	834	15	110	86	884	15	128	88
785	19	124	66	835	15	100	60	885	17	110	70
786	16	126	76	836	19	128	92	886	16	120	80
787	18	140	66	837	19	108	70	887	17	120	80
788	18	120	68	838	16	110	70	888	18	110	70
789	17	136	90	839	18	90	60	889	15	128	80
790	18	136	94	840	17	120	80	890	18	170	110
791	17	122	74	841	16	120	80	891	20	120	80
792	15	102	72	842	15	104	68	892	20	120	70
793	17	130	74	843	16	120	80	893	17	120	70
794	18	138	60	844	17	130	90	894	17	120	70
795	18	118	78	845	17	110	70	895	15	130	80
796	16	116	60	846	17	90	70	896	18	130	100
797	17	122	64	847	18	130	78	897	20	120	70
798	20	140	98	848	17	100	70	898	19	100	70
799	20	122	70	849	15	104	80	899	17	120	70
800	20	98	64	850	15	110	80	900	17	130	80

NIH data: 965 males, ages 15 through 20 (continued)/ NIH data: 945 females, ages 15 through 20

OBS	AGE	SYSBP	DIASBP	OBS	AGE	SYSBP	DIASBP	OBS	AGE	SYSBP	DIASBP
901	16	100	70	951	15	112	76	1001	16	106	66
902	15	170	70	952	16	126	90	1002	18	126	70
903	20	120	70	953	17	104	84	1003	19	110	76
904	20	110	70	954	15	94	74	1004	20	112	50
905	15	110	70	955	18	130	88	1005	20	100	60
906	17	110	80	956	19	132	94	1006	16	150	86
907	17	120	70	957	20	118	88	1007	20	102	68
908	18	120	80	958	20	130	90	1008	19	110	68
909	16	120	90	959	18	100	70	1009	19	110	62
910	19	128	70	960	19	120	70	1010	16	110	70
911	17	118	80	961	19	110	90	1011	18	110	60
912	15	110	70	962	17	120	80	1012	19	114	80
913	19	154	90	963	15	130	70	1013	16	136	76
914	17	120	76	964	17	115	78	1014	18	105	70
915	16	110	70	965	15	80	40	1015	17	126	70
916	16	120	80		FEMALES			1016	17	110	62
917	17	118	70	967	18	114	66	1017	18	96	70
918	17	120	84	968	17	106	80	1018	16	100	58
919	19	110	76	969	15	96	76	1019	20	118	60
920	19	110	80	970	15	110	64	1020	20	114	70
921	16	120	74	971	15	104	76	1021	15	100	64
922	19	120	80	972	17	104	66	1022	17	114	76
923	20	120	80	973	18	102	70	1023	15	122	64
924	18	120	80	974	15	98	60	1024	17	104	74
925	20	140	80	975	18	106	80	1025	17	114	76
926	15	120	70	976	18	86	70	1026	16	102	70
927	16	120	70	977	18	102	70	1027	19	126	68
928	17	110	70	978	19	84	60	1028	19	86	56
929	19	110	70	979	18	102	60	1029	15	94	60
930	17	120	70	980	17	96	54	1030	17	86	60
931	15	120	70	981	15	138	108	1031	16	102	70
932	19	100	70	982	17	128	90	1032	20	108	60
933	16	110	80	983	17	110	72	1033	20	124	86
934	15	120	70	984	17	104	74	1034	17	105	65
935	19	120	80	985	15	124	70	1035	17	105	70
936	20	120	70	986	16	110	70	1036	16	104	80
937	16	124	50	987	18	130	104	1037	19	92	70
938	17	110	60	988	20	138	68	1038	15	104	68
939	16	110	74	989	15	88	68	1039	18	105	70
940	17	114	50	990	16	104	64	1040	18	108	62
941	19	128	80	991	19	106	70	1041	16	78	50
942	16	116	55	992	16	104	66	1042	20	90	60
943	17	105	50	993	16	126	86	1043	18	128	84
944	16	100	70	994	15	122	56	1044	20	134	76
945	15	118	60	995	18	120	82	1045	17	114	72
946	17	115	80	996	16	110	80	1046	15	102	76
947	15	112	60	997	15	100	66	1047	15	98	54
948	17	110	60	998	15	98	64	1048	16	100	56
949	20	130	70	999	19	100	72	1049	18	112	78
950	20	109	70	1000	18	104	66	1050	18	108	80

NIH data: 945 females, ages 15 through 20 (continued)

OBS	AGE	SYSBP	DIASBP	OBS	AGE	SYSBP	DIASBP	OBS	AGE	SYSBP	DIASBP
1051	16	102	52	1101	16	140	86	1151	17	126	80
1052	15	126	66	1102	15	110	76	1152	17	114	76
1053	18	96	64	1103	15	132	74	1153	18	102	62
1054	15	118	72	1104	20	104	74	1154	18	116	64
1055	18	130	88	1105	20	130	78	1155	19	120	74
1056	17	94	68	1106	20	100	70	1156	16	110	58
1057	16	120	76	1107	15	92	60	1157	19	100	60
1058	15	122	80	1108	17	130	94	1158	19	98	66
1059	17	140	94	1109	19	116	76	1159	17	118	76
1060	19	95	70	1110	17	126	88	1160	18	124	80
1061	16	124	78	1111	19	116	74	1161	20	116	76
1062	15	102	70	1112	16	110	62	1162	20	118	76
1063	18	100	60	1113	16	104	70	1163	16	122	78
1064	16	94	68	1114	19	110	68	1164	19	120	86
1065	16	110	75	1115	17	98	68	1165	16	114	78
1066	18	105	50	1116	15	110	80	1166	19	110	68
1067	19	118	60	1117	15	108	70	1167	19	110	82
1068	20	90	68	1118	17	124	80	1168	20	112	68
1069	16	110	88	1119	19	110	70	1169	20	104	60
1070	16	120	70	1120	19	112	70	1170	20	116	84
1071	17	95	70	1121	17	128	70	1171	20	94	70
1072	17	100	80	1122	15	122	68	1172	20	114	82
1073	15	95	68	1123	16	100	70	1173	18	98	68
1074	17	110	70	1124	20	110	70	1174	19	114	78
1075	17	102	70	1125	17	126	68	1175	19	124	72
1076	15	100	65	1126	18	106	82	1176	18	106	90
1077	16	88	60	1127	19	116	86	1177	16	114	80
1078	18	80	48	1128	16	120	74	1178	18	140	80
1079	19	110	70	1129	17	112	86	1179	17	130	80
1080	20	105	55	1130	17	104	76	1180	19	144	80
1081	20	102	70	1131	16	118	66	1181	18	124	70
1082	17	100	60	1132	15	126	76	1182	16	116	78
1083	16	114	68	1133	20	114	86	1183	15	110	70
1084	18	124	78	1134	20	110	70	1184	19	114	80
1085	15	124	76	1135	18	102	64	1185	15	126	74
1086	17	130	68	1136	15	90	70	1186	15	120	74
1087	18	140	76	1137	19	100	60	1187	15	112	70
1088	19	122	80	1138	18	110	70	1188	20	120	78
1089	19	102	72	1139	16	110	78	1189	16	104	66
1090	16	120	74	1140	17	98	70	1190	19	110	60
1091	18	96	70	1141	16	110	70	1191	17	110	76
1092	20	100	66	1142	15	100	70	1192	17	94	66
1093	15	110	70	1143	19	130	80	1193	19	118	76
1094	16	120	70	1144	17	112	70	1194	15	100	72
1095	15	90	64	1145	16	120	73	1195	18	104	74
1096	19	110	76	1146	15	100	60	1196	16	110	72
1097	19	110	84	1147	20	100	70	1197	19	106	64
1098	18	112	100	1148	20	126	90	1198	19	116	80
1099	16	100	60	1149	15	108	78	1199	18	108	46
1100	16	96	60	1150	19	104	60	1200	19	100	64

NIH data: 945 females, ages 15 through 20 (continued)

OBS	AGE	SYSBP	DIASBP	OBS	AGE	SYSBP	DIASBP	OBS	AGE	SYSBP	DIASBP
1201	19	126	88	1251	18	92	58	1301	15	112	74
1202	19	88	60	1252	17	120	68	1302	18	116	76
1203	17	106	74	1253	19	112	70	1303	20	84	64
1204	19	106	76	1254	19	106	68	1304	20	110	64
1205	20	96	70	1255	17	102	70	1305	18	102	60
1206	20	112	70	1256	19	126	74	1306	15	100	80
1207	20	108	66	1257	19	108	80	1307	17	120	98
1208	20	98	64	1258	17	124	74	1308	16	110	80
1209	16	102	66	1259	18	120	80	1309	15	120	50
1210	15	106	70	1260	16	122	84	1310	16	120	92
1211	18	102	64	1261	16	110	70	1311	19	126	70
1212	17	116	84	1262	19	120	70	1312	15	130	70
1213	16	112	78	1263	15	104	64	1313	18	90	70
1214	20	108	70	1264	15	125	80	1314	18	118	70
1215	19	130	60	1265	17	122	65	1315	15	130	82
1216	16	124	78	1266	17	112	80	1316	18	115	64
1217	15	110	76	1267	15	110	80	1317	16	100	60
1218	17	116	60	1268	15	120	80	1318	17	122	80
1219	15	110	78	1269	18	98	70	1319	20	158	70
1220	16	128	82	1270	16	90	68	1320	20	98	70
1221	17	134	84	1271	15	130	95	1321	20	110	70
1222	16	112	76	1272	18	112	70	1322	20	110	80
1223	15	128	82	1273	20	125	78	1323	18	106	58
1224	20	116	94	1274	15	120	84	1324	18	96	50
1225	17	126	88	1275	19	110	78	1325	16	98	64
1226	15	142	64	1276	16	110	80	1326	18	118	80
1227	15	104	80	1277	17	108	78	1327	19	98	64
1228	16	106	70	1278	15	110	70	1328	17	118	82
1229	17	100	72	1279	18	108	76	1329	16	88	64
1230	16	104	70	1280	18	100	68	1330	16	104	56
1231	16	110	72	1281	18	108	·	1331	17	118	78
1232	16	128	76	1282	16	124	86	1332	19	104	72
1233	17	116	82	1283	15	122	92	1333	19	114	80
1234	17	130	88	1284	17	104	68	1334	16	104	76
1235	19	122	84	1285	15	110	86	1335	17	126	90
1236	17	126	74	1286	18	92	60	1336	16	94	58
1237	19	110	76	1287	15	90	60	1337	19	114	80
1238	18	130	60	1288	20	120	90	1338	19	116	70
1239	18	110	84	1289	20	90	70	1339	20	96	68
1240	17	114	64	1290	20	90	60	1340	20	118	78
1241	16	105	70	1291	20	88	70	1341	16	90	60
1242	16	100	65	1292	20	98	74	1342	17	110	70
1243	16	110	74	1293	17	104	68	1343	16	120	70
1244	16	108	60	1294	18	130	76	1344	16	124	86
1245	18	128	80	1295	17	100	70	1345	20	100	60
1246	19	118	75	1296	19	106	64	1346	17	120	74
1247	16	102	70	1297	16	114	74	1347	18	82	58
1248	15	110	54	1298	18	132	74	1348	20	110	70
1249	18	110	78	1299	16	106	80	1349	20	100	70
1250	15	110	84	1300	18	94	58	1350	17	136	84

NIH data: 945 females, ages 15 through 20 (continued)

OBS	AGE	SYSBP	DIASBP	OBS	AGE	SYSBP	DIASBP	OBS	AGE	SYSBP	DIASBP
1351	18	120	68	1401	16	100	60	1451	20	95	55
1352	16	104	68	1402	16	110	70	1452	19	94	70
1353	16	104	76	1403	20	100	70	1453	15	90	68
1354	15	104	72	1404	20	100	60	1454	16	100	70
1355	19	124	68	1405	20	110	70	1455	18	108	80
1356	16	134	80	1406	20	100	60	1456	18	110	70
1357	17	92	62	1407	20	100	70	1457	19	118	88
1358	15	110	74	1408	18	106	54	1458	19	90	60
1359	18	100	68	1409	19	94	70	1459	20	110	90
1360	15	98	68	1410	18	110	76	1460	15	108	70
1361	18	110	62	1411	18	128	76	1461	15	118	80
1362	20	135	100	1412	15	88	60	1462	19	104	60
1363	20	94	68	1413	19	110	70	1463	16	118	74
1364	20	122	72	1414	17	120	78	1464	16	110	78
1365	20	106	72	1415	18	104	70	1465	17	114	60
1366	20	150	90	1416	18	100	66	1466	15	106	74
1367	20	106	70	1417	17	102	66	1467	20	112	78
1368	18	116	74	1418	17	88	70	1468	20	108	70
1369	19	120	78	1419	16	109	80	1469	20	126	70
1370	19	112	66	1420	18	90	60	1470	19	130	100
1371	17	124	74	1421	15	90	60	1471	16	110	70
1372	18	128	74	1422	17	118	76	1472	18	120	80
1373	18	136	72	1423	19	100	60	1473	17	120	70
1374	17	134	74	1424	15	112	68	1474	19	120	90
1375	16	106	56	1425	18	100	60	1475	19	102	60
1376	17	122	76	1426	15	115	70	1476	19	120	74
1377	16	108	68	1427	19	100	70	1477	18	120	84
1378	15	118	74	1428	15	98	68	1478	18	100	80
1379	20	120	90	1429	15	106	76	1479	19	106	74
1380	18	112	75	1430	18	86	64	1480	16	110	70
1381	19	102	50	1431	18	120	68	1481	19	120	70
1382	19	125	75	1432	19	150	70	1482	15	98	80
1383	16	108	80	1433	15	112	64	1483	19	108	80
1384	17	130	95	1434	17	90	60	1484	20	110	80
1385	19	135	80	1435	20	112	60	1485	20	112	70
1386	19	120	70	1436	20	132	102	1486	20	120	70
1387	16	130	80	1437	20	106	68	1487	20	120	80
1388	15	105	70	1438	16	110	70	1488	20	114	74
1389	16	98	70	1439	16	102	60	1489	20	114	70
1390	15	100	60	1440	15	98	40	1490	15	108	70
1391	20	98	70	1441	16	110	60	1491	15	106	60
1392	20	135	88	1442	18	112	60	1492	16	114	66
1393	18	110	60	1443	16	102	60	1493	19	128	66
1394	18	110	70	1444	16	140	70	1494	18	118	54
1395	17	90	60	1445	18	110	80	1495	15	108	60
1396	18	110	60	1446	17	102	55	1496	17	94	76
1397	18	90	60	1447	16	88	60	1497	16	116	68
1398	15	120	70	1448	17	110	80	1498	16	106	60
1399	15	110	70	1449	16	100	70	1499	19	104	74
1400	19	130	100	1450	20	120	60	1500	19	122	84

NIH data: 945 females, ages 15 through 20 (continued)

OBS	AGE	SYSBP	DIASBP	OBS	AGE	SYSBP	DIASBP	OBS	AGE	SYSBP	DIASBP
1501	18	114	76	1551	15	100	65	1601	15	90	60
1502	17	114	62	1552	16	108	78	1602	15	80	50
1503	17	110	80	1553	19	120	90	1603	16	100	60
1504	20	128	70	1554	19	110	88	1604	17	100	60
1505	20	110	56	1555	20	98	70	1605	15	120	70
1506	18	110	64	1556	16	110	68	1606	19	90	60
1507	16	118	68	1557	16	110	70	1607	20	100	60
1508	17	110	80	1558	15	136	76	1608	20	110	70
1509	18	100	70	1559	19	140	108	1609	20	110	60
1510	17	110	70	1560	19	112	70	1610	20	120	70
1511	17	110	72	1561	16	110	62	1611	19	112	80
1512	18	128	80	1562	17	112	72	1612	18	100	64
1513	17	120	80	1563	16	116	58	1613	19	94	64
1514	18	120	68	1564	19	100	70	1614	18	110	66
1515	17	120	80	1565	17	120	80	1615	16	122	76
1516	20	114	64	1566	15	110	70	1616	15	110	62
1517	20	124	80	1567	15	108	64	1617	18	94	68
1518	20	134	70	1568	19	98	70	1618	19	90	68
1519	17	95	58	1569	16	110	78	1619	18	86	60
1520	18	100	70	1570	20	118	66	1620	18	100	66
1521	16	108	68	1571	20	110	78	1621	19	106	66
1522	16	102	75	1572	20	130	72	1622	19	108	70
1523	16	104	60	1573	19	102	68	1623	19	92	72
1524	16	100	70	1574	19	116	76	1624	19	98	64
1525	16	92	60	1575	19	140	80	1625	15	104	64
1526	15	102	70	1576	15	96	62	1626	18	108	58
1527	16	85	58	1577	18	106	60	1627	19	112	78
1528	16	112	80	1578	15	106	56	1628	16	124	72
1529	19	100	55	1579	19	112	58	1629	19	166	90
1530	19	115	70	1580	17	108	66	1630	19	104	74
1531	17	120	70	1581	15	108	84	1631	19	100	68
1532	18	110	70	1582	19	112	66	1632	15	100	66
1533	15	112	70	1583	19	122	84	1633	18	96	70
1534	17	115	80	1584	15	110	76	1634	19	96	64
1535	15	95	70	1585	18	108	70	1635	15	116	78
1536	17	110	80	1586	18	114	68	1636	19	114	76
1537	17	108	70	1587	15	114	80	1637	18	90	58
1538	17	102	60	1588	19	100	64	1638	20	96	70
1539	17	110	60	1589	17	104	66	1639	20	160	116
1540	17	110	78	1590	17	108	70	1640	20	136	88
1541	16	100	65	1591	18	122	70	1641	20	98	68
1542	16	115	78	1592	17	110	76	1642	20	104	70
1543	20	100	60	1593	19	130	74	1643	20	104	68
1544	18	98	78	1594	20	120	64	1644	20	106	68
1545	15	92	62	1595	20	114	80	1645	20	104	66
1546	15	120	70	1596	20	120	96	1646	20	102	66
1547	17	130	88	1597	19	100	70	1647	17	120	70
1548	16	110	58	1598	19	120	70	1648	16	110	70
1549	20	100	80	1599	18	110	50	1649	18	100	60
1550	19	100	70	1600	17	120	70	1650	19	110	60

NIH data: 945 females, ages 15 through 20 (continued)

OBS	AGE	SYSBP	DIASBP	OBS	AGE	SYSBP	DIASBP	OBS	AGE	SYSBP	DIASBP
1651	16	120	70	1701	20	110	86	1751	17	114	70
1652	15	120	70	1702	20	120	70	1752	16	110	80
1653	18	110	70	1703	20	120	68	1753	18	124	74
1654	17	110	70	1704	16	110	65	1754	18	126	90
1655	15	100	60	1705	18	102	80	1755	16	124	86
1656	16	130	80	1706	15	100	80	1756	18	100	70
1657	18	100	60	1707	19	112	58	1757	18	114	70
1658	16	110	70	1708	19	110	70	1758	18	116	80
1659	19	110	80	1709	19	120	65	1759	17	110	80
1660	15	100	60	1710	19	100	65	1760	20	90	70
1661	19	120	70	1711	19	118	72	1761	20	110	68
1662	18	110	70	1712	18	98	70	1762	15	110	80
1663	18	130	80	1713	16	95	78	1763	18	80	50
1664	19	90	60	1714	20	105	72	1764	17	108	80
1665	16	120	70	1715	20	102	60	1765	18	100	70
1666	20	100	60	1716	20	98	70	1766	19	140	80
1667	20	90	60	1717	20	102	80	1767	16	120	80
1668	20	100	60	1718	20	120	85	1768	17	100	70
1669	16	112	80	1719	20	115	70	1769	15	90	70
1670	16	112	68	1720	15	136	86	1770	15	130	90
1671	19	128	76	1721	18	116	70	1771	18	130	70
1672	16	116	64	1722	18	124	70	1772	19	110	80
1673	15	120	64	1723	17	104	66	1773	17	120	80
1674	19	134	92	1724	15	110	78	1774	19	98	68
1675	18	120	78	1725	19	104	66	1775	15	100	60
1676	15	126	70	1726	15	114	82	1776	16	110	80
1677	16	118	70	1727	18	120	76	1777	16	110	70
1678	17	110	76	1728	16	114	70	1778	16	130	80
1679	15	132	98	1729	18	120	64	1779	20	140	70
1680	16	116	82	1730	17	124	86	1780	20	100	80
1681	16	122	72	1731	18	96	76	1781	20	110	80
1682	15	104	70	1732	20	110	82	1782	18	110	70
1683	16	110	74	1733	20	114	74	1783	18	130	80
1684	16	104	70	1734	20	106	64	1784	18	100	70
1685	17	150	70	1735	20	110	56	1785	16	130	70
1686	16	80	·	1736	19	100	74	1786	15	120	70
1687	18	100	70	1737	16	110	90	1787	15	110	70
1688	16	138	80	1738	19	120	76	1788	18	110	60
1689	15	150	80	1739	18	120	74	1789	19	110	70
1690	16	98	50	1740	17	110	60	1790	19	90	60
1691	19	100	72	1741	19	130	90	1791	19	120	70
1692	19	128	60	1742	15	120	76	1792	19	100	70
1693	17	100	78	1743	16	110	80	1793	18	110	70
1694	16	100	64	1744	19	130	80	1794	18	110	70
1695	18	90	60	1745	16	90	60	1795	15	100	70
1696	15	88	70	1746	18	110	70	1796	19	90	60
1697	15	120	80	1747	20	104	80	1797	19	100	60
1698	16	120	78	1748	18	110	60	1798	16	100	60
1699	18	90	70	1749	20	94	64	1799	16	110	60
1700	20	148	120	1750	20	110	76	1800	20	100	70

NIH data: 945 females, ages 15 through 20 (continued)

OBS	AGE	SYSBP	DIASBP	OBS	AGE	SYSBP	DIASBP	OBS	AGE	SYSBP	DIASBP
1801	20	90	60	1838	17	110	70	1875	16	100	70
1802	20	110	70	1839	17	110	70	1876	18	90	60
1803	20	110	70	1840	18	110	70	1877	20	90	60
1804	20	100	70	1841	20	130	80	1878	20	90	60
1805	20	110	70	1842	15	110	60	1879	20	110	70
1806	20	120	70	1843	20	100	60	1880	15	112	60
1807	20	120	70	1844	20	100	60	1881	18	120	82
1808	20	110	70	1845	20	90	60	1882	16	96	70
1809	20	90	60	1846	20	110	70	1883	17	100	58
1810	20	110	70	1847	20	120	80	1884	17	134	88
1811	16	120	80	1848	20	120	70	1885	16	98	50
1812	18	100	70	1849	15	108	70	1886	16	100	50
1813	15	100	70	1850	20	120	80	1887	15	110	50
1814	16	100	60	1851	18	100	70	1888	20	100	65
1815	16	120	80	1852	17	110	80	1889	17	114	90
1816	16	110	70	1853	18	120	80	1890	19	116	80
1817	16	120	70	1854	19	104	60	1891	17	98	64
1818	18	110	80	1855	15	110	80	1892	18	108	70
1819	18	110	70	1856	15	110	60	1893	17	110	60
1820	17	120	80	1857	19	100	70	1894	15	98	80
1821	19	118	82	1858	16	108	60	1895	18	124	68
1822	17	120	80	1859	17	128	80	1896	18	110	78
1823	19	110	80	1860	16	100	60	1897	17	106	78
1824	19	110	70	1861	19	120	60	1898	17	112	76
1825	18	110	70	1862	17	110	70	1899	20	150	96
1826	15	106	60	1863	20	120	80	1900	20	94	68
1827	18	100	70	1864	20	120	70	1901	20	106	68
1828	15	110	70	1865	19	110	70	1902	17	112	86
1829	19	120	70	1866	16	100	70	1903	16	98	70
1830	15	100	60	1867	18	100	60	1904	17	140	94
1831	18	110	60	1868	19	110	60	1905	18	110	90
1832	17	120	70	1869	15	110	70	1906	15	90	70
1833	16	130	60	1870	18	110	70	1907	15	88	68
1834	16	110	70	1871	15	110	70	1908	19	90	78
1835	16	90	60	1872	19	120	70	1909	19	90	60
1836	19	120	70	1873	19	120	68	1910	19	88	68
1837	18	115	70	1874	15	104	70	1911	20	104	78

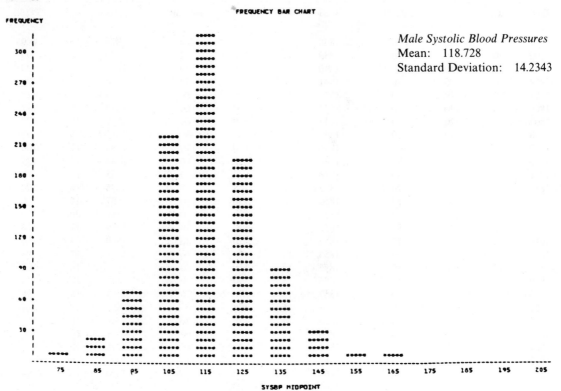

Male Systolic Blood Pressures
Mean: 118.728
Standard Deviation: 14.2343

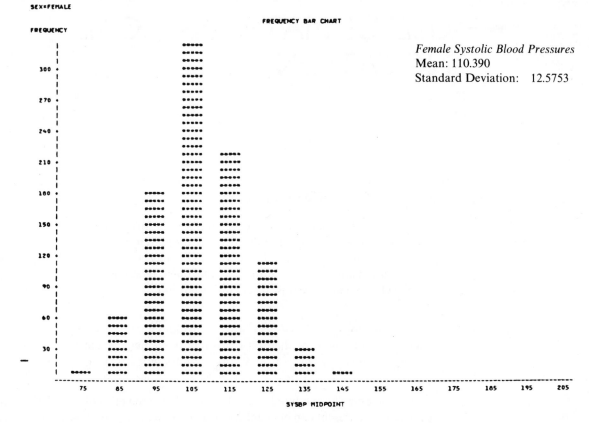

NIH DATA - PERSONS AGED 15 THROUGH 20

SEX=FEMALE

FREQUENCY BAR CHART

Female Systolic Blood Pressures
Mean: 110.390
Standard Deviation: 12.5753

Percentile	0	1	5	10	25	50	75	90	95	99	100
Male	70	90	100	101.2	110	120	128	136.8	142	162	210
Female	78	85.46	90	95	100	110	120	126	130	149.08	166

► USEFUL MATHEMATICAL NOTATION

► **A.I** FUNCTIONAL NOTATION

Consider two sets* of elements (objects, numbers, or anything we want to use) and their relation to one another. Let the symbol x represent an element of the first set and y an element of the second. One could specify a number of different rules defining relationships between x and y. A **functional relationship** between x and y is one such rule, to which we now direct our attention.

Definition

> A **function** consists of two sets of elements and a defined correspondence between an element of the first set x and an element of the second set y such that for each element x there corresponds one, and only one, element y.

A function may be exhibited as a collection of ordered pairs of elements, written (x, y), where x represents an element from the first set and y represents an element from the second. In fact, a function is often defined as a collection of ordered pairs of elements with the property stated in the definition.

Most often, x and y will be variables taking numerical values. The two sets of elements would represent all the possible numerical values that x and y might take, and the rule defining the correspondence between them would be an equation. For example, if we state that x and y are real numbers and that

$$y = x + 2$$

then y is a function of x. Assigning a value to x (that is, choosing an element of the set of all real numbers), there corresponds one, and only one, value of y. When $x = 1$, $y = 3$; when $x = -4$, $y = -2$; and so on.

The area A of a circle is related to the radius r by the formula

$$A = \pi r^2$$

* A set is a collection of specific things.

◁

Note that if we assign a value to r, the value of A can be determined from the formula. Hence, A is a function of r. Likewise, the circumference of a circle C is a function of the radius.

Mathematical writing frequently uses the same phrases or refers to a specific object many times in the course of a discussion. Unnecessary repetition wastes the reader's time, takes up valuable space, and is cumbersome to the writer. Hence, the mathematician resorts to mathematical symbolism that is in some respects a type of mathematical shorthand. Rather than state "the area of a circle A is a function of the radius r," the mathematician would write $A(r)$ or $A = f(r)$. The expression

$$y = f(x)$$

tells us that y is a function of the variable appearing in parentheses, namely, x. Note that this expression does not tell us the specific functional relation that exists between y and x.

Consider the function of x,

$$y = 3x + 2$$

In functional notation, this would be written as

$$f(x) = 3x + 2$$

It is understood that $y = f(x)$.

Functional notation is especially advantageous when we wish to indicate the value of the function y when x takes a specific value, say $x = 2$. For the example above, we see that when $x = 2$,

$$y = 3x + 2 = 3(2) + 2 = 8$$

Rather than write this, we use the simpler notation $f(2) = 8$

To find the value of the function when x equals any value, say $x = c$, we substitute the value of c for x in the equation and obtain

$$f(c) = 3c + 2$$

EXAMPLE A.1   Find $g(4)$ if

$$g(x) = \frac{1}{x} + 3x, \qquad \text{where } x \neq 0$$

(Note that x cannot equal 0 because then $1/x$ is undefined.)

Solution
$$
\begin{aligned}
g(4) &= \frac{1}{4} + 3(4) \\
&= .25 + 12 \\
&= 12.25
\end{aligned}
$$

EXAMPLE A.2 Let $p(y) = (1 - a)a^y$. Find $p(2)$, $p(3)$, and $p(0)$.

Solution
$$p(2) = (1 - a)a^2$$
$$p(3) = (1 - a)a^3$$
$$p(0) = (1 - a)a^0 = 1 - a, \quad \text{since } a^0 = 1$$

EXAMPLE A.3 Let $f(y) = 4$. Find $f(2)$ and $f(3)$.

Solution
$$f(2) = 4$$
$$f(3) = 4$$

Note that $f(y)$ always equals 4, regardless of the value of y. Hence, when a value of y is assigned, the value of the function is determined and will equal 4.

A.2 NUMERICAL SEQUENCES AND SUMMATION NOTATION

A set of objects, $a_1, a_2, a_3, a_4, \ldots$, ordered in the sense that we can identify the first member of the set a_1, the second a_2, and so on, is called a **sequence**. Most often, $a_1, a_2, a_3, \ldots$, called **elements** of the sequence, are numbers, but this is not a requirement. For example, the numbers

$$1, 5, 4, 8, 7, 11, 10, \ldots$$

form a sequence moving from left to right. Note that the elements are ordered only in their position in the sequence and need not be ordered in magnitude.

Although nonnumerical sequences are sometimes of interest, we will be concerned solely with numerical sequences. Specifically, data obtained in a sample from a population will be regarded as a sequence of measurements. For example, if we measure the weights of ten people randomly selected from the population, we could denote the ten observations as

$$w_1, w_2, \ldots, w_{10}$$

More generally, we can denote the n observations in a random sample of size n as

$$x_1, x_2, \ldots, x_n$$

The **typical element** in this sequence is x_i, and the variable i is called the **position variable**.

When we analyze statistical data, we often work with sums of numbers and need a simple notation for indicating a sum. For instance, consider the sequence of numbers

$$1, 2, 3, 4, 5, \ldots$$

and suppose that we wish to discuss the sum of the squares of the first four numbers of the sequence. Using summation notation, this is written as

$$\sum_{y=1}^{4} y^2$$

Interpretation of the summation notation is relatively easy. The Greek letter Σ (capital sigma), corresponding to "S" in the English alphabet (the first letter in the word "Sum"), tells us to sum elements of a sequence. A typical element of the sequence is given to the right of the summation symbol, and the position variable, called the variable of summation, is shown beneath. For our example, y^2 is a typical element, y is the variable of summation, and the implied sequence is

1, 4, 9, 16, 25, 36, . . .

Which elements of the sequence should appear in the sum? The position of the first element in the sum is indicated below the summation sign, the last above. The sum would include all elements proceeding *in order* from the first to last. In our example, the sum would include the sum of the elements commencing with the first and ending with the fourth:

$$\sum_{y=1}^{4} y^2 = (1)^2 + (2)^2 + (3)^2 + (4)^2$$

$$= 1 + 4 + 9 + 16$$

$$= 30$$

EXAMPLE A.4 $\displaystyle\sum_{y=2}^{4} (y - 1) = (2 - 1) + (3 - 1) + (4 - 1)$

$$= 6$$

EXAMPLE A.5 $\displaystyle\sum_{x=2}^{5} 3x = 3(2) + 3(3) + 3(4) + 3(5)$

$$= 42$$

We emphasize that the typical element is a function only of the variable of summation. All other symbols are regarded as constants.

EXAMPLE A.6 $\displaystyle\sum_{i=1}^{3} (x_i - a) = (x_1 - a) + (x_2 - a) + (x_3 - a)$

In Example A.6, note that i is the variable of summation and that it appears as a subscript in the typical element.

EXAMPLE A.7 $\displaystyle\sum_{i=1}^{2} (x - i + 1) = (x - 1 + 1) + (x - 2 + 1)$

$$= 2x - 1$$

EXAMPLE A.8 $\displaystyle\sum_{y=2}^{4} y - 1 = (2 + 3 + 4) - 1$

$$= 8$$

Note the difference between Examples A.4 and A.8. The quantity $(y - 1)$ is the typical element in Example A.4, while y is the typical element in Example A.8.

EXAMPLE A.9   Suppose 122, 129, and 124 represent the blood pressure measurements on three 20-year-old females. Let x_1 represent the measurement 122 and, similarly, let $x_2 = 129$ and $x_3 = 124$. Find

$$\sum_{i=1}^{3} x_i \quad \text{and} \quad \sum_{i=1}^{3} x_i^2$$

Solution   You can see that the expression

$$\sum_{i=1}^{3} x_i = x_1 + x_2 + x_3$$

provides a compact way of writing the sum of the sample measurements. Then,

$$\sum_{i=1}^{3} x_i = x_1 + x_2 + x_3 = 122 + 129 + 124 = 375$$

Similarly,

$$\sum_{i=1}^{3} x_i^2$$

is a compact way of writing the sum of squares of the sample measurements. Thus,

$$\sum_{i=1}^{3} x_i^2 = x_1^2 + x_2^2 + x_3^2 = (122)^2 + (129)^2 + (124)^2$$

$$= 14{,}884 + 16{,}641 + 15{,}376 = 46{,}901$$

A.3   USEFUL THEOREMS RELATING TO SUMS

Consider the summation

$$\sum_{y=1}^{3} 5$$

The typical element is 5 and it does not change. The sequence is, therefore,

5, 5, 5, 5, . . .

and

$$\sum_{y=1}^{3} 5 = 5 + 5 + 5 = 15$$

Theorem A.1   Let c be a constant (an element that does not involve the variable of summation) and y be the variable of summation. Then,

$$\sum_{y=1}^{n} c = nc$$

Proof

$$\sum_{y=1}^{n} c = c + c + c + \cdots + c$$

where the sum involves n elements. Then,

$$\sum_{y=1}^{n} c = nc$$

EXAMPLE A.10

$$\sum_{y=1}^{4} 3a = 4(3a) = 12a$$

(Note that y is the variable of summation and, therefore, that a is a constant.)

EXAMPLE A.11

$$\sum_{i=1}^{3} (3x - 5) = 3(3x - 5)$$

(Note that i is the variable of summation and, therefore, that x is a constant.)

EXAMPLE A.12

$$\sum_{y=4}^{10} 3a = \sum_{y=1}^{10} 3a - \sum_{y=1}^{3} 3a = 10(3a) - 3(3a) = 21a$$

(Note that since the sum of the elements does not commence with $y = 1$, it can be written as the difference between the sum of the elements from 1 to 10 and the sum from 1 to 3.)

A second theorem is illustrated using Example A.5. We note that 3 is a common factor in each term. Therefore,

$$\sum_{x=2}^{5} 3x = 3(2) + 3(3) + 3(4) + 3(5)$$

$$= 3(2 + 3 + 4 + 5)$$

$$= 3 \sum_{x=2}^{5} x$$

Thus, it would appear that the summation of a constant times a variable is equal to the constant times the summation of the variable.

Theorem A.2   Let c be a constant. Then

$$\sum_{i=1}^{n} cx_i = c \sum_{i=1}^{n} x_i$$

Proof

$$\sum_{i=1}^{n} cx_i = cx_1 + cx_2 + cx_3 + \cdots + cx_n$$

$$= c(x_1 + x_2 + \cdots + x_n)$$

$$= c \sum_{i=1}^{n} x_i$$

Theorem A.3
$$\sum_{i=1}^{n} (x_i + y_i + z_i) = \sum_{i=1}^{n} x_i + \sum_{i=1}^{n} y_i + \sum_{i=1}^{n} z_i$$

Proof

$$\sum_{i=1}^{n} (x_i + y_i + z_i) = x_1 + y_1 + z_1 + x_2 + y_2 + z_2$$
$$+ x_3 + y_3 + z_3 + \cdots + x_n + y_n + z_n$$

Regrouping, we have

$$\sum_{i=1}^{n} (x_i + y_i + z_i) = (x_1 + x_2 + \cdots + x_n) + (y_1 + y_2 + \cdots + y_n)$$
$$+ (z_1 + z_2 + \cdots + z_n)$$
$$= \sum_{i=1}^{n} x_i + \sum_{i=1}^{n} y_i + \sum_{i=1}^{n} z_i$$

Theorems A.1, A.2, and A.3 can be used jointly to simplify summations. Consider the following examples:

EXAMPLE A.13
$$\sum_{x=1}^{3} (x^2 + ax + 5) = \sum_{x=1}^{3} x^2 + \sum_{x=1}^{3} ax + \sum_{x=1}^{3} 5$$
$$= \sum_{x=1}^{3} x^2 + a \sum_{x=1}^{3} x + 3(5)$$
$$= (1 + 4 + 9) + a(1 + 2 + 3) + 15$$
$$= 6a + 29$$ ◁

EXAMPLE A.14
$$\sum_{i=1}^{4} (x^2 + 3i) = \sum_{i=1}^{4} x^2 + \sum_{i=1}^{4} 3i$$
$$= 4x^2 + 3 \sum_{i=1}^{4} i$$
$$= 4x^2 + 3(1 + 2 + 3 + 4)$$
$$= 4x^2 + 30$$ ◁

EXAMPLE A.15   Suppose you have a set of sample measurements $x_1, x_2, x_3, \ldots, x_n$, and that you subtract a constant, say c, from each measurement. Then the sum of $x_1 - c$, $x_2 - c, \ldots, x_n - c$, is

$$\sum_{i=1}^{n} (x_i - c) = \sum_{i=1}^{n} x_i - \sum_{i=1}^{n} c = \sum_{i=1}^{n} x_i - nc$$ ◁

APPENDIX III

 TABLES

Table I Cumulative Binomial Probabilities

Tabulated values are $P(x \le a) = \sum\limits_{x=0}^{a} p(x)$. (Computations are rounded at the third decimal place.)

$n = 2$

a	\multicolumn{13}{c}{p}	a
	0.01	0.05	0.10	0.20	0.30	0.40	0.50	0.60	0.70	0.80	0.90	0.95	0.99	
0	.980	.902	.810	.640	.490	.360	.250	.160	.090	.040	.010	.002	.000	0
1	1.000	.998	.990	.960	.910	.840	.750	.640	.510	.360	.190	.098	.020	1
2	1.000	1.000	1.000	1.000	1.000	1.000	1.000	1.000	1.000	1.000	1.000	1.000	1.000	2

$n = 3$

a	\multicolumn{13}{c}{p}	a
	0.01	0.05	0.10	0.20	0.30	0.40	0.50	0.60	0.70	0.80	0.90	0.95	0.99	
0	.970	.857	.729	.512	.343	.216	.125	.064	.027	.008	.001	.000	.000	0
1	1.000	.993	.972	.896	.784	.648	.500	.352	.216	.104	.028	.007	.000	1
2	1.000	1.000	.999	.992	.973	.936	.875	.784	.657	.488	.271	.143	.030	2
3	1.000	1.000	1.000	1.000	1.000	1.000	1.000	1.000	1.000	1.000	1.000	1.000	1.000	3

$n = 4$

a	\multicolumn{13}{c}{p}	a
	0.01	0.05	0.10	0.20	0.30	0.40	0.50	0.60	0.70	0.80	0.90	0.95	0.99	
0	.961	.815	.656	.410	.240	.130	.062	.026	.008	.002	.000	.000	.000	0
1	.999	.986	.948	.819	.652	.475	.312	.179	.084	.027	.004	.000	.000	1
2	1.000	1.000	.996	.973	.916	.821	.688	.525	.348	.181	.052	.014	.001	2
3	1.000	1.000	1.000	.998	.992	.974	.938	.870	.760	.590	.344	.185	.039	3
4	1.000	1.000	1.000	1.000	1.000	1.000	1.000	1.000	1.000	1.000	1.000	1.000	1.000	4

Less than or
= to A

Table I (continued)

$n = 5$

							p							
a	0.01	0.05	0.10	0.20	0.30	0.40	0.50	0.60	0.70	0.80	0.90	0.95	0.99	a
0	.951	.774	.590	.328	.168	.078	.031	.010	.002	.000	.000	.000	.000	0
1	.999	.977	.919	.737	.528	.337	.188	.087	.031	.007	.000	.000	.000	1
2	1.000	.999	.991	.942	.837	.683	.500	.317	.163	.058	.009	.001	.000	2
3	1.000	1.000	1.000	.993	.969	.913	.812	.663	.472	.263	.081	.023	.001	3
4	1.000	1.000	1.000	1.000	.998	.990	.969	.922	.832	.672	.410	.226	.049	4
5	1.000	1.000	1.000	1.000	1.000	1.000	1.000	1.000	1.000	1.000	1.000	1.000	1.000	5

$n = 6$

							p							
a	0.01	0.05	0.10	0.20	0.30	0.40	0.50	0.60	0.70	0.80	0.90	0.95	0.99	a
0	.941	.735	.531	.262	.118	.047	.016	.004	.001	.000	.000	.000	.000	0
1	.999	.967	.886	.655	.420	.233	.109	.041	.011	.002	.000	.000	.000	1
2	1.000	.998	.984	.901	.744	.544	.344	.179	.070	.017	.001	.000	.000	2
3	1.000	1.000	.999	.983	.930	.821	.656	.456	.256	.099	.016	.002	.000	3
4	1.000	1.000	1.000	.998	.989	.959	.891	.767	.780	.345	.114	.033	.001	4
5	1.000	1.000	1.000	1.000	.999	.996	.984	.953	.882	.738	.469	.265	.059	5
6	1.000	1.000	1.000	1.000	1.000	1.000	1.000	1.000	1.000	1.000	1.000	1.000	1.000	6

$n = 7$

							p							
a	0.01	0.05	0.10	0.20	0.30	0.40	0.50	0.60	0.70	0.80	0.90	0.95	0.99	a
0	.932	.698	.478	.210	.082	.028	.008	.002	.000	.000	.000	.000	.000	0
1	.998	.956	.850	.577	.329	.159	.062	.019	.004	.000	.000	.000	.000	1
2	1.000	.996	.974	.852	.647	.420	.227	.096	.029	.005	.000	.000	.000	2
3	1.000	1.000	.997	.967	.874	.710	.500	.290	.126	.033	.003	.000	.000	3
4	1.000	1.000	1.000	.995	.971	.904	.773	.580	.353	.148	.026	.004	.000	4
5	1.000	1.000	1.000	1.000	.996	.981	.938	.841	.671	.423	.150	.044	.002	5
6	1.000	1.000	1.000	1.000	1.000	.998	.992	.972	.918	.790	.522	.302	.068	6
7	1.000	1.000	1.000	1.000	1.000	1.000	1.000	1.000	1.000	1.000	1.000	1.000	1.000	7

Less than or
= to A

Table I   (continued)

$n = 8$

							p							
a	0.01	0.05	0.10	0.20	0.30	0.40	0.50	0.60	0.70	0.80	0.90	0.95	0.99	a
0	.923	.663	.430	.168	.058	.017	.004	.001	.000	.000	.000	.000	.000	0
1	.997	.943	.813	.503	.255	.106	.035	.009	.001	.000	.000	.000	.000	1
2	1.000	.994	.962	.797	.552	.315	.145	.050	.011	.001	.000	.000	.000	2
3	1.000	1.000	.995	.944	.806	.594	.363	.174	.058	.010	.000	.000	.000	3
4	1.000	1.000	1.000	.990	.942	.826	.637	.406	.194	.056	.005	.000	.000	4
5	1.000	1.000	1.000	.999	.989	.950	.855	.685	.448	.203	.038	.006	.000	5
6	1.000	1.000	1.000	1.000	.999	.991	.965	.894	.745	.497	.187	.057	.003	6
7	1.000	1.000	1.000	1.000	1.000	.999	.996	.983	.942	.832	.570	.337	.077	7
8	1.000	1.000	1.000	1.000	1.000	1.000	1.000	1.000	1.000	1.000	1.000	1.000	1.000	8

$n = 9$

							p							
a	0.01	0.05	0.10	0.20	0.30	0.40	0.50	0.60	0.70	0.80	0.90	0.95	0.99	a
0	.914	.630	.387	.134	.040	.010	.002	.000	.000	.000	.000	.000	.000	0
1	.997	.929	.775	.436	.196	.071	.020	.004	.000	.000	.000	.000	.000	1
2	1.000	.992	.947	.738	.463	.232	.090	.025	.004	.000	.000	.000	.000	2
3	1.000	.999	.992	.914	.730	.483	.254	.099	.025	.003	.000	.000	.000	3
4	1.000	1.000	.999	.980	.901	.733	.500	.267	.099	.020	.001	.000	.000	4
5	1.000	1.000	1.000	.997	.975	.901	.746	.517	.270	.086	.008	.001	.000	5
6	1.000	1.000	1.000	1.000	.996	.975	.910	.768	.537	.262	.053	.008	.000	6
7	1.000	1.000	1.000	1.000	1.000	.996	.980	.929	.804	.564	.225	.071	.003	7
8	1.000	1.000	1.000	1.000	1.000	1.000	.998	.990	.960	.866	.613	.370	.086	8
9	1.000	1.000	1.000	1.000	1.000	1.000	1.000	1.000	1.000	1.000	1.000	1.000	1.000	9

$n = 10$

							p							
a	0.01	0.05	0.10	0.20	0.30	0.40	0.50	0.60	0.70	0.80	0.90	0.95	0.99	a
0	.904	.599	.349	.107	.028	.006	.001	.000	.000	.000	.000	.000	.000	0
1	.996	.914	.736	.376	.149	.046	.011	.002	.000	.000	.000	.000	.000	1
2	1.000	.988	.930	.678	.383	.167	.055	.012	.002	.000	.000	.000	.000	2
3	1.000	.999	.987	.879	.650	.382	.172	.055	.011	.001	.000	.000	.000	3
4	1.000	1.000	.998	.967	.850	.633	.377	.166	.047	.006	.000	.000	.000	4
5	1.000	1.000	1.000	.994	.953	.834	.623	.367	.150	.033	.002	.000	.000	5
6	1.000	1.000	1.000	.999	.989	.945	.828	.618	.350	.121	.013	.001	.000	6
7	1.000	1.000	1.000	1.000	.998	.988	.945	.833	.617	.322	.070	.012	.000	7
8	1.000	1.000	1.000	1.000	1.000	.998	.989	.954	.851	.624	.264	.086	.004	8
9	1.000	1.000	1.000	1.000	1.000	1.000	.999	.994	.972	.893	.651	.401	.096	9
10	1.000	1.000	1.000	1.000	1.000	1.000	1.000	1.000	1.000	1.000	1.000	1.000	1.000	10

Table I (continued)

$n = 11$

							p							
a	0.01	0.05	0.10	0.20	0.30	0.40	0.50	0.60	0.70	0.80	0.90	0.95	0.99	a
0	.895	.569	.314	.086	.020	.004	.000	.000	.000	.000	.000	.000	.000	0
1	.995	.898	.697	.322	.113	.030	.006	.001	.000	.000	.000	.000	.000	1
2	1.000	.985	.910	.617	.313	.119	.033	.006	.001	.000	.000	.000	.000	2
3	1.000	.998	.981	.839	.570	.296	.113	.029	.004	.000	.000	.000	.000	3
4	1.000	1.000	.997	.950	.790	.533	.274	.099	.022	.002	.000	.000	.000	4
5	1.000	1.000	1.000	.988	.922	.754	.500	.246	.078	.012	.000	.000	.000	5
6	1.000	1.000	1.000	.998	.978	.901	.726	.467	.210	.050	.003	.000	.000	6
7	1.000	1.000	1.000	1.000	.996	.971	.887	.704	.430	.161	.019	.002	.000	7
8	1.000	1.000	1.000	1.000	.999	.994	.967	.881	.687	.383	.090	.015	.000	8
9	1.000	1.000	1.000	1.000	1.000	.999	.994	.970	.887	.678	.303	.102	.005	9
10	1.000	1.000	1.000	1.000	1.000	1.000	1.000	.996	.980	.914	.686	.431	.105	10
11	1.000	1.000	1.000	1.000	1.000	1.000	1.000	1.000	1.000	1.000	1.000	1.000	1.000	11

$n = 12$

							p							
a	0.01	0.05	0.10	0.20	0.30	0.40	0.50	0.60	0.70	0.80	0.90	0.95	0.99	a
0	.886	.540	.282	.069	.014	.002	.000	.000	.000	.000	.000	.000	.000	0
1	.994	.882	.659	.275	.085	.020	.003	.000	.000	.000	.000	.000	.000	1
2	1.000	.980	.889	.558	.253	.083	.019	.003	.000	.000	.000	.000	.000	2
3	1.000	.998	.974	.795	.493	.225	.073	.015	.002	.000	.000	.000	.000	3
4	1.000	1.000	.996	.927	.724	.438	.194	.057	.009	.001	.000	.000	.000	4
5	1.000	1.000	.999	.981	.882	.665	.387	.158	.039	.004	.000	.000	.000	5
6	1.000	1.000	1.000	.996	.961	.842	.613	.335	.118	.019	.001	.000	.000	6
7	1.000	1.000	1.000	.999	.991	.943	.806	.562	.276	.073	.004	.000	.000	7
8	1.000	1.000	1.000	1.000	.998	.985	.927	.775	.507	.205	.026	.002	.000	8
9	1.000	1.000	1.000	1.000	1.000	.997	.981	.917	.747	.442	.111	.020	.000	9
10	1.000	1.000	1.000	1.000	1.000	1.000	.997	.980	.915	.725	.341	.118	.006	10
11	1.000	1.000	1.000	1.000	1.000	1.000	1.000	.998	.986	.931	.718	.460	.114	11
12	1.000	1.000	1.000	1.000	1.000	1.000	1.000	1.000	1.000	1.000	1.000	1.000	1.000	12

Table I   (continued)

$n = 15$

a	\multicolumn{13}{c}{p}	a
	0.01	0.05	0.10	0.20	0.30	0.40	0.50	0.60	0.70	0.80	0.90	0.95	0.99	
0	.860	.463	.206	.035	.005	.000	.000	.000	.000	.000	.000	.000	.000	0
1	.990	.829	.549	.167	.035	.005	.000	.000	.000	.000	.000	.000	.000	1
2	1.000	.964	.816	.398	.127	.027	.004	.000	.000	.000	.000	.000	.000	2
3	1.000	.995	.944	.648	.297	.091	.018	.002	.000	.000	.000	.000	.000	3
4	1.000	.999	.987	.836	.515	.217	.059	.009	.001	.000	.000	.000	.000	4
5	1.000	1.000	.998	.939	.722	.403	.151	.034	.004	.000	.000	.000	.000	5
6	1.000	1.000	1.000	.982	.869	.610	.304	.095	.015	.001	.000	.000	.000	6
7	1.000	1.000	1.000	.996	.950	.787	.500	.213	.050	.004	.000	.000	.000	7
8	1.000	1.000	1.000	.999	.985	.905	.696	.390	.131	.018	.000	.000	.000	8
9	1.000	1.000	1.000	1.000	.996	.966	.849	.597	.278	.061	.002	.000	.000	9
10	1.000	1.000	1.000	1.000	.999	.991	.941	.783	.485	.164	.013	.001	.000	10
11	1.000	1.000	1.000	1.000	1.000	.998	.982	.909	.703	.352	.056	.005	.000	11
12	1.000	1.000	1.000	1.000	1.000	1.000	.996	.973	.873	.602	.184	.036	.000	12
13	1.000	1.000	1.000	1.000	1.000	1.000	1.000	.995	.965	.833	.451	.171	.010	13
14	1.000	1.000	1.000	1.000	1.000	1.000	1.000	1.000	.995	.965	.794	.537	.140	14
15	1.000	1.000	1.000	1.000	1.000	1.000	1.000	1.000	1.000	1.000	1.000	1.000	1.000	15

$n = 20$

a	\multicolumn{13}{c}{p}	a
	0.01	0.05	0.10	0.20	0.30	0.40	0.50	0.60	0.70	0.80	0.90	0.95	0.99	
0	.818	.358	.122	.012	.001	.000	.000	.000	.000	.000	.000	.000	.000	0
1	.983	.736	.392	.069	.008	.001	.000	.000	.000	.000	.000	.000	.000	1
2	.999	.925	.677	.206	.035	.004	.000	.000	.000	.000	.000	.000	.000	2
3	1.000	.984	.867	.411	.107	.016	.001	.000	.000	.000	.000	.000	.000	3
4	1.000	.997	.957	.630	.238	.051	.006	.000	.000	.000	.000	.000	.000	4
5	1.000	1.000	.989	.804	.416	.126	.021	.002	.000	.000	.000	.000	.000	5
6	1.000	1.000	.998	.913	.608	.250	.058	.006	.000	.000	.000	.000	.000	6
7	1.000	1.000	1.000	.968	.772	.416	.132	.021	.001	.000	.000	.000	.000	7
8	1.000	1.000	1.000	.990	.887	.596	.252	.057	.005	.000	.000	.000	.000	8
9	1.000	1.000	1.000	.997	.952	.755	.412	.128	.017	.001	.000	.000	.000	9
10	1.000	1.000	1.000	.999	.983	.872	.588	.245	.048	.003	.000	.000	.000	10
11	1.000	1.000	1.000	1.000	.995	.943	.748	.404	.113	.010	.000	.000	.000	11
12	1.000	1.000	1.000	1.000	.999	.979	.868	.584	.228	.032	.000	.000	.000	12
13	1.000	1.000	1.000	1.000	1.000	.994	.942	.750	.392	.087	.002	.000	.000	13
14	1.000	1.000	1.000	1.000	1.000	.998	.979	.874	.584	.196	.011	.000	.000	14
15	1.000	1.000	1.000	1.000	1.000	1.000	.994	.949	.762	.370	.043	.003	.000	15
16	1.000	1.000	1.000	1.000	1.000	1.000	.999	.984	.893	.589	.133	.016	.000	16
17	1.000	1.000	1.000	1.000	1.000	1.000	1.000	.996	.965	.794	.323	.075	.001	17
18	1.000	1.000	1.000	1.000	1.000	1.000	1.000	.999	.992	.931	.608	.264	.017	18
19	1.000	1.000	1.000	1.000	1.000	1.000	1.000	1.000	.999	.988	.878	.642	.182	19
20	1.000	1.000	1.000	1.000	1.000	1.000	1.000	1.000	1.000	1.000	1.000	1.000	1.000	20

Table I (continued)

n = 25

							P							
a	0.01	0.05	0.10	0.20	0.30	0.40	0.50	0.60	0.70	0.80	0.90	0.95	0.99	*a*
0	.778	.277	.072	.004	.000	.000	.000	.000	.000	.000	.000	.000	.000	0
1	.974	.642	.271	.027	.002	.000	.000	.000	.000	.000	.000	.000	.000	1
2	.998	.873	.537	.098	.009	.000	.000	.000	.000	.000	.000	.000	.000	2
3	1.000	.966	.764	.234	.033	.002	.000	.000	.000	.000	.000	.000	.000	3
4	1.000	.993	.902	.421	.090	.009	.000	.000	.000	.000	.000	.000	.000	4
5	1.000	.999	.967	.617	.193	.029	.002	.000	.000	.000	.000	.000	.000	5
6	1.000	1.000	.991	.780	.341	.074	.007	.000	.000	.000	.000	.000	.000	6
7	1.000	1.000	.998	.891	.512	.154	.022	.001	.000	.000	.000	.000	.000	7
8	1.000	1.000	1.000	.953	.677	.274	.054	.004	.000	.000	.000	.000	.000	8
9	1.000	1.000	1.000	.983	.811	.425	.115	.013	.000	.000	.000	.000	.000	9
10	1.000	1.000	1.000	.994	.902	.586	.212	.034	.002	.000	.000	.000	.000	10
11	1.000	1.000	1.000	.998	.956	.732	.345	.078	.006	.000	.000	.000	.000	11
12	1.000	1.000	1.000	1.000	.983	.846	.500	.154	.017	.000	.000	.000	.000	12
13	1.000	1.000	1.000	1.000	.994	.922	.655	.268	.044	.002	.000	.000	.000	13
14	1.000	1.000	1.000	1.000	.998	.966	.788	.414	.098	.006	.000	.000	.000	14
15	1.000	1.000	1.000	1.000	1.000	.987	.885	.575	.189	.017	.000	.000	.000	15
16	1.000	1.000	1.000	1.000	1.000	.996	.946	.726	.323	.047	.000	.000	.000	16
17	1.000	1.000	1.000	1.000	1.000	.999	.978	.846	.488	.109	.002	.000	.000	17
18	1.000	1.000	1.000	1.000	1.000	1.000	.993	.926	.659	.220	.009	.000	.000	18
19	1.000	1.000	1.000	1.000	1.000	1.000	.998	.971	.807	.383	.033	.001	.000	19
20	1.000	1.000	1.000	1.000	1.000	1.000	1.000	.991	.910	.579	.098	.007	.000	20
21	1.000	1.000	1.000	1.000	1.000	1.000	1.000	.998	.967	.766	.236	.034	.000	21
22	1.000	1.000	1.000	1.000	1.000	1.000	1.000	1.000	.991	.902	.463	.127	.002	22
23	1.000	1.000	1.000	1.000	1.000	1.000	1.000	1.000	.998	.973	.729	.358	.026	23
24	1.000	1.000	1.000	1.000	1.000	1.000	1.000	1.000	1.000	.996	.928	.723	.222	24
25	1.000	1.000	1.000	1.000	1.000	1.000	1.000	1.000	1.000	1.000	1.000	1.000	1.000	25

Table 2(a) Cumulative Poisson Probabilities

Tabulated values are $P(x \le a) = \sum_{x=0}^{a} p(x)$. (Computations are rounded at the third decimal place.)

					μ						
a	0.1	0.2	0.3	0.4	0.5	0.6	0.7	0.8	0.9	1.0	1.5
0	.905	.819	.741	.670	.607	.549	.497	.449	.407	.368	.223
1	.995	.982	.963	.938	.910	.878	.844	.809	.772	.736	.558
2	1.000	.999	.996	.992	.986	.977	.966	.953	.937	.920	.809
3		1.000	1.000	.999	.998	.997	.994	.991	.987	.981	.934
4				1.000	1.000	1.000	.999	.999	.998	.996	.981
5							1.000	1.000	1.000	.999	.996
6										1.000	.999
7											1.000

Table 2(a) (continued)

	μ										
a	2.0	2.5	3.0	3.5	4.0	4.5	5.0	5.5	6.0	6.5	7.0
0	.135	.082	.050	.030	.018	.011	.007	.004	.003	.002	.001
1	.406	.287	.199	.136	.092	.061	.040	.027	.017	.011	.007
2	.677	.544	.423	.321	.238	.174	.125	.088	.062	.043	.030
3	.857	.758	.647	.537	.433	.342	.265	.202	.151	.112	.082
4	.947	.891	.815	.725	.629	.532	.440	.358	.285	.224	.173
5	.983	.958	.916	.858	.785	.703	.616	.529	.446	.369	.301
6	.995	.986	.966	.935	.889	.831	.762	.686	.606	.563	.450
7	.999	.996	.988	.973	.949	.913	.867	.809	.744	.673	.599
8	1.000	.999	.996	.990	.979	.960	.932	.894	.847	.792	.729
9		1.000	.999	.997	.992	.983	.968	.946	.916	.877	.830
10			1.000	.999	.997	.993	.986	.975	.957	.933	.901
11				1.000	.999	.998	.995	.989	.980	.966	.947
12					1.000	.999	.998	.996	.991	.984	.973
13						1.000	.999	.998	.996	.993	.987
14							1.000	.999	.999	.997	.994
15								1.000	.999	.999	.998
16									1.000	1.000	.999
17											1.000

	μ								
a	7.5	8.0	8.5	9.0	9.5	10.0	12.0	15.0	20.0
0	.001	.000	.000	.000	.000	.000	.000	.000	.000
1	.005	.003	.002	.001	.001	.000	.000	.000	.000
2	.020	.014	.009	.006	.004	.003	.001	.000	.000
3	.059	.042	.030	.021	.015	.010	.002	.000	.000
4	.132	.100	.074	.055	.040	.029	.008	.001	.000
5	.241	.191	.150	.116	.089	.067	.020	.003	.000
6	.378	.313	.256	.207	.165	.130	.046	.008	.000
7	.525	.453	.386	.324	.269	.220	.090	.018	.001
8	.662	.593	.523	.456	.392	.333	.155	.037	.002
9	.776	.717	.653	.587	.522	.458	.242	.070	.005
10	.862	.816	.763	.706	.645	.583	.347	.118	.011
11	.921	.888	.849	.803	.752	.697	.462	.185	.021
12	.957	.936	.909	.876	.836	.792	.576	.268	.039
13	.978	.966	.949	.926	.898	.864	.682	.363	.066
14	.990	.983	.973	.959	.940	.917	.772	.466	.105
15	.995	.992	.986	.978	.967	.951	.844	.568	.157
16	.998	.996	.993	.989	.982	.973	.899	.664	.221
17	.999	.998	.997	.995	.991	.986	.937	.749	.297
18	1.000	.999	.999	.998	.996	.993	.963	.819	.381
19		1.000	.999	.999	.998	.997	.979	.875	.470
20			1.000	1.000	.999	.998	.988	.917	.559
21					1.000	.999	.994	.947	.644
22						1.000	.997	.967	.721
23							.999	.981	.787

Table 2(a) (continued)

	μ								
a	7.5	8.0	8.5	9.0	9.5	10.0	12.0	15.0	20.0
24							.999	.989	.843
25							1.000	.994	.888
26								.997	.922
27								.998	.948
28								.999	.966
29								1.000	.978
30									.987
31									.992
32									.995
33									.997
34									.999
35									.999
36									1.000

Table 2(b)
Values of $e^{-\mu}$

μ	$e^{-\mu}$	μ	$e^{-\mu}$	μ	$e^{-\mu}$
0.00	1.000000	1.40	.246597	2.80	.060810
0.05	.951229	1.45	.234570	2.85	.057844
0.10	.904837	1.50	.223130	2.90	.055023
0.15	.860708	1.55	.212248	2.95	.052340
0.20	.818731	1.60	.201897	3.00	.049787
0.25	.778801	1.65	.192050	3.05	.047359
0.30	.740818	1.70	.182684	3.10	.045049
0.35	.704688	1.75	.173774	3.15	.042852
0.40	.670320	1.80	.165299	3.20	.040762
0.45	.637628	1.85	.157237	3.25	.038774
0.50	.606531	1.90	.149569	3.30	.036883
0.55	.576950	1.95	.142274	3.35	.035084
0.60	.548812	2.00	.135335	3.40	.033373
0.65	.522046	2.05	.128735	3.45	.031746
0.70	.496585	2.10	.122456	3.50	.030197
0.75	.472367	2.15	.116484	3.55	.028725
0.80	.449329	2.20	.110803	3.60	.027324
0.85	.427415	2.25	.105399	3.65	.025991
0.90	.406570	2.30	.100259	3.70	.024724
0.95	.386741	2.35	.095369	3.75	.023518
1.00	.367879	2.40	.090718	3.80	.022371
1.05	.349938	2.45	.086294	3.85	.021280
1.10	.332871	2.50	.082085	3.90	.020242
1.15	.316637	2.55	.078082	3.95	.019255
1.20	.301194	2.60	.074274	4.00	.018316
1.25	.286505	2.65	.070651	4.05	.017422
1.30	.272532	2.70	.067206	4.10	.016573
1.35	.259240	2.75	.063928	4.15	.015764

Table 2(b)
(continued)

μ	$e^{-\mu}$	μ	$e^{-\mu}$	μ	$e^{-\mu}$
4.20	.014996	6.15	.002133	8.10	.000304
4.25	.014264	6.20	.002029	8.15	.000289
4.30	.013569	6.25	.001930	8.20	.000275
4.35	.012907	6.30	.001836	8.25	.000261
4.40	.012277	6.35	.001747	8.30	.000249
4.45	.011679	6.40	.001661	8.35	.000236
4.50	.011109	6.45	.001581	8.40	.000225
4.55	.010567	6.50	.001503	8.45	.000214
4.60	.010052	6.55	.001430	8.50	.000204
4.65	.009562	6.60	.001360	8.55	.000194
4.70	.009095	6.65	.001294	8.60	.000184
4.75	.008652	6.70	.001231	8.65	.000175
4.80	.008230	6.75	.001171	8.70	.000167
4.85	.007828	6.80	.001114	8.75	.000158
4.90	.007447	6.85	.001059	8.80	.000151
4.95	.007083	6.90	.001008	8.85	.000143
5.00	.006738	6.95	.000959	8.90	.000136
5.05	.006409	7.00	.000912	8.95	.000130
5.10	.006097	7.05	.000867	9.00	.000123
5.15	.005799	7.10	.000825	9.05	.000117
5.20	.005517	7.15	.000785	9.10	.000112
5.25	.005248	7.20	.000747	9.15	.000106
5.30	.004992	7.25	.000710	9.20	.000101
5.35	.004748	7.30	.000676	9.25	.000096
5.40	.004517	7.35	.000643	9.30	.000091
5.45	.004296	7.40	.000611	9.35	.000087
5.50	.004087	7.45	.000581	9.40	.000083
5.55	.003887	7.50	.000553	9.45	.000079
5.60	.003698	7.55	.000526	9.50	.000075
5.65	.003518	7.60	.000501	9.55	.000071
5.70	.003346	7.65	.000476	9.60	.000068
5.75	.003183	7.70	.000453	9.65	.000064
5.80	.003028	7.75	.000431	9.70	.000061
5.85	.002880	7.80	.000410	9.75	.000058
5.90	.002739	7.85	.000390	9.80	.000056
5.95	.002606	7.90	.000371	9.85	.000053
6.00	.002479	7.95	.000353	9.90	.000050
6.05	.002358	8.00	.000336	9.95	.000048
6.10	.002243	8.05	.000319	10.00	.000045

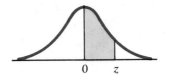

Table 3
Normal curve areas

z	.00	.01	.02	.03	.04	.05	.06	.07	.08	.09
0.0	.0000	.0040	.0080	.0120	.0160	.0199	.0239	.0279	.0319	.0359
0.1	.0398	.0438	.0478	.0517	.0557	.0596	.0636	.0675	.0714	.0753
0.2	.0793	.0832	.0871	.0910	.0948	.0987	.1026	.1064	.1103	.1141
0.3	.1179	.1217	.1255	.1293	.1331	.1368	.1406	.1443	.1480	.1517
0.4	.1554	.1591	.1628	.1664	.1700	.1736	.1772	.1808	.1844	.1879
0.5	.1915	.1950	.1985	.2019	.2054	.2088	.2123	.2157	.2190	.2224
0.6	.2257	.2291	.2324	.2357	.2389	.2422	.2454	.2486	.2517	.2549
0.7	.2580	.2611	.2642	.2673	.2704	.2734	.2764	.2794	.2823	.2852
0.8	.2881	.2910	.2939	.2967	.2995	.3023	.3051	.3078	.3106	.3133
0.9	.3159	.3186	.3212	.3238	.3264	.3289	.3315	.3340	.3365	.3389
1.0	.3413	.3438	.3461	.3485	.3508	.3531	.3554	.3577	.3599	.3621
1.1	.3643	.3665	.3686	.3708	.3729	.3749	.3770	.3790	.3810	.3830
1.2	.3849	.3869	.3888	.3907	.3925	.3944	.3962	.3980	.3997	.4015
1.3	.4032	.4049	.4066	.4082	.4099	.4115	.4131	.4147	.4162	.4177
1.4	.4192	.4207	.4222	.4236	.4251	.4265	.4279	.4292	.4306	.4319
1.5	.4332	.4345	.4357	.4370	.4382	.4394	.4406	.4418	.4429	.4441
1.6	.4452	.4463	.4474	.4484	.4495	.4505	.4515	.4525	.4535	.4545
1.7	.4554	.4564	.4573	.4582	.4591	.4599	.4608	.4616	.4625	.4633
1.8	.4641	.4649	.4656	.4664	.4671	.4678	.4686	.4693	.4699	.4706
1.9	.4713	.4719	.4726	.4732	.4738	.4744	.4750	.4756	.4761	.4767
2.0	.4772	.4778	.4783	.4788	.4793	.4798	.4803	.4808	.4812	.4817
2.1	.4821	.4826	.4830	.4834	.4838	.4842	.4846	.4850	.4854	.4857
2.2	.4861	.4864	.4868	.4871	.4875	.4878	.4881	.4884	.4887	.4890
2.3	.4893	.4896	.4898	.4901	.4904	.4906	.4909	.4911	.4913	.4916
2.4	.4918	.4920	.4922	.4925	.4927	.4929	.4931	.4932	.4934	.4936
2.5	.4938	.4940	.4941	.4943	.4945	.4946	.4948	.4949	.4951	.4952
2.6	.4953	.4955	.4956	.4957	.4959	.4960	.4961	.4962	.4963	.4964
2.7	.4965	.4966	.4967	.4968	.4969	.4970	.4971	.4972	.4973	.4974
2.8	.4974	.4975	.4976	.4977	.4977	.4978	.4979	.4979	.4980	.4981
2.9	.4981	.4982	.4982	.4983	.4984	.4984	.4985	.4985	.4986	.4986
3.0	.4987	.4987	.4987	.4988	.4988	.4989	.4989	.4989	.4990	.4990

Source: This table is abridged from Table 1 of *Statistical Tables and Formulas*, by A. Hald (New York: Wiley, 1952). Reproduced by permission of A. Hald and the publisher, John Wiley & Sons, Inc.

Table 4 Critical values of t	d.f.	$t_{.100}$	$t_{.050}$	$t_{.025}$	$t_{.010}$	$t_{.005}$	d.f.

d.f.	$t_{.100}$	$t_{.050}$	$t_{.025}$	$t_{.010}$	$t_{.005}$	d.f.
1	3.078	6.314	12.706	31.821	63.657	1
2	1.886	2.920	4.303	6.965	9.925	2
3	1.638	2.353	3.182	4.541	5.841	3
4	1.533	2.132	2.776	3.747	4.604	4
5	1.476	2.015	2.571	3.365	4.032	5
6	1.440	1.943	2.447	3.143	3.707	6
7	1.415	1.895	2.365	2.998	3.499	7
8	1.397	1.860	2.306	2.896	3.355	8
9	1.383	1.833	2.262	2.821	3.250	9
10	1.372	1.812	2.228	2.764	3.169	10
11	1.363	1.796	2.201	2.718	3.106	11
12	1.356	1.782	2.179	2.681	3.055	12
13	1.350	1.771	2.160	2.650	3.012	13
14	1.345	1.761	2.145	2.624	2.977	14
15	1.341	1.753	2.131	2.602	2.947	15
16	1.337	1.746	2.120	2.583	2.921	16
17	1.333	1.740	2.110	2.567	2.898	17
18	1.330	1.734	2.101	2.552	2.878	18
19	1.328	1.729	2.093	2.539	2.861	19
20	1.325	1.725	2.086	2.528	2.845	20
21	1.323	1.721	2.080	2.518	2.831	21
22	1.321	1.717	2.074	2.508	2.819	22
23	1.319	1.714	2.069	2.500	2.807	23
24	1.318	1.711	2.064	2.492	2.797	24
25	1.316	1.708	2.060	2.485	2.787	25
26	1.315	1.706	2.056	2.479	2.779	26
27	1.314	1.703	2.052	2.473	2.771	27
28	1.313	1.701	2.048	2.467	2.763	28
29	1.311	1.699	2.045	2.462	2.756	29
inf.	1.282	1.645	1.960	2.326	2.576	inf.

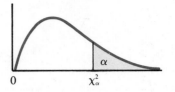

d.f.	$\chi^2_{0.995}$	$\chi^2_{0.990}$	$\chi^2_{0.975}$	$\chi^2_{0.950}$	$\chi^2_{0.900}$
1	0.0000393	0.0001571	0.0009821	0.0039321	0.0157908
2	0.0100251	0.0201007	0.0506356	0.102587	0.210720
3	0.0717212	0.114832	0.215795	0.351846	0.584375
4	0.206990	0.297110	0.484419	0.710721	1.063623
5	0.411740	0.554300	0.831211	1.145476	1.61031
6	0.675727	0.872085	1.237347	1.63539	2.20413
7	0.989265	1.239043	1.68987	2.16735	2.83311
8	1.344419	1.646482	2.17973	2.73264	3.48954
9	1.734926	2.087912	2.70039	3.32511	4.16816
10	2.15585	2.55821	3.24697	3.94030	4.86518
11	2.60321	3.05347	3.81575	4.57481	5.57779
12	3.07382	3.57056	4.40379	5.22603	6.30380
13	3.56503	4.10691	5.00874	5.89186	7.04150
14	4.07468	4.66043	5.62872	6.57063	7.78953
15	4.60094	5.22935	6.26214	7.26094	8.54675
16	5.14224	5.81221	6.90766	7.96164	9.31223
17	5.69724	6.40776	7.56418	8.67176	10.0852
18	6.26481	7.01491	8.23075	9.39046	10.8649
19	6.84398	7.63273	8.90655	10.1170	11.6509
20	7.43386	8.26040	9.59083	10.8508	12.4426
21	8.03366	8.89720	10.28293	11.5913	13.2396
22	8.64272	9.54249	10.9823	12.3380	14.0415
23	9.26042	10.19567	11.6885	13.0905	14.8479
24	9.88623	10.8564	12.4011	13.8484	15.6587
25	10.5197	11.5240	13.1197	14.6114	16.4734
26	11.1603	12.1981	13.8439	15.3791	17.2919
27	11.8076	12.8786	14.5733	16.1513	18.1138
28	12.4613	13.5648	15.3079	16.9279	18.9392
29	13.1211	14.2565	16.0471	17.7083	19.7677
30	13.7867	14.9535	16.7908	18.4926	20.5992
40	20.7065	22.1643	24.4331	26.5093	29.0505
50	27.9907	29.7067	32.3574	34.7642	37.6886
60	35.5346	37.4848	40.4817	43.1879	46.4589
70	43.2752	45.4418	48.7576	51.7393	55.3290
80	51.1720	53.5400	57.1532	60.3915	64.2778
90	59.1963	61.7541	65.6466	69.1260	73.2912
100	67.3276	70.0648	74.2219	77.9295	82.3581

Source: From "Tables of the Percentage Points of the χ^2-Distribution," *Biometrika Tables for Statisticians* 1, 3d ed. (1966). Reproduced by permission of the *Biometrika* Trustees.

Table 5
(continued)

$\chi^2_{0.100}$	$\chi^2_{0.050}$	$\chi^2_{0.025}$	$\chi^2_{0.010}$	$\chi^2_{0.005}$	d.f.
2.70554	3.84146	5.02389	6.63490	7.87944	1
4.60517	5.99147	7.37776	9.21034	10.5966	2
6.25139	7.81473	9.34840	11.3449	12.8381	3
7.77944	9.48773	11.1433	13.2767	14.8602	4
9.23635	11.0705	12.8325	15.0863	16.7496	5
10.6446	12.5916	14.4494	16.8119	18.5476	6
12.0170	14.0671	16.0128	18.4753	20.2777	7
13.3616	15.5073	17.5346	20.0902	21.9550	8
14.6837	16.9190	19.0228	21.6660	23.5893	9
15.9871	18.3070	20.4831	23.2093	25.1882	10
17.2750	19.6751	21.9200	24.7250	26.7569	11
18.5494	21.0261	23.3367	26.2170	28.2995	12
19.8119	22.3621	24.7356	27.6883	29.8194	13
21.0642	23.6848	26.1190	29.1413	31.3193	14
22.3072	24.9958	27.4884	30.5779	32.8013	15
23.5418	26.2962	28.8485	31.9999	34.2672	16
24.7690	27.5871	30.1910	33.4087	35.7185	17
25.9894	28.8693	31.5264	34.8053	37.1564	18
27.2036	30.1435	32.8523	36.1908	38.5822	19
28.4120	31.4104	34.1696	37.5662	39.9968	20
29.6151	32.6705	35.4789	38.9321	41.4010	21
30.8133	33.9244	36.7807	40.2894	42.7956	22
32.0069	35.1725	38.0757	41.6384	44.1813	23
33.1963	36.4151	39.3641	42.9798	45.5585	24
34.3816	37.6525	40.6465	44.3141	46.9278	25
35.5631	38.8852	41.9232	45.6417	48.2899	26
36.7412	40.1133	43.1944	46.9630	49.6449	27
37.9159	41.3372	44.4607	48.2782	50.9933	28
39.0875	42.5569	45.7222	49.5879	52.3356	29
40.2560	43.7729	46.9792	50.8922	53.6720	30
51.8050	55.7585	59.3417	63.6907	66.7659	40
63.1671	67.5048	71.4202	76.1539	79.4900	50
74.3970	79.0819	83.2976	88.3794	91.9517	60
85.5271	90.5312	95.0231	100.425	104.215	70
96.5782	101.879	106.629	112.329	116.321	80
107.565	113.145	118.136	124.116	128.299	90
118.498	124.342	129.561	135.807	140.169	100

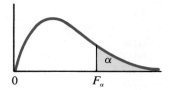

Table 6   Percentage points of the F distribution: $\alpha = .10$

ν_2 (d.f.)	ν_1 (d.f.)								
	1	2	3	4	5	6	7	8	9
1	39.86	49.50	53.59	55.83	57.24	58.20	58.91	59.44	59.86
2	8.53	9.00	9.16	9.24	9.29	9.33	9.35	9.37	9.38
3	5.54	5.46	5.39	5.34	5.31	5.28	5.27	5.25	5.24
4	4.54	4.32	4.19	4.11	4.05	4.01	3.98	3.95	3.94
5	4.06	3.78	3.62	3.52	3.45	3.40	3.37	3.34	3.32
6	3.78	3.46	3.29	3.18	3.11	3.05	3.01	2.98	2.96
7	3.59	3.26	3.07	2.96	2.88	2.83	2.78	2.75	2.72
8	3.46	3.11	2.92	2.81	2.73	2.67	2.62	2.59	2.56
9	3.36	3.01	2.81	2.69	2.61	2.55	2.51	2.47	2.44
10	3.29	2.92	2.73	2.61	2.52	2.46	2.41	2.38	2.35
11	3.23	2.86	2.66	2.54	2.45	2.39	2.34	2.30	2.27
12	3.18	2.81	2.61	2.48	2.39	2.33	2.28	2.24	2.21
13	3.14	2.76	2.56	2.43	2.35	2.28	2.23	2.20	2.16
14	3.10	2.73	2.52	2.39	2.31	2.24	2.19	2.15	2.12
15	3.07	2.70	2.49	2.36	2.27	2.21	2.16	2.12	2.09
16	3.05	2.67	2.46	2.33	2.24	2.18	2.13	2.09	2.06
17	3.03	2.64	2.44	2.31	2.22	2.15	2.10	2.06	2.03
18	3.01	2.62	2.42	2.29	2.20	2.13	2.08	2.04	2.00
19	2.99	2.61	2.40	2.27	2.18	2.11	2.06	2.02	1.98
20	2.97	2.59	2.38	2.25	2.16	2.09	2.04	2.00	1.96
21	2.96	2.57	2.36	2.23	2.14	2.08	2.02	1.98	1.95
22	2.95	2.56	2.35	2.22	2.13	2.06	2.01	1.97	1.93
23	2.94	2.55	2.34	2.21	2.11	2.05	1.99	1.95	1.92
24	2.93	2.54	2.33	2.19	2.10	2.04	1.98	1.94	1.91
25	2.92	2.53	2.32	2.18	2.09	2.02	1.97	1.93	1.89
26	2.91	2.52	2.31	2.17	2.08	2.01	1.96	1.92	1.88
27	2.90	2.51	2.30	2.17	2.07	2.00	1.95	1.91	1.87
28	2.89	2.50	2.29	2.16	2.06	2.00	1.94	1.90	1.87
29	2.89	2.50	2.28	2.15	2.06	1.99	1.93	1.89	1.86
30	2.88	2.49	2.28	2.14	2.05	1.98	1.93	1.88	1.85
40	2.84	2.44	2.23	2.09	2.00	1.93	1.87	1.83	1.79
60	2.79	2.39	2.18	2.04	1.95	1.87	1.82	1.77	1.74
120	2.75	2.35	2.13	1.99	1.90	1.82	1.77	1.72	1.68
∞	2.71	2.30	2.08	1.94	1.85	1.77	1.72	1.67	1.63

Table 6   (continued)

				v_1 (d.f.)							
10	12	15	20	24	30	40	60	120	∞	v_2 (d.f.)
60.19	60.71	61.22	61.74	62.00	62.26	62.53	62.79	63.06	63.33	1
9.39	9.41	9.42	9.44	9.45	9.46	9.47	9.47	9.48	9.49	2
5.23	5.22	5.20	5.18	5.18	5.17	5.16	5.15	5.14	5.13	3
3.92	3.90	3.87	3.84	3.83	3.82	3.80	3.79	3.78	3.76	4
3.30	3.27	3.24	3.21	3.19	3.17	3.16	3.14	3.12	3.10	5
2.94	2.90	2.87	2.84	2.82	2.80	2.78	2.76	2.74	2.72	6
2.70	2.67	2.63	2.59	2.58	2.56	2.54	2.51	2.49	2.47	7
2.54	2.50	2.46	2.42	2.40	2.38	2.36	2.34	2.32	2.29	8
2.42	2.38	2.34	2.30	2.28	2.25	2.23	2.21	2.18	2.16	9
2.32	2.28	2.24	2.20	2.18	2.16	2.13	2.11	2.08	2.06	10
2.25	2.21	2.17	2.12	2.10	2.08	2.05	2.03	2.00	1.97	11
2.19	2.15	2.10	2.06	2.04	2.01	1.99	1.96	1.93	1.90	12
2.14	2.10	2.05	2.01	1.98	1.96	1.93	1.90	1.88	1.85	13
2.10	2.05	2.01	1.96	1.94	1.91	1.89	1.86	1.83	1.80	14
2.06	2.02	1.97	1.92	1.90	1.87	1.85	1.82	1.79	1.76	15
2.03	1.99	1.94	1.89	1.87	1.84	1.81	1.78	1.75	1.72	16
2.00	1.96	1.91	1.86	1.84	1.81	1.78	1.75	1.72	1.69	17
1.98	1.93	1.89	1.84	1.81	1.78	1.75	1.72	1.69	1.66	18
1.96	1.91	1.86	1.81	1.79	1.76	1.73	1.70	1.67	1.63	19
1.94	1.89	1.84	1.79	1.77	1.74	1.71	1.68	1.64	1.61	20
1.92	1.87	1.83	1.78	1.75	1.72	1.69	1.66	1.62	1.59	21
1.90	1.86	1.81	1.76	1.73	1.70	1.67	1.64	1.60	1.57	22
1.89	1.84	1.80	1.74	1.72	1.69	1.66	1.62	1.59	1.55	23
1.88	1.83	1.78	1.73	1.70	1.67	1.64	1.61	1.57	1.53	24
1.87	1.82	1.77	1.72	1.69	1.66	1.63	1.59	1.56	1.52	25
1.86	1.81	1.76	1.71	1.68	1.65	1.61	1.58	1.54	1.50	26
1.85	1.80	1.75	1.70	1.67	1.64	1.60	1.57	1.53	1.49	27
1.84	1.79	1.74	1.69	1.66	1.63	1.59	1.56	1.52	1.48	28
1.83	1.78	1.73	1.68	1.65	1.62	1.58	1.55	1.51	1.47	29
1.82	1.77	1.72	1.67	1.64	1.61	1.57	1.54	1.50	1.46	30
1.76	1.71	1.66	1.61	1.57	1.54	1.51	1.47	1.42	1.38	40
1.71	1.66	1.60	1.54	1.51	1.48	1.44	1.40	1.35	1.29	60
1.65	1.60	1.55	1.48	1.45	1.41	1.37	1.32	1.26	1.19	120
1.60	1.55	1.49	1.42	1.38	1.34	1.30	1.24	1.17	1.00	∞

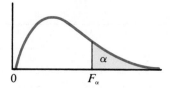

Table 7 Percentage points of the F distribution: $\alpha = .05$

ν_2 (d.f.)	ν_1 (d.f.)								
	1	2	3	4	5	6	7	8	9
1	161.4	199.5	215.7	224.6	230.2	234.0	236.8	238.9	240.5
2	18.51	19.00	19.16	19.25	19.30	19.33	19.35	19.37	19.38
3	10.13	9.55	9.28	9.12	9.01	8.94	8.89	8.85	8.81
4	7.71	6.94	6.59	6.39	6.26	6.16	6.09	6.04	6.00
5	6.61	5.79	5.41	5.19	5.05	4.95	4.88	4.82	4.77
6	5.99	5.14	4.76	4.53	4.39	4.28	4.21	4.15	4.10
7	5.59	4.74	4.35	4.12	3.97	3.87	3.79	3.73	3.68
8	5.32	4.46	4.07	3.84	3.69	3.58	3.50	3.44	3.39
9	5.12	4.26	3.86	3.63	3.48	3.37	3.29	3.23	3.18
10	4.96	4.10	3.71	3.48	3.33	3.22	3.14	3.07	3.02
11	4.84	3.98	3.59	3.36	3.20	3.09	3.01	2.95	2.90
12	4.75	3.89	3.49	3.26	3.11	3.00	2.91	2.85	2.80
13	4.67	3.81	3.41	3.18	3.03	2.92	2.83	2.77	2.71
14	4.60	3.74	3.34	3.11	2.96	2.85	2.76	2.70	2.65
15	4.54	3.68	3.29	3.06	2.90	2.79	2.71	2.64	2.59
16	4.49	3.63	3.24	3.01	2.85	2.74	2.66	2.59	2.54
17	4.45	3.59	3.20	2.96	2.81	2.70	2.61	2.55	2.49
18	4.41	3.55	3.16	2.93	2.77	2.66	2.58	2.51	2.46
19	4.38	3.52	3.13	2.90	2.74	2.63	2.54	2.48	2.42
20	4.35	3.49	3.10	2.87	2.71	2.60	2.51	2.45	2.39
21	4.32	3.47	3.07	2.84	2.68	2.57	2.49	2.42	2.37
22	4.30	3.44	3.05	2.82	2.66	2.55	2.46	2.40	2.34
23	4.28	3.42	3.03	2.80	2.64	2.53	2.44	2.37	2.32
24	4.26	3.40	3.01	2.78	2.62	2.51	2.42	2.36	2.30
25	4.24	3.39	2.99	2.76	2.60	2.49	2.40	2.34	2.28
26	4.23	3.37	2.98	2.74	2.59	2.47	2.39	2.32	2.27
27	4.21	3.35	2.96	2.73	2.57	2.46	2.37	2.31	2.25
28	4.20	3.34	2.95	2.71	2.56	2.45	2.36	2.29	2.24
29	4.18	3.33	2.93	2.70	2.55	2.43	2.35	2.28	2.22
30	4.17	3.32	2.92	2.69	2.53	2.42	2.33	2.27	2.21
40	4.08	3.23	2.84	2.61	2.45	2.34	2.25	2.18	2.12
60	4.00	3.15	2.76	2.53	2.37	2.25	2.17	2.10	2.04
120	3.92	3.07	2.68	2.45	2.29	2.17	2.09	2.02	1.96
∞	3.84	3.00	2.60	2.37	2.21	2.10	2.01	1.94	1.88

Source: From "Tables of Percentage Points of the Inverted Beta (F)-Distribution," *Biometrika* 33 (1943) 73–88, by Maxine Merrington and Catherine M. Thompson. Reproduced by permission of the *Biometrika* Trustees.

Table 7   (continued)

10	12	15	20	24	30	40	60	120	∞	ν_2 (d.f.)
				ν_1 (d.f.)						
241.9	243.9	245.9	248.0	249.1	250.1	251.1	252.2	253.3	254.3	1
19.40	19.41	19.43	19.45	19.45	19.46	19.47	19.48	19.49	19.50	2
8.79	8.74	8.70	8.66	8.64	8.62	8.59	8.57	8.55	8.53	3
5.96	5.91	5.86	5.80	5.77	5.75	5.72	5.69	5.66	5.63	4
4.74	4.68	4.62	4.56	4.53	4.50	4.46	4.43	4.40	4.36	5
4.06	4.00	3.94	3.87	3.84	3.81	3.77	3.74	3.70	3.67	6
3.64	3.57	3.51	3.44	3.41	3.38	3.34	3.30	3.27	3.23	7
3.35	3.28	3.22	3.15	3.12	3.08	3.04	3.01	2.97	2.93	8
3.14	3.07	3.01	2.94	2.90	2.86	2.83	2.79	2.75	2.71	9
2.98	2.91	2.85	2.77	2.74	2.70	2.66	2.62	2.58	2.54	10
2.85	2.79	2.72	2.65	2.61	2.57	2.53	2.49	2.45	2.40	11
2.75	2.69	2.62	2.54	2.51	2.47	2.43	2.38	2.34	2.30	12
2.67	2.60	2.53	2.46	2.42	2.38	2.34	2.30	2.25	2.21	13
2.60	2.53	2.46	2.39	2.35	2.31	2.27	2.22	2.18	2.13	14
2.54	2.48	2.40	2.33	2.29	2.25	2.20	2.16	2.11	2.07	15
2.49	2.42	2.35	2.28	2.24	2.19	2.15	2.11	2.06	2.01	16
2.45	2.38	2.31	2.23	2.19	2.15	2.10	2.06	2.01	1.96	17
2.41	2.34	2.27	2.19	2.15	2.11	2.06	2.02	1.97	1.92	18
2.38	2.31	2.23	2.16	2.11	2.07	2.03	1.98	1.93	1.88	19
2.35	2.28	2.20	2.12	2.08	2.04	1.99	1.95	1.90	1.84	20
2.32	2.25	2.18	2.10	2.05	2.01	1.96	1.92	1.87	1.81	21
2.30	2.23	2.15	2.07	2.03	1.98	1.94	1.89	1.84	1.78	22
2.27	2.20	2.13	2.05	2.01	1.96	1.91	1.86	1.81	1.76	23
2.25	2.18	2.11	2.03	1.98	1.94	1.89	1.84	1.79	1.73	24
2.24	2.16	2.09	2.01	1.96	1.92	1.87	1.82	1.77	1.71	25
2.22	2.15	2.07	1.99	1.95	1.90	1.85	1.80	1.75	1.69	26
2.20	2.13	2.06	1.97	1.93	1.88	1.84	1.79	1.73	1.67	27
2.19	2.12	2.04	1.96	1.91	1.87	1.82	1.77	1.71	1.65	28
2.18	2.10	2.03	1.94	1.90	1.85	1.81	1.75	1.70	1.64	29
2.16	2.09	2.01	1.93	1.89	1.84	1.79	1.74	1.68	1.62	30
2.08	2.00	1.92	1.84	1.79	1.74	1.69	1.64	1.58	1.51	40
1.99	1.92	1.84	1.75	1.70	1.65	1.59	1.53	1.47	1.39	60
1.91	1.83	1.75	1.66	1.61	1.55	1.50	1.43	1.35	1.25	120
1.83	1.75	1.67	1.57	1.52	1.46	1.39	1.32	1.22	1.00	∞

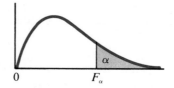

Table 8 Percentage points of the F distribution: $\alpha = .025$

ν_2 (d.f.)	ν_1 (d.f.)								
	1	2	3	4	5	6	7	8	9
1	647.8	799.5	864.2	899.6	921.8	937.1	948.2	956.7	963.3
2	38.51	39.00	39.17	39.25	39.30	39.33	39.36	39.37	39.39
3	17.44	16.04	15.44	15.10	14.88	14.73	14.62	14.54	14.47
4	12.22	10.65	9.98	9.60	9.36	9.20	9.07	8.98	8.90
5	10.01	8.43	7.76	7.39	7.15	6.98	6.85	6.76	6.68
6	8.81	7.26	6.60	6.23	5.99	5.82	5.70	5.60	5.52
7	8.07	6.54	5.89	5.52	5.29	5.12	4.99	4.90	4.82
8	7.57	6.06	5.42	5.05	4.82	4.65	4.53	4.43	4.36
9	7.21	5.71	5.08	4.72	4.48	4.32	4.20	4.10	4.03
10	6.94	5.46	4.83	4.47	4.24	4.07	3.95	3.85	3.78
11	6.72	5.26	4.63	4.28	4.04	3.88	3.76	3.66	3.59
12	6.55	5.10	4.47	4.12	3.89	3.73	3.61	3.51	3.44
13	6.41	4.97	4.35	4.00	3.77	3.60	3.48	3.39	3.31
14	6.30	4.86	4.24	3.89	3.66	3.50	3.38	3.29	3.21
15	6.20	4.77	4.15	3.80	3.58	3.41	3.29	3.20	3.12
16	6.12	4.69	4.08	3.73	3.50	3.34	3.22	3.12	3.05
17	6.04	4.62	4.01	3.66	3.44	3.28	3.16	3.06	2.98
18	5.98	4.56	3.95	3.61	3.38	3.22	3.10	3.01	2.93
19	5.92	4.51	3.90	3.56	3.33	3.17	3.05	2.96	2.88
20	5.87	4.46	3.86	3.51	3.29	3.13	3.01	2.91	2.84
21	5.83	4.42	3.82	3.48	3.25	3.09	2.97	2.87	2.80
22	5.79	4.38	3.78	3.44	3.22	3.05	2.93	2.84	2.76
23	5.75	4.35	3.75	3.41	3.18	3.02	2.90	2.81	2.73
24	5.72	4.32	3.72	3.38	3.15	2.99	2.87	2.78	2.70
25	5.69	4.29	3.69	3.35	3.13	2.97	2.85	2.75	2.68
26	5.66	4.27	3.67	3.33	3.10	2.94	2.82	2.73	2.65
27	5.63	4.24	3.65	3.31	3.08	2.92	2.80	2.71	2.63
28	5.61	4.22	3.63	3.29	3.06	2.90	2.78	2.69	2.61
29	5.59	4.20	3.61	3.27	3.04	2.88	2.76	2.67	2.59
30	5.57	4.18	3.59	3.25	3.03	2.87	2.75	2.65	2.57
40	5.42	4.05	3.46	3.13	2.90	2.74	2.62	2.53	2.45
60	5.29	3.93	3.34	3.01	2.79	2.63	2.51	2.41	2.33
120	5.15	3.80	3.23	2.89	2.67	2.52	2.39	2.30	2.22
∞	5.02	3.69	3.12	2.79	2.57	2.41	2.29	2.19	2.11

Table 8   (continued)

				ν_1 (d.f.)						
10	12	15	20	24	30	40	60	120	∞	ν_2 (d.f.)
968.6	976.7	984.9	993.1	997.2	1001	1006	1010	1014	1018	1
39.40	39.41	39.43	39.45	39.46	39.46	39.47	39.48	39.49	39.50	2
14.42	14.34	14.25	14.17	14.12	14.08	14.04	13.99	13.95	13.90	3
8.84	8.75	8.66	8.56	8.51	8.46	8.41	8.36	8.31	8.26	4
6.62	6.52	6.43	6.33	6.28	6.23	6.18	6.12	6.07	6.02	5
5.46	5.37	5.27	5.17	5.12	5.07	5.01	4.96	4.90	4.85	6
4.76	4.67	4.57	4.47	4.42	4.36	4.31	4.25	4.20	4.14	7
4.30	4.20	4.10	4.00	3.95	3.89	3.84	3.78	3.73	3.67	8
3.96	3.87	3.77	3.67	3.61	3.56	3.51	3.45	3.39	3.33	9
3.72	3.62	3.52	3.42	3.37	3.31	3.26	3.20	3.14	3.08	10
3.53	3.43	3.33	3.23	3.17	3.12	3.06	3.00	2.94	2.88	11
3.37	3.28	3.18	3.07	3.02	2.96	2.91	2.85	2.79	2.72	12
3.25	3.15	3.05	2.95	2.89	2.84	2.78	2.72	2.66	2.60	13
3.15	3.05	2.95	2.84	2.79	2.73	2.67	2.61	2.55	2.49	14
3.06	2.96	2.86	2.76	2.70	2.64	2.59	2.52	2.46	2.40	15
2.99	2.89	2.79	2.68	2.63	2.57	2.51	2.45	2.38	2.32	16
2.92	2.82	2.72	2.62	2.56	2.50	2.44	2.38	2.32	2.25	17
2.87	2.77	2.67	2.56	2.50	2.44	2.38	2.32	2.26	2.19	18
2.82	2.72	2.62	2.51	2.45	2.39	2.33	2.27	2.20	2.13	19
2.77	2.68	2.57	2.46	2.41	2.35	2.29	2.22	2.16	2.09	20
2.73	2.64	2.53	2.42	2.37	2.31	2.25	2.18	2.11	2.04	21
2.70	2.60	2.50	2.39	2.33	2.27	2.21	2.14	2.08	2.00	22
2.67	2.57	2.47	2.36	2.30	2.24	2.18	2.11	2.04	1.97	23
2.64	2.54	2.44	2.33	2.27	2.21	2.15	2.08	2.01	1.94	24
2.61	2.51	2.41	2.30	2.24	2.18	2.12	2.05	1.98	1.91	25
2.59	2.49	2.39	2.28	2.22	2.16	2.09	2.03	1.95	1.88	26
2.57	2.47	2.36	2.25	2.19	2.13	2.07	2.00	1.93	1.85	27
2.55	2.45	2.34	2.23	2.17	2.11	2.05	1.98	1.91	1.83	28
2.53	2.43	2.32	2.21	2.15	2.09	2.03	1.96	1.89	1.81	29
2.51	2.41	2.31	2.20	2.14	2.07	2.01	1.94	1.87	1.79	30
2.39	2.29	2.18	2.07	2.01	1.94	1.88	1.80	1.72	1.64	40
2.27	2.17	2.06	1.94	1.88	1.82	1.74	1.67	1.58	1.48	60
2.16	2.05	1.94	1.82	1.76	1.69	1.61	1.53	1.43	1.31	120
2.05	1.94	1.83	1.71	1.64	1.57	1.48	1.39	1.27	1.00	∞

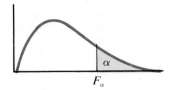

Table 9 Percentage points of the F distribution: $\alpha = .01$

ν_2 (d.f.)	ν_1 (d.f.)								
	1	2	3	4	5	6	7	8	9
1	4052	4999.5	5403	5625	5764	5859	5928	5982	6022
2	98.50	99.00	99.17	99.25	99.30	99.33	99.36	99.37	99.39
3	34.12	30.82	29.46	28.71	28.24	27.91	27.67	27.49	27.35
4	21.20	18.00	16.69	15.98	15.52	15.21	14.98	14.80	14.66
5	16.26	13.27	12.06	11.39	10.97	10.67	10.46	10.29	10.16
6	13.75	10.92	9.78	9.15	8.75	8.47	8.26	8.10	7.98
7	12.25	9.55	8.45	7.85	7.46	7.19	6.99	6.84	6.72
8	11.26	8.65	7.59	7.01	6.63	6.37	6.18	6.03	5.91
9	10.56	8.02	6.99	6.42	6.06	5.80	5.61	5.47	5.35
10	10.04	7.56	6.55	5.99	5.64	5.39	5.20	5.06	4.94
11	9.65	7.21	6.22	5.67	5.32	5.07	4.89	4.74	4.63
12	9.33	6.93	5.95	5.41	5.06	4.82	4.64	4.50	4.39
13	9.07	6.70	5.74	5.21	4.86	4.62	4.44	4.30	4.19
14	8.86	6.51	5.56	5.04	4.69	4.46	4.28	4.14	4.03
15	8.68	6.36	5.42	4.89	4.56	4.32	4.14	4.00	3.89
16	8.53	6.23	5.29	4.77	4.44	4.20	4.03	3.89	3.78
17	8.40	6.11	5.18	4.67	4.34	4.10	3.93	3.79	3.68
18	8.29	6.01	5.09	4.58	4.25	4.01	3.84	3.71	3.60
19	8.18	5.93	5.01	4.50	4.17	3.94	3.77	3.63	3.52
20	8.10	5.85	4.94	4.43	4.10	3.87	3.70	3.56	3.46
21	8.02	5.78	4.87	4.37	4.04	3.81	3.64	3.51	3.40
22	7.95	5.72	4.82	4.31	3.99	3.76	3.59	3.45	3.35
23	7.88	5.66	4.76	4.26	3.94	3.71	3.54	3.41	3.30
24	7.82	5.61	4.72	4.22	3.90	3.67	3.50	3.36	3.26
25	7.77	5.57	4.68	4.18	3.85	3.63	3.46	3.32	3.22
26	7.72	5.53	4.64	4.14	3.82	3.59	3.42	3.29	3.18
27	7.68	5.49	4.60	4.11	3.78	3.56	3.39	3.26	3.15
28	7.64	5.45	4.57	4.07	3.75	3.53	3.36	3.23	3.12
29	7.60	5.42	4.54	4.04	3.73	3.50	3.33	3.20	3.09
30	7.56	5.39	4.51	4.02	3.70	3.47	3.30	3.17	3.07
40	7.31	5.18	4.31	3.83	3.51	3.29	3.12	2.99	2.89
60	7.08	4.98	4.13	3.65	3.34	3.12	2.95	2.82	2.72
120	6.85	4.79	3.95	3.48	3.17	2.96	2.79	2.66	2.56
∞	6.63	4.61	3.78	3.32	3.02	2.80	2.64	2.51	2.41

Source: From "Tables of Percentage Points of the Inverted Beta (F)-Distribution," *Biometrika* 33 (1943) 73–88, by Maxine Merrington and Catherine M. Thompson. Reproduced by permission of the *Biometrika* Trustees.

Table 9   (continued)

10	12	15	20	24	30	40	60	120	∞	v_2 (d.f.)
					v_1 (d.f.)					
6056	6106	6157	6209	6235	6261	6287	6313	6339	6366	1
99.40	99.42	99.43	99.45	99.46	99.47	99.47	99.48	99.49	99.50	2
27.23	27.05	26.87	26.69	26.60	26.50	26.41	.26.32	26.22	26.13	3
14.55	14.37	14.20	14.02	13.93	13.84	13.75	13.65	13.56	13.46	4
10.05	9.89	9.72	9.55	9.47	9.38	9.29	9.20	9.11	9.02	5
7.87	7.72	7.56	7.40	7.31	7.23	7.14	7.06	6.97	6.88	6
6.62	6.47	6.31	6.16	6.07	5.99	5.91	5.82	5.74	5.65	7
5.81	5.67	5.52	5.36	5.28	5.20	5.12	5.03	4.95	4.86	8
5.26	5.11	4.96	4.81	4.73	4.65	4.57	4.48	4.40	4.31	9
4.85	4.71	4.56	4.41	4.33	4.25	4.17	4.08	4.00	3.91	10
4.54	4.40	4.25	4.10	4.02	3.94	3.86	3.78	3.69	3.60	11
4.30	4.16	4.01	3.86	3.78	3.70	3.62	3.54	3.45	3.36	12
4.10	3.96	3.82	3.66	3.59	3.51	3.43	3.34	3.25	3.17	13
3.94	3.80	3.66	3.51	3.43	3.35	3.27	3.18	3.09	3.00	14
3.80	3.67	3.52	3.37	3.29	3.21	3.13	3.05	2.96	2.87	15
3.69	3.55	3.41	3.26	3.18	3.10	3.02	2.93	2.84	2.75	16
3.59	3.46	3.31	3.16	3.08	3.00	2.92	2.83	2.75	2.65	17
3.51	3.37	3.23	3.08	3.00	2.92	2.84	2.75	2.66	2.57	18
3.43	3.30	3.15	3.00	2.92	2.84	2.76	2.67	2.58	2.49	19
3.37	3.23	3.09	2.94	2.86	2.78	2.69	2.61	2.52	2.42	20
3.31	3.17	3.03	2.88	2.80	2.72	2.64	2.55	2.46	2.36	21
3.26	3.12	2.98	2.83	2.75	2.67	2.58	2.50	2.40	2.31	22
3.21	3.07	2.93	2.78	2.70	2.62	2.54	2.45	2.35	2.26	23
3.17	3.03	2.89	2.74	2.66	2.58	2.49	2.40	2.31	2.21	24
3.13	2.99	2.85	2.70	2.62	2.54	2.45	2.36	2.27	2.17	25
3.09	2.96	2.81	2.66	2.58	2.50	2.42	2.33	2.23	2.13	26
3.06	2.93	2.78	2.63	2.55	2.47	2.38	2.29	2.20	2.10	27
3.03	2.90	2.75	2.60	2.52	2.44	2.35	2.26	2.17	2.06	28
3.00	2.87	2.73	2.57	2.49	2.41	2.33	2.23	2.14	2.03	29
2.98	2.84	2.70	2.55	2.47	2.39	2.30	2.21	2.11	2.01	30
2.80	2.66	2.52	2.37	2.29	2.20	2.11	2.02	1.92	1.80	40
2.63	2.50	2.35	2.20	2.12	2.03	1.94	1.84	1.73	1.60	60
2.47	2.34	2.19	2.03	1.95	1.86	1.76	1.66	1.53	1.38	120
2.32	2.18	2.04	1.88	1.79	1.70	1.59	1.47	1.32	1.00	∞

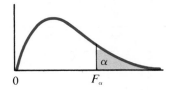

Table 10   Percentage points of the *F* distribution: $\alpha = .005$

ν_2 (d.f.)	ν_1 (d.f.)								
	1	2	3	4	5	6	7	8	9
1	16211	20000	21615	22500	23056	23437	23715	23925	24091
2	198.5	199.0	199.2	199.2	199.3	199.3	199.4	199.4	199.4
3	55.55	49.80	47.47	46.19	45.39	44.84	44.43	44.13	43.88
4	31.33	26.28	24.26	23.15	22.46	21.97	21.62	21.35	21.14
5	22.78	18.31	16.53	15.56	14.94	14.51	14.20	13.96	13.77
6	18.63	14.54	12.92	12.03	11.46	11.07	10.79	10.57	10.39
7	16.24	12.40	10.88	10.05	9.52	9.16	8.89	8.68	8.51
8	14.69	11.04	9.60	8.81	8.30	7.95	7.69	7.50	7.34
9	13.61	10.11	8.72	7.96	7.47	7.13	6.88	6.69	6.54
10	12.83	9.43	8.08	7.34	6.87	6.54	6.30	6.12	5.97
11	12.23	8.91	7.60	6.88	6.42	6.10	5.86	5.68	5.54
12	11.75	8.51	7.23	6.52	6.07	5.76	5.52	5.35	5.20
13	11.37	8.19	6.93	6.23	5.79	5.48	5.25	5.08	4.94
14	11.06	7.92	6.68	6.00	5.56	5.26	5.03	4.86	4.72
15	10.80	7.70	6.48	5.80	5.37	5.07	4.85	4.67	4.54
16	10.58	7.51	6.30	5.64	5.21	4.91	4.69	4.52	4.38
17	10.38	7.35	6.16	5.50	5.07	4.78	4.56	4.39	4.25
18	10.22	7.21	6.03	5.37	4.96	4.66	4.44	4.28	4.14
19	10.17	7.09	5.92	5.27	4.85	4.56	4.34	4.18	4.04
20	9.94	6.99	.5.82	5.17	4.76	4.47	4.26	4.09	3.96
21	9.83	6.89	5.73	5.09	4.68	4.39	4.18	4.01	3.88
22	9.73	6.81	5.65	5.02	4.61	4.32	4.11	3.94	3.81
23	9.63	6.73	5.58	4.95	4.54	4.26	4.05	3.88	3.75
24	9.55	6.66	5.52	4.89	4.49	4.20	3.99	3.83	3.69
25	9.48	6.60	5.46	4.84	4.43	4.15	3.94	3.78	3.64
26	9.41	6.54	5.41	4.79	4.38	4.10	3.89	3.73	3.60
27	9.34	6.49	5.36	4.74	4.34	4.06	3.85	3.69	3.56
28	9.28	6.44	5.32	4.70	4.30	4.02	3.81	3.65	3.52
29	9.23	6.40	5.28	4.66	4.26	3.98	3.77	3.61	3.48
30	9.18	6.35	5.24	4.62	4.23	3.95	3.74	3.58	3.45
40	8.83	6.07	4.98	4.37	3.99	3.71	3.51	3.35	3.22
60	8.49	5.79	4.73	4.14	3.76	3.49	3.29	3.13	3.01
120	8.18	5.54	4.50	3.92	3.55	3.28	3.09	2.93	2.81
∞	7.88	5.30	4.28	3.72	3.35	3.09	2.90	2.74	2.62

Source: From "Tables of Percentage Points of the Inverted Beta (*F*)-Distribution," *Biometrika* 33 (1943) 73–88, by Maxine Merrington and Catherine M. Thompson. Reproduced by permission of the *Biometrika* Trustees.

Table 10 (continued)

10	12	15	20	24	30	40	60	120	∞	ν_2 (d.f.)
24224	24426	24630	24836	24940	25044	25148	25253	25359	25465	1
199.4	199.4	199.4	199.4	199.5	199.5	199.5	199.5	199.5	199.5	2
43.69	43.39	43.08	42.78	42.62	42.47	42.31	42.15	41.99	41.83	3
20.97	20.70	20.44	20.17	20.03	19.89	19.75	19.61	19.47	19.32	4
13.62	13.38	13.15	12.90	12.78	12.66	12.53	12.40	12.27	12.14	5
10.25	10.03	9.81	9.59	9.47	9.36	9.24	9.12	9.00	8.88	6
8.38	8.18	7.97	7.75	7.65	7.53	7.42	7.31	7.19	7.08	7
7.21	7.01	6.81	6.61	6.50	6.40	6.29	6.18	6.06	5.95	8
6.42	6.23	6.03	5.83	5.73	5.62	5.52	5.41	5.30	5.19	9
5.85	5.66	5.47	5.27	5.17	5.07	4.97	4.86	4.75	4.64	10
5.42	5.24	5.05	4.86	4.76	4.65	4.55	4.44	4.34	4.23	11
5.09	4.91	4.72	4.53	4.43	4.33	4.23	4.12	4.01	3.90	12
4.82	4.64	4.46	4.27	4.17	4.07	3.97	3.87	3.76	3.65	13
4.60	4.43	4.25	4.06	3.96	3.86	3.76	3.66	3.55	3.44	14
4.42	4.25	4.07	3.88	3.79	3.69	3.58	3.48	3.37	3.26	15
4.27	4.10	3.92	3.73	3.64	3.54	3.44	3.33	3.22	3.11	16
4.14	3.97	3.79	3.61	3.51	3.41	3.31	3.21	3.10	2.98	17
4.03	3.86	3.68	3.50	3.40	3.30	3.20	3.10	2.99	2.87	18
3.93	3.76	3.59	3.40	3.31	3.21	3.11	3.00	2.89	2.78	19
3.85	3.68	3.50	3.32	3.22	3.12	3.02	2.92	2.81	2.69	20
3.77	3.60	3.43	3.24	3.15	3.05	2.95	2.84	2.73	2.61	21
3.70	3.54	3.36	3.18	3.08	2.98	2.88	2.77	2.66	2.55	22
3.64	3.47	3.30	3.12	3.02	2.92	2.82	2.71	2.60	2.48	23
3.59	3.42	3.25	3.06	2.97	2.87	2.77	2.66	2.55	2.43	24
3.54	3.37	3.20	3.01	2.92	2.82	2.72	2.61	2.50	2.38	25
3.49	3.33	3.15	2.97	2.87	2.77	2.67	2.56	2.45	2.33	26
3.45	3.28	3.11	2.93	2.83	2.73	2.63	2.52	2.41	2.29	27
3.41	3.25	3.07	2.89	2.79	2.69	2.59	2.48	2.37	2.25	28
3.38	3.21	3.04	2.86	2.76	2.66	2.56	2.45	2.33	2.21	29
3.34	3.18	3.01	2.82	2.73	2.63	2.52	2.42	2.30	2.18	30
3.12	2.95	2.78	2.60	2.50	2.40	2.30	2.18	2.06	1.93	40
2.90	2.74	2.57	2.39	2.29	2.19	2.08	1.96	1.83	1.69	60
2.71	2.54	2.37	2.19	2.09	1.98	1.87	1.75	1.61	1.43	120
2.52	2.36	2.19	2.00	1.90	1.79	1.67	1.53	1.36	1.00	∞

ν_1 (d.f.)

Table II
Distribution function of
U, $P(U \leq U_0)$; U_0 is the
argument; $n_1 \leq n_2$;
$3 \leq n_2 \leq 10$

$n_2 = 3$

U_0	n_1		
	1	2	3
0	.25	.10	.05
1	.50	.20	.10
2		.40	.20
3		.60	.35
4			.50

$n_2 = 4$

U_0	n_1			
	1	2	3	4
0	.2000	.0667	.0286	.0143
1	.4000	.1333	.0571	.0286
2	.6000	.2667	.1143	.0571
3		.4000	.2000	.1000
4		.6000	.3143	.1714
5			.4286	.2429
6			.5714	.3429
7				.4429
8				.5571

Note: Computed by M. Pagano, Department of Statistics, University of Florida.

$n_2 = 5$

U_0	n_1				
	1	2	3	4	5
0	.1667	.0476	.0179	.0079	.0040
1	.3333	.0952	.0357	.0159	.0079
2	.5000	.1905	.0714	.0317	.0159
3		.2857	.1250	.0556	.0278
4		.4286	.1964	.0952	.0476
5		.5714	.2857	.1429	.0754
6			.3929	.2063	.1111
7			.5000	.2778	.1548
8				.3651	.2103
9				.4524	.2738
10				.5476	.3452
11					.4206
12					.5000

Table 11
(continued)

$n_2 = 6$

U_0				n_1		
	1	2	3	4	5	6
0	.1429	.0357	.0119	.0048	.0022	.0011
1	.2857	.0714	.0238	.0095	.0043	.0022
2	.4286	.1429	.0476	.0190	.0087	.0043
3	.5714	.2143	.0833	.0333	.0152	.0076
4		.3214	.1310	.0571	.0260	.0130
5		.4286	.1905	.0857	.0411	.0206
6		.5714	.2738	.1286	.0628	.0325
7			.3571	.1762	.0887	.0465
8			.4524	.2381	.1234	.0660
9			.5476	.3048	.1645	.0898
10				.3810	.2143	.1201
11				.4571	.2684	.1548
12				.5429	.3312	.1970
13					.3961	.2424
14					.4654	.2944
15					.5346	.3496
16						.4091
17						.4686
18						.5314

$n_2 = 7$

U_0				n_1			
	1	2	3	4	5	6	7
0	.1250	.0278	.0083	.0030	.0013	.0006	.0003
1	.2500	.0556	.0167	.0061	.0025	.0012	.0006
2	.3750	.1111	.0333	.0121	.0051	.0023	.0012
3	.5000	.1667	.0583	.0212	.0088	.0041	.0020
4		.2500	.0917	.0364	.0152	.0070	.0035
5		.3333	.1333	.0545	.0240	.0111	.0055
6		.4444	.1917	.0818	.0366	.0175	.0087
7		.5556	.2583	.1152	.0530	.0256	.0131
8			.3333	.1576	.0745	.0367	.0189
9			.4167	.2061	.1010	.0507	.0265
10			.5000	.2636	.1338	.0688	.0364
11				.3242	.1717	.0903	.0487
12				.3939	.2159	.1171	.0641
13				.4636	.2652	.1474	.0825
14				.5364	.3194	.1830	.1043
15					.3775	.2226	.1297
16					.4381	.2669	.1588
17					.5000	.3141	.1914
18						.3654	.2279
19						.4178	.2675
20						.4726	.3100
21						.5274	.3552
22							.4024
23							.4508
24							.5000

683

Table 11
(continued)

$n_2 = 8$

U_0	\multicolumn{8}{c}{n_1}
	1	2	3	4	5	6	7	8
0	.1111	.0222	.0061	.0020	.0008	.0003	.0002	.0001
1	.2222	.0444	.0121	.0040	.0016	.0007	.0003	.0002
2	.3333	.0889	.0242	.0081	.0031	.0013	.0006	.0003
3	.4444	.1333	.0424	.0141	.0054	.0023	.0011	.0005
4	.5556	.2000	.0667	.0242	.0093	.0040	.0019	.0009
5		.2667	.0970	.0364	.0148	.0063	.0030	.0015
6		.3556	.1394	.0545	.0225	.0100	.0047	.0023
7		.4444	.1879	.0768	.0326	.0147	.0070	.0035
8		.5556	.2485	.1071	.0466	.0213	.0103	.0052
9			.3152	.1414	.0637	.0296	.0145	.0074
10			.3879	.1838	.0855	.0406	.0200	.0103
11			.4606	.2303	.1111	.0539	.0270	.0141
12			.5394	.2848	.1422	.0709	.0361	.0190
13				.3414	.1772	.0906	.0469	.0249
14				.4040	.2176	.1142	.0603	.0325
15				.4667	.2618	.1412	.0760	.0415
16				.5333	.3108	.1725	.0946	.0524
17					.3621	.2068	.1159	.0652
18					.4165	.2454	.1405	.0803
19					.4716	.2864	.1678	.0974
20					.5284	.3310	.1984	.1172
21						.3773	.2317	.1393
22						.4259	.2679	.1641
23						.4749	.3063	.1911
24						.5251	.3472	.2209
25							.3894	.2527
26							.4333	.2869
27							.4775	.3227
28							.5225	.3605
29								.3992
30								.4392
31								.4796
32								.5204

Table 11
(continued)

$n_2 = 9$

U_0	1	2	3	4	5	6	7	8	9
					n_1				
0	.1000	.0182	.0045	.0014	.0005	.0002	.0001	.0000	.0000
1	.2000	.0364	.0091	.0028	.0010	.0004	.0002	.0001	.0000
2	.3000	.0727	.0182	.0056	.0020	.0008	.0003	.0002	.0001
3	.4000	.1091	.0318	.0098	.0035	.0014	.0006	.0003	.0001
4	.5000	.1636	.0500	.0168	.0060	.0024	.0010	.0005	.0002
5		.2182	.0727	.0252	.0095	.0038	.0017	.0008	.0004
6		.2909	.1045	.0378	.0145	.0060	.0026	.0012	.0006
7		.3636	.1409	.0531	.0210	.0088	.0039	.0019	.0009
8		.4545	.1864	.0741	.0300	.0128	.0058	.0028	.0014
9		.5455	.2409	.0993	.0415	.0180	.0082	.0039	.0020
10			.3000	.1301	.0559	.0248	.0115	.0056	.0028
11			.3636	.1650	.0734	.0332	.0156	.0076	.0039
12			.4318	.2070	.0949	.0440	.0209	.0103	.0053
13			.5000	.2517	.1199	.0567	.0274	.0137	.0071
14				.3021	.1489	.0723	.0356	.0180	.0094
15				.3552	.1818	.0905	.0454	.0232	.0122
16				.4126	.2188	.1119	.0571	.0296	.0157
17				.4699	.2592	.1361	.0708	.0372	.0200
18				.5301	.3032	.1638	.0869	.0464	.0252
19					.3497	.1942	.1052	.0570	.0313
20					.3986	.2280	.1261	.0694	.0385
21					.4491	.2643	.1496	.0836	.0470
22					.5000	.3035	.1755	.0998	.0567
23						.3445	.2039	.1179	.0680
24						.3878	.2349	.1383	.0807
25						.4320	.2680	.1606	.0951
26						.4773	.3032	.1852	.1112
27						.5227	.3403	.2117	.1290
28							.3788	.2404	.1487
29							.4185	.2707	.1701
30							.4591	.3029	.1933
31							.5000	.3365	.2181
32								.3715	.2447
33								.4074	.2729
34								.4442	.3024
35								.4813	.3332
36								.5187	.3652
37									.3981
38									.4317
39									.4657
40									.5000

Table 11
(continued)

$n_2 = 10$

					n_1					
U_0	1	2	3	4	5	6	7	8	9	10
0	.0909	.0152	.0035	.0010	.0003	.0001	.0001	.0000	.0000	.0000
1	.1818	.0303	.0070	.0020	.0007	.0002	.0001	.0000	.0000	.0000
2	.2727	.0606	.0140	.0040	.0013	.0005	.0002	.0001	.0000	.0000
3	.3636	.0909	.0245	.0070	.0023	.0009	.0004	.0002	.0001	.0000
4	.4545	.1364	.0385	.0120	.0040	.0015	.0006	.0003	.0001	.0001
5	.5455	.1818	.0559	.0180	.0063	.0024	.0010	.0004	.0002	.0001
6		.2424	.0804	.0270	.0097	.0037	.0015	.0007	.0003	.0002
7		.3030	.1084	.0380	.0140	.0055	.0023	.0010	.0005	.0002
8		.3788	.1434	.0529	.0200	.0080	.0034	.0015	.0007	.0004
9		.4545	.1853	.0709	.0276	.0112	.0048	.0022	.0011	.0005
10		.5455	.2343	.0939	.0376	.0156	.0068	.0031	.0015	.0008
11			.2867	.1199	.0496	.0210	.0093	.0043	.0021	.0010
12			.3462	.1518	.0646	.0280	.0125	.0058	.0028	.0014
13			.4056	.1868	.0823	.0363	.0165	.0078	.0038	.0019
14			.4685	.2268	.1032	.0467	.0215	.0103	.0051	.0026
15			.5315	.2697	.1272	.0589	.0277	.0133	.0066	.0034
16				.3177	.1548	.0736	.0351	.0171	.0086	.0045
17				.3666	.1855	.0903	.0439	.0217	.0110	.0057
18				.4196	.2198	.1099	.0544	.0273	.0140	.0073
19				.4725	.2567	.1317	.0665	.0338	.0175	.0093
20				.5275	.2970	.1566	.0806	.0416	.0217	.0116
21					.3393	.1838	.0966	.0506	.0267	.0144
22					.3839	.2139	.1148	.0610	.0326	.0177
23					.4296	.2461	.1349	.0729	.0394	.0216
24					.4765	.2811	.1574	.0864	.0474	.0262
25					.5235	.3177	.1819	.1015	.0564	.0315
26						.3564	.2087	.1185	.0667	.0376
27						.3962	.2374	.1371	.0782	.0446
28						.4374	.2681	.1577	.0912	.0526
29						.4789	.3004	.1800	.1055	.0615
30						.5211	.3345	.2041	.1214	.0716
31							.3698	.2299	.1388	.0827
32							.4063	.2574	.1577	.0952
33							.4434	.2863	.1781	.1088
34							.4811	.3167	.2001	.1237
35							.5189	.3482	.2235	.1399
36								.3809	.2483	.1575
37								.4143	.2745	.1763
38								.4484	.3019	.1965
39								.4827	.3304	.2179
40								.5173	.3598	.2406
41									.3901	.2644
42									.4211	.2894
43									.4524	.3153
44									.4841	.3421
45									.5159	.3697
46										.3980
47										.4267
48										.4559
49										.4853
50										.5147

Table 12
Critical values of T in the
Wilcoxon signed-rank
test; $n = 5\,(1)50$

One-sided	Two-sided	$n = 5$	$n = 6$	$n = 7$	$n = 8$	$n = 9$	$n = 10$
$P = .05$	$P = .10$	1	2	4	6	8	11
$P = .025$	$P = .05$		1	2	4	6	8
$P = .01$	$P = .02$			0	2	3	5
$P = .005$	$P = .01$				0	2	3

One-sided	Two-sided	$n = 11$	$n = 12$	$n = 13$	$n = 14$	$n = 15$	$n = 16$
$P = .05$	$P = .10$	14	17	21	26	30	36
$P = .025$	$P = .05$	11	14	17	21	25	30
$P = .01$	$P = .02$	7	10	13	16	20	24
$P = .005$	$P = .01$	5	7	10	13	16	19

One-sided	Two-sided	$n = 17$	$n = 18$	$n = 19$	$n = 20$	$n = 21$	$n = 22$
$P = .05$	$P = .10$	41	47	54	60	68	75
$P = .025$	$P = .05$	35	40	46	52	59	66
$P = .01$	$P = .02$	28	33	38	43	49	56
$P = .005$	$P = .01$	23	28	32	37	43	49

One-sided	Two-sided	$n = 23$	$n = 24$	$n = 25$	$n = 26$	$n = 27$	$n = 28$
$P = .05$	$P = .10$	83	92	101	110	120	130
$P = .025$	$P = .05$	73	81	90	98	107	117
$P = .01$	$P = .02$	62	69	77	85	93	102
$P = .005$	$P = .01$	55	68	68	76	84	92

One-sided	Two-sided	$n = 29$	$n = 30$	$n = 31$	$n = 32$	$n = 33$	$n = 34$
$P = .05$	$P = .10$	141	152	163	175	188	201
$P = .025$	$P = .05$	127	137	148	159	171	183
$P = .01$	$P = .02$	111	120	130	141	151	162
$P = .005$	$P = .01$	100	109	118	128	138	149

One-sided	Two-sided	$n = 35$	$n = 36$	$n = 37$	$n = 38$	$n = 39$	
$P = .05$	$P = .10$	214	228	242	256	271	
$P = .025$	$P = .05$	195	208	222	235	250	
$P = .01$	$P = .02$	174	186	198	211	224	
$P = .005$	$P = .01$	160	171	183	195	208	

Source: From "Some Rapid Approximate Statistical Procedures" (1964) 28, by F. Wilcoxon and R. A. Wilcox.
Reproduced with the kind permission of Lederle Laboratories, a division of American Cyanamid Company.

Table 12
(continued)

One-sided	Two-sided	$n = 40$	$n = 41$	$n = 42$	$n = 43$	$n = 44$	$n = 45$
$P = .05$	$P = .10$	287	303	319	336	353	371
$P = .025$	$P = .05$	264	279	295	311	327	344
$P = .01$	$P = .02$	238	252	267	281	297	313
$P = .005$	$P = .01$	221	234	248	262	277	292

One-sided	Two-sided	$n = 46$	$n = 47$	$n = 48$	$n = 49$	$n = 50$
$P = .05$	$P = .10$	389	408	427	446	466
$P = .025$	$P = .05$	361	379	397	415	434
$P = .01$	$P = .02$	329	345	362	380	398
$P = .005$	$P = .01$	307	323	339	356	373

Table 13
Critical values of
Spearman's rank
correlation coefficient
for a one-tailed test

n	$\alpha = .05$	$\alpha = .025$	$\alpha = .01$	$\alpha = .005$
5	0.900	—	—	—
6	0.829	0.886	0.943	—
7	0.714	0.786	0.893	—
8	0.643	0.738	0.833	0.881
9	0.600	0.683	0.783	0.833
10	0.564	0.648	0.745	0.794
11	0.523	0.623	0.736	0.818
12	0.497	0.591	0.703	0.780
13	0.475	0.566	0.673	0.745
14	0.457	0.545	0.646	0.716
15	0.441	0.525	0.623	0.689
16	0.425	0.507	0.601	0.666
17	0.412	0.490	0.582	0.645
18	0.399	0.476	0.564	0.625
19	0.388	0.462	0.549	0.608
20	0.377	0.450	0.534	0.591
21	0.368	0.438	0.521	0.576
22	0.359	0.428	0.508	0.562
23	0.351	0.418	0.496	0.549
24	0.343	0.409	0.485	0.537
25	0.336	0.400	0.475	0.526
26	0.329	0.392	0.465	0.515
27	0.323	0.385	0.456	0.505
28	0.317	0.377	0.448	0.496
29	0.311	0.370	0.440	0.487
30	0.305	0.364	0.432	0.478

Source: From "Distribution of Sums of Squares of Rank Differences for Small Samples," by E. G. Olds, *Annals of Mathematical Statistics* 9 (1938). Reproduced with the kind permission of the Editor, *Annals of Mathematical Statistics.*

Table 14 Random numbers

Line	1	2	3	4	5	6	7	8	9	10	11	12	13	14
							Column							
1	10480	15011	01536	02011	81647	91646	69179	14194	62590	36207	20969	99570	91291	90700
2	22368	46573	25595	85393	30995	89198	27982	53402	93965	34095	52666	19174	39615	99505
3	24130	48360	22527	97265	76393	64809	15179	24830	49340	32081	30680	19655	63348	58629
4	42167	93093	06243	61680	07856	16376	39440	53537	71341	57004	00849	74917	97758	16379
5	37570	39975	81837	16656	06121	91782	60468	81305	49684	60672	14110	06927	01263	54613
6	77921	06907	11008	42751	27756	53498	18602	70659	90655	15053	21916	81825	44394	42880
7	99562	72905	56420	69994	98872	31016	71194	18738	44013	48840	63213	21069	10634	12952
8	96301	91977	05463	07972	18876	20922	94595	56869	69014	60045	18425	84903	42508	32307
9	89579	14342	63661	10281	17453	18103	57740	84378	25331	12566	58678	44947	05585	56941
10	85475	36857	53342	53988	53060	59533	38867	62300	08158	17983	16439	11458	18593	64952
11	28918	69578	88231	33276	70997	79936	56865	05859	90106	31595	01547	85590	91610	78188
12	63553	40961	48235	03427	49626	69445	18663	72695	52180	20847	12234	90511	33703	90322
13	09429	93969	52636	92737	88974	33488	36320	17617	30015	08272	84115	27156	30613	74952
14	10365	61129	87529	85689	48237	52267	67689	93394	01511	26358	85104	20285	29975	89868
15	07119	97336	71048	08178	77233	13916	47564	81056	97735	85977	29372	74461	28551	90707
16	51085	12765	51821	51259	77452	16308	60756	92144	49442	53900	70960	63990	75601	40719
17	02368	21382	52404	60268	89368	19885	55322	44819	01188	65255	64835	44919	05944	55157
18	01011	54092	33362	94904	31273	04146	18594	29852	71585	85030	51132	01915	92747	64951
19	52162	53916	46369	58586	23216	14513	83149	98736	23495	64350	94738	17752	35156	35749
20	07056	97628	33787	09998	42698	06691	76988	13602	51851	46104	88916	19509	25625	58104
21	48663	91245	85828	14346	09172	30168	90229	04734	59193	22178	30421	61666	99904	32812
22	54164	58492	22421	74103	47070	25306	76468	26384	58151	06646	21524	15227	96909	44592
23	32639	32363	05597	24200	13363	38005	94342	28728	35806	06912	17012	64161	18296	22851
24	29334	27001	87637	87308	58731	00256	45834	15398	46557	41135	10367	07684	36188	18510
25	02488	33062	28834	07351	19731	92420	60952	61280	50001	67658	32586	86679	50720	94953
26	81525	72295	04839	96423	24878	82651	66566	14778	76797	14780	13300	87074	79666	95725
27	29676	20591	68086	26432	46901	20849	89768	81536	86645	12659	92259	57102	80428	25280
28	00742	57392	39064	66432	84673	40027	32832	61362	98947	96067	64760	64585	96096	98253
29	05366	04213	25669	26422	44407	44048	37937	63904	45766	66134	75470	66520	34693	90449
30	91921	26418	64117	94305	26766	25940	39972	22209	71500	64568	91402	42416	07844	69618
31	00582	04711	87917	77341	42206	35126	74087	99547	81817	42607	43808	76655	62028	76630
32	00725	69884	62797	56170	86324	88072	76222	36086	84637	93161	76038	65855	77919	88006
33	69011	65795	95876	55293	18988	27354	26575	08625	40801	59920	29841	80150	12777	48501
34	25976	57948	29888	88604	67917	48708	18912	82271	65424	69774	33611	54262	85963	03547
35	09763	83473	73577	12908	30883	18317	28290	35797	05998	41688	34952	37888	38917	88050
36	91567	42595	27958	30134	04024	86385	29880	99730	55536	84855	29080	09250	79656	73211
37	17955	56349	90999	49127	20044	59931	06115	20542	18059	02008	73708	83517	36103	42791
38	46503	18584	18845	49618	02304	51038	20655	58727	28168	15475	56942	53389	20562	87338
39	92157	89634	94824	78171	84610	82834	09922	25417	44137	48413	25555	21246	35509	20468
40	14577	62765	35605	81263	39667	47358	56873	56307	61607	49518	89656	20103	77490	18062

Source: Abridged from *Handbook of Tables for Probability and Statistics*, 2d ed. Edited by William H. Beyer (Cleveland: The Chemical Rubber Company, 1968). Reproduced by permission of CRC Press, Inc.

Table 14   (continued)

Line	Column													
	1	2	3	4	5	6	7	8	9	10	11	12	13	14
41	98427	07523	33362	64270	01638	92477	66969	98420	04880	45585	46565	04102	46880	45709
42	34914	63976	88720	82765	34476	17032	87589	40836	32427	70002	70663	88863	77775	69348
43	70060	28277	39475	46473	23219	53416	94970	25832	69975	94884	19661	72828	00102	66794
44	53976	54914	06990	67245	68350	82948	11398	42878	80287	88267	47363	46634	06541	97809
45	76072	29515	40980	07391	58745	25774	22987	80059	39911	96189	41151	14222	60697	59583
46	90725	52210	83974	29992	65831	38857	50490	83765	55657	14361	31720	57375	56228	41546
47	64364	67412	33339	31926	14883	24413	59744	92351	97473	89286	35931	04110	23726	51900
48	08962	00358	31662	25388	61642	34072	81249	35648	56891	69352	48373	45578	78547	81788
49	95012	68379	93526	70765	10592	04542	76463	54328	02349	17247	28865	14777	62730	92277
50	15664	10493	20492	38391	91132	21999	59516	81652	27195	48223	46751	22923	32261	85653
51	16408	81899	04153	53381	79401	21438	83035	92350	36693	31238	59649	91754	72772	02338
52	18629	81953	05520	91962	04739	13092	97662	24822	94730	06496	35090	04822	86774	98289
53	73115	35101	47498	87637	99016	71060	88824	71013	18735	20286	23153	72924	35165	43040
54	57491	16703	23167	49323	45021	33132	12544	41035	80780	45393	44812	12515	98931	91202
55	30405	83946	23792	14422	15059	45799	22716	19792	09983	74353	68668	30429	70735	25499
56	16631	35006	85900	98275	32388	52390	16815	69298	82732	38480	73817	32523	41961	44437
57	96773	20206	42559	78985	05300	22164	24369	54224	35033	19687	11052	91491	60383	19746
58	38935	64202	14349	82674	66523	44133	00697	35552	35970	19124	63318	29686	03387	59846
59	31624	76384	17403	53363	44167	64486	64758	75366	76554	31601	12614	33072	60332	92325
60	78919	19474	23632	27889	47914	02584	37680	20801	72152	39339	34806	08930	85001	87820
61	03931	33309	57047	74211	63445	17361	62825	39908	05607	91284	68833	25570	38818	46920
62	74426	33278	43972	10119	89917	15665	52872	73823	73144	88662	88970	74492	51805	99378
63	09066	00903	20795	95452	92648	45454	09552	88815	16553	51125	79375	97596	16296	66092
64	42238	12426	87025	14267	20979	04508	64535	31355	86064	29472	47689	05974	52468	16834
65	16153	08002	26504	41744	81959	65642	74240	56302	00033	67107	77510	70625	28725	34191
66	21457	40742	29820	96783	29400	21840	15035	34537	33310	06116	95240	15957	16572	06004
67	21581	57802	02050	89728	17937	37621	47075	42080	97403	48626	68995	43805	33386	21597
68	55612	78095	83197	33732	05810	24813	86902	60397	16489	03264	88525	42786	05269	92532
69	44657	66999	99324	51281	84463	60563	79312	93454	68876	25471	93911	25650	12682	73572
70	91340	84979	46949	81973	37949	61023	43997	15263	80644	43942	89203	71795	99533	50501
71	91227	21199	31935	27022	84067	05462	35216	14486	29891	68607	41867	14951	91696	85065
72	50001	38140	66321	19924	72163	09538	12151	06878	91903	18749	34405	56087	82790	70925
73	65390	05224	72958	28609	81406	39147	25549	48542	42627	45233	57202	94617	23772	07896
74	27504	96131	83944	41575	10573	08619	64482	73923	36152	05184	94142	25299	84387	34925
75	37169	94851	39117	89632	00959	16487	65536	49071	39782	17095	02330	74301	00275	48280
76	11508	70225	51111	38351	19444	66499	71945	05422	13442	78675	84081	66938	93654	59894
77	37449	30362	06694	54690	04052	53115	62757	95348	78662	11163	81651	50245	34971	52924
78	46515	70331	85922	38329	57015	15765	97161	17869	45349	61796	66345	81073	49106	79860
79	30986	81223	42416	58353	21532	30502	32305	86482	05174	07901	54339	58861	74818	46942
80	63798	64995	46583	09785	44160	78128	83991	42865	92520	83531	80377	35909	81250	54238

Table 14   (continued)

Line								Column						
	1	2	3	4	5	6	7	8	9	10	11	12	13	14
81	82486	84846	99254	67632	43218	50076	21361	64816	51202	88124	41870	52689	51275	83556
82	21885	32906	92431	09060	64297	51674	64126	62570	26123	05155	59194	52799	28225	85762
83	60336	98782	07408	53458	13564	59089	26445	29789	85205	41001	12535	12133	14645	23541
84	43937	46891	24010	25560	86355	33941	25786	54990	71899	15475	95434	98227	21824	19585
85	97656	63175	89303	16275	07100	92063	21942	18611	47348	20203	18534	03862	78095	50136
86	03299	01221	05418	38982	55758	92237	26759	86367	21216	98442	08303	56613	91511	75928
87	79626	06486	03574	17668	07785	76020	79924	25651	83325	88428	85076	72811	22717	50585
88	85636	68335	47539	03129	65651	11977	02510	26113	99447	68645	34327	15152	55230	93448
89	18039	14367	61337	06177	12143	46609	32989	74014	64708	00533	35398	58408	13261	47908
90	08362	15656	60627	36478	65648	16764	53412	09013	07832	41574	17639	82163	60859	75567
91	79556	29068	04142	16268	15387	12856	66227	38358	22478	73373	88732	09443	82558	05250
92	92608	82674	27072	32534	17075	27698	98204	63863	11951	34648	88022	56148	34925	57031
93	23982	25835	40055	67006	12293	02753	14827	23235	35071	99704	37543	11601	35503	85171
94	09915	96306	05908	97901	28395	14186	00821	80703	70426	75647	76310	88717	37890	40129
95	59037	33300	26695	62247	69927	76123	50842	43834	86654	70959	79725	93872	28117	19233
96	42488	78077	69882	61657	34136	79180	97526	43092	04098	73571	80799	76536	71255	64239
97	46764	86273	63003	93017	31204	36692	40202	35275	57306	55543	53203	18098	47625	88684
98	03237	45430	55417	63282	90816	17349	88298	90183	36600	78406	06216	95787	42579	90730
99	86591	81482	52667	61582	14972	90053	89534	76036	49199	43716	97548	04379	46370	28672
100	38534	01715	94964	87288	65680	43772	39560	12918	86737	62738	19636	51132	25739	56947

◁ ▬▶ ANSWERS

Chapter 2

2.1   (a) 8–10 classes   (c) 43/50   (d) 33/50
2.2   (b) 36/50
2.5   (b) 12.8%
2.6   (b) 25/60
2.9   (b) 1.6; 4.9
2.10   (b) yes
2.16   (a) 2   (b) 1   (c) 2
2.17   (a) 5   (b) 3.875   (c) 2.4107
2.18   (a) 5.8   (b) 5.5   (c) 2.573
2.19   (a) approximately 68% of measurements in interval 33–39; approximately 95% of measurements in interval 30–42; almost all of measurements in interval 27–45
    (b) at least 3/4 of measurements in interval 30–42; at least 8/9 of measurements in interval 27–45
2.20   2387 to 26,387
2.21   (a) skewed to the right
    (b) choose the median
2.22   (a) skewed to the right
    (b) mean greater than the median
2.23   (a) 1352 to 6152   (b) 152 to 7352
2.24   (a) skewed to the right
    (b) approximately 68% of measurements in interval 2552–4952; approximately 95% of measurements in interval 1352–6152; almost all of measurements in interval 152–7352
2.26   (a) .68   (b) .95   (c) ≈0
2.27   (a) at least 3/4 of measurements in interval .15–.19; at least 8/9 of measurements in interval .14–.20
    (b) approximately 68% of measurements in interval .16–.18; approximately 95% of measurements in interval .15–.19; almost all of measurements in interval .14–.20
    (c) no, sample is too small for the data to have a mound-shaped distribution

2.28   at least 3/4 of measurements in interval 0–140; at least 8/9 of measurements in interval 0–193
2.29   (a) 972.74   (b) 872.0   (c) yes
2.30   (a) 2.125   (b) $s^2 = 1.2679$; $s = 1.126$
2.31   (a) 2.4   (b) $s^2 = 2.8$; $s = 1.673$
    (c) $s \approx 1$
2.32 $\bar{x} = 3.858$, $m = 3.85$, $s = 1.09$
2.33   (a) $s \approx .15$ (b) $\bar{x} = .76$, $s = .165$
2.34   (a) $\bar{x} = 4.164$, $s = .922$
    (b) 77%, 95%, 97%
2.35   (a) $s \approx 6274.75$ (b) 81.5%
2.36   (a) $s \approx 3.5$ (b) $\bar{x} = 6.22$, $s = 3.497$
    (c) 96%
2.37   (b) 7.729   (c) 1.985

Interval	Actual	Tchebysheff's Theorem	Empirical Rule
$\bar{x} \pm s$	.714	≥ 0	$\approx .68$
$\bar{x} \pm 2s$	.957	$\geq 3/4$	$\approx .95$
$\bar{x} \pm 3s$	1.000	$\geq 8/9$	≈ 1.00

2.38   (a) 32.1   (b) $s \approx 8.025$ (c) 7.671
2.39   (a) 37, 47, 49   (b) yes; yes
2.40   (a) 1.4, 1.4   (b) 1.4, 1.4
2.41 $\bar{x} = 4.65$, $s = 1.27$
2.42   (a) 2.04, 2.806   (b)–(c)

k	$\bar{x} \pm ks$	Actual	Tchebysheff's Theorem	Empirical Rule
1	(−.77, 4.85)	.84	≥ 0	$\approx .68$
2	(−3.57, 7.65)	.92	$\geq 3/4$	$\approx .95$
3	(−6.38, 10.46)	1.00	$\geq 8/9$	≈ 1.00

2.44 $m = 5.5$, $Q_L = 3.25$, $Q_U = 7.75$
2.45   (a) $\bar{x} = 4.75$, $s = 2.454$ (b) −1.94, 1.32, no
2.46 $m = 10$, $Q_L = 6$, $Q_U = 14$

◁

2.47 (a) $m = 872$, $Q_L = 766.25$, $Q_U = 1087$
(b) Alaska—z-score $= 5.92$

2.48 $m = 6.35$, $Q_L = 2.325$, $Q_U = 12.825$

2.49 69% scored lower than you

2.50 $20 is the 10th percentile; $55 is the 90th percentile

2.51 Los Angeles: $35,000 is the 50th percentile; $75,000 is approximately the 90th percentile
San Francisco: $75,000 is the 93rd percentile

2.52 84th; 87th

2.53 (a) $\bar{x} = 900.3$, $s = 1515.005$ (b) 2.84, yes
(c) yes

2.54 inner fences: 17.25 and 31.25; outer fences: 12 and 36.5; 12 is a mild outlier

2.55 inner fences: -1 and 15; outer fences: -7 and 21; 22 is an extreme outlier

2.56 (a) lower hinge $= 770$, upper hinge $= 1084$
(b) inner fences: 299 and 1555; outer fences: -172 and 2026
(c) 1569 is a mild outlier; 3490 is an extreme outlier

2.57 (a) inner fences: -118 and 578; outer fences: -379 and 839
(b) no

2.58 (a) inner fences: -49 and 959; outer fences: -427 and 1337; 5200 is an extreme outlier
(b) yes

2.60 (a) $\bar{x} = 49$, $s^2 = 16.5$, $s = 4.062$
(b) 11; 2.7; no
(c) no, z-score $= 1.48$

2.61 (b) Range $= 147,300$; $s \approx 36,825$
(c) 90,957.69; 2,263,911,338; 47,580.577
(d) .81, .92, 1.00

2.62 (b) $Q_L = 60,475$; $Q_U = 101,700$
(c) mild outliers: 186,600; 183,800; 178,800; 159,900, extreme outliers: 198,400

2.64 $s \approx 4.4$

2.65 .68, .95

2.66 (a) 7.75   (b) 59.2, 10.369 vs. 7.75

2.68 (2, 26)

2.69 45

2.70

k	$\bar{x} \pm ks$	Approximate Fraction in Interval
1	(415, 425)	$\approx .68$
2	(410, 430)	$\approx .95$
3	(405, 435)	≈ 1.00

2.71 (a) skewed right   (b) at most 1/4

2.72 $\sigma \approx 100$

2.73 (a) .025   (b) .84

2.74 (b) .66, 1.387   (c) .95; .96; yes; yes

2.75 (a) at least 3/4 have between 145 and 205 teachers
(b) .16

2.76 $.5/4 = .125$ vs. .132

2.77 (b) approximately 90%

Chapter 3

3.1 $P(D) = 0$, $P(E) = 2/3$, $P(F) = 1/6$

3.2 $P(E_4) = .2$, $P(E_5) = .1$

3.3 $P(E_1) = .45$, $P(E_2) = .15$, $P(E_i) = .05$ for $i = 3, 4, \ldots, 10$

3.4 (a) .21   (b) .91

3.5 (b) $P(E_i) = 1/38$ (c) $P(A) = 1/19$
(d) 9/19

3.6 (a) $(NDQ), (NDH), (NQH), (DQH)$ (b) 3/4
(c) 3/4

3.7 1/6, 1/18

3.8 (a) .58   (b) .14   (c) .46

3.9 (a) choose 2 of 4 cans without replacement
(b) $(N_1N_2), (N_1W_1), (N_1W_2), (N_2W_1), (N_2W_2)$, (W_1W_2) where N = no water, W = water
(c) 1/6

3.10 (a) choose 2 of 5 cans without replacement
(b) $(N_1N_2), (N_1N_3), (N_1W_1), (N_1W_2), (N_2N_3)$, $(N_2W_1), (N_2W_2), (N_3W_1), (N_3W_2), (W_1W_2)$ where N = no water, W = water
(c) 1/10

3.11 (a) rank A, B, C
(b) $(ABC), (ACB), (BAC), (BCA), (CAB), (CBA)$
(c) 1/3, 1/3

3.12 (a) assign four men, one to each of the four jobs
(b) (1234), (1243), (1324), (1342), (1423), (1432), (2134), (2143), (2314), (2341), (2413), (2431), (3124), (3142), (3214), (3241), (3412), (3421), (4123), (4132), (4213), (4231), (4312), (4321); (1 and 2 minority)
(c) 1/6

3.13 (a) 2/27   (b) 20/27

3.14 3/10, 6/10

3.15 (a) 16
(b) $(EEEE), (EEEN), (EENE), (ENEE)$, $(NEEE), (EENN), (ENEN), (NEEN), (ENNE)$, $(NENE), (NNEE), (ENNN), (NENN)$, $(NNEN), (NNNE), (NNNN)$
(d) 11/16, 3/8, 1/4

3.16 (a) select 2 from 5 without replacement
(b) (12), (13), (14), (15), (23), (24), (25), (34), (35), (45)
(c) 1/10

3.17 (a) 3/5 (b) 1/5 (c) 1/5 (d) 1/5
(e) 1/5 (f) 1 (g) 3/5 (h) 1/2
(i) 1/4 (j) 1 (k) 4/5 (l) 0

3.19 (a) 1/4 (b) 1/2

3.20 (a) 1 (b) 1/5 (c) 1/5

3.21 (a) no (b) no

3.22 (a) 1 (b) 1 (c) 1/3 (d) 0 (e) 1/3
(f) 0 (g) 0 (h) 1 (i) 5/6

3.23 (a) no, no (b) no, yes

3.24 .0118

3.25 (a) .0004 (b) .9996 (c) .0004

3.26 (a) .4 (b) .37 (c) .10 (d) .67
(e) .6 (f) .33 (g) .90 (h) .27
(i) .25

3.28 (a) 5.9% (b) .385 (c) .378; .852

3.29 (a) .413 (b) .400 (c) .091

3.30 (a) .14 (b) .56 (c) .30

3.31 .2222

3.32 .05

3.33 .0214

3.34 (a) .99 (b) .01

3.35 .255

3.36 (a) .015625 (b) .421875 (c) .25

3.37 (a) 406 (b) .0025 (c) .0690
(d) .0027

3.38 (a) simple events correspond to a pair of birth dates (d_1, d_2), where d_i, $i = 1, 2$, is the birth date for person i; d_i can assume values $1, 2, 3, \ldots, 365$
(b) $1/(365)^2$
(c) A contains simple events (1, 1), (2, 2), (3, 3), $\ldots$, (365, 365)
(d) 1/365 (e) 364/365

3.39 (a) $P(A) = .9918$, $P(B) = .0082$
(b) $P(A) = .9836$, $P(B) = .0164$

3.40 .012

3.41 .6087, .3913

3.42 .2222, .2778, .5000

3.43 .2432

3.44 .6618, .1176

3.45 (a) $P(M) = .9444$ (b) .9581; .0419

3.46 .25

3.47 .3130

3.48 .9412

3.49 .6667

3.50 .6585

3.51 80

3.52 84

3.53 45

3.54 90

3.55 1320

3.56 1260

3.57 720

3.58 12,144

3.59 8

3.60 (a) 1/1000 (b) 1/8000

3.61 $5.720645(10^{12})$

3.62 $380,204,032 = (52)^5$

3.63 2,598,960

3.64 1/56

3.65 (a) 5^{20} (b) 5^{-20}

3.66 1/120; 7/24

3.67 (a) D (b) C (c) C (d) C (e) D

3.68 (a) C (b) C (c) D (d) D (e) C

3.69 (a) $p(4) = .05$

3.70 (a) .2

3.71 (a) $(B_1 B_2)$, $(B_1 W_1)$, $(B_1 W_2)$, $(B_2 W_1)$, $(B_2 W_2)$, $(W_1 W_2)$; $P(E_i) = 1/6$, $i = 1, 2, \ldots, 6$
(c) $p(0) = 1/6$; $p(1) = 2/3$; $p(2) = 1/6$

3.72 (a)

T	$p(T)$	T	$p(T)$
2	1/36	8	5/36
3	2/36	9	4/36
4	3/36	10	3/36
5	4/36	11	2/36
6	5/36	12	1/36
7	6/36		

3.73 (a) (S), (FS), (FFS), $(FFFS)$
(b) $p(1) = p(2) = p(3) = p(4) = 1/4$

3.74 (a) $p(0) = .5314$, $p(1) = .3740$, $p(2) = .0877$, $p(3) = .0069$
(c) .0946

3.75 (a) $p(x) = \dfrac{C_x^2 C_{2-x}^3}{C_2^5}$ for $x = 0, 1, 2$

3.76 $p(0) = 1/5$, $p(1) = 3/5$, $p(2) = 1/5$

3.77 (a) .1; .09; .081 (b) $p(x) = (.9)^{x-1}(.1)$

3.78 $p(3) = .28$; $p(4) = .3744$; $p(5) = .3456$

3.79 $p(x) = (.6)^{x-1}(.4)$ for $x = 1, 2, \ldots$

3.80 (a) 2.75, 1.3875, 1.1779 (c) (.39, 5.11); 1
(d) yes

3.81 (a) 3.45, 2.0475, 1.4309 (c) (.59, 6.31); .95
(d) yes

3.82 $E(x) = 1.5$

3.83 $E(\text{gain}) = -\$0.26$

3.84 (a) 3.25 (b) 1.3875, 1.1779

3.85 (a) 2.98 (b) 3.94 (c) 5.16

3.86 \$1500

3.87 (a) 7.9 (b) 2.1749 (c) .96

3.88 $E(\text{gain}) = \$0.39$

3.89 \$2050

3.90 70.74

3.91 (a) 4.0656 (b) 4.125 (c) 3.3186

3.92 (a) 10 (b) 210 (c) 3003

3.93 (a) 21 (b) 792 (c) 161,700

3.94 $\approx 2.5524(10^{11})$

3.95 1/3921225

3.100 (a) D (b) C (c) D (d) D (e) C

3.101 (a) C (b) C (c) D (d) D (e) D

3.102 (a) D (b) C (c) C (d) D

3.103 (a) 2.5 (b) 7.5 (c) 1.25; 1.118

3.104 $\sigma^2 = 2.9167$; 1.0

3.106 .0713

3.107 (a) $(C_1C_2), (C_1A_1), (C_1A_2), (C_2C_1), (C_2A_1),$
$(C_2A_2), (A_1C_1), (A_1C_2), (A_1A_2), (A_2C_1),$
$(A_2C_2), (A_2A_1)$
(b) $(C_1C_2), (C_1A_1), (C_1A_2), (C_2C_1), (C_2A_1),$
(C_2A_2)
(c) $(C_1A_1), (C_1A_2), (C_2A_1), (C_2A_2), (A_1C_1),$
$(A_1C_2), (A_2C_1), (A_2C_2)$
(d) $(A_1A_2), (A_2A_1)$

3.108 1/2; 2/3; 1/3; 5/6; 1/6; 0; 2/3

3.109 (b) $(HHHT), (HHTH), (HTHH), (THHH)$
(c) 1/16, 1/4

3.110 identical to coin-tossing problem, Exercise 3.13; H represents customer preference for style 1, T corresponds to style 2
(c) $(HHHH), (TTTT)$ (d) 1/8

3.111 (a) S contains 21 simple events
(b) A contains 6 simple events (c) 2/7

3.112 8333.333

3.113 $p(0) = .0256, p(1) = .1536, p(2) = .3456,$
$p(3) = .3456, p(4) = .1296; .4752$

3.114 $p(0) = .008, p(1) = .096, p(2) = .384,$
$p(3) = .512; .992$

3.115 2.4; .48

3.116 (a) 7 (b) 5.8333 (d) $34/36 = .944$

3.117 1/6

3.118 1/4; 1/12; 0; 1/3

3.119 0; 1/3

3.120 (a) .73 (b) .27

3.121 .3913

3.122 $P(AB) = P(A)P(B) > 0$

3.123 .999999

3.124 (a) .8 (b) .64 (c) .36

3.125 1; .4783; .00781

3.126 .0256; .1296

3.127 .6

3.128 .0625; yes

3.129 .2; .4

3.130 4

3.131 (a) .128 (b) .488

3.132 (a) 1/5 (b) 2/5 (c) 1/2

3.133 (a) 1/8 (b) 1/64
(c) not necessarily; they could have studied together, etc.

3.134 .5; no

3.135 (a) 5/6 (b) 25/36 (c) 11/36

3.136 \$561.60

3.137 1023.7; 40.9%

3.140 20,000

3.141 (a) 6 (b) 15

3.142 .63

3.143 1/42

3.144 yes; $4!(3!)^4/12!$

Chapter 4

4.1 not binomial; dependent trials, p varies from trial to trial

4.2 binomial; $n = 2, p = 3/5$

4.3 (a) .2965 (b) .8145 (c) .1172
(d) .3670

4.4

| | | $p(x)$ | |
x	$p = .2$	$p = .5$	$p = .8$
0	.2097	.0078	.0000
1	.3670	.0547	.0004
2	.2753	.1641	.0043
3	.1147	.2734	.0287
4	.0287	.2734	.1147
5	.0043	.1641	.2753
6	.0004	.0547	.3670
7	.0000	.0078	.2097

4.5 (a) .3670, .0547, .0004
(b) .7903, .9922, 1.0000
(c) .4233, .9375, .9996
(d) .5767, .0625, .0004
(e) 1.4, 3.5, 5.6
(f) 1.0583, 1.3229, 1.0583

4.6 (**a**) .987   (**b**) .213   (**c**) .788
4.7 (**a**) .967   (**b**) .005   (**c**) .000
       (**d**) 1.000
4.8 (**a**) .925   (**b**) .131   (**c**) .651
4.9 (**a**) .748   (**b**) .610   (**c**) .367
       (**d**) .966   (**e**) .656
4.10 (**a**) $\mu = 300, \sigma = 14.49$
       (**b**) $\mu = 4, \sigma = 1.99$
       (**c**) $\mu = 250, \sigma = 11.18$
       (**d**) $\mu = 1280, \sigma = 16$
4.11 (**a**) 1; .99   (**b**) 90; 3   (**c**) 30; 4.58
       (**d**) 70; 4.58   (**e**) 50; 5
4.12 $p = .5$
4.13 (**a**) .9568   (**b**) .957   (**c**) .9569
       (**d**) $\mu = 2; \sigma = 1.342$
       (**e**) .7455, .9569, .9977   (**f**) yes
4.14 binomial; $n = 7, p = 1/6$
4.15 binomial, assuming random sampling;
       $n = 1162, p = $ P(adequate security)
4.16 binomial; $n = 501, p = .17$ (assuming random
       sampling of executives)
4.17 not binomial; experiment does not result in
       one of two outcomes.
4.18 (**a**) .3277   (**b**) .4096   (**c**) .0067
4.19 their performance would seem to be below the
       80% rate of success
4.20 (**a**) .0081   (**b**) .4116   (**c**) 2401
4.21 600; 420; 20.4939; $P(x > 700) \leq .04$
4.22 (**a**) .0429   (**b**) .9997   (**c**) .0176
4.23 (**a**) .9606   (**b**) .9994
4.24 .0037; .3439; $p < .9$ because $P(x \geq 2) = .0523$
4.25 (**a**) $\mu + 3\sigma \approx 256$ (**b**) $\mu - 3\sigma \approx 320$
4.26 (**a**) .983
       (**b**) P(accept lot when $p = .10) = .392$, no
4.27 (**b**) 3   (**c**) 1
4.28 $a = 1$
4.29 $n = 25, a = 1$
4.30 $\mu = 900, \sigma = 21.2132$; no, $x = 1260$ lies more
       than 16 standard deviations above the mean
4.31 (**a**) 200, 11.5470
       (**b**) no, $x = 270$ lies more than 6 standard
       deviations above the mean
4.32 (**a**) 32
       (**b**) no, $x = 50$ lies more than 3 standard
       deviations above the mean
4.33 (**a**) .135335   (**b**) .270671   (**c**) .593994
       (**d**) .036089
4.34 (**a**) .109   (**b**) .958   (**c**) .257   (**d**) .809
4.35 (**a**) .677   (**b**) .6767

4.36 $p(0) \approx .2865, p(1) \approx .3581$; from Table 1,
       $p(0) = .277, p(1) = .365$
4.37 (**a**) .0067   (**b**) .1755   (**c**) .560
4.38 (**a**) .08422, .12465
       (**b**) no, $x = 10$ lies 2.236 standard deviations
       above the mean
4.39 (**a**) .271   (**b**) .594   (**c**) .406
4.40 (**a**) $\mu = 2, \sigma = 1.414$ (**b**) 0 to 4
4.41 no
4.42 (**a**) binomial, $n = 100$ and $p = .05; \mu = 5$
       (**b**) .440
4.43 (**a**) .6   (**b**) .5143   (**c**) .0714
4.44 (**a**) 0   (**b**) .9762   (**c**) .2381
4.45 (**a**) $p(0) = 165/455, p(1) = 220/455$,
       $p(2) = 66/455, p(3) = 4/455$
       (**c**) $\mu = .8, \sigma^2 = .50286$
       (**d**) .99 fall between $-.618$ and 2.218; .99 fall
       between -1.327 and 2.927; yes
4.46 (**a**) $p(0) = 1/6, p(1) = 2/3, p(2) = 1/6$
       (**b**) $\mu = 1, \sigma^2 = 1/3$
4.47 $p(0) = 1/5, p(1) = 3/5, p(2) = 1/5$
4.48 (**a**) $p(x) = \dfrac{C_x^2 C_{2-x}^3}{C_2^5}$ for $x = 0, 1, 2$
       (**b**) $\mu = .8, \sigma^2 = .36$
4.49 .4165; .7834
4.50 $\mu = 3/5, \sigma = .655$; 0 to 1
4.51 .0035; event is very unlikely, possibly
       nonrandom interviews
4.55 (**a**) $p(0) = .125, p(1) = .375, p(2) = .375$,
       $p(3) = .125$
       (**c**) 1.5; .866   (**d**) .75; 1.00
4.56 (**a**) $p(0) = .729, p(1) = .243, p(2) = .027$,
       $p(3) = .001$
       (**c**) .3; .520   (**d**) .729; .972
4.57 (**a**) .9606   (**b**) .9994
4.58 (**a**) .590   (**b**) .168   (**c**) .031   (**d**) 1
       (**e**) 0
4.59 (**a**) .919   (**b**) .528   (**c**) .188   (**d**) 1
       (**e**) 0
4.60 (**a**) .349   (**b**) .028   (**c**) .001   (**d**) 1
       (**e**) 0
4.61 (**a**) .736   (**b**) .149   (**c**) .011   (**d**) 1
       (**e**) 0
4.62 for a given value of n and p, P(accept)
       increases as a increases; for a given value of a
       and p, P(accept) decreases as n increases
4.63 (**a**) 1.000   (**b**) .588   (**c**) .021
4.64 (**a**) .234   (**b**) .136
       (**c**) claim is not unlikely

4.65 (a) $p(x) = 1/10$ for $x = 0, 1, \ldots, 9$
(b) 3/10 (c) 7/10
4.66 (a) .228
(b) no indication that people are more likely to choose middle numbers
4.67 3.5, 1.735; 1 to 6
4.68 .6384; $\mu = 3$
4.69 no; $x = 12$ lies more than 5 standard deviations above the mean
4.70 (a) yes; $x = 13$ lies more than 3 standard deviations below the mean
(b) yes
4.71 (a) (25, 5) (b) (25, 5)
4.72 (a) .983 (b) .736 (c) .392 (d) .069
4.73 (a) 30,000 (b) 122.474
(c) highly unlikely if $p = .5$ (d) yes
4.74 (a) 20 (b) 4 (c) .006
(d) theory is incorrect
4.75 (a) .128 (b) .488
4.76 .017; do not support
4.77 (a) .3 (b) 7.5, 2.291 (c) 1.09 (d) no
4.78 (a) .5 (b) 12.5, 2.5
(c) z-score = 3; there is a preference for the second design
4.79 (a) .0038 (b) .9560
4.80 (a) 3.325, 1.0447 (b) 2 to 5
4.81 (a) binomial; $n = 1000$, $p = .00019$
(b) .1731 (c) $\mu = .19$, $\sigma = .4358$
(d) yes
4.82 :1730
4.83 $\mu + 3\sigma = 19.49$; 20 cases per 100,000 would be unusual
4.84 (a) hypergeometric, or approximately binomial
(b) Poisson (c) approximately .85, .72, .61
4.86 (a) binomial (b) Poisson
(c) no; $x = 9$ lies more than 5 standard deviations above the mean
4.87 (a) 10 (b) 1/10 (c) yes; no
4.88 (a) geometric, $p = 1/2$
(c) $\mu = 2$, $\sigma^2 = 2$; 0 to 6
(d) coin biased towards tails
4.89 (a) 9.09 (b) 2.875 (c) no
4.90 (a) $\mu = 6$, $\sigma = 2.1909$
(b) no; $x = 24$ lies more than 8 standard deviations above the mean

Chapter 5

5.1 (b) .25 (c) $\mu = .5$, $\sigma^2 = .0833$
5.2 (a) 0 (b) .2212 (c) .2149 (d) .6873

5.3 (a) .4452 (b) .4664 (c) .3159
(d) .3159
5.4 (a) .8384 (b) .9544 (c) .9974
5.5 (a) .6753 (b) .2401 (c) .2694
5.6 (a) 1.96 (b) 1.44
5.7 (a) -1.96 (b) .36
5.8 1.36
5.9 (a) 1.65 (b) -1.645
5.10 (a) 1.645 (b) 2.575
5.11 (a) .0401 (b) .1841 (c) .2358
5.12 (a) .1596 (b) .1151 (c) .1359
5.13 $\mu = 9.2$
5.14 13.464
5.15 $\mu = 8$, $\sigma = 2$
5.16 not normally distributed
5.17 (a) .25 (b) .25 (c) .625
(d) $\mu = 30,000$, $\sigma = 11,547.005$ (e) .577
5.18 (a) .135335 (b) .63212 (c) .23254
5.19 (a) .1056; .5000 (b) .8944
5.20 .1562; .0012
5.21 .0505
5.22 (a) .1335 (b) .8665 (c) .1335
5.23 (a) .4721 (b) .0582 (c) .4697
5.24 .0668
5.25 (a) .0104 (b) .2206
5.26 (a) 1.27 (b) .1020
5.27 (a) .0174 (b) .1112
5.28 .0475
5.29 $\mu = 1940.119$
5.30 63,550
5.31 (a) .2177 (b) 52.6 lbs
5.32 set 3.29 psi above customer specifications
5.33 .0175
5.34 (a) .390 (b) .4049
5.35 .3745
5.36 .1711
5.37 .2676
5.38 .3520
5.39 (a) .546 (b) .5468
5.40 (a) .245 (b) .2483
5.41 .3531
5.42 (a) .7611 (b) ≈ 0 (c) .9838
5.43 .1251
5.44 (a) $\mu = 64$, $\sigma = 4.8$ (b) .0047 (c) yes
5.45 .0901; no
5.46 (a) .0054
(b) no; media coverage influences improvement
5.47 .9441
5.48 z-score = 2.882; yes

5.49 z-score $= 3.98$; yes
5.50 (a) yes　(b) yes
5.51 (a) 3.5, .577　(c) .577, 1, 1; yes; no
5.52 (a) .04, .04　(c) .8647, .9502, .9817; yes; no
5.53 (a) .3849　(b) .3159
5.54 (a) .3227　(b) .1586　(c) .4279
　　　(d) .1628
5.55 (a) .7734　(b) .9115
5.56 $z_0 = 0$
5.57 .8612
5.58 $z_0 = .6745$
5.59 (a) .1056　(b) .8944　(c) .1056
5.60 no
5.61 .16
5.62 5.065
5.63 (a) .0778　(b) .0274
5.64 .0401
5.65 3.44%
5.66 (a) .5780　(b) .5752
5.67 85.36
5.68 .3859
5.69 z-score $= -1.26$; no
5.70 .2119
5.71 (a) no　(b) .0179
5.72 7.301
5.73 (a) .0951　(b) .0681
5.74 every 383.5 hours
5.75 (a) 141　(b) .0401
5.76 .8980
5.77 .9474
5.78 (a) .178　(b) .392
5.79 (a) .1725　(b) .4052
5.80 .3557
5.81 z-score $= 1.461$; no
5.82 (a) .1587　(b) 7.935
　　　(c) z-score $= 2.73$, no
5.83 (a) ≈ 0　(b) .6026
　　　(c) survey not reliable
5.84 (a) .2636　(b) .2835
　　　(c) 1898.84 and 276,665.96

Chapter 6

6.1 (a) $\mu = 10, \sigma_{\bar{x}} = .6$　(b) $\mu = 5, \sigma_{\bar{x}} = .2$
　　　(c) $\mu = 120, \sigma_{\bar{x}} = .3536$
6.2 (a) normal
　　　(b) approximately normal for parts (a) and (b)
6.3 (c) .5468
6.4 (c) $\mu_{\bar{x}} = 3.5; \sigma_{\bar{x}} = .765$

6.5 (a) $\mu_{\bar{x}} = 3.5; \sigma_{\bar{x}} = .541$
　　　(b) $\mu_{\bar{x}} = 3.5; \sigma_{\bar{x}} = .442$
　　　(c) $\mu_{\bar{x}} = 3.5; \sigma_{\bar{x}} = .342$
6.6 decreases the standard error

6.7

n	$\sigma_{\bar{x}}$	n	$\sigma_{\bar{x}}$
1	1	16	.250
2	.707	25	.200
4	.500	100	.100
9	.333		

6.8 (a) 1; .161　(b) .0314　(c) ≈ 0
　　　(d) .0132
6.9 (a) 106; 2.4　(b) .0475　(c) .9050
6.10 (b) $\mu = 110.39, \sigma_{\bar{x}} = 6.288$
6.14 (b) ≈ 0　(c) no
6.15 (a) .3758　(b) no
6.16 (a) normal　(b) .0571　(c) $\mu = 21.948$
6.17 (a) 1890; 69.282　(b) .0559
6.18 (a) approximately normal
　　　(b) $\mu_x = 255, \sigma_x = 13.693$
6.19 (a) .3, .0458　(b) .1, .015　(c) .6, .0310
6.21 (b) .9198
6.22 (a) yes　(b) .0681　(c) .5　(d) .8638
6.23 (a) .0099　(b) .03　(c) .0458
　　　(d) .0500　(e) .0458　(f) .03
　　　(g) .0099
6.24 (a) yes　(b) .0668　(c) .9544
6.25 (b) .9556
6.26 (a) approximately normal; $\mu = .4, \sigma = .0152$
　　　(b) .8098　(c) .9372
6.27 (a) approximately normal; $\mu = .51, \sigma = .0158$
　　　(b) .7960　(c) .8740
6.28 (b) .7114
6.29 (a) $-10, .8062$　(b) $-2, .2$
6.30 (a) normal
　　　(b) approximately normal for part (b)
6.31 (c) .8664
6.32 no
6.33 (a) .1357　(b) .2714
6.34 $\mu_1 - \mu_2 = -8.338; \sigma_{(\bar{x}_1 - \bar{x}_2)} = 9.4968$
6.35 (a) .1; .0574　(b) $-.5$; .0287
6.36 (c) .7016
6.37 (a) .1149　(b) .0949　(c) .03
6.39 .0238
6.40 (a) ≈ 0　(b) they are not accurate
6.41 .1894
6.42 .0068

6.43 (a) approximately normal; $\sigma = .00911$
(b) .7286   (c) .0292
6.44 (b) .8384
6.45 (b) 3.5; .987
6.47 (a) 7; 1.449   (b) 1; 1.449
6.48 (a) 64; 12.49   (b) 16; 12.49
6.49 .9591
6.50 approximately normal
6.51 .6170
6.52 (a) .0114
6.53 (a) 6
(d) $p(\bar{x}) = 1/6$ for $\bar{x} = 1.5, 2, 2.5, 3.5, 4, 4.5$
(e) $\mu = 3$; no
6.56 (a) $\sigma \approx 12.5$ (b) ≈ 1
(c) estimate is too high
6.57 .1922
6.58 limit total number of passengers to $n = 11$
6.59 (a) 131.2; 3.677   (b) yes   (c) .1515
6.60 (a) 1312; 11.628   (b) .003
6.61 (b) .8502
6.62 (b) ≈ 1
(c) there is a difference in the two methods of testing
6.63 (a) .0793   (b) ≈ 0

Chapter 7

7.2 (a) .877   (b) .186   (c) .960
7.3 (a) (12.496, 13.704)   (b) (2.651, 2.809)
7.4 (a) (.797, .883)   (b) (21.469, 22.331)
7.5 (a) (32.550, 35.450)
(b) (1047.543, 1050.457)
(c) (65.973, 66.627)
7.6 (a) .8098   (b) .9356
7.7 (a) 3.92   (b) 2.772   (c) 1.96
7.8 (a) decreases by $1/\sqrt{2}$ (b) decreases by 1/2
7.9 (a) 3.29   (b) 5.16
7.11 $39.8^0 \pm 4.768^0$
7.12 $4.2 \pm .339$
7.13 $7.2\% \pm .776\%$
7.14 (a) $\bar{x} \pm 1.96\, s/\sqrt{n}$ (b) (1.922, 1.984)
7.15 (b) (1.756, 1.924)   (c) (486.610, 619.390)
(d) no
7.16 (.1435, .1465)
7.17 (b) yes; bound on error = .0061
7.18 (3.496, 3.904); random sampling
7.19 .690
7.20 (a) $(-2.525, -1.875)$ (b) $(-2.710, -1.690)$
7.21 (15.463, 36.937)

7.22 $(-1.032, -0.368)$
7.23 (a) (6.449, 7.951)   (b) (1.762, 3.238)
7.24 $(-1.030, 8.030)$
7.25 $(-5.785, -0.215)$
7.26 (6.748, 9.052)
7.27 (6.753, 9.923)
7.28 (a) $(-.061, .081)$ (b) $(-39.852, 109.852)$
(d) yes; implies no difference
7.29 (.690, .766)
7.30 (.846, .908)
7.31 (.034, .074)
7.32 (.0405; .0429)
7.33 (a) no   (b) nothing; no
7.34 (a) (.708, .752)   (b) random sampling
7.35 (.66, .68)
7.36 (.039, .097)
7.37 .062
7.38 (a) (.668, .672)   (b) (.708, .712)
7.39 bound on error = .002; claim is incorrect
7.40 (a) .012
7.41 (a) random sampling   (b) yes
(c) (.713, .767)
7.42 .055
7.43 (a) $(-.203, -.117)$ (b) random samples
7.44 (a) (.086, .178)   (b) random samples
7.45 $(-.216, -.144)$
7.46 (a) $(-.082, .002)$ (b) .032
7.47 (a) $(-.172, -.028)$ (b) $(-.097, .065)$
7.48 (.063, .157)
7.49 (.061, .259)
7.50 243
7.51 505
7.52 5207
7.53 1086
7.55 (b) 9604
7.56 97
7.57 (a) (2.837, 3.087)   (b) 276
7.58 (a) (5.543, 5.963)   (b) 530
7.59 2873
7.60 97
7.61 136
7.62 70
7.63 $n_1 = n_2 = 98$
7.65 $29.1 \pm .9555$
7.66 (28.298, 29.902)
7.67 234
7.68 (3.867, 4.533)
7.69 $n_1 = n_2 = 224$
7.70 (a) $.48 \pm .044$ (b) (.443, .517)

7.71 1083

7.72 $(-.0157, .2907)$

7.73 $n_1 = n_2 = 376$

7.74 9.7 ± 2.274

7.75 $(7.792, 11.608)$

7.76 yes; .031, .044, .052

7.77 .032

7.78 97

7.79 68

7.80 $(.754, .847)$

7.81 (a) $.0625 \pm .0237$ (b) 563

7.82 $(20.447, 23.553)$

7.83 (a) $(.042, .618)$ (b) $.67 \pm .017$

7.84 2401; yes

7.85 (a)

	Trt	$\hat{p}$	Standard Error
Male	1	.928	.010687
	2	.903	.012310
	3	.898	.012699
Female	1	.886	.012642
	2	.914	.012415
	3	.927	.010719

(b) $(-.041, .051)$ (c) $(-.055, .029)$

(d) $(-.001, .085)$

7.86 9604

7.87 65

7.88 193

7.89 $(-.026, .166)$

7.90 $(33.41, 34.59)$

7.91 6147

7.92 $(.059, .141)$

7.93 .387; .651

7.94 (a) $(-2.874, 5.074)$

(b) yes; implies no difference

(c) -3 ± 3.350

Chapter 8

8.1 (a) $H_0: \mu = 2.3; H_a: \mu > 2.3$ (c) $z = 1.645$

(d) yes, $z = 2.04$

8.2 $H_0: \mu = 2.9; H_a: \mu < 2.9$; one-tailed

8.3 $H_0: \mu = 2.9; H_a: \mu \neq 2.9$; two-tailed

8.4 (a) 2.38 (b) .3409; .9484; .0071; ≈ 0

8.5 (a) $H_0: \mu = 60; H_a: \mu < 60$ (b) one-tailed

(c) yes; $z = -1.992$

8.6 (a) $H_0: \mu = 7.5; H_a: \mu < 7.5$

(b) one-tailed (d) $z = -5.477$; reject H_0

8.7 (a) $H_0: \mu = 80$ (b) $H_a: \mu \neq 80$

(c) $z = -3.75$; reject H_0

8.8 yes; $z = 11.49$

8.9 no; $z = -.992$

8.10 (a) $H_0: \mu_1 - \mu_2 = 0; H_a: \mu_1 - \mu_2 > 0$

(b) one-tailed (c) $z > 1.28$

(d) reject H_0; $z = 2.087$

8.11 (a) $H_0: \mu_1 - \mu_2 = 0; H_a: \mu_1 - \mu_2 \neq 0$

(b) two-tailed (c) $|z| > 1.96$

(e) do not reject H_0; $z = -1.334$

8.12 $\alpha = .20$

8.13 (a) $H_0: \mu_1 - \mu_2 = 0; H_a: \mu_1 - \mu_2 > 0$;

one-tailed

(b) reject H_0; $z = 2.074$

8.14 (a) $H_0: \mu_1 - \mu_2 = 0; H_a: \mu_1 - \mu_2 > 0$

(b) $z > 1.645$ (d) reject H_0; $z = 1.993$

8.15 no, $z = -2.779$

8.16 (a) no; $z = .46$ (b) no; $z = -1.76$

(c) yes; $z = 3.13$

8.17 (a) $H_0: \mu_1 - \mu_2 = 0; H_a: \mu_1 - \mu_2 \neq 0$

(b) two-tailed (c) no; $z = -.954$

8.18 reject H_0; $z = 13.574$

8.19 (a) $H_0: \mu_1 - \mu_2 = 0; H_a: \mu_1 - \mu_2 < 0$

(b) no; $z = -.426$

8.20 (a) $H_0: p = .3; H_a: p < .3$ (b) one-tailed

(c) do not reject H_0; $z = -1.449$

8.21 (a) $H_0: p = .4; H_a: p \neq .4$ (b) two-tailed

(c) reject H_0; $z = -1.680$

8.22 yes; $z = 2.191$

8.23 yes; $z = 6.390$

8.24 do not reject H_0; $z = 1.43$

8.25 (a) $H_0: p = .55; H_a: p > .55$ (b) $z > 1.645$

(d) reject H_0; $z = 2.697$

8.26 (a) $H_0: p = .054; H_a: p > .054$

(b) $z > 2.33$

(c) do not reject H_0; $z = 1.03$

8.27 (a) do not reject H_0; $z = 1.77$

(b) $(.17, .43)$

8.28 yes; $z = -21.033$

8.29 (a) $H_0: p = .15; H_a: p > .15$

(b) reject H_0 if $z > 2.33$

(c) reject H_0; $z = 9.996$

8.30 (a) $H_0: p_1 - p_2 = 0; H_a: p_1 - p_2 \neq 0$

(b) two-tailed

(c) do not reject H_0; $z = -.84$

8.31 (a) $H_0: p_1 - p_2 = 0; H_a: p_1 - p_2 < 0$

(b) one-tailed

(c) do not reject H_0; $z = -.84$

8.32 (a) $H_0: p_1 - p_2 = 0; H_a: p_1 - p_2 < 0$

(b) one-tailed
(c) do not reject H_0; $z = -.948$

8.33 **(a)** yes; $z = -2.38$

8.34 **(a)** $H_0: p_1 - p_2 = 0$; $H_a: p_1 - p_2 \neq 0$
(b) do not reject H_0; $z = -.38$

8.35 **(a)** yes; $z = 2.39$ **(b)** no

8.36 **(a)** $H_0: p_1 - p_2 = 0$; $H_a: p_1 - p_2 \neq 0$
(c) no; $z = 1.621$

8.37 yes, $z = 2.72$

8.38 **(a)** .0268 **(b)** reject H_0

8.39 **(a)** p-value $< .002$ **(b)** reject H_0

8.40 **(a)** .0718 **(b)** do not reject H_0

8.41 yes

8.42 p-value $< .002$

8.43 p-value $< .001$

8.44 p-value $= .0174$

8.45 **(a)** $H_0: \mu = 3$; $H_a: \mu > 3$ **(b)** one-tailed
(c) p-value $= .4247$ **(d)** no

8.46 **(a)** yes; $z = 3.23$ **(b)** p-value $< .001$

8.47 **(a)** no; p-value $= .1587$ **(b)** no
(c) p-value $= .0336$ **(d)** yes

8.48 **(a)** $H_0: p = 2/3$ **(b)** $H_a: p > 2/3$
(c) yes; $z = 4.6$ **(d)** p-value $< .001$

8.51 **(a)** p-value $= .0025$ **(b)** reject H_0

8.52 yes; $z = -25.298$; reject H_0

8.53 **(a)** p-value $< .001$
(b) reject H_0; conclude that more than 1/3
prefer A

8.54 random sample; either σ known or sample size
30 or more

8.55 **(a)** no; $z = .283$ **(b)** no; $z = -.80$

8.56 no; $z = 1.57$

8.57 yes, $z = 2.858$

8.58 no, $z = 1.684$

8.59 .1151, .0228

8.60 **(a)** $H_0: \mu = 35$; $H_a: \mu < 35$ **(b)** $z < -2.33$
(c) $z = -6.19$; mean is less than 35

8.61 **(a)** p-value $= .0668$ **(b)** do not reject H_0

8.62 309

8.63 yes, $z = 4.00$

8.64 yes, $z = 4.00$

8.65 yes, $z = 5.237$

8.66 **(a)** yes, $z = 4.333$ **(b)** (7.12, 18.88)

8.67

μ	β
873	.3446
875	.6103
877	.8274

8.68

μ	β
873	.6141
875	.7794
877	.8906

8.69 **(a)** no, $z = 1.00$ **(b)** no, $z = .92$ **(c)** no

8.70 **(a)** p-value $< .002$ **(b)** reject H_0

Chapter 9

9.1 **(a)** 2.015 **(b)** 2.306 **(c)** 1.330
(d) ≈ 1.960

9.2 **(a)** 1.356 **(b)** 2.485 **(c)** 1.746
(d) 2.365

9.3 **(a)** $t = -1.438$; do not reject H_0
(b) $.10 < p$-value $< .20$
(c) (45.976, 48.224)

9.4 **(a)** 7.05, .4994 **(b)** (6.537, 7.563)
(c) $t = -2.849$; reject H_0
(d) no; different value of α

9.5 **(a)** (16.412, 19.824) **(b)** (.026, .042)
(c) (26.964, 43.218)

9.6 **(a)** $s \approx 3.86$ **(b)** (8.085, 11.595)

9.7 **(a)** no, $t = -.942$ **(b)** p-value $> .10$
(c) no, $t = -1.296$ **(d)** p-value $> .10$

9.8 no, $t = -1.195$; do not reject H_0

9.9 (1.034, 1.206)

9.10 (55.099, 66.501)

9.11 yes, $t = -3.044$

9.12 (82.88, 96.83)

9.13 (3.652, 3.912)

9.14 $s = .1812$; $n = 1273$

9.15 **(a)** (−183.394, 1983.994) **(b)** (20.287, 379.173)

9.16 **(a)** 22 **(b)** 20 **(c)** 16

9.17 **(a)** 3.775 **(b)** 21.2258

9.18 **(a)** 12.4571 **(b)** (−6.087, 2.887)
(c) $t = -.6761$; do not reject H_0

9.19 **(a)** $H_0: \mu_1 - \mu_2 = 0$; $H_a: \mu_1 - \mu_2 \neq 0$
(b) $|t| > 1.703$ **(c)** $t = 2.795$
(d) p-value $< .01$ **(e)** reject H_0

9.20 (.938, 3.862)

9.21 **(a)** $H_0: \mu_1 - \mu_2 = 0$; $H_a: \mu_1 - \mu_2 > 0$
(b) yes, $t = 2.806$ **(c)** $.005 < p$-value $< .01$

9.22 (.107, .853)

9.23 **(a)** yes, $t = -2.165$
(b) $.02 < p$-value $< .05$

9.24 **(a)** $H_0: \mu_1 - \mu_2 = 0$; $H_a: \mu_1 - \mu_2 > 0$
(b) $t > 1.895$ **(d)** $t = 16.752$; reject H_0

9.25 (a) yes; $t = 6.035$ (b) p-value $< .005$
9.26 yes; $t = 6.661$; p-value $< .005$
9.27 assumption of equal variances is violated
9.28 (a) no, $t = 1.606$; do not reject H_0
(b) $(-.061, .341)$
9.29 (a) two-sided
(b) $H_0: \mu_1 - \mu_2 = 0$; $H_a: \mu_1 - \mu_2 \neq 0$
(c) $|t| > 2.056$ (d) yes, $t = -3.6$
(e) p-value $< .01$
9.30 $t = 2.372$; reject H_0; $.02 < p$-value $< .05$
9.31 $(.014, .586)$
9.32 62
9.33 (a) $H_0: \mu_d = 0$; $H_a: \mu_d > 0$
(b) $t = 1.511$; do not reject H_0
9.34 (a) no, $t = 1.177$ (b) p-value $> .20$
(c) $(-.082, .202)$
9.35 (b) yes, $t = -1.834$
(c) $.025 < p$-value $< .05$
(d) $(-.125, -.001)$
9.36 $(18.196, 206.804)$
9.37 (b) yes, $t = 9.150$ (c) p-value $< .01$
(d) $(80.472, 133.328)$
9.38 (a) no, $t = -1.772$
(b) $.10 < p$-value $< .20$
(c) $(-.053, .009)$
9.40 paired
9.41 (a) yes, $t = -4.326$; reject H_0
(b) $(-2.594, -.566)$
9.42 approximately 65
9.43 (a) yes, $t = 2.821$; reject H_0
(b) 95% confidence interval—$(.2405, 2.7345)$
(d) yes
9.44 (a) paired difference test
(c) p-value $> .20$; yes
9.45 no, $\chi^2 = 34.24$
9.46 $(.190, .685)$
9.47 (a) .699 (b) .291, 3.390)
(c) $\chi^2 = 5.24$; do not reject H_0
(d) p-value $> .20$
9.48 $(.00703, .03264)$
9.49 (a) no, $\chi^2 = 12.8618$ (b) p-value $> .10$
9.50 $\chi^2 = 22.449$, reject H_0: $\sigma^2 = .49$
9.51 $(1.408, 31.264)$
9.52 (a) no, $t = -.232$; do not reject H_0
(b) yes, $\chi^2 = 19.98$; reject H_0
9.53 $(.00476, .00980)$
9.54 (a) no, 1000 is 5 standard deviations away
from 800
(b) yes, $z = 3.262$; reject H_0
(c) yes, $\chi^2 = 57.281$; for $\alpha = .05$, reject H_0

9.55 (a) $(80.207, 372.186)$ (b) yes
9.56 no, $\chi^2 = 29.433$
9.57 (a) no, $F = 1.774$ (b) p-value $> .20$
9.58 $(.677, 4.896)$
9.59 (a) no, $F = 2.316$ (b) $.05 < p$-value $< .10$
9.60 (a) no, $F = 1.394$ (b) $(.47, 4.11)$ (c) yes
9.61 (a) yes, $F = 2.486$; reject H_0
(b) p-value $< .005$
9.62 $(1.544, 4.003)$ (Using $F_{49,49} \approx 1.61$)
9.63 (a) $\sigma_1^2 = \sigma_2^2$ (b) no, $F = 1.64$
9.64 (a) $\sigma_1^2 = \sigma_2^2$ (b) no, $F = 5.897$
9.65 (a) no, $F = 2.904$ (b) $(.050, .254)$
9.66 $F = 4$; do not reject H_0
9.67 Eastern vs. Western: $F = 17.46$; Eastern vs.
Crude: $F = 1219.05$; Western vs. Crude:
$F = 69.82$; Constant warm vs. constant cool:
$F = 13.52$
9.71 For $\alpha = .05$, yes; $t = 2.108$;
$.025 < p$-value $< .05$
9.72 $(9.860, 12.740)$
9.73 $(71.426, 84.100)$
9.74 yes, $t = 5.985$; reject H_0; $(28.375, 33.625)$
9.75 $(3.545, 4.975)$
9.76 yes, $F = 3.268$
9.77 (a) $(169.1, 199.9)$ (b) $(2.28, 2.80)$
(c) $(69.43, 76.57)$
9.78 $n = 72$
9.79 (a) yes, $t = -4.535$; reject H_0
(b) p-value $< .01$
9.80 yes, $F = 2.865$; $.01 < p$-value $< .025$
9.81 (a) yes, $\chi^2 = 52.4$ (b) no, $z = -.79$
9.82 yes, $F = 2.836$
9.83 (a) yes, $t = -3.354$; reject H_0
(b) p-value $< .01$
9.84 $(-10.246, -2.354)$
9.85 (a) approximately 136
9.86 (b) yes, $F = 19.52$ (c) p-value $< .01$
(d) assumption of equal variances is violated
9.87 (a) no, $t = -1.082$
(b) assumption of equal variances is violated
9.88 (a) no (b) yes, $t = 3.237$
(c) yes, $t = 59.342$; p-value $< .01$
9.89 (a) $(.381, .419)$ (b) $(1.226, 2.434)$
9.90 $(13.25, 28.95)$
9.91 (a) no, $t = -.306$ (b) p-value $> .20$
9.92 $(33.465, 34.107)$
9.93 yes, $t = 2.2$
9.94 (a) $H_0: \mu_1 - \mu_2 = 0$; $H_a: \mu_1 - \mu_2 > 0$
(b) $t > 2.552$ (c) no, $t = 1.69$
(d) paired difference test
(e) no, the analysis was incorrect.

9.95 (35.845, 48.405)
9.96 no, $t = -1.712$
9.97 yes, $t = 2.497$
9.98 $(-1.522, .022)$
9.99 $(-1.704, -.046)$
9.100 $(-1.842, .092)$; yes
9.101 $t = 9.568$; reject H_0
9.102 $.259 \pm .047$ unpaired $.259 \pm .050$ paired
9.103 when within-group variation is less than between-group variation
9.104 (a) no, $t = 2.571$ (b) $(.000, .002)$
9.106 (a) two-tailed; $H_a: \sigma_1^2 \neq \sigma_2^2$
　　　(b) lower-tailed; $H_a: \sigma_1^2 < \sigma_2^2$
　　　(c) upper-tailed; $H_a: \sigma_1^2 > \sigma_2^2$
9.107 (a) $F = 1.922$; p-value $> .20$
　　　(b) $(.781, 4.728)$
9.108 yes, $F = 2.407$
9.109 no, $\chi^2 = 7.008$
9.110 $(.185, 2.465)$
9.111 no, $t = -1.8$
9.112 yes, $t = 2.425$
9.113 $(-11.414, -8.958)$
9.114 yes, $t = -2.945$
9.115 no, population distribution not normal
9.116 yes, $t = 3.038$; reject H_0
9.117 no, $t = 1.862$; do not reject H_0
9.118 $(24.582, 73.243)$
9.119 (b) p-value $\approx P(|z| > 1.94) = .0524$
9.120 (a) no, $t = 1.120$ (b) no, $t = -1.519$
9.121 (a) no, $F = 1.331$
　　　(b) $t = -4.537$, p-value $< .01$; reject H_0
9.122 (a) $t = -.402$; do not reject H_0
　　　(b) no, $F = 1.730$ (c) $(-6.964, 9.364)$

Chapter 10

10.1 y intercept $= 1$, slope $= 2$
10.2 y intercept $= 1$, slope $= -2$
10.3 $y = 3 - x$
10.4 $y = x - 3$
10.6 (a) $\hat{y} = 3.0 + 1.2x$
10.7 (a) $\hat{y} = 6 - .557x$
10.8 (a) $\hat{y} = 38.12 - .802x$
10.9 (a) $\hat{y} = 29.611 + .0868x$
10.10 (a) $\hat{y} = 115.333 + 78.848x$
10.11 SSE $= 1.6$; $s^2 = .5333$
10.12 SSE $= .1429$; $s^2 = .0357$
10.13 SSE $= 6.4261$; $s^2 = 6.4261$
10.14 SSE $= 1.5789$; $s^2 = .2256$
10.15 SSE $= 830.60606$; $s^2 = 103.8258$

10.16 yes, $t = 5.196$
10.17 $(.657, 1.743)$
10.18 yes, $t = -12.333$
10.19 $(-.682, -.432)$
10.20 no, $t = -4.142$
10.21 (a) no, $t = 1.584$ (b) $.10 < p$-value $< .20$
10.22 (a) $H_0: \beta_1 = 0$; $H_a: \beta_1 > 0$
　　　(b) $t = 70.29$, p-value $< .005$, reject H_0
10.23 (a) $\hat{y} = 41.3 + .69x$ (c) $s^2 = 9.445$
10.24 yes, $t = 5.020$
10.25 (b) $z = 3.44$; reject $H_0: p_1 - p_2 = 0$
10.26 (a) $\hat{y} = 3 + .475x$ (c) 5.025
10.27 yes, $t = 3.791$; no
10.28 $(.186, .764)$
10.29 (a) $\hat{y} = 21.4 - .0063x$ (b) 122
　　　(c) yes, $t = -9.00$
10.30 (b) $\hat{y} = 9.569 + 2.325x$
　　　(c) $(1.221, 3.429)$; no; yes
10.31 $(3.259, 5.141)$
10.32 $(4.667, 5.105)$
10.33 (a) $\hat{y} = 3.714 + 1.25x$
　　　(b) SSE $= 1.678571$; $s^2 = .3357$
　　　(c) $(4.471, 5.457)$
10.34 $(4.80, 23.32)$
10.35 (a) $\hat{y} = 4.3 + 1.5x$ (c) 1.531
　　　(d) yes, $t = 3.834$ (e) p-value $< .01$
　　　(f) $(8.11, 9.49)$
10.36 (a) $(44.697, 47.989)$
10.37 (a) $(24.093, 30.719)$
　　　(b) $(104,675.991, 107,359.749)$
10.38 (b) $\hat{y} = 80.852 + 270.816x$
　　　(c) yes, $t = 3.96$ (d) p-value $< .01$
　　　(e) $(112.14, 157.90)$
10.39 $(2.318, 6.082)$
10.40 $(4.427, 5.345)$
10.41 $(3.697, 6.231)$
10.42 $(42.138, 50.548)$
10.43 $(96,603.261, 103,373.949)$
10.44 $(-4.43, 32.55)$
10.45 $(9.436, 37.602)$
10.46 (a) $\hat{y} = -.627 + 1.080x$
　　　(b) $(7.843, 12.495)$
10.47 (a) $\hat{y} = .72 + .118x$ (c) $(.110, .126)$
　　　(d) $t = 4.587$, reject H_0 (e) $(6.888, 7.572)$
10.48 (a) $\hat{y} = 40.548 - .161x$
　　　(b) SSE $= .26287$; $s^2 = .03755$
　　　(c) yes, $t = -7.080$ (d) $(33.705, 34.497)$
10.49 (a) 6 (b) yes, $t = -14.49$
　　　(c) $\hat{y} = 3393.93 - 1.319x$
　　　(d) $\hat{y} = 2305.755$; no

10.52 (a) $+1$ (b) -1
10.53 (a) $\hat{y} = 3.0 + 6x$ (b) positive
(c) $r = .9487; r^2 = .9$
10.54 (a) $\hat{y} = 8.267 - 1.314x$ (c) $-.982$
(d) 96.5%
10.55 (a) $\hat{y} = -.933 + 1.314x$ (c) $.982$
(d) 96.5%
10.56 (a) .5137 (b) .2639
10.57 (a) negatively related (b) $r = -.5519$
(c) $H_a: \rho < 0$ (d) reject $H_0, t = -1.872$
10.58 (a) positive (b) no, $t = 1.580$
10.60 (a) yes, $t = -3.260$ (b) p-value $< .01$
10.61 (a) yes, $t = 3.158$ (b) p-value $< .01$
10.62 (a) $\hat{y} = 307.917 + 34.583x$ (b) $.874; .764$
(c) (488.405, 542.425)
(d) (418.031, 612.799) (e) 76.4
10.63 (a) $\hat{y} = 198.925 + .0385x$ (c) $.3297$
(d) 10.9%
10.65 y intercept $= -5/3$, slope $= 2/3$
10.66 all points fall on a straight line
10.67 σ^2
10.68 (a) $\hat{y} = 2.643 + .554x$ (c) $3.277; .6554$
(d) yes, $t = 3.618$ (e) .161, .947)
(f) (1.209, 2.969) (g) (1.900, 5.600)
10.69 (a) .851; 72.36%
10.70 (a) $\hat{y} = 46 - .317x$ (c) 19.033
(d) $(-.419, -.215)$ (e) (27.667, 32.367)
(f) (21.876, 38.458)
10.71 (b) $r = .9188$ (c) .844
(d) yes, $t = 5.204$
10.72 (a) $r = .740$ (b) yes, $t = 7.622$
10.73 (a) $\hat{y} = 1.159 + .100x$ (b) yes, $t = 12.84$
(c) (.086, .114) (d) (2.79, 3.73)
10.74 (a) $\hat{y} = -5.497 + .066x$ (c) .1800
(d) yes, $t = 3.38$ (e) .730
(f) (2.108, 2.656) (g) (1.566, 3.198)
10.75 (a) $\hat{y} = .067 + .517x$ (b) (1.851, 2.415)
(c) no, observations are not independent
10.76 (a) $\hat{y} = -.629 + 1.075x$
(b) yes, $t = 67.83$
10.77 (a) .7738 (b) .5988
(c) $\hat{y} = 11.238 + 1.309x$
10.78 (a) .9803 (b) .9610
(c) $\hat{y} = 21.867 + 14.967x$ (d) yes
10.79 (a) no; $t = 2.066; .10 < p$-value $< .20$
(b) .2992
10.80 (b) $r = -.148$ (c) no, $t = -.423$
(d) .022
10.81 (b) no, $t = -1.08$
10.83 no; variances not stable

Chapter 11

11.2 (e) changes direction of parabola
11.3 (b) shifts the parabola to the right along the
x-axis
(c) shifts the parabola to the left along the
x-axis by the same amount
11.4 (b) parallel lines
11.5 (b) parallel lines
11.6 (b) changes the slopes
(c) allows for interaction
11.7 (b) yes, $F = 57.44$
11.8 (a) $\hat{y} = 1.04 + 1.29x_1 + 2.72x_2 + .41x_3$
(b) all three variables contribute information;
$t_1 = 3.07, t_2 = 4.185, t_3 = 2.412$
(c) parallel lines (e) (.536, 2.044)
11.9 when $v_1 = 1$
11.10 (a) $x_2 = 1: \hat{y} = 1.83 + .83x_1; x_2 = 2:$
$\hat{y} = 0.19 + 1.46x_1; x_2 = 3:$
$\hat{y} = -1.45 + 2.09x_1$
(c) straight lines; yes
11.11 (d) yes, $F = 15.11$
11.12 (a) negative (b) SSE $= 1.0599; s^2 = .3533$
(c) 3 (d) yes, $F = 332.53$
(e) p-value $= .0003$ (f) yes, $t = -4.49$
(g) p-value $= .0206$
(h) $\hat{y} = -44.192 + 16.334x - .820x^2$
(i) $R^2 = .9955$ (j) $\hat{y} = 27.3424$
11.13 $R^2 = .9955$
11.15 (a) $\hat{y} = 8.59 + 3.82x - .217x^2$ (b) .944
(c) yes, $F = 33.44$ (d) yes, $t = -4.93$
11.16 (a) SSE $= .05368; s^2 = .00596$
(b) $\hat{y} = .4381 + .0053x_1 - .0302x_2$
$+ .0007x_1x_2 - .00038x_1^2 + .0004x_2^2$
(c) $R^2 = .9888$ (d) yes, $F = 159.23$
(e) .0001 (f) no
11.17 (a) parabola (b) positive
(c) straight line (d) negative
(e) allows for different curves for different
values of x_2
11.18 (a) SSE $= 152.17748; s^2 = 8.4543$ (b) 18
(c) yes, $F = 31.85$
(d) p-value $< .005$; from printout,
p-value $= .0001$
(e) $R^2 = .898455$; TSS $= S_{yy} = 1498.625$
(f) $\hat{y} = 41.262$
11.19 (a) yes
(b) (i) SSE$_1 = 465.1343$; SSE$_2 = 152.1775$
(ii) $v_1 = 3, v_2 = 18$ (iii) $F = 12.35$
(iv) $F_{.05} = 3.16$ (v) reject H_0

11.20 (a) $E(y) = \beta_0 + \beta_1 x_1$
(b) $E(y) = (\beta_0 + \beta_2) + (\beta_1 + \beta_4)x_1$
(c) $E(y) = (\beta_0 + \beta_3) + (\beta_1 + \beta_5)x_1$
(d) β_2 (e) β_5
(f) $H_0: \beta_4 = \beta_5 = 0; H_a: \beta_i \neq 0$ for some $i = 4, 5$

11.21 (b) 9   (d) $R^2 = .98397$
(f) $\hat{y} = 4.10 + 1.04x_1 + 3.53x_2 + 4.76x_1 - .43x_1x_2 - .08x_1x_3$
(g) yes, $t = -2.613$ (using $s_{\hat{\beta}_4} = .16459$)
(h) $(-.802, -.058)$
(i) no, $t = -.486$ (using $s_{\hat{\beta}_5} = .16459$)

11.22 (c) yes, $F = 48.86$ (d) .0011

11.23 reject H_0, $F = 60.506$

11.25 (a) 123   (c) $F = 33.40$
(d) reject H_0; 3-variable model provides a good fit

11.26 (a) $y = \beta_0 + \beta_1 x_1 + \beta_2 x_2 + \beta_3 x_1 x_2 + \beta_4 x_1^2 x_2 + \epsilon$
(b) $\hat{y} = -17.65 + 10.737x_1 - 28.69x_2 + 12.722x_1x_2 - .3707x_1^2x_2$
(c) .765   (d) yes, $F = 32.53$
(e) $\hat{y} = -17.65 + 10.737x_1$
(f) $\hat{y} = -46.34 + 23.459x_1 - .3707x_1^2$
(g) yes, $t = -3.52$

Chapter 12

12.2 (a) 7.81473   (b) 20.0902   (c) 22.3072
(d) 17.2750

12.3 (a) 10.6446   (b) 21.6660   (c) 22.3621
(d) 5.99147

12.4 (a) 4   (b) $X^2 > 9.4877$
(c) $H_a: p_i \neq p_j$ for some pair $i, j\ (i \neq j)$
(d) $X^2 = 8.00$; do not reject H_0
(e) $.05 < p\text{-value} < .10$

12.5 (a) 2   (b) $X^2 > 5.99147$
(c) $H_a: p_1 \neq p_j$ for some pair $i, j\ (i \neq j)$
(d) $X^2 = 16.88$; reject H_0
(e) $p\text{-value} < .005$

12.6 yes, $X^2 = 24.48\ (\chi^2_{.05} = 7.81)$

12.7 no, $X^2 = .658\ (\chi^2_{.05} = 7.81)$

12.8 (a) $H_0: p_1 = p_2 = \cdots = p_{12}$
(b) $H_a: p_i \neq p_j$ for some pair $i, j\ (i \neq j)$
(c) $X^2 > 19.6751$
(d) $X^2 = 18.66$; do not reject H_0

12.9 yes, $X^2 = 31.77\ (\chi^2_{.05} = 9.49)$

12.10 no, $X^2 = 13.58\ (\chi^2_{.10} = 17.275)$

12.11 no, $X^2 = .2995$, $p\text{-value} > .10$

12.12 (a) $X^2 = 211.71$; d.f. $= 6$
(b) $X^2 = 8.93$; d.f. $= 2$

12.13 8

12.14 (a) 2   (b) $X^2 = 3.059$ (c) $X^2 > 4.605$
(d) do not reject H_0 (e) $p\text{-value} > .10$

12.15 (c) $X^2 > 3.84$ (f) $X^2 = 22.87$; reject H_0

12.16 (a) no, $X^2 = 10.639$ (b) $p\text{-value} > .10$

12.17 (a) no, $X^2 = 0.639$
(b) $H_0: p_1 = p_2; H_a: p_1 \neq p_2$
(c) $H_0: p_1 = p_2; H_a: p_1 < p_2$
(d) z test for the difference between two population proportions

12.18 (a) no, $X^2 = 4.17$ (b) $p\text{-value} > .10$

12.19 (a) $X^2 = 2.56$ (b) yes
(c) insufficient evidence to indicate a dependence

12.20 (a) no, $X^2 = 3.02$ (b) $p\text{-value} > .10$
(c) yes

12.21 (a) $X^2 = 10.597$ (b) $X^2 > 7.779$
(c) reject H_0 (d) $.025 < p\text{-value} < .05$

12.22 (a) $X^2 = 5.322$ (b) $X^2 > 5.99$
(c) do not reject H_0
(d) $.05 < p\text{-value} < .10$

12.23 reject H_0; $X^2 = 6.617$

12.24 (a) yes, $X^2 = 18.53\ (\chi^2_{.05} = 3.84)$

12.25 yes, $X^2 = 8.75\ (\chi^2_{.05} = 5.99)$

12.26 yes, for $\alpha = .10$, $X^2 = 5.49\ (\chi^2_{.10} = 4.61)$

12.27 yes, $X^2 = 38.43\ (\chi^2_{.05} = 12.59)$

12.28 yes, $X^2 = 22.97\ (\chi^2_{.01} = 11.3449)$

12.29 no, $X^2 = 4.4\ (\chi^2_{.10} = 7.78)$

12.30 yes, $X^2 = 21.51\ (\chi^2_{.01} = 11.34)$

12.31 no, $X^2 = 1.89\ (\chi^2_{.01} = 4.61)$

12.32 yes, $X^2 = 6.75\ (\chi^2_{.01} = 6.63)$

12.33 yes, $X^2 = 10.26\ (\chi^2_{.05} = 7.81)$

12.34 yes, $X^2 = 17.54\ (\chi^2_{.05} = 3.84)$

12.35 (a) yes, $X^2 = 22.699\ (\chi^2_{.05} = 5.99)$
(b) $p\text{-value} < .005$

12.36 no, $X^2 = 2.87\ (\chi^2_{.10} = 4.61)$

12.37 yes, $X^2 = 6.18\ (\chi^2_{.05} = 5.99)$

12.39 $\hat{p} = .5$; 6.25, 25, 37.5, 25, 6.25; reject H_0: $X^2 = 8.56\ (\chi^2_{.05} = 7.81)$

12.40 yes, $X^2 = 7.48\ (\chi^2_{.10} = 2.71)$

12.41 (a) yes, $X^2 = 10.27\ (\chi^2_{.05} = 3.84)$
(b) yes, $z = 3.205\ (z_{.05} = 1.645)$

12.42 yes, $X^2 = 6.19\ (\chi^2_{.05} = 5.99)$; $.415 \pm .057$

12.43 yes, $X^2 = 15.28$, $p\text{-value} < .005$

12.44 yes, $X^2 = 23.78\ (\chi^2_{.05} = 5.99)$

12.45 yes, $X^2 = 8.91\ (\chi^2_{.05} = 5.99)$

12.46 (a) $X^2 = 20.51$ (b) $.005 < p\text{-value} < .01$
(c) reject H_0

12.47 yes, $X^2 = 141.03 \, (\chi^2_{.01} = 6.63)$
12.48 yes, $X^2 = 135.62 \, (\chi^2_{.05} = 11.07)$
12.49 no, $X^2 = 6.50 \, (\chi^2_{.01} = 13.28)$
12.50 (a) no, $X^2 = 1.19$ (b) p-value $> .10$
(c) yes

Chapter 13

13.4 $n_1 = n_2 = n/2$, signal amplification
13.5 $n_1 = 34, n_2 = 56$
13.6 $n_1 = n_2 = 48$
13.8 three observations each at $x = 2$ and $x = 5$; signal amplification
13.9 1.46 times as large; slightly more than twice (2.14 times) as many observations
13.10 test for curvature
13.11 (a) CM $= 103.142857$; Total SS $= 26.8571$
(b) SST $= 14.5071$; MST $= 7.2536$
(c) SSE $= 12.3500$; MSE $= 1.1227$
(f) $F > 3.98$ (g) $F = 6.46$; reject H_0
13.12 yes, $t = 3.59$
13.13 (a) (1.95, 3.65) (b) (.27, 2.83)
13.14 (a) 5 (b) no (c) 30
(e) $F > 2.76$ for $\alpha = .05$ (f) $F = 3.11$; yes
13.15 SSE $= 10.75$; $F = 2.93$; do not reject H_0

13.16 (a)

Source	d.f.	SS	MS	F
Method	2	641.88	320.94	5.15
Error	8	498.67	62.33	
Total	10	1140.55		

(b) yes, $F = 5.15$ $(F_{.05} = 4.46)$
13.17 (a) (67.86, 84.14) (b) (55.82, 76.84)
(c) $(-3.628, 22.962)$
(d) no, they are not independent

13.18 (a)

Source	d.f.	SS	MS
Regimens	2	64.31	32.155
Error	25	338.02	13.5208
Total	27	402.33	

(b) no, $F = 2.38$ $(F_{.05} = 3.39)$
(c) $(-8.51, -1.55)$ (d) (1.165, 6.215)
13.19 (a) $T_1 = .54$; $T_2 = .30$; $T_3 = 2.16$
(b) CM $= .3333$ (c) SST $= .004467$
(e) SSE $= .0655$ (g) no, $F = .818$

13.20 (a) yes, $F = 5.7$; p-value $= .0251$
(b) (.422, 11.244) (c) (56.957, 64.043)
(d) yes (note that the observations are the means of four assembly times)
13.21 (a) $T_1 = 348$; $T_2 = 329$; $T_3 = 48$; $T_4 = 48$
(b) CM $= 31,448.89474$
(c) SST $= 5734.105263$ (e) SSE $= 202$
(g) yes, $F = 141.93$
13.22 (a) each observation is the mean length of 10 leaves
(b) yes, $F = 57.38$
(c) reject H_0, $t = 12.105$
(d) (2.030, 2.704)
13.23 (a) CM $= 1,567,504$; $T_1 = 2172$, $T_2 = 3096$, $T_3 = 2244$
(b) SSE $= 4378$

(c)

Source	d.f.	SS	MS	F
Group	2	44024	22012	165.92
Error	33	4378	132.667	
Total	35	48402		

(d) yes, $F = 165.92$; p-value $< .005$
(e) (61.78, 80.22)
13.24 (a) yes, $F = 63.66$ (b) (1.387, 1.933)
13.25 (a) CM $= 1064.0833$; Total SS $= 148.9167$
(b) SST $= 25.5833$; MST $= 8.5278$
(c) SSB $= 120.6667$; MSB $= 60.3333$
(d) SSE $= 2.6667$; MSE $= .4444$
(f) yes, $F = 19.19$ (g) $F = 135.75$
(h) yes
13.26 $(-5.058, -2.942)$
13.27 (a) CM $= 159.414$; Total SS $= 6.496$
(b) SST $= 4.476$; MST $= 2.238$
(c) SSB $= 1.796$; MSB $= .449$
(d) SSE $= .224$; MSE $= .028$
(f) yes, $F = 79.93$ (g) yes, $F = 16.04$
(h) yes (i) (.665, 1.375)
13.28 (a) 7 (b) 7 (c) 5 (e) yes, $F = 9.68$
(f) yes, $F = 8.59$
13.29 (a) yes, $F = 6.46$ $(F_{.05} = 5.14)$
(b) no, $F = 3.75$ $(F_{.05} = 4.76)$
(c) $(-1.85, -.55)$
13.30 (a) yes, $F = 19.44$ $(F_{.05} = 4.76)$
(b) yes, $F = 40.21$ $(F_{.05} = 5.14)$
(c) $(-1.997, -.803)$
13.31 (b) yes, $F = 6.96$ (c) yes, $F = 3.74$

13.32 **(a)**

Source	d.f.	SS	MS	F
Treatments	2	38	19	10.06
Blocks	3	61.667	20.556	10.88
Error	6	11.333	1.889	
Total	11	111.000		

yes, $F = 10.06$ ($F_{.05} = 5.14$)
(b) yes, $F = 10.88$ ($F_{.05} = 4.76$)
(c) (1.612, 5.388)

13.33 **(a)**

Source	d.f.	SS	MS	F
Treatments	2	524,177.167	262,088.58	258.19
Blocks	3	173,415	57,805	56.95
Error	6	6,090.5	1,015.083	
Total	11	703,681.67		

(b) 6   **(c)** yes, $F = 258.19$ ($F_{.05} = 5.14$)
(d) yes, $F = 56.95$ ($F_{.05} = 4.76$)   **(e)** 22.529
(f) $(-292.38, -182.12)$
13.34 **(a)** yes, $F = 5.77$ ($F_{.05} = 3.74$)
(b) $(-.251, 2.751)$
13.35 **(a)** yes, $F = 7.20$ ($F_{.05} = 5.14$)
(b) $(-3.469, -1.081)$
(c) yes, $F = 16.61$ ($F_{.05} = 4.76$)
13.36 **(a)** randomized block design
(b) no, cannot calculate SSB
13.37 $n \geq 32$
13.38 $n \geq 64$
13.39 **(a)** 35   **(b)** 140
13.40 $n_i = 11$
13.41 $b = 16$
13.42 $b = 21$
13.43 **(a)** randomized block design   **(b)** 32
13.44 $b = 18$
13.45 assume $\sigma^2 \approx .03$; $b = 48$
13.47 SSE = 1196.6
13.48 **(a)** no   **(b)** completely randomized design
13.49 **(a)** 16   **(b)** 135
(c) for a 95% confidence interval and $z \approx 2$, half width is 14.14

13.50 **(a)**

Source	d.f.	SS	MS	F
Treatments	4	1.212	.303	11.67
Error	22	.571	.026	
Total	26	1.783		

reject H_0, $F = 11.67$ ($F_{.05} = 2.82$)
(b) $t = -2.73$, reject $H_0: \mu_D - \mu_A = 0$

13.51

Source	d.f.	SS	MS	F
Treatments	4	.787	.197	27.78
Blocks	3	.140	.047	6.59
Error	12	.085	.007	
Total	19	1.012		

$F = 27.78$ ($F_{.05} = 3.26$);
reject $H_0: \mu_A = \mu_B = \mu_C = \mu_D = \mu_E$
13.52 **(a)** yes, $F = 13.03$ **(b)** $(-12.17, -2.83)$
13.53 **(a)** no, $F = .87$ **(b)** $(-1.050, 6.450)$
(c) (24.848, 30.152)   **(d)** 26
13.54 no, $F = .63$ ($F_{.05} = 3.89$)
13.55 **(a)** yes, $F = 9.84$ ($F_{.05} = 3.03$)
(b) (.492, 13.174)   **(c)** (87.724, 96.026)

13.56 **(g)**

Source	d.f.	SS	MS
Treatments	2	1207.4667	603.7333
Error	27	1880.3250	69.6417

(h) yes, $F = 8.67$
(i) no, the variances are unequal
13.57 **(a)** yes, $F = 3.43$ **(c)** $(-22.56, -5.84)$
13.58 **(a)** yes, $F = 5.20$ ($F_{.05} = 3.24$)
(b) no, $t = .88$ ($t_{.05} = 1.746$)
(c) $(-.579, -.117)$
13.59 **(a)** yes, $F = 18.61$ ($F_{.05} = 2.87$)
(b) yes, $F = 7.11$ ($F_{.05} = 2.71$)
(c) $(-6.767, -3.433)$
13.60 **(a)** yes, $F = 7.34$ **(c)** $(-2.31, 2.31)$

Chapter 14

14.1 **(b)** $\alpha = .002, .007, .022, .054, .115$
14.2 **(b)** $\alpha = .004, .014, .044, .108$
14.3 One-tailed: $n = 10$, $\alpha = .001, .011, .055$; $n = 15$, $\alpha = .004, .018, .059$; $n = 20$, $\alpha = .001, .006, .021, .058, .132$

Two-tailed: $n = 10$, $\alpha = .002, .022, .110$; $n = 15$, $\alpha = .008, .036, .118$; $n = 20$, $\alpha = .002, .012, .042, .116$.

14.4 (a) $H_0: p = 1/2$; $H_a: p \neq 1/2$
(b) $x = 0$ or 7 ($\alpha = .016$)
(c) $x = 6$, do not reject H_0 (d) no

14.5 (a) $H_0: p = 1/2$; $H_a: p \neq 1/2$; rejection region: $\{0, 1, 7, 8\}$; $x = 6$; do not reject H_0 at $\alpha = .07$; p-value $= .290$

14.6 (a) Rejection region: $\{x \leq 5$ or $x \geq 15\}$ with $\alpha = .042$; $x = 11$, do not reject H_0
(b) $z = .45$, do not reject H_0

14.7 $z = 3.15$, reject H_0

14.8 (a) yes, $x = 2$; rejection region: $\{0, 1, 2, 8, 9, 10\}$ for $\alpha = .110$
(b) variance not constant

14.9 (a) U_1 (b) $U \leq 13$ (c) $\alpha = .0906$

14.10 (a) the smaller of U_1 and U_2 (b) $U \leq 10$
(c) $\alpha = .0812$

14.11 (a) H_0: populations 1 and 2 are identical; H_a: population 1 is shifted to the left of population 2
(b) $U \leq 4$ with $\alpha = .0476$
(c) $T_1 = 16$, $T_2 = 39$, $U_1 = 24$, $U_2 = 1$
(d) $U = 1$ (e) yes

14.12 $z = -4.75$, p-value $< .002$, reject H_0

14.13 do not reject H_0; $z = -1.59$

14.14 (a) reject H_0, $U = 0$; rejection region: $U \leq 3$ with $\alpha = .0666$
(b) p-value $= .0096$

14.15 (a) $U_1 = 6$; do not reject H_0: rejection region: $\{0, 1, \ldots, 4\}$ (one-tailed) for $\alpha = .0476$
(b) do not reject $H_0: \mu_A - \mu_B = 0$; $t = 1.606$ ($t_{.05} = 1.86$)

14.16 yes, $U_1 = 0$; rejection region: $\{0, 1, \ldots, 18\}$ for $\alpha = .0504$

14.17 $U_2 = 12.5$; reject H_0; rejection region: $\{0, 1, \ldots, 21\}$ for $\alpha = .1012$

14.18 yes, $U_1 = 8$; rejection region: $\{0, 1, \ldots, 18\}$ for $\alpha = .0546$

14.19 yes, $z = 3.49$

14.20 (a) H_0: the two population distributions are identical; H_a: the two population distributions differ in location
(b) $\min(T^+, T^-)$ (c) $T \leq 137$
(d) $T^- = 216$; do not reject H_0

14.21 (a) H_0: the two population distributions are identical; H_a: population distribution 1 is shifted to the right of distribution 2

(b) T^- (c) $T \leq 152$
(d) $T^- = 216$; do not reject H_0

14.22 $z = -.34$; do not reject H_0

14.23 $z = -.34$; do not reject H_0

14.24 (a) $T^- = 1.5$, reject H_0 (b) yes

14.25 (a) $T = 3$; rejection region: $\{T \leq 4\}$; reject H_0; consistent with results of Exercise 9.43

14.26 (a) no, $T = 6.5$; rejection region: $\{T \leq 4\}$

14.27 no, $T = 73.5$; rejection region: $\{T \leq 35\}$; consistent with results of Exercise 14.6

14.28 (a) no, $x = 8$, rejection region: $\{0, 1, 2, 9, 10, 11\}$ for $\alpha = .066$
(b) no, $T = 14.5$; rejection region: $\{T \leq 11\}$ for $\alpha = .05$

14.29 (a) do not reject H_0; $T = 35$; rejection region: $\{T \leq 14\}$ for $\alpha = .05$
(b) p-value $> .10$

14.30 no, $x = 5$ with rejection region $\{x \geq 9\}$ and $\alpha = .073$

14.31 yes, $T = 3.5$; rejection region: $\{T \leq 4\}$ for $\alpha = .05$

14.32 yes, $H = 13.39$ ($\chi^2_{.05} = 5.99$)

14.33 yes, $H = 13.90$ ($\chi^2_{.05} = 7.81$)

14.34 (a) yes, $H = 17.075$ ($\chi^2_{.05} = 7.81$)
(b) p-value $< .005$ (c) p-value $< .005$

14.35 (a) no, $H = 2.63$ ($\chi^2_{.10} = 6.25$)
(b) p-value $> .10$ (c) p-value $> .10$

14.36 (a) yes, $H = 12.48$ ($\chi^2_{.05} = 5.99$)
(b) p-value $< .005$

14.37 (a) reject H_0; $F_r = 7.58$ ($\chi^2_{.05} = 5.99$)
(b) $.01 < p$-value $< .025$ (d) $F = 10.83$
(e) p-value $< .005$

14.38 (a) reject H_0; $F_r = 21.19$ ($\chi^2_{.05} = 7.81$)
(b) p-value $< .005$ (d) $F = 75.43$
(e) p-value $< .005$

14.39 (a) yes, $F_r = 19.57$ ($\chi^2_{.05} = 9.49$)
(b) p-value $< .005$

14.40 (a) no, $F_r = 5.81$ ($\chi^2_{.05} = 5.99$)
(b) $.05 < p$-value $< .10$
(c) no; the more powerful parametric test allows us to reject H_0, as long as parametric assumptions are satisfied

14.41 (a) $r_s \geq .425$ (b) $r_s \geq .601$

14.42 (a) $r_s \leq -.497$ (b) $r_s \leq -.703$

14.43 (a) $|r_s| \geq .400$ (b) $|r_s| \geq .526$

14.44 (a) $r_s = -1$
(b) yes; reject H_0 for $|r_s| \geq .886$

14.45 (a) $-.593$ (b) yes

14.46 (a) $-.745$ (b) yes, reject H_0 for $|r_s| \geq .600$

14.47 (a) $-.465$; do not reject $H_0: \rho_s = 0$
(b) .708; reject $H_0: \rho_s = 0$
(c) .819; reject $H_0: \rho_s = 0$

14.48 .867

14.49 yes

14.50 $r_s = .903$; yes

14.51 yes, $r_s = .9118$

14.52 (a) negative   (b) $-.75$
(c) reject H_0 $(r_s \leq -.714)$

14.53 (a) $-.75$ (b) yes   (c) no

14.54 (a) do not reject H_0; $x = 2$; rejection region: $\{0, 1, 8, 9\}$ for $\alpha = .040$
(b) do not reject H_0; $t = -1.65$; rejection region: $|t| > 2.306$ for $\alpha = .05$

14.55 do not reject H_0; $T = 10.5$; rejection region: $\{T \leq 6\}$ for $\alpha = .05$

14.56 (a) do not reject H_0; $x = 7$; rejection region: $\{0, 1, 9, 10\}$ for $\alpha = .022$
(b) do not reject H_0; $x = 7$; rejection region: $\{8, 9, 10\}$ for $\alpha = .055$

14.57 (a) reject H_0; $T = 6$; rejection region: $T \leq 8$ for $\alpha = .05$
(b) reject H_0; $T = 6$; rejection region: $T \leq 11$ for $\alpha = .05$

14.58 (a) no, $U = 32$; rejection region: $U \leq 21$ for $\alpha = .094$
(b) no, $t = .30$

14.59 (a) $U_B = 3$; reject H_0; rejection region: $U \leq 3$ for $\alpha = .0556$
(b) $F = 4.91$; do not reject H_0; rejection region: $F \geq 9.12$

14.60 no, $x = 2$; rejection region: $\{0, 1, 7, 8\}$ for $\alpha = .0703$

14.61 $-.845$

14.62 yes; yes

14.63 $U = 17.5$; no; rejection region: $U \leq 13$ for $\alpha = .0498$

14.64 $T = 14$; reject H_0; rejection region: $T \leq 16$ for $\alpha = .01$ or $T \leq 25$ for $\alpha = .05$

14.65 .67684, reject H_0; rejection region: $|r_s| \geq .441$ for $\alpha = .10$

14.66 (a) reject H_0; $F_r = 20.13$
(b) results are identical

14.67 (a) yes, $H = 14.52$ ($\chi^2_{.05} = 7.81$)
(b) results are identical

14.68 (a) yes, $H = 9.08$ ($\chi^2_{.05} = 7.81$)
(b) $.025 < p\text{-value} < .05$ (c) yes

14.69 yes, $z = 3.31$, $p\text{-value} < .001$

14.70 no, $x = 2$ with rejection region $\{x \geq 5\}$ and $\alpha = .109$

14.71 $z = -3.45$, reject H_0; black uniforms receive higher malevolence scores.

INDEX